RADIO-FREQUENCY AND MICROWAVE COMMUNICATION CIRCUITS

RADIO-FREQUENCY AND MICROWAVE COMMUNICATION CIRCUITS

Analysis and Design

Second Edition

Devendra K. Misra
University of Wisconsin–Milwaukee

A JOHN WILEY & SONS, INC., PUBLICATION

Published by John Wiley & Sons, Inc., Hoboken, New Jersey.
Published simultaneously in Canada.

For general information on our other products and services please contact our Customer Care Department within the U.S. at 877-762-2974, outside the U.S. at 317-572-3993 or fax 317-572-4002.

Wiley also publishes its books in a variety of electronic formats. Some content that appears in print, however, may not be available in electronic format.

Library of Congress Cataloging-in-Publication Data:

Misra, Devendra, 1949–
Radio-frequency and microwave communication circuits : analysis and design / Devendra K. Misra.—2nd ed.
p. cm.
Includes bibliographical references and index.
ISBN 0-471-47873-3 (Cloth)
1. Radar circuits–Design and construction. 2. Microwave circuits–Design and construction. 3. Electronic circuit design. 4. Radio frequency. I. Title.
TK6560 .M54 2004
621.384′12–dc22

2003026691

Printed in the United States of America.

10 9 8 7 6 5 4 3 2 1

CONTENTS

PREFACE

Wireless technology continues to grow at a tremendous rate, with new applications still reported almost daily. In addition to the traditional applications in communications, such as radio and television, radio-frequency (RF) and microwaves are being used in cordless phones, cellular communication, local area networks, and personal communication systems. Keyless door entry, radio-frequency identification, monitoring of patients in a hospital or a nursing home, cordless mice or keyboards for computers, and wireless networking of home appliances are some of the other areas where RF technology is being employed. Although some of these applications have traditionally used infrared technology, RF circuits are taking over, because of their superior performance. The present rate of growth in RF technology is expected to continue in the foreseeable future. These advances require the addition of personnel in the areas of radio-frequency and microwave engineering. Therefore, in addition to regular courses as a part of electrical engineering curriculums, short courses and workshops are regularly conducted in these areas for practicing engineers.

This edition of the book maintains the earlier approach of a presentation based on a basic course in electronic circuits. At the same time, a new chapter on electromagnetic fields has been added, following several constructive suggestions from those who used the first edition. It provides the added option of using the book for a traditional microwave engineering course with electromagnetic fields and waves. Or, this chapter can be bypassed so as to follow the approach used in the first edition: that is, instead of using electromagnetic fields as most microwave engineering books do, the subject is introduced via circuit concepts. Further, an overview of communication systems is presented in the beginning to provide the reader with an overall perspective of various building blocks involved.

This edition of the book is organized into thirteen chapters and nine appendixes, using a top-down approach. It begins with an introduction to frequency bands, RF and microwave devices, and applications in communication, radar, industrial, and biomedical areas. The introduction includes a brief description of microwave transmission lines: waveguides, strip lines, and microstrip line. An overview of transmitters and receivers is included, along with digital modulation and demodulation techniques. Modern wireless communication systems, such as terrestrial and satellite communication systems and RF wireless services, are discussed briefly in Chapter 2. After introducing antenna terminology, effective isotropic radiated power, the Friis transmission formula, and the radar range equation are presented. In the final section of the chapter, noise and distortion associated with communication systems are introduced.

Chapter 3 begins with a discussion of distributed circuits and construction of a solution to the transmission line equation. Topics presented in this chapter include RF circuit analysis, phase and group velocities, sending-end impedance, reflection coefficient, return loss, insertion loss, experimental determination of characteristic impedance and the propagation constant, the voltage standing wave ratio, and impedance measurement. The final section in Chapter 3 includes a description of the Smith chart and its application in the analysis of transmission line circuits. Fundamental laws of electromagnetic fields are introduced in Chapter 4 along with wave equations and uniform plane wave solutions. Boundary conditions and potential functions that lead to the construction of solutions are then introduced. The chapter concludes with analyses of various metallic waveguides.

Resonant circuits are discussed in Chapter 5, which begins with series and parallel resistance–inductance–capacitance circuits. This is followed by a section on transformer-coupled circuits. The final two sections of the chapter are devoted to transmission line resonant circuits and microwave resonators. Chapters 6 and 7 deal with impedance-matching techniques. Single reactive element or stub, double-stub, and lumped-element matching techniques are discussed in Chapter 6. Chapter 7 is devoted to multisection transmission line impedance transformers, binomial and Chebyshev sections, and impedance tapers.

Chapter 8 introduces circuit parameters associated with two-port networks. Impedance, admittance, hybrid, transmission, scattering, and chain scattering parameters are presented, along with examples that illustrate their characteristic behaviors. Chapter 9 begins with the image parameter method for the design of passive filter circuits. The insertion-loss technique is introduced next to synthesize Butterworth and Chebyshev low-pass filters. The chapter includes descriptions of impedance and frequency scaling techniques to realize high-pass, bandpass, and bandstop networks. The chapter concludes with a section on microwave transmission line filter design.

Concepts of signal-flow-graph analysis are introduced in Chapter 10, along with a representation of voltage source and passive devices, which facilitates formulation of the power gain relations that are needed in the amplifier design discussed in Chapter 11. The chapter begins with stability considerations using scattering parameters of a two-port network followed by design techniques of various amplifiers.

Chapter 12 presents basic concepts and design of various oscillator circuits. The phase-locked loop and its application in the design of frequency synthesizers are also summarized. The final section of the chapter includes the analysis and design of microwave transistor oscillators using S-parameters. Chapter 13 includes the fundamentals of frequency-division multiplexing, amplitude modulation, radio-frequency detection, frequency-modulated signals, and mixer circuits. The book ends with nine appendixes, which include a discussion of logarithmic units (dB, dBm, dBW, dBc, and neper), design equations for selected transmission lines (coaxial line, strip line, and microstrip line), and a list of abbreviations used in the communication area.

Some of the highlights of the book are as follows:

- The presentation begins with an overview of frequency bands, RF and microwave devices, and their applications in various areas. Communication systems are presented in Chapter 2, including terrestrial and satellite systems, wireless services, antenna terminology, the Friis transmission formula, the radar equation, and Doppler radar. Thus, students learn about the systems using blocks of amplifiers, oscillators, mixers, filters, and so on. Students' response has strongly supported this *top-down approach.*
- Since students are assumed to have only one semester of electrical circuits, resonant circuits and two-port networks are included in the book. Concepts of network parameters (impedance, admittance, hybrid, transmission, and scattering) and their characteristics are introduced via examples.
- A separate chapter on oscillator design includes concepts of feedback, the Hartley oscillator, the Colpitts oscillator, the Clapp oscillator, crystal oscillators, phased-locked-loop and frequency synthesizers, transistor oscillator design using S-parameters, and three-port S-parameter description of transistors and their use in feedback network design.
- A separate chapter on detectors and mixers includes amplitude- and frequency-modulated signal characteristics and their detection schemes, single-diode mixers, RF detectors, double-balanced mixers, conversion loss, intermodulation distortion in diode-ring mixers, and field-effect-transistor mixers.
- Appendixes include logarithmic units, design equations for selected transmission lines, and a list of abbreviations used in the communication area.
- There are 153 solved examples with a step-by-step explanation. Therefore, practicing engineers will find the book useful for self-study as well.
- There are 275 class-tested problems at the ends of chapters. Supplementary material is available to instructors adopting the book.

Acknowledgments

I learned this subject from engineers and authors who are too many to include in this short space, but I gratefully acknowledge their contributions. I would like to thank my anonymous reviewers, instructors here and abroad who used the first edition of this book and provided a number of constructive suggestions, and my former students, who made useful suggestions to improve the presentation. I deeply appreciate the support I received from my wife, Ila, and son, Shashank, during the course of this project. The first edition of this book became a reality only because of enthusiastic support from then-senior editor Philip Meyler and his staff at Wiley. I feel fortunate to continue getting the same kind of support from current editor Val Moliere and from Kirsten Rohstedt.

DEVENDRA K. MISRA

1

INTRODUCTION

Scientists and mathematicians of the nineteenth century laid the foundation of telecommunication and wireless technology, which has affected all facets of modern society. In 1864, James C. Maxwell put forth fundamental relations of electromagnetic fields that not only summed up the research findings of Laplace, Poisson, Faraday, Gauss, and others but also predicted the propagation of electrical signals through space. Heinrich Hertz subsequently verified this in 1887, and Guglielmo Marconi transmitted wireless signals across the Atlantic Ocean successfully in 1900. Interested readers may find an excellent discussion of the historical developments of radio frequencies (RFs) and microwaves in the *IEEE Transactions on Microwave Theory and Technique* (Vol. MTT-32, September 1984).

Wireless communication systems require high-frequency signals for the efficient transmission of information. Several factors lead to this requirement. For example, an antenna radiates efficiently if its size is comparable to the signal wavelength. Since the signal frequency is inversely related to its wavelength, antennas operating at RFs and microwaves have higher radiation efficiencies. Further, their size is relatively small and hence convenient for mobile communication. Another factor that favors RFs and microwaves is that the transmission of broadband information signals requires a high-frequency carrier signal. In the case of a single audio channel, the information bandwidth is about 20 kHz. If amplitude modulation (AM) is used to superimpose this information on a carrier, it requires at least this much bandwidth on one side of the spectrum. Further,

Radio-Frequency and Microwave Communication Circuits: Analysis and Design, Second Edition,
By Devendra K. Misra
ISBN 0-471-47873-3

TABLE 1.1 Frequency Bands Used in Commercial Broadcasting

	Channels	Frequency Range	Wavelength Range
AM	107	535–1605 kHz	186.92–560.75 m
TV	2–4	54–72 MHz	4.17–5.56 m
	5–6	76–88 MHz	3.41–3.95 m
FM	100	88–108 MHz	2.78–3.41 m
TV	7–13	174–216 MHz	1.39–1.72 m
	14–83	470–890 MHz	33.7–63.83 cm

commercial AM transmission requires a separation of 10-kHz between the two transmitters. On the other hand, the required bandwidth increases significantly if frequency modulation (FM) is used. Each FM transmitter typically needs a bandwidth of 200 kHz for audio transmission. Similarly, each television channel requires about 6 MHz of bandwidth to carry the video information as well. Table 1.1 shows the frequency bands used for commercial radio and television broadcasts.

In the case of digital transmission, a standard monochrome television picture is sampled over a grid of 512 × 480 elements called *pixels*. Eight bits are required to represent 256 shades of the gray display. To display motion, 30 frames are sampled per second; thus, it requires about 59 Mb/s (512 × 480 × 8 × 30 = 58,982,400). Color transmission requires even higher bandwidth (on the order of 90 Mb/s).

Wireless technology has been expanding very fast, with new applications reported every day. In addition to the traditional applications in communication, such as radio and television, RF and microwave signals are being used in cordless phones, cellular communication, local, wide, and metropolitan area networks and personal communication service. Keyless door entry, radio-frequency identification (RFID), monitoring of patients in a hospital or a nursing home, and cordless mice or keyboards for computers are some of the other areas where RF technology is being used. Although some of these applications have traditionally used infrared (IR) technology, current trends favor RF, because RF is superior to infrared technology in many ways. Unlike RF, infrared technology requires unobstructed line-of-sight connection. Although RF devices have been more expensive than IR, the current trend is downward because of an increase in their production and use.

The electromagnetic frequency spectrum is divided into bands as shown in Table 1.2. Hence, AM radio transmission operates in the medium-frequency (MF) band, television channels 2 to 12 operate in the very high frequency (VHF) band, and channels 18 to 90 operate in the ultrahigh-frequency (UHF) band. Table 1.3 shows the band designations in the microwave frequency range.

In addition to natural and human-made changes, electrical characteristics of the atmosphere affect the propagation of electrical signals. Figure 1.1 shows

TABLE 1.2 IEEE Frequency Band Designations

Band Designation	Frequency Range	Wavelength Range (in Free Space)
VLF	3–30 kHz	10–100 km
LF	30–300 kHz	1–10 km
MF	300–3000 kHz	100 m–1 km
HF	3–30 MHz	10–100 m
VHF	30–300 MHz	1–10 m
UHF	300–3000 MHz	10 cm–1 m
SHF	3–30 GHz	1–10 cm
EHF	30–300 GHz	0.1–1 cm

TABLE 1.3 Microwave Frequency Band Designations

Frequency Bands	Old (Still Widely Used)	New (Not So Commonly Used)
500–1000 MHz	UHF	C
1–2 GHz	L	D
2–4 GHz	S	E
3–4 GHz	S	F
4–6 GHz	C	G
6–8 GHz	C	H
8–10 GHz	X	I
10–12.4 GHz	X	J
12.4–18 GHz	Ku	J
18–20 GHz	K	J
20–26.5 GHz	K	K
26.5–40 GHz	Ka	K

various layers of the ionosphere and the troposphere that are formed due to the ionization of atmospheric air. As illustrated in Figure 1.2(*a*) and (*b*), an RF signal can reach the receiver by propagating along the ground or after reflection from the ionosphere. These signals may be classified as *ground* and *sky waves*, respectively. The behavior of a sky wave depends on the season, day or night, and solar radiation. The ionosphere does not reflect microwaves, and the signals propagate line of sight, as shown in Figure 1.2(*c*). Hence, curvature of the Earth limits the range of a microwave communication link to less than 50 km. One way to increase the range is to place a human-made reflector up in the sky. This type of arrangement is called a *satellite communication system*. Another way to increase the range of a microwave link is to place repeaters at periodic intervals. This is known as a *terrestrial communication system*.

Figures 1.3 and 1.4 list selected devices used at RF and microwave frequencies. Solid-state devices as well as vacuum tubes are used as active elements in RF and

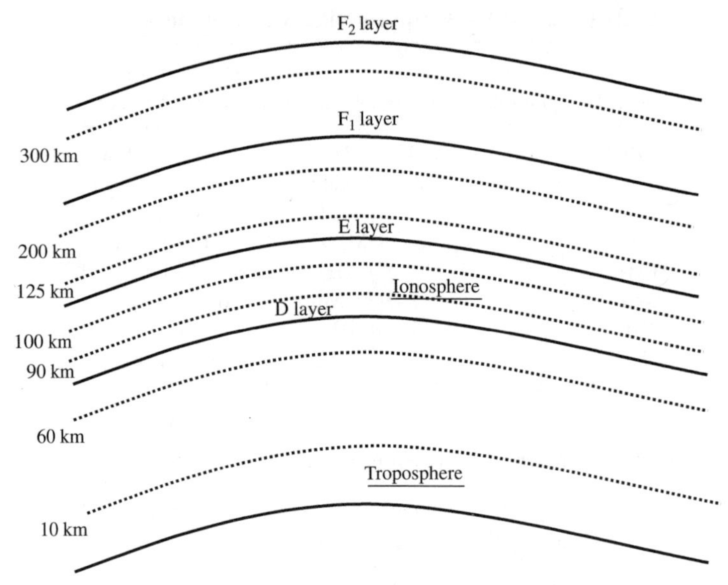

Figure 1.1 Atmosphere surrounding Earth.

microwave circuits. Predominant applications for microwave tubes are in radar, communications, electronic countermeasures (ECMs), and microwave cooking. They are also used in particle accelerators, plasma heating, material processing, and power transmission. Solid-state devices are employed primarily in the RF region and in low-power microwave circuits such as low-power transmitters for local area networks and receiver circuits. Some applications of solid-state devices are listed in Table-1.4.

Figure 1.5 lists some applications of microwaves. In addition to terrestrial and satellite communications, microwaves are used in radar systems as well as in various industrial and medical applications. Civilian applications of radar include air traffic control, navigation, remote sensing, and law enforcement. Its military uses include surveillance; guidance of weapons; and command, control, and communication (C^3). Radio-frequency and microwave energy are also used in industrial heating and for household cooking. Since this process does not use a conduction mechanism for the heat transfer, it can improve the quality of certain products significantly. For example, the hot air used in a printing press to dry ink affects the paper adversely and shortens the product's life span. By contrast, in microwave drying only the ink portion is heated, and the paper is barely affected. Microwaves are also used in material processing, telemetry, imaging, and hyperthermia.

1.1 MICROWAVE TRANSMISSION LINES

Figure 1.6 shows selected transmission lines used in RF and microwave circuits. The most common transmission line used in the RF and microwave range is the

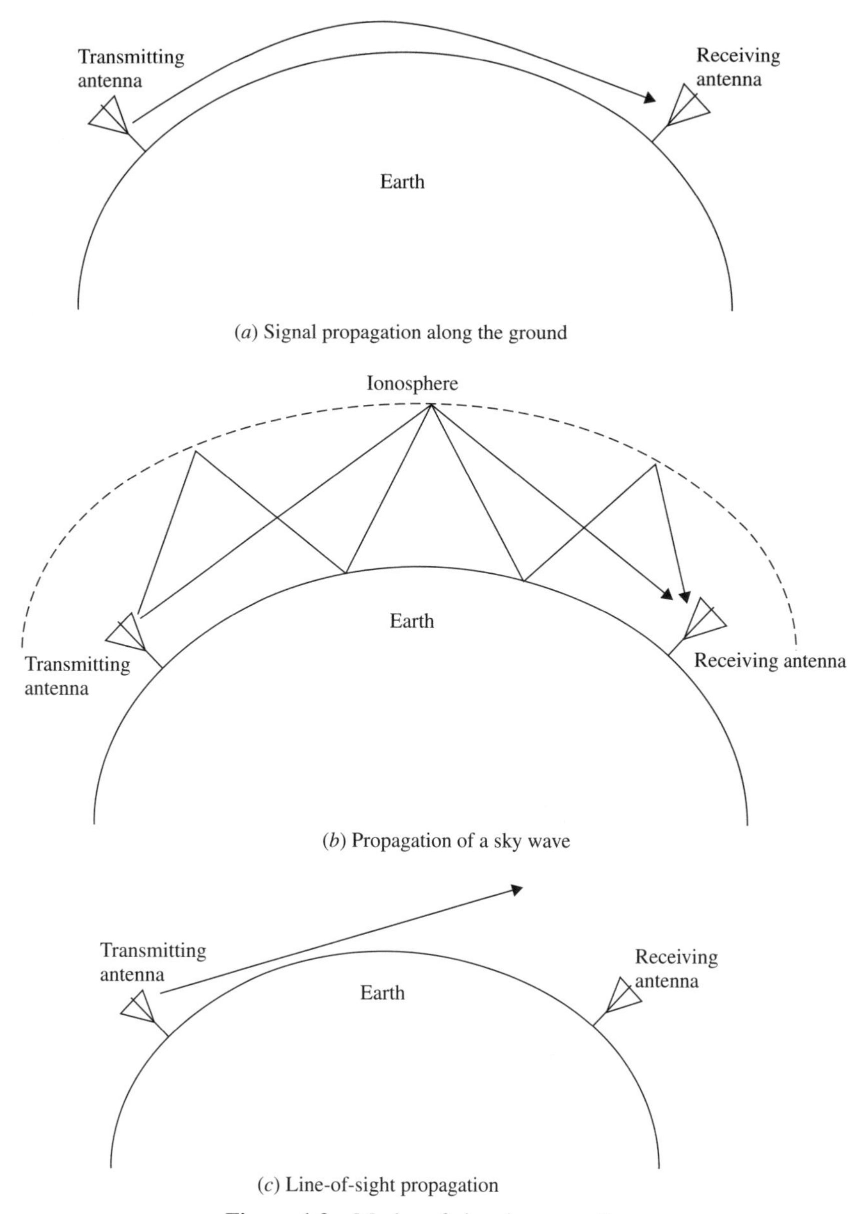

Figure 1.2 Modes of signal propagation.

coaxial line. A low-loss dielectric material is used in these transmission lines to minimize signal loss. Semirigid coaxial lines with continuous cylindrical conductors outside perform well in the microwave range. To ensure single-mode transmission, the cross section of a coaxial line must be much smaller than the signal wavelength. However, this limits the power capacity of these lines. In

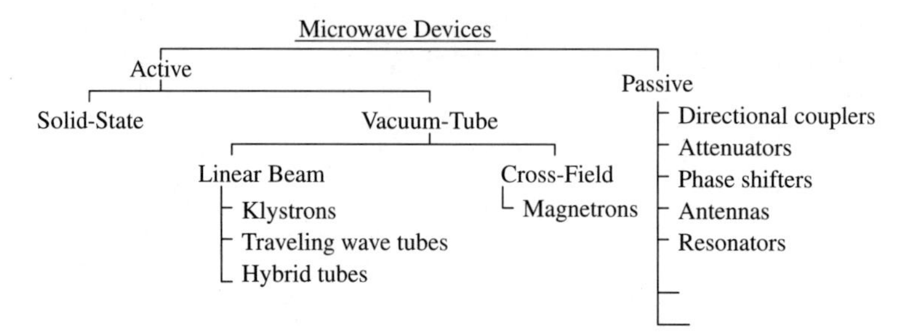

Figure 1.3 Microwave devices.

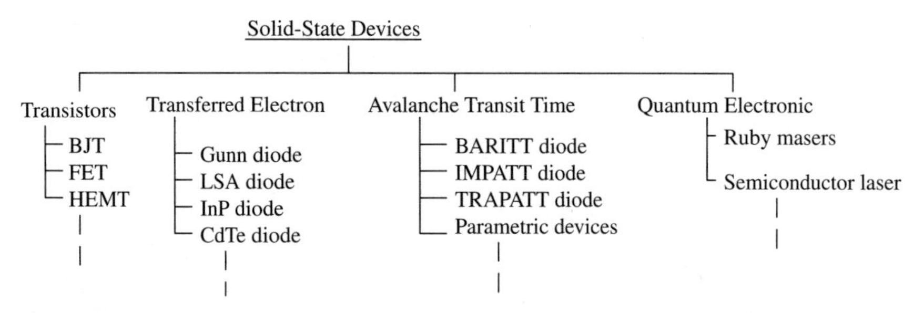

Figure 1.4 Solid-state devices used at RF and microwave frequencies.

TABLE 1.4 Selected Applications of Microwave Solid-State Devices

Devices	Applications	Advantages
Transistors	L-band transmitters for telemetry systems and phased-array radar systems; transmitters for communication systems	Low cost, low power supply, reliable, high-continuous-wave (CW) power output, lightweight
Transferred electron devices (TED)	C-, X-, and Ku-band ECM amplifiers for wideband systems; X- and Ku-band transmitters for radar systems, such as traffic control	Low power supply (12 V), low cost, lightweight, reliable, low noise, high gain
IMPATT diode	Transmitters for millimeter-wave communication	Low power supply, low cost, reliable, high-CW power, lightweight
TRAPATT diode	S-band pulsed transmitter for phased-array radar systems	High peak and average power, reliable, low power supply, low cost
BARITT diode	Local oscillators in communication and radar receivers	Low power supply, low cost, low noise, reliable

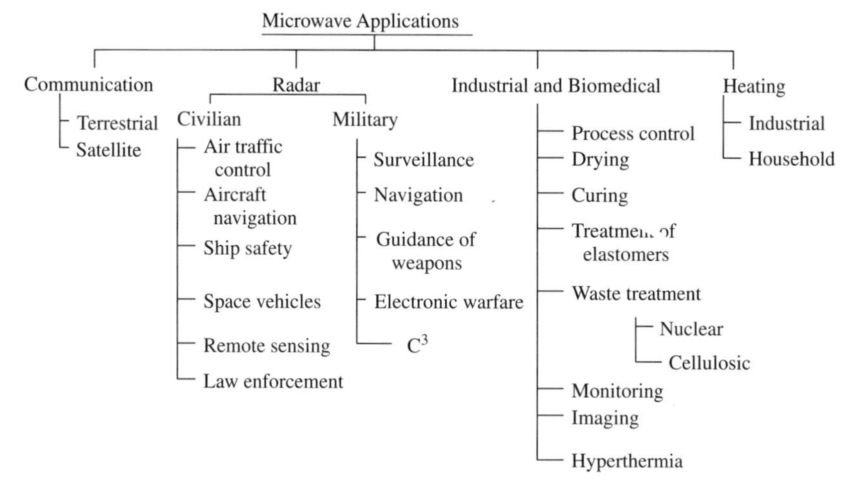

Figure 1.5 Some applications of microwaves.

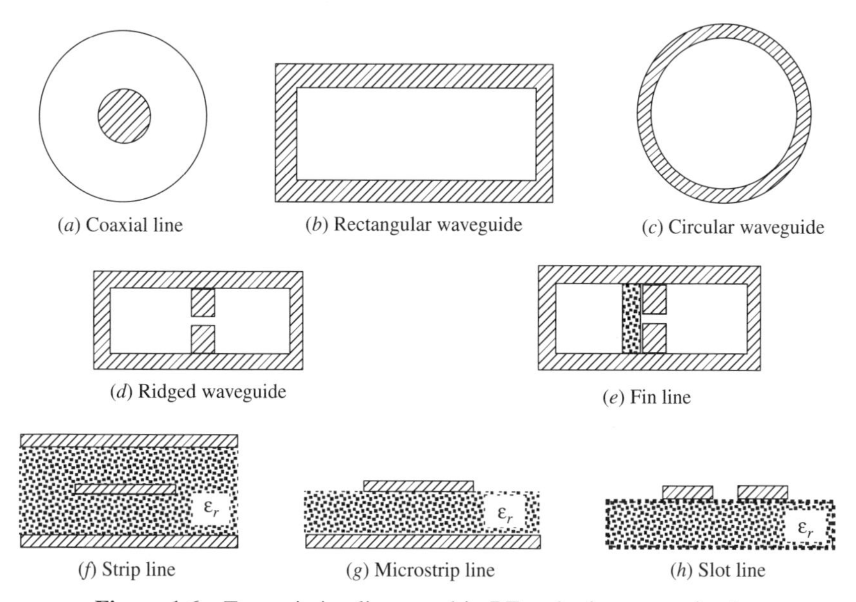

Figure 1.6 Transmission lines used in RF and microwave circuits.

high-power microwave circuits, waveguides are used in place of coaxial lines. Rectangular waveguides are commonly employed for connecting high-power microwave devices because these are easy to manufacture compared with circular waveguides. However, certain devices (e.g., rotary joints) require a circular cross section. In comparison with a rectangular waveguide, a ridged waveguide provides broadband operation. The fin line shown in Figure 1.6(*e*) is commonly used in the millimeter-wave band. Physically, it resembles a slot line enclosed in a rectangular waveguide.

The transmission lines illustrated in Figure 1.6(*f*) to (*h*) are most convenient in connecting the circuit components on a printed circuit board (PCB). The physical dimensions of these transmission lines depend on the dielectric constant ε_r of insulating material and on the operating frequency band. The characteristics and design formulas of selected transmission lines are given in the appendixes.

1.2 TRANSMITTER AND RECEIVER ARCHITECTURES

Wireless communication systems require a transmitter at one end to send the information signal and a receiver at the other to retrieve it. In one-way communication (such as a commercial broadcast), a transmitting antenna radiates the signal according to its radiation pattern. The receiver, located at the other end, receives this signal via its antenna and extracts the information, as illustrated in Figure 1.7. Thus, the transmitting station does not require a receiver, and vice versa. On the other hand, a transceiver (a transmitter and a receiver) is needed at both ends to establish a two-way communication link.

Figure 1.7 is a simplified block diagram of a one-way communication link. At the transmitting end, an information signal is modulated and mixed with a local oscillator to up-convert the carrier frequency. Bandpass filters are used before

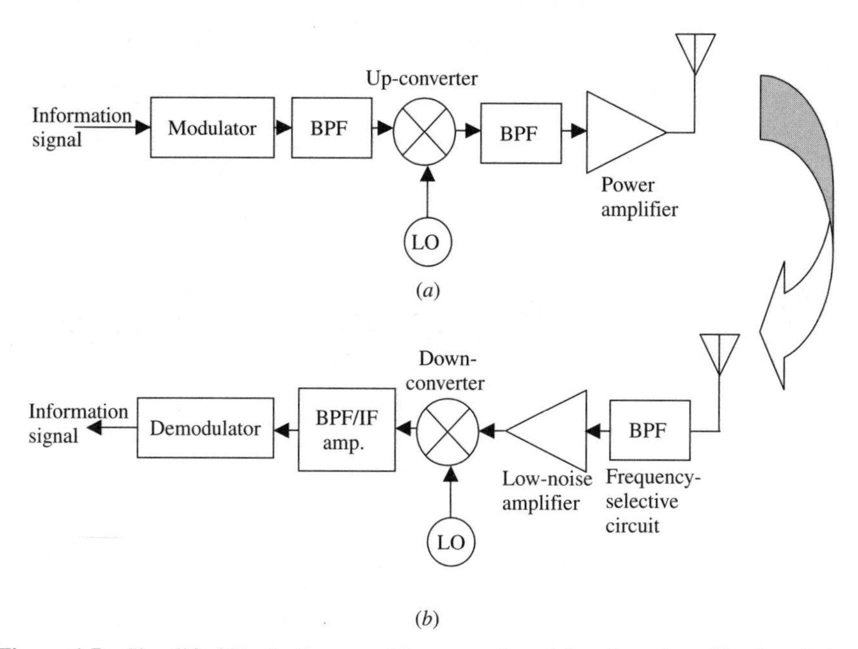

Figure 1.7 Simplified block diagram of the transmitter (*a*) and receiver (*b*) of a wireless communication system.

and after the mixer to stop undesired harmonics. The signal power is amplified before feeding it to the antenna. At the receiving end the entire process is reversed to recover the information. Signal received by the antenna is filtered and amplified to improve the signal-to-noise ratio before feeding it to the mixer for down-converting the frequency. A frequency-selective amplifier (tuned amplifier) amplifies it further, before feeding it to a suitable demodulator, which extracts the information signal.

In an analog communication system, the amplitude or angle (frequency or phase) of the carrier signal is varied according to the information signal. These modulations are known as amplitude modulation (AM), frequency modulation (FM), and phase modulation (PM), respectively. In a digital communication system, the input passes through channel coding, interleaving, and other processing before it is fed to the modulator. Various modulation schemes are available, including on–off keying (OOK), frequency-shift keying (FSK), and phase-shift keying (PSK). As these names suggest, a high-frequency signal is turned on and off in OOK to represent the logic states 1 and 0 of a digital signal. Similarly, two different signal frequencies are employed in FSK to represent the two signal states. In PSK, the phase of the high-frequency carrier is changed according to the 1 or 0 state of the signal. If only two phase states (0° and 180°) of the carrier are used, it is called binary phase-shift keying (BPSK). On the other hand, a phase shift of 90° gives four possible states, each representing 2 bits of information (known as a *dibit*). This type of digital modulation, known as quadrature phase-shift keying (QPSK), is explained further below.

Consider a sinusoidal signal S_{mod}, as given by equation (1.2.1). Its angular frequency and phase are ω radians per second and ϕ, respectively:

$$S_{\text{mod}}(t) = \sqrt{2}\cos(\omega t - \phi) = \sqrt{2}(\cos\omega t\cos\phi + \sin\omega t\sin\phi) \tag{1.2.1}$$

This can be simplified further as follows:

$$S_{\text{mod}}(t) = S_I\cos\omega t + S_Q\sin\omega t \tag{1.2.2}$$

where

$$S_I = \sqrt{2}\cos\phi \tag{1.2.3}$$

and

$$S_Q = \sqrt{2}\sin\phi \tag{1.2.4}$$

The subscripts I and Q represent in-phase and quadrature-phase components of S_{mod}. Table 1.5 shows the values of S_I and S_Q for four different phase states together with the corresponding dibit representations. This scheme can be implemented easily for a polar digital signal (positive peak value representing logic 1 and the negative peak logic 0). It is illustrated in Figure 1.8. The demultiplexer simultaneously feeds one bit of the digital input S_{BB} to the top and the other to the bottom branch of this circuit. The top branch multiplies the signal by $\cos\omega t$

TABLE 1.5 QPSK Scheme

ϕ	S_I	S_Q	Dibit Representation
$\pi/4$	1	1	11
$3\pi/4$	−1	1	01
$5\pi/4$	−1	−1	00
$7\pi/4$	1	−1	10

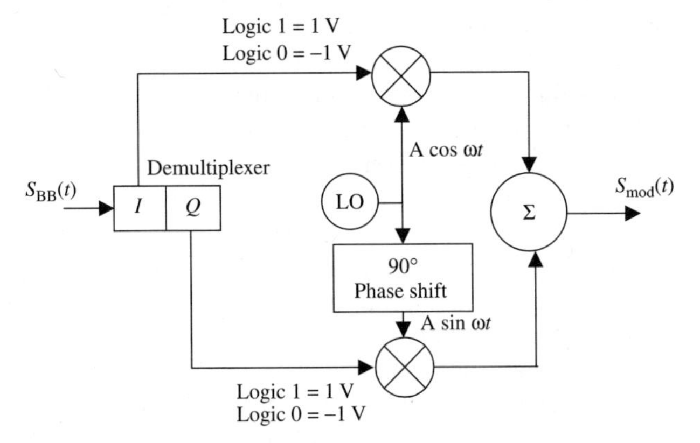

Figure 1.8 Block diagram of a QPSK modulation scheme.

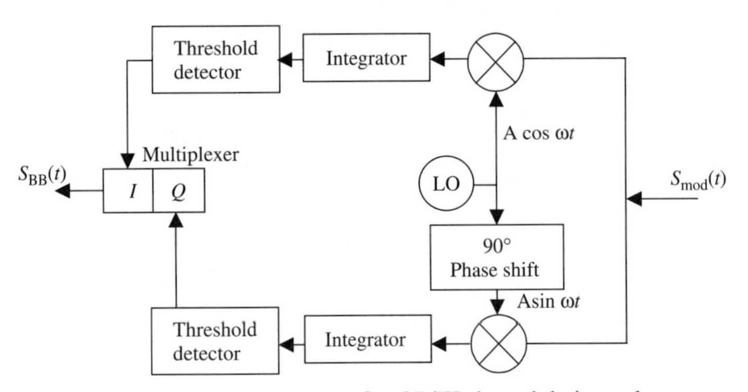

Figure 1.9 Block diagram of a QPSK demodulation scheme.

and the bottom signal by sin ωt. The outputs of the two mixers are then added to generate S_{mod}.

Demodulation inverts the modulation to retrieve S_{BB}. As illustrated in Figure 1.9, S_{mod} is multiplied by cos ωt in the top branch and by sin ωt in the bottom branch of the demodulator. The top integrator stops S_Q, and the bottom integrator, S_I. Two threshold detectors generate the corresponding logic states that are multiplexed by the multiplexer unit to recover S_{BB}. Chapter 2 provides an overview of wireless communication systems and their characteristics.

2

COMMUNICATION SYSTEMS

Modern communication systems require RF and microwave signals for the wireless transmission of information. These systems employ oscillators, mixers, filters, and amplifiers to generate and process various types of signals. The transmitter communicates with the receiver via antennas placed on each side. Electrical noise associated with the systems and the channel affects the performance. A system designer needs to know about the channel characteristics and system noise in order to estimate the required power levels. The chapter begins with an overview of microwave communication systems and RF wireless services to illustrate the applications of circuits and devices that are described in the following chapters. It also covers the placement of various building blocks in a given system.

A short discussion on antennas is included to help in understanding signal behavior when it propagates from a transmitter to a receiver. The Friis transmission formula and the radar range equation are important to understanding the effects of frequency, range, and operating power levels on the performance of a communication system. Note that radar concepts now find many other applications, such as proximity or level sensing in an industrial environment. Therefore, a brief discussion of Doppler radar is also included. Noise and distortion characteristics play a significant role in analysis and design of these systems. Minimum detectable signal (MDS), gain compression, intercept point, and the dynamic range of an amplifier (or receiver) are introduced later. Other concepts associated with noise and distortion characteristics are also introduced in this chapter.

Radio-Frequency and Microwave Communication Circuits: Analysis and Design, Second Edition,
By Devendra K. Misra
ISBN 0-471-47873-3

2.1 TERRESTRIAL COMMUNICATION

As mentioned in Chapter 1, microwave signals propagate along the line of sight. Therefore, the Earth's curvature limits the range over which a microwave communication link can be established. A transmitting antenna sitting on a 25-ft-high tower can typically communicate only up to a distance of about 50 km. Repeaters can be placed at regular intervals to extend the range. Figure 2.1 is a block diagram of a typical repeater.

A repeater system operates as follows. A microwave signal arriving at antenna A works as input to port 1 of a circulator. It is directed to port 2 without loss, assuming that the circulator is ideal. Then it passes through a receiver protection circuit that limits the magnitude of large signals but passes those of low intensity with negligible attenuation. The purpose of this circuit is to block excessively large signals from reaching the receiver input. The mixer following it works as a down-converter that transforms a high-frequency signal to a low-frequency signal, typically in the range of 70 MHz. A Schottky diode is generally employed in the mixer because of its superior noise characteristics. This frequency conversion facilitates amplification of the signal economically. A bandpass filter is used at the output of the mixer to stop undesired harmonics. An intermediate-frequency (IF) amplifier is then used to amplify the signal. It is generally a low-noise solid-state

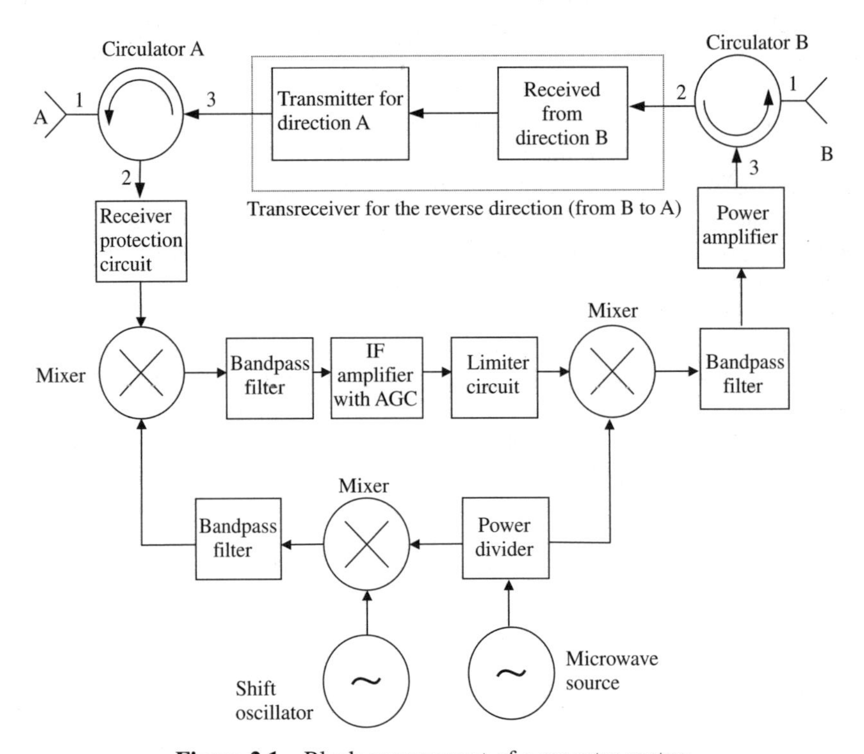

Figure 2.1 Block arrangement of a repeater system.

amplifier with ultralinear characteristics over a broadband. The amplified signal is mixed with another signal for up-conversion of frequency. After filtering out undesired harmonics introduced by the mixer, it is fed to a power amplifier stage that feeds circulator B for onward transmission through antenna B. This up-converting mixer circuit generally employs a varactor diode. Circulator B directs the signal entering at port 3 to the antenna connected at its port 1. Similarly, the signal propagating upstream is received by antenna B and the circulator directs it toward port 2. It then goes through the processing as described for the downstream signal and is radiated by antenna A for onward transmission. Hence, the downstream signal is received by antenna A and transmitted in the forward direction by antenna B. Similarly, the upstream signal is received by antenna B and forwarded to the next station by antenna A. The two circulators help channel the signal in the correct direction.

A parabolic antenna with tapered horn as primary feeder is generally used in microwave links. This type of composite antenna system, known as a *hog horn*, is fairly common in high-density links because of its broadband characteristics. These microwave links operate in the frequency range 4 to 6 GHz, and signals propagating in two directions are separated by a few hundred megahertz. Since this frequency range overlaps with C-band satellite communication, the interference of these signals needs to be taken into design consideration. A single frequency can be used twice for transmission of information using vertical and horizontal polarization.

2.2 SATELLITE COMMUNICATION

The ionosphere does not reflect microwaves as it does RF signals. However, one can place a conducting object (satellite) up in the sky that reflects them back to Earth. A satellite can even improve the signal quality using on-board electronics before transmitting it back. The gravitational force needs to be balanced somehow if this object is to stay in position. An orbital motion provides this balancing force. If a satellite is placed at low altitude, greater orbital force will be needed to keep it in position. These low- and medium-altitude satellites are visible from a ground station only for short periods. On the other hand, satellites placed at an altitude of about 36,000 km over the equator, called *geosynchronous* or *geostationary satellites* are visible from their shadows at all times.

C-band geosynchronous satellites use between 5725 and 7075 MHz for their uplinks. The corresponding downlinks are between 3400 and 5250 MHz. Table 2.1 lists the downlink center frequencies of a 24-channel transponder. Each channel has a total bandwidth of 40 MHz; 36 MHz of that carries the information, and the remaining 4 MHz is used as a guard band. It is accomplished with a 500-MHz bandwidth using different polarization for the overlapping frequencies. The uplink frequency plan may be found easily after adding 2225 MHz to these downlink frequencies. Figure 2.2 illustrates the simplified block diagram of a C-band satellite transponder. A 6-GHz signal received from the Earth station

TABLE 2.1 C-Band Downlink Transponder Frequencies

Horizontal Polarization		Vertical Polarization	
Channel	Center Frequency (MHz)	Channel	Center Frequency (MHz)
1	3720	2	3740
3	3760	4	3780
5	3800	6	3820
7	3840	8	3860
9	3880	10	3900
11	3920	12	3940
13	3960	14	3980
15	4000	16	4020
17	4040	18	4060
19	4080	20	4100
21	4120	22	4140
23	4160	24	4180

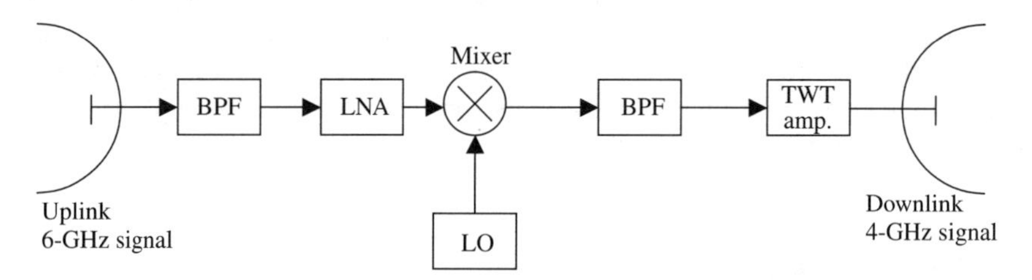

Figure 2.2 Simplified block diagram of a transponder.

is passed through a bandpass filter before amplifying it through a low-noise amplifier (LNA). It is then mixed with a local oscillator (LO) signal to bring down its frequency. A bandpass filter that is connected right after the mixer filters out the unwanted frequency components. This signal is then amplified by a traveling wave tube (TWT) amplifier and transmitted back to Earth.

Another frequency band in which satellite communication has been growing continuously is the Ku-band. The geosynchronous Fixed Satellite Service (FSS) generally operates between 10.7 and 12.75 GHz (space to Earth) and 13.75 to 14.5 GHz (Earth to space). It offers the following advantages over the C-band:

- The size of the antenna can be smaller (3 ft or even smaller, with higher-power satellites against 8 to 10 ft for C-band).
- Because of higher frequencies used in the up- and downlinks, there is no interference with C-band terrestrial systems.

Since higher-frequency signals attenuate faster while propagating through adverse weather (rain, fog, etc.), Ku-band satellites suffer from this major drawback. Signals with higher powers may be used to compensate for this loss. Generally, this power is on the order of 40 to 60 W. The high-power direct broadcast satellite (DBS) system uses power amplifiers in the range 100 to 120 W.

The National Broadcasting Company (NBC) has been using the Ku-band to distribute the programming to its affiliates. Also, various news-gathering agencies have used this frequency band for some time. Convenience stores, auto parts distributors, banks, and other businesses have used the very small aperture terminal (VSAT) because of its small antenna size (typically, on the order of 3 ft in diameter). It offers two-way satellite communication; usually back to hub or headquarters. The Public Broadcasting Service (PBS) uses VSATs for exchanging information among public schools.

Direct broadcast satellites (DBSs) have been around since 1980, but early DBS ventures failed for various reasons. In 1991, Hughes Communications entered into the direct-to-home (DTH) television business. DirecTV was formed as a unit of GM Hughes, with *DBS-1* launched in December 1993. Its longitudinal orbit is at 101.2°W, and it employs a left-handed circular polarization. *DBS-2*, launched in August 1994 uses a right-handed circular polarization, and its orbital longitude is at 100.8°W. DirecTV employs a digital architecture that can utilize video and audio compression techniques. It complies with Motion Picture Experts Group (MPEG)-2. By using compression ratios of 5 to 7, over 150 channels of programs are available from the two satellites. These satellites include 120-W TWT amplifiers that can be combined to form eight pairs at 240 W of power. This higher power can also be utilized for high-definition television (HDTV) transmission. Earth-to-satellite link frequency is 17.3 to 17.8 GHz; satellite-to-Earth link frequency uses the 12.2- to 12.7-GHz band. Circular polarization is used because it is less affected by rain than is linear orthogonal polarization.

Several communication services are now available that use low-Earth-orbit satellites (LEOSs) and medium-Earth-orbit satellites (MEOSs). LEOS altitudes range from 750 to 1500 km; MEOS systems have an altitude around 10,350 km. These services compete with or supplement cellular systems and geosynchronous Earth-orbit satellites (GEOSs). The GEOS systems have some drawbacks, due to the large distances involved. They require relatively large powers, and the propagation time delay creates problems in voice and data transmissions. The LEOS and MEOS systems orbit Earth faster because of being at lower altitudes, and these are therefore visible only for short periods. As Table 2.2 indicates, several satellites are used in a personal communication system to solve this problem.

Three classes of service can be identified for mobile satellite services:

1. Data transmission and messaging from very small, inexpensive satellites
2. Voice and data communications from big LEOSs
3. Wideband data transmission

Another application of L-band microwave frequencies (1227.60 and 1575.42 MHz) is in global positioning systems (GPSs). A constellation of 24 satellites is used to determine a user's geographical location. Two services

TABLE 2.2 Specifications of Certain Personal Communication Satellites

	Iridium (LEO)	Globalstar (LEO)	Odyssey (MEO)
Number of satellites	66	48	12
Altitude (km)	755	1,390	10,370
Uplink (GHz)	1.616–1.6265	1.610–1.6265	1.610–1.6265
Downlink (GHz)	1.616–1.6265	2.4835–2.500	2.4835–2.500
Gateway terminal uplink (GHz)	27.5–30.0	C-band	29.5–30.0
Gateway terminal downlink (GHz)	18.8–20.2	C-band	19.7–20.2
Average satellite connect time (min)	9	10–12	120
Features of handset			
Modulation	QPSK	FQPSK	QPSK
BER	1E-2 (voice) 1E-5 (data)	1E-3 (voice) 1E-5 (data)	1E-3 (voice) 1E-5 (data)
Supportable data rate (kb/s)	4.8 (voice) 2.4 (data)	1.2–9.6 (voice and data)	4.8 (voice) 1.2–9.6 (data)

are available: the standard positioning service (SPS) for civilian use, utilizing a single-frequency course/acquisition (C/A) code, and the precise positioning service (PPS) for the military, utilizing a dual-frequency P-code (protected). These satellites are at an altitude of 10,900 miles above the Earth, with an orbital period of 12 hours.

2.3 RADIO-FREQUENCY WIRELESS SERVICES

A lot of exciting wireless applications have been reported that use voice and data communication technologies. Wireless communication networks consist of microcells that connect people with truly global, pocket-size communication devices, telephones, pagers, personal digital assistants, and modems. Typically, a cellular system employs a 100-W transmitter to cover a cell 0.5 to 10 miles in radius. The handheld transmitter has a power of less than 3 W. Personal communication networks (PCN/PCS) operate with a 0.01- to 1-W transmitter to cover a cell radius of less than 450 yards. The handheld transmitter power is typically less than 10 mW. Table 2.3 shows the cellular telephone standards of selected systems.

There have been no universal standards set for wireless personal communication. In North America, cordless has been CT-0 (an analog 46/49-MHz standard) and cellular AMPS (Advanced Mobile Phone Service) operating at 800 MHz. The situation in Europe has been far more complex; every country has had its own standard. Although cordless was nominally CT-0, different countries used their own frequency plans. This led to a plethora of new standards. These include,

TABLE 2.3 Selected Cellular Telephones

	Analog Cellular Phone Standard		Digital Cellular Phone Standard			
	AMP	ETACS	NADC (IS-54)	NADC (IS-95)	GSM	PDC
Freq. range (MHz)						
Tx	824–849	871–904	824–849	824–849	880–915	940–956 1477–1501
Rx	869–894	916–949	869–894	869–894	925–960	810–826 1429–1453
Transmitter power (max.)			600 mW	200 mW	1 W	
Multiple access	FDMA	FDMA	TDMA/FDM	CDMA/FDM	TDMA/FDM	TDMA/FDM
Number of channels	832	1000	832	20	124	1600
Channel spacing (kHz)	30	25	30	1250	200	25
Modulation	FM	FM	π/4 DQPSK	QPSK/BPSK	GMSK	π/4 DQPSK
Bit rate (kb/s)	—	—	48.6	1228.8	270.833	42

TABLE 2.4 Selected Cordless Telephones

	Analog Cordless Phone Standard		Digital Cordless Phone Standard		
	CT-0	CT-1 and CT-1+	CT-2 and CT-2+	DECT	PHS (Formerly PHP)
Frequency range (MHz)	46/49	CT-1: 914/960; CT-1+: 885–932	CT-1: 864–868; CT-2+: 930/931; 940/941	1880–1900	1895–1918
Transmitter power (max.) (mW)	—	—	10 and 80	250	80
Multiple access	FDMA	FDMA	TDMA/FDM	TDMA/FDM	TDMA/FDM
Number of channels	10–20	CT-1: 40; CT-1+: 80	40	10 (12 users per channel)	300 (four users per channels)
Channel spacing (kHz)	40	25	100	1728	300
Modulation	FM	FM	GFSK	GFSK	π/4 DQPSK
Bit rate (kb/s)	—	—	72	1152	384

but are not limited to, CT-1, CT-1+, DECT (Digital European Cordless Telephone), PHP (Personal Handy Phone in Japan), E-TACS (Extended Total Access Communication System in the UK), NADC (North American Digital Cellular), GSM (Global System for Mobile Communication), and PDC (Personal Digital Cellular). Specifications for selected cordless telephones are given in Table 2.4.

2.4 ANTENNA SYSTEMS

Figure 2.3 illustrates some of the antennas that are used in communication systems. These can be categorized into two groups: wire antennas and aperture-type antennas. Electric dipole, monopole, and the loop antennas belong to the former group; horn, reflector, and lens belong to the latter category. The aperture antennas can be further subdivided into primary and secondary (or passive) antennas. Primary antennas are directly excited by the source and can be used independently for transmission or reception of signals. On the other hand, a secondary antenna requires another antenna as its feeder. Horn antennas fall in the first category, whereas the reflector and lens belong to the second. Various types of horn antennas are commonly used as feeders in reflector and lens antennas.

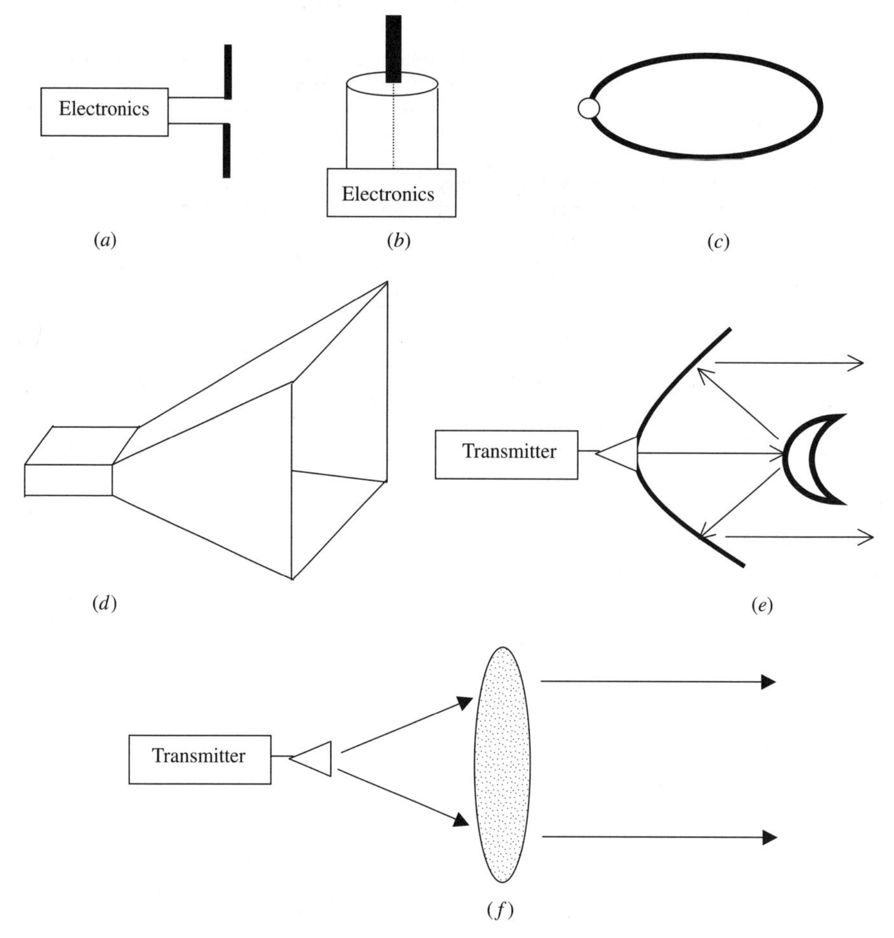

Figure 2.3 Some commonly used antennas: (*a*) electric dipole; (*b*) monopole; (*c*) loop; (*d*) pyramidal horn; (*e*) Cassegrain reflector; (*f*) lens.

When an antenna is energized, it generates two types of electromagnetic fields. Part of the energy stays nearby and part propagates outward. The propagating signal represents the radiation fields, while the nonpropagating is reactive (capacitive or inductive) in nature. Space surrounding the antenna can be divided into three regions. The reactive fields dominate in the nearby region but are reduced in strength at a faster rate than those associated with the propagating signal. If the largest dimension of an antenna is D and the signal wavelength is λ, reactive fields dominate up to about $0.62\sqrt{D^3/\lambda}$ and diminish after $2D^2/\lambda$. The region beyond $2D^2/\lambda$ is called the *far-field* (or *radiation field*) *region*.

Power radiated by an antenna per unit solid angle is known as the *radiation intensity* U. It is a far-field parameter that is related to power density (power per unit area) W_{rad} and distance r as follows:

$$U = r^2 W_{\text{rad}} \tag{2.4.1}$$

Directive Gain and Directivity

If an antenna radiates uniformly in all directions, it is called an *isotropic antenna*. This is a hypothetical antenna that helps in defining the characteristics of a real one. The directive gain D_G is defined as the ratio of radiation intensity due to the test antenna to that of an isotropic antenna. It is assumed that total radiated power remains the same in the two cases. Hence,

$$D_G = \frac{U}{U_0} = \frac{4\pi U}{P_{\text{rad}}} \tag{2.4.2}$$

where U is the radiation intensity due to the test antenna in watts per unit solid angle, U_0 the radiation intensity due to the isotropic antenna in watts per unit solid angle, and P_{rad} the total power radiated in watts. Since U is a directional-dependent quantity, the directive gain of an antenna depends on the angles θ and ϕ. If the radiation intensity assumes its maximum value, the directive gain is called the *directivity* D_0. That is,

$$D_0 = \frac{U_{\max}}{U_0} = \frac{4\pi U_{\max}}{P_{\text{rad}}} \tag{2.4.3}$$

Gain of an Antenna

The *power gain* of an antenna is defined as the ratio of its radiation intensity at a point to the radiation intensity that results from a uniform radiation of the same input power. Hence,

$$\text{gain} = 4\pi \frac{\text{radiation intensity}}{\text{total input power}} = 4\pi \frac{U(\theta, \phi)}{P_{\text{in}}} \tag{2.4.4}$$

Most of the time we deal with relative gain. It is defined as a ratio of the power gain of the test antenna in a given direction to the power gain of a reference antenna. Both antennas must have the same input power. The reference antenna is usually a dipole, horn, or any other antenna whose gain can be calculated or is known. However, the reference antenna is a lossless isotropic radiator in most cases. Hence,

$$\text{gain} = 4\pi \frac{U(\theta, \phi)}{P_{\text{in}}(\text{lossless isotropic antenna})} \tag{2.4.5}$$

When the direction is not stated, the power gain is usually taken in the direction of maximum radiation.

Radiation Patterns and Half-Power Beam Width

Far-field power distribution at a distance r from the antenna depends on the spatial coordinates θ and ϕ. Graphical representations of these distributions on the orthogonal plane (θ-plane or ϕ-plane) at a constant distance r from the antenna are called its *radiation patterns*. Figure 2.4 illustrates the radiation pattern of the vertical dipole antenna with θ. Its ϕ-plane pattern can be found after rotating it about the vertical axis. Thus, a three-dimensional picture of the radiation pattern of a dipole is doughnut shaped. Similarly, the power distributions of other antennas generally show peaks and valleys in the radiation zone. The highest peak between the two valleys is known as the *main lobe*; the others are called *side lobes*. The total angle about the main peak over which power is reduced by 50% of its maximum value is called the *half-power beam width* on that plane.

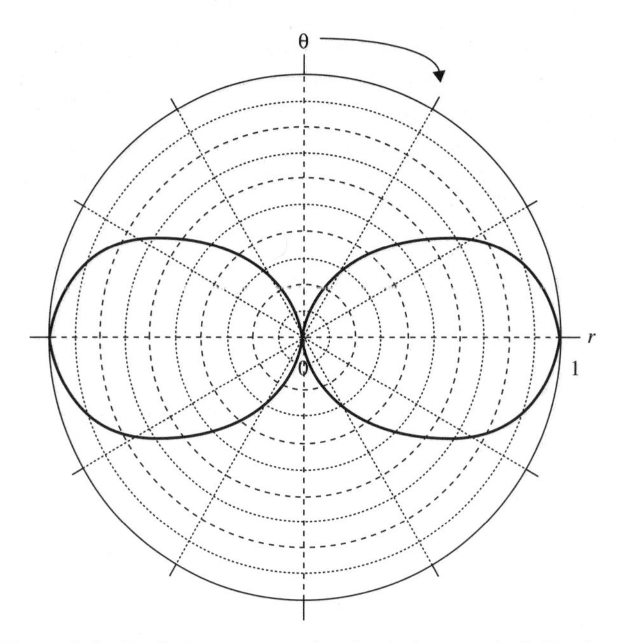

Figure 2.4 Radiation pattern of a dipole in a vertical (θ) plane.

The following relations are used to estimate the power gain G and the half-power beam width (HPBW, or simply BW) of an aperture antenna:

$$G = \frac{4\pi}{\lambda^2} A_e = \frac{4\pi}{\lambda^2} A\kappa \tag{2.4.6}$$

and

$$\text{BW(degrees)} = \frac{65\lambda}{d} \tag{2.4.7}$$

where A_e is the effective area of radiating aperture in square meters, A its physical area ($\pi d^2/4$ for a reflector antenna dish with diameter d), κ the antenna efficiency (ranges from 0.6 to 0.65), and λ the signal wavelength in meters.

Example 2.1 Calculate the power gain (in decibels) and the half-power beam width of a parabolic dish antenna 30 m in diameter that is radiating at 4 GHz.

SOLUTION The signal wavelength and area of the aperture are

$$\lambda = \frac{3 \times 10^8}{4 \times 10^9} = 0.075\,\text{m}$$

and

$$A = \frac{\pi d^2}{4} = \pi \frac{30^2}{4} = 706.8584\,\text{m}^2$$

Assuming that the aperture efficiency is 0.6, the antenna gain and half-power beam width are found as follows:

$$G = \frac{4\pi}{(0.075)^2} \times 706.8584 \times 0.6 = 947{,}482.09$$

$$= 10\log_{10}(947{,}482.09) = 59.76 \approx 60\,\text{dB}$$

$$\text{BW} = \frac{65 \times 0.075}{30} = 0.1625^\circ$$

Antenna Efficiency

If an antenna is not matched with its feeder, a part of the signal available from the source is reflected back. It is considered to be the reflection (or mismatch) loss. The *reflection* (or *mismatch*) *efficiency* is defined as a ratio of power input to the antenna to that of power available from the source. Since the ratio of reflected power to that of power available from the source is equal to the square of the magnitude of the voltage reflection coefficient, the reflection efficiency e_r is given by

$$e_r = 1 - |\Gamma|^2$$

where

$$\Gamma = \text{voltage reflection coefficient} = \frac{Z_A - Z_0}{Z_A + Z_0}$$

Here Z_A is the antenna impedance and Z_0 is the characteristic impedance of the feeding line.

In addition to mismatch, the signal energy may dissipate in an antenna due to imperfect conductor or dielectric material. These efficiencies are hard to compute. However, the combined conductor and dielectric efficiency e_{cd} can be determined experimentally after measuring the input power P_{in} and the radiated power P_{rad}. It is given as

$$e_{cd} = \frac{P_{\text{rad}}}{P_{\text{in}}}$$

The overall efficiency e_o is a product of the efficiencies above. That is,

$$e_o = e_r e_{cd} \tag{2.4.8}$$

Example 2.2 A 50-Ω transmission line feeds a lossless one-half-wavelength-long dipole antenna. The antenna impedance is 73 Ω. If its radiation intensity, $U(\theta, \phi)$, is given as

$$U = B_0 \sin^3 \theta$$

find the maximum overall gain.

SOLUTION The maximum radiation intensity, U_{max}, is the B_0 value that occurs at $\theta = \pi/2$. Its total radiated power is found as follows:

$$P_{\text{rad}} = \int_0^{2\pi} \int_0^{\pi} B_0 \sin^3 \theta \,\sin \theta \, d\theta \, d\phi = \tfrac{3}{4}\pi^2 B_0$$

Hence,

$$D_0 = 4\pi \frac{U_{\text{max}}}{P_{\text{rad}}} = \frac{4\pi B_0}{\frac{3}{4}\pi^2 B_0} = \frac{16}{3\pi} = 1.6977$$

or

$$D_0(\text{dB}) = 10 \log_{10}(1.6977)\text{dB} = 2.2985 \text{ dB}$$

Since the antenna is lossless, the radiation efficiency e_{cd} is unity (0 dB). Its mismatch efficiency is computed as follows.

The voltage reflection coefficient at its input (it is formulated in Chapter 2) is

$$\Gamma = \frac{Z_A - Z_0}{Z_A + Z_0} = \frac{73 - 50}{73 + 50} = \frac{23}{123}$$

Therefore, the mismatch efficiency of the antenna is

$$e_r = 1 - (23/123)^2 = 0.9650 = 10\log_{10}(0.9650)\text{dB} = -0.1546\,\text{dB}$$

The overall gain G_0 (in decibels) is found as follows:

$$G_0(\text{dB}) = 2.2985 - 0 - 0.1546 = 2.1439\,\text{dB}$$

Bandwidth

Antenna characteristics such as gain, radiation pattern, impedance, and so on are frequency dependent. The bandwidth of an antenna is defined as the frequency band over which its performance with respect to some characteristic (HPBW, directivity, etc.) conforms to a specified standard.

Polarization

Polarization of an antenna is the same as polarization of its radiating wave. It is a property of the electromagnetic wave describing the time-varying direction and relative magnitude of an electric field vector. The curve traced by the instantaneous electric field vector with time is the polarization of that wave. The polarization is classified as follows:

- *Linear polarization.* If the tip of the electric field intensity traces a straight line in some direction with time, the wave is linearly polarized.
- *Circular polarization.* If the end of the electric field traces a circle in space as time passes, that electromagnetic wave is circularly polarized. Further, it may be right-handed circularly polarized (RHCP) or left-handed circularly polarized (LHCP), depending on whether the electric field vector rotates clockwise or counterclockwise.
- *Elliptical polarization.* If the tip of the electric field intensity traces an ellipse in space as time lapses, the wave is elliptically polarized. As in the preceding case, it may be right- or left-handed elliptical polarization (RHEP and LHEP).

In a receiving system, the polarization of the antenna and the incoming wave need to be matched for maximum response. If this is not the case, there will be some signal loss, known as *polarization loss*. For example, if there is a vertically polarized wave incident on a horizontally polarized antenna, the induced voltage available across its terminals will be zero. In this case, the antenna is *cross-polarized with an incident wave*. The square of the cosine of the angle between wave polarization and antenna polarization is a measure of the polarization loss. It can be determined by squaring the scalar product of unit vectors representing the two polarizations.

Example 2.3 The electric field intensity of an electromagnetic wave propagating in a lossless medium in the z-direction is given by

$$\mathbf{E}(\mathbf{r}, t) = \hat{x} E_0(x, y) \cos(\omega t - kz) \qquad \text{V/m}$$

It is incident upon an antenna that is linearly polarized as follows:

$$\mathbf{E}_a(\mathbf{r}) = (\hat{x} + \hat{y}) E(x, y, z) \qquad \text{V/m}$$

Find the polarization loss factor.

SOLUTION In this case, the incident wave is linearly polarized along the x-axis while the receiving antenna is linearly polarized at 45° from it. Therefore, one-half of the incident signal is cross-polarized with the antenna. It is determined mathematically as follows. The unit vector along the polarization of incident wave is

$$\hat{u}_i = \hat{x}$$

The unit vector along the antenna polarization may be found as

$$\hat{u}_a = \frac{1}{\sqrt{2}}(\hat{x} + \hat{y})$$

Hence, the polarization loss factor is

$$|\hat{u}_i \bullet \hat{u}_a|^2 = 0.5 = -3.01 \text{ dB}$$

Effective Isotropic Radiated Power

The effective isotropic radiated power (EIRP) is a measure of the power gain of the antenna. It is equal to the power needed by an isotropic antenna that provides the same radiation intensity at a given point as that of the directional antenna. If power input to the feeding line is P_t and the antenna gain is G_t, the EIRP is defined as

$$\text{EIRP} = \frac{P_t G_t}{L} \tag{2.4.9}$$

where L is the input/output power ratio of the transmission line that is connected between the output of the final power amplifier stage of the transmitter and the antenna. It is given by

$$L = \frac{P_t}{P_{\text{ant}}} \tag{2.4.10}$$

Alternatively, the EIRP can be expressed in dBw as

$$\text{EIRP(dBw)} = P_t(\text{dBw}) - L(\text{dB}) + G(\text{dB}) \tag{2.4.11}$$

Example 2.4 In a transmitting system, the output of the final high-power amplifier is 500 W, and the line feeding its antenna has an attenuation of 20%. If the gain of the transmitting antenna is 60 dB, find the EIRP in dBw.

SOLUTION

$$P_t = 500\,\text{W} = 26.9897\,\text{dBw}$$

$$P_{\text{ant}} = 0.8 \times 500 = 400\,\text{W}$$

$$G = 60\,\text{dB} = 10^6$$

and

$$L = \frac{500}{400} = 1.25 = 10\log_{10}(1.25) = 0.9691\,\text{dB}$$

Hence,

$$\text{EIRP(dBw)} = 26.9897 - 0.9691 + 60 = 86.0206\,\text{dBw}$$

or

$$\text{EIRP} = \frac{500 \times 10^6}{1.25} = 400 \times 10^6\,\text{W}$$

Space Loss

The transmitting antenna radiates in all directions, depending on its radiation characteristics. However, the receiving antenna receives only the power that is incident on it. Hence, the rest of the power is not used and is lost in space. It is represented by the space loss. It can be determined as follows.

The power density w_t of a signal transmitted by an isotropic antenna is given by

$$w_t = \frac{P_t}{4\pi R^2} \qquad \text{W/m}^2 \tag{2.4.12}$$

where P_t is the transmitted power in watts and R is the distance from the antenna in meters. The power received by a unity-gain antenna located at R is found to be

$$P_r = w_t A_{\text{eu}} \tag{2.4.13}$$

where A_{eu} is the effective area of an isotropic antenna.

From (2.4.6), for an isotropic antenna,

$$G = \frac{4\pi}{\lambda^2} A_{\text{eu}} = 1$$

or

$$A_{\text{eu}} = \frac{\lambda^2}{4\pi}$$

Hence, (2.4.12) can be written as

$$P_r = \frac{P_t}{4\pi R^2}\frac{\lambda^2}{4\pi} \tag{2.4.14}$$

and the space loss ratio is found to be

$$\frac{P_r}{P_t} = \left(\frac{\lambda}{4\pi R}\right)^2 \tag{2.4.15}$$

It is usually expressed in decibels as follows:

$$\text{space loss ratio} = 20\log_{10}\frac{\lambda}{4\pi R} \quad \text{dB} \tag{2.4.16}$$

Example 2.5 A geostationary satellite is 35,860 km away from Earth's surface. Find the space loss ratio if it is operating at 4 GHz.

SOLUTION

$$R = 35{,}860{,}000\,\text{m} \quad \text{and} \quad \lambda = \frac{3\times 10^8}{4\times 10^9} = 0.075\,\text{m}$$

Hence,

$$\text{space loss ratio} = \left(\frac{4\pi \times 35{,}860{,}000}{0.075}\right)^2 = 2.77\times 10^{-20} = -195.5752\,\text{dB}$$

Friis Transmission Formula and Radar Range Equation

Analysis and design of communication and monitoring systems often require an estimation of transmitted and received powers. The Friis transmission formula and radar range equation provide the means for such calculations. The former is applicable to a one-way communication system where the signal is transmitted at one end and is received at the other end of the link. The radar range equation is applicable when the signal transmitted hits a target and the signal reflected is generally received at the location of the transmitter. We consider these two formulations next.

Friis Transmission Equation

Consider a simplified communication link as illustrated in Figure 2.5. A distance R separates the transmitter and the receiver. Effective apertures of transmitting and receiving antennas are A_{et} and A_{er}, respectively. Further, the two antennas

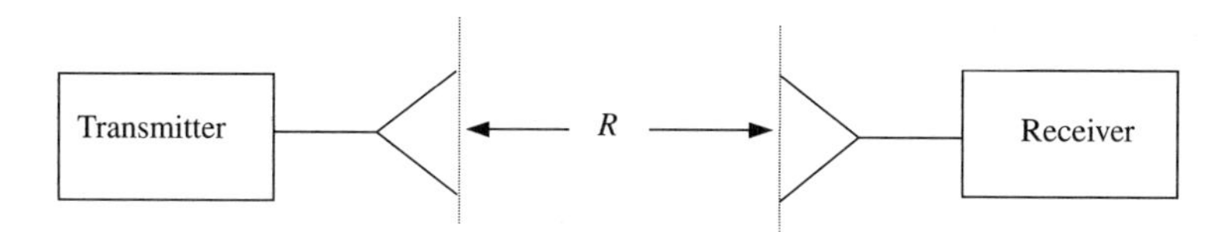

Figure 2.5 Simplified block diagram of a communication link.

are assumed to be polarization matched. If power input to the transmitting antenna is P_t, isotropic power density w_0 at a distance R from the antenna is given as

$$w_0 = \frac{P_t e_t}{4\pi R^2} \tag{2.4.17}$$

where e_t is the radiation efficiency of the transmitting antenna. For a directional transmitting antenna, the power density w_t can be written as

$$w_t = \frac{P_t G_t}{4\pi R^2} = \frac{P_t e_t D_t}{4\pi R^2} \tag{2.4.18}$$

where G_t is the gain and D_t is the directivity of transmitting antenna. The power collected by the receiving antenna is

$$P_r = A_{er} w_t \tag{2.4.19}$$

From (2.4.6),

$$A_{er} = \frac{\lambda^2}{4\pi} G_r \tag{2.4.20}$$

where the receiving antenna gain is G_r. Therefore, we find that

$$P_r = \frac{\lambda^2}{4\pi} G_r w_t = \frac{\lambda^2}{4\pi} G_r \frac{P_t G_t}{4\pi R^2}$$

or

$$\frac{P_r}{P_t} = \left(\frac{\lambda}{4\pi R}\right)^2 G_r G_t = e_r e_t \left(\frac{\lambda}{4\pi R}\right)^2 D_r D_t \tag{2.4.21}$$

If signal frequency is f, for a free-space link,

$$\frac{\lambda}{4\pi R} = \frac{3 \times 10^8}{4\pi f R} \tag{2.4.22}$$

where f is in hertz and R is in meters.

Generally, the link distance is long and the signal frequency is high, such that kilometer and megahertz will be more convenient units than the usual meter and hertz, respectively. For R in kilometers and f in megahertz, we find that

$$\frac{\lambda}{4\pi R} = \frac{3 \times 10^8}{4\pi \times 10^6 f_{\text{MHz}} \times 10^3 R_{\text{km}}} = \frac{0.3}{4\pi} \frac{1}{f_{\text{MHz}} R_{\text{km}}} \tag{2.4.23}$$

Hence, from (2.4.21),

$$P_r(\text{dBm}) = P_t(\text{dBm}) + 20 \log_{10} \frac{0.3}{4\pi} - 20 \log_{10}(f_{\text{MHz}} R_{\text{km}}) + G_t(\text{dB}) + G_r(\text{dB})$$

or

$$P_r(\text{dBm}) = P_t(\text{dBm}) + G_t(\text{dB}) + G_r(\text{dB}) - 20 \log_{10}(f_{\text{MHz}} R_{\text{km}}) - 32.4418 \tag{2.4.24}$$

where the power transmitted and the power received are in dBm while the two antenna gains are in decibels.

Example 2.6 A 20-GHz transmitter on board the satellite uses a parabolic antenna that is 45.7 cm in diameter. The antenna gain is 37 dB and its radiated power is 2 W. The ground station that is 36,941.031 km away from it has an antenna gain of 45.8 dB. Find the power collected by the ground station. How much power would be collected at the ground station if there were isotropic antennas on both sides?

SOLUTION The power transmitted, $P_t(\text{dBm}) = 10 \log_{10}(2000) = 33.0103$ dBm and

$$20 \log_{10}(f_{\text{MHz}} R_{\text{km}}) = 20 \log_{10}(20 \times 10^3 \times 36{,}941.031) = 177.3708 \text{ dB}$$

Hence, the power received at the Earth station is found as follows:

$$P_r(\text{dBm}) = 33.0103 + 37 + 45.8 - 177.3708 - 32.4418 = -94.0023 \text{ dBm}$$

or

$$P_r = 3.979 \times 10^{-10} \text{ mW}$$

If the two antennas are isotropic, $G_t = G_r = 1$ (or 0 dB) and therefore

$$\text{P}_r(\text{dBm}) = 33.0103 + 0 + 0 - 177.3708 - 32.4418 = -176.8023 \text{ dBm}$$

or

$$P_r = 2.0882 \times 10^{-18} \text{ mW}$$

Radar Equation

In the case of a radar system, the signal transmitted is scattered by the target in all possible directions. The receiving antenna collects part of the energy that is scattered back toward it. Generally, a single antenna is employed for both the transmitter and the receiver, as shown in Fig. 2.6. If power input to the transmitting antenna is P_t and its gain is G_t, the power density w_{inc} incident on the target is

$$w_{\text{inc}} = \frac{P_t G_t}{4\pi R^2} = \frac{P_t A_{et}}{\lambda^2 R^2} \tag{2.4.25}$$

where A_{et} is the effective aperture of the transmitting antenna.

The radar cross section σ of an object is defined as the area intercepting that amount of power that when scattered isotropically produces at the receiver a power density that is equal to that scattered by the actual target. Hence,

$$\text{radar cross section} = \frac{\text{scattered power}}{\text{incident power density}} \qquad \text{square meters}$$

or

$$\sigma = \frac{4\pi r^2 w_r}{w_{\text{inc}}} \tag{2.4.26}$$

where w_r is isotropically backscattered power density at a distance r and w_{inc} is power density incident on the object. Hence, the radar cross section of an object is its effective area that intercepts an incident power density w_{inc} and gives an isotropically scattered power of $4\pi r^2 w_r$ for a backscattered power density. Radar cross sections of selected objects are listed in Table 2.5.

Using the radar cross section of a target, the power intercepted by it can be found as follows:

$$P_{\text{inc}} = \sigma w_{\text{inc}} = \frac{\sigma P_t G_t}{4\pi R^2} \tag{2.4.27}$$

Power density arriving back at the receiver is

$$w_{\text{scatter}} = \frac{P_{\text{inc}}}{4\pi R^2} \tag{2.4.28}$$

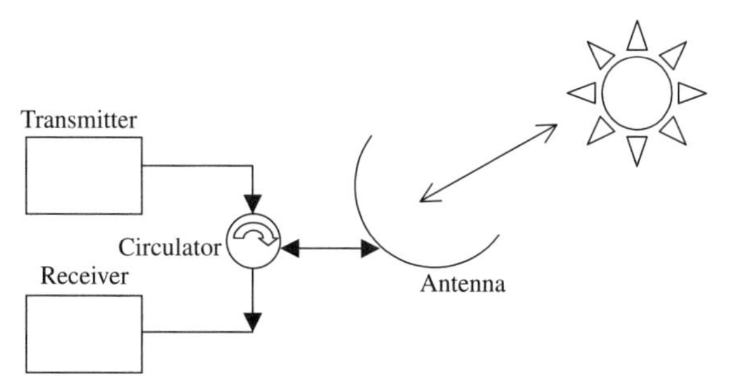

Figure 2.6 Radar system.

TABLE 2.5 Radar Cross Sections of Selected Objects

Object	Radar Cross Section (m^2)
Pickup truck	200
Automobile	100
Jumbo-jet airliner	100
Large bomber	40
Large fighter aircraft	6
Small fighter aircraft	2
Adult male	1
Conventional winged missile	0.5
Bird	0.01
Insect	0.00001
Advanced tactical fighter	0.000001

and power available at the receiver input is

$$P_r = A_{er} w_{\text{scatter}} = \frac{G_r \lambda^2 \sigma P_t G_t}{4\pi(4\pi R^2)^2} = \frac{\sigma A_{er} A_{et} P_t}{4\pi \lambda^2 R^4} \tag{2.4.29}$$

Example 2.7 A distance of 100λ separates two lossless X-band horn antennas (Figure 2.7). Reflection coefficients at the terminals of transmitting and receiving antennas are 0.1 and 0.2, respectively. Maximum directivities of the transmitting and receiving antennas are 16 and 20 dB, respectively. Assuming that the input power in a lossless transmission line connected to the transmitting antenna is 2 W and that the two antennas are aligned for maximum radiation between them and are polarization matched, find the power input to the receiver.

SOLUTION As discussed in Chapter 3, impedance discontinuity generates an echo signal very similar to that of an acoustical echo. Hence, signal power available beyond the discontinuity is reduced. The ratio of the reflected signal voltage to that of the incident is called the *reflection coefficient*. Since the power is proportional to the square of the voltage, the power reflected from the discontinuity is equal to the square of the reflection coefficient times the incident power.

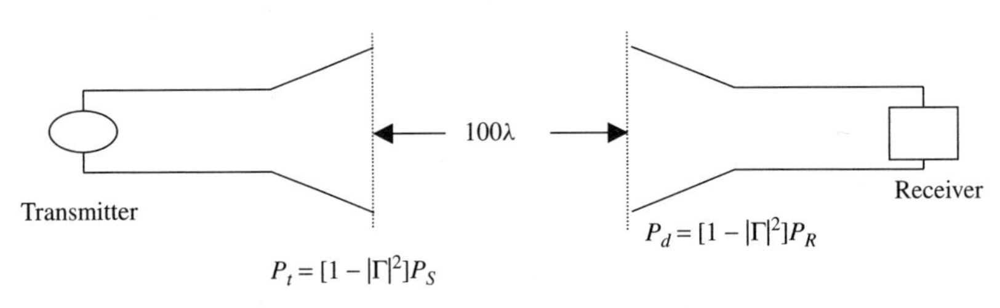

Figure 2.7 Setup for Example 2.7.

Therefore, power transmitted in the forward direction will be given by

$$P_t = (1 - |\Gamma|^2) P_{\text{in}}$$

and the power radiated by the transmitting antenna is found to be

$$P_t = (1 - 0.1^2)2 = 1.98\,\text{W}$$

Since the Friis transmission equation requires the antenna gain as a ratio instead of in decibels, G_t and G_r are calculated as follows:

$$G_t = 16\,\text{dB} = 10^{1.6} = 39.8107 \quad \text{and} \quad G_r = 20\,\text{dB} = 10^{2.0} = 100$$

Hence, from (2.4.21),

$$P_r = \left(\frac{\lambda}{4\pi \times 100\lambda}\right)^2 \times 100 \times 39.8107 \times 1.98$$

or

$$P_r = 5\,\text{mW}$$

and power delivered to the receiver, P_d, is

$$P_d = (1 - 0.2^2)5 = 4.8\,\text{mW}$$

Example 2.8 Radar operating at 12 GHz transmits 25 kW through an antenna of 25 dB gain. A target with its radar cross section at 8 m^2 is located at 10 km from the radar. If the same antenna is used for the receiver, determine the power received.

SOLUTION

$$P_t = 25\,\text{kW}$$

$$f = 12\,\text{GHz} \rightarrow \lambda = \frac{3 \times 10^8}{12 \times 10^9} = 0.025\,\text{m}$$

$$G_r = G_t = 25\,\text{dB} \rightarrow 10^{2.5} = 316.2278$$

$$R = 10\,\text{km} \qquad \sigma = 8\,\text{m}^2$$

Hence,

$$P_r = \frac{G_r G_t P_t \sigma \lambda^2}{4\pi(4\pi R^2)^2} = \frac{316.2278^2 \times 25,000 \times 8 \times 0.025^2}{(4\pi)^3(10^4)^4} = 6.3 \times 10^{-13}\,\text{W}$$

or

$$P_r = 0.63\,\text{pW}$$

Doppler Radar

An electrical signal propagating in free space can be represented by a simple expression as follows:

$$v(z,t) = A\cos(\omega t - kz) \tag{2.4.30}$$

The signal frequency is ω radians per second and k is its wave number (equal to ω/c, where c is the speed of light in free space) in radians per meter. Assume that there is a receiver located at $z = R$, as shown in Figure 2.5 and R is changing with time (the receiver may be moving toward or away from the transmitter). In this situation, the receiver response $v_o(t)$ is given as follows:

$$v_o(t) = V\cos(\omega t - kR) \tag{2.4.31}$$

The angular frequency, ω_0, of v_o(t) can be determined easily after differentiating the argument of the cosine function with respect to time. Hence,

$$\omega_0 = \frac{d}{dt}(\omega t - kR) = \omega - k\frac{dR}{dt} \tag{2.4.32}$$

Note that k is time independent and that the time derivative of R represents the velocity, v_r, of the receiver with respect to the transmitter. Hence, (2.4.32) can be written

$$\omega_0 = \omega - \frac{\omega v_r}{c} = \omega\left(1 - \frac{v_r}{c}\right) \tag{2.4.33}$$

If the receiver is closing in, v_r will be negative (negative slope of R), and therefore the signal received will indicate a signal frequency higher than ω. On the other hand, it will show a lower frequency if R is increasing with time. It is the Doppler frequency shift that is employed to design the Doppler radar.

Consider the simplified block diagram shown in Figure 2.8. A microwave signal generated by the oscillator is split into two parts via a power divider. The circulator feeds one part of this power to an antenna that illuminates a target while a mixer uses the remaining fraction as its reference signal. Further, the antenna intercepts a part of the signal that is scattered by an object. It is then directed to the mixer through a circulator. The mixer output includes a difference frequency signal that can be filtered out for further processing. Two inputs to the mixer will have the same frequency if the target is stationary, and therefore the Doppler shift $\delta\omega$ will be zero. On the other hand, the mixer output will have Doppler frequency if the target is moving. Note that the signal travels twice over the same distance, and therefore the Doppler frequency shift in this case will be twice that found via (2.4.33). Mathematically,

$$\omega_0 = \omega\left(1 - \frac{2v_r}{c}\right) \tag{2.4.34}$$

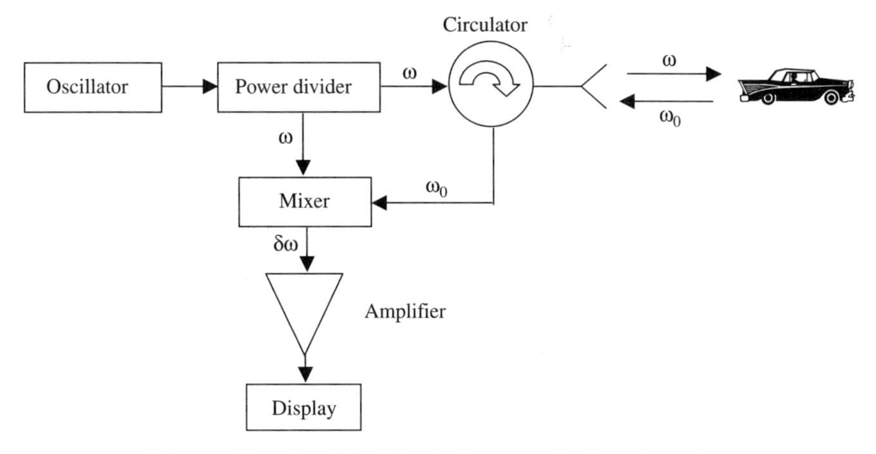

Figure 2.8 Simplified block diagram of a Doppler radar.

and

$$\delta\omega = \frac{2\omega v_r}{c} \tag{2.4.35}$$

2.5 NOISE AND DISTORTION

Random movement of charges or charge carriers in an electronic device generates currents and voltages that vary randomly with time. In other words, the amplitude of these electrical signals cannot be predicted at any time. However, it can be expressed in terms of probability density functions. These signals are termed *noise*. For most applications it suffices to know the mean-square or root-mean-square value. Since the square of the current or the voltage is proportional to the power, mean-square noise voltage and current values are generally called *noise power*. Further, noise power is normally a function of frequency and the power per unit frequency (watts per hertz) is defined as the power spectral density of noise. If the noise power is the same over the entire frequency band of interest, it is called *white noise*. There are several mechanisms that can cause noise in an electronic device. Some of these are as follows:

- *Thermal noise.* This is the most basic type of noise, which is caused by thermal vibration of bound charges. Johnson studied this phenomenon in 1928, and Nyquist formulated an expression for spectral density around the same time. Therefore, it is also known as *Johnson noise* or *Nyquist noise.* In most electronic circuits, thermal noise dominates; therefore, it will be described further because of its importance.
- *Shot noise.* This is due to random fluctuations of charge carriers that pass through the potential barrier in an electronic device. For example, electrons

emitted from the cathode of thermionic devices or charge carriers in Schottky diodes produce a current that fluctuates about the average value I. The mean-square current due to shot noise is generally given by

$$\langle i_{\mathrm{Sh}}^2 \rangle = 2eIB \tag{2.5.1}$$

where e is electronic charge (1.602×10^{-19} C) and B is the bandwidth in hertz.

- *Flicker noise.* This occurs in solid-state devices and vacuum tubes operating at low frequencies. Its magnitude decreases with an increase in frequency. It is generally attributed to chaos in the dynamics of a system. Since the flicker noise power varies inversely with frequency, it is often called *1/f noise*. Sometimes it is referred to as *pink noise*.

Thermal Noise

Consider a resistor R that is at a temperature of T Kelvin. Electrons in this resistor are in random motion with a kinetic energy that is proportional to the temperature T. These random motions produce small, random voltage fluctuations across its terminals. This voltage has a zero average value but a nonzero mean-square value $\langle v_n^2 \rangle$. It is given by Planck's blackbody radiation law as

$$\langle v_n^2 \rangle = \frac{4hfRB}{\exp(hf/kT) - 1} \tag{2.5.2}$$

where h is Planck's constant (6.546×10^{-34} J · s), k the Boltzmann constant (1.38×10^{-23} J/K), T the temperature in kelvin, B the system bandwidth in hertz, and f the center frequency of the bandwidth in hertz.

For frequencies below 100 GHz, the product hf will be smaller than 6.546×10^{-23} J and kT will be greater than 1.38×10^{-22} J if T stays above 10 K. Therefore, kT will be larger than hf for such cases. Hence, the exponential term in equation (2.5.2) can be approximated as follows:

$$\exp\left(\frac{hf}{kT}\right) \approx 1 + \frac{hf}{kT}$$

Therefore,

$$\langle v_n^2 \rangle \approx \frac{4hfRB}{hf/kT} = 4BRkT \tag{2.5.3}$$

This is known as the *Rayleigh–Jeans approximation.*

A Thévenin-equivalent circuit can replace the noisy resistor as shown in Figure 2.9. As illustrated, it consists of a noise-equivalent voltage source in series with a noise-free resistor. This source will supply a maximum power to a load

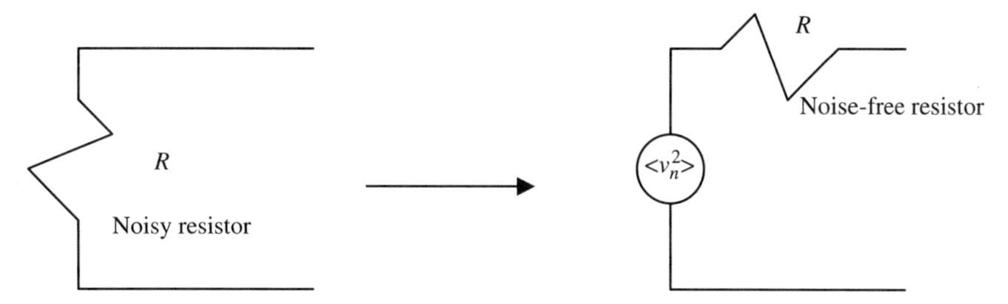

Figure 2.9 Noise-equivalent circuit of a resistor.

of resistance R. The power delivered to that load in a bandwidth B is found as follows:

$$P_n = \frac{\langle v_n^2 \rangle}{4R} = kTB \tag{2.5.4}$$

Conversely, if an arbitrary white noise source with its driving point impedance R delivers a noise power P_s to a load R, it can be represented by a noisy resistor of value R that is at temperature T_e. Hence,

$$T_e = \frac{P_s}{kB} \tag{2.5.5}$$

where T_e is an equivalent temperature selected so that the same noise power is delivered to the load.

Consider the noisy amplifier as shown in Figure 2.10. Its gain is G over the bandwidth B. Let the amplifier be matched to the noiseless source and the load resistors. If the source resistor is at a hypothetical temperature of $T_s = 0\,\text{K}$, the power input to the amplifier P_i will be zero and the output noise power P_o will only be due to noise generated by the amplifier. We can obtain the same noise power at the output of an ideal noiseless amplifier by raising the temperature T_s of the source resistor to T_e:

$$T_e = \frac{P_o}{GkB} \tag{2.5.6}$$

Hence, the output power in both cases is $P_o = GkT_eB$. The temperature T_e is known as the equivalent noise temperature of the amplifier.

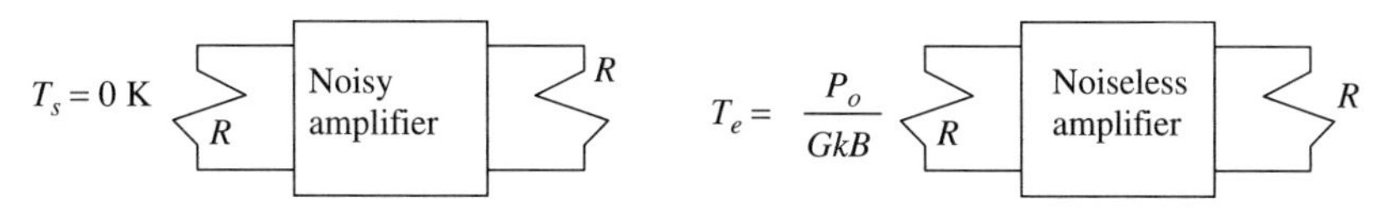

Figure 2.10 Noise-equivalent representation of an amplifier.

Measurement of Noise Temperature by the Y-Factor Method

According to the definition, the noise temperature of an amplifier (or any other two-port network) can be determined by setting the source resistance R at 0 K and then measuring the output noise power. However, a temperature of 0 K cannot be achieved in practice. We can circumvent this problem by repeating the experiment at two different temperatures. This procedure is known as the *Y-factor method.*

Consider an amplifier with a power gain of G over a frequency band of B hertz. Further, its equivalent noise temperature is T_e(K). The input port of the amplifier is terminated by a matched resistor R while a matched power meter is connected at its output, as illustrated in Figure 2.11. With R at temperature T_h, the power meter measures the noise output as P_1. Similarly, the noise power is found to be P_2 when the temperature of R is set at T_c. Hence,

$$P_1 = GkT_hB + GkT_eB$$

and

$$P_2 = GkT_cB + GkT_eB$$

For T_h higher than T_c, the noise power P_1 will be larger than P_2. Therefore,

$$\frac{P_1}{P_2} = Y = \frac{T_h + T_e}{T_c + T_e}$$

or

$$T_e = \frac{T_h - YT_c}{Y - 1} \tag{2.5.7}$$

For T_h larger than T_c, Y will be greater than unity. Further, measurement accuracy is improved by selecting two temperature settings that are far apart. Therefore, T_h and T_c represent hot and cold temperatures, respectively.

Example 2.9 An amplifier has a power gain of 10 dB in the 500-MHz to 1.5-GHz frequency band. The following data are obtained for this amplifier using the

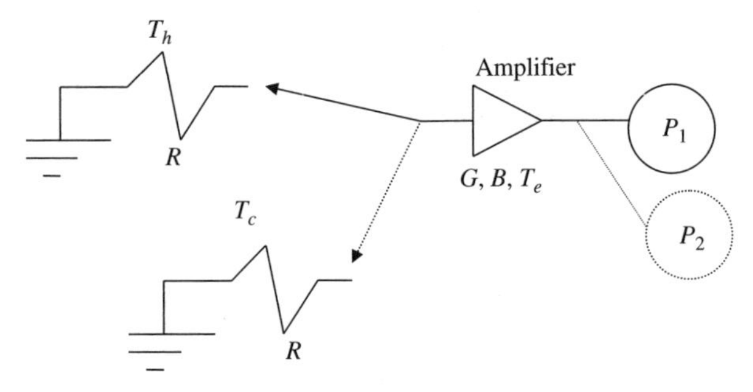

Figure 2.11 Experimental setup for measurement of noise temperature.

Y-factor method: At $T_h = 290\,\text{K}$, $P_1 = -70\,\text{dBm}$; at $T_c = 77\,\text{K}$, $P_2 = -75\,\text{dBm}$. Determine its equivalent noise temperature. If this amplifier is used with a source that has an equivalent noise temperature of 450 K, find the output noise power in dBm.

SOLUTION Since P_1 and P_2 are given in dBm, the difference of these two values will give Y in decibels. Hence,

$$Y = (P_1 - P_2)\ \text{dBm} = (-70) - (-75) = 5\,\text{dB}$$

or

$$Y = 10^{0.5} = 3.1623$$

Therefore,

$$T_e = \frac{290 - (3.1623)(77)}{3.1623 - 1} = 21.51\,\text{K}$$

If a source with an equivalent noise temperature of $T_s = 450\,\text{K}$ drives the amplifier, the noise power input to this will be kT_sB. The total noise power at the output of the amplifier will be

$$P_o = GkT_sB + GkT_eB = 10 \times 1.38 \times 10^{-23} \times 10^9 \times (450 + 21.51)$$
$$= 6.5068 \times 10^{-11}\,\text{W}$$

Therefore,

$$P_o = 10\log(6.5068 \times 10^{-8}) = -71.8663\,\text{dBm}$$

Noise Factor and Noise Figure

The noise factor of a two-port network is obtained by dividing the signal-to-noise ratio at its input port by the signal-to-noise ratio at its output. Hence,

$$\text{noise factor, } F = \frac{S_i/N_i}{S_o/N_o}$$

where S_i, N_i, S_o, and N_o represent the power in input signal, input noise, output signal, and output noise, respectively. If the two-port network is noise-free, the signal-to-noise ratio at its output will be the same as its input, resulting in a noise factor of unity. In reality, the network will add its own noise, while the input signal and noise will be altered by the same factor (gain or loss). It will lower the output signal/noise ratio, resulting in a higher noise factor. Therefore, the noise factor of a two-port network is generally greater than unity. By definition, the input noise power is assumed to be the noise power resulting from a matched resistor at $T_0 = 290\,\text{K}$ (i.e., $N_i = kT_0B$). Using the circuit arrangement illustrated

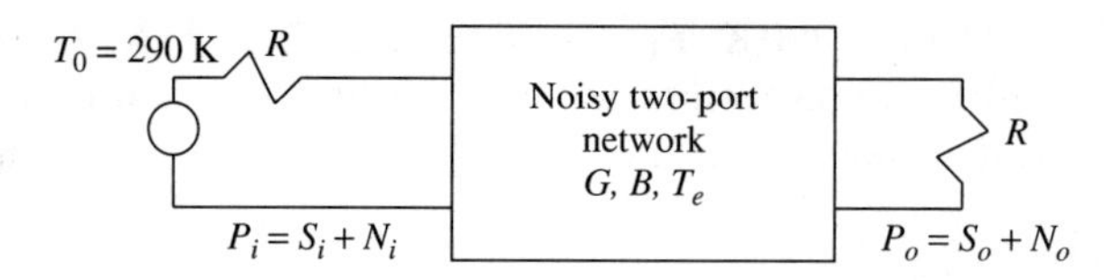

Figure 2.12 Circuit arrangement for determination of the noise factor of a noisy two-port network.

in Figure 2.12, the noise factor of a noisy two-port network can be defined as follows:

$$\text{noise factor} = \frac{\text{total output noise in } B \text{ when input source temperature is } 290\,\text{K}}{\text{output noise of source (only) at } 290\,\text{K}}$$

or

$$F = \frac{kT_0BG + P_{\text{int}}}{kT_0BG} = 1 + \frac{P_{\text{int}}}{kT_0BG} \tag{2.5.8}$$

where P_{int} represents the output noise power that is generated by the two-port network internally. It can be expressed in terms of noise factor as follows:

$$P_{\text{int}} = k(F-1)T_0BG = kT_eBG \tag{2.5.9}$$

where T_e is known as the equivalent noise temperature of a two-port network. It is related to the noise factor as follows:

$$T_e = (F-1)T_0 \tag{2.5.10}$$

When the noise factor is expressed in decibels, it is commonly called the *noise figure* (NF). Hence,

$$\text{NF} = 10\log_{10} F \qquad \text{dB} \tag{2.5.11}$$

Example 2.10 The power gain of an amplifier is 20 dB in the frequency band 10 to 12 GHz. If its noise figure is 3.5 dB, find the output noise power in dBm.

SOLUTION

$$\text{Output noise power} = kT_0BG + P_{\text{int}} = FkT_0BG = N_o$$

$$F = 3.5\,\text{dB} = 10^{0.35} = 2.2387$$

$$G = 20\,\text{dB} = 10^2 = 100$$

Therefore,

$$N_o = 2.2387 \times 1.38 \times 10^{-23} \times 290 \times 2 \times 10^9 \times 100 = 1.7919 \times 10^{-9}\,\text{W}$$

or

$$N_o = 10\log_{10}(1.7919 \times 10^{-6}\,\text{mW})\,\text{dBm} = -57.4669\,\text{dBm} \approx -57.5\,\text{dBm}$$

Noise in Two-Port Networks

Consider the noisy two-port network shown in Figure 2.13(*a*). $Y_s = G_s + jB_s$ is the source admittance connected at port 1 of the network. The noise it generates is represented by the current source i_s with a root-mean-square value of I_s. This noisy two-port can be replaced by a noise-free network with a current source i_n and a voltage source v_n connected at its input, as shown in Figure 2.13(*b*). I_n and V_n represent the corresponding root-mean-square current and voltage of the noise. It is assumed that the noise represented by i_s is uncorrelated with that represented by i_n and v_n. However, a part of i_n, i_{nc}, is assumed to be correlated with v_n via the correlation admittance $Y_c = G_c + jX_c$ while the remaining part i_{nu} is uncorrelated. Hence,

$$I_s^2 = \langle i_s^2 \rangle = 4kTBG_S \tag{2.5.12}$$

$$V_n^2 = \langle v_n^2 \rangle = 4kTBR_n \tag{2.5.13}$$

and

$$I_{nu}^2 = \langle i_{nu}^2 \rangle = 4kTBG_{nu} \tag{2.5.14}$$

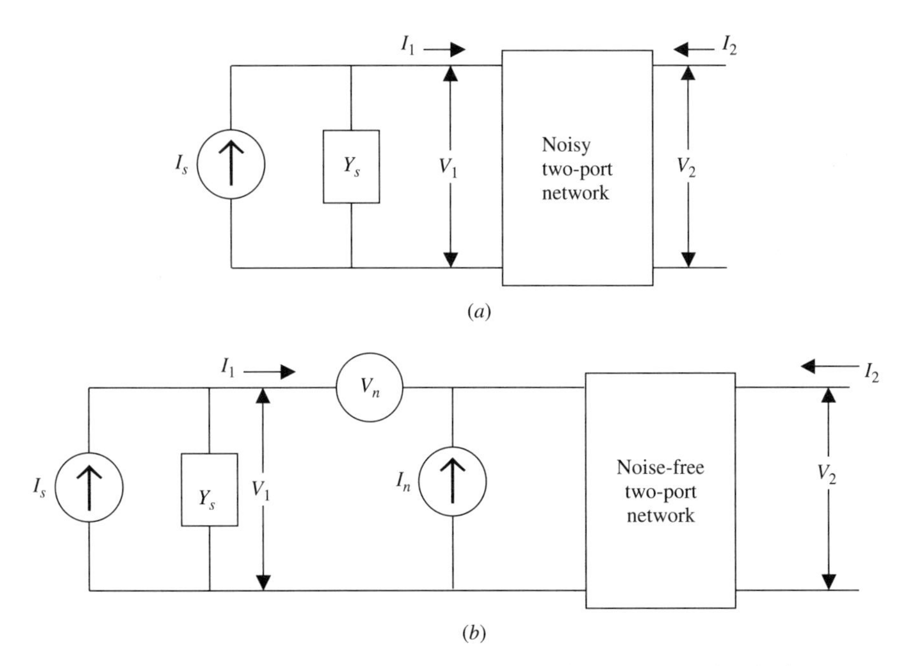

Figure 2.13 Noisy two-port network (*a*) and its equivalent circuit (*b*).

Now we find a Norton equivalent for the circuit that is connected at the input of the noise-free two-port network shown in Figure 2.13(*b*). Since these are random variables, the mean-square noise current $< i_{\text{eq}}^2 >$ is found as follows:

$$\langle i_{\text{eq}}^2 \rangle = \langle i_s^2 \rangle + \langle |i_n + Y_s v_n|^2 \rangle = \langle i_s^2 \rangle + \langle |i_{nu} + i_{nc} + Y_s v_n|^2 \rangle$$
$$= \langle i_s^2 \rangle + \langle |i_{nu} + (Y_c + Y_s) v_n|^2 \rangle$$

or

$$\langle i_{\text{eq}}^2 \rangle = \langle i_s^2 \rangle + \langle i_{nu}^2 \rangle + |Y_c + Y_n|^2 \langle v_n^2 \rangle \tag{2.5.15}$$

Hence, the noise factor F is

$$F = \frac{\langle i_{\text{eq}}^2 \rangle}{\langle i_s^2 \rangle} = 1 + \frac{\langle i_{nu}^2 \rangle}{\langle i_s^2 \rangle} + |Y_c + Y_s|^2 \frac{\langle v_n^2 \rangle}{\langle i_s^2 \rangle} = 1 + \frac{G_{nu}}{G_s} + |Y_c + Y_s|^2 \frac{R_n}{G_s}$$

or

$$F = 1 + \frac{G_{nu}}{G_s} + \frac{R_n}{G_s}[(G_s + G_c)^2 + (X_s + X_c)^2] \tag{2.5.16}$$

For a minimum noise factor, F_{min},

$$\frac{\partial F}{\partial G_s} = 0 \Rightarrow G_s^2 = G_c^2 + \frac{G_{nu}}{R_n} = G_{\text{opt}}^2 \tag{2.5.17}$$

and

$$\frac{\partial F}{\partial X_s} = 0 \Rightarrow X_s = -X_c = X_{\text{opt}} \tag{2.5.18}$$

From (2.5.17),

$$G_{nu} = R_n (G_{\text{opt}}^2 - G_c^2) \tag{2.5.19}$$

Substituting (2.5.18) and (2.5.19) into (2.5.16), the minimum noise factor is found as

$$F_{\text{min}} = 1 + 2R_n (G_{\text{opt}} + G_c) \tag{2.5.20}$$

Using (2.5.18) to (2.5.20), (2.5.16) can be expressed as

$$F = F_{\text{min}} + \frac{R_n}{G_s} |Y_s - Y_{\text{opt}}|^2 \tag{2.5.21}$$

Noise Figure of a Cascaded System

Consider a two-port network with gain G_1, noise factor F_1, and equivalent noise temperature T_{e1}. It is connected in cascade with another two-port network, as shown in Figure 2.14(*a*). The second two-port network has gain G_2, noise factor

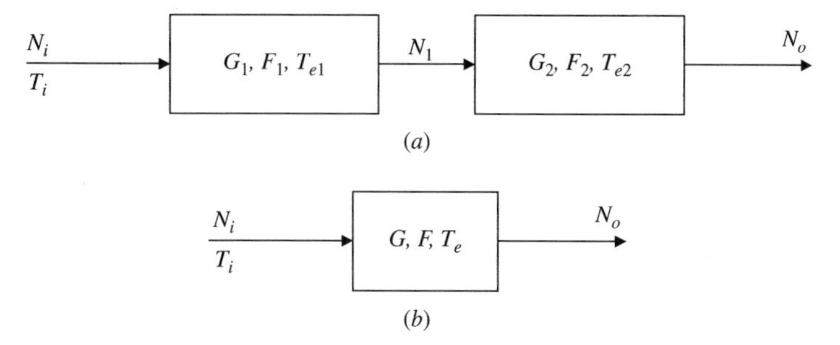

Figure 2.14 Two networks connected in cascade (*a*) and its equivalent system (*b*).

F_2, and noise temperature T_{e2}. Our goal is to find the noise factor F and equivalent noise temperature T_e of the overall system illustrated in Figure 2.14(*b*).

Assume that the noise power input to the first two-port network is N_i. Its equivalent noise temperature is T_i. The output noise power of the first system is N_1, whereas it is N_o after the second system. Hence,

$$N_1 = G_1 k T_i B + G_1 k T_{e1} B \tag{2.5.22}$$

and

$$\begin{aligned} N_o &= G_2 N_1 + G_2 k T_{e2} B = G_1 G_2 k B (T_i + T_{e1} + T_{e2}/G_1) \\ &= G_2[G_1 k B (T_i + T_{e1})] + G_2 k T_{e2} B \end{aligned}$$

or

$$N_o = G_1 G_2 k B (T_e + T_i) = G k B (T_e + T_i) \tag{2.5.23}$$

Therefore, the noise temperature of a cascaded system is

$$T_e = T_{e1} + \frac{T_{e2}}{G_1}$$

and from $T_{e1} = (F_1 - 1)T_i$, $T_{e2} = (F_2 - 1)T_i$, and $T_e = (F - 1)T_i$, we get

$$F = F_1 + \frac{F_2 - 1}{G_1}$$

The equations for T_e and F above can be generalized as follows:

$$T_e = T_{e1} + \frac{T_{e2}}{G_1} + \frac{T_{e3}}{G_1 G_2} + \frac{T_{e4}}{G_1 G_2 G_3} + \cdots \tag{2.5.24}$$

and

$$F = F_1 + \frac{F_2 - 1}{G_1} + \frac{F_3 - 1}{G_1 G_2} + \frac{F_4 - 1}{G_1 G_2 G_3} + \cdots \tag{2.5.25}$$

Example 2.11 A receiving antenna is connected to an amplifier through a transmission line that has an attenuation of 2 dB. The gain of the amplifier is 15 dB, and its noise temperature is 150 K over a bandwidth of 100 MHz. All the components are at an ambient temperature of 300 K.

(**a**) Find the noise figure of this cascaded system.

(**b**) What would be the noise figure if the amplifier were placed before the transmission line?

SOLUTION First we need to determine the noise factor of the transmission line alone. The formulas derived in the preceding section can then provide the noise figures desired for the two cases. Consider a transmission line that is match-terminated at both its ends by resistors R_0, as illustrated in Figure 2.15. Since the entire system is in thermal equilibrium at T(K), noise powers delivered to the transmission line and available at its output are kTB. Mathematically,

$$P_o = kTB = GkTB + GN_{\text{added}}$$

where N_{added} is noise generated by the line as it appcars at its input terminals. G is an output-to-input power ratio and B is the bandwidth of the transmission line. Note that input noise is attenuated in a lossy transmission line, but there is noise generated by it as well. Hence,

$$N_{\text{added}} = \frac{1}{G}(1 - G)kTB = \left(\frac{1}{G} - 1\right)kTB = kT_eB$$

An expression for the equivalent noise temperature T_e of the transmission line is found from it as follows:

$$T_e = \left(\frac{1}{G} - 1\right)T$$

and

$$F = 1 + \frac{T_e}{T_0} = 1 + \left(\frac{1}{G} - 1\right)\frac{T}{T_0}$$

The gain of the amplifier, $G_{\text{amp}} = 15\,\text{dB} = 10^{1.5} = 31.6228$. For the transmission line, $1/G = 10^{0.2} = 1.5849$. Hence, the noise factor of the line is $1 + (1.5849 - 1)300/290 = 1.6051$. The corresponding noise figure is 2.05 dB. Similarly, the

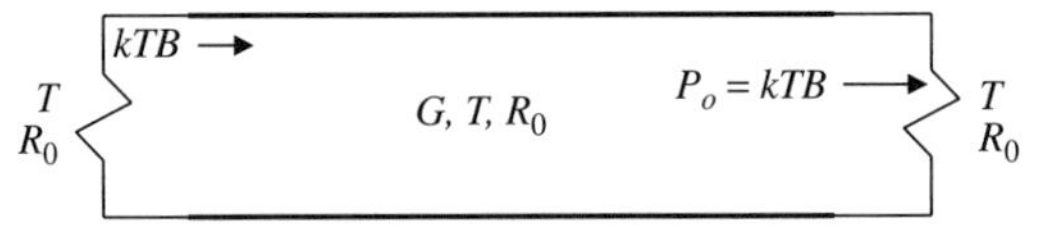

Figure 2.15 Lossy transmission line that is matched terminated at its ends.

noise factor of the amplifier is found to be $1 + 150/290 = 1.5172$. Its noise figure is 1.81 dB.

(**a**) In this case the noise figure of the cascaded system, F_{cascaded}, is found as

$$F_{\text{cascaded}} = F_{\text{line}} + \frac{F_{\text{amp}} - 1}{G_{\text{line}}} = 1.6051 + \frac{1.5172 - 1}{1/1.5849} = 2.4248 = 3.8468\,\text{dB}$$

(**b**) If the amplifier is connected before the line (i.e., the amplifier is placed right at the antenna),

$$F_{\text{cascaded}} = F_{\text{amp}} + \frac{F_{\text{line}} - 1}{G_{\text{amp}}} = 1.5172 + \frac{1.6051 - 1}{31.6228} = 1.5363 = 1.8649\,\text{dB}$$

Note that the amplifier alone has a noise figure of 1.81 dB. Hence, the noisy transmission line connected after it does not alter the noise figure significantly.

Example 2.12 Two amplifiers, each with a 20-dB gain, are connected in cascade as shown in Figure 2.16. The noise figure of amplifier A_1 is 3 dB, while that of A_2 is 5 dB. Calculate the overall gain and noise figure for this arrangement. If the order of two amplifiers is changed in the system, find its resulting noise figure.

SOLUTION The noise factors and gains of two amplifiers are

$$F_1 = 3\,\text{dB} = 10^{0.3} = 2$$
$$F_2 = 5\,\text{dB} = 10^{0.5} = 3.1623$$
$$G_1 = G_2 = 20\,\text{dB} = 10^2 = 100$$

Therefore, the overall gain and noise figure of the cascaded system is found as follows:

$$G = \frac{P_3}{P_1} = \frac{P_3}{P_2} \times \frac{P_2}{P_1} = 100 \times 100 = 10,000 = 40\,\text{dB}$$

and

$$F = F_1 + \frac{F_2 - 1}{G_1} = 2 + \frac{3.1623 - 1}{100} = 2.021623 = 3.057\,\text{dB}$$

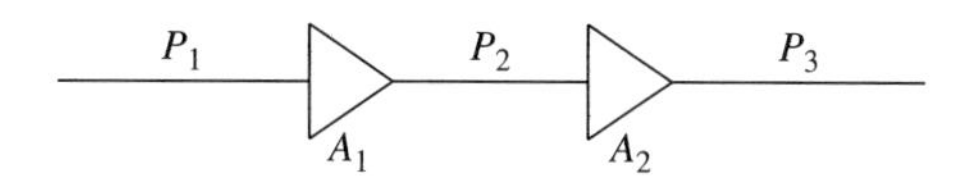

Figure 2.16 Setup for Example 2.12.

If the order of amplifiers is changed, the overall gain will stay the same. However, the noise figure of the new arrangement will change as follows:

$$F = F_2 + \frac{F_1 - 1}{G_2} = 3.1623 + \frac{2-1}{100} = 3.1723 = 5.013743\,\text{dB}$$

Minimum Detectable Signal

Consider a receiver circuit with gain G over a bandwidth B. Assume that its noise factor is F. P_I and P_o represent the signal power at its input and output ports, respectively. N_I is the input noise power and N_o is the total noise power at its output, as illustrated in Figure 2.17. Hence,

$$N_o = kT_0FBG \tag{2.5.26}$$

This constitutes the noise floor of the receiver. A signal weaker than this will be lost in noise. N_o can be expressed in dBW as follows:

$$N_o(\text{dBW}) = 10\log_{10} kT_o + F(\text{dB}) + 10\log_{10} B + G(\text{dB}) \tag{2.5.27}$$

The minimum detectable signal must have power higher than this. Generally, it is taken as 3 dB above this noise floor. Further,

$$10\log_{10} kT_0 = 10\log_{10}(1.38 \times 10^{-23} \times 290) \approx -204\,\text{dBW/Hz}$$

Hence, the minimum detectable signal $P_{o\text{MDS}}$ at the output is

$$P_{o\text{MDS}} = -201 + F(\text{dB}) + 10\log_{10} B_{\text{Hz}} + G(\text{dB}) \tag{2.5.28}$$

The corresponding signal power $P_{I\text{MDS}}$ at its input is

$$P_{I\text{MDS}}(\text{dBW}) = -201 + F(\text{dB}) + 10\log_{10} B_{\text{Hz}} \tag{2.5.29}$$

Alternatively, $P_{I\text{MDS}}$ can be expressed in dBm as follows:

$$P_{I\text{MDS}}(\text{dBm}) = -111 + F(\text{dB}) + 10\log_{10} B_{\text{MHz}} \tag{2.5.30}$$

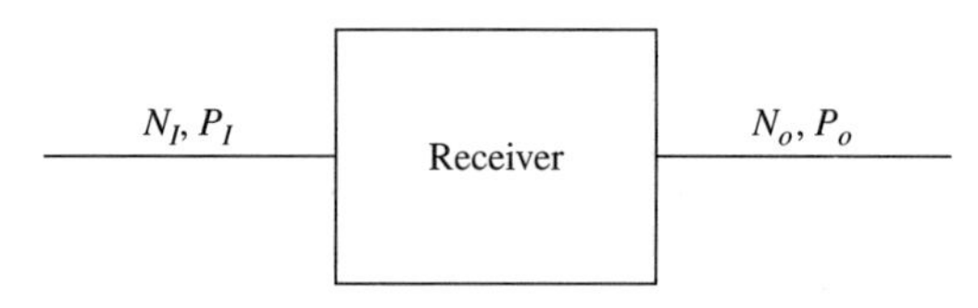

Figure 2.17 Signals at the two ports of a receiver with a noise figure of F decibels.

Example 2.13 The noise figure of a communication receiver is found as 10 dB at room temperature (290 K). Determine the minimum detectable signal power if (**a**) $B = 1$ MHz, (**b**) $B = 1$ GHz, (**c**) $B = 10$ GHz, and (**d**) $B = 1$ kHz.

SOLUTION From (2.5.30),

$$P_{I\text{MDS}} = -111 + F(\text{dB}) + 10\log_{10} B_{\text{MHz}} \text{ dBm}$$

Hence,

(**a**) $P_{I\text{MDS}} = -111 + 10 + 10\log_{10}(1) = -101 \text{ dBm} = 7.94 \times 10^{-11} \text{ mW}$

(**b**) $P_{I\text{MDS}} = -111 + 10 + 10\log_{10}(10^3) = -71 \text{ dBm} = 7.94 \times 10^{-8} \text{ mW}$

(**c**) $P_{I\text{MDS}} = -111 + 10 + 10\log_{10}(10^4) = -61 \text{ dBm} = 7.94 \times 10^{-7} \text{ mW}$

(**d**) $P_{I\text{MDS}} = -111 + 10 + 10\log_{10}(10^{-3}) = -131 \text{ dBm} = 7.94 \times 10^{-14} \text{ mW}$

These results show that the receiver can detect a relatively weak signal when its bandwidth is narrow.

Intermodulation Distortion

The electrical noise of a system determines the minimum signal level that it can detect. On the other hand, the signal will be distorted if its level is too high. This occurs because of the nonlinear characteristics of electrical devices, such as diodes, transistors, and so on. In this section we analyze the distortion characteristics and introduce the associated terminology.

Consider the nonlinear system illustrated in Figure 2.18. Assume that its nonlinearity is frequency independent and can be represented by the following power series:

$$v_o = k_1 v_i + k_2 v_i^2 + k_3 v_i^3 + \cdots \tag{2.5.31}$$

For simplicity, we assume that the k_i are real and the first three terms of this series are sufficient to represent its output signal. Further, it is assumed that the input signal has two different frequency components that can be expressed as follows:

$$v_i = a\cos\omega_1 t + b\cos\omega_2 t \tag{2.5.32}$$

Therefore, the corresponding output signal can be written as

$$v_o \approx k_1(a\cos\omega_1 t + b\cos\omega_2 t) + k_2(a\cos\omega_1 t + b\cos\omega_2 t)^2 + k_3(a\cos\omega_1 t + b\cos\omega_2 t)^3$$

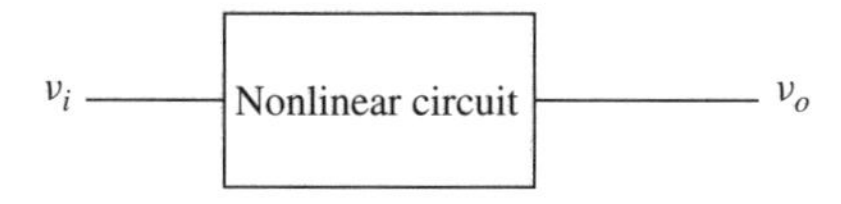

Figure 2.18 Nonlinear circuits with input signal v_i that produces v_o at its output.

After simplifying and rearranging it, we get

$$v_o = k_1(a\cos\omega_1 t + b\cos\omega_2 t) + k_2\left[\frac{a^2}{2}(1+\cos 2\omega_1 t) + \frac{b^2}{2}(1+\cos 2\omega_2 t) + \frac{ab}{2}\{\cos(\omega_1+\omega_2)t + \cos(\omega_1-\omega_2 t)\}\right]$$
$$+ k_3\left[\begin{array}{l} \frac{3}{4}a^3\cos\omega_1 t + \frac{3}{2}ab^2\cos\omega_1 t + \frac{a^3}{4}\cos 3\omega_1 t \\ \quad + \frac{3}{4}ab^2\cos(\omega_1 - 2\omega_2)t + \frac{3}{4}a^2 b\cos(2\omega_1-\omega_2)t \\ \quad + \frac{3}{2}a^2 b\cos\omega_2 t + \frac{3}{4}b^3\cos\omega_2 t + \frac{b^3}{4}\cos 3\omega_2 t \\ \quad + \frac{3}{4}a^2 b\cos(2\omega_1+\omega_2)t + \frac{3}{4}ab^2\cos(\omega_1+2\omega_2)t \end{array}\right] \tag{2.5.33}$$

Therefore, the output signal has several frequency components in its spectrum. Amplitudes of various components are listed in Table 2.6.

Figure 2.19 illustrates the input–output characteristic of an amplifier. If the input signal is too low, it may be submerged under the noise. The output power rises linearly above the noise as the input is increased. However, it deviates from the linear characteristic after a certain level of input power. In the linear region, output power can be expressed in dBm as follows:

$$P_{\text{out}}(\text{dBm}) = P_{\text{in}}(\text{dBm}) + G(\text{dB})$$

The input power for which output deviates by 1 dB below its linear characteristic is known as 1-dB compression point. In this figure, it occurs at an input power

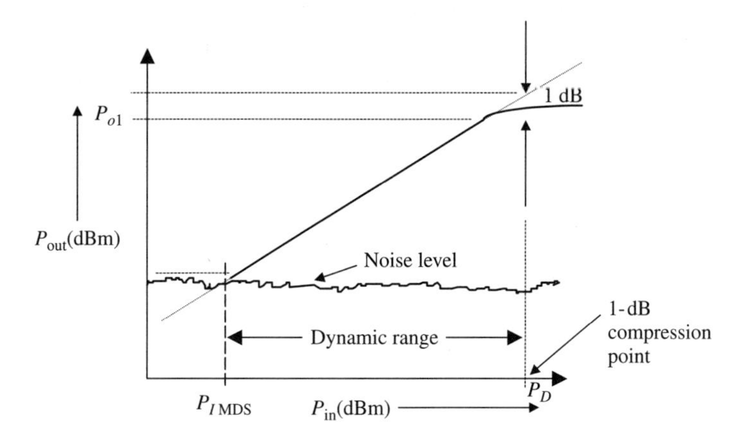

Figure 2.19 Gain characteristics of an amplifier.

TABLE 2.6 Amplitudes of Various Harmonics in the Output

Harmonic Components	Amplitude
ω_1	$k_1 a + k_3\left(\frac{3}{4}a^3 + \frac{3}{2}ab^2\right)$
ω_2	$k_1 b + k_3\left(\frac{3}{4}b^3 + \frac{3}{2}a^2 b\right)$
$\omega_1 - \omega_2$	$k_2 \frac{ab}{2}$
$\omega_1 + \omega_2$	$k_2 \frac{ab}{2}$
$2\omega_1$	$k_2 \frac{a^2}{2}$
$2\omega_2$	$k_2 \frac{b^2}{2}$
$3\omega_1$	$k_3 \frac{a^3}{4}$
$3\omega_2$	$k_3 \frac{b^3}{4}$
$2\omega_1 - \omega_2$	$\frac{3}{4}k_3 a^2 b$
$\omega_1 - 2\omega_2$	$\frac{3}{4}k_3 ab^2$
$2\omega_1 + \omega_2$	$\frac{3}{4}k_3 a^2 b$
$\omega_1 + 2\omega_2$	$\frac{3}{4}k_3 ab^2$

of P_D(dBm) that produces an output of P_{o1}(dBm). From the relation above, we find that

$$P_{o1}(\text{dBm}) + 1 = P_D(\text{dBm}) + G(\text{dB})$$

or

$$P_D(\text{dBm}) = P_{o1}(\text{dBm}) + 1 - G(\text{dB}) \tag{2.5.34}$$

The difference between the input power at 1-dB compression point and the minimum detectable signal defines the *dynamic range* (DR). Hence,

$$\text{DR} = P_D(\text{dBm}) - P_{I\text{MDS}}$$

From (2.5.30) and (2.5.34), we find that

$$\text{DR} = P_{o1}(\text{dBm}) + 112 - G(\text{dB}) - F(\text{dB}) - 10\log_{10} B_{\text{MHz}} \tag{2.5.35}$$

Gain Compression

Nonlinear characteristics of the circuit (amplifier, mixer, etc.) compress its gain. If there is only one input signal [i.e., b is zero in (2.5.32)], the amplitude a_1 of $\cos \omega_1 t$ in its output is found from Table 2.6 to be

$$a_1 = k_1 a + \tfrac{3}{4} k_3 a^3 \tag{2.5.36}$$

The first term of a_1 represents the linear (ideal) case, while its second term results from the nonlinearity. At the 1-dB compression point,

$$\left(k_1 a + \tfrac{3}{4} k_3 a^3\right)\Big|_{\text{dB}} = k_1 a|_{\text{dB}} - 1 \rightarrow 20 \log \frac{k_1 a + \frac{3}{4} k_3 a^3}{k_1 a}$$
$$= -1 = -20 \log (1.122)$$

or

$$20 \log \left(\frac{k_1 + \frac{3}{4} k_3 a^2}{k_1} \times 1.122 \right) = 0 \rightarrow \frac{k_1 + \frac{3}{4} k_3 a^2}{k_1} \times 1.122 = 1$$

Therefore,

$$k_3 = -0.145 \frac{k_1}{a^2} \tag{2.5.37}$$

This indicates that k_3 is a negative constant for a positive k_1, or vice versa. Therefore, it tends to reduce a_1, resulting in a lower gain. The single-tone gain compression factor may be defined as follows:

$$A_{1C} = \frac{a_1}{k_1 a} = 1 + \frac{3k_3}{4k_1} a^2 \tag{2.5.38}$$

Let us now consider the case when both of the input signals are present in (2.5.32). The amplitude of $\cos \omega_1 t$ in the output now becomes $k_1 a + k_3 \left(\frac{3}{4} a^3 + \frac{3}{2} a b^2\right)$. If b is large compared with a, the term with k_3 may dominate (undesired) over the first (desired) one. Since k_3 and k_1 have opposite signs, the desired output $k_1 a$ may be *blocked* completely for a certain value of b.

Second Harmonic Distortion

Second harmonic distortion occurs due to k_2. If b is zero, the amplitude of the second harmonic will be $k_2 \left(\dfrac{a^2}{2}\right)$. Since power is proportional to the square of the voltage, the desired term in the output can be expressed as

$$P_1 = 10 \log_{10} \left(\zeta k_1 \frac{a}{2}\right)^2 = 20 \log_{10} a + C_1 \tag{2.5.39}$$

where ζ is the proportionality constant and

$$C_1 = 20 \log \frac{\zeta k_1}{2}$$

Similarly, power in the second harmonic component can be expressed as

$$P_2 = 10 \log_{10} \left(\varsigma k_2 \frac{a^2}{2} \right)^2 = 40 \log_{10} a + C_2 \tag{2.5.40}$$

and the input power is

$$P_{\text{in}} = 10 \log_{10} \left(\varsigma \frac{a}{2} \right)^2 = 20 \log_{10} a + C_3 \tag{2.5.41}$$

Proportionality constants ς and k_2 are embedded in C_2 and C_3. From (2.5.39)–(2.5.41) we find that

$$P_1 = P_{\text{in}} + D_1 \tag{2.5.42}$$

and

$$P_2 = 2P_{\text{in}} + D_2 \tag{2.5.43}$$

where D_1 and D_2 replace $C_1 - C_3$ and $C_2 - 2C_3$, respectively. Equations (2.5.42) and (2.5.43) indicate that both the fundamental and second harmonic signals in the output are linearly related with input power. However, the second harmonic power increases at twice the rate of the fundamental (the desired) component.

Intermodulation Distortion Ratio

From Table 2.6 we find that the cubic term produces intermodulation frequencies $2\omega_1 \pm \omega_2$ and $2\omega_2 \pm \omega_1$. If ω_1 and ω_2 are very close, $2\omega_1 + \omega_2$ and $2\omega_2 + \omega_1$ will be far away from the desired signals, and therefore, these can be filtered out easily. However, the other two terms, $2\omega_1 - \omega_2$ and $2\omega_2 - \omega_1$, will be so close to ω_1 and ω_2 that these components may be within the passband of the system. It will distort the output. This characteristic of a nonlinear circuit is specified via the intermodulation distortion. It is obtained after dividing the amplitude of one of the intermodulation terms by the desired output signal. For an input signal with both ω_1 and ω_2 (i.e., a two-tone input), the intermodulation distortion ratio (IMR) may be found as

$$\text{IMR} = \frac{\frac{3}{4} k_3 a^2 b}{k_1 a} = \frac{3k_3}{4k_1} ab \tag{2.5.44}$$

Intercept Point

Since power is proportional to the square of the voltage, the intermodulation distortion power may be defined as

$$P_{\text{IMD}} = \varsigma \frac{\left(\frac{3}{4}k_3 a^2 b\right)^2}{2} \tag{2.5.45}$$

where ς is the proportionality constant. If the two input signals are equal in amplitude, $a = b$ and the expression for intermodulation distortion power simplifies to

$$P_{\text{IMD}} = \varsigma \frac{\left(\frac{3}{4}k_3 a^3\right)^2}{2} \tag{2.5.46}$$

Similarly, the power in one of the input signal components can be expressed as

$$P_{\text{in}} = \varsigma \frac{(a)^2}{2}$$

Therefore, the intermodulation distortion power can be expressed as

$$P_{\text{IMD}} = \alpha P_{\text{in}}^3$$

where α is another constant.

The ratio of intermodulation distortion power (P_{IMD}) to the desired output power P_o ($P_o = k_1^2 P_{\text{in}}$) for the case where two input signal amplitudes are the same is known as the *intermodulation distortion ratio*. It is found to be

$$P_{\text{IMR}} = \frac{P_{\text{IMD}}}{P_o} = \alpha_1 P_{\text{in}}^2 \tag{2.5.47}$$

Note that P_{IMD} increases as the cube of input power and the desired signal power P_o is linearly related with P_{in}. Hence, P_{IMD} increases three times as fast as P_o on a log-log plot (or both of them are expressed in dBm before displaying on a linear graph). In other words, for a change of 1 dBm in P_{in}, P_o changes by 1 dBm, whereas P_{IMD} changes by 3 dBm. The value of the input power for which P_{IMD} is equal to P_o is referred to as the *intercept point* (IP). It is also referred to as IIP_3 and the corresponding output power as OIP_3. Hence, P_{IMR} is unity at the intercept point. If P_{IP} is input power at IP,

$$P_{\text{IMR}} = 1 = \alpha_1 P_{\text{IP}}^2 \Rightarrow \alpha_1 = \frac{1}{P_{\text{IP}}^2} \tag{2.5.48}$$

Therefore, the intermodulation distortion ratio P_{IMR} is related to P_{IP} as

$$P_{\text{IMR}} = \left(\frac{P_{\text{in}}}{P_{\text{IP}}}\right)^2 \tag{2.5.49}$$

Example 2.14 The intercept point in the transfer characteristic of a nonlinear system is found to be at 25 dBm. If a −15-dBm signal is applied to this system, find the intermodulation ratio.

SOLUTION

$$P_{\text{IMR}} = \left(\frac{P_{\text{in}}}{P_{\text{IP}}}\right)^2 \rightarrow P_{\text{IMR}}(\text{dB}) = 2[P_{\text{in}}(\text{dBm}) - P_{\text{IP}}(\text{dBm})]$$
$$= 2(-15 - 25) = -80\,\text{dB}$$

Dynamic Range

As mentioned earlier, noise at one end and distortion at the other limit the range of detectable signals of a system. The amount of distortion that can be tolerated depends somewhat on the type of application. If we set the upper limit that a system can detect as the signal level at which intermodulation distortion is equal to the minimum detectable signal, we can formulate an expression for its dynamic range. Thus, the ratio of the signal power that causes distortion power (in one frequency component) to be equal to the noise floor and the minimum detectable signal is the dynamic range (DR) of the system (amplifier, mixer, or receiver). It is also known as the *spurious-free dynamic range* (SFDR).

Since the ideal power output P_o is linearly related to input as

$$P_o = k_1^2 P_{\text{in}} \tag{2.5.50}$$

the distortion power with reference to input can be expressed as

$$P_{di} = \frac{P_{\text{IMD}}}{k_1^2} \tag{2.5.51}$$

Therefore,

$$P_{\text{IMR}} = \frac{P_{\text{IMD}}}{P_o} = \frac{k_1^2 P_{di}}{k_1^2 P_{\text{in}}} = \frac{P_{di}}{P_{\text{in}}} = \left(\frac{P_{\text{in}}}{P_{\text{IP}}}\right)^2 \tag{2.5.52}$$

If $P_{\text{di}} = N_f$ (noise floor at the input),

$$\frac{N_f}{P_{\text{in}}} = \left(\frac{P_{\text{in}}}{P_{\text{IP}}}\right)^2 \Rightarrow P_{\text{in}}^3 = P_{\text{IP}}^2 N_f \Rightarrow P_{\text{in}} = (P_{\text{IP}}^2 N_f)^{1/3} \tag{2.5.53}$$

and the dynamic range is

$$\text{DR} = \frac{(P_{\text{IP}}^2 N_f)^{1/3}}{N_f} = \left(\frac{P_{\text{IP}}}{N_f}\right)^{2/3} \tag{2.5.54}$$

or

$$\text{DR(dB)} = \tfrac{2}{3}[P_{\text{IP}}(\text{dBm}) - N_f(\text{dBm})] \tag{2.5.55}$$

Example 2.15 A receiver is operating at 900 MHz with its bandwidth at 500 kHz and the noise figure at 8 dB. If its input impedance is 50 Ω and IP is 10 dBm, find its dynamic range.

SOLUTION

$$N_f = kT_0BF \Rightarrow N_f(\text{dBm}) = 10\log_{10}(1.38 \times 10^{-23} \times 290 \times 500 \times 10^3 \times 10^3) + 8 = -108.99\,\text{dBm}$$

Therefore,

$$\text{DR} = \tfrac{2}{3}(10 + 108.99) = 79.32\,\text{dB}$$

SUGGESTED READING

Balanis, C. A. , *Antenna Theory*. New York: Wiley, 1997.

Cheah, J. Y. C., Introduction to wireless communications applications and circuit design, in L. E. Larsen (ed.), *RF and Microwave Circuit Design for Wireless Communications*. Boston: Artech House, 1996.

Razavi, B., *RF Microelectronics*. Upper Saddle River, NJ: Prentice Hall, 1998.

Schiller, J. H., *Mobile Communication*. Reading, MA: Addison-Wesley, 2000.

Smith, J. R., *Modern Communication Circuits*. New York: McGraw-Hill, 1997.

Stutzman, W. L., and G. A. Thiele, *Antenna Theory and Design*. New York: Wiley, 1998.

PROBLEMS

2.1. **(a)** What is the gain in decibels of an amplifier with a power gain of 4?

(b) What is the power gain ratio of a 5-dB amplifier?

(c) Express 2 kW of power in terms of dBm and dBw.

2.2. The maximum radiation intensity of a 90% efficient antenna is 200 mW per unit solid angle. Find the directivity and gain (dimensionless and in dB) when **(a)** the input power is 40π milliwatts and **(b)** the radiated power is 40π milliwatts.

2.3. The normalized far-zone field pattern of an antenna is given by

$$E = \begin{cases} \sqrt{\sin\theta\cos^2\phi} & 0 \leqslant \theta \leqslant \pi \text{ and } 0 \leqslant \phi \leqslant \pi/2,\ 3\pi/2 \leqslant \phi \leqslant 2\pi \\ 0 & \text{elsewhere} \end{cases}$$

Find the directivity of this antenna.

2.4. The radiation intensity pattern of an antenna is given by

$$U(\theta, \phi) = \begin{cases} 10 \sin\theta \cos^2\phi & 0 \leqslant \theta \leqslant \pi \text{ and } 0 \leqslant \phi \leqslant \pi/2, 3\pi/2 \leqslant \phi \leqslant 2\pi \\ 0 & \text{elsewhere} \end{cases}$$

Find the directivity of this antenna.

2.5. For an X-band (8.2 to 12.4 GHz) rectangular antenna with aperture dimensions of 5.5 and 7.4 cm, find its maximum effective aperture (in cm^2) when its gain is **(a)** 14.8 dB at 8.2 GHz, **(b)** 16.5 dB at 10.3 GHz, and **(c)** 18.0 dB at 12.4 GHz.

2.6. Transmitting and receiving antennas operating at 1 GHz with gains of 20 and 15 dB, respectively, are separated by a distance of 1 km. Find the maximum power delivered to the load when the input power is 150 W, assuming that the antennas are polarization matched.

2.7. An antenna with a total loss resistance of $1\,\Omega$ is connected to a generator whose internal impedance is $50 + j25\,\Omega$. Assuming that the peak voltage of the generator is 2 V and the impedance of antenna is $74 + j42.5\,\Omega$, find the power **(a)** supplied by the source (real power), **(b)** radiated by the antenna, and **(c)** dissipated by the antenna.

2.8. The antenna connected to a radio receiver induces $9\,\mu\text{V}$ of root-mean-square voltage into its input impedance, which is $50\,\Omega$. Calculate the input power in watts, dBm, and dBW. If the signal is amplified by 127 dB before it is fed to an 8-Ω speaker, find the output power.

2.9. The electric field radiated by a rectangular aperture, mounted on an infinite ground plane with z perpendicular to the aperture, is given by

$$\mathbf{E} = \lfloor \hat{\theta} \cos\phi - \hat{\phi} \sin\phi \cos\theta \rfloor f(r, \theta, \phi)$$

where $f(r, \theta, \phi)$ is a scalar function that describes the field variation of the antenna. Assuming that the receiving antenna is linearly polarized along the x-axis, find the polarization loss factor.

2.10. A radar receiver has a sensitivity of 10^{-12} W. If the radar antenna's effective aperture is $1\,\text{m}^2$ and the wavelength is 10 cm, find the transmitter power required to detect an object at a distance of 3 km with a radar cross section of $5\,\text{m}^2$.

2.11. Two space vehicles are separated by 10^8 m. Each has an antenna with $D = 1000$ operating at 2.5 GHz. If vehicle A's receiver requires 1 pW for a 20-dB signal-to-noise ratio, what transmitter power is required on vehicle B to achieve this signal-to-noise ratio?

2.12. **(a)** Design an Earth-based radar system that receives 10^{-15} W of peak echo power from Venus. It is to operate at 10 GHz with a single antenna to be used for both transmitting and receiving. Specify the effective aperture of the antenna and the peak transmitter power. Assume that the Earth to

Venus distance is 3 light-minutes; the diameter of Venus is 13×10^6 m; and the radar cross section of Venus is 15% of its physical cross section.

(b) If the system of part (a) is used to observe the moon, determine the power received. Assume that the moon diameter is 3.5×10^6 m, its radar cross section is 15% of the physical cross section, and the Earth-to-moon distance is 1.2 light-seconds.

2.13. Consider an imaging satellite that sends close-up pictures of the planet Neptune. It uses a 10-W transmitter operating at 18 GHz and a 2.5-m-diameter parabolic dish antenna. What Earth-station system temperature is required to provide a signal-to-noise ratio of 3 dB for reception of a picture with 3×10^6 pixels (picture elements) in 2 min if the Earth-station antenna diameter is 75 m? Assume aperture efficiencies of 70% and the Earth–Neptune distance as 4 light-hours. One pixel is equal to one bit, and two bits per second has a bandwidth of 1 Hz.

2.14. A Doppler radar operating at 10 GHz employs an antenna with its gain at 20 dB. If the signal reflected back from a target is a uniform plane wave with a power density of 10^{-8} W/m^2 at the antenna, find the power delivered to a matched receiver. Also, if this radar is intended to detect target velocities ranging from 90 to 130 km/h, find the required passband of the Doppler filter.

2.15. A satellite transmitter is 36,500 km away from the receiver. It radiates a power of 10 W through an antenna of 19 dB gain toward the receiver. The satellite operates at a frequency of 11.5 GHz. If the receiving antenna has a gain of 60.5 dB, find the received power in dBm.

2.16. Consider the WLAN receiver front end shown in Figure P2.16, in which the bandpass filter has a bandwidth of 250 MHz centered at 5.4 GHz. If the system is operating at room temperature (290 K), calculate **(a)** the overall noise figure in decibels and **(b)** the minimum detectable signal in milliwatts. **(c)** Suggest if there is a better arrangement to increase the signal to noise ratio at the output.

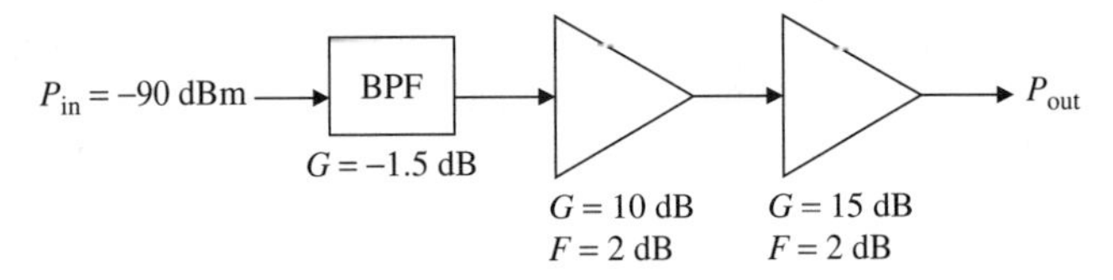

Figure P2.16

2.17. Various units of a 4-GHz receiver have the gains and noise temperatures indicated in Figure P2.17.

(a) If the transmission line has a loss of 2 dB and the system is at an ambient temperature of 300 K, find the noise figure (in decibels) of this receiver.

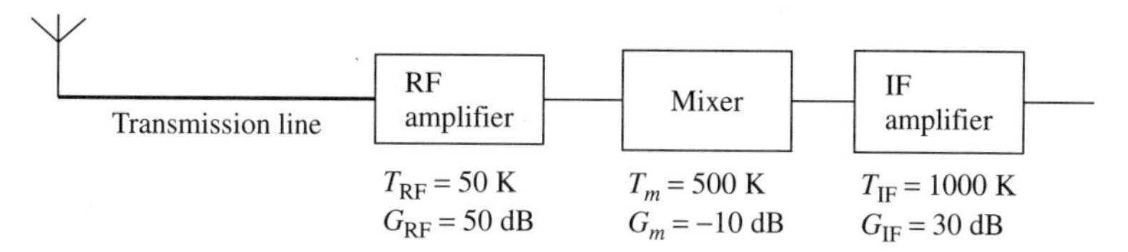

Figure P2.17

(b) Suggest a better receiver architecture (if it exists) that improves the noise figure. Also, find the noise temperature of the new arrangement.

2.18. Two receivers for a design trade-off study are shown in Figure P2.18(*a*) and (*b*).

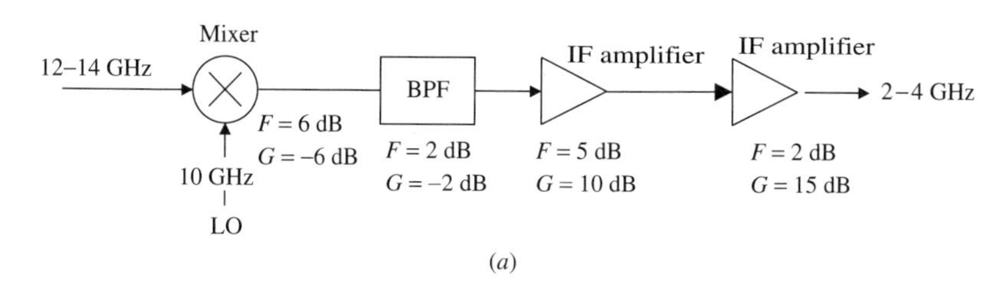

(*a*)

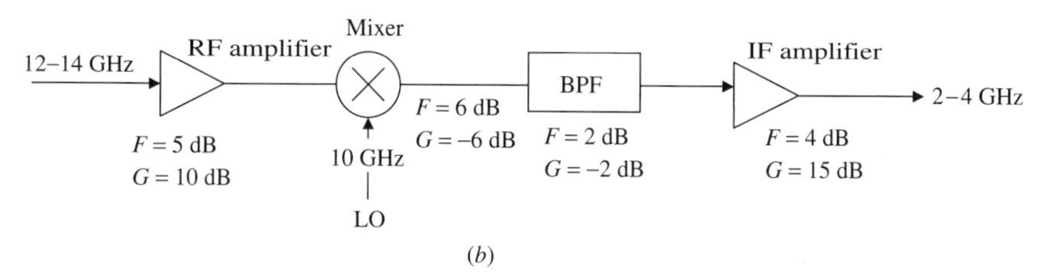

(*b*)

Figure P2.18

(a) Calculate the noise figure for the two systems.

(b) Calculate the overall gain for the two systems.

(c) Calculate the minimum detectable signal at the input for the two systems.

(d) If the output power at the 1-dB compression point is 10 mW for both systems, calculate the dynamic range for the two systems.

2.19. Calculate the input minimum detectable signal in milliwatts at room temperature for a receiver with **(a)** BW = 1 GHz, $F = 5$ dB and **(b)** BW = 100 MHz, $F = 10$ dB.

2.20. (a) What is the root-mean-square (rms) noise voltage produced in a 10-kΩ resistance when its temperature is 45°C and the effective bandwidth is 100 MHz?

(b) A 50-kΩ resistance is connected in parallel with the 10-kΩ resistance of part (a). What is the resulting rms noise voltage?

2.21. The intercept point in the transfer characteristic of a nonlinear system is found to be 33 dBm. If a −18-dBm signal is applied to this system, find the intermodulation ratio.

2.22. A receiver is operating at 2455 MHz with its bandwidth at 500 kHz and the noise figure at 15 dB. If its input impedance is 50 Ω and IP is 18 dBm, find its dynamic range.

2.23. An amplifier has a gain of 6 dB. Its intermodulation ratio is found to be −40 dB at an input power level of −4 dBm. Calculate the values of its input and output intercept points. What input signal level is needed if we want an intermodulation ratio of −50 dB?

3

TRANSMISSION LINES

Transmission lines are needed for connecting various circuit elements and systems. Open wire and coaxial lines are commonly used for circuits operating at low frequencies. On the other hand, coaxial line, strip line, microstrip line, and waveguides are employed at radio and microwave frequencies. Generally, low-frequency signal characteristics are not affected as the signal propagates through the line. However, radio frequency and microwave signals are affected significantly because of the fact that the circuit size is comparable to the wavelength. A comprehensive understanding of signal propagation requires the analysis of electromagnetic fields in a given line. On the other hand, a generalized formulation can be obtained using circuit concepts on the basis of line parameters.

This chapter begins with an introduction to line parameters and a distributed model of the transmission line. Solutions to the transmission line equation are then constructed to explain the behavior of the propagating signal. This is followed by discussions on sending-end impedance, reflection coefficient, return loss, and insertion loss. A quarter-wave impedance transformer is also presented, along with a few examples of matching resistive loads. Impedance measurement using the voltage standing wave ratio is then discussed. Finally, the Smith chart is introduced to facilitate graphical analysis and design of transmission line circuits.

3.1 DISTRIBUTED CIRCUIT ANALYSIS OF TRANSMISSION LINES

Any transmission line can be represented by a distributed electrical network (Figure 3.1) comprised of series inductors and resistors and shunt capacitors and

Radio-Frequency and Microwave Communication Circuits: Analysis and Design, Second Edition,
By Devendra K. Misra
ISBN 0-471-47873-3

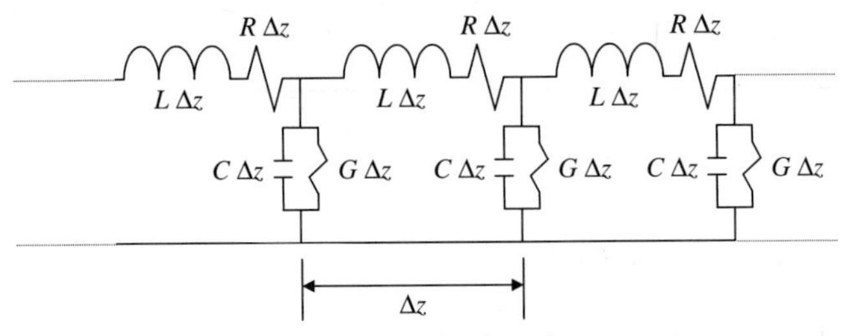

Figure 3.1 Distributed network model of a transmission line.

resistors. These distributed elements are defined as follows:

$$L = \text{inductance per unit length (H/m)}$$

$$R = \text{resistance per unit length } (\Omega/\text{m})$$

$$C = \text{capacitance per unit length (F/m)}$$

$$G = \text{conductance per unit length (S/m)}$$

L, R, C, and G, the *line parameters*, are determined theoretically by electromagnetic field analysis of a transmission line. These parameters are influenced by their cross-section geometry and the electrical characteristics of their constituents. For example, if a line is made up of an ideal dielectric and a perfect conductor, its R and G values will be zero. If it is a coaxial cable with inner and outer radii a and b, respectively, as shown in Figure 3.2, then

$$C = \frac{55.63\varepsilon_r}{\ln(b/a)} \qquad \text{pF/m} \tag{3.1.1}$$

and

$$L = 200\ln(b/a) \qquad \text{nH/m} \tag{3.1.2}$$

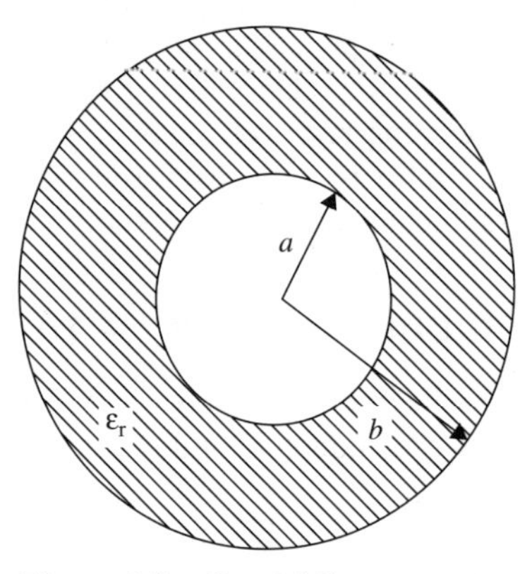

Figure 3.2 Coaxial line geometry.

where ε_r is the dielectric constant of the material between two coaxial conductors of the line.

If the coaxial line has small losses due to an imperfect conductor and insulator, its resistance and conductance parameters can be calculated as follows:

$$R \approx 10\left(\frac{1}{a}+\frac{1}{b}\right)\sqrt{\frac{f_{\mathrm{GHz}}}{\sigma}} \qquad \Omega/\mathrm{m} \tag{3.1.3}$$

and

$$G = \frac{0.3495\varepsilon_r f_{\mathrm{GHz}} \tan\delta}{\ln(b/a)} \qquad \mathrm{S/m} \tag{3.1.4}$$

where $\tan\delta$ is the loss tangent of the dielectric material, σ the conductivity (in S/m) of the conductors, and f_{GHz} is the signal frequency in gigahertz.

Characteristic Impedance of a Transmission Line

Consider a transmission line that extends to infinity, as shown in Figure 3.3. The voltages and currents at several points on the line are as indicated. When a voltage is divided by the current through that point, the ratio is found to remain constant. This ratio is called the *characteristic impedance* of the transmission line. Mathematically,

$$\text{characteristic impedance} = Z_0 = V_1/I_1 = V_2/I_2 = V_3/I_3 = \cdots = V_n/I_n$$

In actual electrical circuits, the length of the transmission lines is always finite. Hence, it seems that the characteristic impedance has no significance in the real world. However, that is not the case. When the line extends to infinity, an electrical signal continues propagating in the forward direction without reflection. On the other hand, it may be reflected back by the load that terminates a transmission line of finite length. If one varies this termination, the strength of the reflected signal changes. When the transmission line is terminated by a load impedance that absorbs all the incident signal, the voltage source sees an infinite electrical length. The voltage-to-current ratio at any point on this line is a constant equal

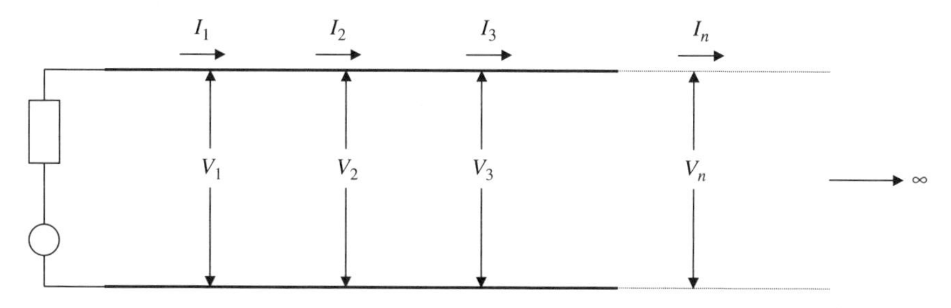

Figure 3.3 Infinitely long transmission line and voltage source.

to the terminating impedance. In other words, there is a unique impedance for every transmission line that does not produce an echo signal when the line is terminated by it. The terminating impedance that does not produce an echo on the line is equal to its characteristic impedance.

If

$$Z = R + j\omega L = \text{impedance per unit length}$$

$$Y = G + j\omega C = \text{admittance per unit length}$$

then using the definition of characteristic impedance and the distributed model shown in Figure 3.1, we can write

$$Z_0 = \frac{(Z_0 + Z\ \Delta z)(1/Y\ \Delta z)}{Z_0 + Z\ \Delta z + 1/Y\ \Delta z}$$

$$= \frac{Z_0 + Z\ \Delta z}{1 + Y\ \Delta z(Z_0 + Z\ \Delta z)} \Rightarrow Z_0 Y(Z_0 + Z\ \Delta z) = Z$$

For $\Delta z \to 0$,

$$Z_0 = \sqrt{\frac{Z}{Y}} = \sqrt{\frac{R + j\omega L}{G + j\omega C}} \tag{3.1.5}$$

Some special cases are as follows:

1. For a dc signal, $Z_{0\ \mathrm{dc}} = \sqrt{R/G}$.
2. For $\omega \to \infty$, $\omega L \gg R$ and $\omega C \gg G$; therefore, $Z_{0(\omega \to \mathrm{large})} = \sqrt{L/C}$.
3. For a lossless line, $R \to 0$ and $G \to 0$; therefore, $Z_0 = \sqrt{L/C}$.

Thus, a lossless semirigid coaxial line with $2a = 0.036$ in., $2b = 0.119$ in., and an ε_r value of 2.1(Teflon-filled) will have $C = 97.71$ pF/m and $L = 239.12$ nH/m. Its characteristic impedance will be 49.5 Ω. Since the conductivity of copper is 5.8×10^7 S/m and the loss tangent of Teflon is 0.00015, $Z = 3.74 + j1.5 \times 10^3$ Ω/m and $Y = 0.092 + j613.92$ mS/m at 1 GHz. The corresponding characteristic impedance is $49.5 - j0.058$ Ω, which is very close to the approximate value of 49.5 Ω.

Example 3.1 Calculate the equivalent impedance and admittance of a 1-m-long line that is operating at 1.6 GHz. The line parameters are $L = 0.002\ \mu$ H/m, $C = 0.012$ pF/m, $R = 0.015$ Ω/m, and $G = 0.1$ mS/m. What is the characteristic impedance of this line?

SOLUTION

$$Z = R + j\omega L = 0.015 + j2\pi \times 1.6 \times 10^9 \times 0.002 \times 10^{-6}\ \Omega/\mathrm{m}$$

$$= 0.015 + j20.11\ \Omega/\mathrm{m}$$

$$Y = G + j\omega C = 0.0001 + j2\pi \times 1.6 \times 10^9 \times 0.012 \times 10^{-12}\,\mathrm{S/m}$$
$$= 0.1 + j0.1206\,\mathrm{mS/m}$$

$$Z_0 = \sqrt{\frac{Z}{Y}} = 337.02 + j121.38\,\Omega$$

Transmission Line Equations

Consider the equivalent distributed circuit of a transmission line that is terminated by a load impedance Z_L, as shown in Figure 3.4. The line is excited by a voltage source $v(t)$ with its internal impedance Z_S. We apply Kirchhoff's voltage and current laws over a small length, Δz, of this line as follows: For the loop,

$$v(z,t) = L\ \Delta z \frac{\partial i(z,t)}{\partial t} + R\ \Delta z\ i(z,t) + v(z + \Delta z, t)$$

or

$$\frac{v(z + \Delta z, t) - v(z,t)}{\Delta z} = -Ri(z,t) - L\frac{\partial i(z,t)}{\partial t}$$

Under the limit $\Delta z \to 0$, the equation above reduces to

$$\frac{\partial v(z,t)}{\partial z} = -\left[R \times i(z,t) + L\frac{\partial i(z,t)}{\partial t}\right] \tag{3.1.6}$$

Similarly, at node A,

$$i(z,t) = i(z + \Delta z, t) + G\ \Delta z\ v(z + \Delta z, t) + C\ \Delta z \frac{\partial v(z + \Delta z, t)}{\partial t}$$

or

$$\frac{i(z + \Delta z, t) - i(z,t)}{\Delta z} = -\left[G \times v(z + \Delta z, t) + C\frac{\partial v(z + \Delta z, t)}{\partial t}\right]$$

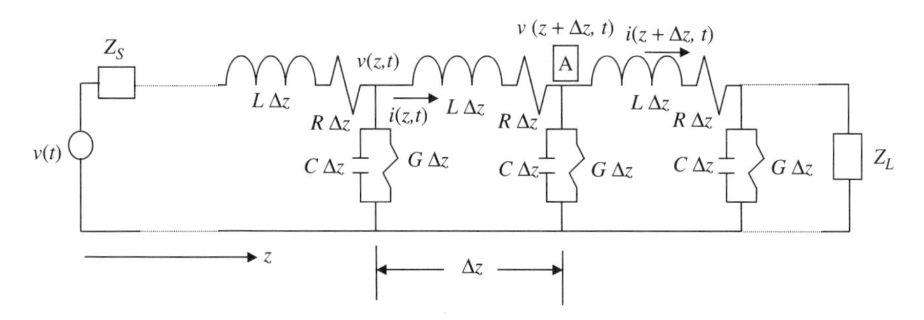

Figure 3.4 Distributed circuit model of a transmission line.

Again, under the limit $\Delta z \to 0$, it reduces to

$$\frac{\partial i(z,t)}{\partial z} = -Gv(z,t) - C\frac{\partial v(z,t)}{\partial t} \tag{3.1.7}$$

Now, from equations (3.1.6) and (3.1.7), $v(z,t)$ or $i(z,t)$ can be eliminated to formulate the following:

$$\frac{\partial^2 v(z,t)}{\partial z^2} = RGv(z,t) + (RC+LG)\frac{\partial v(z,t)}{\partial t} + LC\frac{\partial^2 v(z,t)}{\partial t^2} \tag{3.1.8}$$

and

$$\frac{\partial^2 i(z,t)}{\partial z^2} = RGi(z,t) + (RC+LG)\frac{\partial i(z,t)}{\partial t} + LC\frac{\partial^2 i(z,t)}{\partial t^2} \tag{3.1.9}$$

Some special cases are as follows:

1. For a lossless line, R and G will be zero, and these equations reduce to well-known homogeneous scalar wave equations,

$$\frac{\partial^2 v(z,t)}{\partial z^2} = LC\frac{\partial^2 v(z,t)}{\partial t^2} \tag{3.1.10}$$

and

$$\frac{\partial^2 i(z,t)}{\partial z^2} = LC\frac{\partial^2 i(z,t)}{\partial t^2} \tag{3.1.11}$$

It is to be noted that the velocity of these waves is $1/\sqrt{LC}$.

2. If the source is sinusoidal with time (i.e., time-harmonic), we can switch to phasor voltages and currents. In that case, equations (3.1.8) and (3.1.9) can be simplified as follows:

$$\frac{d^2 V(z)}{dz^2} = ZYV(z) = \gamma^2 V(z) \tag{3.1.12}$$

and

$$\frac{d^2 I(z)}{dz^2} = ZYI(z) = \gamma^2 I(z) \tag{3.1.13}$$

where $V(z)$ and $I(z)$ are phasor quantities; Z and Y are impedance per unit length and admittance per unit length, respectively, as defined earlier. $\gamma = \sqrt{ZY} = \alpha + j\beta$ is known as the *propagation constant* of the line. α and β are called the *attenuation constant* and the *phase constant*, respectively. Equations (3.1.12) and (3.1.13) are referred to as *homogeneous Helmholtz equations*.

Solution of Helmholtz Equations

Note that both differential equations have the same general format. Therefore, we consider the solution to the following generic equation here. Expressions for voltage and current on the line can be constructed on the basis of that.

$$\frac{d^2 f(z)}{dz^2} - \gamma^2 f(z) = 0 \tag{3.1.14}$$

Assume that $f(z) = Ce^{\kappa z}$, where C and κ are arbitrary constants. Substituting it into (3.1.14), we find that $\kappa = \pm\gamma$. Therefore, a complete solution to this equation may be written as follows:

$$f(z) = C_1 e^{-\gamma z} + C_2 e^{\gamma z} \tag{3.1.15}$$

where C_1 and C_2 are integration constants that are evaluated through the boundary conditions.

Hence, complete solutions to equations (3.1.12) and (3.1.13) can be written as follows:

$$V(z) = V_{\text{in}} e^{-\gamma z} + V_{\text{ref}} e^{\gamma z} \tag{3.1.16}$$

and

$$I(z) = I_{\text{in}} e^{-\gamma z} + I_{\text{ref}} e^{\gamma z} \tag{3.1.17}$$

where V_{in}, V_{ref}, I_{in}, and I_{ref} are integration constants that may be complex, in general. These constants can be evaluated from the known values of voltages and currents at two different locations on the transmission line. If we express the first two of these constants in polar form as

$$V_{\text{in}} = v_{\text{in}} e^{j\phi} \quad \text{and} \quad V_{\text{ref}} = v_{\text{ref}} e^{j\varphi}$$

the line voltage, in time domain, can be evaluated as follows:

$$\begin{aligned} v(z,t) &= \text{Re}[V(z)e^{j\omega t}] \\ &= \text{Re}[V_{\text{in}} e^{-(\alpha+j\beta)z} e^{j\omega t} + V_{\text{ref}} e^{+(\alpha+j\beta)z} e^{j\omega t}] \end{aligned}$$

or

$$v(z,t) = v_{\text{in}} e^{-\alpha z} \cos(\omega t - \beta z + \phi) + v_{\text{ref}} e^{+\alpha z} \cos(\omega t + \beta z + \varphi) \tag{3.1.18}$$

At this point, it is important to analyze and understand the behavior of each term on the right-hand side of this equation. At a given time, the first term changes sinusoidally with distance, z, while its amplitude decreases exponentially. It is illustrated in Figure 3.5(a). On the other hand, the amplitude of the second sinusoidal term increases exponentially. It is shown in Figure 3.5(b). Further, the argument of the cosine function decreases with distance in the former case and

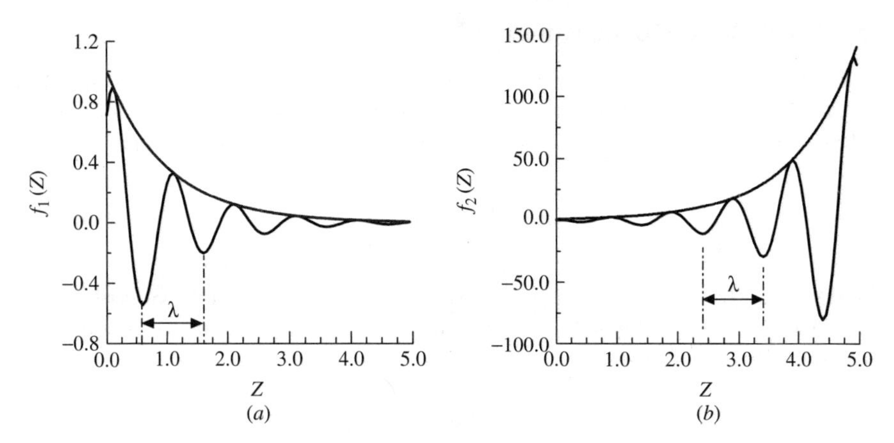

Figure 3.5 Behavior of two solutions to the Helmholtz equation with distance.

increases in the latter case. When a signal is propagating away from the source along $+z$-axis, its phase should be delayed. Further, if it is propagating in a lossy medium, its amplitude should decrease with distance z. Thus, the first term on the right-hand side of equation (3.1.16) represents a wave traveling along the $+z$-axis (an incident or outgoing wave). Similarly, the second term represents a wave traveling in the opposite direction (a reflected or incoming wave). This analysis is also applied to equation (3.1.17). Note that I_{ref} is reflected current that will be 180° out of phase with incident current I_{in}. Hence,

$$\frac{V_{\text{in}}}{I_{\text{in}}} = -\frac{V_{\text{ref}}}{I_{\text{ref}}} = Z_0$$

and therefore equation (3.1.17) may be written as follows:

$$I(z) = \frac{V_{\text{in}}}{Z_0}e^{-\gamma z} - \frac{V_{\text{ref}}}{Z_0}e^{\gamma z} \tag{3.1.19}$$

Incident and reflected waves change sinusoidally with both space and time. Time duration over which the phase angle of a wave goes through a change of 360° (2π radians) is known as its *time period*. The inverse of the time period in seconds is the signal frequency in hertz. Similarly, the distance over which the phase angle of the wave changes by 360° (2π radians) is known as its *wavelength* (λ). Therefore, the phase constant β is equal to 2π divided by the wavelength in meters.

Phase and Group Velocities

The velocity with which the phase of a time-harmonic signal moves is known as its *phase velocity*. In other words, if we tag a phase point of the sinusoidal wave and monitor its velocity, we obtain the *phase velocity*, v_p, of this wave. Mathematically,

$$v_p = \frac{\omega}{\beta}$$

A transmission line has no dispersion if the phase velocity of a propagating signal is independent of frequency. Hence, a graphical plot of ω versus β will be a straight line passing through the origin. This type of plot is called the *dispersion diagram* of a transmission line. An information-carrying signal is composed of many sinusoidal waves. If the line is dispersive, each of these harmonics will travel at a different velocity. Therefore, the information will be distorted at the receiving end. The velocity with which a group of waves travels is called the *group velocity*, v_g. It is equal to the slope of the dispersion curve of the transmission line.

Consider two sinusoidal signals with angular frequencies $\omega + \delta\omega$ and $\omega - \delta\omega$, respectively. Assume that these waves of equal amplitudes are propagating in the z-direction with corresponding phase constants $\beta + \delta\beta$ and $\beta - \delta\beta$. The resulting wave can be found as follows:

$$\begin{aligned} f(z,t) &= \mathrm{Re}[Ae^{j[(\omega+\delta\omega)t-(\beta+\delta\beta)z]} + Ae^{j[(\omega-\delta\omega)t-(\beta-\delta\beta)z]}] \\ &= 2A\cos(\delta\omega t - \delta\beta z)\cos(\omega t - \beta z) \end{aligned}$$

Hence, the resulting wave, $f(z,t)$, is amplitude modulated. The envelope of this signal moves with the group velocity,

$$v_g = \frac{\delta\omega}{\delta\beta}$$

Example 3.2 A signal generator has an internal resistance of 50 Ω and an open-circuit voltage $v(t) = 3\cos(2\pi \times 10^8 t)$ V. It is connected to a 75-Ω lossless transmission line that is 4 m long and terminated by a matched load at the other end. If the signal propagation velocity on this line is 2.5×10^8 m/s, find the instantaneous voltage and current at an arbitrary location on the line.

SOLUTION Since the transmission line is terminated by a load that is equal to its characteristic impedance, there will be no echo signal. Further, an equivalent circuit at its input end may be drawn as shown in Figure 3.6. Using the voltage-division rule and Ohm's law, incident voltage and current can be determined as follows:

$$\text{incident voltage at the input end, } V_{\text{in}}(z=0) = \frac{75}{50+75} 3\,\angle 0^\circ = 1.8\,\angle 0^\circ \text{ V}$$

$$\text{incident current at the input end, } I_{\text{in}}(z=0) = \frac{3\,\angle 0^\circ}{50+75} = 0.024\,\angle 0^\circ \text{ A}$$

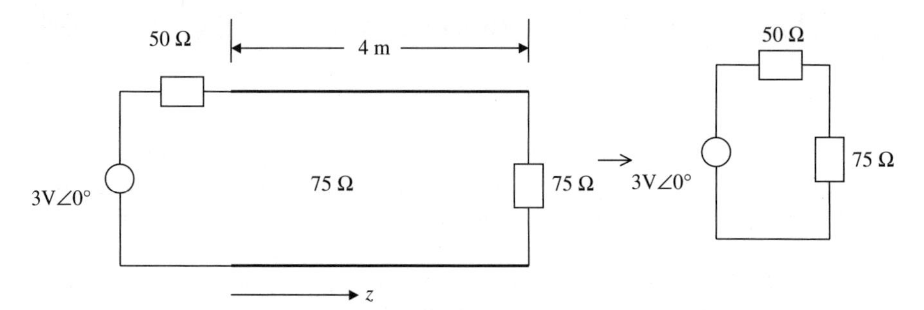

Figure 3.6 Circuit arrangement for Example 3.2.

and

$$\beta = \frac{\omega}{v_p} = \frac{2\pi \times 10^8}{2.5 \times 10^8} = 0.8\pi \text{ rad/m}$$

Therefore, $V(z) = 1.8e^{-j0.8\pi z}$ V and $I(z) = 0.024e^{-j0.8\pi z}$ A. Hence,

$$v(z,t) = 1.8\cos(2\pi \times 10^8 t - 0.8\pi z) \text{ V} \quad \text{and}$$
$$i(z,t) = 0.024\cos(2\pi \times 10^8 t - 0.8\pi z) \text{ A}$$

Example 3.3 The parameters of a transmission line are $R = 2\,\Omega$/m, $G = 0.5$ mS/m, $L = 8$ nH/m, and $C = 0.23$ pF/m. If the signal frequency is 1 GHz, calculate its characteristic impedance (Z_0) and the propagation constant (γ).

SOLUTION

$$Z_0 = \sqrt{\frac{R + j\omega L}{G + j\omega C}} = \sqrt{\frac{2 + j2\pi \times 10^9 \times 8 \times 10^{-9}}{0.5 \times 10^{-3} + j2\pi \times 10^9 \times 0.23 \times 10^{-12}}}\ \Omega$$
$$= \sqrt{\frac{2 + j50.2655}{0.5 \times 10^{-3} + j1.4451 \times 10^{-3}}}\ \Omega = \sqrt{\frac{50.31\ \angle 1.531 \text{ rad}}{15.29 \times 10^{-4}\ \angle 1.2377 \text{ rad}}}\ \Omega$$
$$= 181.39\,\Omega\ \angle 8.4^\circ = 179.44 + j26.51\,\Omega$$

and

$$\gamma = \sqrt{ZY} = \sqrt{(50.31\ \angle 1.531 \text{ rad}) \times (15.29 \times 10^{-4}\ \angle 1.2377 \text{ rad})}$$
$$= 0.2774\ \angle 79.31^\circ \text{m}^{-1} = 0.0514 + j0.2726 \text{ m}^{-1} = \alpha + j\beta$$

Therefore, $\alpha = 0.0514$ Np/m and $\beta = 0.2726$ rad/m.

Example 3.4 Two antennas are connected through a quarter-wavelength-long lossless transmission line, as shown in Figure 3.7. However, the characteristic

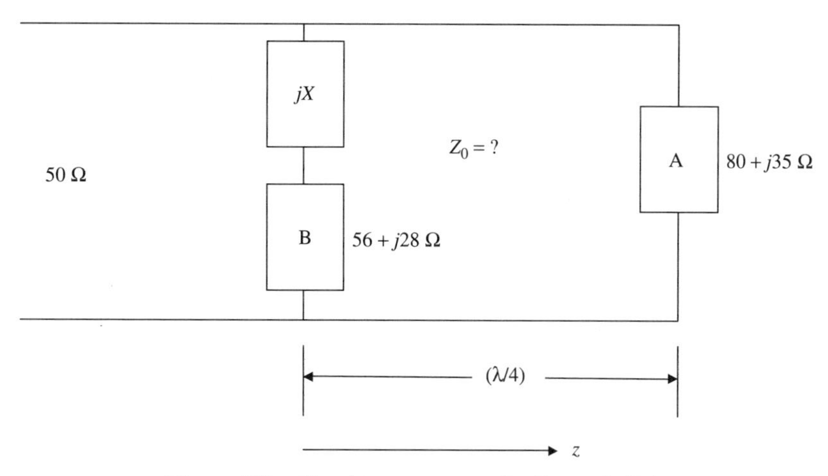

Figure 3.7 Circuit arrangement for Example 3.4.

impedance of this line is unknown. The array is excited through a 50-Ω line. Antenna A has an impedance of $80 + j35\ \Omega$ and antenna B has $56 + j28\ \Omega$. Currents (peak values) through these antennas are found to be $1.5\ \angle 0^\circ$ and $1.5\ \angle 90^\circ$ A, respectively. Determine the characteristic impedance of the line connecting these two antennas and the value of a reactance connected in series with antenna B.

SOLUTION Assume that V_{in} and V_{ref} are the incident and reflected phasor voltages, respectively, at antenna A. Therefore, the current, I_{A}, through this antenna is

$$I_A = \frac{V_{\text{in}} - V_{\text{ref}}}{Z_0} = 1.5\ \angle 0^\circ\ \text{A} \Rightarrow V_{\text{in}} - V_{\text{ref}} = Z_0 I_{\text{A}} = Z_0 1.5\ \angle 0^\circ \text{V}$$

Since the connecting transmission line is a quarter-wavelength long, the incident and reflected voltages across the transmission line at the location of B will be jV_{in} and $-jV_{\text{ref}}$, respectively. Therefore, the total voltage, V_{TBX}, appearing across the antenna B and the reactance jX combined will be equal to $j(V_{\text{in}} - V_{\text{ref}})$.

$$V_{\text{TBX}} = j(V_{\text{in}} - V_{\text{ref}}) = \text{j}Z_0 I_{\text{A}} = j1.5Z_0 = 1.5Z_0\ \angle 90^\circ$$

Also, Ohm's law can be used to find this voltage as follows:

$$V_{\text{TBX}} = (56 + j28 + jX)1.5\ \angle 90^\circ$$

Therefore, $X = -28\ \Omega$ and $Z_0 = 56\ \Omega$. Note that the unknown characteristic impedance is a real quantity because the transmission line is lossless.

3.2 SENDING-END IMPEDANCE

Consider a transmission line of length l and characteristic impedance Z_0. It is terminated by a load impedance Z_L, as shown in Figure 3.8. Assume that the

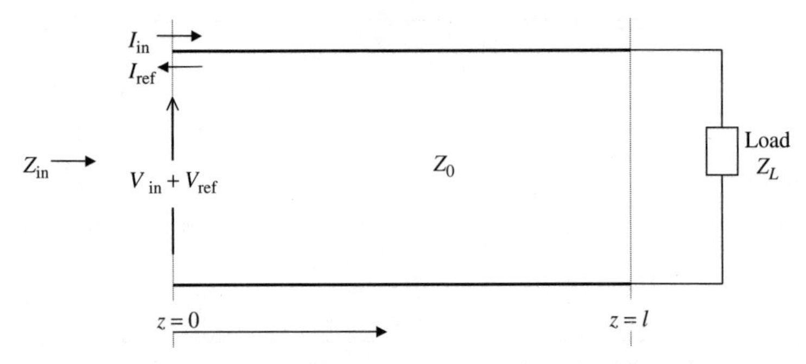

Figure 3.8 Transmission line terminated by a load impedance.

incident and reflected voltages at its input ($z = 0$) are V_{in} and V_{ref}, respectively. The corresponding currents are represented by I_{in} and I_{ref}.

If $V(z)$ represents the total phasor voltage at point z on the line and $I(z)$ is the total current at that point,

$$V(z) = V_{in}e^{-\gamma z} + V_{ref}e^{\gamma z} \tag{3.2.1}$$

and

$$I(z) = I_{in}e^{-\gamma z} + I_{ref}e^{\gamma z} \tag{3.2.2}$$

where V_{in}, V_{ref}, I_{in}, and I_{ref} are incident voltage, reflected voltage, incident current, and reflected current at $z = 0$, respectively.

Impedance at the input of this transmission line, Z_{in}, can be found after dividing total voltage by the total current at $z = 0$. Thus,

$$Z_{in} = \frac{V(z=0)}{I(z=0)} = \frac{V_{in} + V_{ref}}{I_{in} + I_{ref}} = \frac{V_{in} + V_{ref}}{V_{in}/Z_0 - V_{ref}/Z_0} = Z_0\frac{V_{in} + V_{ref}}{V_{in} - V_{ref}}$$

or

$$Z_{in} = Z_0\frac{1 + V_{ref}/V_{in}}{1 - V_{ref}/V_{in}} = Z_0\frac{1 + \Gamma_0}{1 - \Gamma_0} \tag{3.2.3}$$

where $\Gamma_0 = \rho e^{j\phi}$ is known as the *input reflection coefficient*. Further,

$$Z_{in} = Z_0\frac{1 + \Gamma_0}{1 - \Gamma_0} \Rightarrow \frac{Z_{in}}{Z_0} = \overline{Z}_{in} = \frac{1 + \Gamma_0}{1 - \Gamma_0}$$

where $\overline{Z}_{in}$ is called the *normalized input impedance*. Similarly, voltage and current at $z = l$ are related through load impedance as follows:

$$\begin{aligned} Z_L &= \frac{V(z=l)}{I(z=l)} = \frac{V_{in}e^{-\gamma l} + V_{ref}e^{+\gamma l}}{I_{in}e^{-\gamma l} + I_{ref}e^{+\gamma l}} \\ &= Z_0\frac{V_{in}e^{-\gamma l} + V_{ref}e^{+\gamma l}}{V_{in}e^{-\gamma l} - V_{ref}e^{+\gamma l}} = Z_0\frac{e^{-\gamma l} + \Gamma_0 e^{+\gamma l}}{e^{-\gamma l} - \Gamma_0 e^{+\gamma l}} \end{aligned}$$

Therefore,

$$\overline{Z_L} = \frac{e^{-\gamma l} + \Gamma_0 e^{\gamma l}}{e^{-\gamma l} - \Gamma_0 e^{+\gamma l}} \Rightarrow \Gamma_0 = \frac{\overline{Z_L} - 1}{\overline{Z_L} + 1} e^{-2\gamma l} \tag{3.2.4}$$

and equation (3.2.3) can be written as follows:

$$\overline{Z_{\text{in}}} = \frac{1 + [(\overline{Z_L} - 1)/(\overline{Z_L} + 1)]e^{-2\gamma l}}{1 - [(\overline{Z_L} - 1)/(\overline{Z_L} + 1)]e^{-2\gamma l}} = \frac{\overline{Z_L}(1 + e^{-2\gamma l}) + (1 - e^{-2\gamma l})}{\overline{Z_L}(1 - e^{-2\gamma l}) + (1 + e^{-2\gamma l})}$$

since

$$\frac{1 - e^{-2\gamma l}}{1 + e^{-2\gamma l}} = \frac{e^{+\gamma l} - e^{-\gamma l}}{e^{+\gamma l} + e^{-\gamma l}} = \frac{\sinh \gamma l}{\cosh \gamma l} = \tanh \gamma l$$

$$\overline{Z}_{\text{in}} = \frac{\overline{Z}_L + \tanh \gamma l}{1 + \overline{Z}_L \tanh \gamma l}$$

or

$$Z_{\text{in}} = Z_0 \frac{Z_L + Z_0 \tanh \gamma l}{Z_0 + Z_L \tanh \gamma l} \tag{3.2.5}$$

For a lossless line, $\gamma = \alpha + j\beta = j\beta$, and therefore $\tanh \gamma l = \tanh j\beta l = j \tan \beta l$. Hence, equation (3.2.5) simplifies as follows:

$$Z_{\text{in}} = Z_0 \frac{Z_L + jZ_0 \tan \beta l}{Z_0 + jZ_L \tan \beta l} \tag{3.2.6}$$

It is to be noted from this equation that Z_{in} repeats periodically every one-half wavelength on the transmission line. In other words, the input impedance on a lossless transmission line will be the same at points $d \pm n\lambda/2$, where n is an integer, due to the fact that

$$\beta l = \frac{2\pi}{\lambda}\left(d \pm \frac{n\lambda}{2}\right) = \frac{2\pi d}{\lambda} \pm n\pi$$

and

$$\tan \beta l = \tan\left(\frac{2\pi d}{\lambda} \pm n\pi\right) = \tan \frac{2\pi d}{\lambda}$$

Some special cases are as follows:

1. $Z_L = 0$ (i.e., a lossless line is short circuited) $\Rightarrow Z_{\text{in}} = jZ_0 \tan \beta l$.
2. $Z_L = \infty$ (i.e., a lossless line has an open circuit at the load) $\Rightarrow Z_{\text{in}} = -jZ_0 \cot \beta l$.
3. $l = \lambda/4$, and therefore $\beta l = \pi/2 \Rightarrow Z_{\text{in}} = Z_0^2/Z_L$.

According to the first two cases, a lossless line can be used to synthesize an arbitrary reactance. The third case indicates that a quarter-wavelength-long line of suitable characteristic impedance can be used to transform a load impedance Z_L to a new value of Z_{in}. This type of transmission line, called an *impedance transformer*, is useful in impedance-matching applications. Further, this equation can be rearranged as follows:

$$\overline{Z}_{in} = \frac{1}{\overline{Z}_L} = \overline{Y}_L$$

Hence, normalized impedance at a point a quarter-wavelength away from the load is equal to the normalized load admittance.

Example 3.5 A transmission line of length d and characteristic impedance Z_0 acts as an impedance transformer to match a 150-Ω load to a 300-Ω line (see Figure 3.9). If the signal wavelength is 1 m, find **(a)** d, **(b)** Z_0, and **(c)** the reflection coefficient at the load.

SOLUTION

(a) $d = \lambda/4 \rightarrow d = 0.25\,\text{m}$

(b) $Z_0 = \sqrt{Z_{in}Z_L} \rightarrow Z_0 = (150 \times 300)^{1/2} = 212.132\ \Omega$

(c) $\Gamma_L = \dfrac{\overline{Z}_L - 1}{\overline{Z}_L + 1} = \dfrac{Z_L - Z_0}{Z_L + Z_0} = \dfrac{150 - 212.132}{150 + 212.132} = 0.1716$

Example 3.6 Design a quarter-wavelength transformer to match a 20-Ω load to the 45-Ω line at 3 GHz. If this transformer is made from a Teflon-filled ($\varepsilon_r = 2.1$) coaxial line, calculate its length (in centimeters). Also, determine the diameter of its inner conductor if the inner diameter of the outer conductor is 0.5 cm. Assume that the impedance transformer is lossless.

SOLUTION

$$Z_0 = \sqrt{Z_{in}Z_L} = \sqrt{45 \times 20} = 30\ \Omega$$

and

$$Z_0 = \sqrt{\frac{L}{C}} = \sqrt{\frac{200 \times 10^{-9}}{55.63 \times \varepsilon_r \times 10^{-12}}}\,\ln\frac{b}{a} = 30$$

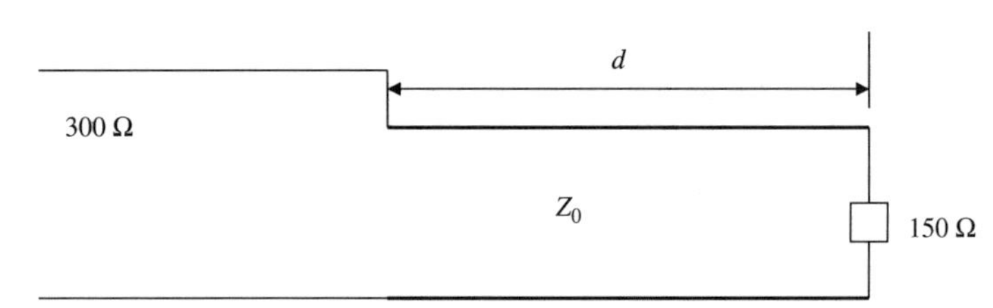

Figure 3.9 Circuit arrangement for Example 3.5.

Therefore,

$$\ln \frac{b}{a} = \frac{30}{41.3762} = 0.7251$$

and

$$\frac{b}{a} = \frac{2b}{2a} = e^{0.7251} = 2.0649 \Rightarrow 2a = \frac{2b}{2.0649} = \frac{0.5}{2.0649} = 0.2421 \text{ cm}$$

Phase constant $\beta = \omega\sqrt{LC}$:

$$\beta = \frac{2\pi}{\lambda} \Rightarrow \frac{1}{\lambda} = \frac{\omega}{2\pi}\sqrt{LC} = f \times \sqrt{200 \times 55.63 \times \varepsilon_r \times 10^{-21}}$$
$$= 3 \times 10^9 \times 10^{-10} \times \sqrt{200 \times 55.63 \times 2.1 \times 0.1}$$
$$= 14.5011 \text{ m}^{-1}$$

Therefore, $\lambda = 0.06896$ m and $d = \lambda/4 = 0.01724$ m $= 1.724$ cm.

Example 3.7 Design a quarter-wavelength microstrip impedance transformer to match a patch antenna of 80 Ω with a 50-Ω line. The system is to be fabricated on a 1.6-mm-thick substrate ($\varepsilon_r = 2.3$) that operates at 2 GHz (see Figure 3.10).

SOLUTION The characteristic impedance of the microstrip line impedance transformer must be

$$Z_0 = \sqrt{Z_{01} Z_L} = 63.2456\ \Omega$$

Design formulas for microstrip line are given in Appendix 2. Assume that the strip thickness t is less than 0.096 mm and that the dispersion is negligible for the time being at the operating frequency.

$$A = \frac{63.2456}{60}\left(\frac{2.3+1}{2}\right)^{1/2} + \frac{2.3-1}{2.3+1} \times \left(0.23 + \frac{0.11}{2.3}\right) = 1.4635$$

$$B = \frac{60\pi^2}{63.2456\sqrt{2.3}} = 6.1739$$

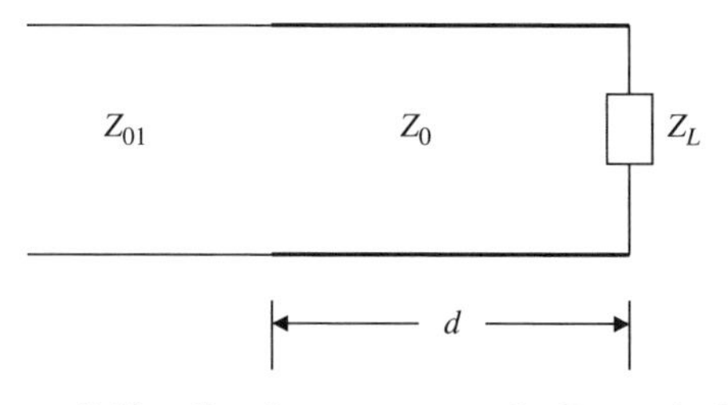

Figure 3.10 Circuit arrangement for Example 3.7.

Since $A \leq 1.52$,

$$\frac{w}{h} = \frac{2}{\pi}\left\{6.1739 - 1 - \ln(2 \times 6.1739 - 1) + \frac{2.3 - 1}{2 \times 2.3}\left[\ln(6.1739 - 1) + 0.39 - \frac{0.61}{2.3}\right]\right\}$$

$$= \frac{2 \times 3.2446}{\pi} = 2.0656 \Rightarrow w = 3.3\,\text{mm}$$

At this point, we can check if the dispersion in the line is really negligible. For that, we determine the effective dielectric constant as follows:

$$F\left(\frac{w}{h}\right) = \left(1 + 12\frac{h}{w}\right)^{-1/2} = \left(1 + \frac{12}{2.0656}\right)^{-1/2} = 0.383216$$

Assuming that $t/h = 0.005$,

$$\varepsilon_e = \frac{2.3 + 1}{2} + \frac{2.3 - 1}{2} \times 0.383216 - \frac{2.3 - 1}{4.6} \times \frac{0.005}{\sqrt{2.0656}} = 1.8981 \approx 1.9$$

and

$$F = \frac{4h\sqrt{\varepsilon_r - 1}}{\lambda_0}\left\{0.5 + \left[1 + 2 \times \log\left(1 + \frac{w}{h}\right)\right]^2\right\} = 0.213712$$

$$\varepsilon_e(f) = \left(\frac{\sqrt{2.3} - \sqrt{1.9}}{1 + 4 \times F^{-1.5}} + \sqrt{1.9}\right)^2 = 1.909192$$

Since $\varepsilon_e(f)$ is very close to ε_e, dispersion in the line can be neglected. Therefore,

$$\lambda = \frac{3 \times 10^8}{2 \times 10^9 \times \sqrt{1.9}}\text{m} = 10.8821\,\text{cm} \Rightarrow \text{length of line} = \frac{10.8821}{4} = 2.72\,\text{cm}$$

Reflection Coefficient, Return Loss, and Insertion Loss

The *voltage reflection coefficient* is defined as the ratio of reflected to incident phasor voltages at a location in the circuit. In the case of a transmission line terminated by load Z_L, the voltage reflection coefficient is given by equation (3.2.4). Hence,

$$\Gamma = \frac{V_{\text{ref}}}{V_{\text{in}}} = \frac{Z_L - Z_0}{Z_L + Z_0}e^{-2\gamma l} = \rho_L e^{j\theta}e^{-2(\alpha + j\beta)l} = \rho_L e^{-2\alpha l}e^{-j(2\beta l - \theta)} \qquad (3.2.7)$$

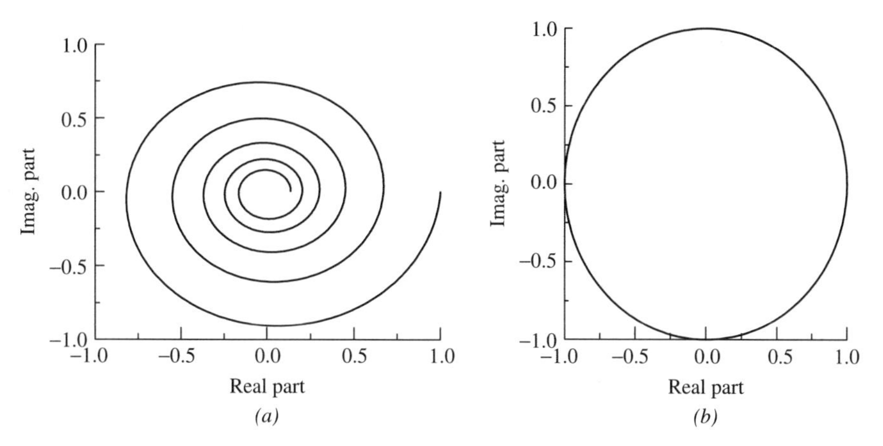

Figure 3.11 Reflection coefficient on (*a*) a lossy and (*b*) a lossless transmission line.

where $\rho_L e^{j\theta} = (Z_L - Z_0)/(Z_L + Z_0)$ is called the *load reflection coefficient*. Equation (3.2.7) indicates that the magnitude of reflection coefficient decreases by a factor of $e^{-2\alpha l}$ as the observation point moves away from the load. Further, its phase angle changes by $-2\beta l$. A polar (magnitude and phase) plot of it will look like a spiral, as shown in Figure 3.11(*a*). However, the magnitude of reflection coefficient will not change if the line is lossless. Therefore, the reflection coefficient point will be moving clockwise on a circle of radius equal to its magnitude as the line length is increased. As illustrated in Figure 3.11 (*b*), it makes one complete revolution for each half-wavelength distance away from the load (because $-2\beta\lambda/2 = -2\pi$).

Similarly, the *current reflection coefficient*, Γ_c, is defined as a ratio of reflected to incident signal-current phasors. It is related to the voltage reflection coefficient as follows:

$$\Gamma_c = \frac{I_{\text{ref}}}{I_{\text{in}}} = \frac{-V_{\text{ref}}/Z_0}{V_{\text{in}}/Z_0} = -\Gamma$$

The *return loss* of a device is defined as the ratio of reflected power to incident power at its input. Since the power is proportional to the square of the voltage at that point, it may be found as

$$\text{return loss} = \frac{\text{reflected power}}{\text{incident power}} = \rho^2$$

Generally, it is expressed in decibels as follows:

$$\text{return loss} = 20\log_{10}\rho \qquad \text{dB} \tag{3.2.8}$$

The *insertion loss* of a device is defined as the ratio of power transmitted (power available at the output port) to that of the power incident at its input. Since the power transmitted is equal to the difference of the incident and reflected powers for a lossless device, the insertion loss can be expressed as follows:

$$\text{insertion lossof a lossless device} = 10\log_{10}(1-\rho^2) \quad \text{dB} \tag{3.2.9}$$

Low-Loss Transmission Lines

Most practical transmission lines possess a very small loss of propagating signal. Therefore, expressions for the propagation constant and the characteristic impedance can be approximated for such lines as follows:

$$\gamma = \sqrt{ZY} = \sqrt{(R+j\omega L)(G+j\omega C)} = \sqrt{-\omega^2 LC\left(1+\frac{R}{j\omega L}\right)\left(1+\frac{G}{j\omega C}\right)}$$

For $R \ll \omega L$ and $G \ll \omega C$, a first-order approximation is

$$\gamma = \alpha + j\beta \approx j\omega\sqrt{LC}\left(1+\frac{R}{j2\omega L}\right)\left(1+\frac{G}{j2\omega C}\right)$$

$$\approx j\omega\sqrt{LC}\left(1+\frac{R}{j2\omega L}+\frac{G}{j2\omega C}\right)$$

Therefore,

$$\alpha \approx \frac{1}{2}\left(R\sqrt{\frac{C}{L}}+G\sqrt{\frac{L}{C}}\right) \quad \text{Np/m} \tag{3.2.10}$$

and

$$\beta \approx \omega\sqrt{LC} \quad \text{rad/m} \tag{3.2.11}$$

$$Z_0 = \sqrt{\frac{R+j\omega L}{G+j\omega C}} = \sqrt{\frac{L}{C}}\left(1+\frac{R}{j\omega L}\right)^{1/2}\left(1+\frac{G}{j\omega C}\right)^{-1/2}$$

Hence,

$$Z_0 \approx \sqrt{\frac{L}{C}}\left(1+\frac{R}{j2\omega L}\right)\left(1-\frac{G}{j2\omega C}\right) \approx \sqrt{\frac{L}{C}}\left(1+\frac{R}{j2\omega L}-\frac{G}{j2\omega C}\right)$$

$$= \sqrt{\frac{L}{C}}\left[1+\frac{1}{j2\omega}\left(\frac{R}{L}-\frac{G}{C}\right)\right] \tag{3.2.12}$$

Thus, the attenuation constant of a low-loss line is independent of frequency, while its phase constant is linear, as in the case of a lossless line. However, the frequency dependency of its characteristic impedance is of concern to communication engineers, because it will distort the signal. If $RC = GL$, the frequency-dependent term will go to zero. This type of low-loss line is called a *distortionless line*. Hence,

$$\begin{aligned}\gamma &= \sqrt{ZY} = \sqrt{(R + j\omega L)(G + j\omega C)} \\ &= (R + j\omega L)\sqrt{\frac{C}{L}} = \alpha + j\beta \qquad \left(\because G = \frac{RC}{L}\right)\end{aligned} \tag{3.2.13}$$

Example 3.8 A signal propagating through a 50-Ω distortionless transmission line attenuates at the rate of 0.01 dB/m. If this line has a capacitance of 100 pF/m, find **(a)** R, **(b)** L, **(c)** G, and **(d)** v_p.

SOLUTION Since the line is distortionless,

$$Z_0 = \sqrt{\frac{L}{C}} = 50$$

and

$$\alpha = 0.01\,\text{dB/m} \approx 0.01/8.69\,\text{Np/m} = 1.15 \times 10^{-3}\,\text{Np/m}$$

Hence,

(a) $R = \alpha\sqrt{L/C} = 1.15 \times 10^{-3} \times 50 = 0.057\,\Omega/\text{ m}$
(b) $L = CZ_0^2 = 10^{-10} \times 50^2\text{ H/m} = 0.25\ \mu\text{H/m}$
(c) $G = RC/L = R/Z_0^2 = 0.057/50^2\text{ S/m} = 22.8\mu\text{S/ m}$
(d) $v_p = 1/\sqrt{LC} = 2 \times 10^8\text{ m/s}$

Experimental Determination of Characteristic Impedance and Propagation Constant of a Transmission Line

A transmission line of length d is kept open at one end, and the impedance at its other end is measured using an impedance bridge. Assume that it is Z_{oc}. The process is repeated after placing a short circuit at its open end, and this impedance is recorded as Z_{sc}. Using equation (3.2.5), one can write

$$Z_{oc} = Z_0 \coth \gamma d \tag{3.2.14}$$

and

$$Z_{sc} = Z_0 \tanh \gamma d \tag{3.2.15}$$

Therefore,

$$Z_0 = \sqrt{Z_{oc}Z_{sc}} \tag{3.2.16}$$

and

$$\tanh \gamma d = \sqrt{\frac{Z_{sc}}{Z_{oc}}} \Rightarrow \gamma = \frac{1}{d}\tanh^{-1}\sqrt{\frac{Z_{sc}}{Z_{oc}}} \tag{3.2.17}$$

These two equations can be solved to determine Z_0 and γ. The following identity can be used to facilitate the evaluation of the propagation constant:

$$\tanh^{-1} Z = \frac{1}{2} \ln \frac{1+Z}{1-Z}$$

The following examples illustrate this procedure.

Example 3.9 Impedance at one end of the transmission line is measured to be $Z_{\text{in}} = 30 + j60\,\Omega$ using an impedance bridge, while its other end is terminated by a load Z_L. The experiment is repeated twice, with the load replaced first by a short circuit and then by an open circuit. These data are recorded as $j53.1\,\Omega$ and $-j48.3\,\Omega$, respectively. Find the characteristic resistance of this line and the load impedance.

SOLUTION

$$Z_0 = \sqrt{Z_{\text{sc}} \times Z_{\text{oc}}} = \sqrt{j53.1 \times (-j48.3)} = 50.6432\,\Omega$$

$$Z_{\text{in}} = Z_0 \frac{Z_L + jZ_0 \tan\beta l}{Z_0 + jZ_L \tan\beta l} = Z_0 \frac{Z_L + Z_{\text{sc}}}{Z_0 + jZ_L \tan\beta l}$$

$$= \frac{Z_0}{j\tan\beta l} \frac{Z_L + Z_{\text{sc}}}{Z_L + Z_0/(j\tan\beta l)}$$

Therefore,

$$Z_{\text{in}} = Z_{\text{oc}} \frac{Z_L + Z_{\text{sc}}}{Z_L + Z_{\text{oc}}} \Rightarrow Z_L = Z_{\text{oc}} \frac{Z_{\text{sc}} - Z_{\text{in}}}{Z_{\text{in}} - Z_{\text{oc}}}$$

$$Z_L = -j48.3 \times \frac{j53.1 - (30 + j60)}{30 + j60 - (-j48.3)}$$

Therefore,

$$Z_L = 11.6343 + j6.3\,\Omega$$

Example 3.10 Measurements are made on a 1.5-m-long transmission line using an impedance bridge. After short-circuiting at one of its ends, impedance at the other end is found to be $j103\,\Omega$. Repeating the experiment with the short circuit now replaced by an open circuit gives $-j54.6\,\Omega$. Determine the propagation constant and the characteristic impedance of this line.

SOLUTION

$$Z_0 = \sqrt{Z_{\text{oc}} Z_{\text{sc}}} = \sqrt{-j54.6 \times j103} = 74.99 \approx 75\,\Omega$$

$$\tanh 1.5\gamma = \sqrt{\frac{j103}{-j54.3}} = j1.8969 \Rightarrow 1.5\gamma = \tanh^{-1} j1.8969$$

$$= \frac{1}{2} \ln \frac{1 + j1.8969}{1 - j1.8969}$$

$$1.5\gamma = \frac{1}{2} \ln \frac{1 + j1.8969}{1 - j1.8969} = \frac{1}{2} \ln(1 \angle 2.1713 \text{ rad}) = \frac{1}{2} \ln(e^{j2.1713}) = j1.08565$$

and $\gamma = j0.7238\,\text{m}^{-1}$.

Example 3.11 A 10-m-long 50-Ω lossless transmission line is terminated by a load, $Z_L = 100 + j50\,\Omega$. It is driven by a signal generator that has an open-circuit voltage V_s at $100 \angle 0^\circ$ V and source impedance Z_s at 50 Ω. The propagating signal has a phase velocity of 200 m/μS at 26 MHz. Determine the impedance at its input end and the phasor voltages at both its ends (see Figure 3.12).

SOLUTION

$$Z_{\text{in}} = Z_0 \frac{Z_L + jZ_0 \tan \beta l}{Z_0 + jZ_L \tan \beta l} \quad \text{and} \quad \beta = \frac{\omega}{v_p} = \frac{2\pi \times 26 \times 10^6}{200 \times 10^6} = 0.8168 \text{ rad/m}$$

$$= 50 \frac{100 + j50 + j50 \tan(8.168)}{50 + j(100 + j50) \tan(8.168)} = 19.5278\,\Omega \angle 0.181 \text{ rad}$$

$$= 19.21 + j3.52\,\Omega$$

The equivalent circuits at its input and at the load can be drawn as in Figure 3.13.

$$\text{Voltage at the input end, } V_A = \frac{V_s}{Z_s + Z_{\text{in}}} Z_{\text{in}} = \frac{100 \times 19.5278 \angle 0.181 \text{ rad}}{50 + 19.21 + j3.52} \text{ V}$$

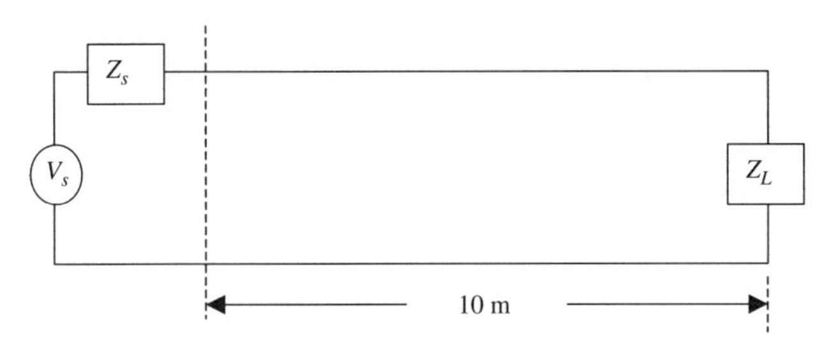

Figure 3.12 Circuit arrangement for Example 3.11.

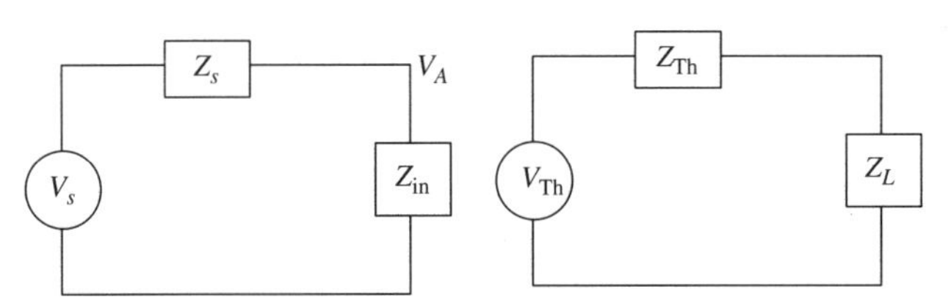

Figure 3.13 Equivalent circuits at the two ends of the transmission line shown in Figure 3.12.

or

$$V_A = \frac{1952.78\ \angle 0.181\,\text{rad}}{69.2995\ \angle 0.0508\,\text{rad}} = 28.1788\,\text{V}\ \angle 0.1302\,\text{rad} = 28.1788\,\text{V}\ \angle 7.46^\circ$$

For determining voltage at the load, a Thévenin equivalent circuit can be used as follows:

$$Z_{\text{Th}} = Z_0 \frac{Z_s + jZ_0 \tan \beta l}{Z_0 + jZ_s \tan \beta l} = Z_0 \quad (\because Z_s = Z_0 = 50\,\Omega)$$

and

$$V_{\text{Th}} = (V_{\text{in}} + V_{\text{ref}})_{\text{at o.c.}} = (2 \times V_{\text{in}})_{\text{at o.c.}} = 100\,\text{V}\ \angle -8.168\,\text{rad}$$

Therefore,

$$V_L = \frac{V_{\text{Th}}}{Z_{\text{Th}} + Z_L} Z_L = \frac{100\ \angle -8.168\,\text{rad}}{50 + 100 + j50} \times (100 + j50)\ \text{V}$$

or

$$V_L = 70.71\,\text{V}\ \angle -8.0261\,\text{rad} = 70.71\,\text{V}\ \angle -459.86^\circ = 70.71\,\text{V}\ \angle -99.86^\circ$$

Alternatively,

$$V(z) = V_{\text{in}} e^{-j\beta z} + V_{\text{ref}} e^{j\beta z} = V_{in}(e^{-j\beta z} + \Gamma e^{j\beta z})$$

where V_{in} is incident voltage at $z = 0$ and Γ is the input reflection coefficient.

$$V_{\text{in}} = 50\,\text{V}\ \angle 0^\circ$$

and

$$\Gamma = \frac{19.21 + j3.52 - 50}{19.21 + j3.52 + 50} = 0.4472\ \angle 2.9769\,\text{rad}$$

Hence,

$$\begin{aligned} V_L = V(z = 10\,\text{m}) &= 50(e^{-j8.168} + 0.4472 e^{j(2.9769+8.168)}) \\ &= 50 \times 1.4142\,\text{V}\ \angle -1.743\,\text{rad} \end{aligned}$$

or

$$V_L = 70.7117\,\text{V}\ \angle -99.8664^\circ$$

Example 3.12 Two identical signal generators are connected in parallel through a quarter-wavelength-long lossless 50-Ω transmission line (Figure 3.14). Each of these generators has an open-circuit voltage of 12 $\angle 0^\circ$ V and a source impedance

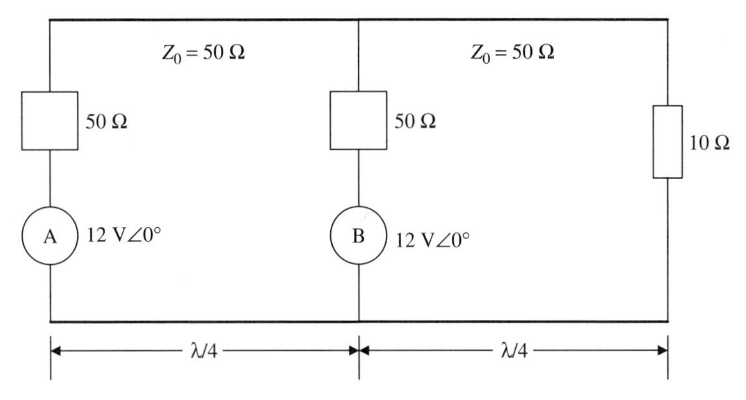

Figure 3.14 Circuit arrangement for Example 3.12.

of 50 Ω. It drives a 10-Ω load through another quarter-wavelength-long similar transmission line. Determine the power dissipated by the load.

SOLUTION The signal generator connected at the left (source A) and the quarter-wavelength-long line connecting it can be replaced by a Thévenin equivalent voltage source of $12\,\text{V}\angle{-90^\circ}$ with its internal resistance at 50 Ω. This voltage source can be combined with the $12\ \angle 0^\circ$ V (source B) already there, which also has an internal resistance of 50 Ω. The resulting source will have a Thévenin voltage of $6 - j6\,\text{V} = 6\sqrt{2}\,\text{V}\ \angle{-45^\circ}$ and an internal resistance of 25 Ω (i.e., two 50-Ω resistances connected in parallel). The load will transform to $50 \times 50/10 = 250\ \Omega$ at the location of source B. Therefore, a simplified circuit can be drawn as in Figure 3.15.

The voltage across 250 Ω will be equal to

$$250 \times \frac{6 - j6}{275\,\text{V}} = 60\sqrt{2}/11\ \angle{-45^\circ}\text{V}$$

Thus, the dissipated power, P_d, can be calculated as follows:

$$P_d = \frac{60^2 \times 2}{11^2 \times 2 \times 250} = 0.119008\,\text{W} = 119.008\,\text{mW}$$

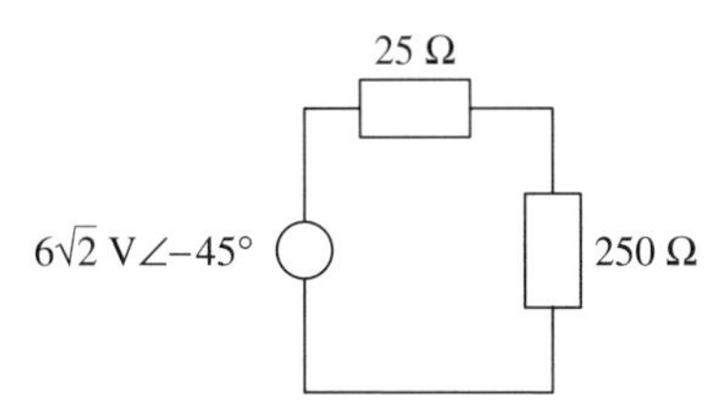

Figure 3.15 Simplification of the circuit shown in Figure 3.14.

3.3 STANDING WAVE AND STANDING WAVE RATIO

Consider a lossless transmission line that is terminated by a load impedance Z_L, as shown in Figure 3.16. Incident and reflected voltage phasors at its input (i.e., at $z = 0$) are assumed to be V_{in} and V_{ref}, respectively. Therefore, total voltage $V(z)$ can be expressed as follows:

$$V(z) = V_{\text{in}}e^{-j\beta z} + V_{\text{ref}}e^{+j\beta z}$$

Alternatively, $V(x)$ can be written as

$$V(x) = V_+e^{+j\beta x} + V_-e^{-j\beta x} = V_+(e^{+j\beta x} + \Gamma e^{-j\beta x})$$

or

$$V(x) = V_+[e^{+j\beta x} + \rho e^{-j(\beta x - \phi)}] \tag{3.3.1}$$

where

$$\Gamma = \rho e^{j\phi} = \frac{V_-}{V_+}$$

V_+ and V_- represent incident and reflected wave voltage phasors, respectively, at the load point (i.e., at $x = 0$).

Let us first consider two extreme conditions at the load. In one case, the transmission line has an open circuit, whereas in the other, it has a short-circuit termination. Therefore, the magnitude of the reflection coefficient, ρ, is unity in both cases. However, the phase angle, ϕ, is zero for the former case and π for the latter. Phasor diagrams for these two cases are depicted in Figure 3.17(*a*) and (*b*), respectively. As distance x from the load increases, the phasor $e^{j\beta x}$ rotates counterclockwise because of the increase in its phase angle βx. On the other hand, the phasor $e^{-j\beta x}$ rotates clockwise by the same amount. Therefore, the phase angle of the resulting voltage, $V(x)$, remains constant with space coordinate x while its magnitude varies sinusoidally between $\pm 2V_+$. Since the phase angle of the resulting signal $V(x)$ does not change with distance, it does not represent

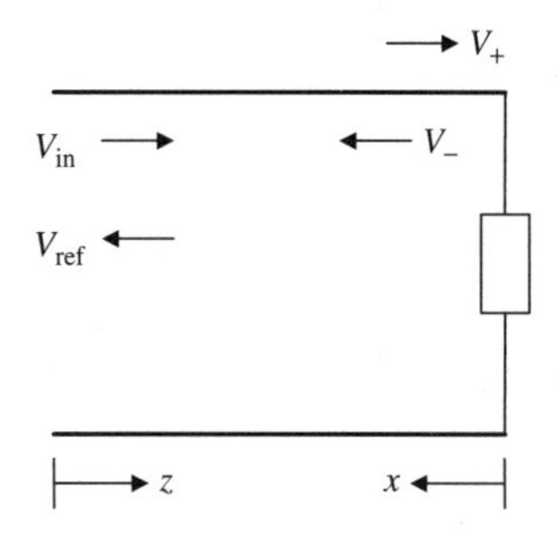

Figure 3.16 Lossless transmission line with termination.

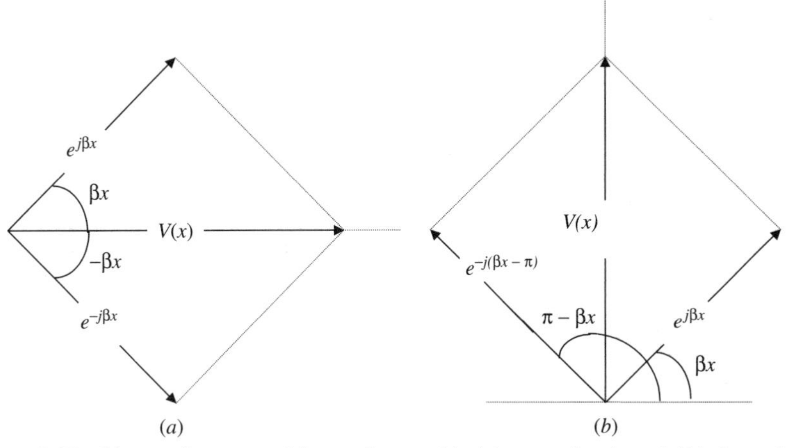

Figure 3.17 Phasor diagrams of line voltage with *(a)* open-circuit and *(b)* short-circuit termination.

a propagating wave. Note that there are two waves propagating on this line in opposite directions. However, the resulting signal represents a *standing wave*.

When the terminating load is of an arbitrary value that is different from the characteristic impedance of the line, there are still two waves propagating in opposite directions. The interference pattern of these two signals is stationary with time. Assuming that V_+ is unity, a phasor diagram for this case is drawn as shown in Figure 3.18. The magnitude of the resulting signal, $V(x)$, can be determined using the law of parallelograms as follows:

$$|V(x)| = |V_+|[1 + \rho^2 + 2\rho\cos(2\beta x - \phi)]^{1/2} \tag{3.3.2}$$

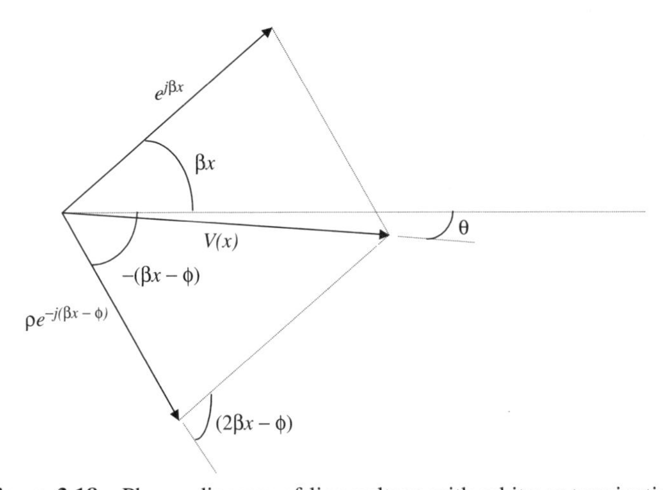

Figure 3.18 Phasor diagram of line voltage with arbitrary termination.

The reflection coefficient, $\Gamma(x)$, can be expressed as

$$\Gamma(x) = \frac{V_{-}e^{-j\beta x}}{V_{+}e^{j\beta x}} = \rho e^{-j(2\beta x-\phi)} \tag{3.3.3}$$

Hence, the magnitude of this standing wave changes with location on the transmission line. Since x appears only in the argument of cosine function, the voltage magnitude has an extreme value whenever this argument is an integer multiple of π. It has a maximum whenever the reflected wave is in phase with the incident signal, which requires that the following condition be satisfied:

$$2\beta x - \phi = \pm 2n\pi, \qquad n = 0, 1, 2, \ldots \tag{3.3.4}$$

On the other hand, $V(x)$ has a minimum value where the reflected and incident signals are out of phase. Hence, x satisfies the following condition at a minimum of the interference pattern:

$$2\beta x - \phi = \pm(2m+1)\pi, \qquad m = 0, 1, 2, \ldots \tag{3.3.5}$$

Further, these extreme values of the standing waves are

$$|V(x)|_{\max} = |V_{+}|(1+\rho) \tag{3.3.6}$$

and

$$|V(x)|_{\min} = |V_{+}|(1-\rho) \tag{3.3.7}$$

The ratio of maximum to minimum values of voltage $V(x)$ is called the *voltage standing wave ratio* (VSWR). Therefore,

$$\text{VSWR} = S = \frac{|V(x)|_{\max}}{|V(x)|_{\min}} = \frac{1+\rho}{1-\rho} \tag{3.3.8}$$

Since $0 \le \rho \le 1$ for a passive load, the minimum value of the VSWR will be unity (for a matched load) while its maximum value can be infinity (for total reflection, with a short circuit or an open circuit as the load).

Assume that there is a voltage minimum at x_1 from the load, and if one keeps moving toward the source, the next minimum occurs at x_2. In other words, there are two consecutive minimums at x_1 and x_2 with $x_2 > x_1$. Hence,

$$2(\beta x_1 - \phi) = (2m_1 + 1)\pi$$

and

$$2(\beta x_2 - \phi) = [2(m_1+1)+1]\pi$$

Subtracting the former equation from the latter, one finds that

$$2\beta(x_2 - x_1) = 2\pi \Rightarrow x_2 - x_1 = \frac{\lambda}{2}$$

where $x_2 - x_1$ is the separation between the two consecutive minimums.

Similarly, it can be proved that two consecutive maximums are a half-wavelength apart and also that separation between the consecutive maximum and minimum is a quarter-wavelength. This information can be used to measure the wavelength of a propagating signal. In practice, the location of a minimum is preferred over that of a maximum. This is because minimums are sharper in comparison with maximums, as illustrated in Figure 3.19. Further, a short (or open) circuit must be used as the load for the best measurement accuracy.

Measurement of Impedance

Impedance of a one-port microwave device can be determined from measurement on the standing wave at its input. Those required parameters are the VSWR and the location of the first minimum (or maximum) from the load. A slotted line that is equipped with a detector probe is connected before the load to facilitate measurement. Since the output of a detector is proportional to power, the square root of the ratio is taken to find the VSWR. Since it may not be possible in most cases to probe up to the input terminals of the load, the location of the first minimum is determined as follows. An arbitrary minimum is located on the slotted line with unknown load. The load is then replaced by a short circuit. As a result, there is a shift in minimum, as shown in Figure 3.19. The shift of the original minimum away from the generator is equal to the location of the first minimum from the load.

Since the reflected voltage is out of phase with that of the incident signal at the minimum of the standing wave pattern, a relation between the reflection coefficient, $\Gamma_1 = -\rho$, and the impedance Z_1 at this point may be written as

$$\Gamma_1 = -\rho = \frac{\overline{Z}_1 - 1}{\overline{Z}_1 + 1}$$

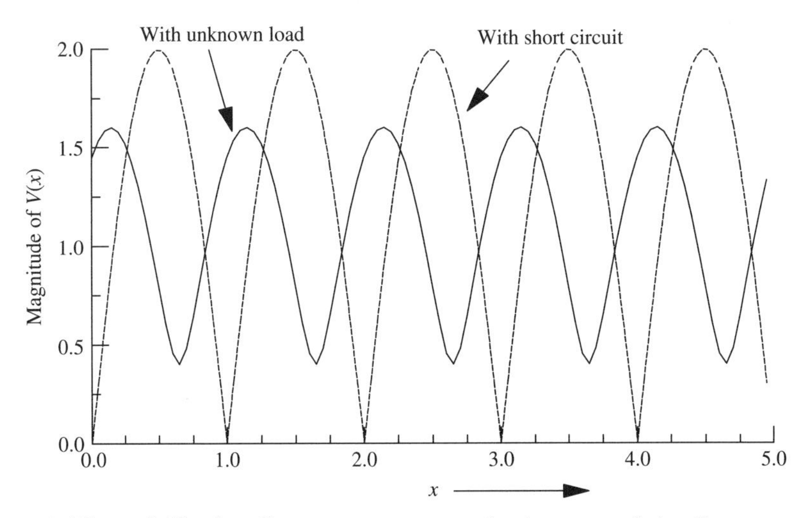

Figure 3.19 Standing wave pattern on a lossless transmission line.

where $\overline{Z}_1$ is the normalized impedance at the location of the first minimum of the voltage standing wave. Since

$$S = \frac{1+\rho}{1-\rho} \quad \text{and} \quad \rho = \frac{S-1}{S+1}$$

and therefore,

$$-\frac{S-1}{S+1} = \frac{\overline{Z}_1 - 1}{\overline{Z}_1 + 1} \Rightarrow \overline{Z}_1 = \frac{1}{S}$$

This normalized impedance is equal to the input impedance of a line that is terminated by load Z_L and has a length of d_1. Hence,

$$\overline{Z}_1 = \frac{1}{S} = \frac{\overline{Z}_L + j\tan\beta d_1}{1 + j\overline{Z}_L \tan\beta d_1} \Rightarrow \overline{Z}_L = \frac{1 - jS\tan\beta d_1}{S - j\tan\beta d_1}$$

Similarly, the reflected voltage is in phase with the incident signal at the maximum of the standing wave. Assume that the first maximum is located at a distance d_2 from the unknown load and that the impedance at this point is Z_2. One can write

$$\Gamma_2 = \rho = \frac{\overline{Z}_2 - 1}{\overline{Z}_2 + 1} = \frac{S-1}{S+1} \Rightarrow \overline{Z}_2 = S$$

and

$$\overline{Z}_L = \frac{\overline{Z}_2 - j\tan\beta d_2}{1 - j\overline{Z}_2 \tan\beta d_2}$$

Since the maximum and minimum are measured on a lossless line that feeds the unknown load, magnitude of the reflection coefficient ρ does not change at different points on the line.

Example 3.13 A load impedance of $73 - j42.5\ \Omega$ terminates a transmission line of characteristic impedance $50 + j0.01\ \Omega$. Determine its reflection coefficient and the voltage standing wave ratio.

SOLUTION

$$\Gamma = \frac{Z_L - Z_0}{Z_L + Z_0} = \frac{73 - j42.5 - (50 + j0.01)}{73 - j42.5 + (50 + j0.01)}$$

$$= \frac{23 - j42.51}{123 - j42.49} = \frac{48.3332\ \angle -1.0749\,\text{rad}}{130.1322\ \angle -0.3326\,\text{rad}}$$

$$= 0.3714\ \angle -0.7423\,\text{rad} = 0.3714\ \angle -42.53^\circ$$

and

$$\text{VSWR} = \frac{1+\rho}{1-\rho} = \frac{1+0.3714}{1-0.3714} = 2.1817$$

Note that this line is lossy, and therefore the reflection coefficient and VSWR will change with distance from the load.

Example 3.14 A 100-Ω transmission line is terminated by the load Z_L. Measurements indicate that it has a VSWR of 2.4 and the standing wave minimums are 100 cm apart. The scale reading at one of these minimums is found to be 275 cm. When the load is replaced by a short circuit, the minimum moves away from the generator to a point where the scale shows 235 cm. Find the signal wavelength and the load impedance.

SOLUTION Since the minimums are 100 cm apart, the signal wavelength $\lambda = 2 \times 100 = 200\,\text{cm} = 2\,\text{m}$, $\beta = \pi$ rad/m, and $d_1 = 275 - 235 = 40\,\text{cm}$. Hence,

$$\overline{Z}_L = \frac{1 - j2.4\tan(0.4\pi)}{2.4 - j\tan(0.4\pi)} = 1.65 - j0.9618$$

or

$$Z_L = 165 - j96.18\,\Omega$$

3.4 SMITH CHART

Normalized impedance at a point in the circuit is related to its reflection coefficient as

$$\overline{Z} = \overline{R} + j\overline{X} = \frac{1 + \Gamma}{1 - \Gamma} = \frac{1 + \Gamma_r + j\Gamma_i}{1 - \Gamma_r - j\Gamma_i} \tag{3.4.1}$$

where Γ_r and Γ_i represent real and imaginary parts of the reflection coefficient, while $\overline{R}$ and $\overline{X}$ are real and imaginary parts of the normalized impedance, respectively. This complex equation can be split into two, after equating real and imaginary components on its two sides. Hence,

$$\left(\Gamma_r - \frac{\overline{R}}{1 + \overline{R}}\right)^2 + \Gamma_i^2 = \left(\frac{1}{1 + \overline{R}}\right)^2 \tag{3.4.2}$$

and

$$(\Gamma_r - 1)^2 + \left(\Gamma_i - \frac{1}{\overline{X}}\right)^2 = \left(\frac{1}{\overline{X}}\right)^2 \tag{3.4.3}$$

These two equations represent a family of circles on the complex Γ-plane. The circle in (3.4.2) has its center at $(\overline{R}/(1 + \overline{R}), 0)$ and a radius of $1/(1 + \overline{R})$. For $\overline{R} = 0$ it is centered at the origin with unity radius. As $\overline{R}$ increases, the center of the constant-resistance circle moves on a positive real axis and its radius decreases. When $\overline{R} = \infty$, the radius reduces to zero and the center of the circle moves to (1,0). These plots are shown in Figure 3.20. Note that for passive impedance, $0 \le \overline{R} \le \infty$ and $-\infty \le \overline{X} \le +\infty$.

Similarly, (3.4.3) represents a circle that is centered at $(1, 1/\overline{X})$ with a radius of $1/\overline{X}$. For $\overline{X} = 0$, its center lies at $(1,\infty)$ with an infinite radius. Hence, it is a straight line along the Γ_r-axis. As $\overline{X}$ increases on the positive side (i.e., $0 \leq \overline{X} \leq \infty$), the center of the circle moves toward point (1,0) along a vertical line defined by $\Gamma_r = 1$, and its radius becomes smaller and smaller in size. For $\overline{X} = \infty$, it becomes a point that is located at (1,0). Similar characteristics are observed for $0 \geq \overline{X} \geq -\infty$. As shown in Figure 3.20, a graphical representation of these two equations for all possible normalized resistance and reactance values is known as the *Smith chart*. Thus, a normalized impedance point on the Smith chart represents the corresponding reflection coefficient in polar coordinates on the complex Γ-plane. According to (3.2.7), the magnitude of the reflection coefficient on a lossless transmission line remains constant as ρ_L while its phase angle decreases as $-2\beta l$. Hence, it represents a circle of radius ρ_L. As one moves away from the load (i.e., toward the generator), the reflection coefficient point moves clockwise on this circle. Since the reflection coefficient repeats periodically at every half-wavelength, the circumference of the circle is equal to $\lambda/2$. For a given reflection coefficient, normalized impedance may be found using the impedance scale of the Smith chart. Further, $\overline{R}$ in (3.4.1) is equal to the VSWR for $\Gamma_r > 0$ and $\Gamma_i = 0$. On the other hand, it equals the inverse of VSWR for $\Gamma_r < 0$ and $\Gamma_i = 0$. Hence, $\overline{R}$ values on the positive Γ_r-axis also represent the VSWR.

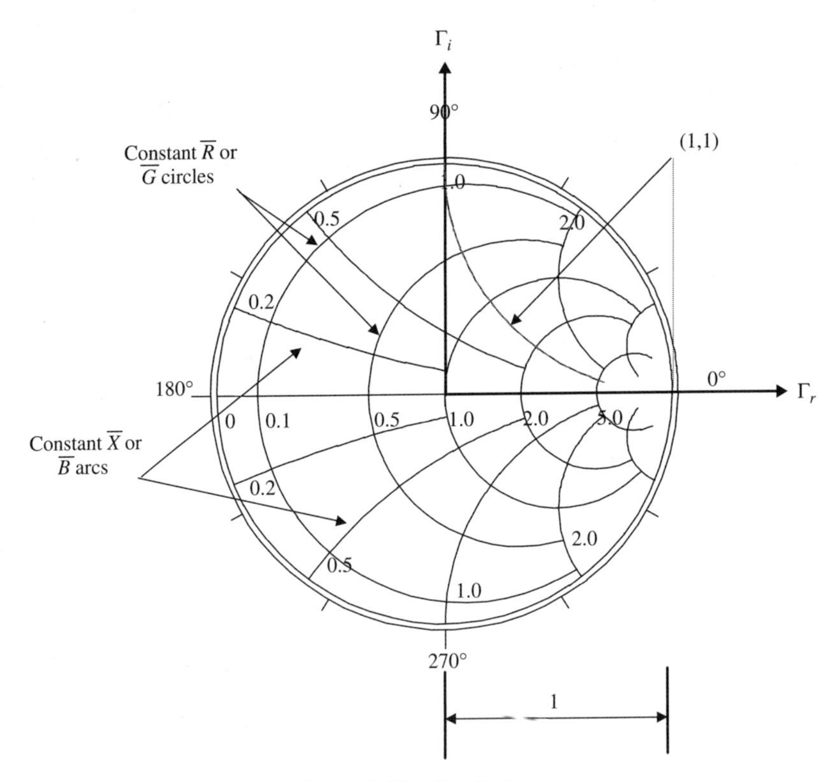

Figure 3.20 Smith chart.

Since admittance represents the inverse of impedance, we can write

$$\overline{Y} = \frac{1}{\overline{Z}} = \frac{1-\Gamma}{1+\Gamma}$$

Hence, a similar analysis can be performed with the normalized admittance instead of impedance. It results in the same kind of chart except that the normalized conductance circles replace normalized resistance circles, while the normalized susceptance arcs replace normalized reactance.

Normalized resistance (or conductance) of each circle is indicated on the Γ_r -axis of the Smith chart. Normalized positive reactance (or susceptance) arcs are shown on the upper half, while negative reactance (or susceptance) arcs are seen in the lower half. The Smith chart in conjunction with equation (3.2.7) facilitates the analysis and design of transmission line circuits.

Example 3.15 A load impedance of $50 + j100\ \Omega$ terminates a lossless, quarter-wavelength-long transmission line. If characteristic impedance of the line is $50\ \Omega$, find the impedance at its input end, the load reflection coefficient, and the VSWR on this transmission line.

SOLUTION This problem can be solved using equations (3.2.6), (3.2.7), and (3.3.8) or the Smith chart. Let us try it both ways.

$$\beta l = \frac{2\pi}{\lambda} \times \frac{\lambda}{4} = \frac{\pi}{2} = 90^\circ$$

$$Z_{\text{in}} = Z_0 \frac{Z_L + jZ_0 \tan\beta l}{Z_0 + jZ_L \tan\beta l} = 50 \frac{(50 + j100) + j50\tan(90^\circ)}{50 + j(50 + j100)\tan(90^\circ)}$$

or

$$Z_{\text{in}} = 50 \frac{j50}{j(50 + j100)} = \frac{2500}{50 + j100} = 10 - j20\ \Omega$$

$$\Gamma = \frac{Z_L - Z_0}{Z_L + Z_0} = \frac{50 + j100 - 50}{50 + j100 + 50} = \frac{j100}{100 + j100} = \frac{100\ \angle 90^\circ}{100 \times \sqrt{2}\ \angle 45^\circ}$$

$$= 0.7071\ \angle 45^\circ$$

$$\text{VSWR} = \frac{1 + |\Gamma|}{1 - |\Gamma|} = \frac{1 + 0.7071}{1 - 0.7071} = \frac{1.7071}{0.2929} = 5.8283$$

To solve this problem graphically using the Smith chart, normalized load impedance is determined as follows:

$$\overline{Z_L} = \frac{Z_L}{Z_0} = \frac{50 + j100}{50} = 1 + j2$$

This point is located on the Smith chart as shown in Figure 3.21. A circle that passes through $1 + j2$ is then drawn with point $1 + j0$ as its center. Since the

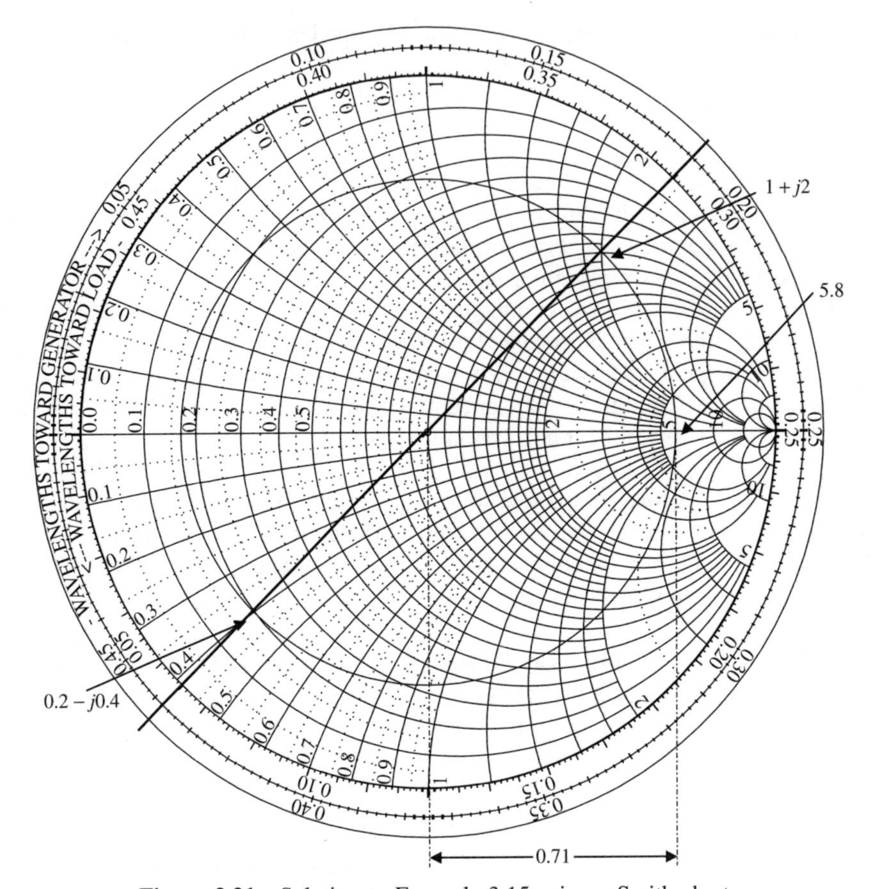

Figure 3.21 Solution to Example 3.15 using a Smith chart.

radius of the Smith chart represents unity magnitude, the radius of this circle is equal to the magnitude of the reflection coefficient ρ_L. In other words, the normalized radius of this circle (i.e., the radius of this circle divided by the radius of the Smith chart) is equal to ρ_L. Note that a clockwise movement on this circle corresponds to a movement away from the load on the transmission line. Hence, a point d meters away from the load is located at $-2\beta d$ on the chart. Therefore, the input port of the line that is a quarter-wavelength away from the load (i.e., $d = \lambda/4$) can be located on this circle after moving by $-\pi$ to a point at 0.438λ on the "wavelengths toward generator" scale. Thus, from the Smith chart, VSWR = 5.8, $\Gamma_L = 0.71\ \angle 45^\circ$, and $\overline{Z}_{\text{in}} = 0.2 - j0.4$:

$$Z_{\text{in}} = 50(0.2 - j0.4) = 10 - j20\ \Omega$$

Example 3.16 A lossless 75-Ω transmission line is terminated by an impedance of $150 + j150\ \Omega$. Using the Smith chart, find **(a)** Γ_L, **(b)** VSWR, **(c)** Z_{in} at a distance of 0.375λ from the load, (**d**) the shortest length of the line for which impedance is purely resistive, and (**e**) the value of this resistance.

SOLUTION

$$\overline{Z}_L = \frac{150 + j150}{75} = 2 + j2$$

After locating this normalized impedance point on the Smith chart, the constant VSWR circle is drawn as shown in Figure 3.22.

(**a**) The magnitude of the reflection coefficient is equal to the radius of the VSWR circle (with the radius of the Smith chart as unity). The angle made by the radial line that connects the load impedance point with the center of the chart is equal to the phase angle of reflection coefficient. Hence,

$$\Gamma_L = 0.62 \angle 30^\circ$$

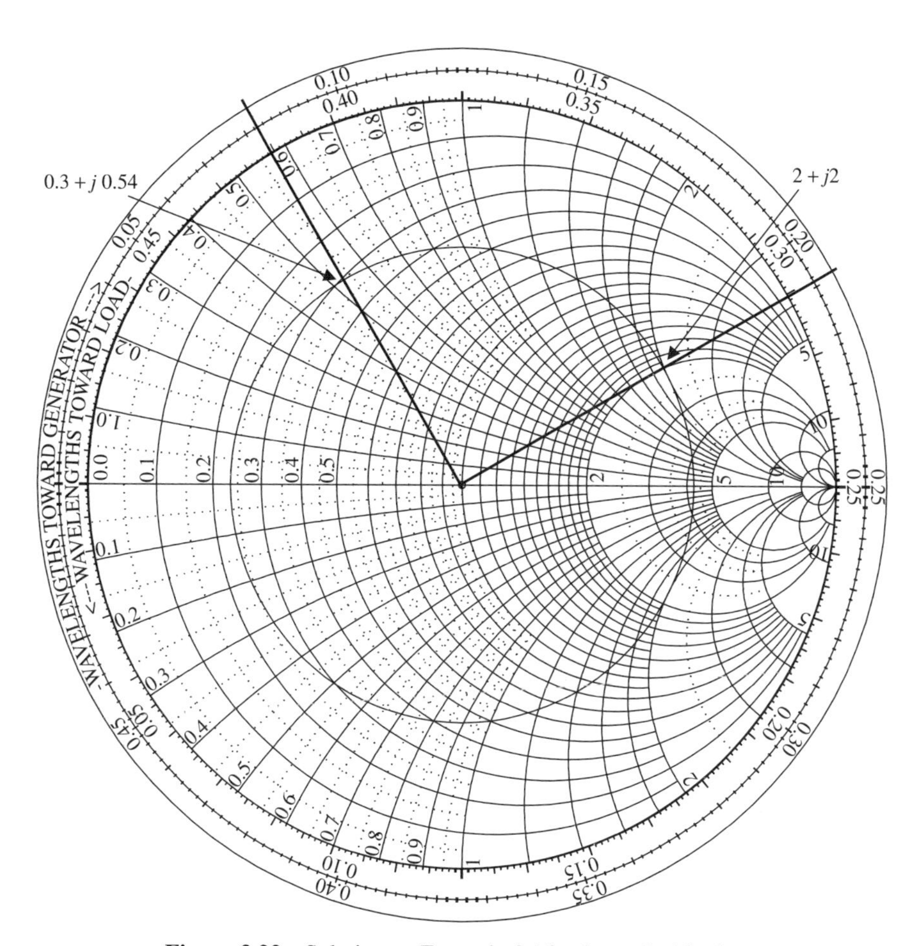

Figure 3.22 Solution to Example 3.16 using a Smith chart.

(**b**) VSWR is found to be 4.25 from the scale reading for the point where the circle intersects the $+\Gamma_r$ axis.

(**c**) For $d = 0.375\lambda$, $-2\beta d = -4.7124\,\text{rad} = -270^\circ$ (clockwise from the load). This point is located after moving on the VSWR circle by 0.375λ from the load (at 0.084λ on the "wavelengths toward generator" scale). The corresponding normalized impedance is found to be $0.3 + j0.54$. Therefore,

$$Z_{\text{in}}(l = 0.375\lambda) = 75(0.3 + j0.54) = 22.5 + j40.5\,\Omega$$

(**d**) While moving clockwise from the load point, the VSWR circle crosses the Γ_r-axis for the first time at 0.25λ. The imaginary part of impedance is zero at this intersection point. Therefore, $d = (0.25 - 0.208)\lambda = 0.042\lambda$. Normalized impedance at this point is 4.25. The next point on the transmission line where the impedance is purely real occurs a quarter-wavelength from it (i.e., 0.292λ from load). Normalized impedance at this point is 0.23.

(**e**) The normalized resistance and VSWR are the same at this point. Therefore,

$$R = 75 \times 4.25 = 318.75\,\Omega$$

Example 3.17 A lossless 100-Ω transmission line is terminated by an impedance of $100 + j100\,\Omega$ as illustrated in Figure 3.23. Find the location of the first V_{max}, the first V_{min}, and the VSWR if the operating wavelength is 5 cm.

SOLUTION

$$\overline{Z}_L = \frac{100 + j100}{100} = 1 + j1$$

As shown in Figure 3.24, this point is located on the Smith chart and the VSWR circle is drawn. From the chart, VSWR = 2.6. The scale reading on the "wavelengths toward generator" is 0.162λ at the load point. When one moves away from this point clockwise (toward generator) on this VSWR circle, the voltage maximum is found first at 0.25λ and then a minimum at 0.5λ. If the first voltage maximum is at d_{max} from the load, $d_{\text{max}} = (0.25 - 0.162)\lambda = 0.088\lambda = 0.44\,\text{cm}$

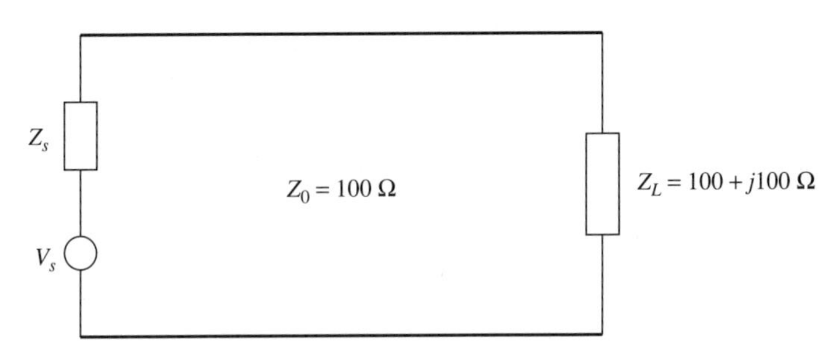

Figure 3.23 Circuit arrangement for Example 3.17.

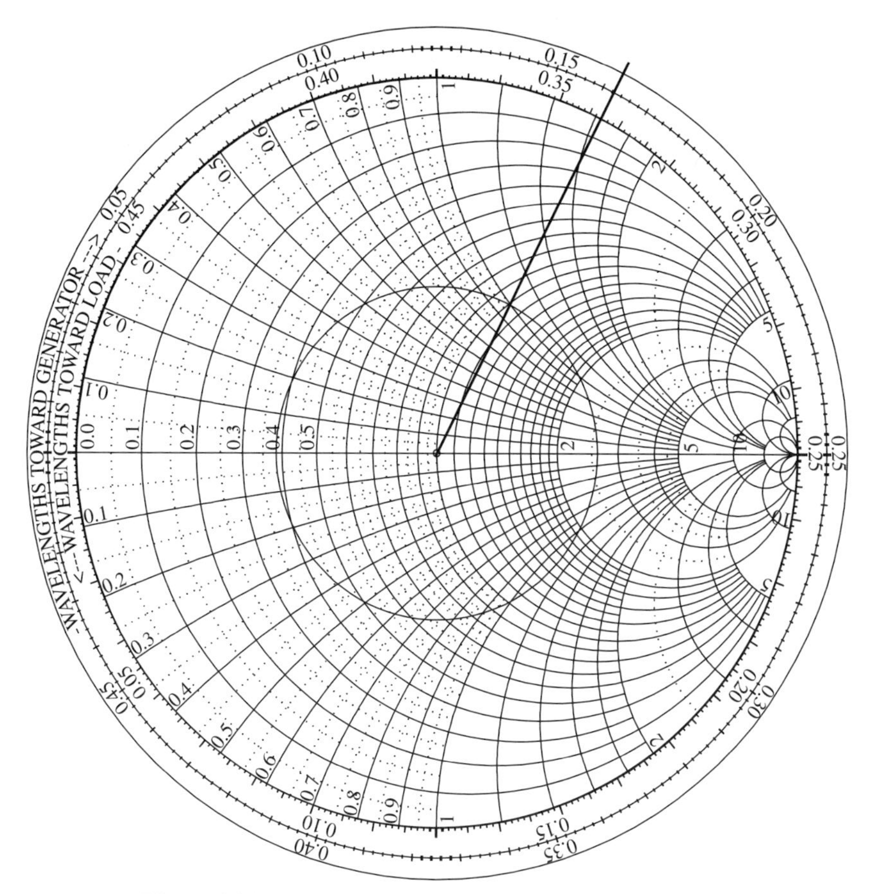

Figure 3.24 Solution to Example 3.17 using a Smith chart.

from the load. The first minimum is a quarter-wavelength away from the point of voltage maximum. Hence, $d_{\min} = (0.5 - 0.162)\lambda = 0.338\lambda = 1.69\,\text{cm}$.

Example 3.18 A 150-Ω load terminates a 75-Ω line. Find the impedance at points 2.15λ and 3.75λ from the termination:

$$\overline{Z}_L = \frac{150}{75} = 2$$

SOLUTION As illustrated in Figure 3.25, this point is located on the Smith chart and the VSWR circle is drawn. Note that the VSWR on this line is 2 and the load reflection coefficient is about $0.33\ \angle 0^\circ$. As one moves on the transmission line toward the generator, the phase angle of reflection coefficient changes by $-2\beta d$, where d is the distance away from the load. Hence, one revolution around the VSWR circle is completed for every half-wavelength. Therefore, the

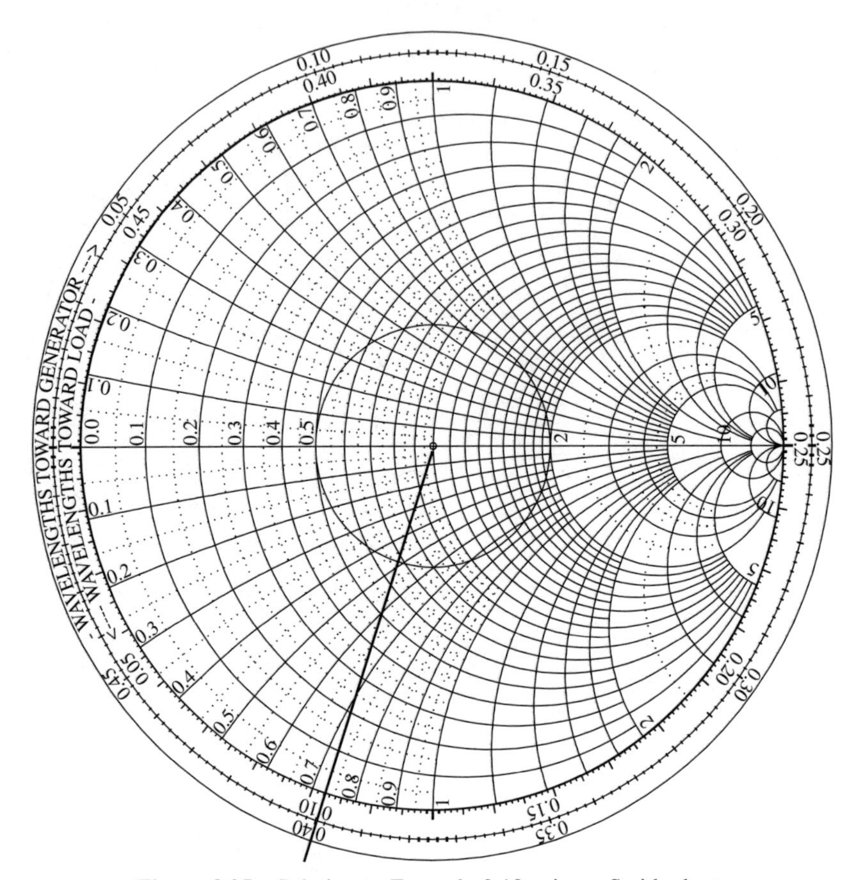

Figure 3.25 Solution to Example 3.18 using a Smith chart.

normalized impedance will be 2 at every integer multiple of a half-wavelength from the load. It will be true for a point located at 2λ as well as at 3.5λ. For the remaining 0.15λ, the impedance point is located on the VSWR circle at 0.40λ (i.e., 0.25 + 0.15) on the "wavelengths toward generator" scale. Similarly, the point corresponding to 3.75λ from the load is found at 0.5λ. From the Smith chart, the normalized impedance at 2.15λ is 0.68 − j0.48, while it is 0.5 at 3.75λ. Therefore,

$$\text{impedance at } 2.15\lambda,\ Z_1 = (0.68 - j0.48) \times 75 = 51 - j36\,\Omega$$

and

$$\text{impedance at } 3.75\lambda,\ Z_2 = (0.5) \times 75 = 37.5\,\Omega.$$

Example 3.19 A lossless 100-Ω transmission line is terminated by an admittance of 0.0025 − j0.0025 S. Find the impedance at a point 3.15λ away from the load and the VSWR on this line.

SOLUTION

$$\overline{Y}_L = \frac{Y_L}{Y_0} = Y_L \times Z_0 = 0.25 - j0.25$$

As before, this normalized admittance point is located on the Smith chart and the VSWR circle is drawn. It is shown in Figure 3.26. There are two choices available at this point. The given normalized load admittance is converted to corresponding impedance by moving to a point on the diametrically opposite side of the VSWR circle. It shows a normalized load impedance as $2 + j2$. Moving from this point by 3.15λ toward the generator, the normalized impedance is found as $0.55 - j1.08$. Alternatively, we can first move from the normalized admittance point by 3.15λ toward the generator to a normalized admittance point $0.37 + j0.75$. This is then converted to normalized impedance by moving to the diametric opposite point on the VSWR circle. Thus, the normalized impedance at a point 3.15λ away from the load is $0.55 - j1.08$. The impedance at this point is $100 \times (0.55 - j1.08) = 55 - j108\ \Omega$, and the VSWR is approximately 4.3.

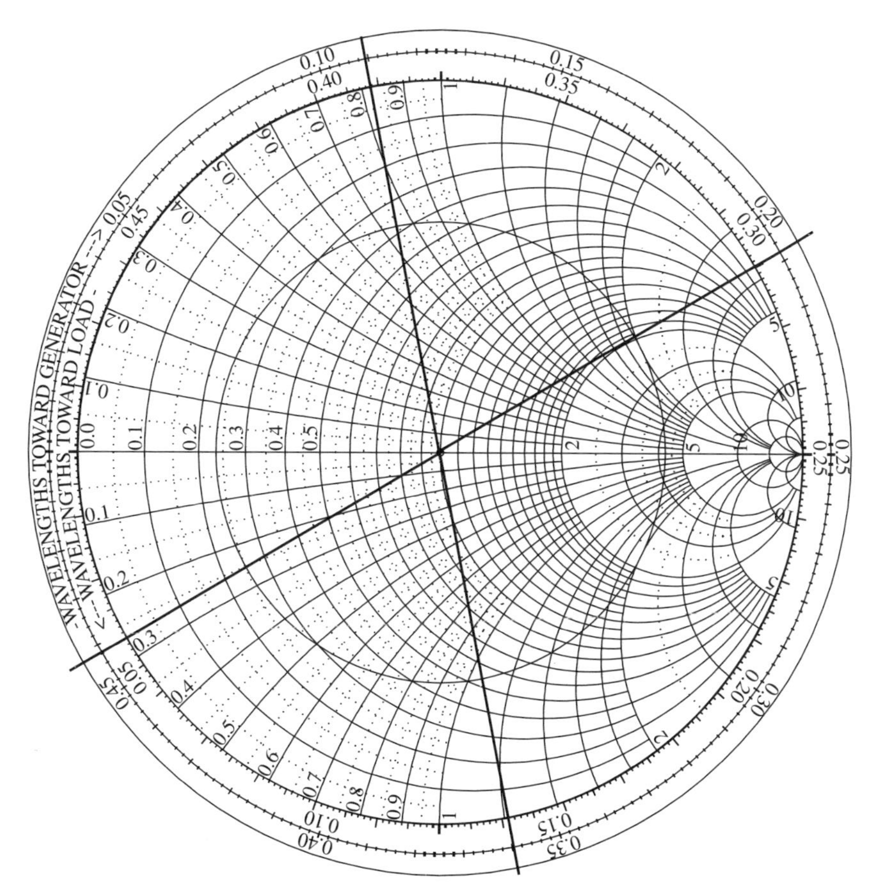

Figure 3.26 Solution to Example 3.19 using a Smith chart.

Example 3.20 An experiment is performed using the circuit illustrated in Figure 3.27. First, a load Z_L is connected at the end of a 100-Ω transmission line and its VSWR is found to be 2. After that, the detector probe is placed at one of the minimums on the line. It is found that this minimum shifts toward the load by 15 cm when the load is replaced by a short circuit. Further, two consecutive minimums are found to be 50 cm apart. Determine the load impedance.

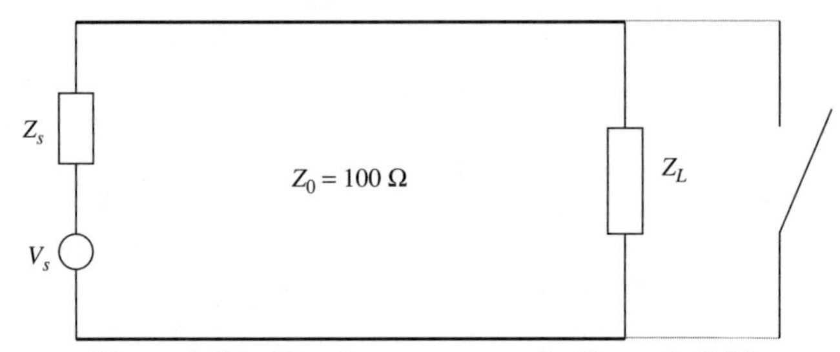

Figure 3.27 Circuit arrangement for Example 3.20.

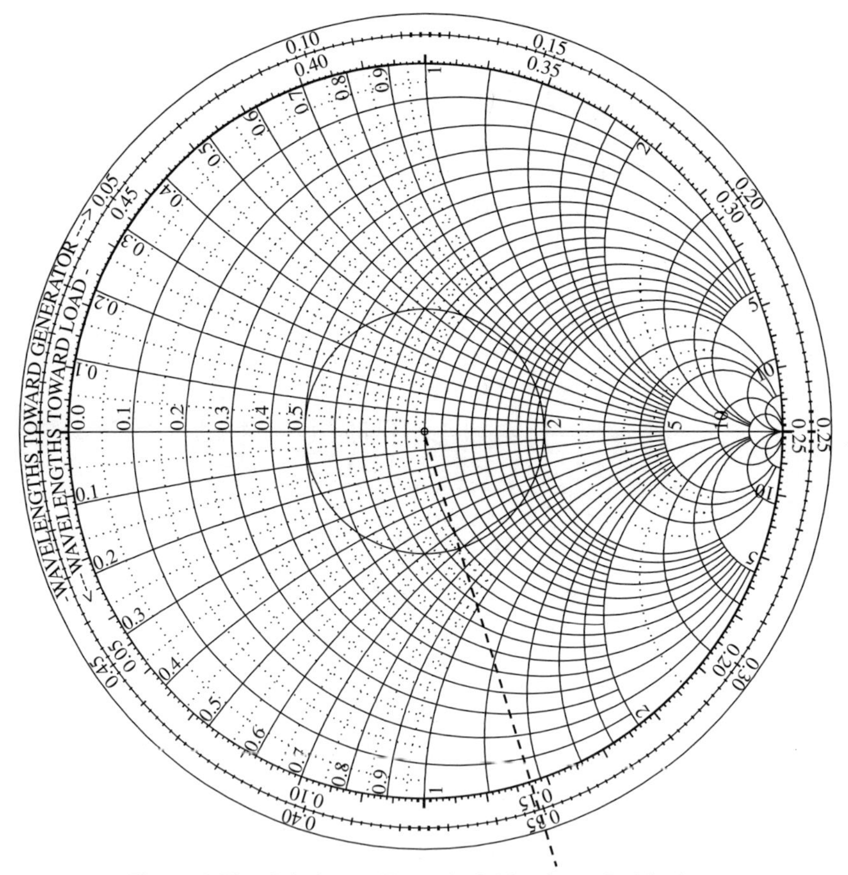

Figure 3.28 Solution to Example 3.20 using a Smith chart.

SOLUTION Since the separation between consecutive minimums is 50 cm, the signal wavelength on the line is 100 cm. Therefore, the first minimum of the standing wave pattern occurs at 0.15λ from the load. Since the VSWR on the line is measured as 2, this circle is drawn on the Smith chart as illustrated in Figure 3.28. As explained in the preceding section, a minimum in a voltage standing wave occurs when the phase angle of the reflection coefficient is 180°. It is located on the Smith chart at a point where the VSWR circle intersects the $-\Gamma_r$ axis. From this point we move toward the load by 0.15λ (i.e., counterclockwise) to locate the normalized load impedance point. It is found to be $1 - j0.7$. Therefore,

$$Z_L = 100 \times (1 - j0.7) = 100 - j70\ \Omega$$

SUGGESTED READING

Collin, R. E., *Foundations for Microwave Engineering*. New York: McGraw-Hill, 1992.

Edwards, T., *Foundations for Microstrip Circuit Design*. New York: Wiley, 1992.

Gupta, K. C., and I. J. Bahl, *Microstrip Lines and Slotlines*. Boston: Artech House, 1979.

Magnusson, P. C., G. C. Alexander, and V. K. Tripathi, *Transmission Lines and Wave Propagation*. Boca Raton, FL: CRC Press, 1992.

Pozar, D. M., *Microwave Engineering*. New York: Wiley, 1998.

Ramo, S., J. R. Whinnery, and T. Van Duzer. *Fields and Waves in Communication Electronics*. New York: Wiley, 1994.

Rizzi, P. A., *Microwave Engineering*. Englewood Cliffs, NJ: Prentice Hall, 1988.

Sinnema, W., *Electronic Transmission Technology*. Englewood Cliffs, NJ: Prentice Hall, 1988.

PROBLEMS

3.1. A lossless semirigid coaxial line has its inner and outer conductor radii as 1.325 and 4.16 mm, respectively. Find the line parameters, characteristic impedance, and the propagation constant for a signal frequency of 500 MHz ($\varepsilon_r = 2.1$).

3.2. Calculate the magnitude of the characteristic impedance and the propagation constant for a coaxial line at 2 GHz. Assume that $b = 3$ cm, $a = 0.5$ cm, and $\varepsilon = \varepsilon_0\ (2.56 - j0.005)$.

3.3. A certain telephone line has the following electrical parameters: $R = 40\ \Omega$/mile, $L = 1.1$ mH/mile, $G \approx 0$, and $C = 0.062\ \mu$F/mile. Loading coils are added which provide an additional inductance of 30 mH/mile as well as an additional resistance of 8 Ω/mile. Obtain the attenuation constant and phase velocities at frequencies of 300 Hz and 3.3 kHz.

3.4. A given transmission line has the following parameters: $Z_0 = 600\ \angle{-6^\circ}\ \Omega$, $\alpha = 2.0 \times 10^{-5}$ dB/m, $v_p = 2.97 \times 10^8$ m/s, and $f = 1.0$ kHz. Write phasor

$V(z)$ and $I(z)$ and the corresponding instantaneous values for a wave traveling in the z-direction if the maximum value of the current wave at $z = 0$ is 0.3 mA and it has a maximum positive value with respect to time at $t = 0$.

3.5. A $100 + j200$-Ω load is connected to a 75-Ω transmission line that is 1 m long. If signal wavelength on the line is 8 m, find the input impedance and the signal frequency. Assume that the phase velocity is 70% of the speed of light in free space.

3.6. A 30-km-long transmission line is terminated by a $100 + j200$-Ω load. A sinusoidal source with output voltage $v(t) = 15\cos(8000\pi t)$ V and internal resistance 75 Ω is connected at its other end. The characteristic impedance of the line is 75 Ω and the phase velocity of the signal is 2.5×10^8 m/s. Find the total voltage across its input end and the load.

3.7. A 2.5-m-long transmission line is short-circuited at one end and the impedance at its other end is found to be $j5$ Ω. When the short is replaced by an open circuit, the impedance at the other end changes to $-j500$ Ω. A 1.9-MHz sinusoidal source is used in this experiment, and the transmission line is less than a quarter-wavelength long. Determine the characteristic impedance of the line and phase velocity of the signal.

3.8. A lossless 75-Ω transmission line is connected between the load impedance of $37.5 - j15$ Ω and a signal generator with internal impedance of 75 Ω. Find (**a**) the reflection coefficient at 0.15λ from the load, (**b**) the VSWR on the line, and (**c**) the input impedance at 1.3λ from the load.

3.9. Two antennas are connected through a quarter-wavelength-long lossless transmission line, as shown in the Figure P3.9. However, the characteristic impedance of this line is unknown. The array is excited through a 50-Ω line. Antenna A has an impedance of $80 - j80$ Ω and antenna B, $75 - j55$ Ω. The peak voltage across antenna A is found to be $113.14\ \angle{-45^\circ}$ V and the peak current through antenna B is $1.0\ \angle 90^\circ$ A. Determine the characteristic impedance of the line connecting the two antennas and the value of a reactance connected in series with antenna B.

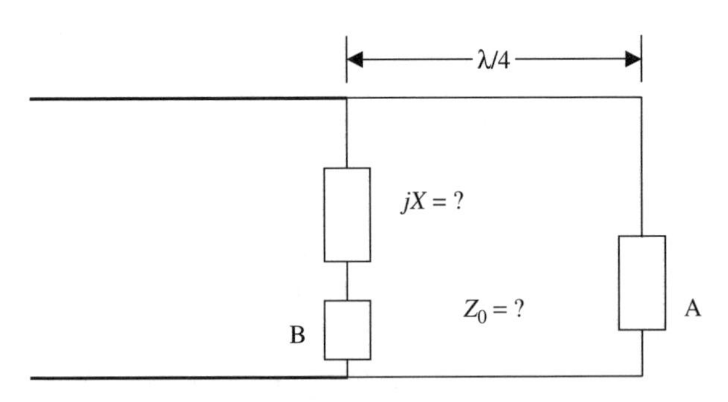

Figure P3.9

3.10. Measurements are made on a 1-m-long coaxial line (with negligible loss) using an impedance bridge that operates at 100 MHz. First, one of its ends is short-circuited and the impedance at the other end is measured as $-j86.6025\,\Omega$. Repeating the measurements with an open circuit in place of a short circuit shows an impedance of $j28.8675\,\Omega$. Find the characteristic impedance of the line and the propagation constant. If the coaxial line is Teflon filled ($\varepsilon_r = 2.1$) and the inner diameter of its outer conductor is 0.066 cm, determine the diameter of its inner conductor.

3.11. A lossless 75-Ω transmission line is terminated by a load impedance of $150 + j150\,\Omega$. Find the shortest length of line that results in (**a**) $Z_{\text{in}} = 75 - j120\,\Omega$, (**b**) $Z_{\text{in}} = 75 - j75\,\Omega$, and (**c**) $Z_{\text{in}} = 17.6\,\Omega$.

3.12. The open- and short-circuit impedances measured at the input terminals of a lossless transmission line of length 1.5 m, which is less than a quarter-wavelength, are $-j54.6\,\Omega$ and $j103\,\Omega$, respectively.

(**a**) Find the characteristic impedance Z_0 and propagation constant γ of the line.

(**b**) Without changing the operating frequency, find the input impedance of a short-circuited line that is twice the given length.

(**c**) How long should the short-circuited line be for it to appear as an open circuit at the input terminals?

3.13. At a frequency of 100 MHz, the following values are appropriate for a certain transmission line: $L = 0.25\,\mu\text{H/m}$, $C = 80\,\text{pF/m}$, $R = 0.15\,\Omega\text{/m}$, and $G = 8\,\mu\text{S/m}$. Calculate (**a**) the propagation constant, $\gamma = \alpha + j\beta$, (**b**) the signal wavelength, (**c**) the phase velocity, and (**d**) the characteristic impedance.

3.14. The open- and short-circuit impedances measured at the input terminals of a 3-m-long (i.e., greater than a quarter-wavelength but less than one-half wavelength) lossless transmission line are $j24.2\,\Omega$ and $-j232.4\,\Omega$, respectively.

(**a**) Find the characteristic impedance and propagation constant of this line.

(**b**) How long should the open-circuited line be for it to appear as a short circuit at the input terminals?

3.15. A 2.25λ-long lossless transmission line with its characteristic impedance at 75 Ω is terminated by a 300-Ω load. It is energized at the other port by a signal generator that has an open-circuit voltage of 20 V (peak value) and internal impedance of 75 Ω.

(**a**) Find the impedance at the input end of the line.

(**b**) Determine the total voltage across the load.

3.16. A 22.5-m-long lossless transmission line with $Z_0 = 50\,\Omega$ is short-circuited at one end, and a voltage source $v_S = 20\cos(4\pi \times 10^6 t - 30^\circ)$ V is connected at its input terminals, as shown in Figure P3.16. If the source impedance is 50 Ω and the phase velocity on the line is 1.8×10^8 m/s, find the total currents at its input and through the short circuit.

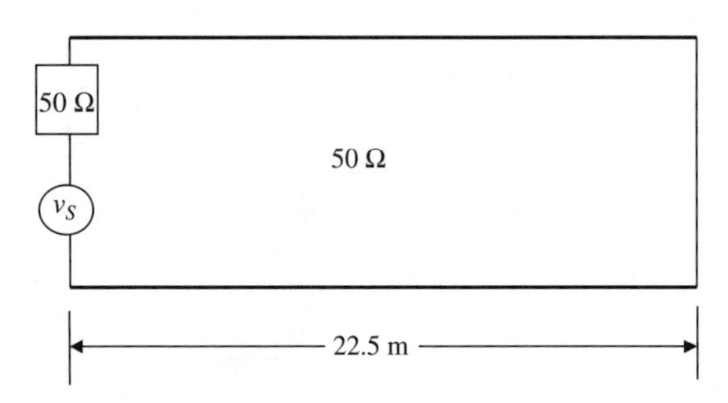

Figure P3.16

3.17. A 15-cm-long lossless transmission line with $Z_0 = 75\,\Omega$ is open at one end, and a voltage source $v_s = 10\cos(2\pi \times 10^9 t - 45°)$ V is connected at its other end, as shown in Figure P3.17. The source impedance is 75 Ω and the phase velocity on the line is 2×10^8 m/s. Compute the total voltage and current at its input as well as at its open end.

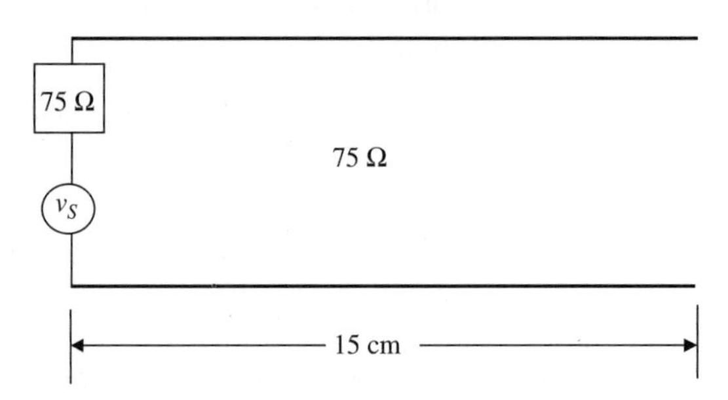

Figure P3.17

3.18. Determine the characteristic impedance and the phase velocity of a 25-cm-long loss-free transmission line from the following experimental data: $Z_{\text{s.c.}} = -j90\,\Omega$, $Z_{\text{o.c.}} = j40\,\Omega$, and $f = 300$ MHz. Assume that βl is less than π radians.

3.19. A lossless line is terminated in a load resistance of 50 Ω. Calculate the two possible values of characteristic impedance Z_0 if one-fourth of the incident voltage wave is reflected back.

3.20. Measurements on a 0.6-m-long lossless coaxial cable at 100 kHz show a capacitance of 54 pF when the cable is open-circuited and an inductance of 0.3 μH when it is short-circuited. Determine the characteristic impedance of the line and the phase velocity of this signal on the line. (Assume that the line length is less than a quarter-wavelength.)

3.21. For the reflection coefficients and characteristic impedances given, find the reflecting impedance in each case: (**a**) $\Gamma = 0.7\ \angle 30^\circ$, $Z_0 = 50\ \Omega$; (**b**) $\Gamma = 0.9\ \angle{-35^\circ}$, $Z_0 = 100\ \Omega$; (**c**) $\Gamma = 0.1 - j0.2$, $Z_0 = 50\ \Omega$; (**d**) $\Gamma = 0.5 - j0$, $Z_0 = 600\ \Omega$.

3.22. A 75-Ω line lossless line is terminated by a load impedance $Z_L = 100 + j150\ \Omega$. Using a Smith chart, determine (**a**) the load reflection coefficient, (**b**) the VSWR, (**c**) the load admittance, (**d**) Z_{in} at 0.4λ from the load, (**e**) the location of $V_{\max}$ and $V_{\min}$ with respect to the load if the line is 0.6λ long, and (**f**) Z_{in} at the input end.

3.23. A lossless 50-Ω line is terminated by a $25 - j60$-Ω load. Find (**a**) the reflection coefficient at the load, (**b**) the VSWR on the line, (**c**) the impedance at 3.85λ from the load, (**d**) the shortest length of the line for which impedance is purely resistive, and (**e**) the value of this resistance.

3.24. An antenna of input impedance $Z_L = 75 + j150\ \Omega$ at 2 MHz is connected to a transmitter through a 100-m-long section of coaxial line, which has the following distributed constants: $R = 153\ \Omega/\text{km}$, $L = 1.4\ \text{mH/km}$, $G = 0.8\ \mu\text{S/km}$, and $C = 0.088\ \mu\text{F/km}$. Determine the characteristic impedance, the propagation constant, and the impedance at the input end of this line.

3.25. A lossless 50-Ω transmission line is terminated in $25 + j50\ \Omega$. Find (**a**) the voltage reflection coefficient, (**b**) the impedance at 0.3λ from the load, (**c**) the shortest length of the line for which impedance is purely resistive, and (**d**) the value of this resistance.

3.26. A uniform transmission line has constants $R = 15\ \text{m}\Omega/\text{m}$, $L = 2\ \mu\text{H/m}$, $G = 1.2\ \mu\text{S/m}$, and $C = 1.1\ \text{nF/m}$. The line is 1 km long and it is terminated in a resistive load of 80 Ω. At 8 kHz, find (**a**) the input impedance, (**b**) the attenuation in decibels, and (**c**) the characteristic impedance.

3.27. A 50-Ω lossless line is terminated by a load impedance $Z_L = 50 + j100\ \Omega$. Using a Smith chart, determine (**a**) the load reflection coefficient, (**b**) the VSWR, (**c**) the load admittance, (**d**) Z_{in} at 0.4λ from the load, (**e**) the location of $V_{\max}$ and $V_{\min}$ with respect to load if the line is 0.6λ long, and (**f**) Z_{in} at the input end.

3.28. A 50-Ω lossless line is terminated by a load admittance $Y_L = 100 + j150\ \text{S}$. Using a Smith chart, determine (**a**) the load reflection coefficient, (**b**) the VSWR, (**c**) the load admittance, (**d**) Z_{in} at 0.4λ from the load, (**e**) the locations of $V_{\max}$ and $V_{\min}$ with respect to load if the line is 0.6λ long, and (**f**) Z_{in} at the input end.

3.29. Design a quarter-wavelength stripline impedance transformer to match a 150-Ω load with a 75-Ω line. The ground plane separation is 2.6 mm, and the dielectric constant of the filling material is 3.1. It operates at 2 GHz.

3.30. Design a quarter-wavelength microstrip impedance transformer to match a 250-Ω load with a 50-Ω line. The system is to be fabricated on a 1.6-mm-thick substrate ($\varepsilon_r = 2.3$) that operates at 2 GHz.

3.31. A microwave transmitter is fabricated on a GaAs substrate. An antenna used in the system offers a resistive load of 40 Ω. The electronic circuit has an output impedance of 1 kΩ. Design a microstrip $\lambda/4$ impedance transformer to match the system. The operating frequency is 4 GHz and the substrate thickness is 0.05 mm. The ε_r value of the GaAs substrate is 12.3.

3.32. A microstrip line is designed on a 0.1-mm-thick GaAs substrate ($\varepsilon_r = 14$). The strip thickness is 0.0001 mm and its width is 0.01 mm. Compute (**a**) the characteristic impedance and (**b**) the effective dielectric constant.

3.33. For the transmission line shown in Figure 3.33, find (**a**) the reflection coefficient (Γ_{in}) at the input end, (**b**) the VSWR on the line, and (**c**) the input admittance (Y_{in}).

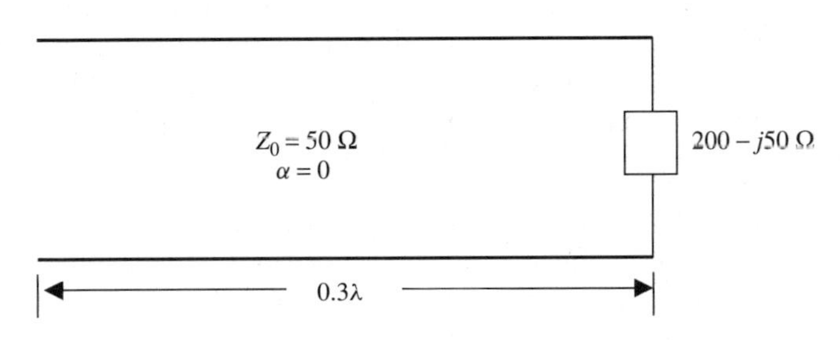

Figure P3.33

3.34. A lossless transmission line has a characteristic impedance of 50 Ω and is terminated with an unknown load. The standing wave ratio on the line is 3.0. Successive voltage minimums are 20 cm apart, and the first minimum is 5 cm from the load.

(**a**) Determine the value of terminating load.

(**b**) Find the position and value of a reactance that might be added in series with the line at some point to eliminate reflections of waves incident from the source end.

3.35. A lossless 100-Ω transmission line is terminated with an unknown load. The standing wave ratio on the line is 2.5. Successive voltage minima are 15 cm apart, and the first maximum is 7.5 cm from the load.

(**a**) Determine the value of terminating load.

(**b**) Find the position closest to the load and value of a reactance that might be added in series with the line to eliminate reflections of waves incident from the source end.

3.36. A lossless transmission line has a characteristic impedance of 50 Ω and is terminated with an unknown load. The standing wave ratio on the line is 3.0. Successive voltage minimums are 2 cm apart, and the first minimum is 0.5 cm from the load.

(a) Determine the admittance of terminating load.

(b) Find the shortest length of the line from the load for which the admittance is purely conductive.

(c) Determine the value of this conductance in siemens.

(d) Find the shortest length of the line from the load for which the impedance is purely resistive.

(e) Determine the value of this resistance in ohms.

3.37. A lossless 50-Ω line is connected to a load as shown in Figure P3.37. The input reflection coefficient Γ_{in} at 1.45λ from the load is found at $0.3\ \angle{-60^\circ}$. Find **(a)** the load impedance in ohms, **(b)** the VSWR, **(c)** the shortest length of the line from the load for which the impedance is purely resistive, and **(d)** the value of this resistance in ohms.

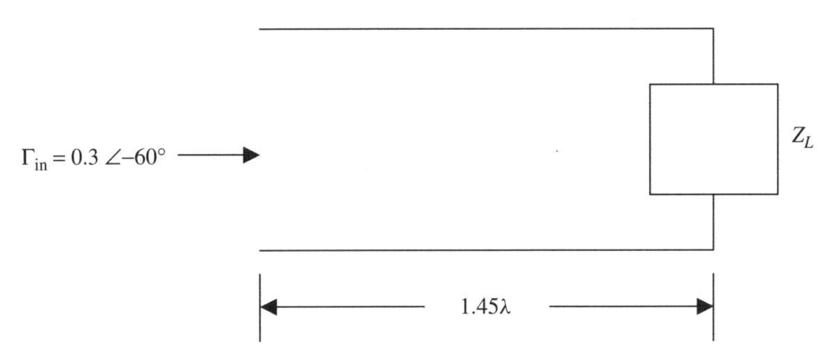

Figure P3.37

3.38. A lossless 100-Ω transmission line is terminated in a $50 + j150$-Ω load. Find **(a)** the reflection coefficient at the load, **(b)** the VSWR, and **(c)** Z_{in} $(z = 0.35\lambda)$.

3.39. A lossless 50-Ω transmission line is terminated by a $25 - j50$-Ω load. Find **(a)** Γ_L, **(b)** the VSWR, **(c)** Z_{in} at 2.35λ from the load, **(d)** the shortest length of the line where impedance is purely resistive, and **(e)** the value of this resistance.

3.40. For the transmission line circuit shown in Figure P3.40, find **(a)** the load reflection coefficient, **(b)** the impedance seen by the generator, **(c)** the VSWR on the transmission line, and **(d)** the fraction of the input power delivered to the load.

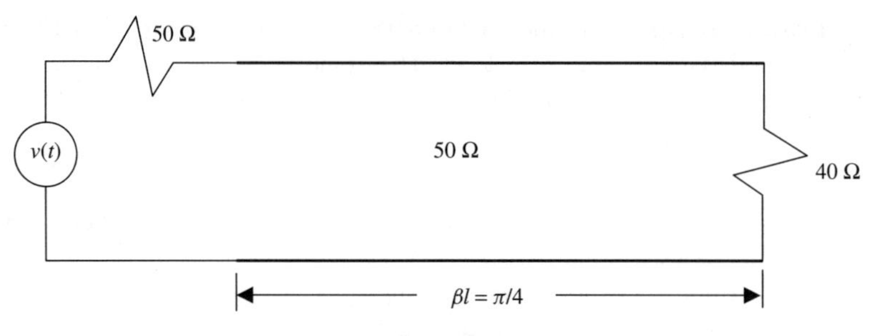

Figure P3.40

3.41. For the transmission line circuit shown in Figure P3.41, find the required value of Z_0 that will match the 20-Ω load resistance to the generator. The generator internal resistance is 60 Ω. Find the VSWR on the transmission line. Is the load resistance matched to the transmission line?

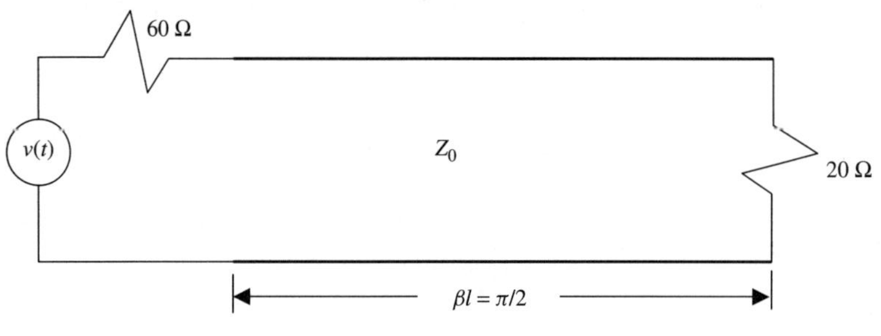

Figure P3.41

3.42. Use a Smith chart to find the following for the transmission line circuit shown in Figure P3.42: (**a**) the reflection coefficient at the load, (**b**) the reflection coefficient at the input, (**c**) the VSWR on the line, and (**d**) the input impedance.

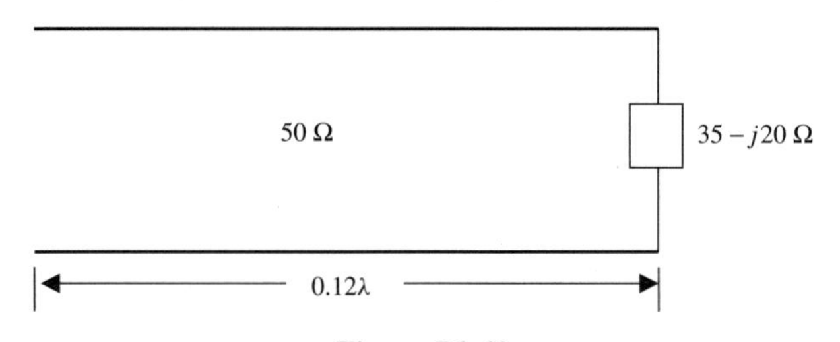

Figure P3.42

3.43. A lossless 40-Ω line is terminated in a 20-Ω load. Find the shortest line length that results in (**a**) $Z_{in} = 40 - j28\ \Omega$ and (**b**) $Z_{in} = 48 + j16\ \Omega$. Use a Smith chart rather than the impedance transformation equation.

3.44. A 30-m-long lossless transmission line with $Z_0 = 50\,\Omega$ operating at 2 MHz is terminated with a load $Z_L = 60 + j40\,\Omega$. If the phase velocity on the line is 1.8×10^8 m/s, find (**a**) the load reflection coefficient, (**b**) the standing wave ratio, and (**c**) the input impedance.

3.45. A 50-Ω transmission line is terminated by an unknown load. The total voltage at various points of the line is measured and found to be as displayed in Figure P3.45. Determine (**a**) the magnitude of reflection coefficient, (**b**) the VSWR, and (**c**) the signal wavelength in meters.

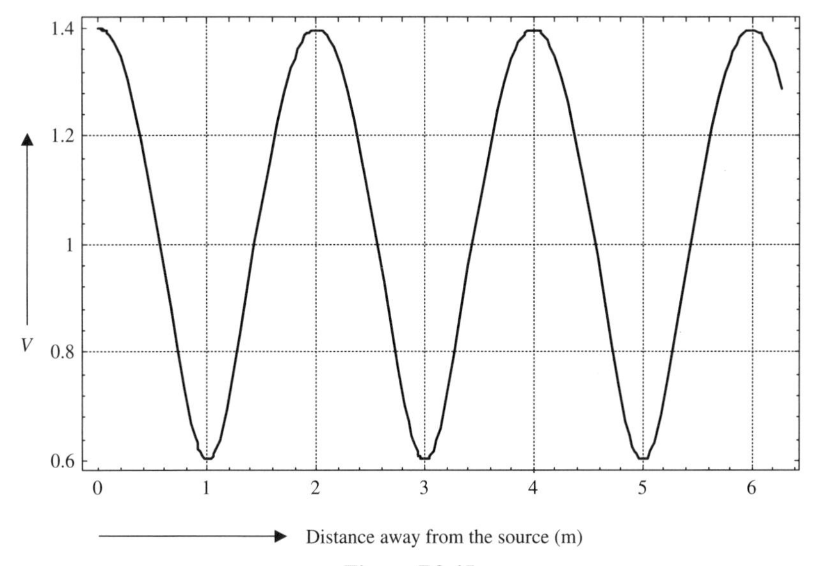

Figure P3.45

3.46. A 100-Ω lossless line is terminated by a series *RLC* circuit, as shown in Figure P3.46.

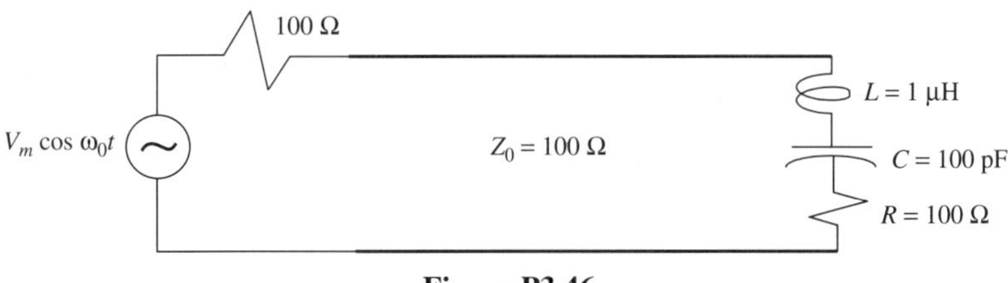

Figure P3.46

(**a**) Find the source frequency f_0 for which VSWR on the line will be unity.

(**b**) Find the source frequencies f_1 and f_2 on either side of f_0 for which the VSWR on the line will be 2.

4

ELECTROMAGNETIC FIELDS AND WAVES

This chapter begins with an introduction to the fundamental laws of electromagnetic fields. Maxwell's equations in integral (or large-scale) form are introduced initially because these are relatively easier to correlate with the experimental observations. The corresponding differential expressions (or point forms) are obtained via the Stokes and Gauss theorems. After a brief discussion on time-harmonic fields and the constitutive relations, uniform plane waves are introduced. The boundary conditions are summarized next along with their applications in reflection and transmission of uniform plane waves at planar interfaces.

A short presentation on the magnetic sources (current and charge) is included following uniform plane wave propagation. After introducing modified Maxwell's equations, potential functions are presented. This facilitates the construction of solutions in different coordinate systems. The chapter ends with a presentation on hollow metallic cylindrical waveguides.

4.1 FUNDAMENTAL LAWS OF ELECTROMAGNETIC FIELDS

In general, electromagnetic fields and sources vary with space coordinates and time. Using vector notations, these may be represented as follows:

$\mathcal{E}(\mathbf{r}, t)$ electric field intensity in V/m
$\mathcal{H}(\mathbf{r}, t)$ magnetic field intensity in A/m

Radio-Frequency and Microwave Communication Circuits: Analysis and Design, Second Edition,
By Devendra K. Misra
ISBN 0-471-47873-3

$\mathcal{D}(\mathbf{r}, t)$	electric flux density in C/m^2
$\mathcal{B}(\mathbf{r}, t)$	magnetic flux density in T
$\mathcal{J}(\mathbf{r}, t)$	electric current density in A/m^2
$\rho(\mathbf{r}, t)$	electric charge density in C/m^3

Using this notation, we present several laws as they appear in Maxwell's equations. For simplicity, the space and time dependence of these field quantities is not shown explicitly below but is implied.

Faraday's Law of Induction

Faraday discovered that if a conducting wire loop is exposed to the magnetic flux, it might induce an *electromotive force* (emf). The induced emf depends on the time rate of change of magnetic flux that leaves the surface bound by the loop. Induction is possible only when the magnetic flux is changing with time or the loop is moving through a nonuniform magnetic field. Consider a loop c that bounds the surface s, as illustrated in Figure 4.1. Magnetic flux leaving the surface can be evaluated by integrating the normal component of magnetic flux density at every point on it. Mathematically,

$$\oint_c \mathcal{E} \cdot dl = -\frac{\partial}{\partial t} \int_s \mathcal{B} \cdot d\mathbf{s} \tag{4.1.1}$$

Generalized Ampère's Law

The generalized Ampère law is based on experiments conducted by Oersted and the theoretical reasoning of Maxwell. Oersted found that electric current flowing through a conductor generates an encircling magnetic field. Maxwell introduced the displacement current term along with the conduction current. This was a

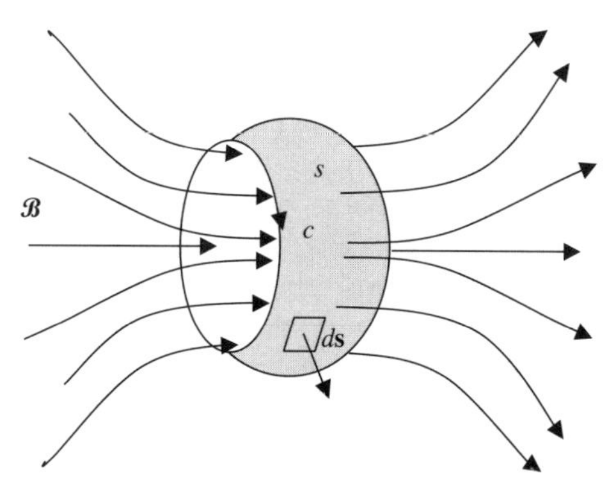

Figure 4.1 Magnetic flux density $\mathcal{B}$ passing through an area s bounded by a curve c.

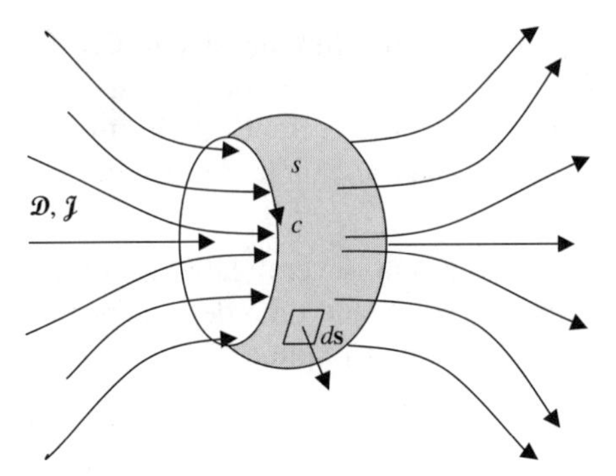

Figure 4.2 Electric flux density $\mathcal{D}$ and current density $\mathcal{J}$ passing through an area s bounded by a curve c.

significant contribution to efforts to explain current flow through capacitors in a circuit as well as to predict the propagation of electromagnetic waves in space.

Consider a closed path c that bounds a surface s as illustrated in Figure 4.2. Integration of the tangential component of the magnetic field intensity along this closed path (the line integral of vector magnetic field intensity along c) gives the circulation of the magnetic field intensity, called the *magnetomotive force* (mmf), found to be equal to the net current enclosed by the closed path c. Mathematically,

$$\oint_c \mathcal{H} \cdot dl = \int_s \mathcal{J} \cdot d\mathbf{s} + \frac{\partial}{\partial t} \int_s \mathcal{D} \cdot d\mathbf{s} \tag{4.1.2}$$

Thus, it is analogous to Faraday's law. Since the electric field intensity is expressed in volts per meter, its line integral (the emf) is in volts. Similarly, the unit of magnetic field intensity is amperes per meter, and therefore the mmf has the unit amperes. The first term on the right-hand side of (4.1.2) represents the conduction current (the net transfer of electric charge) and its second term is the time rate of change of electric flux leaving the surface s. There is no term analogous to conduction current in Faraday's law because there is never a net transfer of magnetic charge.

Gauss's Law for Electric Fields

Gauss's law for electric fields is based on the experimental observations of Gauss and Faraday. Consider a closed surface s that bounds a volume v, as shown in Figure 4.3. There is an electric charge distributed in volume v with a charge density of ρ. This electrical charge sets up a displacement flux with its density as $\mathcal{D}$. Gauss's law for electric fields relates the enclosed charge with the total displacement flux that emanates from the closed surface s. According to this law, the electric displacement flux that emanates from a closed surface s is equal to the net charge contained within the volume v. Mathematically,

$$\oint_s \mathcal{D} \cdot d\mathbf{s} = \int_v \rho \, dv \tag{4.1.3}$$

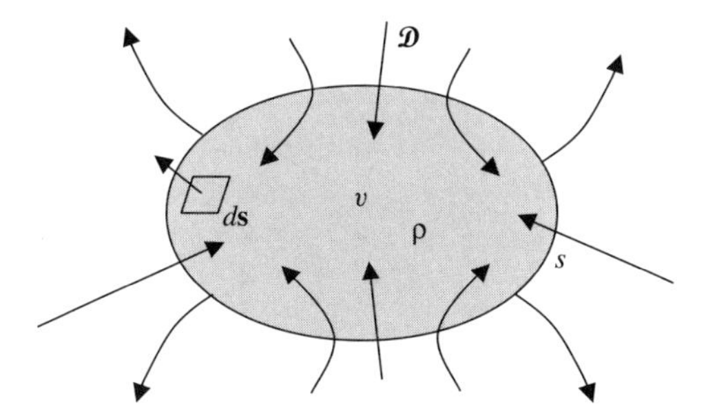

Figure 4.3 Electric flux density $\mathcal{D}$ emanating from a volume v bounded by a surface s.

Gauss's Law for Magnetic Fields

This law is analogous to Gauss's law for electric fields. Since the magnetic flux lines are always closed and magnetic charges do not exist, unlike positive or negative electric charges, net magnetic charge in a volume v has to be zero. As illustrated in Figure 4.4, the magnetic flux emanating from the closed surface s is therefore equal to zero. Mathematically,

$$\oint_s \mathcal{B} \cdot d\mathbf{s} = 0 \tag{4.1.4}$$

Equation of Continuity/Conservation of Charge

Consider a closed surface s that bounds a volume v as shown in Figure 4.5. There is an electrical charge distributed in this volume that has a charge density ρ. If there is a current flow across the closed surface s, there should be an effect on the charge distribution. If the net current flowing through the surface s is zero, there is no change in total charge. The net current emanating from the surface s is

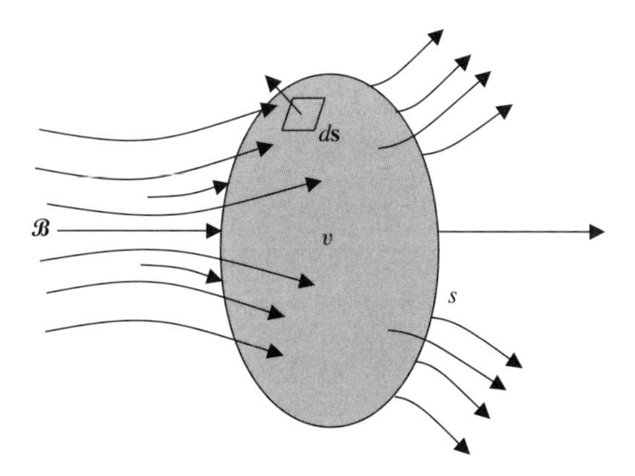

Figure 4.4 Electric flux density $\mathcal{B}$ emanating from a volume v bounded by a surface s.

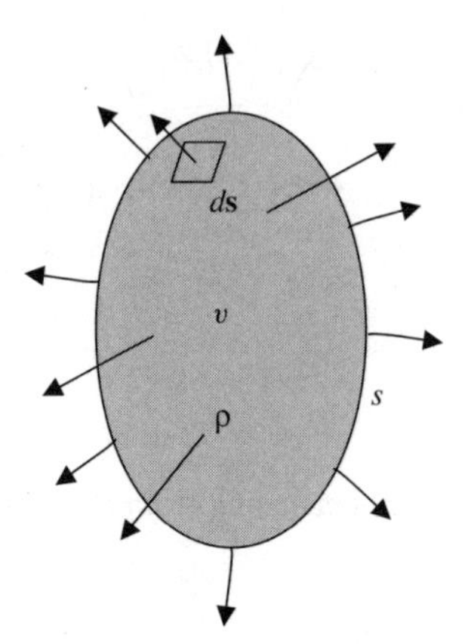

Figure 4.5 Electric charges leaving a volume v bounded by a surface s.

equal to the time rate of decrease in the charge enclosed. Stated mathematically,

$$\oint_s \mathcal{J} \cdot d\mathbf{s} = -\frac{\partial}{\partial t} \int_v \rho \, dv \tag{4.1.5}$$

Example 4.1 As shown in Figure 4.6, three arms of a rectangular conducting loop are fixed and the fourth is sliding in the x-direction with a velocity of 500 m/s. The loop lies on the $z = 0$ plane, and there is a magnetic flux density of 5 T in the z-direction. Find the emf induced in the loop. What happens if the magnetic flux varies with time as follows?

$$\mathcal{B} = \hat{z} \cdot 5 \cos 10^9 t \quad \text{T}$$

SOLUTION

$$\oint_c \mathcal{E} \cdot d\boldsymbol{l} = -\frac{\partial}{\partial t} \int_0^x \int_0^2 \mathcal{B} \cdot \hat{z} \, dx \, dy = -\frac{\partial}{\partial t}(5 \times 2 \times x) = -10\frac{dx}{dt} = -10 \times 500$$
$$= -5000 \text{ V}$$

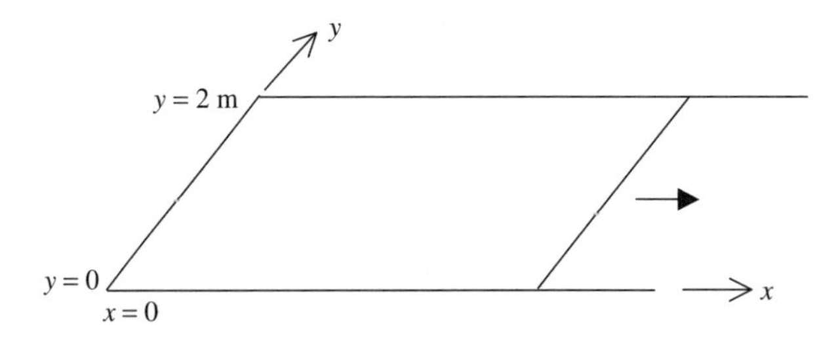

Figure 4.6

For the case of a magnetic field changing with time:

$$\oint_c \mathcal{E} \cdot dl = -\frac{\partial}{\partial t} \int_0^x \int_0^2 \mathcal{B} \cdot \hat{z}\, dx\, dy = -\frac{\partial}{\partial t}(2 \times x \times 5 \cos 10^9 t) \quad \text{V}$$

$$= -2\frac{dx}{dt} 5 \cos 10^9 t + 2 \times x \times 10^9 \times 5 \sin 10^9 t \quad \text{V}$$

$$= x \times 10^{10} \times \sin 10^9 t - 5000 \cos 10^9 t \quad \text{V}$$

Example 4.2 An infinitely long solid cylindrical wire with its radius at 2 m is laying along the z-axis. It carries a current with its density given as $\mathbf{J}(\rho) = J_0(1 - 0.5\rho)\hat{z}\text{A/m}^2$. Find the magnetic field intensity inside as well as outside the wire.

SOLUTION From the condition given, it is clear that only the encircling magnetic field H_ϕ exists. Hence, Ampère's law for inside the wire gives

$$\oint_c \mathbf{H} \cdot dl = \int_s \mathbf{J} \cdot ds \to \int_0^{2\pi} H_\phi \rho\, d\phi = \int_0^\rho \int_0^{2\pi} J_0(1 - 0.5\rho'\, d\rho'\, d\phi)$$

or

$$2\pi\rho H_\phi = 2\pi J_0 \left(\frac{\rho^2}{2} - 0.5\frac{\rho^3}{3}\right) \to H_\phi = J_0 \left(\frac{\rho}{2} - \frac{\rho^2}{6}\right) \quad \text{A/m} \qquad 0 \le \rho \le 2$$

Similarly, for outside the wire,

$$\oint_c \mathbf{H} \cdot dl = \int_s \mathbf{J} \cdot ds \to \int_0^{2\pi} H_\phi \rho\, d\phi = \int_0^2 \int_0^{2\pi} J_0(1 - 0.5\rho'\, d\rho'\, d\phi)$$

or

$$2\pi\rho H_\phi = 2\pi J_0 \left(\frac{2^2}{2} - 0.5\frac{2^3}{3}\right) \to H_\phi = \frac{2J_0}{3\rho} \quad \text{A/m} \qquad 2 \le \rho$$

Equations (4.1.1)–(4.1.4) are known as *Maxwell's equations in integral* (or *large-scale*) *form*. A differential (or point) form of these equations can be obtained via the application of the Stokes and Gauss theorems as follows. From (4.1.1), (4.1.2), and the Stokes theorem,

$$\oint_c \mathcal{E} \cdot dl = \int_s (\nabla \times \mathcal{E}) \cdot ds = -\frac{\partial}{\partial t} \int_s \mathcal{B} \cdot ds \to \nabla \times \mathcal{E} = -\frac{\partial \mathcal{B}}{\partial t}$$

and

$$\oint_c \mathcal{H} \cdot dl = \int_s (\nabla \times \mathcal{H}) \cdot ds = \int_s \mathcal{J} \cdot ds + \frac{\partial}{\partial t} \int_s \mathcal{D} \cdot ds \to \nabla \times \mathcal{H} = \mathcal{J} + \frac{\partial \mathcal{D}}{\partial t}$$

Similarly, from (4.1.3), (4.1.4), and the Gauss theorem,

$$\oint_s \mathcal{D} \cdot d\mathbf{s} = \int_v \nabla \cdot \mathcal{D}\, dv = \int_v \rho\, dv \rightarrow \nabla \cdot \mathcal{D} = \rho$$

and

$$\oint_s \mathcal{B} \cdot d\mathbf{s} = \int_v \nabla \cdot \mathcal{B}\, dv = 0 \rightarrow \nabla \cdot \mathcal{B} = 0$$

Application of the Gauss theorem to (4.1.5) gives

$$\oint_s \mathcal{J} \cdot d\mathbf{s} = \int_v \nabla \cdot \mathcal{J}\, dv = -\frac{\partial}{\partial t}\int_v \rho\, dv \rightarrow \nabla \cdot \mathcal{J} = -\frac{\partial \rho}{\partial t}$$

These results may be summarized as follows:

$$\nabla \times \mathcal{E} = -\frac{\partial \mathcal{B}}{\partial t} \tag{4.1.6}$$

$$\nabla \times \mathcal{H} = \mathcal{J} + \frac{\partial \mathcal{D}}{\partial t} \tag{4.1.7}$$

$$\nabla \cdot \mathcal{D} = \rho \tag{4.1.8}$$

$$\nabla \cdot \mathcal{B} = 0 \tag{4.1.9}$$

and

$$\nabla \cdot \mathcal{J} = -\frac{\partial \rho}{\partial t} \tag{4.1.10}$$

Equations (4.1.6) to (4.1.9) are known as *Maxwell's equations in point* (or *differential*) *form*; (4.1.10) is the equation of continuity in differential form.

Time-Harmonic Fields

Electromagnetic fields and sources considered up to this point were assumed to be an arbitrary function of time and space. The analysis can be simplified significantly with the assumption that these vary sinusoidally with time. It is to be noted that this assumption includes a large number of cases. Further, this formulation can be extended to include nonsinusoidal cases via Fourier series or Fourier integrals, as necessary.

Consider a vector field $\mathcal{A}(\mathbf{r}, t)$ that is sinusoidal with time. Using complex notations, it can be expressed as

$$\mathcal{A}(\mathbf{r}, t) = \mathbf{A}(\mathbf{r}) \cos(\omega t + \theta) = \mathrm{Re}\{[\mathbf{A}(\mathbf{r})e^{j\theta}]e^{j\omega t}\} \tag{4.1.11}$$

Therefore, the time derivative of this vector can be written as

$$\frac{\partial}{\partial t}\mathcal{A}(\mathbf{r}, t) = \mathrm{Re}\{[j\omega \mathbf{A}(\mathbf{r})e^{j\theta}]e^{j\omega t}\} \tag{4.1.12}$$

This indicates that a complex vector $\mathbf{A}(\mathbf{r})e^{j\theta}$ (a phasor quantity) can be used in place of $\mathcal{A}(\mathbf{r}, t)$, $j\omega$ can replace the time derivative, and $e^{j\omega t}$ can be suppressed during the field analysis. Further, it can be proved that division by $j\omega$ will replace the time integral. The time dependence can be recovered after multiplying the given complex vector by $e^{j\omega t}$ and then extracting the real part of that. This is a familiar process used in ac circuit analysis to define the reactance of an inductor or a capacitor.

Using phasor representations for the field quantities, equations (4.1.6) to (4.1.10) can be written as follows:

$$\nabla \times \mathbf{E}(\mathbf{r}) = -j\omega\mathbf{B}(\mathbf{r}) \tag{4.1.13}$$

$$\nabla \times \mathbf{H}(\mathbf{r}) = \mathbf{J}(\mathbf{r}) + j\omega\mathbf{D}(\mathbf{r}) \tag{4.1.14}$$

$$\nabla \bullet \mathbf{D}(\mathbf{r}) = \rho(\mathbf{r}) \tag{4.1.15}$$

$$\nabla \bullet \mathbf{B}(\mathbf{r}) = 0 \tag{4.1.16}$$

and

$$\nabla \bullet \mathbf{J}(\mathbf{r}) = -j\omega\rho(\mathbf{r}) \tag{4.1.17}$$

Bold roman symbols are now employed instead of bold script to distinguish the phasor field quantities. Further, to simplify the notation, space dependence of the field quantities will no longer be included explicitly.

Constitutive Relations

A general analysis involves the evaluation of **E, D, H**, and **B** for given sources (current and charge densities). Maxwell's equations represent only eight scalar equations, whereas unknown fields have 12 scalar components. Further, the two Gauss laws can be found from the other two Maxwell equations and the equation of continuity. Therefore, Maxwell's equations give only six independent scalar equations, whereas 12 are needed to find 12 unknown field components. The constitutive relations provide the remaining six independent scalar equations as follows:

$$\mathbf{D} = \varepsilon_0\mathbf{E} + \mathbf{P} \tag{4.1.18}$$

where

$$\mathbf{P} = \begin{cases} \overline{\overline{\chi_e}}\varepsilon_0\mathbf{E} & \text{for a linear medium} \\ \chi_e\varepsilon_0\mathbf{E} & \text{for a linear and isotropic medium} \end{cases} \tag{4.1.19}$$

P is the polarization density vector; ε_0 the permittivity of the free space (8.854×10^{-12} F/m), $\overline{\overline{\chi_e}}$ a dimensionless electrical susceptibility tensor, and χ_e a dimensionless electrical susceptibility scalar. Therefore, for a linear and isotropic medium,

$$\mathbf{D} = \varepsilon_0(1 + \chi_e)\mathbf{E} = \varepsilon\mathbf{E} = \varepsilon_0\varepsilon_r\mathbf{E} \tag{4.1.20}$$

where ε_r is called the *relative permittivity* (or *dielectric constant*) of the medium and ε is its permittivity in F/m.

Similarly,

$$\mathbf{B} = \mu_0(\mathbf{H} + \mathbf{M}) \tag{4.1.21}$$

where

$$\mathbf{M} = \begin{cases} \overline{\overline{\chi_m}}\mathbf{H} & \text{for a linear medium} \\ \chi_m\mathbf{H} & \text{for a linear and isotropic medium} \end{cases} \tag{4.1.22}$$

$\mathbf{M}$ is the magnetization density vector, μ_0 the permeability of the free space ($4\pi \times 10^{-7}$ H/m), $\overline{\overline{\chi_m}}$ a dimensionless magnetic susceptibility tensor, and χ_m a dimensionless magnetic susceptibility scalar. Therefore, for a linear and isotropic medium,

$$\mathbf{B} = \mu_0(1 + \chi_m)\mathbf{H} = \mu\mathbf{H} = \mu_0\mu_r\mathbf{H} \tag{4.1.23}$$

where μ is the permeability of the medium and μ_r is its relative permeability.

Also, the current density term in (4.1.14) may have two parts, one due to an impressed source and the other due to conduction. Conduction current density can be found via Ohm's law as follows:

$$\mathbf{J} = \sigma\mathbf{E} \tag{4.1.24}$$

where σ is the conductivity of the medium in S/m. Therefore, (4.1.14) may be expressed as follows:

$$\nabla \times \mathbf{H} = \mathbf{J}^e + \sigma\mathbf{E} + j\omega\varepsilon\,\mathbf{E} = \mathbf{J}^e + j\omega\varepsilon\left(1 - j\frac{\sigma}{\omega\varepsilon}\right)\mathbf{E} = \mathbf{J}^e + j\omega\varepsilon^*\mathbf{E} \tag{4.1.25}$$

where

$$\varepsilon^* = \left(\varepsilon - j\frac{\sigma}{\omega}\right) = \varepsilon(1 - j\tan\delta) \rightarrow \varepsilon_r^* = \frac{\varepsilon}{\varepsilon_0} - j\frac{\sigma}{\omega\varepsilon_0} = \varepsilon' - j\varepsilon'' \tag{4.1.26}$$

$$\tan\delta = \frac{\sigma}{\omega\varepsilon} \tag{4.1.27}$$

J^e is the impressed current source density, ε^* the complex permittivity of the material, ε_r^* the complex relative permittivity, and $\tan\delta$ is known as the loss tangent. The loss tangent represents a ratio of the magnitude of conduction current density to the magnitude of displacement current density in the medium.

Example 4.3 The electric field intensity in a source-free region is given as follows:

$$\mathbf{E} = \hat{z} \cdot 4e^{-j(x-3y)} \qquad \text{V/m}$$

If $\varepsilon = \varepsilon_0$ and $\mu = \mu_0$ in the region, find the signal frequency,

SOLUTION Since all electromagnetic fields must satisfy Maxwell's equations, we first determine the corresponding magnetic field and then try to find the electric field.

$$\nabla \times \mathbf{E} = -j\omega\mu_0\mathbf{H} \rightarrow \mathbf{H} = -\frac{1}{j\omega\mu_0}\begin{vmatrix} \hat{x} & \hat{y} & \hat{z} \\ \frac{\partial}{\partial x} & \frac{\partial}{\partial y} & \frac{\partial}{\partial z} \\ 0 & 0 & 4e^{-j(x-3y)} \end{vmatrix}$$

$$= \frac{12\hat{x} + 4\hat{y}}{\omega\mu_0} e^{-j(x-3y)}$$

and

$$\nabla \times \mathbf{H} = j\omega\varepsilon_0\mathbf{E} \rightarrow \mathbf{E} = \frac{1}{j\omega\varepsilon_0}\begin{vmatrix} \hat{x} & \hat{y} & \hat{z} \\ \frac{\partial}{\partial x} & \frac{\partial}{\partial y} & \frac{\partial}{\partial z} \\ \frac{12}{\omega\mu_0}e^{-j(x-3y)} & \frac{4}{\omega\mu_0}e^{-j(x-3y)} & 0 \end{vmatrix}$$

$$= \hat{x}0 + \hat{y}0 + \hat{z}\frac{16}{\omega^2\mu_0\varepsilon_0}e^{-j(x-3y)}$$

Therefore,

$$\frac{16}{\omega^2\mu_0\varepsilon_0} = 4 \rightarrow \omega = \frac{2}{\sqrt{\mu_0\varepsilon_0}} = 6 \times 10^8 \text{ rad/s}$$

Poynting's Vector and the Power Flow

Poynting's vector represents the electromagnetic power flow through a surface. The magnitude of Poynting's vector represents the power per unit area leaving that surface in the direction along the power flow (normal to that surface). An instantaneous Poynting vector can be evaluated for given instantaneous electric and magnetic field intensities as follows:

$$\mathcal{P} = \mathcal{E} \times \mathcal{H} \tag{4.1.28}$$

A time-averaged Poynting's vector can be determined from given phasor electric and magnetic field intensities as follows:

$$\mathbf{P} = \tfrac{1}{2}\,\text{Re}(\mathbf{E} \times \mathbf{H}^*) \tag{4.1.29}$$

The asterisk on the magnetic field intensity represents its complex conjugate.

Example 4.4 The electromagnetic fields of an antenna at a large distance are found as follows:

$$\mathbf{E}(r,\theta) = \hat{\theta} j \frac{120\pi}{r\sin\theta}\cos\left(\frac{\pi}{2}\cos\theta\right)e^{-jk_0 r} \qquad \text{V/m}$$

and

$$\mathbf{H}(r, \theta) = \hat{\phi} j \frac{1}{r \sin\theta} \cos\left(\frac{\pi}{2}\cos\theta\right) e^{-jk_0 r} \qquad \text{A/m}$$

Find the power radiated by this antenna.

SOLUTION

$$\mathbf{P} = \frac{1}{2}\operatorname{Re}(\mathbf{E} \times \mathbf{H}^*) = \hat{r}\frac{60\pi}{r^2 \sin^2\theta}\cos^2\left(\frac{\pi}{2}\cos\theta\right) \qquad \text{W/m}^2$$

and

$$P_{\text{radiated}} = \int_0^{\pi}\int_0^{2\pi} \mathbf{P}\cdot\hat{r}r^2 \sin\theta\, d\theta\, d\phi = \int_0^{\pi}\int_0^{2\pi} \frac{60\pi}{\sin\theta}\cos^2\left(\frac{\pi}{2}\cos\theta\right) d\theta\, d\phi$$

$$= 1443.5\,\text{W} = 1.4435\,\text{kW}$$

4.2 THE WAVE EQUATION AND UNIFORM PLANE WAVE SOLUTIONS

In a source-free, linear, isotropic, homogeneous region, Maxwell's curl equations in phasor form are

$$\nabla \times \mathbf{E} = -j\omega\mu\mathbf{H} \tag{4.2.1}$$

$$\nabla \times \mathbf{H} = -j\omega\varepsilon_0\varepsilon^*\mathbf{E} \tag{4.2.2}$$

After taking the curl of (4.2.1) and substituting (4.2.2) on its right-hand side, we get

$$\nabla \times \nabla \times \mathbf{E} = -j\omega\mu\nabla \times \mathbf{H} = -j\omega\mu(j\omega\varepsilon_0\varepsilon^*\mathbf{E}) = k^2\mathbf{E} \tag{4.2.3}$$

where

$$k = \omega\sqrt{\mu\varepsilon_0\varepsilon^*} = \beta - j\alpha \tag{4.2.4}$$

k, α, and β being the wave number, attenuation constant, and phase constant, respectively.

Further,

$$\nabla \times \nabla \times \mathbf{E} = \nabla(\nabla\cdot\mathbf{E}) - \nabla^2\mathbf{E} = k^2\mathbf{E} \rightarrow \nabla^2\mathbf{E} + k^2\mathbf{E} = 0 \tag{4.2.5}$$

Similarly, starting from (4.2.2), we find that

$$\nabla^2\mathbf{H} + k^2\mathbf{H} = 0 \tag{4.2.6}$$

Equations (4.2.5) and (4.2.6) are known as the *Helmholtz equation* or *vector wave equation*.

Consider an electric field that is directed along the x-axis (x-polarized). Further, it is uniform in the x and y directions (i.e., no variations with x or y). Therefore, the wave equation (4.2.5) for this case reduces to

$$\frac{d^2E_x(z)}{dz^2} + k^2E_x(z) = 0 \tag{4.2.7}$$

This is an ordinary differential equation of second order, similar to (3.1.14). Its solution can be found easily as follows:

$$E_x(z) = E_+e^{-jkz} + E_-e^{jkz} = E_+e^{-\alpha z}e^{-j\beta z} + E_-e^{\alpha z}e^{j\beta z} \tag{4.2.8}$$

E_+ and E_- are integration constants. As noted in (3.1.18), the first term represents a wave traveling along $+z$ (outgoing wave), and the second term represents a wave traveling along $-z$ (incoming wave).

Associated magnetic field intensity can be found from (4.2.2) instead of solving (4.2.6). It can be proved that only the y-component of the magnetic field exists in this case. Therefore, the electric field, magnetic field, and direction of wave propagation are orthogonal to each other. As before, the wavelength and phase velocity of the wave are

$$\lambda = \frac{2\pi}{\beta} \tag{4.2.9}$$

and

$$v_p = \frac{\omega}{\beta} \tag{4.2.10}$$

The penetration depth of the wave is defined as a distance d_p over which the amplitude of the electric field decays to $1/e$ of its value at $z = 0$. Hence,

$$d_p = \frac{1}{\alpha} \tag{4.2.11}$$

When the medium is highly conductive, the depth of penetration is called the *skin depth*, δ. It is found to be

$$\delta = \sqrt{\frac{2}{\omega\mu\sigma}} \tag{4.2.12}$$

The intrinsic impedance η of the medium, defined as

$$\eta = \sqrt{\frac{\mu}{\varepsilon^*}} = \frac{E_x}{H_y} = \frac{\omega\mu}{k} \tag{4.2.13}$$

represents the ratio of electric field intensity component to magnetic field intensity component transverse to the direction of wave propagation such that it has a positive real part.

Example 4.5 The electric field intensity of a time-harmonic field in free space is given as follows:

$$\mathbf{E} = \hat{z}(1 + j)e^{-j(2\pi x)} \qquad \text{mV/m}$$

Assuming that the distance x is measured in meters, find (**a**) the wavelength, (**b**) the frequency, and (**c**) the associated magnetic field intensity.

SOLUTION

(**a**) $\lambda = \dfrac{2\pi}{\beta} = \dfrac{2\pi}{2\pi} = 1\,\text{m}$

(**b**) $f = \dfrac{3 \times 10^8}{\lambda} = 3 \times 10^8\,\text{Hz} = 300\,\text{MHz}$

(**c**) $\mathbf{H} = -\dfrac{1}{j\omega\mu_0}\nabla \times \mathbf{E} = -\dfrac{1}{j\omega\mu_0}\begin{vmatrix} \hat{x} & \hat{y} & \hat{z} \\ \dfrac{\partial}{\partial x} & \dfrac{\partial}{\partial y} & \dfrac{\partial}{\partial z} \\ 0 & 0 & E_z \end{vmatrix}$

$= -\hat{y}\dfrac{2\pi}{\omega\mu_0}(1 + j)e^{-j2\pi x} \qquad \text{mA/m}$

Alternatively, the intrinsic impedance can be used in conjunction with Poynting's vector to determine the associated magnetic field:

$$\eta = \frac{E_z}{-H_y} = \frac{\omega\mu_0}{k} = \frac{2\pi \times 3 \times 10^8 \times 4\pi \times 10^{-7}}{2\pi} = 376.9911\,\Omega$$

Example 4.6 A 10-kHz electromagnetic wave is being used to communicate between two submerged submarines. If the conductivity and dielectric constant of the seawater are 4 S/m and 81, respectively, find (**a**) the propagation constant, (**b**) the wavelength, (**c**) the attenuation constant, (**d**) the phase velocity, and (**e**) the skin depth of the wave.

SOLUTION Since the seawater is nonmagnetic, $\mu = \mu_0$ and

$$k = \omega\sqrt{\mu_0\varepsilon_0\varepsilon^*} = \omega\sqrt{\mu_0\varepsilon_0\left(\varepsilon_r - j\frac{\sigma}{\omega\varepsilon_0}\right)}$$

Further,

$$\frac{\sigma}{\omega\varepsilon_0} = \frac{4}{2\pi \times 10^4 \times 8.854 \times 10^{-12}} = 7190.2 \gg 81$$

Therefore,

$$k \approx \omega\sqrt{\mu_0\varepsilon_0\left(-j\frac{\sigma}{\omega\varepsilon_0}\right)} = \sqrt{\frac{\omega\mu_0\sigma}{2}} - j\sqrt{\frac{\omega\mu_0\sigma}{2}} = 0.3974 - j0.3974\ \mathrm{m}^{-1}$$

Hence, $\beta = 0.3974$ rad/m and $\alpha = 0.3974$ Np/m.

$$\lambda = \frac{2\pi}{\beta} = \frac{2\pi}{0.3974} = 15.8107\ \mathrm{m}$$

$$v_p = \frac{\omega}{\beta} = \frac{2\pi \times 10^4}{0.3974} = 15.8107 \times 10^4\ \mathrm{m/s}$$

and

$$\delta = \frac{1}{\alpha} = \frac{1}{0.3974} = 2.5163\ \mathrm{m}$$

Uniform Plane Wave Propagating Along *r*

The preceding analysis can be extended easily for the case of a wave that propagates at an angle from a given axis. Consider a uniform plane wave propagating along r at an angle θ from the z-axis, as illustrated in Figure 4.7. This wave has an x-component of the electric field intensity that can be written as

$$\mathbf{E} = \hat{x}E_0e^{-jk_r r} \tag{4.2.13}$$

Since the magnetic field intensity must be orthogonal to both its electric field and the direction of propagation, it will have a y as well as a z component. We select a point B at r on the propagation path and drop a perpendicular from that on the y-axis. Further, we drop a perpendicular from point C to A in the direction

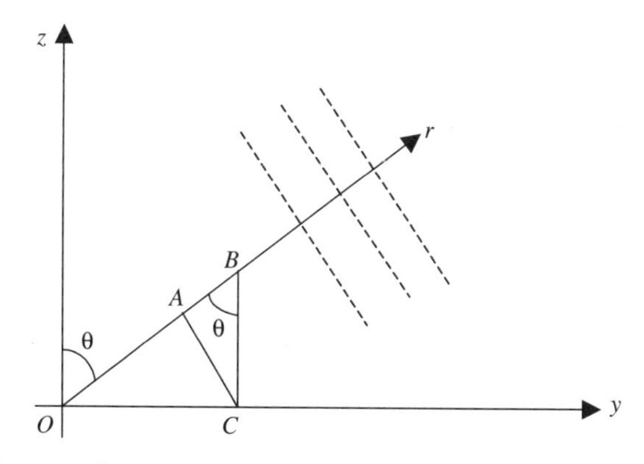

Figure 4.7 Geometry for a uniform plane wave propagating at an angle θ from the z-axis.

of propagation. As indicated in Figure 4.7, the angle ABC is equal to θ as well. Assume that the Cartesian coordinates of point B are y and z. Hence,

$$r = OA + AB = OC \sin\theta + BC \cos\theta = y \sin\theta + z \cos\theta \quad (4.2.14)$$

and

$$k_r r = k_r \sin\theta \; y + k_r \cos\theta \; z = k_y y + k_z z \quad (4.2.15)$$

where

$$k_y = k_r \sin\theta$$

and

$$k_z = k_r \cos\theta$$

Therefore, (4.2.13) may be written as

$$\mathbf{E} = \hat{x} E_0 e^{-j(k_y y + k_z z)} \quad (4.2.16)$$

Phase velocities v_{py} and v_{pz} of this wave with respect to the y and z axes can be found as

$$v_{py} = \frac{\omega}{\mathrm{Re}(k_y)} \quad (4.2.17)$$

and

$$v_{pz} = \frac{\omega}{\mathrm{Re}(k_z)} \quad (4.2.18)$$

Example 4.7 The electric field intensity of a time-harmonic wave traveling in a source-free free space is given as follows:

$$\mathbf{E} = (4\hat{y} + 3\hat{z}) e^{-j(6y - 8z)} \qquad \mathrm{mV/m}$$

Assuming that y and z represent their respective distances in meters, determine (**a**) the angle of the propagation direction relative to the z-axis, (**b**) the wavelengths of the wave along the r, y, and z directions, (**c**) the phase velocities along the r, y, and z directions, (**d**) the energy velocities along the r, y, and z directions, (**e**) the frequency of the wave, and (**f**) the associated magnetic field intensity.

SOLUTION From the information given, $k_y = k \sin\theta = 6$ and $k_z = k \cos\theta = 8$. Therefore,

$$k_r = \sqrt{k_y^2 + k_z^2} = \sqrt{6^2 + 8^2} = 10\,\mathrm{m}^{-1}$$

(**a**) $\sin\theta = \dfrac{k_y}{k_r} = 0.6 \rightarrow \theta = 36.87^\circ$

An analysis of the given electric field indicates that this angle is from the $-z$ direction, and therefore it should be subtracted from 180° to obtain the angle measured from $+z$. Hence, $\theta = 180° - 36.87° = 143.13°$.

(b) $\lambda_r = \frac{2\pi}{k_r} = \frac{2\pi}{10} = 0.6283\,\text{m}$, $\lambda_y = \frac{2\pi}{k_y} = \frac{2\pi}{6} = 1.0472\,\text{m}$, and $\lambda_z = \frac{2\pi}{k_z} = \frac{2\pi}{8} = 0.7854\,\text{m}$

(c) Since the wave is propagating in free space, its phase velocity along r is 3×10^8 m/s. The frequency of this wave may be found as follows:

$$\omega = v_r k_r = 3 \times 10^9\,\text{rad/s} \quad \text{or} \quad f = \frac{3 \times 10^8}{\lambda_r}\text{Hz} = 477.47\,\text{MHz}$$

$$v_{py} = \frac{\omega}{k_y} = \frac{3 \times 10^9}{6} = 5 \times 10^8 \text{ m/s}$$

and

$$v_{pz} = \frac{\omega}{k_z} = \frac{3 \times 10^9}{8} = 3.75 \times 10^8 \text{ m/s}$$

(d) $v_{er} = 3 \times 10^8$ m/s, $v_{ey} = v_{er} \sin\theta = 1.8 \times 10^8$ m/s, and $v_{ez} = v_{er} \cos\theta = 2.4 \times 10^8$ m/s

(e) The frequency is 477.47 MHz, as found in part (c).

(f) $$\mathbf{H} = -\frac{1}{j\omega\mu_0}\nabla \times \mathbf{E} = -\frac{1}{j\omega\mu_0}\begin{vmatrix} \hat{x} & \hat{y} & \hat{z} \\ \frac{\partial}{\partial x} & \frac{\partial}{\partial y} & \frac{\partial}{\partial z} \\ 0 & E_y & E_z \end{vmatrix}$$

$$= \hat{x} \bullet 13.2629 \times 10^{-6} e^{-j(6y-8z)} \qquad \text{A/m}$$

4.3 BOUNDARY CONDITIONS

The electromagnetic field problems considered so far have involved a single medium of infinite extent. However, there are numerous real-world problems that require analysis of electromagnetic fields in the presence of more than one medium. Therefore, it is important to know the behavior of electromagnetic fields across the interface of two media.

Consider an interface of the two media, as illustrated in Figure 4.8. The electrical characteristics of these media are as indicated. When (4.1.2) is applied to an infinitesimal area across the interface and its side Δh is reduced to zero under the limit, it results in the relation

$$(\mathcal{H}_2 - \mathcal{H}_1) \times \hat{n} = \mathcal{J}_s \tag{4.3.1}$$

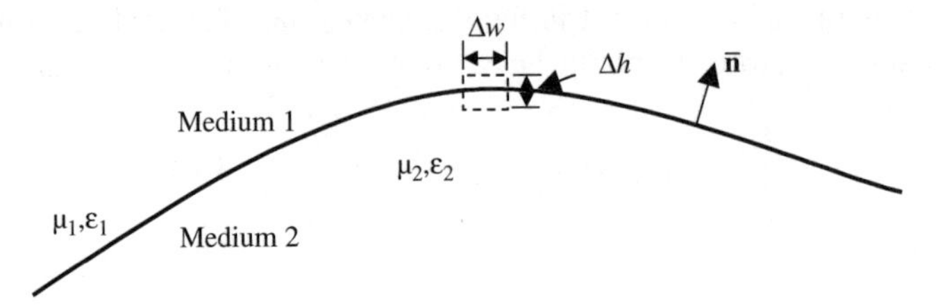

Figure 4.8 Infinitesimal area across the interface of medium 1 and medium 2.

where subscripts 1 and 2 are used to indicate the magnetic field intensity in the corresponding medium. $\mathcal{J}_s$ represents the surface current density in A/m on the boundary, and the unit vector **n** is directed into medium 1.

Similarly, when (4.1.1) is applied to this infinitesimal area across the interface and its side Δh is reduced to zero under the limit, it results in the relation

$$\hat{n} \times (\mathcal{E}_2 - \mathcal{E}_1) = 0 \tag{4.3.2}$$

Subscripts 1 and 2 are used to indicate the electric field intensity in the corresponding medium. Now, consider a pillbox volume across the interface of the two media, as shown in Figure 4.9. When (4.1.3) is applied to this volume and the side Δh is reduced to zero under the limit, it results in the relation

$$\hat{n} \bullet (\mathcal{D}_1 - \mathcal{D}_2) = \rho_s \tag{4.3.3}$$

where subscripts 1 and 2 are used to indicate the electric flux density in the corresponding medium ρ_s represents the surface charge density in C/m^2 on the boundary, and the unit vector **n** is directed into medium 1.

Similarly, the following relation is found via (4.1.4) for the magnetic flux densities in the two media:

$$\hat{n} \bullet (\mathcal{B}_1 - \mathcal{B}_2) = 0 \tag{4.3.4}$$

Equations (4.3.1) to (4.3.4) can be used to determine the electromagnetic fields from one to other medium across the boundary. It is to be noted that (4.3.3)

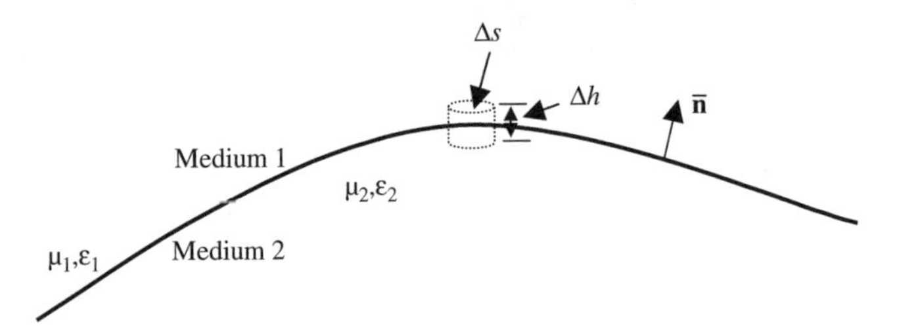

Figure 4.9 Pillbox volume across the interface of medium 1 and medium 2.

and (4.3.4) are formulated via Gauss's laws that are not independent, and therefore these conditions are satisfied automatically if (4.3.1) and (4.3.2) are met. Further, if medium 2 is a perfect conductor, the fields inside it vanish, and therefore the boundary conditions (4.3.1) to (4.3.4) reduce to

$$\hat{n} \times \mathcal{H}_1 = \mathcal{J}_s \tag{4.3.5}$$

$$\hat{n} \times \mathcal{E}_1 = 0 \tag{4.3.6}$$

$$\hat{n} \cdot \mathcal{D}_1 = \rho_s \tag{4.3.7}$$

and

$$\hat{n} \cdot \mathcal{B}_1 = 0 \tag{4.3.8}$$

If the second medium is a good conductor (finite conductivity), the fields penetrate mostly up to its skin depth. In such a case, the following Leontovich impedance boundary condition can be used provided that the radius of curvature of the surface is significantly greater than the skin depth $(2/\omega\mu\sigma)^{1/2}$.

$$\mathcal{E}_t = Z_s(\hat{n} \times \mathcal{H}) \tag{4.3.9}$$

where

$$Z_s = \sqrt{\frac{\mu}{\varepsilon}} \approx \sqrt{\frac{j\mu\omega}{\sigma}} \tag{4.3.10}$$

Example 4.8 The region $x > 0$ is a perfect dielectric with $\varepsilon_r = 2.25$, and the region $x < 0$ is a free space. At the interface, subscript 1 denotes field components on the $+x$ side of the boundary and subscript 2 those on the $-x$ side. If $\mathbf{D}_1 = \hat{x} + 2\hat{y}\mathrm{C/m^2}$, find $\mathbf{D}_2$, $\mathbf{E}_1$, and $\mathbf{E}_2$.

SOLUTION The conditions stated in the problem are illustrated in Figure 4.10. The interface is $x = 0$ plane, and therefore the x component, will be normal, with

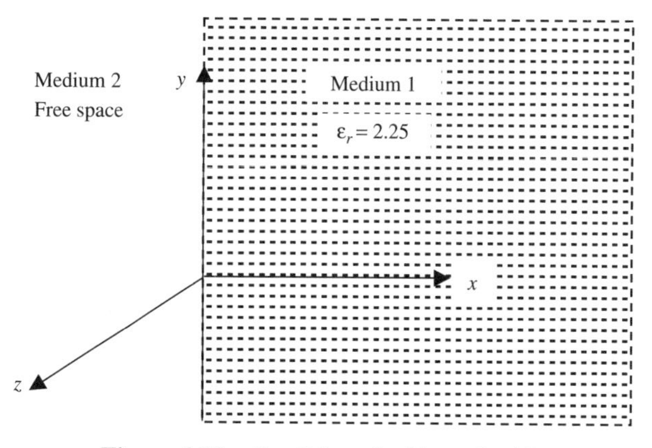

Figure 4.10 Conditions for Example 4.8.

the y component tangential to this plane.

$$\mathbf{D}_1 = \varepsilon_1 \mathbf{E}_1 \rightarrow \mathbf{E}_1 = \frac{\mathbf{D}_1}{\varepsilon_1} = \frac{1}{2.25\varepsilon_0}\hat{x} + \frac{2}{2.25\varepsilon_0}\hat{y} \qquad \text{V/m}$$

From the boundary conditions, the x component of electric flux density and y component of electric field intensity in medium 2 will be same as in medium 1. Hence,

$$D_{2x} = 1 \quad \text{and} \quad E_{2y} = \frac{2}{2.25\varepsilon_0}$$

Therefore,

$$\mathbf{D}_2 = \hat{x} + \frac{2}{2.25}\hat{y} \qquad \text{C/m}^2$$

and

$$\mathbf{E}_2 = \frac{\mathbf{D}_2}{\varepsilon_0} = \frac{1}{\varepsilon_0}\hat{x} + \frac{2}{2.25\varepsilon_0}\hat{y} \qquad \text{V/m}$$

Example 4.9 A sphere of 2 m radius is made of a perfect dielectric material (medium 1). It is surrounded by free space (medium 2). Electric field intensities in the two media are given as follows:

$$\mathbf{E}_1 = E_{01}(\cos\theta\hat{r} - \sin\theta\hat{\theta}) \qquad r \le 2\,\text{m}$$

and

$$\mathbf{E}_2 = E_{02}\left[\left(1 + \frac{8}{r^3}\right)\cos\theta\hat{r} - \left(1 - \frac{4}{r^3}\right)\sin\theta\hat{\theta}\right] \qquad r \ge 2\,\text{m}$$

Find the permittivity of the spherical medium.

SOLUTION As indicated in Figure 4.11, the r components of the electric fields are normal and the θ components are tangential at the interface. Therefore,

$$-\sin\theta E_{01} = -\left(1 - \frac{4}{8}\right)\sin\theta E_{02} \rightarrow E_{02} = 2E_{01}$$

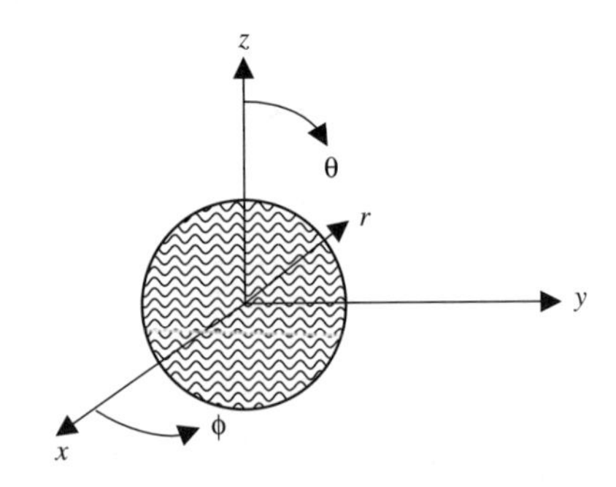

Figure 4.11 Conditions for Example 4.9.

and

$$\varepsilon_1 E_{01} \cos\theta = \varepsilon_0 E_{02} \cdot 2\cos\theta \rightarrow \varepsilon_1 = 2\varepsilon_0 \frac{E_{02}}{E_{01}} = 4\varepsilon_0 \qquad \text{F/m}$$

4.4 UNIFORM PLANE WAVE INCIDENT NORMALLY ON AN INTERFACE

Consider a uniform plane electromagnetic wave propagating in medium 1 as illustrated in Figure 4.12. The electric field $\mathbf{E}^i$ of this wave is y directed, and its magnetic field $\mathbf{H}^i$ is in the $-x$ direction. A wave propagating in the $+z$ direction is incident normally on the interface of medium 2 with medium 1. The electrical characteristics of the two media are as indicated in Figure 4.12. Since the wave experiences a change in electrical characteristics in its path, a part of the incident signal is reflected back, while the remaining signal is transmitted into medium 2. Thus, there are two waves in medium 1 propagating in opposite directions, and one in medium 2 that continues along $+z$.

The electromagnetic fields associated with the incident wave can be expressed as follows:

$$\mathbf{E}^i = \hat{y} E_0 e^{-jk_1 z} \tag{4.4.1}$$

and

$$\mathbf{H}^i = -\hat{x} H_0 e^{-jk_1 z} = -\hat{x}\frac{E_0}{\eta_1} e^{-jk_1 z} \tag{4.4.2}$$

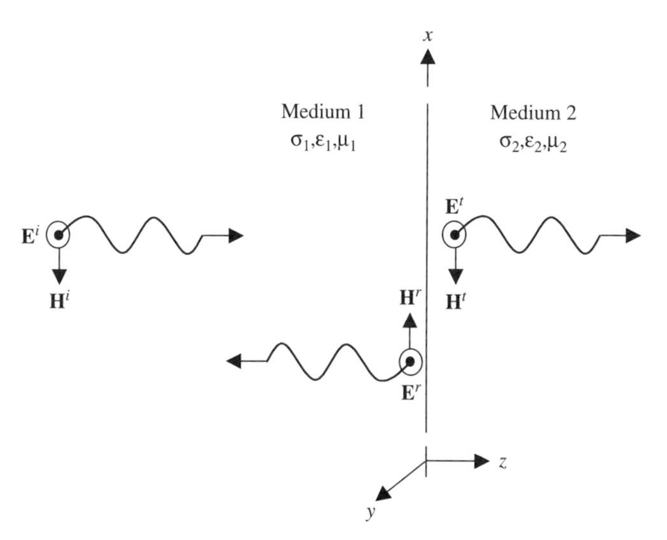

Figure 4.12 Geometry of a uniform plane wave incident normally on the interface of two media.

where E_0 and H_0 are arbitrary constants, k_1 the complex wave number, and η_1 the complex intrinsic impedance of medium 1.

Similarly, electromagnetic fields associated with the reflected wave ($\mathbf{E}^r$ and $\mathbf{H}^r$) as well as the transmitted wave ($\mathbf{E}^t$ and $\mathbf{H}^t$) can be expressed as follows:

$$\mathbf{E}^r = \hat{y} R E_0 e^{+jk_1 z} \tag{4.4.3}$$

$$\mathbf{H}^r = \hat{x} R H_0 e^{-jk_1 z} = \hat{x} R \frac{E_0}{\eta_1} e^{+jk_1 z} \tag{4.4.4}$$

$$\mathbf{E}^t = \hat{y} T E_0 e^{-jk_2 z} \tag{4.4.5}$$

and

$$\mathbf{H}^t = -\hat{x} T H_0 e^{-jk_1 z} = -\hat{x} T \frac{E_0}{\eta_2} e^{-jk_2 z} \tag{4.4.6}$$

where R and T are complex reflection and transmission coefficients, respectively, k_2 the complex wave number in medium 2, and η_2 the complex intrinsic impedance of medium 2.

Now, applying the boundary conditions (tangential components of the fields are continuous across the boundary) at $z = 0$ gives

$$E_0 + R E_0 = T E_0 \rightarrow 1 + R = T \tag{4.4.7}$$

and

$$-\frac{E_0}{\eta_1} + R \frac{E_0}{\eta_1} = -T \frac{E_0}{\eta_1} \rightarrow 1 - R = \frac{\eta_1}{\eta_2} T \tag{4.4.8}$$

Equations (4.4.7) and (4.4.8) can be solved for R and T as follows:

$$T = \frac{2\eta_2}{\eta_1 + \eta_2} \tag{4.4.9}$$

and

$$R = \frac{\eta_2 - \eta_1}{\eta_1 + \eta_2} \tag{4.4.10}$$

Therefore, the total electric field at any point in medium 1 is found to be

$$\mathbf{E}_1 = \hat{y} E_0 e^{-jk_1 z} + \hat{y} R E_0 e^{+jk_1 z}$$

$$|\mathbf{E}_1|^2 = |E_0|^2 [e^{-2\alpha_1 z} + \rho^2 e^{2\alpha_1 z} + 2\rho \cos(2\beta_1 z + \theta)] \tag{4.4.11}$$

where

$$R = \rho e^{j\theta} \tag{4.4.12}$$

$$k_1 = \beta_1 - j\alpha_1 \tag{4.4.13}$$

β_1 is the phase constant in rad/s and α_1 is the attenuation constant in Np/m for medium 1.

It is to be noted that the propagation characteristics of this wave are very much similar to that of a wave propagating on the transmission line, as discussed in Chapter 3.

Example 4.10 The electric field intensity of a uniform plane wave propagating in free space is given as

$$\mathcal{E}^i(z,t) = \hat{y} \bullet 4.8 \cos(10^9 t - k_0 z) \qquad \text{V/m}$$

If this wave impinges normally on a lossless medium ($\mu_r = 1$ and $\varepsilon_r = 2.25$) in the $z \geq 0$ region, find k_0, k_2, R, T, the VSWR, and the average power transmitted into the medium.

SOLUTION The wave numbers and intrinsic impedance of the wave in the two media are found as follows:

$$k_0 = \omega\sqrt{\mu_0\varepsilon_0} = 10^9\sqrt{4\pi \times 10^{-7} \times 8.854 \times 10^{-12}} = 3.3356\,\text{rad/m}$$

$$k_2 = \omega\sqrt{\mu_0\varepsilon_0\varepsilon_r} = k_0\sqrt{2.25} = 5.0034\,\text{rad/m}$$

$$\eta_1 = \sqrt{\frac{\mu_0}{\varepsilon_0}} = \sqrt{\frac{4\pi \times 10^{-7}}{8.854 \times 10^{-12}}} = 376.7343\,\Omega$$

and

$$\eta_2 = \sqrt{\frac{\mu_0}{\varepsilon_0\varepsilon_r}} = \sqrt{\frac{4\pi \times 10^{-7}}{8.854 \times 10^{-12} \times 2.25}} = 251.1562\,\Omega$$

Now R and T can be found via (4.4.9) and (4.4.10) as

$$R = \frac{\eta_2 - \eta_1}{\eta_1 + \eta_2} \approx -0.2$$

and

$$T = \frac{2\eta_2}{\eta_1 + \eta_2} \approx 0.8$$

The VSWR in medium 1 is found to be

$$\text{VSWR} = \frac{1+\rho}{1-\rho} = \frac{1+0.2}{1-0.2} = 1.5$$

The electric field intensity in medium 2 is

$$\mathcal{E}^t(z,t) = \hat{y} \bullet 3.84 \cos(10^9 t - k_2 z) \qquad \text{V/m}$$

Therefore, the time-averaged power transmitted into medium 2 may be found as follows:

$$P_{\text{av}} = \frac{1}{2}\,\text{Re}(\mathbf{E} \times \mathbf{H}^*) = \frac{1}{2}\frac{|\mathbf{E}^t|^2}{\eta_2} = \frac{1}{2}\left(\frac{3.84^2}{251.1562}\right) = 0.0294\ \text{W/m}^2$$

4.5 MODIFIED MAXWELL'S EQUATIONS AND POTENTIAL FUNCTIONS

Equation (4.1.9) implies that the magnetic flux does not start from or terminate at a magnetic charge the way that electric flux does with electric charge. This is because there is no real magnetic charge (or magnetic monopole). In fact, the smallest unit of the magnetic source is a magnetic dipole that is actually an infinitesimal current loop.

Consider a general case where the source consists of two types of currents: a linear current that flows or oscillates in the linear direction, and an infinitesimal circulatory current that flows or oscillates around an infinitesimal loop. For this case it is rather hard to describe mathematically the electric current density J that includes both parts. Mathematically, it is sometimes advantageous to recognize its linear part as electric current J_e and imagine its infinitesimal circulatory current as an equivalent magnetic dipole source. If the circulatory current varies in magnitude and direction or oscillates around the loop with time, the equivalent magnetic dipole strength also oscillates with time. When the magnetic dipole oscillates with time, it can be considered as an imaginary magnetic current flowing up and down. This is completely analogous to a time-varying electric dipole. Therefore, a time-varying circulatory current can be represented by an equivalent magnetic current and charges, J_m and ρ_m, respectively.

Imagine now that the magnetic charge ρ_m produces a magnetic flux, analogous to the electric charge ρ_e producing the electric flux. Then equation (4.1.9) should be modified to

$$\nabla \cdot \mathcal{B} = \rho_m$$

If (4.1.9) is modified, (4.1.6) also needs modification because divergence of the curl of a vector is always zero, which is not true because ρ_m is a function of time:

$$\nabla \cdot (\nabla \times \mathcal{E}) = -\nabla \cdot \frac{\partial \mathcal{B}}{\partial t} = -\frac{\partial}{\partial t}(\nabla \cdot \mathcal{B}) = -\frac{\partial}{\partial t}\rho_m \neq 0$$

If it is assumed that the usual continuity equation should hold between ρ_m and J_m,

$$\nabla \cdot \mathcal{J}_m = -\frac{\partial \rho_m}{\partial t}$$

Hence, equation (4.1.6) should be modified to

$$\nabla \times \mathcal{E} = -\mathcal{J}_m - \frac{\partial \mathcal{B}}{\partial t}$$

Therefore, the modified Maxwell's equations for the time-harmonic field and the two continuity equations can be expressed as:

$$\nabla \times \mathbf{E}(\mathbf{r}) = -\mathbf{J}_m(\mathbf{r}) - j\omega\mathbf{B}(\mathbf{r}) \tag{4.5.1}$$

$$\nabla \times \mathbf{H}(\mathbf{r}) = \mathbf{J}_e(\mathbf{r}) + j\omega\,\mathbf{D}(\mathbf{r}) \tag{4.5.2}$$

$$\nabla \bullet \mathbf{D}(\mathbf{r}) = \rho_e(\mathbf{r}) \tag{4.5.3}$$

$$\nabla \bullet \mathbf{B}(\mathbf{r}) = \rho_m(\mathbf{r}) \tag{4.5.4}$$

$$\nabla \bullet \mathbf{J}_e(\mathbf{r}) = -j\omega\rho_e(\mathbf{r}) \tag{4.5.5}$$

and

$$\nabla \bullet \mathbf{J}_m(\mathbf{r}) = -j\omega\rho_m(\mathbf{r}) \tag{4.5.6}$$

Magnetic Vector and Electric Scalar Potentials

Potential functions are introduced to transform Maxwell's equations mathematically and so facilitate solutions to electromagnetic problems. Consider that there are only electric sources (current and charge) present in a volume under consideration (i.e., $\mathbf{J}_e \neq 0$, $\rho_e \neq 0$, $\mathbf{J}_m = 0$, and $\rho_m = 0$). Therefore, Maxwell's equations for time-harmonic fields and the equation of continuity can be written as follows:

$$\nabla \times \mathbf{E} = -j\omega\mathbf{B} \tag{4.5.7}$$

$$\nabla \times \mathbf{H} = \mathbf{J}_e + j\omega\mathbf{D} \tag{4.5.8}$$

$$\nabla \bullet \mathbf{D} = \rho_e \tag{4.5.9}$$

$$\nabla \bullet \mathbf{B} = 0 \tag{4.5.10}$$

and

$$\nabla \bullet \mathbf{J}_e = -j\omega\rho_e \tag{4.5.11}$$

As defined earlier, the constitutive relations are

$$\mathbf{B} = \mu\mathbf{H} \tag{4.5.12}$$

and

$$\mathbf{D} = \varepsilon\mathbf{E} \tag{4.5.13}$$

Since divergence of the curl of a vector is always zero, the magnetic flux density in (4.5.10) may be assumed to be the curl of vector **A**. Hence,

$$\nabla \bullet \mathbf{B} = 0 \Rightarrow \mathbf{B} = \nabla \times \mathbf{A} \tag{4.5.14}$$

Substitution of **B** from (4.5.14) into (4.5.7), and recognizing the fact that the curl of the gradient of a scalar is always zero gives

$$\nabla \times \mathbf{E} = -j\omega \nabla \times \mathbf{A} \Rightarrow \nabla \times (\mathbf{E} + j\omega \mathbf{A}) = 0 \Rightarrow \mathbf{E} + j\omega \mathbf{A} = -\nabla \phi_e$$

Therefore,

$$\mathbf{E} = -\nabla \phi_e - j\omega \mathbf{A} \tag{4.5.15}$$

where **A** is called the *magnetic vector potential* and ϕ_e is known as the *electric scalar potential.*

Using (4.5.12), (4.5.13), and (4.5.15), (4.5.8) may be written as

$$\nabla \times \frac{\nabla \times \mathbf{A}}{\mu} = \mathbf{J}_e + j\omega\varepsilon \mathbf{E} = \mathbf{J}_e + j\omega\varepsilon(-\nabla \phi_e - j\omega \mathbf{A})$$

If μ is not changing with space coordinates, this relation simplifies further as follows:

$$\nabla \times \nabla \times \mathbf{A} = \nabla(\nabla \bullet \mathbf{A}) - \nabla^2 \mathbf{A} = \mu \mathbf{J}_e - j\omega\varepsilon\mu \nabla \phi_e + \omega^2 \mu\varepsilon \mathbf{A}$$

or

$$\nabla^2 \mathbf{A} + \omega^2 \mu\varepsilon \mathbf{A} = -\mu \mathbf{J}_e + j\omega\varepsilon\mu \nabla \phi_e + \nabla(\nabla \bullet \mathbf{A})$$

or

$$\nabla^2 \mathbf{A} + \omega^2 \mu\varepsilon \mathbf{A} = -\mu \mathbf{J}_e + \nabla(\nabla \bullet \mathbf{A} + j\omega\varepsilon\mu\phi_e) \tag{4.5.16}$$

Similarly for ε not changing with space coordinates, (4.5.9), (4.5.13), and (4.5.15) give

$$\nabla \bullet (-\nabla \phi_e - j\omega \mathbf{A}) = \frac{\rho_e}{\varepsilon} \rightarrow \nabla^2 \phi_e + j\omega(\nabla \bullet \mathbf{A}) = -\frac{\rho_e}{\varepsilon} \tag{4.5.17}$$

Equations (4.5.16) and (4.5.17) represent a pair of coupled partial differential equations for ϕ and **A**. These may be uncoupled via *Helmholtz's theorem*, which states that a vector is specified completely by its curl and divergence. Since only the curl of **A** is defined via (4.5.14), we are at liberty to specify its divergence. The Lorentz condition may be used to define it as

$$\nabla \bullet \mathbf{A} = -j\omega\varepsilon\mu\phi_e \tag{4.5.18}$$

Thus, (4.5.16) and (4.5.17) simplify to

$$\nabla^2 \mathbf{A} + k^2 \mathbf{A} = -\mu \mathbf{J} \tag{4.5.19}$$

and

$$\nabla^2 \phi_e + k^2 \phi_e = -\frac{\rho_e}{\varepsilon} \tag{4.5.20}$$

where $k^2 = \omega^2\mu\varepsilon$ and k is the wave number. It should be noted at this point that the continuity equation is implied by the Lorentz condition.

Electric Vector and Magnetic Scalar Potentials

Electric vector and magnetic scalar potentials are introduced when the existence of only magnetic current and charge is assumed in the region (i.e, $\rho_e = 0$ and $\mathbf{J}_e = 0$). Therefore, Maxwell's equations and the equation of continuity for this case are found to be

$$\nabla \times \mathbf{E} = -\mathbf{J}_m - j\omega\mathbf{B} \tag{4.5.21}$$

$$\nabla \times \mathbf{H} = j\omega\mathbf{D} \tag{4.5.22}$$

$$\nabla \cdot \mathbf{D} = 0 \tag{4.5.23}$$

$$\nabla \cdot \mathbf{B} = \rho_m \tag{4.5.24}$$

and

$$\nabla \cdot \mathbf{J}_m = -j\omega\rho_m \tag{4.5.25}$$

As before, $\mathbf{D} = \varepsilon\mathbf{E}$ and $\mathbf{B} = \mu\mathbf{H}$. Because of (4.5.23), it is assumed that

$$\mathbf{D} = -\nabla \times \mathbf{F} \tag{4.5.26}$$

where $\mathbf{F}$ is called the *electric vector potential.*

After substituting (4.5.26) into (4.5.22) and noting that curl of the gradient of a scalar is always zero, we get

$$\nabla \times \mathbf{H} = -j\omega\nabla \times \mathbf{F} \rightarrow \nabla \times (\mathbf{H} + j\omega\mathbf{F}) = 0 \rightarrow \mathbf{H} + j\omega\mathbf{F} = -\nabla\phi_m$$

Therefore,

$$\mathbf{H} = -\nabla\phi_m - j\omega\mathbf{F} \tag{4.5.27}$$

where ϕ_m is called the *magnetic scalar potential.* Now, from (4.5.21), (4.5.26), and (4.5.27),

$$\nabla \times \left(-\frac{\nabla \times \mathbf{F}}{\varepsilon}\right) = -\mathbf{J}_m - j\omega\mu(-\nabla\phi_m - j\omega\mathbf{F})$$

or

$$-\nabla \times \nabla \times \mathbf{F} = -\varepsilon\mathbf{J}_m + j\omega\mu\varepsilon\nabla\phi_m - \omega^2\mu\varepsilon\mathbf{F}$$

or

$$-[\nabla(\nabla \cdot \mathbf{F}) - \nabla^2\mathbf{F}] + \omega^2\mu\varepsilon\mathbf{F} = -\varepsilon\mathbf{J}_m + j\omega\mu\varepsilon\nabla\phi_m \tag{4.5.28}$$

As in the preceding case, if the Lorentz condition

$$\nabla \cdot \mathbf{F} = -j\omega\mu\varepsilon\phi_m \tag{4.5.29}$$

is introduced, then

$$\nabla^2\mathbf{F} + \omega^2\mu\varepsilon\mathbf{F} = -\varepsilon\mathbf{J}_m \tag{4.5.30}$$

Further, (4.5.24), (4.5.27), and (4.5.29) give

$$\nabla \cdot [\mu(-\nabla\phi_m - j\omega\mathbf{F})] = \rho_m \Rightarrow -\nabla^2\phi_m - j\omega(-j\omega\mu\varepsilon\phi_m) = \frac{\rho_m}{\mu}$$

or

$$\nabla^2\phi_m + \omega^2\mu\varepsilon\phi_m = -\frac{\rho_m}{\mu} \tag{4.5.31}$$

An electromagnetic field problem can be divided into these two cases, solved for each case, and the individual solutions can be superimposed to find the complete solution. Thus, the electromagnetic fields may be found easily after determining the vector potentials via (4.5.19) and (4.5.30) and the corresponding Lorentz conditions. This procedure is used in the following section to further formulate general solution techniques for source-free cases that are specialized subsequently for a few waveguides.

4.6 CONSTRUCTION OF SOLUTIONS

The analysis presented in Section 4.5 is specialized here for a source-free region with rectangular as well as cylindrical coordinates. It is further divided into two cases; the electric vector potential is assumed to be zero in one, and the magnetic vector potential, zero in the other.

Rectangular Coordinate System

1. Assume that $\mathbf{A} = \hat{z}A_z$ and $\mathbf{F} = 0$; then (4.5.14), (4.5.15), and (4.5.18) give

$$\mathbf{H} = \frac{1}{\mu}\nabla \times \hat{z}A_z = \frac{1}{\mu}\left[\hat{x}\frac{\partial A_z}{\partial y} + \hat{y}\left(-\frac{\partial A_z}{\partial x}\right) + \hat{z}0\right] \tag{4.6.1}$$

and

$$\mathbf{E} = -j\omega\hat{z}A_z + \frac{1}{j\omega\mu\varepsilon}\nabla\frac{\partial A_z}{\partial z} \tag{4.6.2}$$

For this case, (4.5.19) simplifies to

$$\nabla^2 A_z + k^2 A_z = 0 \rightarrow \frac{\partial^2 A_z}{\partial x^2} + \frac{\partial^2 A_z}{\partial y^2} + \frac{\partial^2 A_z}{\partial z^2} + k^2 A_z = 0 \tag{4.6.3}$$

After (4.6.3) is solved for a given problem, field components are found from (4.6.1) and (4.6.2) as follows:

$$H_x = \frac{1}{\mu}\frac{\partial A_z}{\partial y} \tag{4.6.4}$$

$$H_y = -\frac{1}{\mu}\frac{\partial A_z}{\partial x} \tag{4.6.5}$$

$$H_z = 0 \tag{4.6.6}$$

$$E_x = \frac{1}{j\omega\mu\varepsilon}\frac{\partial^2 A_z}{\partial x \partial z} \tag{4.6.7}$$

$$E_y = \frac{1}{j\omega\mu\varepsilon}\frac{\partial^2 A_z}{\partial y \partial z} \tag{4.6.8}$$

and

$$E_z = \frac{1}{j\omega\mu\varepsilon}\left(\frac{\partial^2}{\partial z^2} + k^2\right) A_z \tag{4.6.9}$$

Since the magnetic fields are transverse to the z-axis ($H_z = 0$), these fields are known as TM^z (*transverse magnetic to* z) *mode fields*.

2. Assume that $\mathbf{F} = \hat{z}F_z$ and $\mathbf{A} = 0$; then (4.5.26) and (4.5.27) give

$$\mathbf{E} = -\frac{1}{\varepsilon}\nabla \times \hat{z}F_z = -\frac{1}{\varepsilon}\left[\hat{x}\frac{\partial F_z}{\partial y} + \hat{y}\left(-\frac{\partial F_z}{\partial x}\right) + \hat{z}0\right] \tag{4.6.10}$$

and

$$\mathbf{H} = -j\omega\hat{z}F_z + \frac{1}{j\omega\mu\varepsilon}\nabla\frac{\partial F_z}{\partial z} \tag{4.6.11}$$

For this case, (4.5.30) simplifies to

$$\nabla^2 F_z + k^2 F_z = 0 \rightarrow \frac{\partial^2 F_z}{\partial x^2} + \frac{\partial^2 F_z}{\partial y^2} + \frac{\partial^2 F_z}{\partial z^2} + k^2 F_z = 0 \tag{4.6.12}$$

After (4.6.12) is solved for a given problem, field components are found from (4.6.10) and (4.6.11) as follows:

$$E_x = -\frac{1}{\varepsilon}\frac{\partial F_z}{\partial y} \tag{4.6.13}$$

$$E_y = \frac{1}{\varepsilon}\frac{\partial F_z}{\partial x} \tag{4.6.14}$$

$$E_z = 0 \tag{4.6.15}$$

$$H_x = \frac{1}{j\omega\mu\varepsilon}\frac{\partial^2 F_z}{\partial x\, \partial z} \tag{4.6.16}$$

$$H_y = \frac{1}{j\omega\mu\varepsilon} \frac{\partial^2 F_z}{\partial y \partial z} \tag{4.6.17}$$

and

$$H_z = \frac{1}{j\omega\mu\varepsilon} \left(\frac{\partial^2}{\partial z^2} + k^2 \right) F_z \tag{4.6.18}$$

Since the electric fields are transverse to the z-axis in this case, these fields are known as TE^z (*transverse electric to* z) *mode fields.*

Cylindrical Coordinate System

1. Assume that $\mathbf{A} = \hat{z} A_z$ and $\mathbf{F} = 0$; then from (4.5.14), (4.5.15), and (4.5.18), we get

$$\mathbf{H} = \frac{1}{\mu} \nabla \times \hat{z} A_z = \frac{1}{\mu} \left[\hat{\rho} \frac{1}{\rho} \frac{\partial A_z}{\partial \phi} + \hat{\phi} \left(-\frac{\partial A_z}{\partial \rho} \right) + \hat{z} 0 \right] \tag{4.6.19}$$

and

$$\mathbf{E} = -j\omega \hat{z} A_z + \frac{1}{j\omega\mu\varepsilon} \nabla \frac{\partial A_z}{\partial z} \tag{4.6.20}$$

For this case, (4.5.19) in cylindrical coordinates may be written as follows:

$$\nabla^2 A_z + k^2 A_z = 0 \rightarrow \frac{1}{\rho} \frac{\partial}{\partial \rho} \left(\rho \frac{\partial A_z}{\partial \rho} \right) + \frac{1}{\rho^2} \frac{\partial^2 A_z}{\partial \phi^2} + \frac{\partial^2 A_z}{\partial z^z} + k^2 A_z = 0 \tag{4.6.21}$$

Once solutions to (4.6.21) are found, electromagnetic fields may be determined from (4.6.19) and (4.6.20) as follows:

$$H_\rho = \frac{1}{\mu\rho} \frac{\partial A_z}{\partial \phi} \tag{4.6.22}$$

$$H_\phi = -\frac{1}{\mu} \frac{\partial A_z}{\partial \rho} \tag{4.6.23}$$

$$H_z = 0 \tag{4.6.24}$$

$$E_\rho = \frac{1}{j\omega\mu\varepsilon} \frac{\partial^2 A_z}{\partial \rho \partial z} \tag{4.6.25}$$

$$E_\phi = \frac{1}{j\omega\mu\varepsilon\rho} \frac{\partial^2 A_z}{\partial \phi \partial z} \tag{4.6.26}$$

and

$$E_z = \frac{1}{j\omega\mu\varepsilon} \left(\frac{\partial^2}{\partial z^2} + k^2 \right) A_z \tag{4.6.27}$$

This is the TM^z mode (magnetic field transverse to the z-axis) in a cylindrical coordinate system.

2. For $\mathbf{F} = \hat{z}F_z$ and $\mathbf{A} = 0$, (4.5.26) and (4.5.27) in cylindrical coordinates give

$$\mathbf{E} = -\frac{1}{\varepsilon}\nabla \times \hat{z}F_z = -\frac{1}{\varepsilon}\left[\hat{\rho}\frac{1}{\rho}\frac{\partial F_z}{\partial \phi} + \hat{\phi}\left(-\frac{\partial F_z}{\partial \rho}\right) + \hat{z}0\right] \quad (4.6.28)$$

and

$$\mathbf{H} = -j\omega\hat{z}F_z + \frac{1}{j\omega\mu\varepsilon}\nabla\left(\frac{\partial F_z}{\partial z}\right) \quad (4.6.29)$$

Further, (4.5.30) reduces to

$$\nabla^2 F_z + k^2 F_z = 0 \rightarrow \frac{1}{\rho}\frac{\partial}{\partial \rho}\left(\rho\frac{\partial F_z}{\partial \rho}\right) + \frac{1}{\rho^2}\frac{\partial^2 F_z}{\partial \phi^2} + \frac{\partial^2 F_z}{\partial z^2} + k^2 F_z = 0 \quad (4.6.30)$$

After (4.6.30) is solved, the field components are determined from (4.6.28) and (4.6.29) as follows:

$$E_\rho = -\frac{1}{\varepsilon\rho}\frac{\partial F_z}{\partial \phi} \quad (4.6.31)$$

$$E_\phi = \frac{1}{\varepsilon}\frac{\partial F_z}{\partial \rho} \quad (4.6.32)$$

$$E_z = 0 \quad (4.6.33)$$

$$H_\rho = \frac{1}{j\omega\mu\varepsilon}\frac{\partial^2 F_z}{\partial \rho \partial z} \quad (4.6.34)$$

$$H_\phi = \frac{1}{j\omega\mu\varepsilon\rho}\frac{\partial^2 F_z}{\partial \phi\, \partial z} \quad (4.6.35)$$

and

$$H_z = \frac{1}{j\omega\mu\varepsilon}\left(\frac{\partial^2}{\partial z^2} + k^2\right)F_z \quad (4.6.36)$$

This is the TE^z mode (electric field transverse to the z-axis) in a cylindrical coordinate system.

These formulations are used in the following sections to analyze propagation of electromagnetic signals through the waveguide.

4.7 METALLIC PARALLEL-PLATE WAVEGUIDE

Consider a parallel-plate waveguide that consists of two perfectly conductive plates extending to infinity on the y–z plane, as illustrated in Figure 4.13. Separation between the two plates is assumed to be a. There is an electromagnetic

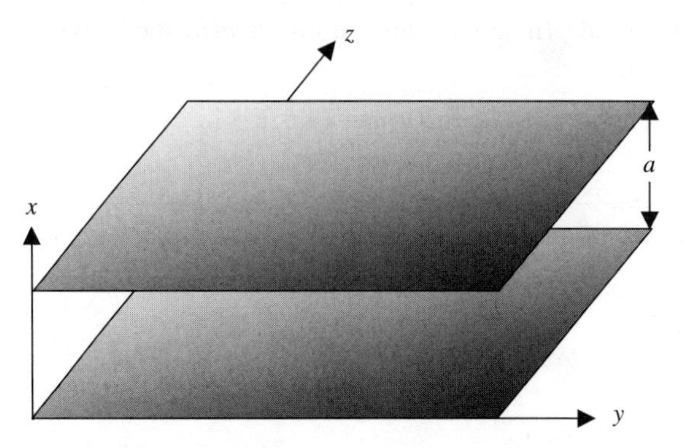

Figure 4.13 Metallic parallel-plate waveguide.

signal that propagates along the z-axis with a complex propagation constant γ. Therefore, the field variation along the z-axis is assumed to be $e^{-\gamma z}$, whereas it stays constant along the y-axis (i.e., $\partial/\partial y \to 0$). For analyzing TM modes between the two plates, partial differential equation (4.6.3) simplifies to the following ordinary differential equation:

$$\frac{d^2 A_z(x)}{dx^2} + (k^2 + \gamma^2) A_z(x) = 0 \tag{4.7.1}$$

A solution to this equation can be found easily as follows:

$$A_z(x) = c_1 \cos k_x x + c_2 \sin k_x x \tag{4.7.2}$$

where c_1 and c_2 are integration constants to be evaluated from the boundary conditions, and

$$k_x^2 = k^2 + \gamma^2 \tag{4.7.3}$$

Therefore,

$$A_z(x, y, z) = (c_1 \cos k_x x + c_2 \sin k_x x) e^{-\gamma z} \tag{4.7.4}$$

Since the boundary conditions require that the tangential electric fields must be zero on the conducting surfaces at $x = 0$ and at $x = a$, E_z must be zero on these surfaces. It is to be noted that E_y is zero everywhere because A_z is independent of y. For a nontrivial solution,

$$E_z|_{\substack{x=0\\x=a}} = 0 \to A_z(x, y, z)|_{\substack{x=0\\x=a}} = 0 \tag{4.7.5}$$

Equation (4.7.5) is satisfied only if

$$c_1 = 0 \tag{4.7.6}$$

and

$$k_x = \frac{m\pi}{a}, \qquad m = 0, 1, 2, \ldots \tag{4.7.7}$$

Therefore,

$$A_z = c_2 \sin\left(\frac{m\pi}{a}x\right) e^{-\gamma z} \tag{4.7.8}$$

Further, the propagation constant can be found from (4.7.3) as follows:

$$\gamma = \sqrt{\left(\frac{m\pi}{a}\right)^2 - k^2} \tag{4.7.9}$$

Therefore, the signal will propagate without attenuation if $k > m\pi/a$, and it will attenuate for $k < m\pi/a$. The cutoff occurs at $k = m\pi/a$.

Once A_z is determined, the field components may be found from (4.6.4)–(4.6.9) as follows:

$$H_x = 0 \tag{4.7.10}$$

$$H_y = -\frac{c_2}{\mu}\frac{m\pi}{a}\cos\left(\frac{m\pi}{a}x\right) e^{-\gamma z} \tag{4.7.11}$$

$$H_z = 0 \tag{4.7.12}$$

$$E_x = -c_2 \frac{\gamma}{j\omega\mu\varepsilon}\frac{m\pi}{a}\cos\left(\frac{m\pi}{a}x\right) e^{-\gamma z} \tag{4.7.13}$$

$$E_y = 0 \tag{4.7.14}$$

and

$$E_z = \frac{c_2}{j\omega\mu\varepsilon}\left(\frac{m\pi}{a}\right)^2 \sin\left(\frac{m\pi}{a}x\right) e^{-\gamma z} \tag{4.7.15}$$

These results are summarized in Table 4.1 assuming that $-c_2(k_c/\mu) = E_0$.

A similar procedure may be used to determine the characteristics of TE mode fields propagating through this waveguide. Final results are summarized in Table 4.1.

Example 4.11 A metallic parallel-plate waveguide is air-filled, and the separation between its plates is 4.5 cm. Investigate the characteristics of a 12-GHz signal propagating in TM_{m0} modes.

SOLUTION

$$k_c = \frac{m\pi}{a} \rightarrow \lambda_c = \frac{2\pi}{k_c} = \frac{2a}{m} = \frac{2 \times 4.5 \times 10^{-2}}{m} \quad \text{m}$$

and

$$f_c = \frac{3 \times 10^8}{\lambda_c} = \frac{3 \times 10^8 \times m}{9 \times 10^{-2}}\text{Hz} = 3.3333m \quad \text{GHz}$$

TABLE 4.1 Signal Propagation in Parallel-Plate Waveguides

	TE_{m0} Modes	TM_{m0} Modes
k_c	$\frac{m\pi}{a}$	$\frac{m\pi}{a}$
γ_{m0}	$\sqrt{k_c^2 - k_0^2}$	$\sqrt{k_c^2 - k_0^2}$
$H_z(x, z)$	$-\frac{k_c}{j\omega\mu} H_0 \cos(k_c x) e^{-\gamma_{m0} z}$	0
$E_z(x, z)$	0	$jE_0 \frac{k_c}{\omega\varepsilon} \sin(k_c a) e^{-\gamma_{m0} z}$
$H_x(x, z)$	$-\frac{\gamma_{m0}}{j\omega\mu} H_0 \sin(k_c x) e^{-\gamma_{m0} z}$	0
$H_y(x, z)$	0	$E_0 \cos(k_c x) e^{-\gamma_{m0} z}$
$E_x(x, z)$	0	$\frac{\gamma_{m0}}{j\omega\varepsilon} E_0 \cos(k_c x) e^{-\gamma_{m0} z}$
$E_y(x, z)$	$H_0 \sin(k_c x) e^{-\gamma_{m0} z}$	0

Hence, the cutoff frequencies for first four TM modes are found as follows:

$$TM_{10} \rightarrow f_c = 3.3333\,\text{GHz}$$

$$TM_{20} \rightarrow f_c = 6.6667\,\text{GHz}$$

$$TM_{30} \rightarrow f_c = 10\,\text{GHz}$$

$$TM_{40} \rightarrow f_c = 13.3333\,\text{GHz}$$

Since the cutoff frequency for the TM_{40} mode is higher than the signal frequency of 12 GHz, only TM_{10}, TM_{20}, and TM_{30} modes will exist for this signal. The corresponding cutoff wavelengths are 9, 4.5, and 3 cm, respectively. The signal wavelength inside the guide for each mode can be found as follows:

$$TM_{10} \rightarrow \lambda_g = \frac{\lambda_0}{\sqrt{1 - (\lambda_0/\lambda_c)^2}} = \frac{2.5}{\sqrt{1 - (2.5/9)^2}}\,\text{cm} = 2.6024\,\text{cm}$$

$$TM_{20} \rightarrow \lambda_g = \frac{\lambda_0}{\sqrt{1 - (\lambda_0/\lambda_c)^2}} = \frac{2.5}{\sqrt{1 - (2.5/4.5)^2}}\,\text{cm} = 3.0067\,\text{cm}$$

and

$$TM_{30} \rightarrow \lambda_g = \frac{\lambda_0}{\sqrt{1 - (\lambda_0/\lambda_c)^2}} = \frac{2.5}{\sqrt{1 - (2.5/3)^2}}\,\text{cm} = 4.5227\,\text{cm}$$

4.8 METALLIC RECTANGULAR WAVEGUIDE

In this section we present another application of the formulation described in Section 4.6 to analyze the electromagnetic signals propagating through a hollow metallic cylindrical waveguide of rectangular cross section, as shown in Figure 4.14. Both the geometry of a rectangular waveguide and the coordinate system are shown. It is assumed that its cross section is $a \times b$, and the electromagnetic signal is propagating along the z-axis. Therefore, the field variation along the z-axis is given as $e^{-jk_z z}$. TM and TE modes propagating through this waveguide can be analyzed using equations (4.6.1) to (4.6.18). For TE modes, (4.6.12) reduces to

$$\frac{\partial^2 F_z(x, y)}{\partial x^2} + \frac{\partial^2 F_z(x, y)}{\partial y^2} + (k^2 - k_z^2) F_z(x, y) = 0 \tag{4.8.1}$$

It is a second-order partial differential equation that can be solved using the separation-of-variables technique as follows.

Assume that $F_z(x, y)$ is a product of two functions, one dependent on x only and the other on y only. Hence,

$$F_z(x, y) = f_1(x) f_2(y) \tag{4.8.2}$$

Therefore,

$$\frac{\partial^2 F_z(x, y)}{\partial x^2} = f_2(y) \frac{d^2 f_1(x)}{dx^2} \tag{4.8.3}$$

and

$$\frac{\partial^2 F_z(x, y)}{\partial y^2} = f_1(x) \frac{d^2 f_2(y)}{dy^2} \tag{4.8.4}$$

Substituting (4.8.2) to (4.8.4) into (4.8.1) and rearranging results in

$$\frac{1}{f_1(x)} \frac{d^2 f_1(x)}{dx^2} + \frac{1}{f_2(y)} \frac{d^2 f_2(y)}{dy^2} = -(k^2 - k_z^2) \tag{4.8.5}$$

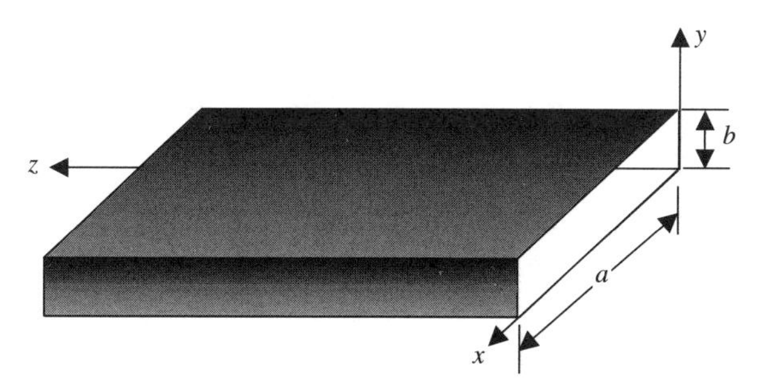

Figure 4.14 Metallic rectangular waveguide.

In (4.8.5), the first term is only x dependent, the second is dependent on y only, and the sum of these two is always a constant, as its right-hand side indicates. Therefore, each term on the left-hand side must be independently constant as follows:

$$\frac{1}{f_1(x)}\frac{d^2 f_1(x)}{dx^2} = -k_x^2 \tag{4.8.6}$$

and

$$\frac{1}{f_2(y)}\frac{d^2 f_2(y)}{dy^2} = -k_y^2 \tag{4.8.7}$$

where k_x and k_y are constants.

Equations (4.8.6) and (4.8.7) are in the form of the harmonic equation considered earlier, and its solutions are the harmonic functions. Complete solutions to each of these can be written as

$$f_1(x) = c_1 \cos k_x x + c_2 \sin k_x x \tag{4.8.8}$$

and

$$f_2(y) = c_3 \cos k_y y + c_4 \sin k_y y \tag{4.8.9}$$

Therefore, the solution to (4.8.1) is found to be

$$F_z(x, y) = (c_1 \cos k_x x + c_2 \sin k_x x)(c_3 \cos k_y y + c_4 \sin k_y y) \tag{4.8.10}$$

Further, equations (4.8.1), (4.8.6), and (4.8.7) give the separation equation,

$$k_x^2 + k_y^2 + k_z^2 = k^2 \tag{4.8.11}$$

Since tangential electric fields must be zero on the conducting surfaces, the boundary conditions for (4.8.1) are found to be

$$E_x\big|_{\substack{y=0\\y=b}} = 0 \rightarrow \left.\frac{\partial F_z}{\partial y}\right|_{\substack{y=0\\y=b}} = 0 \tag{4.8.12}$$

and

$$E_y\big|_{\substack{x=0\\x=a}} = 0 \rightarrow \left.\frac{\partial F_z}{\partial x}\right|_{\substack{x=0\\x=a}} = 0 \tag{4.8.13}$$

To satisfy these boundary conditions, (4.8.10) in conjunction with (4.8.12) and (4.8.14) gives

$$c_4 = 0 \tag{4.8.14}$$

$$k_y = \frac{n\pi}{b} \tag{4.8.15}$$

$$c_2 = 0 \tag{4.8.16}$$

and

$$k_x = \frac{m\pi}{a} \tag{4.8.17}$$

where m and n are integers, including zero, except that both cannot be zero at the same time (which is a trivial case). Therefore,

$$F_z(x, y) = c_1 c_3 \cos\left(\frac{m\pi}{a}x\right) \cos\left(\frac{n\pi}{b}y\right) \tag{4.8.18}$$

Further, k_z may be found from (4.8.11) as

$$k_z = \sqrt{k^2 - \left(\frac{m\pi}{a}\right)^2 - \left(\frac{n\pi}{b}\right)^2} \tag{4.8.19}$$

Consider a case when the waveguide is air-filled and therefore $k = k_0$ is a pure real number. The following conditions are possible in this situation.

(1)

$$k_0^2 < \left(\frac{m\pi}{a}\right)^2 + \left(\frac{n\pi}{b}\right)^2 \tag{4.8.20}$$

In this case, the quantity under the square root in (4.8.19) becomes negative. As a result, k_z is imaginary and the signal attenuates without propagating along the waveguide. This may be used to design attenuators.

(2)

$$k_0^2 > \left(\frac{m\pi}{a}\right)^2 + \left(\frac{n\pi}{b}\right)^2 \tag{4.8.21}$$

In this case, the quantity under the square root in (4.8.19) is positive. Therefore, k_z is real and the signal propagates through the waveguide without attenuation (ideally). This may be used to transmit the signal.

(3)

$$k_0^2 = \left(\frac{m\pi}{a}\right)^2 + \left(\frac{n\pi}{b}\right)^2 = k_c^2 \tag{4.8.22}$$

In this case, the quantity under the square root in (4.8.19) goes to zero. As a result, k_z is zero (i.e., the cutoff condition). This wave number is known as the cutoff wave number k_c, and the corresponding wavelength of the signal is called the *cutoff wavelength* λ_c. The corresponding frequency f_c is called the *cutoff frequency* of the waveguide. Thus, the waveguide works as a high-pass filter. A signal propagates only when its frequency is greater than the cutoff frequency of the waveguide.

The wavelength and phase velocity of a signal propagating on the waveguide are found as follows:

$$\lambda_g = \frac{2\pi}{\sqrt{k_0^2 - k_c^2}} = \frac{2\pi}{k_0\sqrt{1-(k_c/k_0)^2}} = \frac{\lambda_0}{\sqrt{1-(\lambda_0/\lambda_c)^2}}$$
$$= \frac{\lambda_0}{\sqrt{1-(f_c/f)^2}} \tag{4.8.23}$$

and

$$v_p = \frac{\omega}{k_z} = \frac{f\lambda_0}{\sqrt{1-(\lambda_0/\lambda_c)^2}} = \frac{3\times 10^8}{\sqrt{1-(\lambda_0/\lambda_c)^2}} \quad \text{m/s} \tag{4.8.24}$$

where f is the signal frequency, k_0 its wave number in free space, and λ_0 the signal wavelength in free space.

Fields inside the waveguide are found after substituting (4.8.18) into (4.6.13) to (4.6.18) as follows:

$$E_x(x, y, z) = jH_0\omega\mu k_y \cos k_x x \sin k_y y\, e^{-jk_z z} \tag{4.8.25}$$

$$E_y(x, y, z) = -jH_0\omega\mu k_x \sin k_x x \cos k_y y\, e^{-jk_z z} \tag{4.8.26}$$

$$E_z(x, y, z) = 0 \tag{4.8.27}$$

$$H_x(x, y, z) = jH_0 k_z k_x \sin\left(\frac{m\pi}{a}x\right)\cos\left(\frac{n\pi}{b}y\right) e^{-jk_z z} \tag{4.8.28}$$

$$H_y(x, y, z) = jH_0 k_z k_y \cos k_x x \sin k_y y e^{-jk_z z} \tag{4.8.29}$$

and

$$H_z(x, y, z) = H_0 k_c^2 \cos k_x x \cos k_y y e^{-jk_z z} \tag{4.8.30}$$

where

$$H_0 = -jc_1c_3\frac{1}{\omega\mu\varepsilon} \tag{4.8.31}$$

It is to be noted that if m and n are both zero, the fields do not exist. Therefore, for a nontrivial case, m or n has to be a nonzero integer. For $a > b$, the lowest-order mode that propagates is TE_{10} (i.e., for $m = 1$ and $n = 0$). It is known as the dominant mode for rectangular waveguides. From the expressions given in Table 4.2, it may be found easily that the lowest-order TM mode that can propagate through these waveguides is TM_{11} (i.e., for $m = 1$ and $n = 1$).

For the TE_{10} mode, only the H_z, H_x, and E_y field components will be nonzero. Further, $\lambda_c = 2a$ and

$$\lambda_{g(TE_{10})} = \frac{\lambda_0}{\sqrt{1-(\lambda_0/2a)^2}} \tag{4.8.32}$$

Further,

$$E_y(x, y, z) = -jH_0\omega\mu_0\frac{\pi}{a}\sin\left(\frac{\pi}{a}x\right)e^{-jk_z z}$$
$$= -H_0\omega\mu_0\frac{\pi}{2a}\left(e^{j[(\pi x/a)+k_z z]} - e^{-j[(\pi x/a)+k_z z]}\right) \tag{4.8.33}$$

TABLE 4.2 Signal Propagation in Metallic Waveguides of Rectangular Cross Section

	TE_{mn} Modes	TM_{mn} Modes
k_c	$\sqrt{\left(\frac{m\pi}{a}\right)^2 + \left(\frac{n\pi}{b}\right)^2}$	$\sqrt{\left(\frac{m\pi}{a}\right)^2 + \left(\frac{n\pi}{b}\right)^2}$
$k_z = -j\gamma$	$\sqrt{k_0^2 - k_c^2}$	$\sqrt{k_0^2 - k_c^2}$
λ_g	$\frac{\lambda_0}{\sqrt{1 - (k_c/k_0)^2}}$	$\frac{\lambda_0}{\sqrt{1 - (k_c/k_0)^2}}$
$H_z(x, y)$	$H_0 k_c^2 \cos\left(\frac{m\pi}{a}x\right)\cos\left(\frac{n\pi}{b}y\right)$	0
$E_z(x, y)$	0	$E_0 k_c^2 \sin\left(\frac{m\pi}{a}x\right)\sin\left(\frac{n\pi}{b}y\right)$
$H_x(x, y)$	$jH_0 k_z \frac{m\pi}{a} \sin\left(\frac{m\pi}{a}x\right)\cos\left(\frac{n\pi}{b}y\right)$	$jE_0\omega\varepsilon \frac{n\pi}{b} \sin\left(\frac{m\pi}{a}x\right)\cos\left(\frac{n\pi}{b}y\right)$
$H_y(x, y)$	$jH_0 k_z \frac{n\pi}{b} \cos\left(\frac{m\pi}{a}x\right)\sin\left(\frac{n\pi}{b}y\right)$	$-jE_0\omega\varepsilon \frac{m\pi}{a} \cos\left(\frac{m\pi}{a}x\right)\sin\left(\frac{n\pi}{b}y\right)$
$E_x(x, y)$	$jH_0\omega\mu_0 \frac{n\pi}{b} \cos\left(\frac{m\pi}{a}x\right)\sin\left(\frac{n\pi}{b}y\right)$	$-j\frac{m\pi k_z}{a} E_0 \cos\left(\frac{m\pi}{a}x\right)\sin\left(\frac{n\pi}{b}y\right)$
$E_y(x, y)$	$-jH_0\omega\mu_0 \frac{m\pi}{a} \sin\left(\frac{m\pi}{a}x\right)\cos\left(\frac{n\pi}{b}y\right)$	$-j\frac{n\pi k_z}{b} E_0 \sin\left(\frac{m\pi}{a}x\right)\cos\left(\frac{n\pi}{b}y\right)$

On comparing this with (4.2.16), it may be found that this represents two plane electromagnetic waves propagating at angles $\pm\theta$ after reflection from the sidewalls of the waveguide, as shown in Figure 4.15. This angle is found to be

$$\theta = \sin^{-1}\frac{\lambda_0}{2a} \tag{4.8.34}$$

Therefore, $\theta \to 90^\circ$ as $\lambda_0 \to 2a$ (i.e., λ_c), and the wave ceases to propagate.

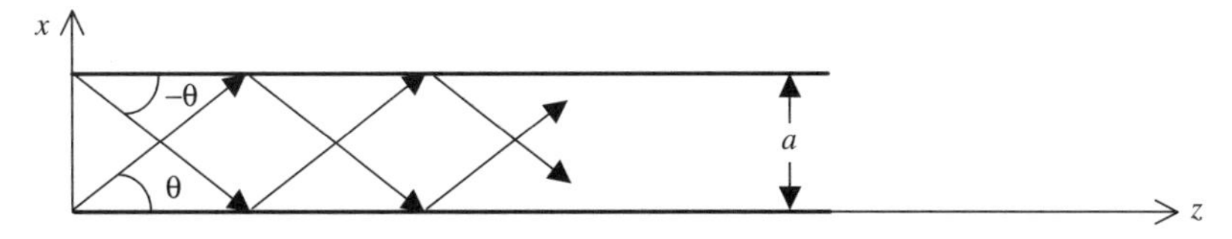

Figure 4.15 Two uniform plane waves bouncing from the conducting sidewalls of a rectangular waveguide.

Example 4.12 The inside cross section of an air-filled rectangular metallic waveguide is $1.58\,\text{cm} \times 0.79\,\text{cm}$. **(a)** Determine its cutoff frequencies for TE_{10}, TE_{20}, TE_{01}, and TE_{11} modes. **(b)** Find the mode(s) that will propagate through this waveguide if the signal frequency is anywhere between 12 and 18 GHz. **(c)** If this waveguide is being used to send a 15.8-GHz signal in TE_{10} mode, find the phase velocity.

SOLUTION

(a) Since

$$k_c = \sqrt{\left(\frac{m\pi}{a}\right)^2 + \left(\frac{n\pi}{b}\right)^2} \rightarrow f_c = 3 \times 10^8 \sqrt{\left(\frac{m}{2a}\right)^2 + \left(\frac{n}{2b}\right)^2}$$

cutoff frequencies for various modes are found to be

$$TE_{10} \rightarrow f_c = 3 \times 10^8 \sqrt{\left(\frac{1}{2 \times 0.0158}\right)^2 + \left(\frac{0}{2 \times 0.0079}\right)^2}\ \text{Hz} = 9.4937\,\text{GHz}$$

$$TE_{20} \rightarrow f_c = 3 \times 10^8 \sqrt{\left(\frac{2}{2 \times 0.0158}\right)^2 + \left(\frac{0}{2 \times 0.0079}\right)^2}\ \text{Hz} = 18.9873\,\text{GHz}$$

$$TE_{01} \rightarrow f_c = 3 \times 10^8 \sqrt{\left(\frac{0}{2 \times 0.0158}\right)^2 + \left(\frac{1}{2 \times 0.0079}\right)^2}\ \text{Hz} = 18.9873\,\text{GHz}$$

and

$$TE_{11} \rightarrow f_c = 3 \times 10^8 \sqrt{\left(\frac{1}{2 \times 0.0158}\right)^2 + \left(\frac{1}{2 \times 0.0079}\right)^2}\ \text{Hz} = 21.2285\,\text{GHz}$$

(b) From part (a), the cutoff frequency for the TE_{10} mode is 9.4937 GHz. Since the next-higher modes have cutoff at 18.9873 GHz, only the TE_{10} mode will exist for the signal frequency band 12 to 18 GHz.

(c) $$v_p = \frac{3 \times 10^8}{\sqrt{1 - (\lambda_0/\lambda_c)^2}} = \frac{3 \times 10^8}{\sqrt{1 - (f_c/f)^2}} = \frac{3 \times 10^8}{\sqrt{1 - (9.4937/15.8)^2}}$$

$$= 3.7531 \times 10^8\ \text{m/s}$$

4.9 METALLIC CIRCULAR WAVEGUIDE

Electromagnetic signal propagating through a hollow conducting cylindrical waveguide of circular cross section can be analyzed using (4.6.19) to (4.6.36). Figure 4.16 shows the geometry and coordinate systems for such analysis.

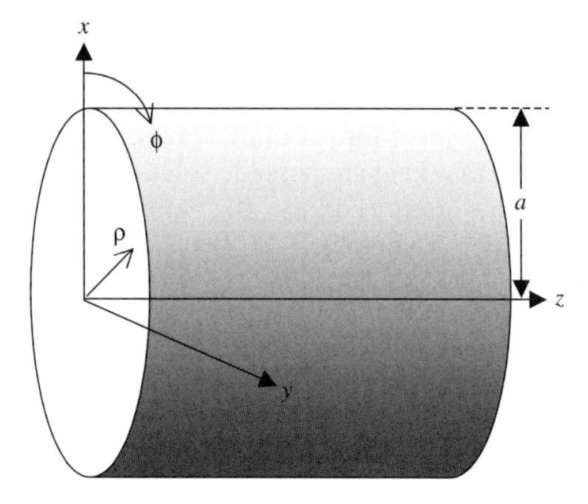

Figure 4.16 Metallic circular waveguide geometry.

Assume that the waveguide radius is a and the field variation in the z-direction is $e^{-jk_z z}$ (because the signal is propagating along z). For a TM mode, (4.6.21) gives

$$\rho\frac{\partial}{\partial\rho}\left(\rho\frac{\partial A_z(\rho,\phi)}{\partial\rho}\right)+\frac{\partial^2 A_z(\rho,\phi)}{\partial\phi^2}+(k^2-k_z^2)\rho^2 A_z(\rho,\phi)=0 \tag{4.9.1}$$

Following the technique used in Section 4.8, assume that $A_z(\rho, \phi)$ is a product of two functions (one dependent on ρ only and the other on ϕ only) and solve (4.9.1) using the separation of variables. Hence,

$$A_z(\rho,\phi)=f_1(\rho)f_2(\phi) \tag{4.9.2}$$

Therefore, (4.9.1) may be written as

$$\frac{\rho}{f_1(\rho)}\frac{d}{d\rho}\left(\rho\frac{df_1(\rho)}{d\rho}\right)+(k^2-k_z^2)\rho^2=-\frac{1}{f_2(\phi)}\frac{d^2 f_2(\phi)}{d\phi^2}=n^2 \tag{4.9.3}$$

where n is a constant that turns out to be an integer so that $f_2(\phi)$ is a single-valued harmonic function.

After defining $k^2 - k_z^2 = k_\rho^2$ for simplicity, (4.9.3) gives

$$\rho\frac{d}{d\rho}\left(\rho\frac{df_1(\rho)}{d\rho}\right)+(k_\rho^2\rho^2-n^2)f_1(\rho)=0 \tag{4.9.4}$$

and

$$\frac{d^2 f_2(\phi)}{d\phi^2}+n^2 f_2(\phi)=0 \tag{4.9.5}$$

Equation (4.9.4) is known as *Bessel's equation of order n*. Its solutions are the Bessel's function of the first kind, the Neumann function (also known as the

Bessel's function of the second kind), and the Hankel functions of the first and second kinds. Any two of these functions are linearly independent solutions to (4.9.4). Further, the first two of these represent standing waves, whereas the Hankel functions are convenient for traveling waves. For the present case, only Bessel's functions of the first kind are possible solutions because Neumann functions go to infinity at the origin (i.e., $\rho = 0$). Equation (4.9.5) is the harmonic equation considered earlier. Therefore,

$$A_z(\rho, \phi, z) = cJ_n(k_\rho \rho) \begin{cases} \sin n\phi \\ \cos n\phi \end{cases} \tag{4.9.6}$$

Since tangential electric fields must be zero on the conducting wall at $\rho = a$,

$$J_n(k_\rho a) = 0 \rightarrow k_\rho = \frac{p_{nm}}{a} \tag{4.9.7}$$

where p_{nm} is the mth zero of $J_n(k_\rho a)$.

TABLE 4.3 Signal Propagation in Circular-Cross-Section Waveguides

	TE$_{nm}$ Modes	TM$_{nm}$ Modes
k_c	$\frac{p'_{nm}}{a}$	$\frac{p_{nm}}{a}$
$k_z = -j\gamma_{mn}$	$\sqrt{k_0^2 - k_c^2}$	$\sqrt{k_0^2 - k_c^2}$
$H_z(\rho, \phi)$	$H_0 J_n(k_c\rho) \begin{cases} \cos n\phi \\ \sin n\phi \end{cases}$	0
$E_z(\rho, \phi)$	0	$E_0 J_n(k_c\rho) \begin{cases} \cos n\phi \\ \sin n\phi \end{cases}$
$H_\rho(\rho, \phi)$	$-j\frac{k_z}{k_c} H_0 J'_n(k_c\rho) \begin{cases} \cos n\phi \\ \sin n\phi \end{cases}$	$j\frac{n\omega\varepsilon}{\rho k_c^2} E_0 J_n(k_c\rho) \begin{cases} -\sin n\phi \\ \cos n\phi \end{cases}$
$H_\phi(\rho, \phi)$	$-j\frac{nk_z}{\rho k_c^2} H_0 J_n(k_c\rho) \begin{cases} -\sin n\phi \\ \cos n\phi \end{cases}$	$-j\frac{\omega\varepsilon}{k_c} E_0 J'_n(k_c\rho) \begin{cases} \cos n\phi \\ \sin n\phi \end{cases}$
$E_\rho(\rho, \phi)$	$-j\frac{n\omega\mu}{\rho k_c^2} H_0 J_n(k_c\rho) \begin{cases} -\sin n\phi \\ \cos n\phi \end{cases}$	$-j\frac{k_z}{k_c} E_0 J'_n(k_c\rho) \begin{cases} \cos n\phi \\ \sin n\phi \end{cases}$
$E_\phi(\rho, \phi)$	$j\frac{\omega\mu}{k_c} H_0 J'_n(k_c\rho) \begin{cases} \cos n\phi \\ \sin n\phi \end{cases}$	$-j\frac{nk_z}{\rho k_c^2} E_0 J_n(k_c\rho) \begin{cases} -\sin n\phi \\ \cos n\phi \end{cases}$

Similarly, (4.6.30) is solved to analyze TE modes in the circular waveguide. In this case, enforcement of the boundary condition ($E_\phi = 0$ at $\rho = a$) gives

$$\left.\frac{\partial J_n(k_\rho \rho)}{\partial (k_\rho \rho)}\right|_{\rho=a} = J_n'(k_\rho a) = 0 \tag{4.9.8}$$

and

$$k_\rho = \frac{p'_{nm}}{a} \tag{4.9.9}$$

where p'_{nm} is the mth root of (4.9.8).

Characteristics of the TM and TE modes in a circular waveguide are listed in Table 4.3. Detailed derivations are deferred to Problems 4.34 and 4.35.

SUGGESTED READING

Arfken, G., *Mathematical Methods for Physicists*. San Diego, CA: Academic Press, 1985.

Balanis, C. A., *Advanced Engineering Electromagnetics*. New York: Wiley, 1989.

Collin, R. E., *Foundations for Microwave Engineering*. New York: McGraw-Hill, 1992.

Harrington, R. F., *Time-Harmonic Electromagnetic Fields*. New York: McGraw-Hill, 1961.

Inan, U. S., and A. S. Inan, *Electromagnetic Waves*. Upper Saddle River, NJ: Prentice Hall, 2000.

PROBLEMS

4.1. The magnetic flux density on the $y = 0$ plane is given by

$$\mathcal{B} = 5 \cos 3x \cos 10^5 t\, \hat{y} \qquad \text{T}$$

A square loop of conducting wire is placed on this plane with its vertices at $(x, 0, 2)$, $(x, 0, 3)$, $(x + 1, 0, 3)$, and $(x + 1, 0, 2)$.

(a) Find the emf induced around the loop assuming that the loop is stationary.

(b) If the loop is moving with a velocity of $\mathbf{v} = \hat{x} \cdot 10^4$ m/s, determinc the new induced emf.

4.2. If $\mathcal{B} = \hat{z} \cdot 5 \cos(10^5 t - 0.5\pi x - \pi y)$T, find the emf induced around a closed loop formed by connecting successively the points (0, 0, 0), (2, 0, 0), (2, 1, 0), (0, 1, 0), and (0, 0, 0).

4.3. For the following charge distribution, find the displacement flux that emanates from a cubical surface bounded by $x = \pm 2$, $y = \pm 2$, and $z = \pm 2$.

$$\rho(x, y, z) = 5(3 - x^2 - y^2 - z^2) \qquad \text{C/m}^3$$

4.4. Find the charge densities that will produce the following electric flux densities: **(a)** $\mathbf{D} = xy\hat{x} + yz\hat{y} + zx\hat{z}$ C/m^2 and **(b)** $\mathbf{D} = \rho \sin\phi\hat{\phi}$ C/m^2.

4.5. In cylindrical coordinates, charge is distributed with uniform density 5 C/m^3 within the region $1 < \rho < 2$. Find the electric flux density in the region **(a)** $1 < \rho < 2$ and **(b)** $\rho > 2$.

4.6. If the electric field intensity in free space is $\mathcal{E} = 5\cos(10^9 t - x + ky)\hat{z}$ V/m, find the value(s) of k.

4.7. The magnetic field intensity of an electromagnetic wave is given as

$$\mathcal{H}(x, t) = \hat{z} \bullet 0.6 \cos\beta x \cos 10^8 t \qquad \text{A/m}$$

If the region is nonmagnetic with $\sigma = 0$ and $\varepsilon = 6.25\varepsilon_0$, find the corresponding electric field intensity and β.

4.8. A uniform plane electromagnetic wave is propagating in free space. Its instantaneous magnetic field intensity is given by

$$\mathcal{H} = \hat{z} \bullet 5\cos\left(10^8 t - \beta y + \frac{\pi}{4}\right) \qquad \mu\text{A/m}$$

(a) Determine β and the signal wavelength.

(b) Find the corresponding instantaneous electric field intensity.

4.9. A uniform plane wave of 50 MHz propagates in a lossless nonmagnetic medium in the $+x$ direction. A probe located at $x = 0$ measures the phase angle of the wave to be 95°. An identical probe located at $x = 2.5$ m measures the phase to be $-12°$. What is the relative permittivity of the medium?

4.10. The magnetic field intensity of a uniform plane wave of 50 MHz is given as

$$\mathbf{H}(z) = (5\hat{x} + j10\hat{y})e^{-j2z} \qquad \text{A/m}$$

If it is in a lossless and nonmagnetic medium, find v_p, η, and the instantaneous electric field intensity.

4.11. The electric field intensity of a uniform plane wave propagating in a nonmagnetic medium of zero conductivity is given as follows:

$$\mathcal{E}(y, t) = \hat{x} \bullet 12\pi \cos(9\pi \times 10^7 t + 0.3\pi y) \qquad \text{V/m}$$

Find **(a)** the frequency, **(b)** the wavelength, **(c)** the direction of wave propagation, **(d)** the dielectric constant of the medium, and **(e)** the associated magnetic field intensity.

4.12. A uniform plane wave of 2 GHz is propagating in the $+x$ direction in a nonmagnetic medium that has a dielectric constant of 2.25 and a loss tangent of 0.1. Its electric field is in the y-direction.

(a) Find the distance over which the amplitude of the propagating wave will reduce by 50 percent.

(b) Find the intrinsic impedance, the wavelength, and the phase velocity of the wave in the medium.

4.13. A uniform plane wave propagates in the $+z$ (downward) direction into the ocean ($\epsilon_r = 80$, $\mu_r = 1$, and $\sigma = 4$ S/m). Its magnetic field at the ocean surface ($z = 0$) is give as

$$\mathcal{H}(0, t) = \hat{y} \cdot 2 \cos 10^9 t \qquad \text{A/m}$$

(a) Determine the skin depth and intrinsic impedance of the ocean water.

(b) Find the corresponding electric field intensity in the ocean.

4.14. A uniform plane wave of frequency 100 kHz is propagating in a material medium. It is given that (1) the fields attenuate by the factor e^{-1} over a distance of 1205 m, (2) the fields undergo a change in phase by 2π radians in a distance of 1321 m, and (3) the ratio of the amplitudes of its electric and magnetic field intensities at a point in the medium is 163.54 Ω. Find the propagation constant γ and the intrinsic impedance η.

4.15. The electric field intensity of a uniform plane wave propagating through a lossless nonmagnetic medium is given by

$$\mathcal{E}(z, t) = \hat{x} \cdot 0.2 \cos(10^9 t + \pi z) \qquad \text{V/m}$$

Find **(a)** the direction of propagation, **(b)** the velocity of the signal, **(c)** the wavelength, and **(d)** the associated magnetic field intensity.

4.16. A material has an intrinsic impedance of $120\angle 24°\,\Omega$ at 320 MHz. If its relative permeability is 4, find its **(a)** loss tangent, **(b)** dielectric constant, **(c)** conductivity, and **(d)** complex permittivity.

4.17. The electric field intensity of a uniform plane wave traveling through a material medium ($\epsilon_r = 4, \sigma = 0.1$ S/m, $\mu_r = 1$) is given by $\hat{y} \cdot 10e^{\gamma z}$ V/m. If the signal frequency is 2.45 GHz, find the associated magnetic field intensity and propagation constant γ.

4.18. For a uniform plane wave propagating in the $+z$ direction in a nonmagnetic material medium with $\sigma = 10$ S/m and $\epsilon_r = 9$, the magnetic field intensity in the $z = 0$ plane is given by

$$\mathcal{H}(z = 0, t) = \hat{y} \cdot 0.1 \cos(2\pi \times 10^8 t) \qquad \text{A/m}$$

Find the associated electric field intensity in the medium.

4.19. The magnetic field intensity of a uniform plane wave propagating through a certain nonmagnetic medium is given by

$$\mathcal{H}(x, t) = \hat{z} \cdot 5 \cos(10^9 t - 6x) \qquad \text{A/m}$$

Find **(a)** the direction of wave travel, **(b)** the velocity of wave, **(c)** the wavelength, and **(d)** the associated electric field intensity.

4.20. A uniform plane wave propagating through a lossless nonmagnetic medium has a power density of 8.05 W/m^2 and an rms electric field intensity at 40.2 V/m. Find **(a)** the intrinsic impedance of the medium, **(b)** the rms magnetic field intensity, and **(c)** the velocity of the wave.

4.21. The electric field intensity of a signal transmitted through a coaxial line is given as follows:

$$\mathbf{E} = \hat{\rho}\frac{E_0}{\rho}e^{-j\beta z} \qquad \text{V/m} \qquad a \leq \rho \leq b$$

where a and b are inner and outer conductor radii, respectively, and $\beta = \omega(\mu\varepsilon)^{0.5}$. Find **(a)** the associated magnetic field intensity, **(b)** the surface current density J_s, **(c)** the surface charge density ρ_s, and **(d)** the displacement current per unit length.

4.22. The region $x > 0$ is a perfect dielectric with $\varepsilon_r = 6.25$, while the region $x < 0$ is a perfect dielectric of $\varepsilon_r = 2.25$. At the interface, subscript 1 denotes field components on the $+x$ side of the boundary and subscript 2 on the $-x$ side. For $\mathbf{D}_1 = 4\hat{x} + 2\hat{y}$ C/m^2, find $\mathbf{D}_2$, $\mathbf{E}_1$, and $\mathbf{E}_2$.

4.23. A sphere of 1 m radius is made of perfect dielectric material (medium 1). It is surrounded by free space (medium 2). The electric field intensities in the two media are given as follows:

$$\mathbf{E}_1 = E_{01}(\cos\theta\,\hat{r} - \sin\theta\,\hat{\theta}) \qquad r \leq 1\,\text{m}$$

and

$$\mathbf{E}_2 = E_{02}\left[\left(1 + \frac{8}{r^3}\right)\cos\theta\,\hat{r} - \left(1 - \frac{4}{r^3}\right)\sin\theta\,\hat{\theta}\right] \qquad r \geq 1\,\text{m}$$

Find the permittivity of the spherical medium.

4.24. A 150-MHz plane wave is normally incident from air (region 1) onto a semi-infinite nonmagnetic dielectric slab (region 2). If the voltage standing wave ratio in front of the slab is 2.5 and E_{min} is at the boundary, find **(a)** η_2, **(b)** ε_{r2}, **(c)** the reflection coefficient, and **(d)** the distance d from the boundary to the nearest E_{max} of the standing wave pattern.

4.25. A uniform plane wave traveling through a nonmagnetic dielectric medium with $\varepsilon_r = 4$ is incident normally upon its interface with free space. If its electric field intensity is given by

$$\mathbf{E}_{\text{in}} = \hat{y} \cdot 3e^{-j200\pi z} \qquad \text{mV/m}$$

Find **(a)** the corresponding magnetic field intensity, **(b)** the reflected electric field intensity, **(c)** the reflected magnetic field intensity, **(d)** the transmitted electric field intensity, and **(e)** the transmitted magnetic field intensity.

4.26. A uniform plane wave of 3 GHz propagating through free space is incident normally on a lossless dielectric medium with $\varepsilon_r = 4$ and $\mu_r = 1$. The incident electric field at the interface has a value of 2 mV/m right before it strikes the interface. In free space, find **(a)** the reflection coefficient, **(b)** the VSWR, **(c)** the positions (in meters) of the maxima and minima of the electric field, and **(d)** the maximum and minimum values of the electric field.

4.27. The electric field intensity of a uniform plane wave propagating in air is $\mathbf{E}_i(z) = \hat{x} \bullet 5e^{-j6z}$ V/m. The wave is incident normally on an interface at $z = 0$ with a medium that has a dielectric constant at 6.25 and a loss tangent at 0.3. Find **(a)** the phasor expressions for reflected and transmitted electric and magnetic fields, and **(b)** the time-averaged power flow per unit area in the lossy medium.

4.28. Region 1 ($x < 0$) is air, whereas region 2 ($x > 0$) is a nonmagnetic medium characterized by $\sigma = 10^{-4}$ S/m and $\varepsilon_r = 4$. A uniform plane wave with its electric field intensity as given below is incident on the interface at $x = 0$ from region 1.

$$\mathcal{E}_{\text{in}} = 5\cos\left(2\pi \times 10^8 t - \frac{2\pi}{3}x\right) \qquad \text{V/m}$$

Obtain the reflected and transmitted wave electric fields.

4.29. Verify the characteristics of the TE mode fields given in Table 4.1 for an electromagnetic signal that propagates in a metallic parallel-plate waveguide.

4.30. Separation between the plates of a parallel-plate waveguide is 0.625 cm, filled with a nonmagnetic dielectric material of $\varepsilon_r = 4$. Find the cutoff frequencies for the TM_{00}, TM_{10}, TM_{20}, TE_{10}, and TE_{20} modes. Can a signal propagate in the TE_{00} mode? Determine the phase velocities for all possible TE and TM modes that exist at 30 GHz.

4.31. Verify the characteristics of TM mode fields given in Table 4.2 for an electromagnetic signal that propagates in a metallic waveguide of rectangular cross section.

4.32. The wavelength of a signal propagating through an air-filled WR-340 rectangular waveguide is found to be 20 cm. Determine its frequency in gigahertz.

4.33. The inside cross section of an air-filled rectangular waveguide is 2.286 cm × 1.016 cm.

(a) Determine its cutoff frequencies for the TE_{10}, TE_{20}, TE_{01}, and TE_{11} modes.

(b) Which mode(s) will propagate through this waveguide if the signal frequency is anywhere between 8 and 12 GHz?

(c) Determine the phase velocity of a 9.375 GHz signal propagating in TE_{10} mode.

4.34. Verify the characteristics of TE mode fields given in Table 4.3 for an electromagnetic signal that propagates in a hollow metallic cylindrical waveguide of circular cross section.

4.35. Verify the characteristics of the TM mode fields given in Table 4.3 for an electromagnetic signal that propagates in a hollow metallic cylindrical waveguide of circular cross section.

5

RESONANT CIRCUITS

A communication circuit designer frequently requires means to select (or reject) a band of frequencies from a wide signal spectrum. Resonant circuits provide such filtering. There are well-developed, sophisticated methodologies to meet virtually any specification. However, a simple circuit suffices in many cases. Further, resonant circuits are an integral part of the frequency-selective amplifier as well as of the oscillator designs. These networks are also used for impedance transformation and matching.

In this chapter we describe the analysis and design of these simple frequency-selective circuits and present the characteristic behavior of series and parallel resonant circuits. Related parameters, such as quality factor, bandwidth, and input impedance, are introduced that will be used in several subsequent chapters. Transmission lines with an open or short circuit at their ends are considered next, and their relationships with the resonant circuits are established. Transformer-coupled parallel resonant circuits are discussed briefly because of their significance in the radio-frequency range. In the final section we summarize the design procedure for rectangular and circular cylindrical cavities, and the dielectric resonator.

5.1 SERIES RESONANT CIRCUITS

Consider the series R–L–C circuit shown in Figure 5.1. Since the inductive reactance is directly proportional to signal frequency, it tries to block the high-frequency contents of the signal. On the other hand, capacitive reactance is

Radio-Frequency and Microwave Communication Circuits: Analysis and Design, Second Edition,
By Devendra K. Misra
ISBN 0-471-47873-3

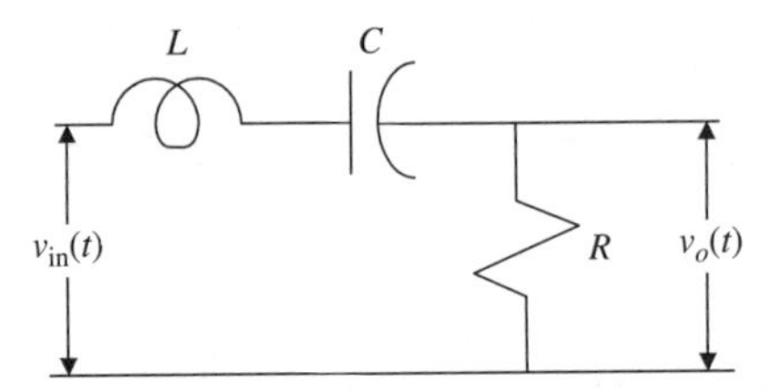

Figure 5.1 Series R–L–C circuit with input–output terminals.

inversely proportional to the frequency. Therefore, it tries to stop its lower frequencies. Note that the voltage across an ideal inductor leads the current by 90° (i.e., the phase angle of an inductive reactance is 90°). In the case of a capacitor, voltage across its terminals lags behind the current by 90° (i.e., the phase angle of a capacitive reactance is −90°). This means it is possible that the inductive reactance will be canceled out by the capacitive reactance at some intermediate frequency. This frequency is called the *resonant frequency* of the circuit. If the input signal frequency is equal to the resonant frequency, maximum current will flow through the resistor and it will be in phase with the input voltage. In this case, the output voltage V_o will be equal to the input voltage V_{in}. It can be analyzed as follows:

From Kirchhoff's voltage law,

$$\frac{L}{R}\frac{dv_o(t)}{dt} + \frac{1}{RC}\int_{-\infty}^{t} v_o(t)\,dt + v_o(t) = v_{\text{in}}(t) \tag{5.1.1}$$

Taking the Laplace transform of this equation with initial conditions as zero (i.e., no energy storage initially), we get

$$\left(\frac{sL}{R} + \frac{1}{sRC} + 1\right) V_o(s) = V_{\text{in}}(s) \tag{5.1.2}$$

where s is the complex frequency (Laplace variable). The transfer function of this circuit, $T(s)$, is given by

$$T(s) = \frac{V_o(s)}{V_{\text{in}}(s)} = \frac{1}{sL/R + 1/sRC + 1} = \frac{sR}{s^2L + sR + 1/C} \tag{5.1.3}$$

Therefore, the transfer function of this circuit has a zero at the origin of the complex s-plane and also has two poles. The location of these poles can be determined by solving the quadratic equation

$$s^2L + sR + \frac{1}{C} = 0 \tag{5.1.4}$$

Two possible solutions to this equation are as follows:

$$s_{1,2} = -\frac{R}{2L} \pm \sqrt{\left(\frac{R}{2L}\right)^2 - \frac{1}{LC}} \tag{5.1.5}$$

The circuit response will be influenced by the location of these poles. Therefore, these networks can be characterized as follows:

1. If $R/2L > 1/\sqrt{LC}$ (i.e., $R > 2\sqrt{L/C}$), both of these poles will be real and distinct, and the circuit is *overdamped.*
2. If $R/2L = 1/\sqrt{LC}$ (i.e., $R = 2\sqrt{L/C}$), the transfer function will have double poles at $s = -R/2L = -1/\sqrt{LC}$. The circuit is *critically damped.*
3. If $R/2L < 1/\sqrt{LC}$ (i.e., $R < 2\sqrt{L/C}$), the two poles of $T(s)$ will be complex conjugate of each other. The circuit is *underdamped.*

Alternatively, the transfer function may be rearranged as follows:

$$T(s) = \frac{sCR}{s^2LC + sRC + 1} = \frac{sCR\omega_0^2}{s^2 + 2\zeta\omega_0 s + \omega_0^2} \tag{5.1.6}$$

where

$$\zeta = \frac{R}{2}\sqrt{\frac{C}{L}} \tag{5.1.7}$$

$$\omega_0 = \frac{1}{\sqrt{LC}} \tag{5.1.8}$$

ζ is called the *damping ratio* and ω_0 is the *undamped natural frequency.*

Poles of $T(s)$ are determined by solving the equation

$$s^2 + 2\zeta\omega_0 s + \omega_0^2 = 0 \tag{5.1.9}$$

For $\zeta < 1$, $s_{1,2} = -\zeta\omega_0 \pm j\omega_0\sqrt{1-\zeta^2}$. As shown in Figure 5.2, the two poles are complex conjugates of each other. The output transient response will be oscillatory with a ringing frequency of $\omega_0(1-\zeta^2)$ and an exponentially decaying amplitude. This circuit is underdamped.

For $\zeta = 0$, the two poles move on the imaginary axis. Transient response will be oscillatory. It is a critically damped case. For $\zeta = 1$, the poles are on the negative real axis. Transient response decays exponentially. In this case, the circuit is overdamped.

Consider the unit step function shown in Figure 5.3. It is like a direct voltage source of 1 V that is turned on at time $t = 0$. If it represents input voltage $v_{\text{in}}(t)$, the corresponding output $v_o(t)$ can be determined via Laplace transform technique. The Laplace transform of a unit step at the origin is equal to $1/s$. Hence, the output voltage, $v_o(t)$, is found as follows:

$$v_o(t) = L^{-1}V_o(s) = L^{-1}\frac{sCR\omega_0^2}{s^2 + 2\zeta\omega_0 s + \omega_0^2}\frac{1}{s} = L^{-1}\frac{CR\omega_0^2}{(s+\zeta\omega_0)^2 + (1-\zeta^2)\omega_0^2}$$

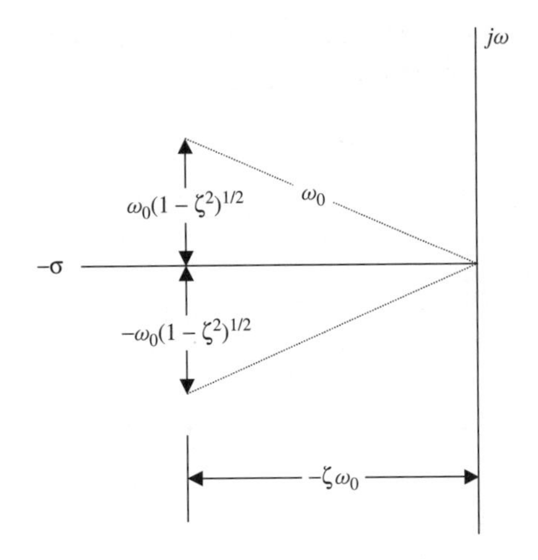

Figure 5.2 Pole–zero plot of the transfer function.

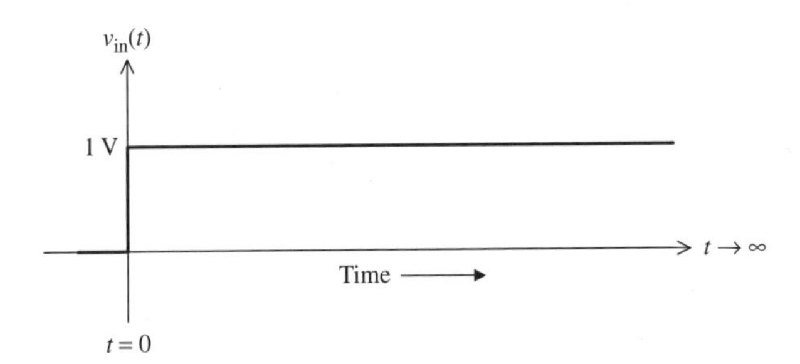

Figure 5.3 Unit-step input voltage.

where L^{-1} represents the inverse Laplace transform operator. Therefore,

$$v_o(t) = \frac{2\zeta}{\sqrt{1-\zeta^2}} e^{-\zeta\omega_0 t} \sin(\omega_0 t\sqrt{1-\zeta^2})u(t)$$

This response is illustrated in Figure 5.4 for three different damping factors. As can be seen, initial ringing lasts longer for a lower damping factor.

A sinusoidal steady-state response of the circuit can be determined easily after replacing s by $j\omega$, as follows:

$$V_o(j\omega) = \frac{V_{\text{in}}(j\omega)}{j\omega L/R + 1/j\omega RC + 1} = \frac{V_{\text{in}}(j\omega)}{(1 + j/RC)(LC\omega - 1/\omega)}$$

or

$$V_o(j\omega) = \frac{V_{\text{in}}(j\omega)}{(1 + j/RC)(\omega/\omega_0^2 - 1/\omega)} = \frac{V_{\text{in}}(j\omega)}{(1 + j/\omega_0 RC)(\omega/\omega_0 - \omega_0/\omega)}$$

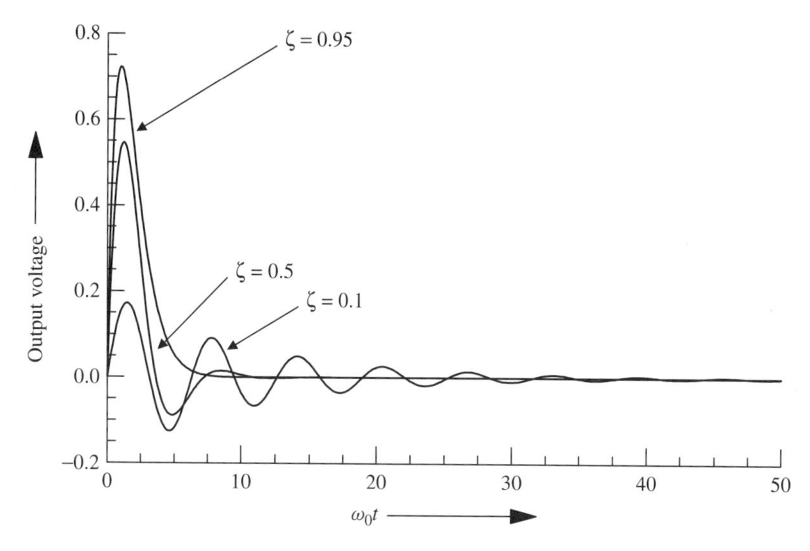

Figure 5.4 Response of a series R–L–C circuit to a unit step input for three different damping factors.

The *quality factor*, Q, of the resonant circuit is a measure of its frequency selectivity. It is defined as

$$Q = \omega_0 \frac{\text{average stored energy}}{\text{power loss}} \tag{5.1.10}$$

Hence,

$$Q = \omega_0 \frac{\frac{1}{2}LI^2}{\frac{1}{2}I^2R} = \frac{\omega_0 L}{R}$$

Since $\omega_0 L = 1/\omega_0 C$,

$$Q = \frac{\omega_0 L}{R} = \frac{1}{\omega_0 RC} = \frac{\sqrt{LC}}{RC} = \frac{1}{R}\sqrt{\frac{L}{C}} = \frac{1}{2\zeta} \tag{5.1.11}$$

Therefore,

$$V_o(j\omega) = \frac{V_{\text{in}}(j\omega)}{1 + jQ(\omega/\omega_0 - \omega_0/\omega)} \tag{5.1.12}$$

Alternatively,

$$\frac{V_o(j\omega)}{V_{\text{in}}(j\omega)} = A(j\omega) = \frac{1}{1 + jQ(\omega/\omega_0 - \omega_0/\omega)} \tag{5.1.13}$$

The magnitude and phase angle of (5.1.13) are illustrated in Figures 5.5 and 5.6, respectively. Figure 5.5 shows that the output voltage is equal to the input for

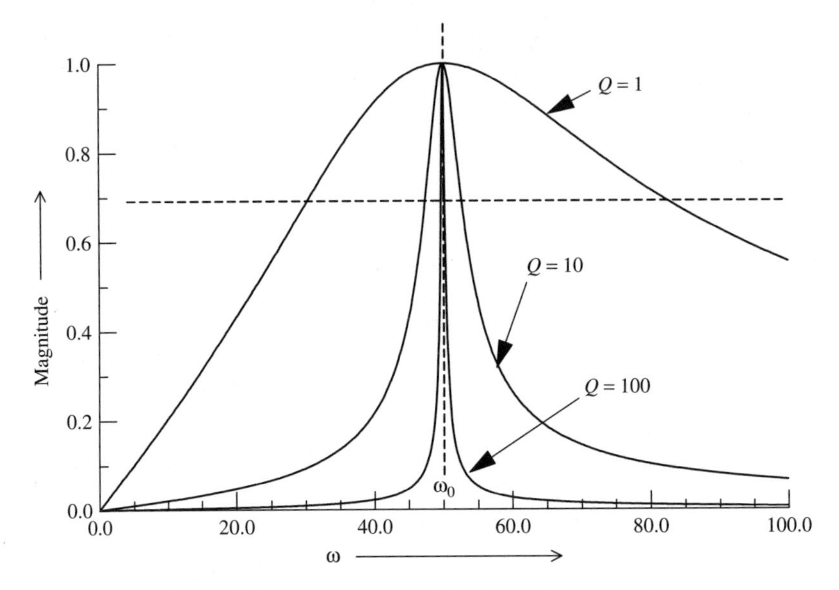

Figure 5.5 Magnitude of $A(j\omega)$ as a function of ω.

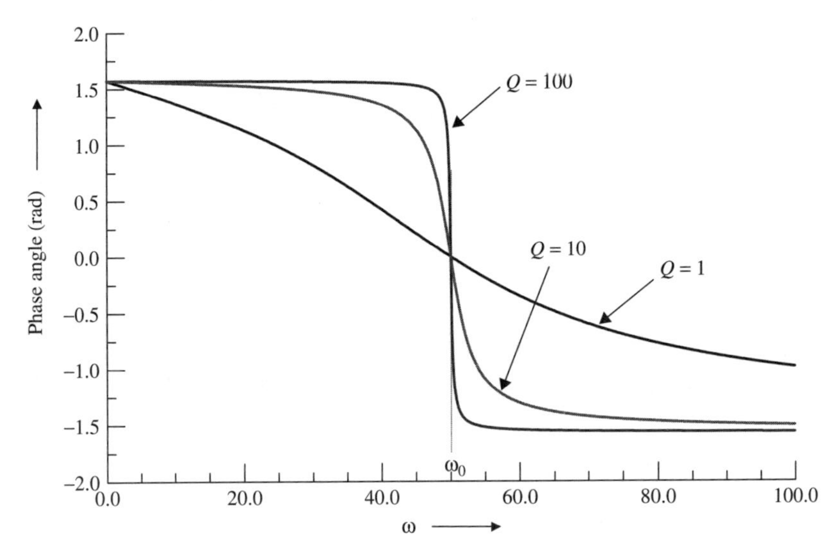

Figure 5.6 Phase angle of $A(j\omega)$ as a function of ω.

a signal frequency equal to the resonant frequency of the circuit. Further, phase angles of the two signals in Figure 5.6 are the same at this frequency, irrespective of the quality factor of the circuit. As signal frequency moves away from this point on either side, the output voltage decreases. The rate of decrease depends on the quality factor of the circuit. For higher Q, the magnitude is sharper, indicating a higher selectivity of the circuit. If signal frequency is below the resonant frequency, the output voltage leads the input. For a signal frequency far below the resonance, the output leads the input by almost 90°. On the other hand, it lags behind the input for higher frequencies. It converges to −90° as the signal frequency moves far beyond the resonant frequency. Thus, the phase angle changes between $\pi/2$ and $-\pi/2$, following a sharper change around the resonance for high-Q circuits. Note that the voltage across the series-connected inductor and capacitor combined has inverse characteristics to those of the voltage across the resistor. Mathematically,

$$V_{LC}(j\omega) = V_{\text{in}}(j\omega) - V_o(j\omega)$$

where $V_{LC}(j\omega)$ is the voltage across the inductor and capacitor combined. In this case, sinusoidal steady-state response can be obtained as follows:

$$\frac{V_{LC}(j\omega)}{V_{\text{in}}(j\omega)} = 1 - \frac{V_o(j\omega)}{V_{\text{in}}(j\omega)} = 1 - \frac{1}{1 + jQ(\omega/\omega_0 - \omega_0/\omega)}$$

$$= \frac{jQ(\omega/\omega_0 - \omega_0/\omega)}{1 + jQ(\omega/\omega_0 - \omega_0/\omega)}$$

Hence, this configuration of the circuit represents a band-rejection filter.

Half-power frequencies ω_1 and ω_2 of a bandpass circuit can be determined from (5.1.13) as

$$\frac{1}{2} = \frac{1}{1 + Q^2(\omega/\omega_0 - \omega_0/\omega)^2} \Rightarrow 2 = 1 + Q^2\left(\frac{\omega}{\omega_0} - \frac{\omega_0}{\omega}\right)^2$$

Therefore,

$$Q\left(\frac{\omega}{\omega_0} - \frac{\omega_0}{\omega}\right) = \pm 1$$

Assuming that $\omega_1 < \omega_0 < \omega_2$,

$$Q\left(\frac{\omega_1}{\omega_0} - \frac{\omega_0}{\omega_1}\right) = -1$$

and

$$Q\left(\frac{\omega_2}{\omega_0} - \frac{\omega_0}{\omega_2}\right) = 1$$

Therefore,

$$\frac{\omega_2}{\omega_0} - \frac{\omega_0}{\omega_2} = -\left(\frac{\omega_1}{\omega_0} - \frac{\omega_0}{\omega_1}\right)$$

or

$$\omega_2 - \frac{\omega_0^2}{\omega_2} = -\omega_1 + \frac{\omega_0^2}{\omega_1} \Rightarrow (\omega_2 + \omega_1) = \frac{\omega_0^2}{\omega_1} + \frac{\omega_0^2}{\omega_1} = \omega_0^2\left(\frac{1}{\omega_1} + \frac{1}{\omega_2}\right)$$

or

$$\omega_0^2 = \omega_1\,\omega_2 \tag{5.1.14}$$

and

$$\frac{\omega_1}{\omega_0} - \frac{\omega_0}{\omega_1} = -\frac{1}{Q} \Rightarrow \omega_1 - \frac{\omega_0^2}{\omega_1} = -\frac{\omega_0}{Q}$$

or

$$\omega_1 - \omega_2 = -\frac{\omega_0}{Q} \Rightarrow Q = \frac{\omega_0}{\omega_2 - \omega_1} \tag{5.1.15}$$

Example 5.1 Determine the element values of a resonant circuit that passes all the sinusoidal signals from 9 to 11 MHz. This circuit is to be connected between a voltage source with negligible internal impedance and a communication system with its input impedance at 50 Ω. Plot its characteristics in a frequency band of 1 to 20 MHz.

SOLUTION From (5.1.14),

$$\omega_0 = \sqrt{\omega_1\omega_2} \rightarrow f_0 = \sqrt{f_1 f_2} = \sqrt{9 \times 11} = 9.949874\,\text{MHz}$$

From (5.1.11) and (5.1.15),

$$Q = \frac{\omega_0 L}{R} = \frac{\omega_0}{\omega_1 - \omega_2} \Rightarrow L = \frac{R}{\omega_1 - \omega_2}$$
$$= \frac{50}{2\pi \times 10^6 \times (11 - 9)} = 3.978874 \times 10^{-6}\ \text{H} \approx 4\,\mu\text{H}$$

From (5.1.8),

$$\omega_0 = \frac{1}{\sqrt{LC}} \Rightarrow C = \frac{1}{L\omega_0^2} = 6.430503 \times 10^{-11}\,\text{F} \approx 64.3\,\text{pF}$$

The circuit arrangement is shown in Figure 5.7. Its magnitude and phase characteristics are displayed in Figure 5.8.

Input Impedance

Impedance across the input terminals of a series R–L–C circuit can be determined as follows:

$$Z_{\text{in}} = R + j\omega L + \frac{1}{j\omega C} = R + j\omega L\left(1 - \frac{\omega_0^2}{\omega^2}\right) \tag{5.1.16}$$

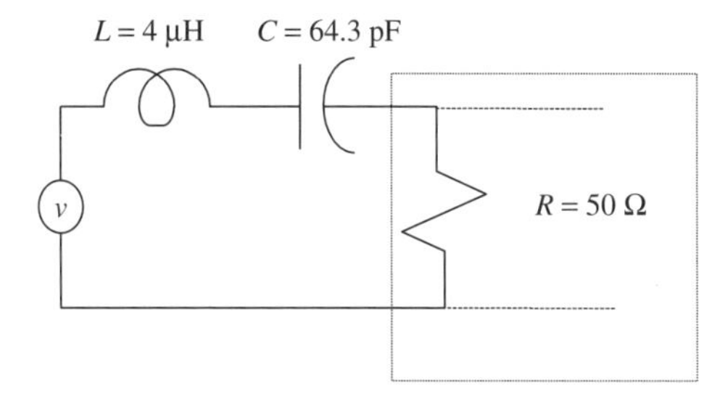

Figure 5.7 Filter circuit arrangement for Example 5.1.

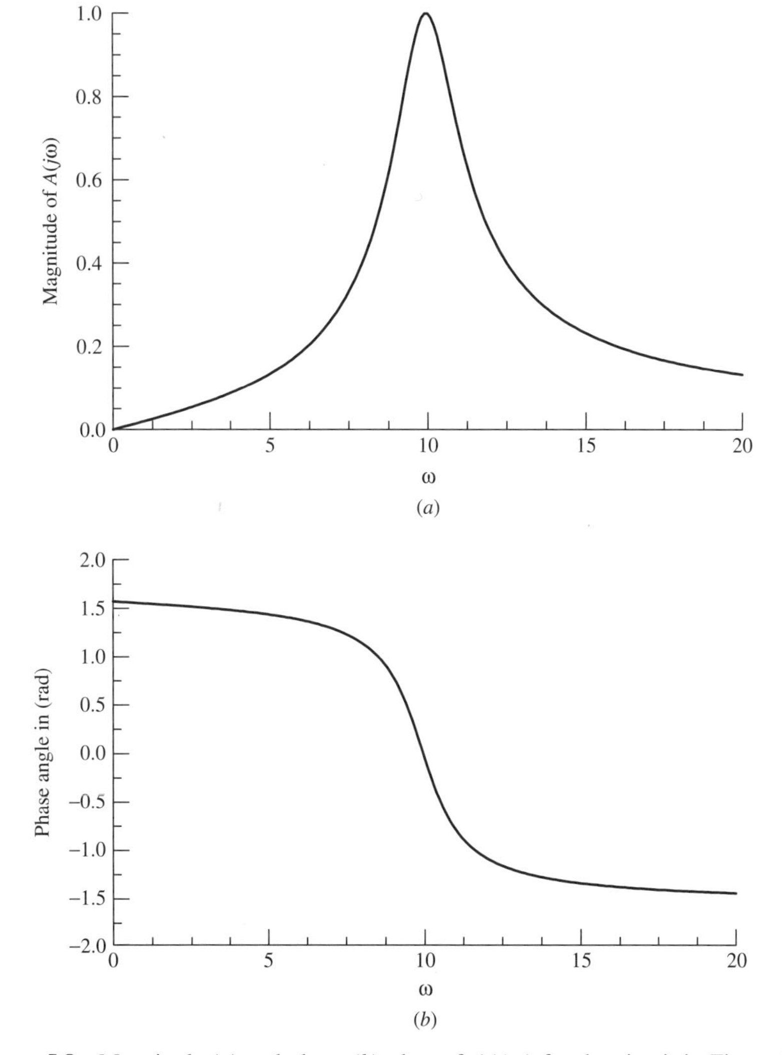

Figure 5.8 Magnitude (*a*) and phase (*b*) plots of $A(j\omega)$ for the circuit in Figure 5.7.

At resonance, the inductive reactance cancels out the capacitive reactance. Therefore, the input impedance reduces to the total resistance of the circuit. If signal frequency changes from the resonant frequency by $\pm\delta\omega$, the input impedance can be approximated as

$$Z_{\text{in}} = R + j\omega L \frac{(\omega + \omega_0)(\omega - \omega_0)}{\omega^2} \approx R + j2\delta\omega L = R + j\frac{2QR\delta\omega}{\omega_0} \quad (5.1.17)$$

Alternatively,

$$Z_{\text{in}} \approx R + j2\delta\omega L = \frac{\omega_0 L}{Q} + j2(\omega - \omega_0)L = j2\left(\omega - \omega_0 + \frac{\omega_0}{j2Q}\right)L$$

$$= j2\left[\omega - \omega_0\left(1 + j\frac{1}{2Q}\right)\right]L \quad (5.1.18)$$

Therefore, a series resonant circuit can be analyzed with R as zero (i.e., assuming that the circuit is lossless). The losses can be included subsequently by replacing a real resonant frequency, ω_0, by the complex frequency, $\omega_0[1 + j(1/2Q)]$.

At resonance, the current through the circuit, I_r, is

$$I_r = \frac{V_{\text{in}}}{R} \quad (5.1.19)$$

Therefore, voltages across the inductor, V_L, and the capacitor, V_C, are

$$V_L = j\omega_0 L \frac{V_{\text{in}}}{R} = jQV_{\text{in}} \quad (5.1.20)$$

and

$$V_C = \frac{1}{j\omega_0 C}\frac{V_{\text{in}}}{R} = -jQV_{\text{in}} \quad (5.1.21)$$

Hence, the magnitude of voltage across the inductor is equal to the quality factor times input voltage, while its phase leads by 90°. The magnitude of the voltage across the capacitor is the same as that across the inductor. However, it is 180° out of phase because it lags behind the input voltage by 90°.

5.2 PARALLEL RESONANT CIRCUITS

Consider an R–L–C circuit in which the three components are connected in parallel, as shown in Figure 5.9. A subscript p is used to differentiate the circuit elements from those used in series circuit of Section 5.1. A current source, $i_{\text{in}}(t)$, is connected across its terminals and $i_o(t)$ is current through the resistor R_p.

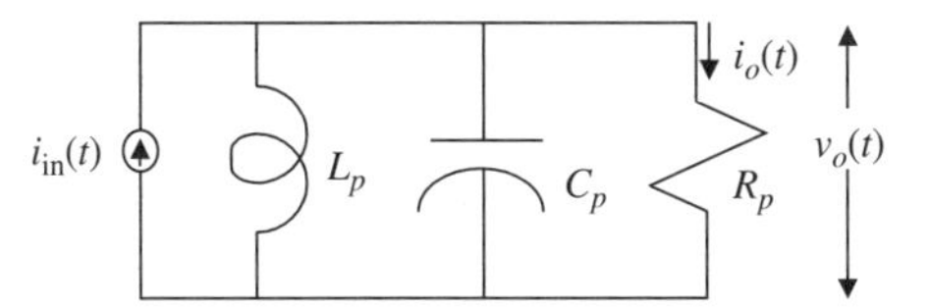

Figure 5.9 Parallel $R-L-C$ circuit.

Voltage across this circuit is $v_o(t)$. From Kirchhoff's current law,

$$i_{\text{in}}(t) = \frac{1}{L_p}\int_{-\infty}^{t} R_p i_o(t)\,dt + C_p\frac{d(R_p i_o(t))}{dt} + i_o(t) \tag{5.2.1}$$

Assuming that no energy was stored in the circuit initially, we take the Laplace transform of (5.2.1). It gives

$$I_{\text{in}}(s) = \left(\frac{R_p}{sL_p} + sR_pC_p + 1\right) I_o(s)$$

Hence,

$$\frac{I_o(s)}{I_{\text{in}}(s)} = \frac{sL_p/R_p}{s^2L_pC_p + s(L_p/R_p) + 1} \tag{5.2.2}$$

Note that this equation is similar to (5.1.6). It changes to $T(s)$ if RC replaces L_p/R_p. Therefore, the results of the series resonant circuit can be used for this parallel resonant circuit, provided that

$$\zeta = \frac{1}{2\,\omega_0 R_p C_p}$$

and

$$\omega_0 = \frac{1}{\sqrt{L_pC_p}} \tag{5.2.3}$$

Hence,

$$\zeta = \frac{1}{2\,\omega_0 R_p C_p} = \frac{1}{2R_p}\sqrt{\frac{L_p}{C_p}} \tag{5.2.4}$$

The quality factor, Q_p, and the impedance, Z_p, of the parallel resonant circuit can be determined as follows:

$$Q_{\text{series}} = \frac{\omega_0 L}{R} = \frac{1}{\omega_0 RC} \Rightarrow Q_p = \omega_0 R_p C_p = \frac{R_p}{\omega_0 L_p} \tag{5.2.5}$$

$$Z_p = \frac{V_o(j\omega)}{I_{\text{in}}(i\omega)} = \frac{I_0(j\omega)R_p}{I_{\text{in}}(j\omega)} = \frac{j\omega L_p}{-\omega^2 L_pC_p + j\omega(L_p/R_p) + 1} \tag{5.2.6}$$

Input Admittance

Admittance across input terminals of the parallel resonant circuit (i.e., the admittance seen by the current source) can be determined as follows:

$$Y_{\text{in}} = \frac{1}{Z_p} = \frac{1}{R_p} + j\omega C_p + \frac{1}{j\omega L_p} = \frac{1}{R_p} + j\omega C_p\left(1 - \frac{\omega_0^2}{\omega^2}\right) \tag{5.2.7}$$

Hence, input admittance will be equal to $1/R_p$ at the resonance. It will become zero (i.e., the impedance will be infinite) for a lossless circuit. It can be approximated around the resonance, $\omega_0 \pm \delta\omega$, as

$$Y_{\text{in}} \approx \frac{1}{R_p} + j2\delta\omega C_p = \frac{1}{R_p} + j\frac{2\delta\omega Q}{\omega_0 R_p} \tag{5.2.8}$$

The corresponding impedance is

$$Z_p \approx \frac{R_p}{1 + j(2Q\delta\omega/\omega_0)} \tag{5.2.9}$$

Current through the capacitor, I_c, at the resonance is

$$I_c = j\omega_0 C_p R_p I_{\text{in}} = jQI_{\text{in}} \tag{5.2.10}$$

and current through the inductor, I_L, is

$$I_L = \frac{1}{j\omega_0 L_p} R_p I_{\text{in}} = -jQI_{\text{in}} \tag{5.2.11}$$

Thus, current through the inductor is equal in magnitude but opposite in phase to that through the capacitor. Further, these currents are larger than the input current by a factor of Q.

Quality Factor of a Resonant Circuit

If resistance R represents losses in the resonant circuit, Q given by the preceding formulas is known as the *unloaded* Q. If the power loss due to external load coupling is included through an additional resistance R_L, the *external* Q_e is defined as follows:

$$Q_e = \begin{cases} \dfrac{\omega_0 L}{R_L} & \text{for series resonant circuit} \\[2ex] \dfrac{R_L}{\omega_0 L_p} & \text{for parallel resonant circuit} \end{cases} \tag{5.2.12}$$

The loaded Q, Q_L, of a resonant circuit includes internal losses as well as the power extracted by the external load. It is defined as follows:

$$Q_L = \begin{cases} \dfrac{\omega_0 L}{R_L + R} & \text{for series resonant circuit} \\ \dfrac{R_L \| R_p}{\omega_0 L_p} & \text{for parallel resonant circuit} \end{cases} \tag{5.2.13}$$

where

$$R_L \| R_p = \frac{R_L R_p}{R_L + R_p}$$

Hence, the following relation holds good for both types of resonant circuit:

$$\frac{1}{Q_L} = \frac{1}{Q_e} + \frac{1}{Q} \tag{5.2.14}$$

See Table 5.1.

Example 5.2 Consider the loaded parallel resonant circuit in Figure 5.10. Compute the resonant frequency in radians per second, unloaded Q, and the loaded Q of this circuit.

SOLUTION

$$\omega_0 = \frac{1}{\sqrt{L_p C_p}} = \frac{1}{\sqrt{10^{-5} \times 10^{-11}}} = 10^8 \text{ rad/s}$$

TABLE 5.1 Relations for Series and Parallel Resonant Circuits

	Series	Parallel
ω_0	$\dfrac{1}{\sqrt{LC}}$	$\dfrac{1}{\sqrt{L_p C_p}}$
Damping factor, ζ	$\dfrac{R}{2}\sqrt{\dfrac{C}{L}}$	$\dfrac{1}{2R_p}\sqrt{\dfrac{L_p}{C_p}}$
Unloaded Q	$\dfrac{\omega_0 L}{R} = \dfrac{1}{\omega_0 RC}$	$\dfrac{R_p}{\omega_0 L_p} = \omega_0 R_p C_p$
External $Q = Q_e$	$\dfrac{\omega_0 L}{R_L} = \dfrac{1}{\omega_0 R_L C}$	$\dfrac{R_L}{\omega_0 L_p} = \omega_0 R_L C_p$
Loaded $Q = Q_L$	$\dfrac{Q Q_e}{Q + Q_e}$	$\dfrac{Q Q_e}{Q + Q_e}$
Input impedance, Z_{in}, around resonance	$R + j\dfrac{2RQ\delta\omega}{\omega_0}$	$\dfrac{R}{1 + j(2Q\delta\omega/\omega_0)}$

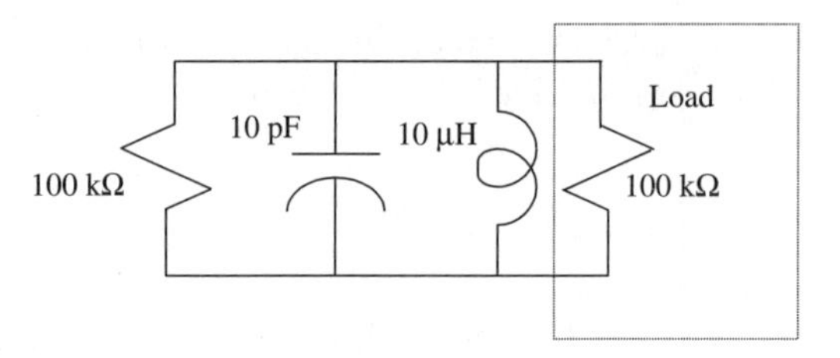

Figure 5.10 Circuit for Example 5.2.

The unloaded

$$Q = \frac{R_p}{\omega_0 L_p} = \frac{10^5}{10^8 \times 10^{-5}} = 100$$

The external Q,

$$Q_e = \frac{R_L}{\omega_0 L_p} = \frac{10^5}{10^8 \times 10^{-5}} = 100$$

The loaded Q,

$$Q_L = \frac{R_p \| R_L}{\omega_0 L_p} = \frac{Q Q_e}{Q + Q_e} = \frac{50 \times 10^3}{10^8 \times 10^{-5}} = 50$$

5.3 TRANSFORMER-COUPLED CIRCUITS

Transformers are used as a means of coupling as well as of impedance transforming in electronic circuits. Transformers with tuned circuits in one or both of their sides are employed in voltage amplifiers and oscillators operating at radio frequencies. In this section we present an equivalent model and an analytical procedure for the transformer-coupled circuits.

Consider a load impedance Z_L that is coupled to the voltage source V_s via a transformer as illustrated in Figure 5.11. Source impedance is assumed to be Z_s. The transformer has a turns ratio of $n:1$ between its primary (the source) and secondary (the load) sides. Using the notations indicated, equations for various voltages and currents can be written in phasor form as follows:

$$V_1 = j\omega L_1 I_1 + j\omega M I_2 \tag{5.3.1}$$

$$V_2 = j\omega M I_1 + j\omega L_2 I_2 \tag{5.3.2}$$

where M is the mutual inductance between the two sides of the transformer. Standard convention with a dot on each side is used. Hence, magnetic fluxes reinforce each other for the case of currents entering this terminal on both sides, and M is positive.

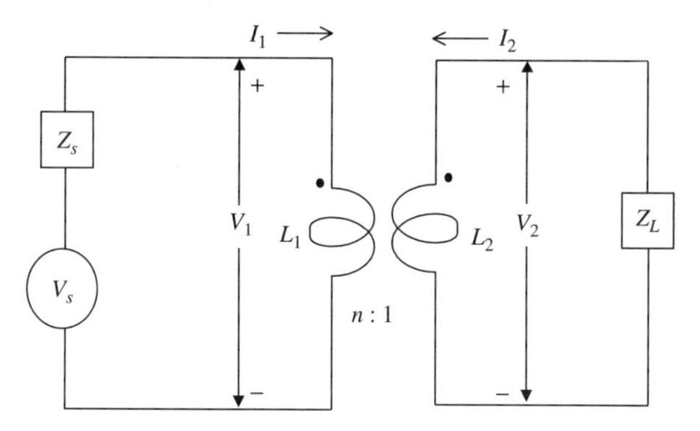

Figure 5.11 Transformer-coupled circuit.

The following relations hold for an ideal transformer operating at any frequency:

$$V_1 = nV_2 \tag{5.3.3}$$

$$I_1 = -\frac{I_2}{n} \tag{5.3.4}$$

and

$$\frac{V_1}{I_1} = Z_1 = \frac{nV_2}{-I_2/n} = n^2\frac{V_2}{-I_2} = n^2 Z_2 \tag{5.3.5}$$

Several equivalent circuits are available for a transformer. We consider one of these that is most useful in analyzing the communication circuits. This equivalent circuit is illustrated in Figure 5.12. The following equations for phasor voltages and currents may be formulated using the notations indicated in the figure.

$$V_1 = j\omega(1-\xi)L_1 I_1 + j\omega\xi L_1\left(I_1 + \frac{I_2}{n}\right) = j\omega L_1 I_1 + j\omega\xi L_1\frac{I_2}{n} \tag{5.3.6}$$

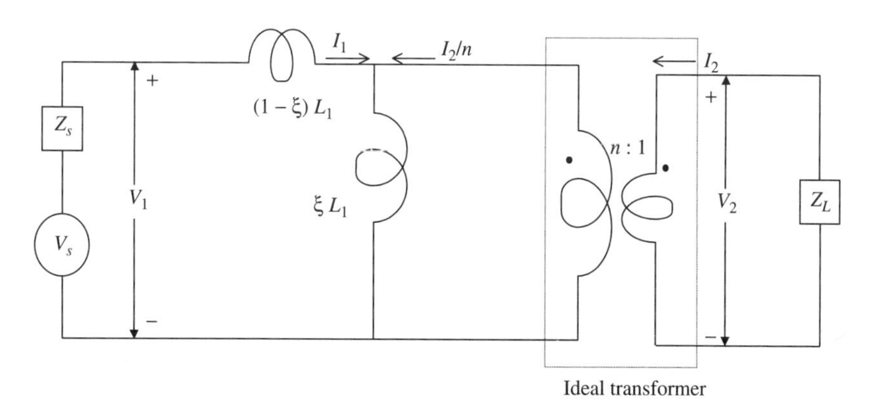

Figure 5.12 Equivalent model of the transformer-coupled circuit shown in Figure 5.11.

and

$$V_2 = \frac{1}{n}\left[j\omega\xi L_1\left(I_1 + \frac{I_2}{n}\right)\right] \tag{5.3.7}$$

If the circuit shown in Figure 5.12 is equivalent to that shown in Figure 5.11, these two equations represent the same voltages as those of (5.3.1) and (5.3.2). Hence,

$$\frac{\xi L_1}{n} = M \tag{5.3.8}$$

and

$$\frac{\xi L_1}{n^2} = L_2 \tag{5.3.9}$$

In other words,

$$n = \sqrt{\frac{\xi L_1}{L_2}} \tag{5.3.10}$$

and

$$\xi = \frac{nM}{L_1} = \frac{M\sqrt{\xi}}{\sqrt{L_1 L_2}} \Rightarrow \sqrt{\xi} = \frac{M}{\sqrt{L_1 L_2}} = \kappa \tag{5.3.11}$$

where κ is called the *coefficient of coupling*. It is close to unity for a tightly coupled transformer, and close to zero for a loose coupling.

Example 5.3 A tightly coupled transformer is used in the circuit shown in Figure 5.13. Inductances of its primary and secondary sides are 320 and 20 nH, respectively. Find its equivalent circuit, the resonant frequency, and the Q of this circuit.

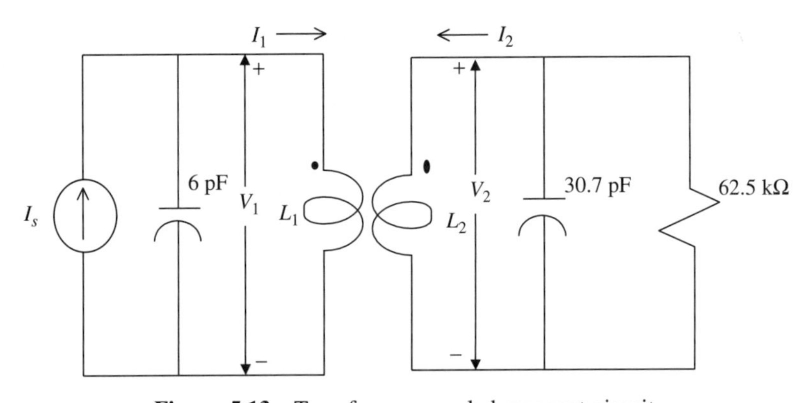

Figure 5.13 Transformer-coupled resonant circuit.

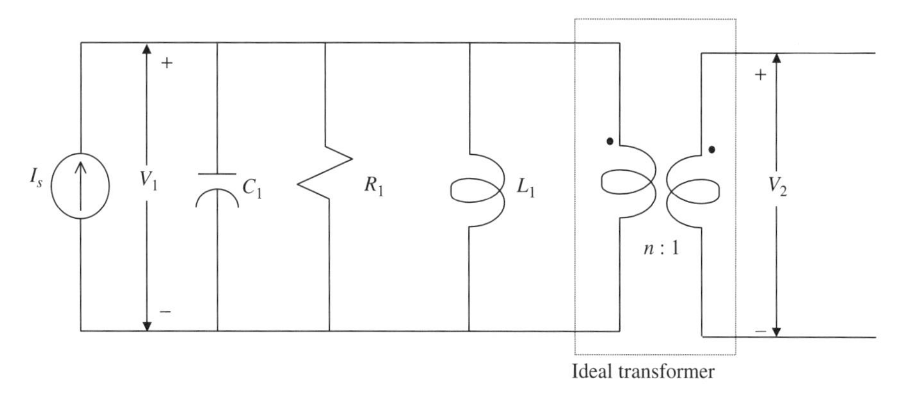

Figure 5.14 Equivalent model of the transformer-coupled circuit shown in Figure 5.13.

SOLUTION Since the transformer is tightly coupled, $\kappa \approx 1$. Therefore, $\xi \approx 1$, $1 - \xi \approx 0$, and its equivalent circuit simplifies as shown in Figure 5.14. From (5.3.10),

$$n = \sqrt{\frac{\xi L_1}{L_2}} \approx \sqrt{\frac{320}{20}} = 4$$

and from (5.3.5),

$$Z_1 = n^2 Z_2 \Rightarrow Y_1 = \frac{Y_2}{n^2} = \frac{1}{n^2}\left(\frac{1}{R_2} + j\omega C_2\right)$$

Therefore,

$$R_1 = n^2 R_2 = 16 \times 62.5\,\text{k}\Omega = 1\,\text{M}\Omega$$

$$C_1 = 6 + \frac{1}{4^2} 30.7\,\text{pF} = 6 + 1.91875 \approx 7.92\,\text{pF}$$

$$\omega_0 = \frac{1}{\sqrt{L_1 C_1}} = \frac{1}{\sqrt{320 \times 10^{-9} \times 7.92 \times 10^{-12}}} = 628.1486 \times 10^6\,\text{rad/s}$$

$$f_0 = \frac{\omega_0}{2\pi} = 99.97\,\text{MHz} \approx 100\,\text{MHz}$$

and

$$Q = \frac{R_1}{\omega_0 L_1} = \frac{10^6}{628.1486 \times 10^6 \times 320 \times 10^{-9}} = 4974.9375$$

Example 5.4 A transformer-coupled circuit is shown in Figure 5.15. Draw its equivalent circuit using an ideal transformer.

(a) If the transformer is tuned only to its secondary side (i.e., R_1 and C_1 are removed), determine its resonant frequency and impedance seen by the current source.

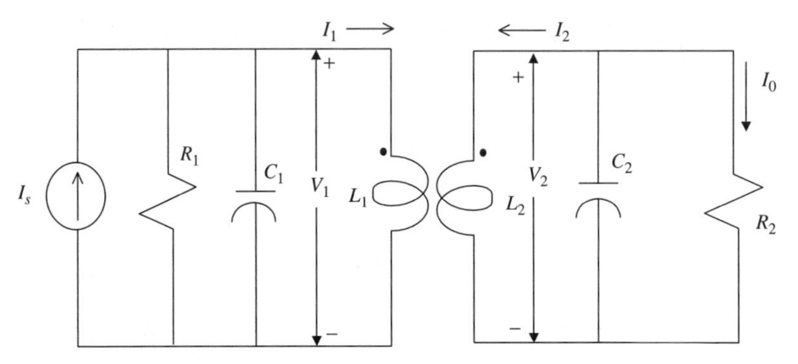

Figure 5.15 Double-tuned transformer-coupled circuit.

(b) Determine the current transfer function (I_0/I_s) for the entire circuit (i.e., R_1 and C_1 are included in the circuit). If the transformer is loosely coupled, and both of the sides have identical quality factors as well as the resonant frequencies, determine the locations of the poles of the current transfer function on the complex ω-plane.

SOLUTION Following the preceding analysis and Figure 5.12, the equivalent circuit can be drawn easily, as shown in Figure 5.16. Using notations as indicated in the figure, circuit voltages and currents can be found as follows:

$$nV_2 = s\xi L_1\left(I_1 + \frac{I_2}{n}\right) \tag{5.3.12}$$

$$V_2 = -I_2\frac{R_2}{1 + sR_2C_2} = R_2I_0 \tag{5.3.13}$$

$$I_s = I_1 + \left(\frac{1}{R_1} + sC_1\right)[nV_2 + s(1 - \xi)L_1I_1] \tag{5.3.14}$$

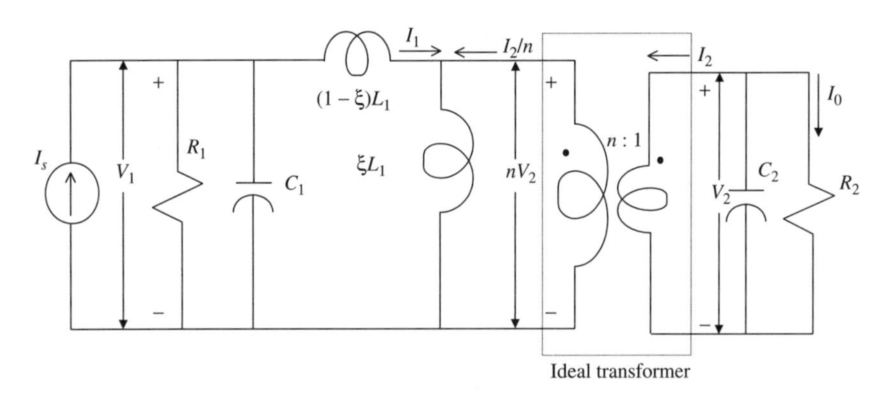

Figure 5.16 Equivalent representation for the circuit shown in Figure 5.15.

From (5.3.12), (5.3.13), and (5.3.10), we find that

$$\left(1+\frac{s^2\xi L_1R_2C_2+s\xi L_1}{n^2R_2}\right)V_2=\left(1+s^2L_2C_2+\frac{sL_2}{R_2}\right)R_2I_0=\frac{s\xi L_1}{n}I_1$$

or

$$\frac{I_0}{I_1}=\frac{s\xi L_1/nR_2}{1+s^2L_2C_2+sL_2/R_2} \tag{5.3.15}$$

When R_1 and C_1 are absent, I_1 will be equal to I_s, and (5.3.15) will represent the current transfer characteristics. This equation is similar to (5.2.2). Hence, the resonant frequency is found as follows:

$$\omega_0=\frac{1}{\sqrt{L_2C_2}} \tag{5.3.16}$$

Note that the resonant frequency in this case depends only on L_2 and C_2. It is independent of inductance L_1 of the primary side. The impedance seen by the current source, with R_1 and C_1 removed, is determined as follows:

$$Z_{\text{in}}(s)=\frac{V_1}{I_1}=sL_1(1-\xi)+\frac{n^2sL_2}{s^2L_2C_2+s(L_2/R_2)+1} \tag{5.3.17}$$

At resonance,

$$Z_{\text{in}}(j\omega_0)=j\omega_0L_1(1-\xi)+n^2R_2 \tag{5.3.18}$$

and if transformer is tightly coupled, $\xi\approx 1$,

$$Z_{\text{in}}(j\omega_0)=n^2R_2 \tag{5.3.19}$$

When R_1 and C_1 are included in the circuit, (5.3.12) to (5.3.14) can be solved to obtain a relationship between I_0 and I_s. The desired current transfer function can be obtained as

$$\frac{I_0}{I_s}=\frac{s\xi L_1/nR_2}{[s^2L_2C_2+s(L_2/R_2)+1]\{1+s(1-\xi)L_1[(1/R_1)+sC_1]\}+s\xi L_1[(1/R_1)+sC_1]} \tag{5.3.20}$$

Note that the denominator of this equation is now a polynomial of the fourth degree. Hence, this current ratio has four poles on the complex ω-plane. If the transformer is loosely coupled, ξ will be negligible. In that case, the denominator of (5.3.20) can be approximated as follows:

$$\frac{I_0}{I_s}\approx\frac{s\xi L_1/nR_2}{[s^2L_2C_2+s(L_2/R_2)+1]\{1+sL_1[(1/R_1)+sC_1]\}+s\xi L_1[(1/R_1)+sC_1]} \tag{5.3.21}$$

For $\omega_1 = \omega_2 = \omega_0$, and $Q_1 = Q_2 = Q_0$,

$$L_1C_1 = L_2C_2 = \frac{1}{\omega_0^2}$$

and

$$\frac{R_1}{L_1} = \frac{R_2}{L_2} = \omega_0 Q_0$$

Hence, (5.3.21) may be written as

$$\frac{I_0}{I_s} \approx \frac{s\xi L_1/nR_2}{(s^2/\omega_0^2 + s/\omega_0 Q + 1)^2 + \xi(s^2/\omega_0^2 + s/\omega_0 Q)} \tag{5.3.22}$$

The poles of (5.3.22) are determined by solving the equation

$$\left(\frac{s^2}{\omega_0^2} + \frac{s}{\omega_0 Q_0} + 1\right)^2 + \xi\left(\frac{s^2}{\omega_0^2} + \frac{s}{\omega_0 Q_0}\right) = 0 \tag{5.3.23}$$

The roots of this equation are found to be

$$s \approx -\frac{\omega_0}{2Q_0} \pm j\omega_0\sqrt{(1 \mp \sqrt{\xi}) - \left(\frac{1}{2Q_0}\right)^2} \tag{5.3.24}$$

5.4 TRANSMISSION LINE RESONANT CIRCUITS

Transmission lines with open- or short-circuited ends are frequently used as resonant circuits in the ultrahigh frequency and microwave range. We consider such networks in this section. Since the quality factor is an important parameter of these circuits, we need to include the finite (even though small) loss in the line. There are four basic types of these networks, as illustrated in Figure 5.17.

It can be found easily by analyzing the input impedance characteristics around the resonant wavelength, λ_r, that the circuits of Figure 5.17 (*a*) and (*d*) behave like a series R–L–C circuit. On the other hand, the other two transmission lines possess the characteristics of a parallel resonant circuit. A quantitative analysis of these circuits is presented below for n as unity.

Short-Circuited Line

Consider a transmission line of length l and characteristic impedance Z_0. It has a propagation constant, $\gamma = \alpha + j\beta$. The line is short circuited at one of its ends, as shown in Figure 5.18. The impedance at its other end, Z_{in}, can be determined

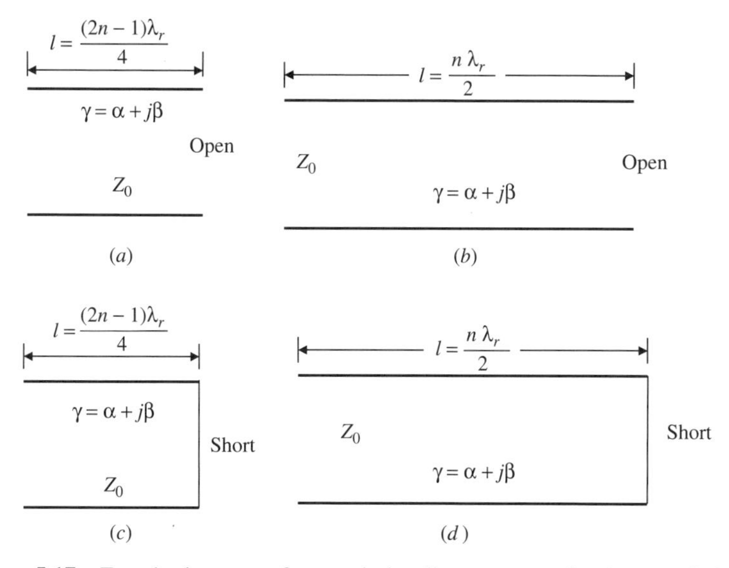

Figure 5.17 Four basic types of transmission line resonant circuits, $n = 1, 2, \ldots$.

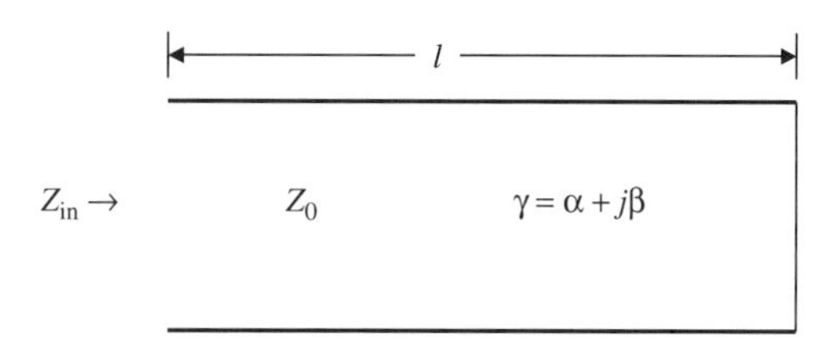

Figure 5.18 Short-circuited lossy transmission line.

from (3.2.5) as follows:

$$Z_{\text{in}} = Z_0 \tanh \gamma l = Z_0 \tanh(\alpha + j\beta)l = Z_0 \frac{\tanh \alpha l + j \tan \beta l}{1 + j \tanh \alpha l \tan \beta l} \tag{5.4.1}$$

For $\alpha l \ll 1, \tanh \alpha l \approx \alpha l$, and assuming that the line supports only the TEM mode, we find that $\beta l = \omega l / v_p$. Hence, it can be simplified around the resonant frequency, ω_0, as follows:

$$\beta l = \frac{\omega l}{v_p} = \frac{\omega_0 l}{v_p} + \frac{\delta \omega l}{v_p} = \beta_r l + \frac{\delta \omega l}{v_p} \tag{5.4.2}$$

where β_r is the phase constant at the resonance.

If the transmission line is one-half wavelength long at the resonant frequency,

$$\beta_r l = \pi \quad \text{and} \quad \frac{l}{v_p} = \frac{\pi}{\omega_0}.$$

Therefore, $\tan \beta l$ can be approximated as follows:

$$\tan \beta l = \tan\left(\pi + \frac{\delta\omega\ l}{v_p}\right) = \tan\left(\pi + \frac{\pi\ \delta\omega}{\omega_0}\right) = \tan\frac{\pi\ \delta\omega}{\omega_0} \approx \frac{\pi\ \delta\omega}{\omega_0}$$

and

$$Z_{\text{in}} \approx Z_0 \frac{\alpha l + j(\pi\ \delta\omega/\omega_0)}{1 + j(\alpha l)(\pi\ \delta\omega/\omega_0)} \approx Z_0\left(\alpha l + j\frac{\pi\ \delta\omega}{\omega_0}\right) \tag{5.4.3}$$

It is assumed in (5.4.3) that $(\alpha l)(\pi\ \delta\omega/\omega_0) \ll 1$.

For a series resonant circuit, $Z_{\text{in}} \approx R + j2\delta\omega L$. Hence, a half-wavelength-long transmission line with short-circuit termination is similar to a series resonant circuit. The equivalent-circuit parameters are found as follows (assuming that losses in the line are small such that the characteristic impedance is a real number):

$$R \approx Z_0 \alpha l = \tfrac{1}{2} Z_0 \alpha \lambda_r \tag{5.4.4}$$

$$L \approx \frac{\pi}{2}\frac{Z_0}{\omega_0} \tag{5.4.5}$$

$$C \approx \frac{2}{\pi \omega_0 Z_0} \tag{5.4.6}$$

and

$$Q = \frac{\omega_0 L}{R} \approx \frac{\pi}{2\alpha l} = \frac{\beta_r}{2\alpha} \tag{5.4.7}$$

On the other hand, if the transmission line is a quarter-wavelength long at the resonant frequency, $\beta_r l = \pi/2$, and $l/v_p = \pi/2\,\omega_0$. Therefore,

$$\tan \beta l = \tan\left(\frac{\pi}{2} + \frac{\delta\omega l}{v_p}\right) = \tan\left(\frac{\pi}{2} + \frac{\pi\delta\omega}{2\,\omega_0}\right) = -\cot\left(\frac{\pi\delta\omega}{2\,\omega_0}\right) \approx -\frac{2\,\omega_0}{\pi\delta\omega} \tag{5.4.8}$$

and

$$Z_{\text{in}} \approx Z_0 \frac{-j(2\,\omega_0/\pi\delta\omega)}{1 - j\alpha l(2\,\omega_0/\pi\delta\omega)} = \frac{Z_0/\alpha l}{1 + j(\pi/\alpha l)(\delta\omega/2\,\omega_0)} \tag{5.4.9}$$

This expression is similar to the one obtained for a parallel resonant circuit in (5.2.9). Therefore, this transmission line is working as a parallel resonant circuit with equivalent parameters as follows:

$$R_p \approx \frac{Z_0}{\alpha l} = \frac{4Z_0}{\alpha \lambda_r} \tag{5.4.10}$$

$$L_p \approx \frac{4Z_0}{\pi \omega_0} \tag{5.4.11}$$

$$C_p \approx \frac{\pi}{4\,\omega_0 Z_0} \tag{5.4.12}$$

and

$$Q = \frac{\pi}{4\alpha l} = \frac{\beta_r}{2\alpha} \tag{5.4.13}$$

Open-Circuited Line

The analysis of the open-circuited transmission line shown in Figure 5.19 can be performed following a similar procedure. These results are summarized below. From (3.2.5),

$$Z_{in}|_{Z_L=\infty} = \frac{Z_0}{\tanh \gamma l} \tag{5.4.14}$$

For a transmission line that is a half-wavelength long, the input impedance can be approximated as follows:

$$Z_{in} \approx Z_0 \frac{1 + j(\alpha l)(\pi\delta\omega/\omega_0)}{\alpha l + j(\pi\delta\omega/\omega_0)} \approx \frac{Z_0}{\alpha l + j(\pi\delta\omega/\omega_0)} \tag{5.4.15}$$

This is similar to (5.2.9). Therefore, a half-wavelength-long open-circuited transmission line is similar to a parallel resonant circuit with equivalent parameters as follows:

$$R_p \approx \frac{Z_0}{\alpha l} = \frac{2Z_0}{\alpha\lambda_r} \tag{5.4.16}$$

$$L_p \approx \frac{2Z_0}{\pi\omega_0} \tag{5.4.17}$$

$$C_p \approx \frac{\pi}{2\,\omega_0 Z_0} \tag{5.4.18}$$

and

$$Q = \frac{\pi}{2\alpha l} = \frac{\beta_r}{2\alpha} \tag{5.4.19}$$

If the line is only a quarter-wavelength long, the input impedance can be approximated as

$$Z_{in} \approx Z_0 \frac{1 - j(\alpha l)(2\,\omega_0/\pi\delta\omega)}{\alpha l - j(2\,\omega_0/\pi\delta\omega)} \approx Z_0 \left(\alpha l + j\frac{\pi\delta\omega}{2\,\omega_0}\right) \tag{5.4.20}$$

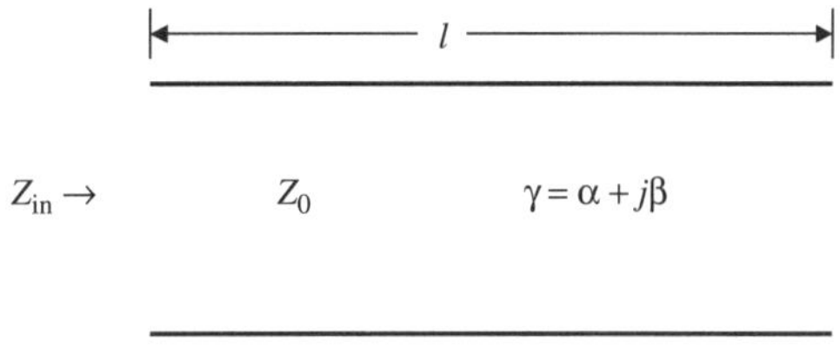

Figure 5.19 Open-circuited transmission line.

TABLE 5.2 Equivalent-Circuit Parameters of the Resonant Lines for $n = 1$

	Quarter-wavelength line		Half-wavelength line	
	Open Circuit	Short Circuit	Open Circuit	Short Circuit
Resonance	Series	Parallel	Parallel	Series
R	$\frac{1}{4}Z_0\alpha\lambda_r$	$\frac{4Z_0}{\alpha\lambda_r}$	$\frac{2Z_0}{\alpha\lambda_r}$	$\frac{1}{2}Z_0\alpha\lambda_r$
L	$\frac{\pi Z_0}{4\,\omega_0}$	$\frac{4Z_0}{\pi\omega_0}$	$\frac{2Z_0}{\pi\omega_0}$	$\frac{\pi Z_0}{2\,\omega_0}$
C	$\frac{4}{\pi Z_0\omega_0}$	$\frac{\pi}{4Z_0\omega_0}$	$\frac{\pi}{2Z_0\omega_0}$	$\frac{2}{\pi Z_0\omega_0}$
Q	$\frac{\beta_r}{2\alpha}$	$\frac{\beta_r}{2\alpha}$	$\frac{\beta_r}{2\alpha}$	$\frac{\beta_r}{2\alpha}$

Since (5.4.20) is similar to (5.1.17), this transmission line works as a series resonant circuit. Its equivalent-circuit parameters are found as follows (see Table 5.2):

$$R \approx Z_0\alpha l = \tfrac{1}{4}Z_0\alpha\lambda_r \tag{5.4.21}$$

$$L \approx \frac{\pi Z_0}{4\,\omega_0} \tag{5.4.22}$$

$$C \approx \frac{4}{\pi\omega_0 Z_0} \tag{5.4.23}$$

and

$$Q = \frac{\pi Z_0}{4R} = \frac{\pi}{4\alpha l} = \frac{\beta_0}{2\alpha} \tag{5.4.24}$$

Example 5.5 Design a half-wavelength-long coaxial line resonator that is short-circuited at its ends. Its inner conductor radius is 0.455 mm and the inner radius of the outer conductor is 1.499 mm. The conductors are made of copper. Compare the Q value of an air-filled resonator to that of a Teflon-filled resonator operating at 5 GHz. The dielectric constant of Teflon is 2.08 and its loss tangent is 0.0004.

SOLUTION From the relations for coaxial lines given in Appendix 2,

$$R_s = \sqrt{\frac{\omega\mu_0}{2\sigma}} = \sqrt{\frac{2\pi \times 5 \times 10^9 \times 4\pi \times 10^{-7}}{2 \times 5.813 \times 10^7}} = 0.018427\ \Omega$$

$$\alpha_c = \frac{R_s}{2\ln(b/a)\sqrt{\mu_0/\varepsilon_0}}\left(\frac{1}{a} + \frac{1}{b}\right)$$

$$= \frac{0.018427}{2 \times 376.7343 \times \ln(1.499/0.455)} \left(\frac{100}{0.455} + \frac{100}{1.499} \right)$$
$$= 0.058768 \text{ Np/m}$$

With Teflon filling,

$$\alpha_c = \frac{R_s}{2\sqrt{\mu_0/\varepsilon_0\varepsilon_r}\,\ln(b/a)} \left(\frac{1}{a} + \frac{1}{b} \right)$$
$$= \frac{0.018427 \times \sqrt{2.08}}{2 \times 376.7343 \times \ln(1.499/0.455)} \left(\frac{100}{0.455} + \frac{100}{1.499} \right)$$
$$= 0.084757 \text{ Np/m}$$
$$\alpha_d = \frac{\omega}{2}\sqrt{\mu_0\varepsilon_0\varepsilon_r}\tan\delta = \pi \times 5 \times 10^9 \times 3 \times 10^8 \times \sqrt{2.08} \times 0.0004$$
$$= 0.030206 \text{ Np/m}$$
$$\beta_0 = \frac{2\pi \times 5 \times 10^9}{3 \times 10^8} = 104.719755 \text{ rad/m}$$
$$\beta_d = \frac{2\pi \times 5 \times 10^9 \times \sqrt{2.08}}{3 \times 10^8} = 151.029 \text{ rad/m}$$
$$Q_{\text{air}} = \frac{\beta_0}{2\alpha} = \frac{104.719755}{2 \times 0.058768} = 890.96$$
$$Q_{\text{Teflon}} = \frac{\beta_d}{2\alpha} = \frac{151.029}{2 \times (0.084757 + 0.030206)} = 656.8629$$

Example 5.6 Design a half-wavelength-long microstrip resonator using a 50-Ω line that is short circuited at its ends. The substrate thickness is 0.159 cm, with its dielectric constant 2.08 and the loss tangent 0.0004. The conductors are made of copper. Compute the length of the line for resonance at 2 GHz and the Q of the resonator. Assume that thickness t of the microstrip is 0.159 μm.

SOLUTION Design equations for the microstrip line (from Appendix 2) are

$$\frac{w}{h} = \begin{cases} \dfrac{8e^A}{e^{2A} - 2} & \text{for } A > 1.52 \\ \dfrac{2}{\pi}\left\{B - 1 - \ln(2B - 1) + \dfrac{\varepsilon_r - 1}{2\varepsilon_r} \times \left[\ln(B - 1) + 0.39 - \dfrac{0.61}{\varepsilon_r}\right]\right\} & \text{for } A \le 1.52 \end{cases}$$

where

$$A = \frac{Z_0}{60}\left(\frac{\varepsilon_r + 1}{2}\right)^{1/2} + \frac{\varepsilon_r - 1}{\varepsilon_r + 1}\left(0.23 + \frac{0.11}{\varepsilon_r}\right)$$

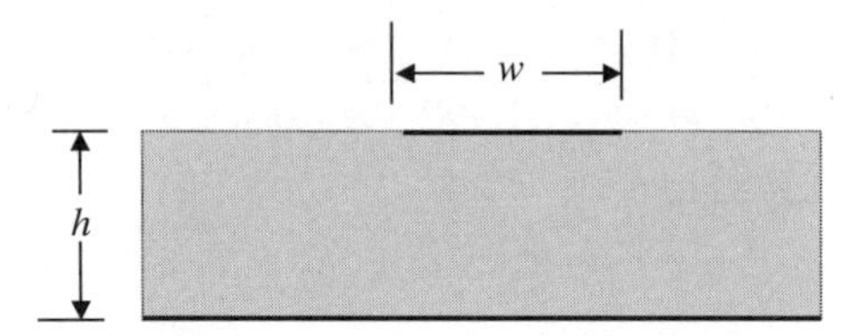

Figure 5.20 Geometry of the microstrip line.

and

$$B = \frac{60\pi^2}{Z_0\sqrt{\varepsilon_r}}$$

See Figure 5.20. Therefore,

$$A = \frac{50}{60}\left(\frac{2.08+1}{2}\right)^{1/2} + \frac{2.08-1}{2.08+1}\left(0.23 + \frac{0.11}{2.08}\right) = 1.1333$$

Since A is smaller than 1.52, we need B to determine the width of microstrip. Hence,

$$B = \frac{60\pi^2}{50 \times \sqrt{2.08}} = 8.212$$

$$\frac{w}{h} = \frac{2}{\pi}\left\{8.212 - 1 - \ln(2 \times 8.212 - 1) + \frac{2.08-1}{2 \times 2.08}\right.$$
$$\left. \times \left[\ln(8.212 - 1) + 0.39 - \frac{0.61}{2.08}\right]\right\} = 3.1921$$

Therefore,

$$w = 0.507543 \text{ cm} \approx 0.51 \text{ cm}$$

Further,

$$\varepsilon_{re} = 0.5\left[2.08 + 1 + (2.08 - 1) \times \frac{1}{\sqrt{1 + 12/3.192094}}\right]$$
$$- \frac{(2.08 - 1)0.0001}{4.6\sqrt{3.1921}} = 1.7875$$

Therefore, the length of the resonator can be determined as follows:

$$l = \frac{\lambda}{2} = \frac{v_p/f}{2} = \frac{c}{2f\sqrt{\varepsilon_e}} = \frac{3 \times 10^{10}}{2 \times 2 \times 10^9 \times \sqrt{1.7875}} \text{ cm} = 5.6096 \text{ cm}$$

Since $\sigma_{copper} = 5.813 \times 10^7$ S/m, the quality factor of this resonator is determined as follows. For $w/h \geq 1/2\pi$,

$$\frac{w_e}{h} = \frac{w}{h} + 0.3979\frac{t}{h}\left[1 + \ln\left(2\frac{h}{t}\right)\right] = 3.1923$$

and

$$\zeta = 1 + \frac{h}{w_e}\left[1 - \frac{1.25t}{\pi h} + \frac{1.25}{\pi}\ln\left(2\frac{h}{t}\right)\right] = 2.5476$$

Therefore, for $w/h \geq 1$, α_c is found to be

$$\alpha_c = 44.1255 \times 10^{-5}\frac{\zeta Z_0 \varepsilon_{re}}{h}\sqrt{\frac{f_{\mathrm{GHz}}}{\sigma}}\left[\frac{w_e}{h} + \frac{0.667\,(w_e/h)}{w_e/h + 1.444}\right] \quad \mathrm{Np/m}$$
$$= 0.0428\ \mathrm{Np/m}$$

Similarly, α_d is found from (A2.28) as follows:

$$\alpha_d = 10.4766\frac{\varepsilon_r}{\varepsilon_r - 1}\frac{\varepsilon_{re} - 1}{\sqrt{\varepsilon_{re}}} f_{\mathrm{GHz}} \tan\delta \quad \mathrm{Np/m} = 0.0095\,\mathrm{Np/m}$$

and

$$\beta = \frac{2\pi}{\lambda} = \frac{2\pi}{2l} = \frac{2\pi}{2 \times 5.6096/100}\ \mathrm{rad/m} = 56.0035\,\mathrm{rad/m}$$

Hence, Q of the resonator is found as

$$Q = \frac{\beta}{2\alpha} = \frac{56.0035}{2 \times (0.0428 + 0.0095)} = 535.4$$

5.5 MICROWAVE RESONATORS

Cavities and dielectric resonators are commonly employed as resonant circuits at frequencies above 1 GHz. These resonators provide much higher Q than the lumped elements and transmission line circuits. However, the characterization of these devices requires analysis of associated electromagnetic fields. Characteristic relations of selected microwave resonators are summarized in this section. Interested readers can find several excellent references and textbooks analyzing these and other resonators.

Rectangular Cavities

Consider a rectangular cavity made of conducting walls with dimensions $a \times b \times c$, as shown in Figure 5.21. It is filled with a dielectric material of dielectric constant ε_r and relative permeability μ_r. In general, it can support both TE_{mnp}

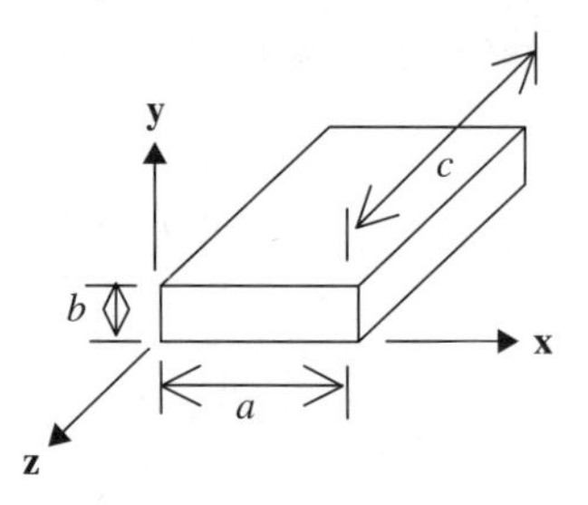

Figure 5.21 Geometry of the rectangular cavity resonator.

and TM_{mnp} modes, which may be degenerate. The cutoff frequency of TE_{mn} and TM_{mn} modes can be determined from the formula

$$f_c(\text{MHz}) = \frac{300}{\sqrt{\mu_r \varepsilon_r}} \sqrt{\left[\left(\frac{m}{2a}\right)^2 + \left(\frac{n}{2b}\right)^2\right]} \quad \text{MHz}$$

The resonant frequency f_r of a rectangular cavity operating in either TE_{mnp} or TM_{mnp} mode can be determined as follows:

$$f_r(\text{MHz}) = \frac{300}{\sqrt{\mu_r \varepsilon_r}} \sqrt{\left[\left(\frac{m}{2a}\right)^2 + \left(\frac{n}{2b}\right)^2 + \left(\frac{p}{2c}\right)^2\right]} \quad \text{MHz} \tag{5.5.1}$$

If the cavity is made of a perfect conductor and filled with a perfect dielectric, it will have infinite Q. However, it is not possible in practice. In the case of cavity walls having a finite conductivity (instead of infinite for the perfect conductor) but filled with a perfect dielectric (no dielectric loss), its quality factor Q_c for TE_{10p} mode is given by

$$Q_c|_{10p} = \frac{60b(ac\omega\sqrt{\mu\varepsilon})^3}{\pi R_s(2p^2a^3b + 2bc^3 + p^2a^3c + ac^3)} \sqrt{\frac{\mu_r}{\varepsilon_r}} \tag{5.5.2}$$

The permittivity and permeability of the dielectric filling are given by ε and μ, respectively, ω is angular frequency, and R_s is the surface resistivity of walls, which is related to the skin depth δ_s and the conductivity σ as follows.

$$R_s = \frac{1}{\sigma\delta_s} = \sqrt{\frac{\omega\mu}{2\sigma}} \tag{5.5.3}$$

On the other hand, if the cavity is filled with a dielectric with its loss tangent as $\tan\delta$ while its walls are made of a perfect conductor, the quality factor Q_d is given as

$$Q_d = \frac{1}{\tan\delta} \tag{5.5.4}$$

When there is power loss in both the cavity walls and the dielectric filling, the quality factor Q is found from Q_c and Q_d as follows:

$$\frac{1}{Q} = \frac{1}{Q_c} + \frac{1}{Q_d} \tag{5.5.5}$$

Example 5.7 An air-filled rectangular cavity is made from a piece of copper WR-90 waveguide. If it resonates at 9.379 GHz in TE_{101} mode, find the required length c and the Q value of this resonator.

SOLUTION Specifications of WR-90 are given in Table A3.3 as follows:

$$a = 0.9\,\text{in.} = 2.286\,\text{cm}$$

and

$$b = 0.4\,\text{in.} = 1.016\,\text{cm}$$

From (5.5.1),

$$\frac{1}{2c} = \sqrt{\left(\frac{9379}{300}\right)^2 - \left(\frac{1}{2a}\right)^2} = 22.3383 \Rightarrow c = 2.238\,\text{cm}$$

Next, the surface resistance, R_s, is determined from (5.5.3) as 0.0253 Ω, and the Q_c is found from (5.5.2) to be about 7858.

Example 5.8 A rectangular cavity made of copper has inner dimensions $a = 1.6\,\text{cm}$, $b = 0.71\,\text{cm}$, and $c = 1.56\,\text{cm}$. It is filled with Teflon ($\varepsilon_r = 2.05$ and $\tan\delta = 2.9268 \times 10^{-4}$). Find the TE_{101} mode resonant frequency and Q value of this cavity.

SOLUTION From (5.5.1),

$$f_r = \frac{300}{\sqrt{2.05}}\sqrt{\left(\frac{1}{2 \times 0.016}\right)^2 + \left(\frac{1}{2 \times 0.0156}\right)^2} \approx 9379\,\text{MHz} = 9.379\,\text{GHz}$$

Since power dissipates both in the dielectric filling as well as in the sidewalls, overall Q will be determined from (5.5.5). From (5.5.2) and (5.5.4), Q_c and Q_d are found to be 5489 and 3417, respectively. Substituting these into (5.5.5), the Q value of this cavity is found to be 2106.

Circular Cylindrical Cavities

Figure 5.22 shows the geometry of a circular cylindrical cavity of radius r and height h. It is filled with a dielectric material of relative permeability μ_r and

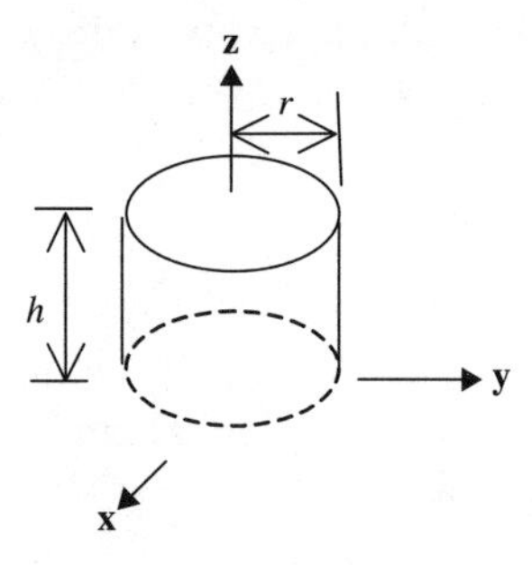

Figure 5.22 Geometry of the circular cylindrical cavity resonator.

dielectric constant ε_r. Its resonant frequency f_r in megahertz is given as

$$f_r(\text{MHz}) = \frac{300}{2\pi\sqrt{\mu_r\varepsilon_r}}\sqrt{\left(\frac{\chi_{nm}}{r}\right)^2 + \left(\frac{p\pi}{h}\right)^2} \quad \text{MHz} \tag{5.5.6}$$

where

$$\chi_{nm} = \begin{cases} p_{nm} & \text{for TM modes} \\ p'_{nm} & \text{for TE modes} \end{cases} \tag{5.5.7}$$

As discussed in Section 4.9, p_{nm} represents the mth zero of the Bessel function of the first kind and order n and p'_{nm} represents the mth zero of the derivative of the Bessel function. In other words, these represent roots of the following equations, respectively (see Table 5.3):

$$J_n(x) = 0 \tag{5.5.8}$$

and

$$\frac{dJ_n(x)}{dx} = 0 \tag{5.5.9}$$

The Q_c of a circular cylindrical cavity operating in TE_{nmp} mode and filled with a lossless dielectric can be found from the following formula:

$$Q_c = \frac{47.7465}{\delta_s f_r(\text{MHz})} \frac{\left[1 - \left(n/p'_{nm}\right)^2\right]\left[(p'_{nm})^2 + (p\pi r/h)^2\right]^{1.5}}{\left\{\left[p'^2_{nm} + (2r/h)(p\pi r/h)^2\right] + \left[1 - (2r/h)\left(np\pi r/p'_{nm}h\right)^2\right]\right\}} \tag{5.5.10}$$

TABLE 5.3 Zeros of $J_n(x)$ and $J'_n(x)$

n	p_{n1}	p_{n2}	p_{n3}	p'_{n1}	p'_{n2}	p'_{n3}
0	2.405	5.520	8.654	3.832	7.016	10.173
1	3.832	7.016	10.173	1.841	5.331	8.536
2	5.136	8.417	11.620	3.054	6.706	9.969

In the case of TM_{nmp} mode, Q_c is given by

$$Q_c = \frac{47.7465}{\delta_s f_r(\text{MHz})} \frac{\sqrt{[p_{nm}^2 + (p\pi r/h)^2]}}{1 + (2r/h)} \qquad p > 0 \tag{5.5.11}$$

For $p = 0$,

$$Q_c \approx \frac{47.7465}{\delta_s f_r(\text{MHz})} \frac{p_{nm}}{1 + (r/h)} \qquad p = 0 \tag{5.5.12}$$

The quality factor Q_d due to dielectric loss and the overall Q are determined as before from (5.5.4) and (5.5.5) with appropriate Q_c.

Example 5.9 Determine the dimensions of an air-filled circular cylindrical cavity that resonates at 5 GHz in TE_{011} mode. It should be made of copper, and its height should be equal to its diameter. Find the Q of this cavity given that $n = 0$, $m = p = 1$, and $h = 2r$.

SOLUTION From (5.5.6), with χ_{01} as 3.832 from Table 5.3, we get

$$r = \sqrt{\frac{3.832^2 + (\pi/2)^2}{(5000 \times 2\pi/300)^2}} = 0.03955\ \text{m}$$

and

$$h = 2r = 0.0791\ \text{m}$$

Since

$$\delta_s = \sqrt{\frac{2}{\omega\mu\sigma}} = \sqrt{\frac{2}{2\pi \times 5 \times 10^9 \times 4\pi \times 10^{-7} \times 5.8 \times 10^7}} = 9.3459 \times 10^{-7}\ \text{m}$$

The Q value of the cavity can be found from (5.5.10) as

$$Q_c = \frac{47.7465 \times 10^6}{9.3459 \times 10^{-7} \times 5 \times 10^9} \times \frac{[3.832^2 + (\pi/2)^2]^{1.5}}{[3.832^2 + (\pi/2)^2] + 1} = 39,984.6$$

Further, there is no loss of power in air ($\tan\delta \approx 0$), Q_d is infinite, and therefore Q is the same as Q_c.

Dielectric Resonators

Dielectric resonators (DRs), made of high-permittivity ceramics, provide high Q values with smaller size. These resonators are generally a solid sphere or cylinder of circular or rectangular cross section. High-purity TiO_2 ($\varepsilon_r \approx 100$, $\tan\delta \approx 0.0001$) was used in early dielectric resonators. However, it was found

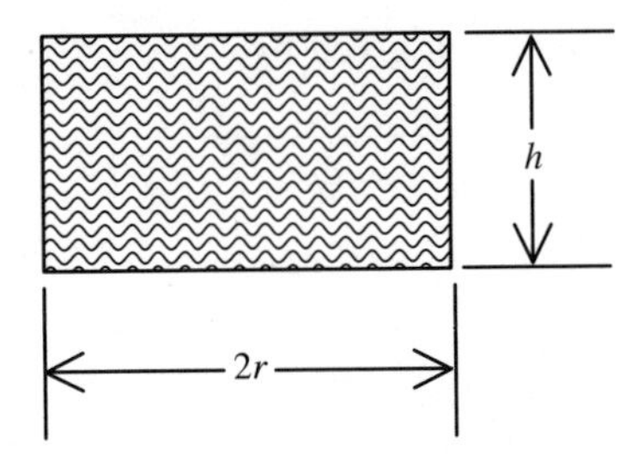

Figure 5.23 Geometry of an isolated circular cylindrical dielectric resonator.

to be intolerably dependent on the temperature. With the development of new ceramics, temperature dependence of DR can be reduced significantly.

Figure 5.23 shows an isolated circular cylindrical DR. Its analysis is beyond the scope of this book. Its $TE_{01\delta}$ -mode resonant frequency, f_r in GHz, can be found from the following formula:

$$f_r(\text{GHz}) = \frac{34}{r\varepsilon_r}\left(3.45 + \frac{r}{h}\right) \qquad \text{GHz} \tag{5.5.13}$$

where r and h are in millimeters. This relation is found to be accurate within $\pm 2\%$ if

$$2 > \frac{r}{h} > 0.5 \tag{5.5.14}$$

and

$$50 > \varepsilon_r > 30 \tag{5.5.15}$$

An isolated DR is of almost no use in practice. A more practical situation is illustrated in Figure 5.24, where it is coupled with a microstrip line in a conducting enclosure. Assume that the dielectric constants of the DR and the substrate are ε_r and ε_s, respectively. With various dimensions as shown (all in meters), the following formulas can be used to determine its radius r and the height h. If r is selected between the following bounds, the formulas presented in this section are found to have a tolerance of no more than 2% (Kajfez and Guillon, 1986):

$$\frac{1.2892 \times 10^8}{f_r\sqrt{\varepsilon_s}} > r > \frac{1.2892 \times 10^8}{f_r\sqrt{\varepsilon_r}} \tag{5.5.16}$$

where f_r is resonant frequency in hertz.

The height of the DR is then found as

$$h = \frac{1}{\beta}\left(\tan^{-1}\frac{\alpha_1}{\beta\tanh\alpha_1 t} + \tan^{-1}\frac{\alpha_2}{\beta\tanh\alpha_2 d}\right) \tag{5.5.17}$$

where

$$\alpha_1 = \sqrt{k'^2 - k_0^2 \varepsilon_s} \tag{5.5.18}$$

$$\alpha_2 = \sqrt{k'^2 - k_0^2} \tag{5.5.19}$$

$$\beta = \sqrt{k_0^2 \varepsilon_r - k'^2} \tag{5.5.20}$$

$$k' = \frac{2.405}{r} + \frac{y}{2.405r[1 + (2.43/y) + 0.291y]} \tag{5.5.21}$$

$$y = \sqrt{(k_0 r)^2(\varepsilon_r - 1) - 5.784} \tag{5.5.22}$$

and

$$k_0 = \omega\sqrt{\mu_0 \varepsilon_0} \tag{5.5.23}$$

Example 5.10 Design a $TE_{01\delta}$-mode cylindrical dielectric resonator for use at 35 GHz in the microstrip circuit shown in Figure 5.24. The substrate is 0.25 mm thick and its dielectric constant is 9.9. The dielectric constant of the material available for DR is 36. Further, the top of the DR should have a clearance of 1 mm from the conducting enclosure.

SOLUTION Since

$$\frac{1.2892 \times 10^8}{f_r\sqrt{\varepsilon_s}} = 1.171 \times 10^{-3}\text{ m}$$

and

$$\frac{1.2892 \times 10^8}{f_r\sqrt{\varepsilon_r}} = 6.139 \times 10^{-4}\text{ m}$$

the radius r of the DR may be selected as 0.835 mm.

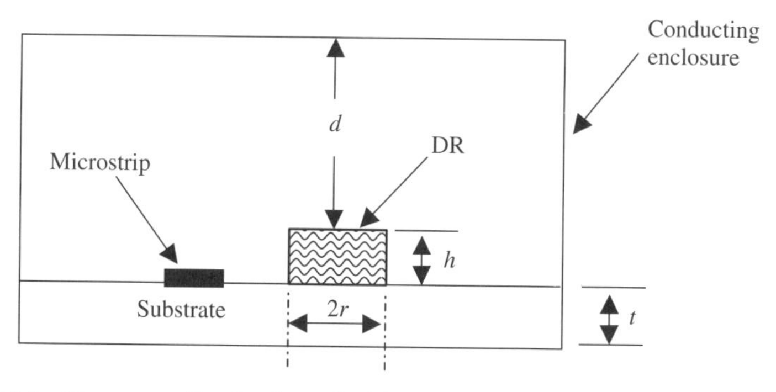

Figure 5.24 Geometry of a circular cylindrical dielectric resonator in MIC configuration.

The height h of the DR is then determined from (5.5.17) to (5.5.22), as follows:

$$y = \sqrt{(k_0 r)^2(\varepsilon_r - 1) - 5.784} = 2.707$$

$$k' = \frac{2.405}{r} + \frac{y}{2.405r[1 + (2.43/y) + 0.291y]} = 3.382 \times 10^3$$

$$\alpha_1 = \sqrt{k'^2 - k_0^2 \varepsilon_s} = 2.474 \times 10^3$$

$$\alpha_2 = \sqrt{k'^2 - k_0^2} = 3.302 \times 10^3$$

and

$$\beta = \sqrt{k_0^2 \varepsilon_r - k'^2} = 2.812 \times 10^3$$

After substituting these numbers into (4.5.17), h is found to be

$$h = \frac{1}{\beta}\left(\tan^{-1}\frac{\alpha_1}{\beta \tanh \alpha_1 t} + \tan^{-1}\frac{\alpha_2}{\beta \tanh \alpha_2 d}\right) = 6.683 \times 10^{-4}\,\text{m} = 0.668\,\text{mm}$$

SUGGESTED READING

Bahl, I., and P. Bhartia, *Microwave Solid State Circuit Design*. New York: Wiley, 1988.

Collin, R. E., *Foundations for Microwave Engineering*. New York: McGraw-Hill, 1992.

Elliott, R. S., *An Introduction to Guided Waves and Microwave Circuits*. Englewood Cliffs, NJ: Prentice Hall, 1993.

Kajfez, D., and P. Guillon, *Dielectric Resonators*. Dedham, MA: Artech House, 1986.

Pozar, D. M., *Microwave Engineering*. New York: Wiley, 1998.

Ramo, S., J. R. Whinnery, and T. Van Duzer, *Fields and Waves in Communication Electronics*. New York: Wiley, 1994.

Rizzi, P.A., *Microwave Engineering*. Englewood Cliffs, NJ: Prentice Hall, 1988.

Smith, J. R., *Modern Communication Circuits*. New York: McGraw-Hill, 1998.

PROBLEMS

5.1. Determine the resonant frequency, bandwidth, and Q of the circuit shown in Figure P5.1.

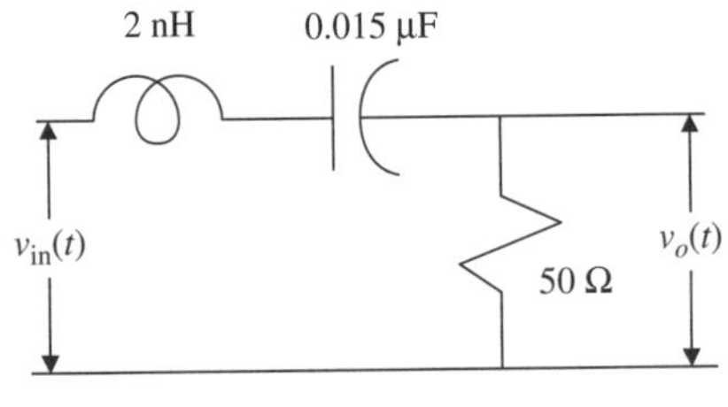

Figure P5.1

5.2. The quality factor of a 100-nH inductor is 150 at 100 MHz. It is used in a series resonant circuit with a 50-Ω load resistor. Find the capacitor required to resonate this circuit at 100 MHz. Find the loaded Q value of the circuit.

5.3. Find the capacitance C in the series $R-L-C$ circuit shown in Figure P5.3 if it is resonating at 1200 rad/s. Compute the output voltage $v_o(t)$ when $v_{\text{in}}(t) = 4\cos(\omega t + 0.2\pi)$ V for ω as 1200, 300, and 4800 rad/s.

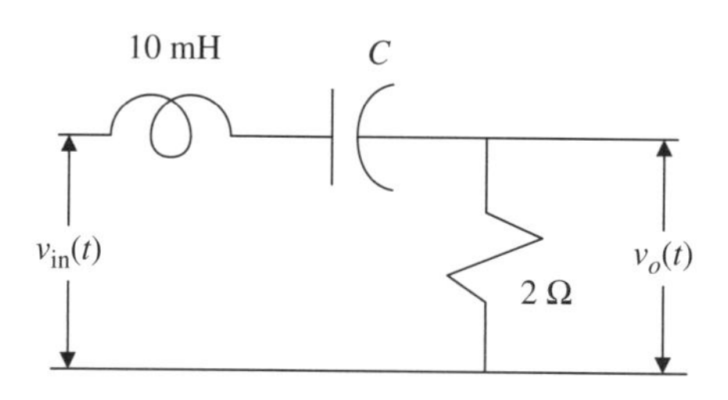

Figure P5.3

5.4. The magnitude of input impedance in the series $R-L-C$ circuit shown in Figure P5.4 is found to be 3 Ω at a resonant frequency of 1600 rad/s. Find the inductance L, Q, and the bandwidth.

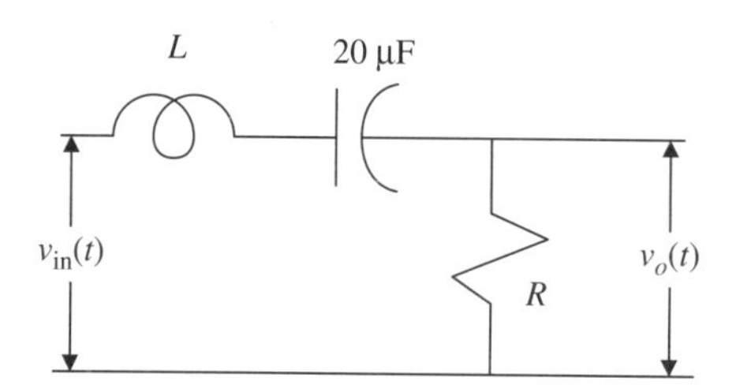

Figure P5.4

5.5. Determine the element values of a resonant circuit to filter all the sinusoidal signals from 100 to 130 MHz. This filter is to be connected between a voltage source with negligible internal impedance and a communication system that has an input impedance of 50 Ω. Plot its response over the frequency range 50 to 200 MHz.

5.6. A series resonant circuit is resonating at 10^5 rad/s and has a bandwidth of 100 rad/s. If the resistance in the circuit is 50 Ω, find its inductance and capacitance.

5.7. For the circuit given in Figure P5.7, the poles are located as illustrated in Figure P7.2. Find **(a)** the Q, **(b)** ω_0, **(c)** $\Delta\omega$, **(d)** ω_1, and **(e)** ω_2.

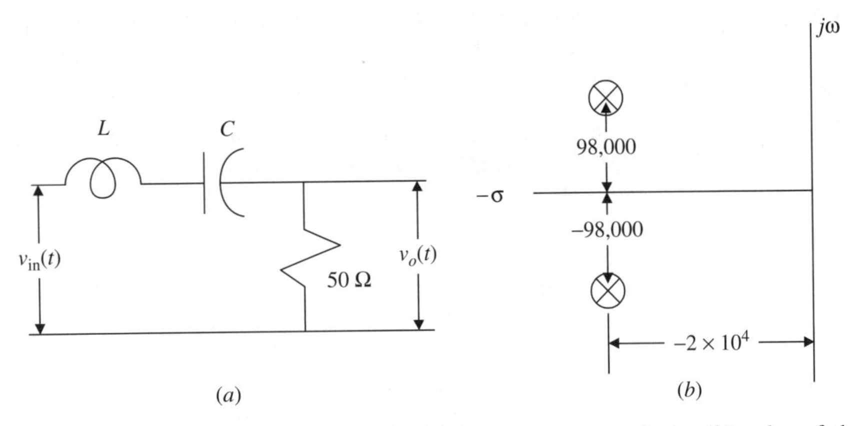

Figure P5.7 (*a*) Series $R-L-C$ circuit with input–output terminals; (*b*) poles of the transfer function.

5.8. Calculate the resonant frequency, Q, and bandwidth of the parallel resonant circuit shown in Figure P5.8.

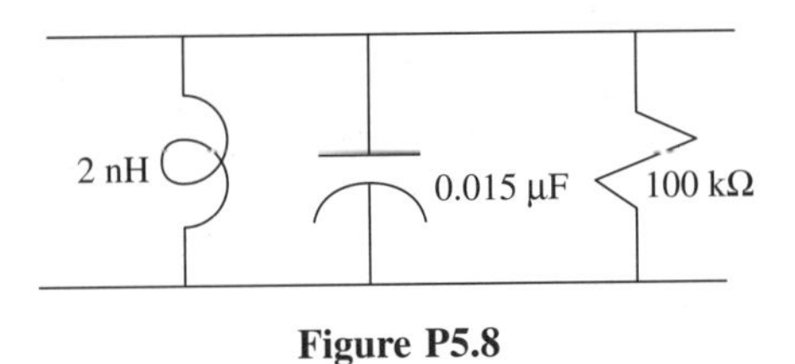

Figure P5.8

5.9. Find the resonant frequency, unloaded Q, and loaded Q of the parallel resonant circuit shown in Figure P5.9.

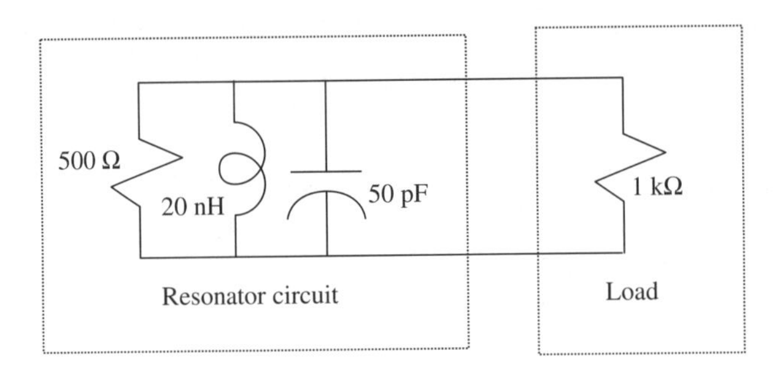

Figure P5.9

5.10. The parallel $R-L-C$ circuit shown in Figure P5.10 is resonant at 20,000 rad/s. If its admittance has a magnitude of 1 mS at the resonance, find the values of R and L.

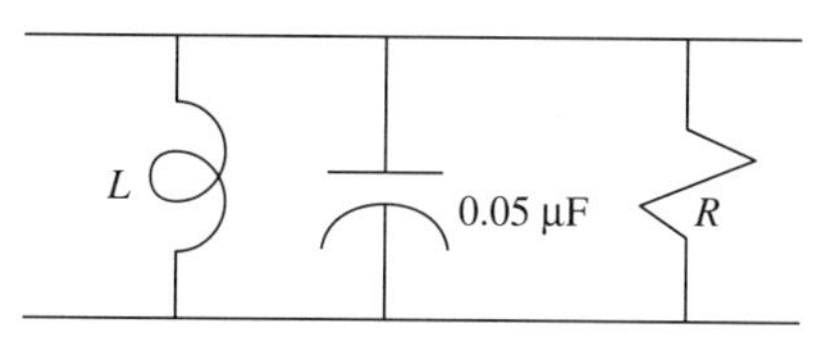

Figure P5.10

5.11. Consider the loaded parallel resonant circuit shown in Figure P5.11. Compute the resonant frequency in radians per second, unloaded Q, and loaded Q of this circuit.

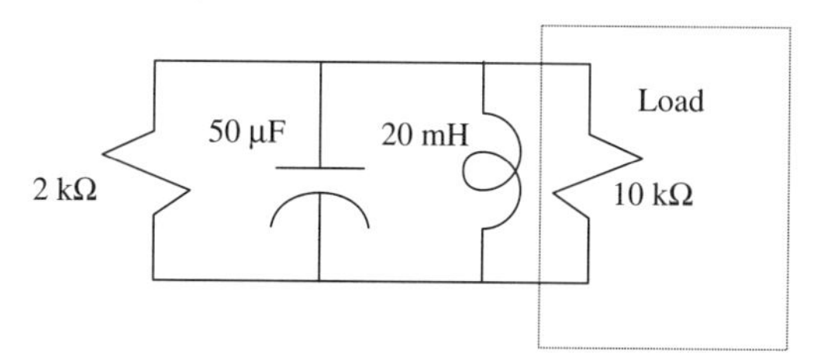

Figure P5.11

5.12. Resistance R in a parallel $R-L-C$ circuit is $200\,\Omega$. The circuit has a bandwidth of 80 rad/s with a lower half-power frequency at 800 rad/s. Find the inductance and capacitance of the circuit.

5.13. A parallel $R-L-C$ circuit is resonant at 2×10^6 rad/s and has a bandwidth of 20,000 rad/s. If its impedance at resonance is $2000\,\Omega$, find the circuit parameters.

5.14. A resonator is fabricated from a 2λ-long transmission line that is short-circuited at its ends. Find its Q.

5.15. A 3-cm-long 100-Ω air-filled coaxial line is used to fabricate a resonator. It is short-circuited at one end while a capacitor is connected at the other end to obtain the resonance at 6 GHz. Find the value of this capacitor.

5.16. A 100-Ω air-filled coaxial line of length l is used to fabricate a resonator. It is terminated by $10 - j5000\ \Omega$ at its ends. If the signal wavelength is 1 m, find the required length for the first resonance and the Q of this resonator.

5.17. Design a half-wavelength-long coaxial line resonator that is short-circuited at its ends. Its inner conductor radius is 0.455 mm and the inner radius of the outer conductor is 1.499 mm. The conductors are made of copper. Compare the Q value of an air-filled resonator to that of a Teflon-filled resonator operating at 800 MHz. The dielectric constant of Teflon is 2.08 and its loss tangent is 0.0004.

5.18. Design a half-wavelength-long microstrip resonator using a 50-Ω line that is short-circuited at its ends. The substrate thickness is 0.12 cm with a dielectric constant of 2.08 and a loss tangent of 0.0004. The conductors are of copper. Compute the length of the line for resonance at 900 MHz and the Q value of the resonator.

5.19. The secondary side of a tightly coupled transformer is terminated by a 2-kΩ load in parallel with capacitor C_2. A current source $I = 10\cos(4 \times 10^6)$ mA is connected on its primary side. The inductance L_1 of its primary is 0.30 μH, the mutual inductance $M = 3$ μH, and the unloaded Q of the secondary coil is 70. Determine the value of C_2 such that this circuit resonates at 4×10^6 rad/s. Find its input impedance and the output voltage at the resonant frequency. What is the circuit bandwidth?

5.20. A 16-Ω resistor terminates the secondary side of a tightly coupled transformer. A 40-μF capacitor is connected in series with its primary coil, which has an inductance L_1 at 100 mH. If the secondary coil has an inductance L_2 at 400 mH, find the resonant frequency and the Q value of the circuit.

5.21. A 16-Ω resistor terminates the secondary side of a tightly coupled transformer. A 30-μF capacitor is connected in series with its primary coil, which has an inductance of 25 mH. If the circuit Q value is 50, find the inductance L_2 of the secondary coil.

5.22. An air-filled rectangular cavity is made from a piece of copper WR-430 waveguide. If it resonates at 2 GHz in TE_{101} mode, find the required length c and the Q of this resonator.

5.23. Design an air-filled circular cylindrical cavity that resonates at 9 GHz in TE_{011} mode. It should be made of copper, and its height should be equal to its diameter. Find the Q value of this cavity.

5.24. Design a $TE_{01\delta}$ mode cylindrical dielectric resonator for use at 4.267 GHz in the microstrip circuit shown in Figure 5.24. The substrate is 0.7 mm thick and its dielectric constant is 9.6. The dielectric constant of the material available for the DR is 34.19. Further, the top of the DR should have a clearance of 0.72 mm from the conducting enclosure.

6

IMPEDANCE-MATCHING NETWORKS

One of the most critical requirements in the design of high-frequency electronic circuits is that the maximum possible signal energy is transferred at each point. In other words, the signal should propagate in a forward direction with a negligible echo (ideally, zero). Echo signal not only reduces the power available but also deteriorates the signal quality due to the presence of multiple reflections. As noted in Chapter 5, impedance can be transformed to a new value by adjusting the turns ratio of a transformer that couples it with the circuit. However, it has several limitations. In this chapter we present a few techniques to design other impedance-transforming networks. These circuits include transmission line stubs and resistive and reactive networks. Further, the techniques introduced are needed in active circuit design at RF and microwave frequencies.

As shown in Figure 6.1, impedance-matching networks are employed at the input and output of an amplifier circuit. These networks may also be needed to perform other tasks, such as filtering the signal and blocking or passing the dc bias voltages. This chapter begins with a section on impedance-matching techniques that use a single reactive element or stub connected in series or in shunt. Theoretical principles behind the technique are explained, and the graphical procedure to design these circuits using a Smith chart is presented. Principles and procedures of double-stub matching are discussed in the following section. The chapter ends with sections on resistive and reactive L-section matching networks. Both analytical and graphical procedures to design these networks using ZY-charts are included.

Radio-Frequency and Microwave Communication Circuits: Analysis and Design, Second Edition,
By Devendra K. Misra
ISBN 0-471-47873-3

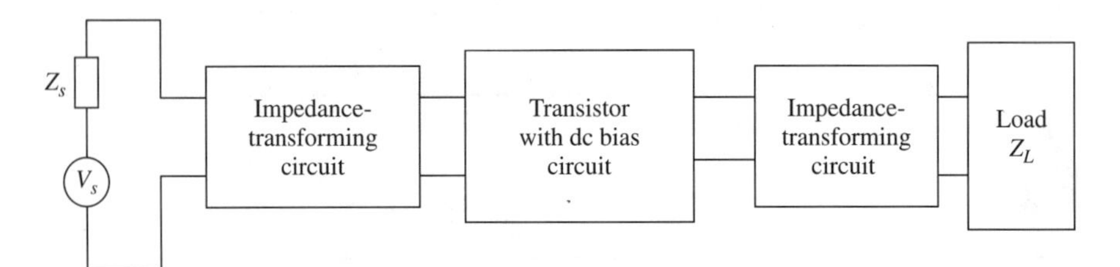

Figure 6.1 Block diagram of an amplifier circuit.

6.1 SINGLE REACTIVE ELEMENT OR STUB MATCHING NETWORKS

When a lossless transmission line is terminated by an impedance Z_L, the magnitude of the reflection coefficient (and hence, the VSWR) on it remains constant, but its phase angle can be anywhere between $+180°$ and $-180°$. As we have seen in Chapter 3, it represents a circle on a Smith chart, and a point on this circle represents the normalized load. As one moves away from the load, the impedance (or the admittance) value changes. This movement is clockwise on the VSWR circle. The real part of the normalized impedance (or normalized admittance) becomes unity at certain points on the line. Addition of a single reactive element or a transmission line stub at this point can eliminate the echo signal and reduce the VSWR to unity beyond this point. A finite-length transmission line with its other end an open or short circuit is called a *stub* and behaves like a reactive element, as explained in Chapter 3.

In this section we discuss the procedure for determining the location on a lossless feeding line where a stub or a reactive element can be connected to eliminate the echo signal. Two different possibilities, a series or a shunt element, are considered. Mathematical equations as well as graphical methods are presented to design the circuits.

Shunt Stub or Reactive Element

Consider a lossless transmission line of characteristic impedance Z_0 that is terminated by a load admittance Y_L, as shown in Figure 6.2. Corresponding normalized input admittance at a point d_s away from the load can be found from (3.2.6) as

$$\overline{Y}_{\text{in}} = \frac{\overline{Y}_L + j\tan\beta d_s}{1 + j\overline{Y}_L\tan\beta d_s} \tag{6.1.1}$$

To obtain a matched condition at d_s, the real part of the input admittance must be equal to the characteristic admittance of the line [i.e., the real part of (6.1.1) must be unity]. This requirement is used to determine d_s. The parallel susceptance B_s is then connected at d_s to cancel out the imaginary part of Y_{in}.

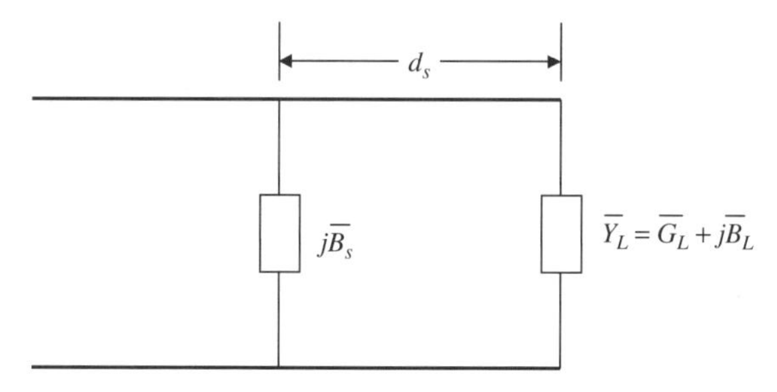

Figure 6.2 Transmission line with a shunt matching element.

Hence,

$$d_s = \frac{1}{\beta}\tan^{-1}\frac{\overline{B}_L \pm \sqrt{\overline{B}_L^2 - A(1-\overline{G}_L)}}{A} \tag{6.1.2}$$

where $A = \overline{G}_L(\overline{G}_L - 1) + \overline{B}_L^2$.

The imaginary part of the normalized input admittance at d_s is found as follows:

$$\overline{B}_{\text{in}} = \frac{(\overline{B}_L + \tan\beta d_s)(1 - \overline{B}_L\tan\beta d_s) - \overline{G}_L^2\tan\beta d_s}{(\overline{G}_L\tan\beta d_s)^2 + (1 - \overline{B}_L\tan\beta d_s)^2} \tag{6.1.3}$$

The other requirement to obtain a matched condition is

$$\overline{B}_s = -\overline{B}_{\text{in}} \tag{6.1.4}$$

Hence, a shunt inductor is needed at d_s if the input admittance is found capacitive (i.e., B_{in} is positive). On the other hand, it will require a capacitor if Y_{in} is inductive at d_s. As mentioned earlier, a lossless transmission line section can be used in place of this inductor or capacitor. The length of this transmission line section is determined according to the susceptance needed by (6.1.4) and the termination (i.e., an open circuit or a short circuit) at its other end. This transmission line section is called a *stub*. If l_s is the stub length that has a short circuit at its other end,

$$l_s = \frac{1}{\beta}\cot^{-1}(-\overline{B}_s) = \frac{1}{\beta}\cot^{-1}\overline{B}_{\text{in}} \tag{6.1.5}$$

On the other hand, if there is an open circuit at the other end of the stub,

$$l_s = \frac{1}{\beta}\tan^{-1}\overline{B}_s = \frac{1}{\beta}\tan^{-1}(-\overline{B}_{\text{in}}) \tag{6.1.6}$$

Series Stub or Reactive Element

If a reactive element (or stub) needs to be connected in series as shown in Figure 6.3, the design procedure can be developed as follows. The normalized

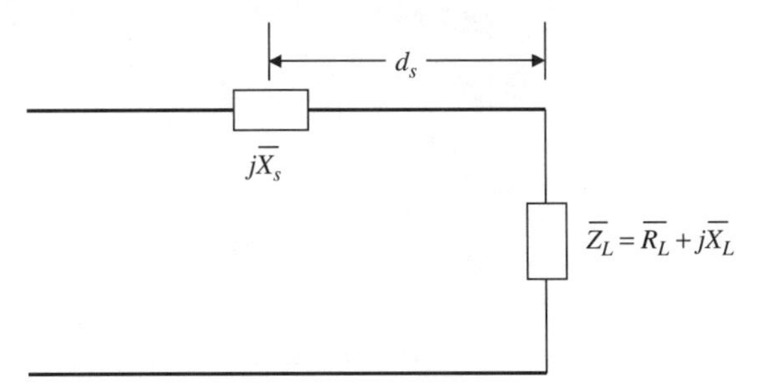

Figure 6.3 Transmission line with a matching element connected in series.

input impedance at d_s is

$$\overline{Z}_{\text{in}} = \frac{\overline{Z}_L + j\tan\beta d_s}{1 + j\overline{Z}_L \tan\beta d_s} \tag{6.1.7}$$

To obtain a matched condition at d_s, the real part of the input impedance must be equal to the characteristic impedance of the line [i.e., the real part of (6.1.7) must be unity]. This condition is used to determine d_s. A reactance X_s is then connected in series at d_s to cancel out the imaginary part of Z_{in}. Hence,

$$d_s = \frac{1}{\beta}\tan^{-1}\frac{\overline{X}_L \pm \sqrt{\overline{X}_L^2 - A_z(1 - \overline{R}_L)}}{A_z} \tag{6.1.8}$$

where $A_z = \overline{R}_L(\overline{R}_L - 1) + \overline{X}_L^2$.

The imaginary part of the normalized input impedance at d_s is found as follows:

$$\overline{X}_{\text{in}} = \frac{(\overline{X}_L + \tan\beta d_s)(1 - \overline{X}_L \tan\beta d_s) - \overline{R}_L^2 \tan\beta d_s}{(\overline{R}_L \tan\beta d_s)^2 + (1 - \overline{X}_L \tan\beta d_s)^2} \tag{6.1.9}$$

To obtain a matched condition at d_s, the reactive part X_{in} must be eliminated by adding an element of opposite nature. Hence,

$$\overline{X}_s = -\overline{X}_{\text{in}} \tag{6.1.10}$$

Therefore, a capacitor will be needed in series if the input impedance is inductive. It will require an inductor if the input reactance is capacitive. As before, a transmission line stub can be used instead of an inductor or a capacitor. The length of this stub with an open circuit at its other end can be determined as follows:

$$l_s = \frac{1}{\beta}\cot(-\overline{X}_s) = \frac{1}{\beta}\cot\overline{X}_{\text{in}} \tag{6.1.11}$$

However, if the stub has a short circuit at its other end, its length will be a quarter-wavelength shorter (or longer, if the resulting number becomes negative) than this value. It can be found as

$$l_s = \frac{1}{\beta} \tan \overline{X}_s = \frac{1}{\beta} \tan(-\overline{X}_{\text{in}}) \tag{6.1.12}$$

Note that the location d_s and the stub length l_s are periodic in nature in both cases. It means that the matching conditions will also be satisfied at points one-half wavelength apart. However, the shortest possible values of d_s and l_s are preferred because those provide the matched condition over a broader frequency band.

Graphical Method

These matching networks can also be designed graphically using a Smith chart. The procedure is similar for both series- and shunt-connected elements, except that the former is based on the normalized impedance, whereas the latter works with normalized admittance. It can be summarized in the following steps:

1. Determine the normalized impedance of the load and locate that point on a Smith chart.
2. Draw the constant-VSWR circle. If the stub needs to be connected in parallel, move a quarter-wavelength away from the load impedance point. This point is located at the other end of the diameter that connects the load point with the center of the circle. For a series stub, stay at the normalized impedance point.
3. From the point found in step 2, move toward the generator (i.e., clockwise) on the VSWR circle until it intersects the unity resistance (or conductance) circle. The distance traveled to this intersection point from the load is equal to d_s. There will be at least two such points within one-half wavelength from the load. A matching element can be placed at either one of these points.
4. If the admittance in step 3 is $1 + j\overline{B}$, a susceptance of $-j\overline{B}$ in shunt is needed for matching. This can be a discrete reactive element (inductor or capacitor, depending on a negative or positive susceptance value) or a transmission line stub.
5. In the case of a stub, the length required is determined as follows. Since its other end will have an open or a short, the VSWR on it will be infinite. It is represented by the outermost circle of the Smith chart. Locate the desired susceptance point (i.e., $0 - j\overline{B}$) on this circle and then move toward load (counterclockwise) until an open circuit (i.e., a zero susceptance) or a short circuit (an infinite susceptance) is found. This distance is equal to the stub length l_s.

For a series reactive element or stub, steps 4 and 5 will be the same except that the normalized reactance replaces the normalized susceptance.

Example 6.1 A uniform, lossless 100-Ω line is connected to a load of $50 - j75\,\Omega$, as illustrated in Figure 6.4. A single stub of 100-Ω characteristic impedance is connected in parallel at a distance d_s from the load. Find the shortest values of d_s and stub length l_s for a match.

SOLUTION As mentioned in the preceding analysis, design equations (6.1.2), (6.1.3), (6.1.5), and (6.1.6) for a shunt stub use admittance parameters. On the other hand, the series-connected stub design uses impedance parameters in (6.1.8), (6.1.9), (6.1.11), and (6.1.12). Therefore, d_s and l_s can be determined theoretically, as follows:

$$\overline{Y}_L = \frac{Y_L}{Y_0} = \frac{Z_0}{Z_L} = \frac{100}{50 - j75} = 0.6154 + j0.9231$$

$$A = \overline{G}_L(\overline{G}_L - 1) + \overline{B}_L^2 = 0.6154(0.6154 - 1) + 0.9231^2 = 0.6154$$

From (6.1.2), the two possible values of d_s are

$$d_s = \frac{\lambda}{2\pi}\tan^{-1}\frac{-0.75 + \sqrt{(-0.75)^2 - 0.6154(1 - 0.5)}}{0.6154} = 0.1949\lambda$$

and

$$d_s = \frac{\lambda}{2\pi}\tan^{-1}\frac{-0.75 - \sqrt{(-0.75)^2 - 0.6154(1 - 0.5)}}{0.6154} = 0.0353\lambda$$

At 0.1949λ from the load, the real part of the normalized admittance is unity and its imaginary part is -1.2748. Hence, the stub should provide $j1.2748$ to cancel it out. Length of a short-circuited stub, l_s, is calculated from (6.1.5) as

$$l_s = \frac{1}{\beta}\cot^{-1}(-1.2748) = 0.3941\lambda$$

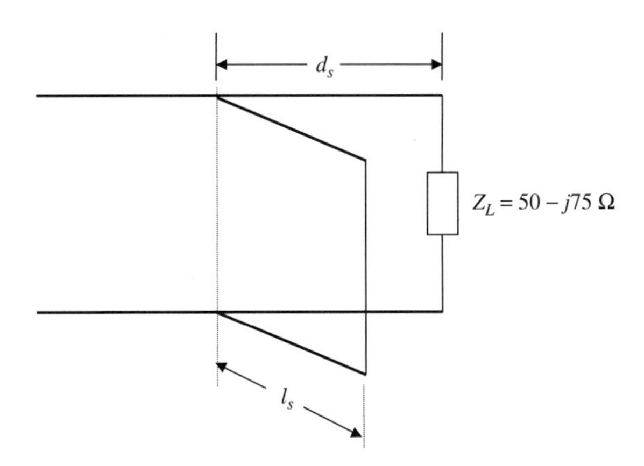

Figure 6.4 Shunt stub matching network.

On the other hand, normalized admittance is $1 + j1.2748$ at 0.0353λ from the load. To obtain a matched condition, the stub at this point must provide a normalized susceptance of $-j1.2748$. Hence,

$$l_s = \frac{1}{\beta}\cot^{-1}(1.2748) = 0.1059\lambda$$

Thus, there are two possible solutions to this problem. In one case, a short-circuited 0.3941λ-long stub is needed at 0.1949λ from the load. The other design requires a 0.1059λ-long short-circuited stub at 0.0353λ from the load. It is preferred over the former design because of its shorter lengths.

The following steps are needed to solve this example graphically with a Smith chart:

1. Determine the normalized load admittance.

$$\overline{Z}_L = \frac{50 - j75}{100} = 0.5 - j0.75$$

2. Locate the normalized load impedance point on the Smith chart. Draw the VSWR circle as shown in Figure 6.5.
3. From the load impedance point, move to the diametrically opposite point and locate the corresponding normalized load admittance. It is the point $0.62 + j0.91$ on the chart.
4. Locate the point on the VSWR circle where the real part of the admittance is unity. There are two such points with normalized admittance values $1 + j1.3$ (say, point A) and $1 - j1.3$ (say, point B), respectively.
5. The distance d_s of $1 + j1.3$ (point A) from the load admittance can be determined as 0.036λ (i.e., $0.170\lambda - 0.134\lambda$) and for point B ($1 - j1.3$) as 0.195λ (i.e., $0.329\lambda - 0.134\lambda$).
6. If a susceptance of $-j1.3$ is added at point A or $j1.3$ at point B, the load will be matched.
7. Locate the point $-j1.3$ along the lower circumference of the chart and from there move toward the load (counterclockwise) until the short circuit (infinity on the chart) is reached. Separation between these two points is $0.25\lambda - 0.146\lambda = 0.104\lambda$. Hence a 0.104λ-long transmission line with a short circuit at its rear end will have the desired susceptance for point A.
8. For a matching stub at point B, locate the point $j1.3$ on the upper circumference of the chart and then move toward the load up to the short circuit (i.e., the right-hand end of the chart). Hence, the stub length l_s for this case is determined as $0.25\lambda + 0.146\lambda = 0.396\lambda$.

Therefore, a 0.104λ-long stub at 0.036λ from the load (point A) or a 0.396λ-long stub at 0.195λ (point B) from the load will match the load. These values are comparable to those obtained earlier.

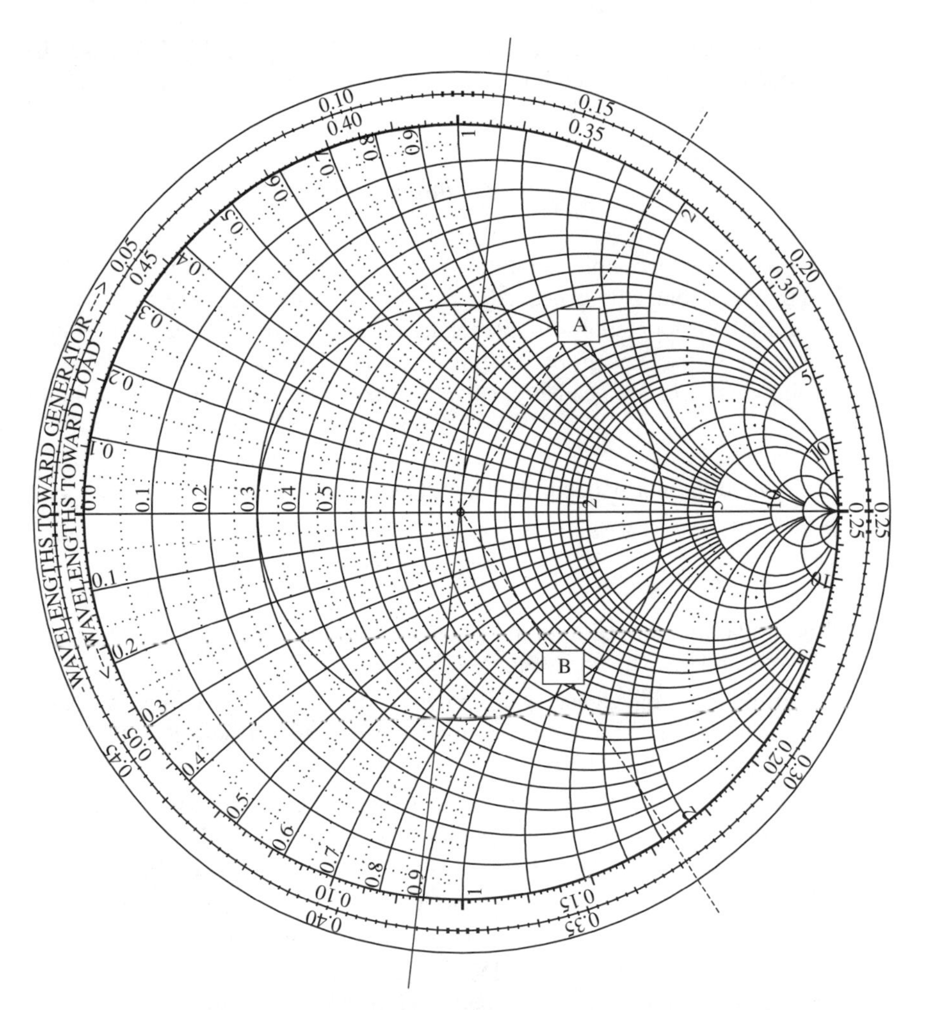

Figure 6.5 Graphical design of matching circuit for Example 6.1.

As mentioned earlier, point A is preferred over point B in matching the network design because it is closer to the load and the stub length is shorter in this case. To compare the frequency response of these two designs, the input reflection coefficient is calculated for the network. Its magnitude plot is shown in Figure 6.6. Since various lengths in the circuit are known in terms of wavelength, it is assumed that the circuit is designed for a signal wavelength of λ_d. As the signal frequency is changed, its wavelength changes to λ. The normalized wavelength used for this plot is equal to λ_d/λ. Since the wavelength is inversely related to the propagation constant, the horizontal scale may also be interpreted as a normalized frequency scale, with 1 being the design frequency.

Plot (a) in Figure 6.6 corresponds to design A (which requires a shorter stub closer to the load) and plot (b) represents design B (a longer stub and away from

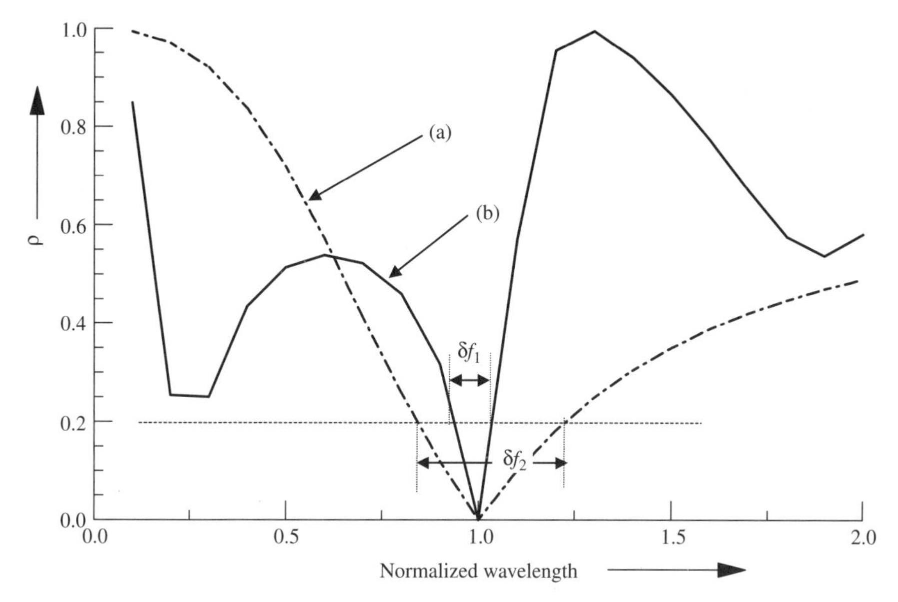

Figure 6.6 Magnitude of the reflection coefficient as a function of signal frequency.

the load). At the normalized wavelength of unity, both of these curves go to zero. As signal frequency is changed on either side (i.e., decreased or increased from this setting), reflection coefficient increases. However, this change in plot (a) is gradual in compared with that in plot (b). In other words, for an allowed reflection coefficient of 0.2, the bandwidth for design A is δf_2, which is much wider than the δf_1 of design B.

Example 6.2 A lossless 100-Ω line is to be matched with a $100/(2 + j3.732)$-Ω load by means of a lossless short-circuited stub, as shown in Figure 6.7. The characteristic impedance of the stub is 200 Ω. Find the position closest to the load and the length of the stub using a Smith chart.

SOLUTION

1. In this example it will be easier to determine the normalized load admittance directly, as follows:

$$\overline{Y}_L = \frac{1}{\overline{Z}_L} = \frac{Z_0}{Z_L} = 2 + j3.732$$

2. Locate this normalized admittance point on the Smith chart and draw the VSWR circle. It is illustrated in Figure 6.8.
3. Move toward the generator (clockwise) on the VSWR circle until the real part of the admittance is unity. One such point is $1 - j2.7$ and the other is

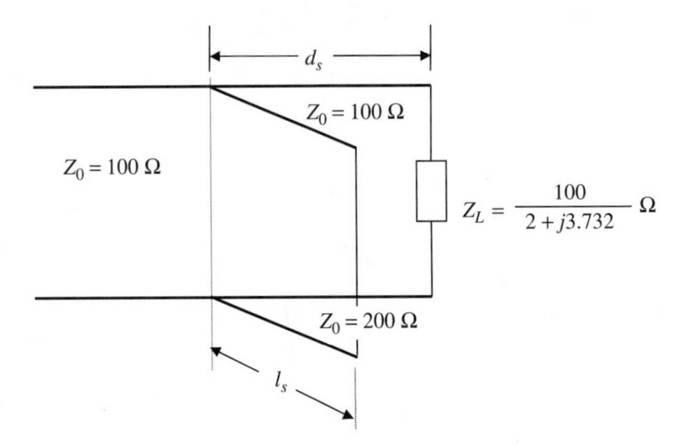

Figure 6.7 Matching circuit for Example 6.2.

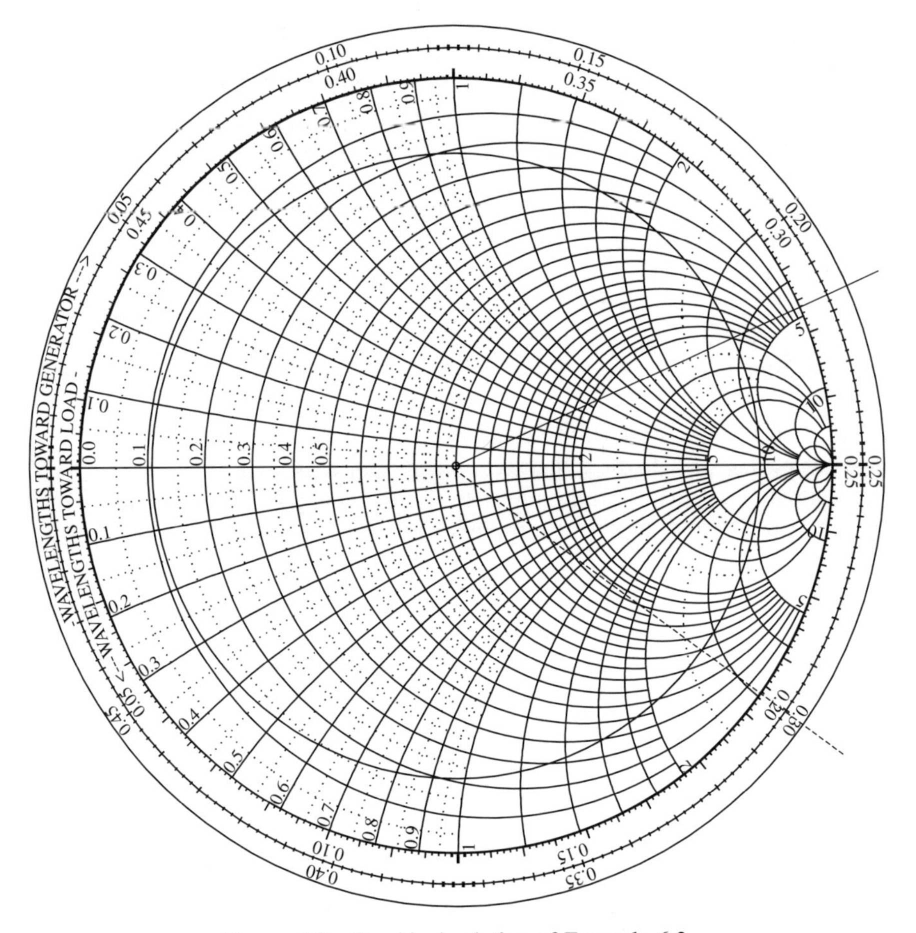

Figure 6.8 Graphical solution of Example 6.2.

$1 + j2.7$. Since the former is closer to the load, the stub must be placed at this point. Hence,

$$d_s = (0.3 - 0.217)\lambda = 0.083\lambda$$

4. The normalized susceptance needed for matching at this point is $j2.7$. However, it is normalized with 100 Ω, while characteristic impedance of the stub is 200 Ω. This means that the normalization must be corrected before determining the stub length l_s. It can be done as follows:

$$j\overline{B}_s = \frac{j2.7 \times 200}{100} = j5.4$$

5. Point $j5.4$ is located on the upper scale of the Smith chart. Moving from this point toward the load (i.e., counterclockwise), open-circuit admittance (zero) is found first. Moving farther in the same direction, the short-circuit admittance point is found next. This means that the stub length will be shorter with an open circuit at its other end. However, a short-circuited stub is used in Figure 6.7. Hence,

$$l_s = 0.22\lambda + 0.25\lambda = 0.47\lambda$$

Example 6.3 A load reflection coefficient is given as $0.4 \angle -30°$. It is desired to get a load with reflection coefficient at $0.2 \angle 45°$. Two different circuits are shown in Figure 6.9. However, the information provided for these circuits is incomplete. Complete or verify the designs at 4 GHz. Assume that the characteristic impedance is 50 Ω.

SOLUTION This example can be solved using equations (3.2.4) and (3.2.6). Alternatively, a graphical procedure can be adopted. Both of these methods are illustrated here. From (3.2.4), the given load and the desired normalized

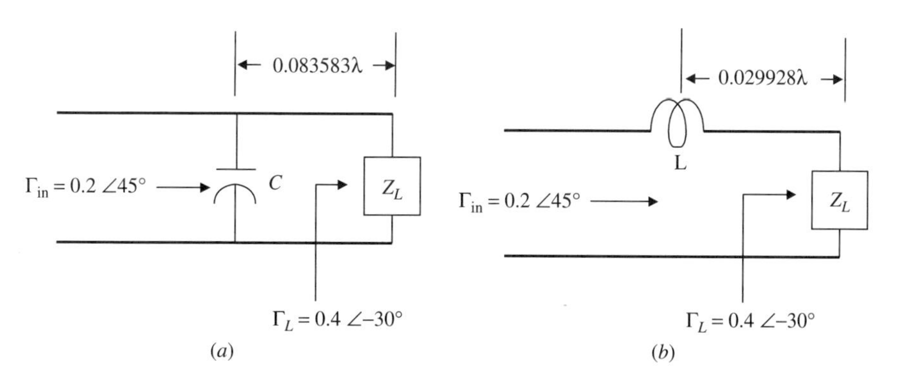

Figure 6.9 Circuit designs for Example 6.3.

impedance or admittance can be calculated as follows. The normalized load impedance is

$$\overline{Z}_L = \frac{1+\Gamma_L}{1-\Gamma_L} = \frac{1+0.4\ \angle{-30^\circ}}{1-0.4\ \angle{-30^\circ}} = 1.7980 - j0.8562$$

The desired normalized input impedance is

$$\overline{Z}_{\text{in}} = \frac{1+\Gamma_{\text{in}}}{1-\Gamma_{\text{in}}} = \frac{1+0.2\ \angle 45^\circ}{1-0.2\ \angle 45^\circ} = 1.2679 + j0.3736$$

and the corresponding normalized input admittance is

$$\overline{Y}_{\text{in}} = \frac{1-\Gamma_{\text{in}}}{1+\Gamma_{\text{in}}} = \frac{1-0.2\ \angle 45^\circ}{1+0.2\ \angle 45^\circ} = 0.7257 - j0.2138$$

From (3.2.6), the normalized input admittance at $l = 0.0836\lambda$ from the load is

$$\overline{Y}_{\text{in}} = \frac{1}{\overline{Z}_{\text{in}}} = \frac{1 + j\overline{Z}_L \tan\beta l}{\overline{Z}_L + j\tan\beta l} = 0.7257 + j0.6911$$

Hence, the real part of this admittance is equal to the value desired. However, its imaginary part is off by $-j0.9049$. A negative susceptance is inductive, while the given circuit has a capacitor that adds a positive susceptance. Therefore, the desired reflection coefficient cannot be obtained by the circuit given in Figure 6.9(a).

In the circuit shown in Figure 6.9(b), components are connected in series. Therefore, it will be easier to solve this problem using impedance than by using admittance. From (3.2.6), the normalized impedance at $l_2 = 0.0299\lambda$ from the load is

$$\overline{Z}_{\text{in}} = \frac{\overline{Z}_L + j\tan\beta l_2}{1 + j\overline{Z}_L \tan\beta l_2} = 1.2679 - j0.9456$$

Hence, its real part is equal to the value desired. However, its imaginary part needs modification by $j1.3192$ to get $j0.3736$. Hence, an inductor is required at this point. The circuit given in Figure 6.9(b) has a series inductor. Therefore, this circuit will have the desired reflection coefficient provided that its value is

$$L = \frac{1.3192 \times 50}{2\pi \times 4 \times 10^9}\text{H} = 2.6245 \times 10^{-9}\text{H} \approx 2.62\,\text{nH}$$

Figure 6.10 illustrates the graphical procedure for solving this example using a Smith chart. VSWR circles are drawn for the reflection coefficient magnitudes given. Using the phase angle of the load reflection coefficient, the normalized load impedance point is identified as $1.8 - j0.85$, which is close to the value calculated earlier. The process is repeated for the desired input reflection coefficient, and

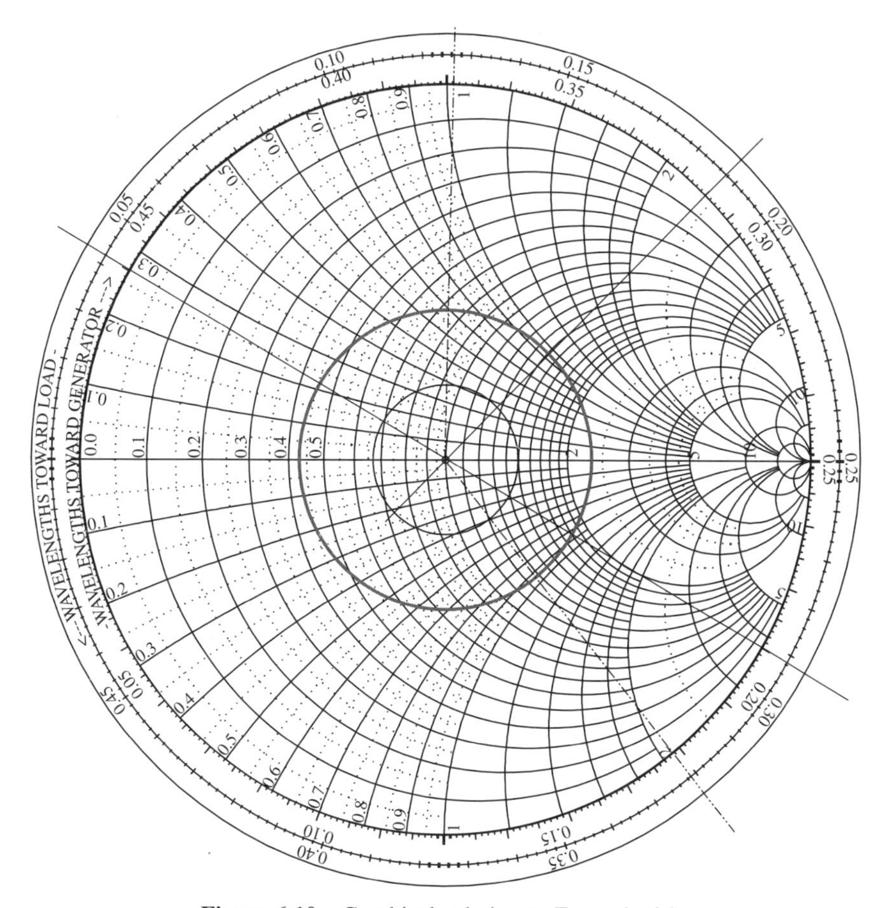

Figure 6.10 Graphical solution to Example 6.3.

the corresponding input impedance point is identified as $1.27 + j0.37$. For the circuit given in Figure 6.9(a), the admittance (normalized) points are found as $0.45 + j0.22$ and $0.73 - j0.21$, respectively. Next, move from the normalized load admittance point toward the generator by a distance of 0.083583λ (i.e., $0.042\lambda + 0.084\lambda = 0.126\lambda$ of the scale "wavelengths toward generator"). The normalized admittance value of this point is found to be $0.73 + j0.69$. Hence, its real part is the same as that of the desired input admittance. However, its susceptance is $j0.69$, whereas the desired value is $-j0.2$. Hence, an inductor will be needed in parallel at this point. Since the given circuit has a capacitor, this design is not possible.

For the circuit shown in Figure 6.9(b), elements are connected in series. Therefore, normalized impedance points need to be used in this case. Move from the load impedance ($1.8 - j0.85$) point toward the generator by a distance of 0.029928λ (i.e., $0.292\lambda + 0.03\lambda = 0.322\lambda$ on the "wavelengths toward generator" scale). Normalized impedance value at this point is $1.27 - j0.95$. Thus, the resistance at this point is found to be equal to the value desired. However, its reactance is $-j0.95$, whereas the required value is $j0.37$. Therefore, a series reactance of $j1.32$ is needed at this point. The given circuit has an inductor that provides a positive reactance. Hence, this circuit will work. The required inductance L is found as

$$L = \frac{1.32 \times 50}{2\pi \times 4 \times 10^9} \text{H} = 2.626\,\text{nH}$$

6.2 DOUBLE-STUB MATCHING NETWORKS

The matching technique presented in Section 5.1 requires that a reactive element or stub be placed at a precise distance from the load. This point will shift with load impedance. Sometimes it may not be feasible to match the load using a single reactive element. Another possible technique to match the circuit employs two stubs with a fixed separation between them. This device can be inserted at a convenient point before the load. The impedance is matched by adjusting the lengths of the two stubs. Of course, it does not provide a universal solution. As will be seen later in this section, separation between the two stubs limits the range of load impedance that can be matched with a given double-stub tuner.

Let l_1 and l_2 be the lengths of two stubs, as shown in Figure 6.11. The first stub is located at a distance l from the load, $Z_L = R_L + jX_L$ ohms. Separation between the two stubs is d, and the characteristic impedance of every transmission line is Z_0. In double-stub matching, the load impedance Z_L is transformed to the normalized admittance at the location of the first stub. Since the stub is connected in parallel, its normalized susceptance is added to that and the resulting normalized admittance is transferred to the location of the second stub. Matching conditions at this point require that the real part of this normalized admittance

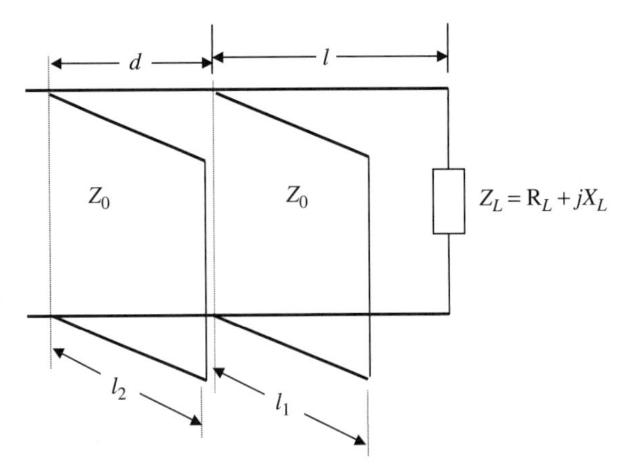

Figure 6.11 Double-stub matching network.

be equal to unity while its imaginary part is canceled by a conjugate susceptance of the second stub. Mathematically,

$$\frac{\overline{Y} + j(\overline{B}_1 + \tan\beta d)}{1 + j(\overline{Y} + j\overline{B}_1)\tan\beta d} + j\overline{B}_2 = 1 \tag{6.2.1}$$

where

$$\overline{Y} = \frac{1 + j\overline{Z}_L \tan\beta l}{\overline{Z}_L + j\tan\beta l} = \frac{\overline{Y}_L + j\tan\beta l}{1 + j\overline{Y}_L \tan\beta l} = \overline{G} + j\overline{B} \tag{6.2.2}$$

jB_1 and jB_2 are the susceptance of the first and second stubs, respectively, and β is the propagation constant over the line. For

$$\text{Re}\left[\frac{\overline{Y} + j(\overline{B}_1 + \tan\beta d)}{1 + j(\overline{Y} + j\overline{B}_1)\tan\beta d}\right] = 1$$

$$\overline{G}^2 \tan^2\beta d - \overline{G}(1 + \tan^2\beta d) + [1 - (\overline{B} + \overline{B}_1)\tan\beta d]^2 = 0 \tag{6.2.3}$$

Since conductance of the passive network must be a positive quantity, (6.2.3) requires that a given double stub can be used for matching only if the following condition is satisfied:

$$0 \le \overline{G} \le \csc^2\beta d \tag{6.2.4}$$

Two possible susceptances of the first stub that can match the load are determined by solving (6.2.3):

$$\overline{B}_1 = \cot\beta d\left[1 - \overline{B}\tan\beta d \pm \sqrt{\overline{G}\sec^2\beta d - (\overline{G}\tan\beta d)^2}\right] \tag{6.2.5}$$

Normalized susceptance of the second stub is determined from (6.2.1) as

$$\overline{B}_2 = \frac{\overline{G}^2 \tan\beta d - (\overline{B} + \overline{B}_1 + \tan\beta d)[1 - (\overline{B} + \overline{B}_1)\tan\beta d]}{(\overline{G}\tan\beta d)^2 + [1 - (\overline{B} + \overline{B}_1)\tan\beta d]^2} \tag{6.2.6}$$

Once the susceptance of a stub is known, its short-circuit length can be determined easily as follows:

$$l_1 = \frac{1}{\beta}\cot^{-1}(-\overline{B}_1) \tag{6.2.7}$$

and

$$l_2 = \frac{1}{\beta}\cot^{-1}(-\overline{B}_2) \tag{6.2.8}$$

Graphical Method

A two-stub matching network can also be designed graphically with the help of a Smith chart. This procedure follows the analytical concepts discussed above and can be summarized as follows:

1. Locate the normalized load-impedance point on a Smith chart and draw the VSWR circle. Move to the corresponding normalized admittance point. If the load is connected right at the first stub, go to the next step; otherwise, move toward the generator (clockwise) by $2\beta l$ on the VSWR circle. Assume that the normalized admittance of this point is $g + jb$.
2. Rotate the unity-conductance circle counterclockwise by $2\beta d$. The conductance circle that touches this circle encloses the "forbidden region." In other words, this tuner can match only those $\overline{Y}$ that lie outside this circle. It is a graphical representation of the condition expressed by (6.2.4).
3. From $g + jb$, move on the constant-conductance circle until it intersects the rotated unity-conductance circle. There are at least two such points, providing two design possibilities. Let the normalized admittance of one of these points be $g + jb_1$.
4. The normalized susceptance required of the first stub is $j(b_1 - b)$.
5. Draw a VSWR circle through point $g + jb_1$ and move toward the generator (clockwise) by $2\beta d$ on it. This point will fall on the unity-conductance circle of the Smith chart. Assume that this point is $1 + jb_2$.
6. The susceptance required from the second stub is $-jb_2$.
7. Once the stub susceptances are known, their lengths can be determined following the procedure used in the technique described earlier.

Example 6.4 For the double-stub tuner shown in Figure 6.12, find the shortest values of l_1 and l_2 to match the load.

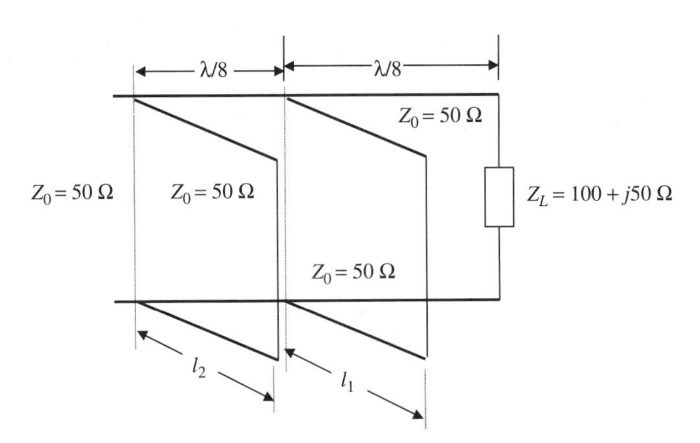

Figure 6.12 Two-stub matching network for Example 6.4.

SOLUTION Since the two stubs are separated by $\lambda/8$, βd is equal to $\pi/4$ and the condition (6.2.4) gives

$$0 \leq \overline{G} \leq 2$$

This means that the real part of the normalized admittance at the first stub (load-side stub) must be less than 2; otherwise, it cannot be matched.

The graphical procedure requires the following steps to find stub settings:

1. $\overline{Z}_L = \dfrac{100 + j50}{50} = 2 + j1$
2. Locate the normalized load impedance on a Smith chart (point A in Figure 6.13) and draw the VSWR circle. Move to the diametrically opposite side and locate the corresponding normalized admittance point B at $0.4 - j0.2$.

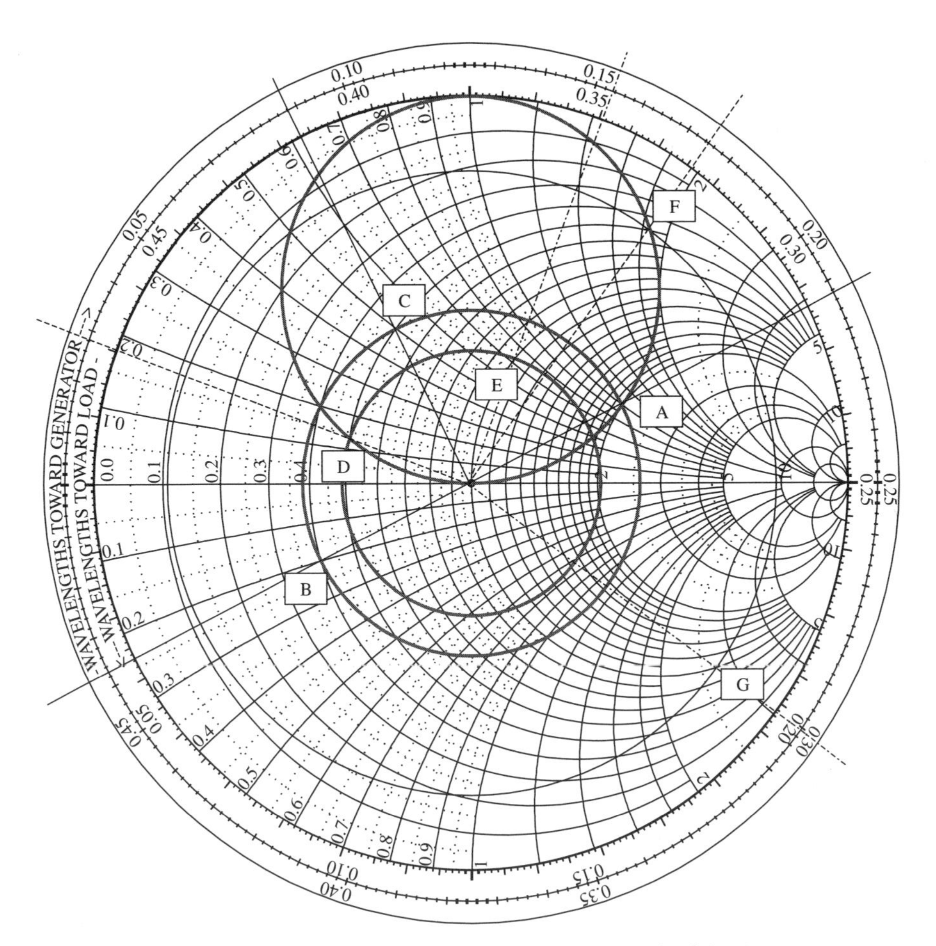

Figure 6.13 Smith chart solution to Example 6.4.

3. Rotate the unity-conductance circle counterclockwise by $2\beta d = \pi/2 = 90^\circ$. This shows that this tuner can only match the admittance with a real part of less than 2 (because it touches the constant-conductance circle of 2).
4. Move clockwise from the point $0.4 - j0.2$ (0.463λ on the "wavelengths toward generator" scale) by a distance of $2\beta l = \pi/2 = 90^\circ$ or $\lambda/8$ on the VSWR circle and locate point C at 0.088λ ($0.463\lambda + 0.125\lambda = 0.588\lambda$) as the normalized admittance $0.5 - j0.5$ of the load transferred to the first stub's location.
5. From point C, move on the constant-conductance circle until it intersects the rotated unity-conductance circle. There are two such points, D and F. If point D is used for the design, the susceptance of the first stub must be equal to $-j0.37$ (i.e., $j0.13 - j0.5$). On the other hand, it will be $j1.36$ (i.e., $j1.86 - j0.5$) for point F.
6. Locate point $-j0.37$ and move toward the load along the circumference of the Smith chart until you reach the short circuit (infinite susceptance). It gives $l_1 = 0.194\lambda$. Similarly, l_1 is found to be 0.4λ for $j1.36$.
7. Draw the VSWR circle through point D and move on it by 0.125λ (i.e., point 0.154λ on the "wavelengths toward generator" scale). The real part of the admittance at this point (point E) is unity, and its susceptance is $j0.72$. Therefore, the second stub must be set for $-j0.72$ if the first one is set for $-j0.37$. Hence, the second stub is required to be 0.15λ long (i.e., $l_2 = 0.15\lambda$).
8. Draw the VSWR circle through point F and move on it by 0.125λ (0.3λ on the "wavelengths toward generator" scale). The real part of the admittance at this point (point G) is unity and its susceptance is $-j2.7$. Therefore, the second stub must be set for $j2.7$ if the first one is set for $j1.36$. In this case, the length of the second stub is found as 0.442λ (i.e., $l_2 = 0.442\lambda$).
9. Hence, the length of the first stub should be equal to 0.194λ and that of the second stub should be 0.15λ. The other possible design, where the respective lengths are found to be 0.4λ and 0.442λ, is not recommended.

Alternatively, the required stub susceptances, and hence lengths l_1 and l_2, can be calculated from (6.2.2) to (6.2.8) as follows:

$$\overline{Y} = \overline{G} + j\overline{B} = 0.5 + j0.5$$

$$\overline{B}_1 = j1.366 \quad \text{or} \quad -j0.366$$

For $\overline{B}_1 = j1.366$, $\overline{B}_2 = j2.7321$, and for $\overline{B}_1 = -j0.366$, $\overline{B}_2 = -j0.7321$. The corresponding lengths are found to be 0.3994λ, 0.4442λ, 0.1942λ, and 0.1494λ. These values are fairly close to those obtained graphically from the Smith chart.

6.3 MATCHING NETWORKS USING LUMPED ELEMENTS

The matching networks described so far may not be useful for certain applications. For example, the wavelength of a 100-MHz signal is 3 m, and therefore it may not be practical to use stub matching in this case because of its size on a printed circuit board. In this section we present matching networks utilizing discrete components that can be useful, especially in such cases. Two different types of L-section matching circuits are described here. The section begins with resistive matching circuits that can be used for broadband applications. However, these networks dissipate signal energy and introduce thermal noise. The section ends with a presentation of the reactive matching networks that are almost lossless, but the design is frequency dependent.

Resistive L-Section Matching Circuits

Consider a signal generator with internal resistance R_s that feeds a load resistance R_L as illustrated in Figure 6.14. Since source resistance is different from the load, a part of the signal is reflected back. Assume that R_s is larger than R_L and there is a resistive L-section introduced between the two. Further, voltages at its input and output ports are assumed to be V_{in} and V_o, respectively. If this circuit is matched at both its ports, the following two conditions must be true:

1. With R_L connected, resistance looking into the input port must be R_s.
2. With R_s terminating the input port, resistance looking into the output port must be R_L.

From the first condition,

$$R_s = R_1 + \frac{R_2 R_L}{R_2 + R_L} = \frac{R_1 R_2 + R_1 R_L + R_2 R_L}{R_2 + R_L} \tag{6.3.1}$$

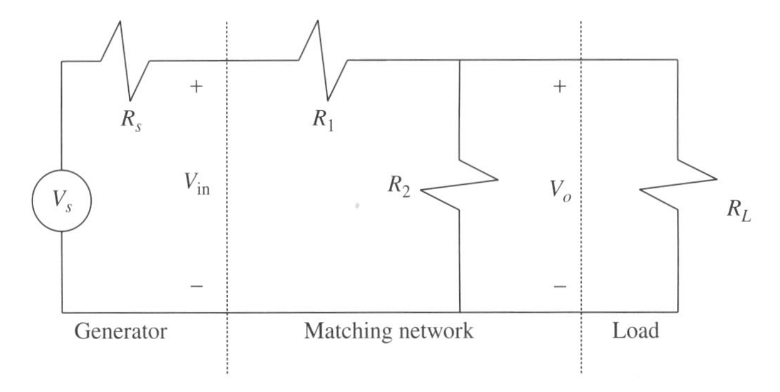

Figure 6.14 Resistive L-section matching circuit.

and from the second,

$$R_L = \frac{R_2(R_1 + R_s)}{R_2 + R_1 + R_s} = \frac{R_1R_2 + R_2R_s}{R_2 + R_1 + R_s} \tag{6.3.2}$$

Equations (6.3.1) and (6.3.2) can be solved for R_1 and R_2 as follows:

$$R_1 = \sqrt{R_s(R_s - R_L)} \tag{6.3.3}$$

and

$$R_2 = \sqrt{\frac{R_L^2 R_s}{R_s - R_L}} \tag{6.3.4}$$

The voltage across the load and the attenuation in the matching network are

$$V_o = \frac{V_{\text{in}}}{R_1 + R_2 \| R_L} R_2 \| R_L \Rightarrow \frac{V_o}{V_{\text{in}}} = \frac{R_2 R_L}{R_1 R_2 + R_1 R_L + R_2 R_L} \tag{6.3.5}$$

and

$$\text{attenuation (dB)} = 20 \log \frac{R_2 R_L}{R_1(R_2 + R_L) + R_2 R_L} \tag{6.3.6}$$

Note that R_1 and R_2 are real only when R_s is greater than R_L. If this condition is not satisfied (i.e., $R_s < R_L$), the circuit shown in Figure 6.14 will require modification. Flipping the L-section the other way around (i.e., R_1 connected in series with R_L and R_2 across the input) will match the circuit. Design equations for that case can easily be obtained.

Example 6.5 Internal resistance of a signal generator is 75 Ω. If it is being used to excite a 50-Ω transmission line, design a resistive network to match the two. Calculate the attenuation in decibels that occurs in the inserted circuit.

SOLUTION Since $R_L = 50\,\Omega$ and $R_s = 75\,\Omega$, R_s is greater than R_L, and therefore the circuit shown in Figure 6.14 can be used.

$$R_1 = \sqrt{75(75 - 50)} = 43.3\,\Omega$$

$$R_2 = \sqrt{\frac{75 \times 50^2}{75 - 50}} = 86.6\,\Omega$$

and

$$\text{attenuation (dB)} = 20 \log \frac{86.6 \times 50}{43.3(86.6 + 50) + 86.6 \times 50}$$
$$= 20 \log(0.4227) = -7.48\,\text{dB}$$

The final circuit arrangement is shown in Figure 6.15.

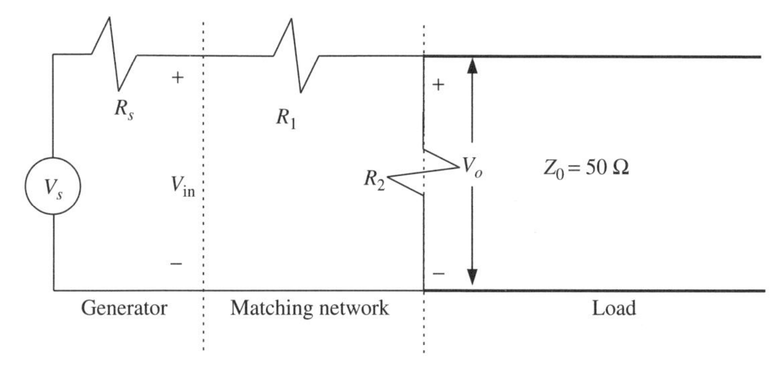

Figure 6.15 Resistive matching circuit for Example 6.5.

Reactive L-Section Matching Circuits

As mentioned earlier, resistive matching circuits are frequency insensitive but dissipate a part of the signal power that affects the signal-to-noise ratio adversely. Here we consider an alternative design using reactive components. In this case, power dissipation is ideally zero, but the matching is frequency dependent.

Consider the two circuits shown in Figure 6.16. In one of these circuits, resistor R_s is connected in series with a reactance X_s, while in the other, resistor R_p is connected in parallel with a reactance X_p. If impedance of one circuit is a complex conjugate of the other, then

$$|R_s + jX_s| = \left| \frac{jX_p R_p}{R_p + jX_p} \right|$$

or

$$\sqrt{R_s^2 + X_s^2} = \frac{X_p R_p}{\sqrt{R_p^2 + X_p^2}} \tag{6.3.7}$$

The quality factor, Q, of a reactive circuit was defined in Chapter 5 as follows:

$$Q = \omega \frac{\text{energy stored in the network}}{\text{average power loss}}$$

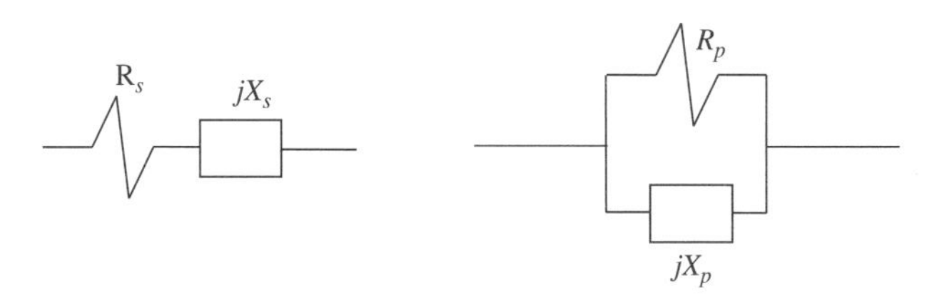

Figure 6.16 Series- and parallel-connected impedance circuits.

where ω is the angular frequency of the signal. For a series circuit,

$$Q = \frac{X_s}{R_s} \tag{6.3.8}$$

and for a parallel circuit,

$$Q = \frac{R_p}{X_p} \tag{6.3.9}$$

Assuming that the quality factors of these two circuits are equal, (6.3.7) can be simplified as follows:

$$\sqrt{R_s^2 + R_s^2 Q^2} = \frac{(R_p/Q)R_p}{\sqrt{R_p^2 + (R_p/Q)^2}} \Rightarrow R_s\sqrt{1+Q^2} = \frac{R_p}{\sqrt{1+Q^2}}$$

Hence,

$$1 + Q^2 = \frac{R_p}{R_s} \tag{6.3.10}$$

The design procedure is based on equations (6.3.8) to (6.3.10). For a resistive load to be matched with another resistor (it may be a transmission line or generator), R_p and R_s are defined such that the former is greater. Q of the circuit is then calculated from (6.3.10). Respective reactances are subsequently determined from (6.3.8) and (6.3.9). If one reactance is selected capacitive, the other must be inductive. X_p will be connected in parallel with R_p, and X_s will be in series with R_s. The following example illustrates the procedure.

Example 6.6 Design a reactive L-section that matches a 600-Ω resistive load to a 50-Ω transmission line. Determine the component values if the matching is desired at 400 MHz.

SOLUTION Since R_p must be larger than R_s, the 600-Ω load is selected as R_p and a 50-Ω line as R_s. Hence, from (6.3.10) we have

$$Q^2 + 1 = \frac{R_p}{R_s} = \frac{600}{50} = 12 \Rightarrow Q = \sqrt{11} = 3.3166$$

Now, from (6.3.8) and (6.3.9),

$$X_s = QR_s = 3.3166 \times 50 = 165.8312\,\Omega$$

$$X_p = \frac{R_p}{Q} = \frac{600}{3.3166} = 180.9068\,\Omega$$

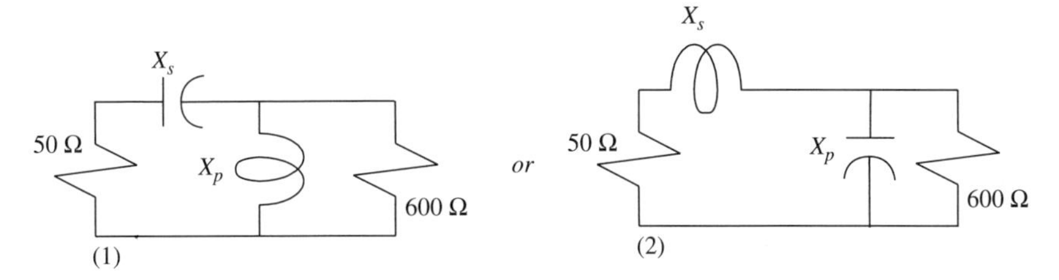

Figure 6.17 Design of two reactive matching circuits for Example 6.6.

Therefore, either one of the two circuits in Figure 6.17 can be used to match the load with the line.

For circuit (1) in Figure 6.17, the component values are determined as follows:

$$X_s = 165.8312 = \frac{1}{\omega C_s}$$

Therefore,

$$C_s = \frac{1}{2\pi \times 400 \times 10^6 \times 165.8312} \approx 2.4 \times 10^{-12}\ \mathrm{F} = 2.4\,\mathrm{pF}$$

and

$$X_p = 180.9068 = \omega L_p$$

Therefore,

$$L_p = \frac{180.9068}{2\pi \times 400 \times 10^6} = 71.9805 \times 10^{-9}\ \mathrm{H} \approx 72\,\mathrm{nH}$$

Similarly, the component values for circuit (2) in Figure 6.17 are

$$X_s = 165.8312 = \omega L_s$$

Therefore,

$$L_s = \frac{165.8312}{2\pi \times 400 \times 10^6} = 65.9821 \times 10^{\ 9}\ \mathrm{H} \approx 66\,\mathrm{nH}$$

and

$$X_p = 180.9068 = \frac{1}{\omega C_p}$$

Therefore,

$$C_p = \frac{1}{2\pi \times 400 \times 10^6 \times 180.9068} = 2.2 \times 10^{-12}\ \mathrm{F} = 2.2\,\mathrm{pF}$$

Example 6.7 Consider Example 6.6 again and transform a 600-Ω resistive load to a 173.2-Ω load by adjusting the Q value of a matching circuit. Continue the transformation process to get 50 Ω from 173.2 Ω. Compare the frequency response (reflection coefficient versus frequency) of your circuits.

SOLUTION For the first part of this example, R_p is 600 Ω and R_s is 173.2 Ω. Hence, from (6.3.8) to (6.3.10),

$$Q = \sqrt{\frac{R_p}{R_s} - 1} = \sqrt{\frac{600}{173.2} - 1} = 1.569778$$

$$X_s = QR_s = 271.8856\,\Omega$$

and

$$X_p = \frac{R_p}{Q} = 382.2196\,\Omega$$

Repeating the calculations with $R_p = 173.2\,\Omega$ and $R_s = 50\,\Omega$, for the second part of the example we get

$$Q = \sqrt{\frac{R_p}{R_s} - 1} = \sqrt{\frac{173.2}{50} - 1} = 1.5697$$

This value of Q is very close to that obtained earlier. This is because 173.2 is close to the geometric mean of 600 and 50. Hence,

$$X_s = QR_s = 78.4857\,\Omega$$

and

$$X_p = \frac{R_p}{Q} = 110.3386\,\Omega$$

Two of the possible circuits (3) and (4) are shown in Figure 6.18 along with their component values. For these circuits, the impedance that the 50-Ω transmission line sees can be determined. The reflection coefficient is then determined for each case. This procedure is repeated for the circuits obtained in Example 6.6 as well. Magnitude of the reflection coefficient for each of these four cases is displayed in Figure 6.19. Curves (a) and (b) are obtained for circuits (1) and (2) of Figure 6.17, respectively. The frequency response of circuit (3) in Figure 6.18 is illustrated by curve (c) of Figure 6.19, while curve (d) displays the frequency response of circuit (4). The horizontal axis of Figure 6.19 represents the normalized frequency (i.e., the signal frequency divided by 400 MHz).

As Figure 6.19 indicates, the reflection coefficient is zero for all four circuits at a normalized frequency of unity. However, it increases if the signal frequency is changed on either side. Further, the rate of increase in reflection is higher for circuits obtained in Example 6.6. For example, if a reflection coefficient of

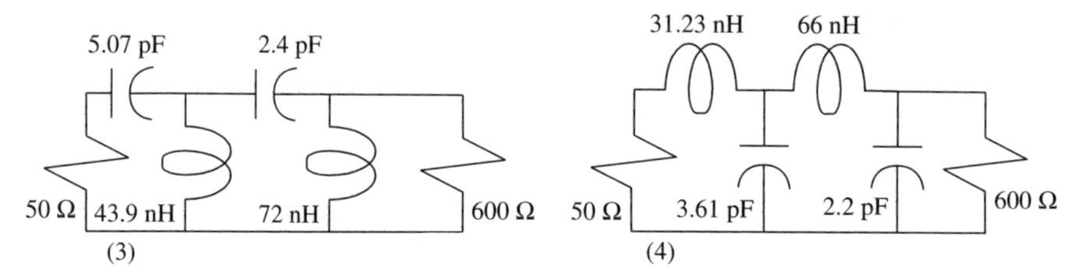

Figure 6.18 Design of reactive matching networks for Example 6.7.

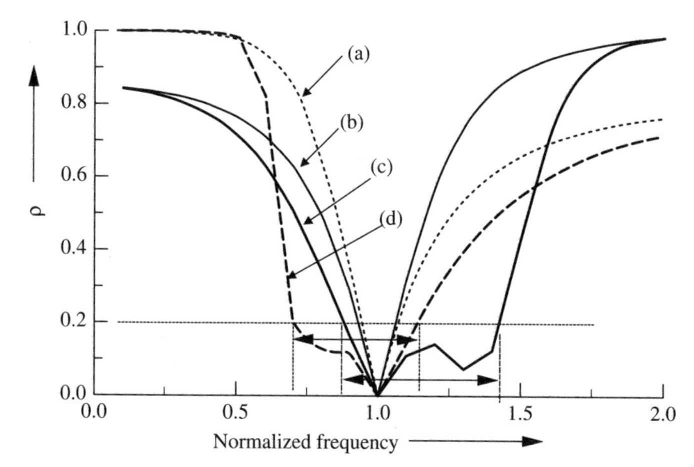

Figure 6.19 Reflection coefficient (magnitude) versus normalized frequency for the matching circuits designed in Examples 6.6 and 6.7.

0.2 is acceptable, circuits (3) and (4) can provide a much wider bandwidth than those of Example 6.6. This concept can be used to shape the reflection coefficient characteristics over the desired frequency band.

Example 6.8 A 50-Ω transmission line is to be matched with a $10 + j10$-Ω load. Design two different L-section reactive circuits and find the component values at 500 MHz.

SOLUTION Unlike the cases considered so far, the load is complex in this example. However, the same design procedure is still applicable. We consider only the real part of load impedance initially and take its imaginary part into account later in the design. Since characteristic impedance of the transmission line is greater than the real part of the load (i.e., $50\,\Omega > 10\,\Omega$), R_p is 50 Ω while R_s is 10 Ω. Therefore,

$$Q^2 = \frac{50}{10} - 1 = 4 \Rightarrow Q = 2$$

X_s and X_p can now be determined from (6.3.8) and (6.3.9):

$$X_s = QR_s = 2 \times 10 = 20\,\Omega$$

and

$$X_p = \frac{R_p}{Q} = \frac{50}{2} = 25\,\Omega$$

Reactance X_p is connected in parallel with the transmission line and X_s is in series with the load. These two matching circuits are shown in Figure 6.20. If X_p is a capacitive reactance of $25\,\Omega$, X_s must be inductive $20\,\Omega$. The reactive part that has not been taken into account so far needs to be included at this point in X_s. Since this reactive part of the load is inductive $10\,\Omega$, another series inductor for the remaining $10\,\Omega$ is needed, as shown in circuit (1) of Figure 6.20. Hence,

$$\omega L_s = 20 - 10$$

Therefore,

$$L_s = \frac{10}{2\pi \times 500 \times 10^6}\,\text{H} = 0.3183 \times 10^{-8}\ \text{H} = 3.183\,\text{nH}$$

and

$$\frac{1}{\omega C_p} = 25$$

Therefore,

$$C_p = \frac{1}{2\pi \times 500 \times 10^6 \times 25} = 0.0127324 \times 10^{-9}\ \text{F} = 12.7324\,\text{pF}$$

In the second case, X_p is assumed inductive and therefore X_s needs to be capacitive. It is circuit (2) in Figure 6.20. Since the load has a 10-Ω inductive reactance,

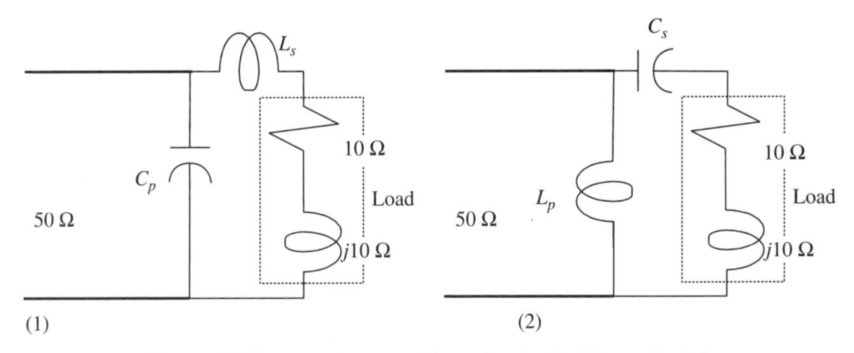

Figure 6.20 Reactive matching circuits for Example 6.8.

it must be taken into account in determining the required capacitance. Therefore,

$$\omega C_s = \frac{1}{20+10} \Rightarrow C_s = \frac{1}{2\pi \times 500 \times 10^6 \times 30} \text{F} = 10.61\,\text{pF}$$

and

$$\omega L_p = 25 \Rightarrow L_p = \frac{25}{2\pi \times 500 \times 10^6} \text{H} = 7.9577\,\text{nH}$$

The matching procedure used so far is based on the transformation of series impedance to a parallel circuit of the same quality factor. Similarly, one can use an admittance transformation, as shown in Figure 6.21.

These design equations can be obtained easily following the procedure used earlier:

$$\frac{G_s}{G_p} = 1 + Q^2 \tag{6.3.11}$$

$$Q = \frac{G_s}{B_s} \tag{6.3.12}$$

and

$$Q = \frac{B_p}{G_p} \tag{6.3.13}$$

Note from (6.3.11) that G_s must be greater than G_p for a real Q.

Example 6.9 A load admittance, $Y_L = 8 - j12$ mS, is to be matched with a 50-Ω line. There are three different L-section matching networks given to you, as shown in Figure 6.22. Complete or verify each of these circuits. Find the element values at 1 GHz.

SOLUTION Since the characteristic impedance of the line is 50 Ω, the corresponding conductance will be 0.02 S. The real part of the load admittance is 0.008 S. Therefore, G_s must be 0.02 S in (6.3.11). However, the circuits in Figure 6.22 have a capacitor or an inductor connected in parallel with the 50-Ω line. Obviously, these design equations cannot be used here. Equations (6.3.8)

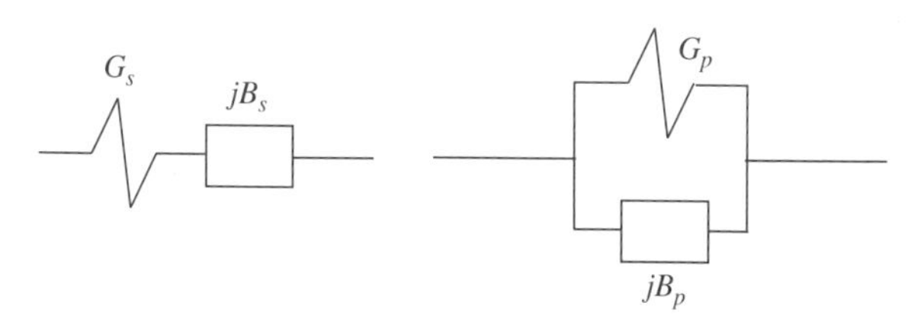

Figure 6.21 Series- and parallel-connected admittance circuits.

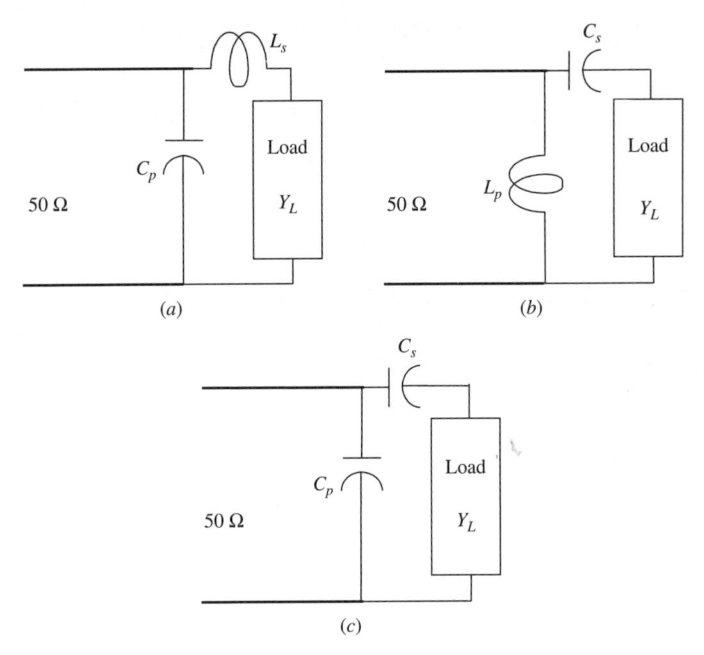

Figure 6.22 Three possible L-section matching circuits given in Example 6.9.

to (6.3.10) use the impedance values. Hence, we follow the procedure outlined below:

$$Y_L = \frac{1}{Z_L} \Rightarrow Z_L = \frac{10^3}{8 - j12}\,\Omega = 38.4615 + j57.6923\,\Omega$$

Therefore,

$$R_p = 50\,\Omega \quad \text{and} \quad R_s = 38.4615\,\Omega \Rightarrow Q = \sqrt{\frac{50}{38.4615} - 1} = 0.5477$$

Now, from (6.3.8) and (6.3.9),

$$X_s = QR_s = 0.5477 \times 38.4615 = 21.0654\,\Omega$$

and

$$X_p = \frac{R_p}{Q} = \frac{50}{0.5477} = 91.2909\,\Omega$$

For circuit (a) in Figure 6.22, capacitor C_p can be selected, which provides 91.29-Ω reactance. It means that the inductive reactance on its right-hand side must be 21.06 Ω. However, an inductive reactance of 57.69 Ω is already present there, due to load. Another series inductor will increase it further, whereas it needs to be reduced to 21.06 Ω. Thus, we conclude that circuit (a) cannot be used.

In circuit (*b*) in Figure 6.22, there is an inductor in parallel with the transmission line. For a match, its reactance must be 91.29 Ω, and overall reactance to the right must be capacitive 21.06 Ω. Inductive reactance of 57.69 Ω of the load needs to be neutralized as well. Hence,

$$X_c = 57.6923 + 21.0654 = 78.7577\,\Omega$$

Therefore,

$$C = \frac{1}{2\pi \times 10^9 \times 78.7577}\,\text{F} = 2.0208\,\text{pF}$$

and

$$X_p = 91.2909\,\Omega \Rightarrow L = \frac{91.2909}{2\pi \times 10^9}\,\text{H} = 14.5294\,\text{nH}$$

Circuit (*c*) in Figure 6.22 has a capacitor across the transmission line terminals. Assuming that its reactance is 91.29 Ω, reactance to the right must be inductive 21.06 Ω. As noted before, the reactive part of the load is an inductive 57.69 Ω. It can be reduced to the desired value by connecting a capacitor in series. Hence, the component values for circuit (*c*) are calculated as follows:

$$X_s = 57.6923 - 21.0654 = 36.6269\,\Omega$$

Therefore, $C_s = 4.3453\,\text{pF}$ and

$$X_p = 91.2909\,\Omega \Rightarrow C_p = 1.7434\,\text{pF}$$

Example 6.10 Reconsider Example 6.9, where $Y_L = 8 - j12\,\text{mS}$ is to be matched with a 50-Ω line. Complete or verify the L-section matching circuits shown in Figure 6.23 at a signal frequency of 1 GHz.

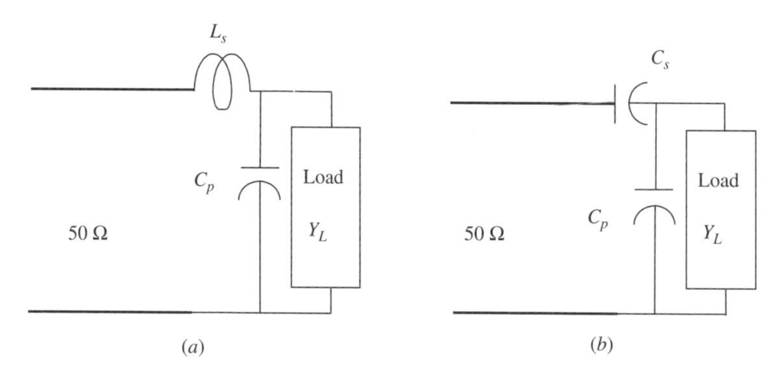

Figure 6.23 L-section matching networks given in Example 6.10.

SOLUTION As noted in Example 6.9, G_s is 0.02 S and G_p is 8 mS. Looking over the given circuit configurations, it seems possible to complete or verify these circuits through (6.3.11) to (6.1.13). Hence,

$$Q = \sqrt{\frac{G_s}{G_p} - 1} = \sqrt{\frac{0.02}{0.008} - 1} = 1.2247$$

$$B_s = \frac{G_s}{Q} = \frac{0.02}{1.2247} = 0.0163\ \text{S}$$

and

$$B_p = QG_p = 1.2247 \times 0.008 = 0.0097978 \approx 0.01\ \text{S}$$

If B_s is selected as an inductive susceptance, B_p must be capacitive. Since load has an inductive susceptance of 12 mS, the capacitor must neutralize that, too. Hence, circuit (*a*) of Figure 6.23 will work provided that

$$B_s = 0.0163\ \text{S} \Rightarrow L_s = 9.746\ \text{nH}$$

and

$$B_p = (0.01 + 0.012)\ \text{S} = 0.022\ \text{S} \Rightarrow C_p = 3.501\ \text{pF}$$

In circuit (*b*) of Figure 6.23, a capacitor is connected in series with G_s. Therefore, B_p must be an inductive susceptance. Since it is 0.012 S whereas only 0.01 S is required for matching, a capacitor in parallel is needed. Hence,

$$B_s = 0.0163\ \text{S} \Rightarrow C_s = 2.599\ \text{pF}$$

and

$$B_p = (0.012 - 0.01) = 0.002\ \text{S (capacitive)} \Rightarrow C_p = 0.3183\ \text{pF}$$

Thus, both circuits can work provided that the component values are as found above.

Graphical Method

As described earlier, L-section reactive networks can be used for matching the impedance. One of these reactive elements appears in series with the load or the desired impedance, while the other one is connected in parallel. Thus, the resistive part stays constant when a reactance is connected in series with impedance. Similarly, a change in the shunt-connected susceptance does not affect the conductive part of the admittance. Consider the normalized impedance point X on the Smith chart shown in Figure 6.24. To distinguish it from others, it may be called the *Z-Smith chart*. If its resistive part needs to be kept constant at 0.5, one must stay on this constant-resistance circle. A clockwise movement from

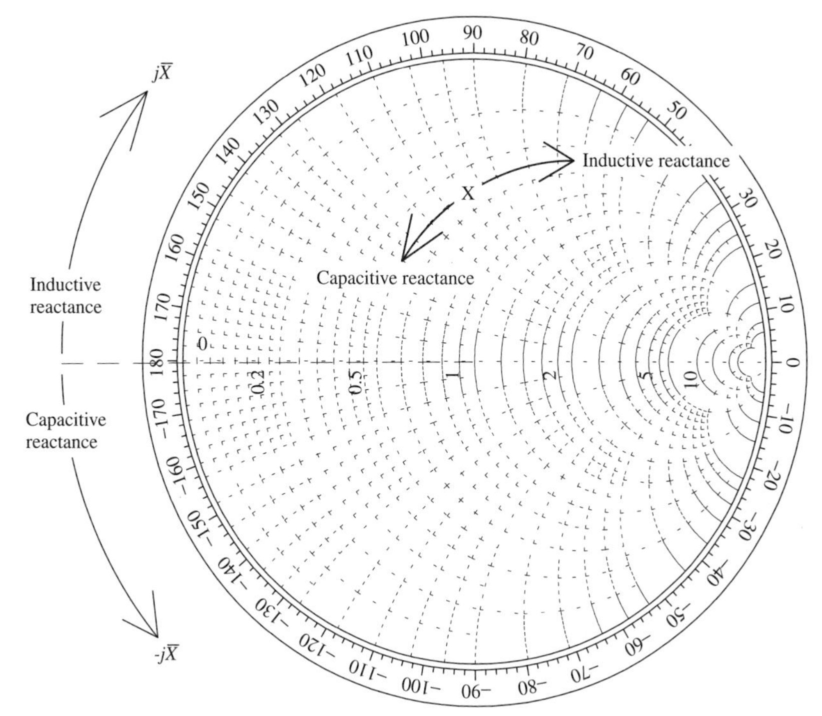

Figure 6.24 Impedance ($\overline{Z}$-) Smith chart.

this point increases the positive reactance. It means that the inductance increases in series with the impedance of X. On the other hand, the positive reactance decreases with a counterclockwise movement. A reduction in positive reactance means a decrease in series inductance or an increase in series capacitance. Note that this movement also represents an increase in negative reactance.

Now consider a Smith chart that is rotated by 180° from its usual position, as shown in Figure 6.25. It may be called a *Y-Smith chart* because it represents the admittance plots. In this case, addition (or subtraction) of a susceptance to admittance does not affect its real part. Hence, it represents a movement on the constant-conductance circle. Assume that a normalized admittance is located at point X. The conductive part of this admittance is 0.5. If a shunt inductance is added, it moves counterclockwise on this circle. On the other hand, a capacitive susceptance moves this point clockwise on the constant-conductance circle.

A superposition of Z- and Y-Smith charts is shown in Figure 6.26. It is generally referred to as a *ZY-Smith chart* because it includes impedance as well as admittance plots at the same time. A short-circuit impedance is zero while the corresponding admittance goes to infinity. A single point on the ZY-Smith chart represents it. Similarly, it may be found that other impedance points of the Z chart coincide with the corresponding admittance points of the Y chart as well. Hence, impedance

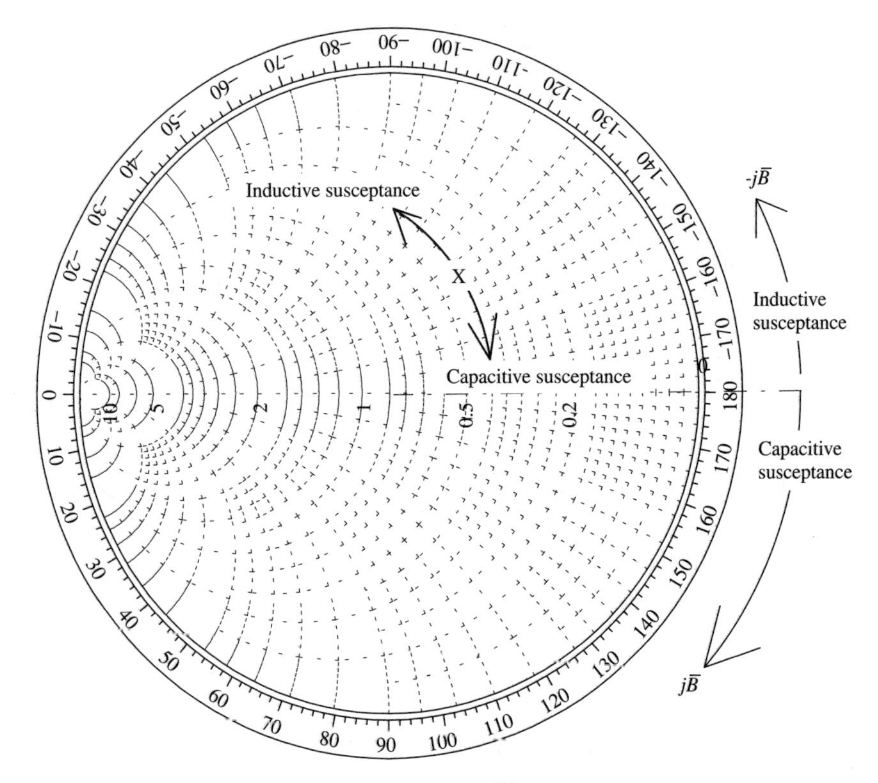

Figure 6.25 Admittance ($\overline{Y}$-) Smith chart..

can be transformed to corresponding admittance just by switching from the Z- to the Y-scales of a ZY-Smith chart. For example, a normalized impedance point of $0.9 - j1$ is located using the Z-chart scales as A in Figure 6.26. The corresponding admittance is read from a Y-chart as $0.5 + j0.55$.

Reactive L-section matching circuits can be designed easily using a ZY-Smith chart. Load and desired impedance points are identified on this chart using Z-scales. Y-scales may be used if either one is given as admittance. Note that the same characteristic impedance is used for normalizing these values. Now, move from the point to be transformed toward the desired point following a constant-resistance or constant-conductance circle. Movement on the constant-conductance circle gives the required susceptance (i.e., a reactive element will be needed in shunt). It is determined by subtracting initial susceptance (the starting point) from the value reached. When moving on the conductance circle, use the Y-scale of the chart. Similarly, moving on a constant-resistance circle will mean that a reactance must be connected in series. It can be determined using the Z-scale of the ZY-Smith chart.

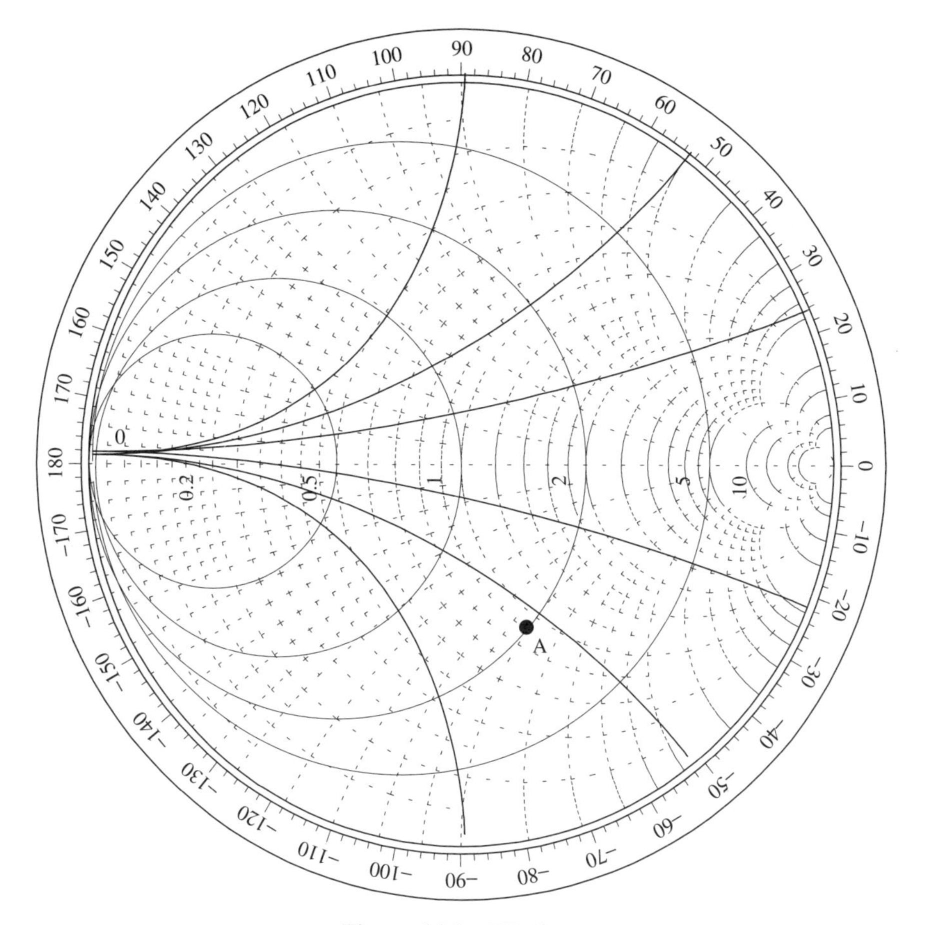

Figure 6.26 ZY-chart.

Note from the ZY-chart that if a normalized load value falls inside the unity resistance circle, it can be matched with the characteristic impedance Z_0 only after shunting it with an inductor or capacitor (a series-connected inductor or capacitor at the load will not work). Similarly, a series inductor or capacitor is required at the load if the normalized value is inside the unity conductance circle. Outside these two unity circles, if the load point falls in the upper half, the first component required for matching would be a capacitor that can be series or parallel connected. On the other hand, an inductor will be needed at the load first if the load point is located in the lower half and outside the unity circles.

Example 6.11 Use a ZY-Smith chart to design the matching circuits of Examples 6.9 and 6.10.

SOLUTION In this case the given load admittance can be normalized by the characteristic admittance of the transmission line. Hence,

$$\overline{Y}_L = \frac{Y_L}{Y_0} = Y_L Z_0 = 0.4 - j0.6$$

This point is found on the ZY-Smith chart as A in Figure 6.27. The matching requires that this admittance must be transformed to 1. This point can be reached through a unity conductance or resistance circle. Hence, a matching circuit can be designed if somehow we can reach one of these circles through constant-resistance and constant-conductance circles only. Since constant-conductance circles intersect constant- resistance circles, one needs to start on a constant-

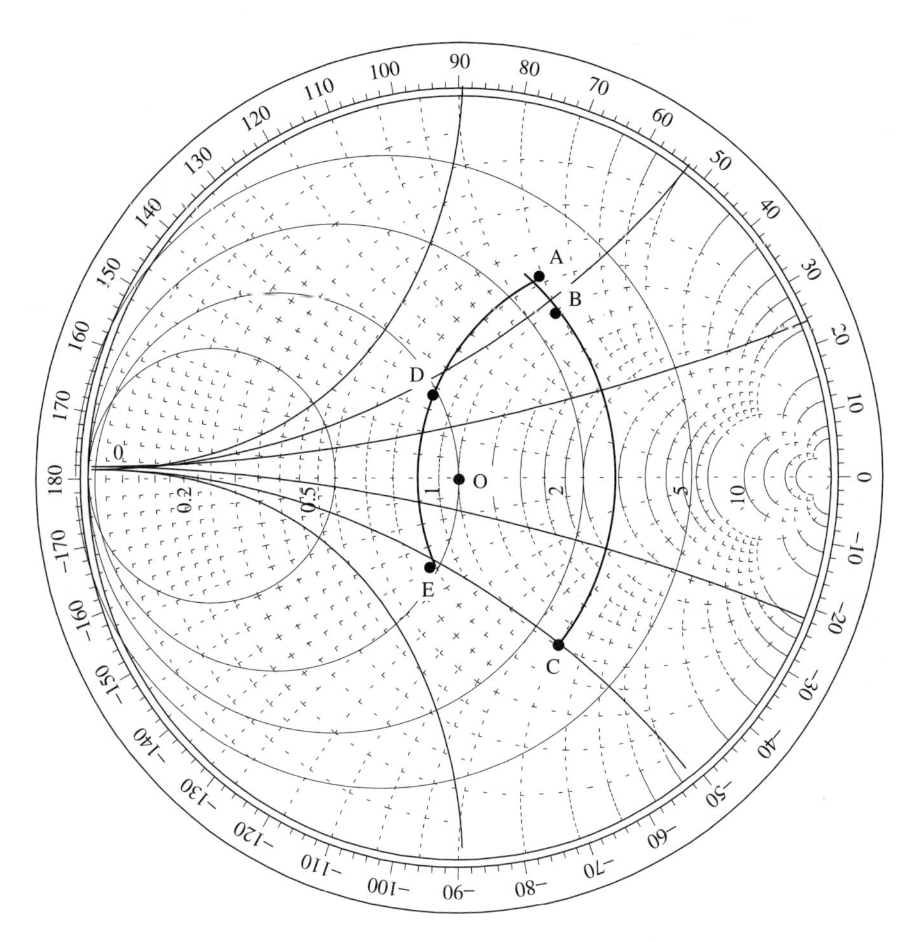

Figure 6.27 Graphical solution to Example 6.11.

resistance circle to get on the unity conductance circle, or vice versa. There are two circles (i.e., 0.4 conductance and 0.78 resistance circles) passing through point A. Hence, it is possible to get on to the unity resistance circle via the 0.4 conductance circle. Alternatively, one can reach on the unity conductance circle through the 0.78 resistance circle. In either case, a circuit can be designed. From point A, one can move on the 0.4 conductance circle (or 0.78 resistance circle) so that the real part of the admittance (or impedance) remains constant. If we start counterclockwise from A on the conductance circle, we end up at infinite susceptance without intersecting the unity resistance circle. Obviously, this does not yield a matching circuit. Similarly, a clockwise movement on the 0.78 resistance circle from A cannot produce a matching network. Hence, an inductor in parallel or in series with the given load cannot be used for the design of a matching circuit. This proves that the circuit shown in Figure 6.22(*a*) cannot be designed.

There are four possible circuits. In one case, move from point A to B on the conductance circle (a shunt capacitor) and then from B to O on the unity resistance circle (a series capacitor). The second possibility is via A to C (a shunt capacitor) and then from C to O (a series inductor). A third circuit can be obtained following the path from A to D (a series capacitor) and then from D to O (a shunt capacitor). The last circuit can be designed by following the path from A to E (a series capacitor) and then from E to O (a shunt inductor). A–E to E–O and A–D to D–O correspond to the circuits shown in Figure 6.22(*b*) and (*c*), respectively. Similarly, A–C to C–O and A–B to B–O correspond to circuits shown in Figure 6.23(*a*) and (*b*), respectively. Component values for each of these circuits are determined following the corresponding susceptance or reactance scales. For example, we move along a resistance circle from A to D. That means that there will be a change in reactance. The required element value is determined after subtracting the reactance at A from the reactance at D. This difference is $j0.42 - j1.15 = -j0.73$. Negative reactance means that it is a series capacitor. Note that $-j0.73$ is a normalized value. The actual reactance is $-j0.73 \times 50 = -j36.5\,\Omega$. It is close to the corresponding value of $36.6269\,\Omega$ obtained earlier. Other components can be determined as well.

Example 6.12 Two types of L-section matching networks are shown in Figure 6.28. Select one that can match the load $Z_L = 25 + j10\,\Omega$ to a 50-Ω transmission line. Find the element values at 500 MHz.

SOLUTION First, let us consider circuit (*a*). R_p must be $50\,\Omega$ because it is required to be greater than R_s. However, the reactive element is connected in series with it. That will be possible only with R_s. Hence, this circuit cannot be designed using (6.3.8) to (6.3.10). The other set of design equations, namely, (6.3.11) to (6.3.13), requires admittance instead of impedance. Therefore, the given impedances are inverted to find the corresponding admittances as

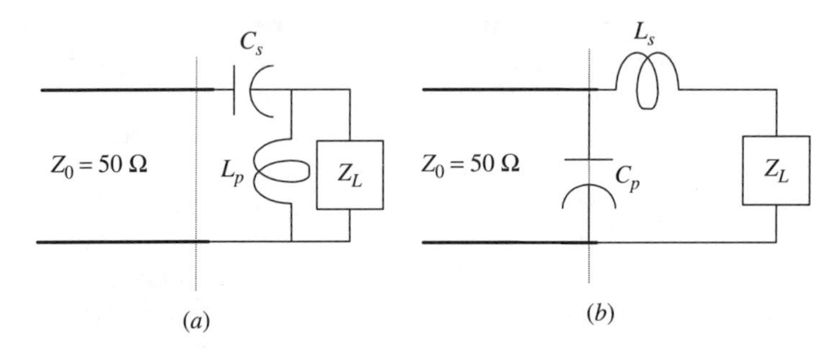

Figure 6.28 L-section matching circuits given for Example 6.12.

follows:

$$Y_L = \frac{1}{Z_L} = \frac{1}{25 + j10} = 0.034483 - j0.01379\,\text{S}$$

and

$$Y_0 = \frac{1}{Z_0} = \frac{1}{50} = 0.02\,\text{S}$$

Since G_s must be larger than G_p, this set of equations produces a matching circuit that has a reactance in series with Y_L. Thus, we conclude that the circuit given in Figure 6.28(*a*) cannot be designed.

Now consider the circuit shown in Figure 6.28(*b*). For (6.3.8) to (6.3.10), R_p is $50\,\Omega$ while R_s is $25\,\Omega$. Hence, one reactance of the matching circuit will be connected across $50\,\Omega$ and the other one will go in series with Z_L. The given circuit has this configuration. Hence, it will work. Its component values are calculated as

$$Q = \sqrt{\frac{50}{25} - 1} = 1$$

Therefore,

$$X_s = R_s = 25\,\Omega$$

and

$$X_p = R_p = 50\,\Omega$$

For X_p to be capacitive, we have

$$\frac{1}{\omega C_p} = 50 \Rightarrow C_p = 6.366\,\text{pF}$$

As 10 Ω of inductive reactance is already included in the load, another inductive 15 Ω will suffice. Thus,

$$\omega L_s = 25 - 10 = 15 \Rightarrow L_s = \frac{15}{\omega}\text{H} = 4.775\,\text{nH}$$

This example can be solved using a ZY-Smith chart as well. To that end, the load impedance is normalized with characteristic impedance of the line. Hence,

$$\overline{Z}_L = \frac{25 + j10}{50} = 0.5 + j0.2$$

This point is located on a ZY-Smith chart as shown in Figure 6.29. It is point A on this chart. If we move on the conductance circle from A, we never reach

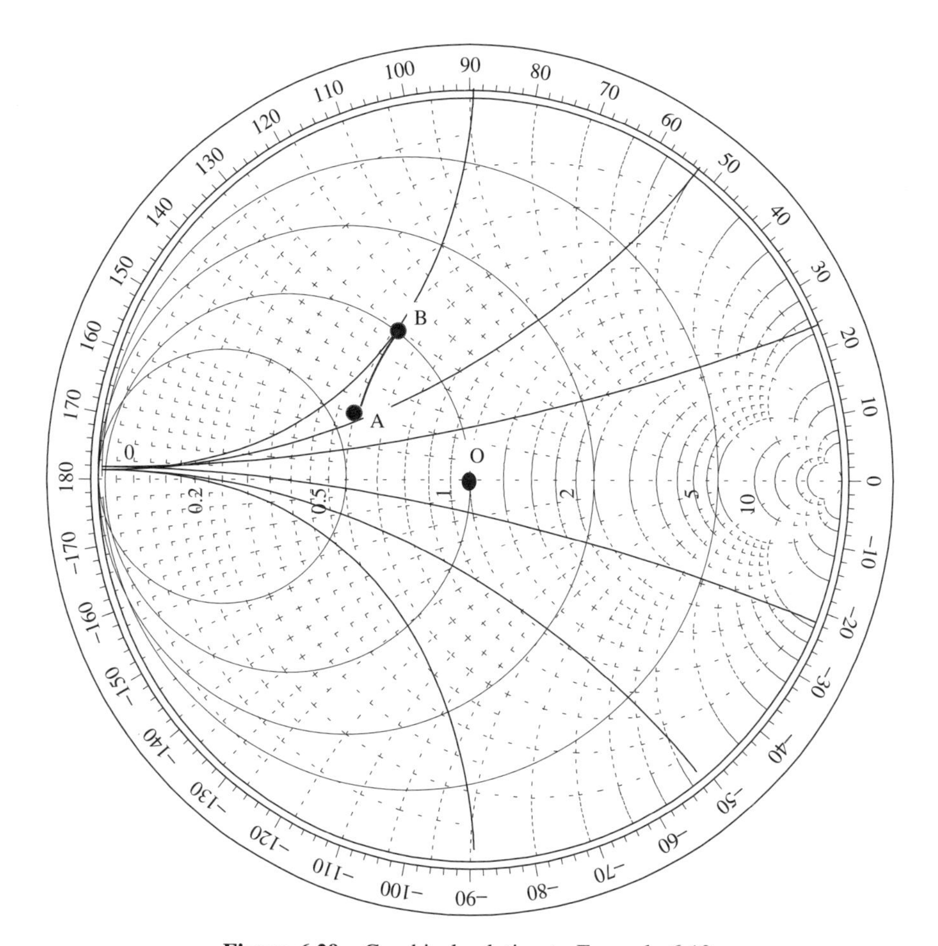

Figure 6.29 Graphical solution to Example 6.12.

the unity-resistance circle. This means that an L-section matching circuit that has a reactance (inductor or capacitor) in parallel with the load cannot be designed. Therefore, circuit (*a*) of Figure 6.28 is not possible.

Next, we try a reactance in series with the load. Moving on the constant-resistance circle from A indicates that there are two possible circuits. Movement from A to B on the resistance circle and then from B to O on the unity conductance circle provides component values of circuit shown in Figure 6.28(*b*). The normalized reactance required is found to be $j0.3$ (i.e., $j0.5 - j0.2$). It is an inductor because of its positive value. Further, it will be connected in series with the load because it is found by moving on a constant-resistance circle from A.

The series inductance required is determined as

$$X_L = 0.3 \times 50 = 15 \Rightarrow L_s = \frac{15}{\omega}\text{H} = 4.775\,\text{nH}$$

Movement from B to O gives a normalized susceptance of $j1$. Positive susceptance is a capacitor in parallel with a 50-Ω line. Hence,

$$\omega C_p = \frac{1}{50} \Rightarrow C_p = 6.366\,\text{pF}$$

These results are found to be exactly equal to those found earlier analytically.

SUGGESTED READING

Bahl, I., and P. Bhartia, *Microwave Solid State Circuit Design*. New York: Wiley, 1988.

Collin, R. E., *Foundations for Microwave Engineering*. New York: McGraw-Hill, 1992.

Fusco, V. F., *Microwave Circuits*. Englewood Cliffs, NJ: Prentice Hall, 1987.

Pozar, D. M., *Microwave Engineering*. New York: Wiley, 1998.

Ramo, S., J. R. Whinnery, and T. Van Duzer, *Fields and Waves in Communication Electronics*. New York: Wiley, 1994.

Rizzi, P. A., *Microwave Engineering*, Englewood Cliffs, NJ: Prentice Hall, 1988.

Sinnema, W., *Electronic Transmission Technology*. Englewood Cliffs, NJ: Prentice Hall, 1988.

PROBLEMS

6.1. A 10-V (rms) voltage source in series with a 50-Ω resistance represents a signal generator. It is to be matched with a 100-Ω load. Design a matching circuit that provides perfect matching over the frequency band 1 kHz to 1 GHz. Determine the power delivered to the load.

6.2. A 1000-Ω load is to be matched with a signal generator that can be represented by a 1-A (rms) current source in parallel with a 100-Ω resistance. Design a resistive circuit to match it. Determine the power dissipated in the matching network.

6.3. Design a single-stub network to match a $800 - j300$-Ω load to a 400-Ω lossless line. The stub should be located as close to the load as possible and it is to be connected in parallel with the transmission line.

6.4. A $140 - j70$-Ω load is terminating a 70-Ω transmission line, as shown in Figure P6.4. Find the location and length of a short-circuited stub of a 40-Ω characteristic impedance that will match the load with the line.

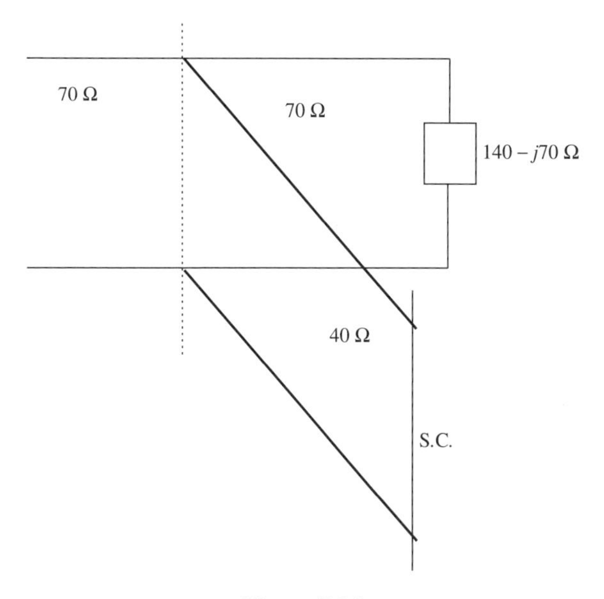

Figure P6.4

6.5. An antenna has impedance of $40 + j30\,\Omega$ at its input. Match it with a 50-Ω line using short-circuited shunt stubs. Determine (**a**) the required stub admittance, (**b**) the distance between the stub and the antenna, (**c**) the stub length, and (**d**) the VSWR on each section of the circuit.

6.6. A lossless 100-Ω transmission line is to be matched with a $100 + j100$-Ω load using a double-stub tuner (Figure P6.6). Separation between the two stubs is $\lambda/8$ and the characteristic impedance is $100\,\Omega$. A load is connected right at the location of the first stub. Determine the shortest possible lengths of the two stubs to obtain the matched condition, and find the VSWR between the two stubs.

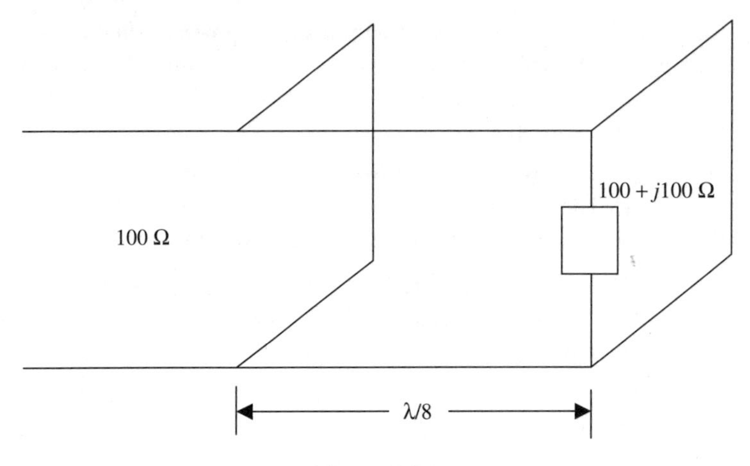

Figure P6.6

6.7. A lossless 75-Ω transmission line is to be matched with a 150 $+j15$-Ω load using a shunt-connected double-stub tuner. Separation between the two stubs is $\lambda/8$ and the characteristic impedance is 75 Ω. The stub closest to the load (first stub) is $\lambda/2$ away from it. Determine the shortest possible lengths of the two stubs to obtain the matched condition, and find the VSWR between the two stubs.

6.8. A lossless 50-Ω transmission line is to be matched with a 25 $+j50$-Ω load using a double-stub tuner. Separation between the two stubs is $3\lambda/8$ and the characteristic impedance is 50 Ω. A load is connected at 0.2λ from the first stub, as shown in Figure P6.8. Determine the shortest possible lengths of the two stubs to obtain the matched condition, and find the VSWR between the two stubs.

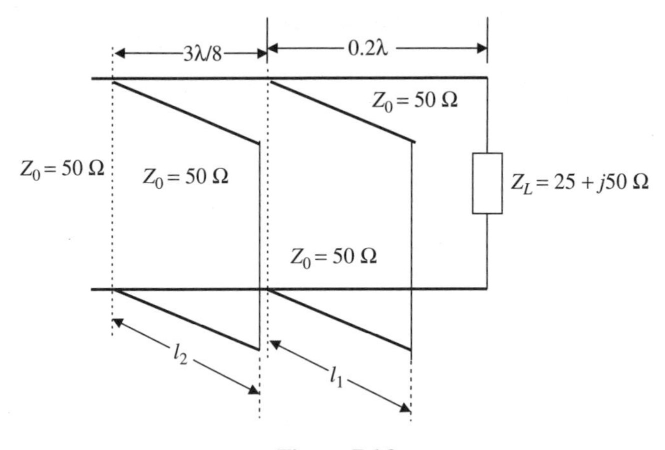

Figure P6.8

6.9. Complete or verify the two interstage designs of 1 GHz shown as (*a*) and (*b*) in Figure P6.9.

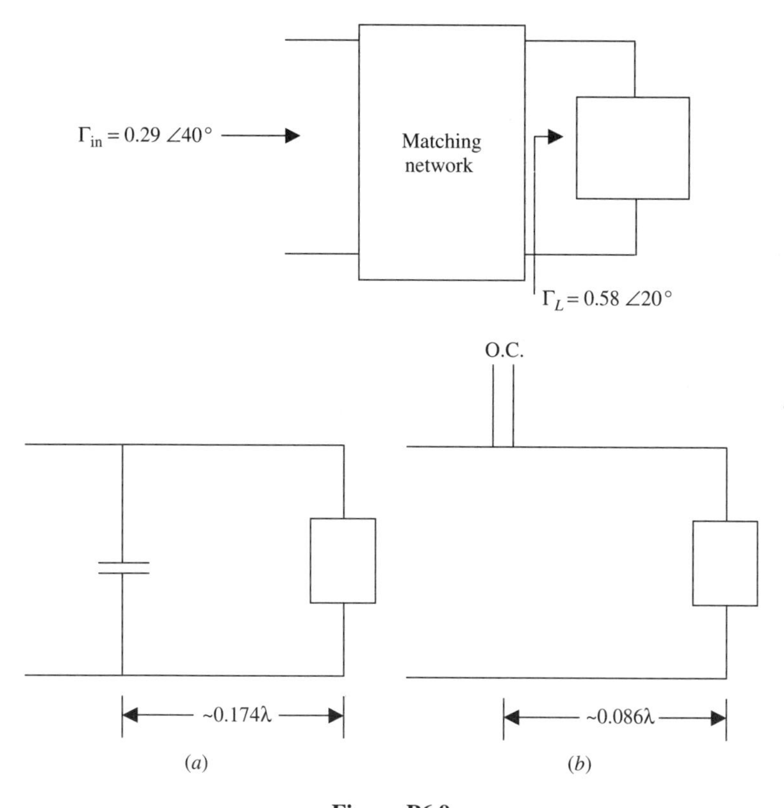

Figure P6.9

6.10. Determine the shortest length d in the two networks in Figure P6.10 to match the load to a 50-Ω lossless line. Also find the stub length l for circuit (*a*) and the required inductance L for circuit (*b*).

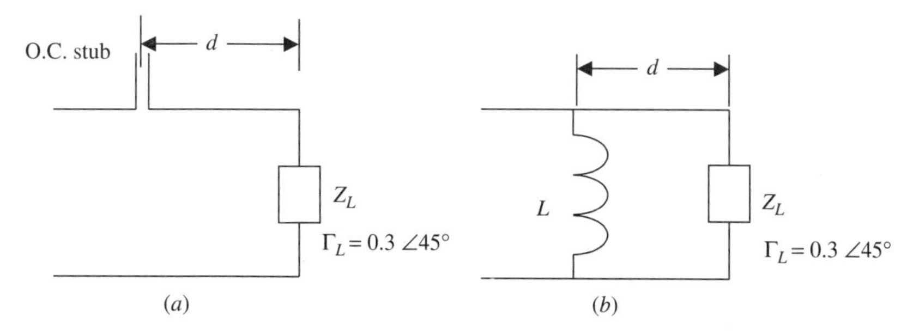

Figure P6.10

6.11. Determine the shortest length d in the two networks in Figure P6.11 to match a $100 - j50$-Ω load to a 50-Ω lossless line. Also find the required capacitance C for circuit (a) and the stub length l for circuit (b).

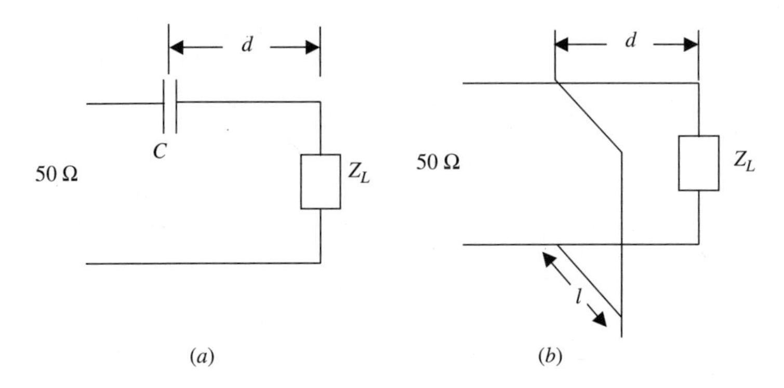

Figure P6.11

6.12. Complete or verify the two interstage designs of 1 GHz shown as (a) and (b) in Figure P6.12. Assume that the characteristic impedance is 50 Ω.

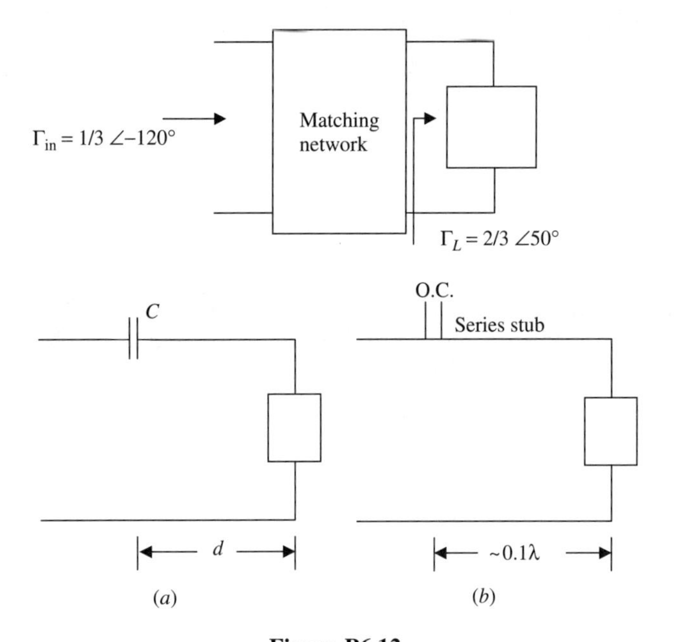

Figure P6.12

6.13. Complete or verify the interstage designs at $f = 4$ GHz, shown as (a) and (b) in Figure P6.13.

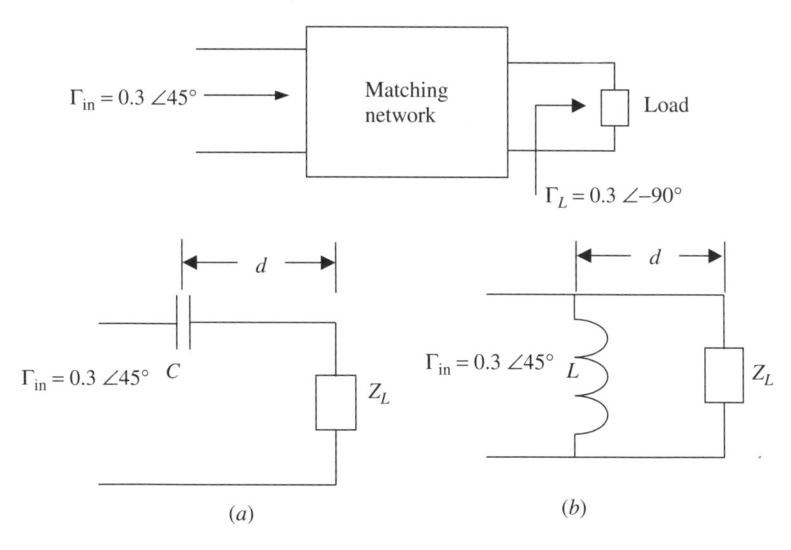

Figure P6.13

6.14. Complete or verify the interstage designs at $f = 4\,\text{GHz}$ shown as (*a*) and (*b*) in Figure P6.14. The characteristic impedance is $50\,\Omega$.

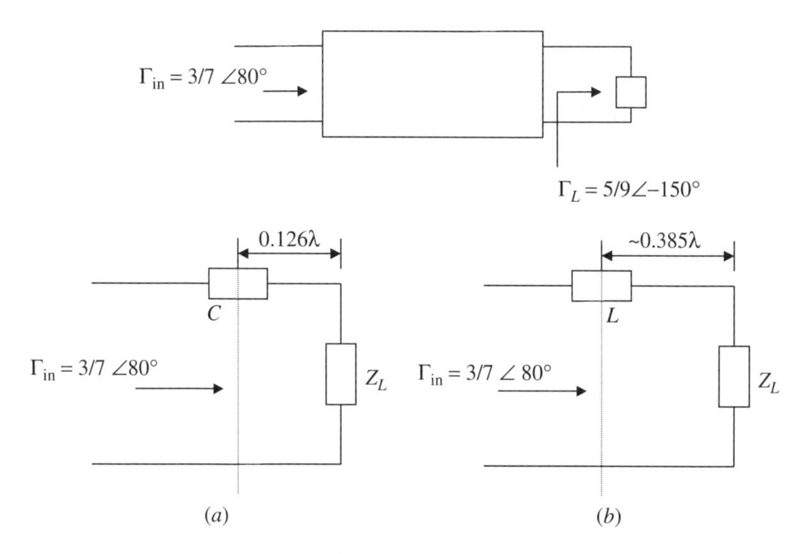

Figure P6.14

6.15. Two types of L-section matching networks are shown in Figure P6.15. Select one that can match the load $Z_L = 20 - j100\,\Omega$ to a 50-Ω transmission line. Find the element values at 500 MHz.

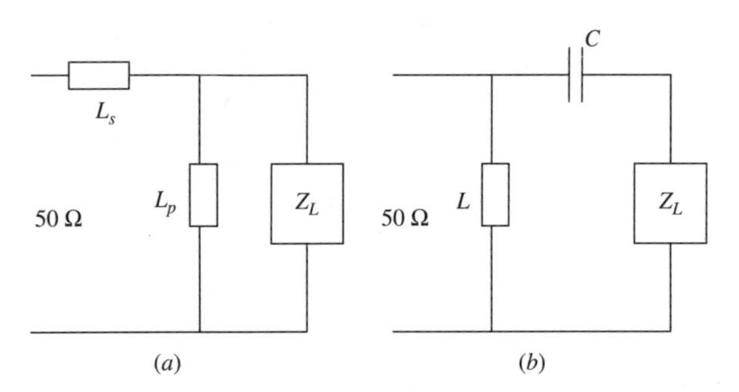

Figure P6.15

6.16. Two types of L-section matching networks are shown in Figure P6.16. Select one that can match a $30 + j50$-Ω load to a 50-Ω line at 1 GHz.

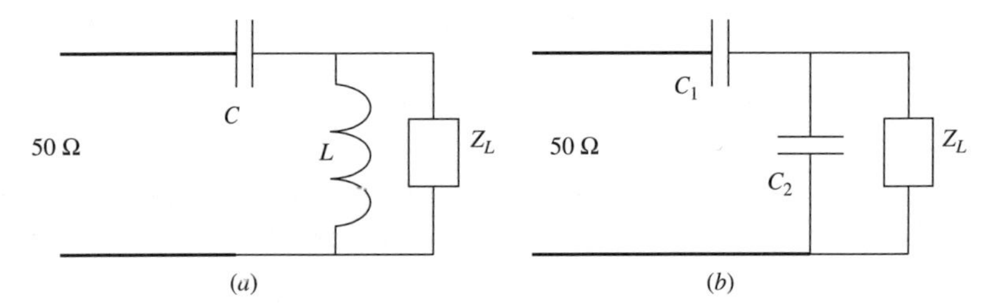

Figure P6.16

6.17. Design a matching network (Figure P6.17) that will transform a $50 - j50$-Ω load to the input impedance of $25 + j25$ Ω.

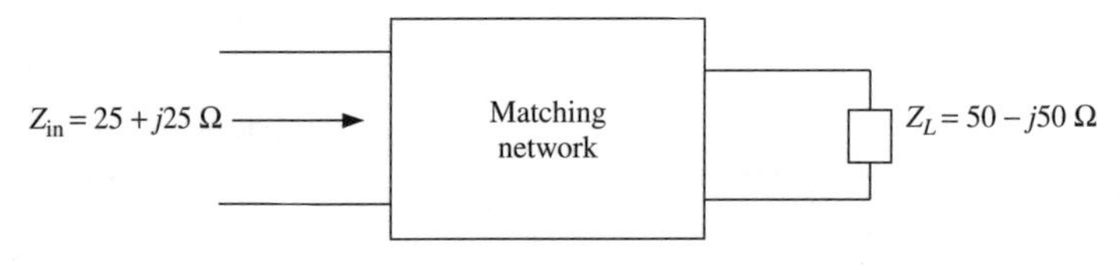

Figure P6.17

6.18. Match the load in Figure P6.18 to a 50-Ω generator using lumped elements. Assume that the signal frequency as 4 GHz.

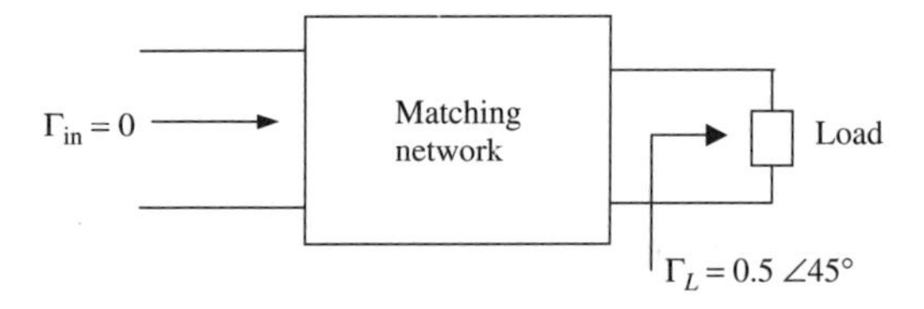

Figure P6.18

6.19. Two types of L-section matching networks are shown in Figure P6.19. Select one that can match the load $Z_L = 60 - j20\,\Omega$ to a 50-Ω transmission line. Find the element values at 500 MHz.

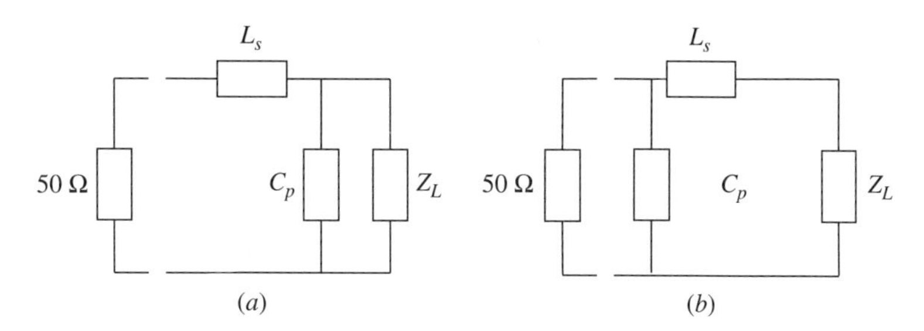

Figure P6.19

6.20. Design a lumped-element network that will provide a load $Z_L = 30 + j50\,\Omega$ for the amplifier shown in Figure P6.20 operating at 2 GHz.

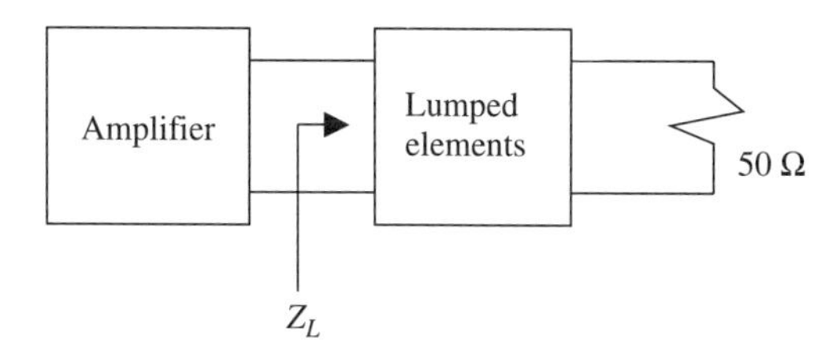

Figure P6.20

6.21. Design a lumped-element network that will provide a load $Z_L = 30 + j50\,\Omega$ for the amplifier shown in Figure P6.21, and stop the biasing voltage from appearing across 50 Ω. The amplifier is operating at 500 MHz.

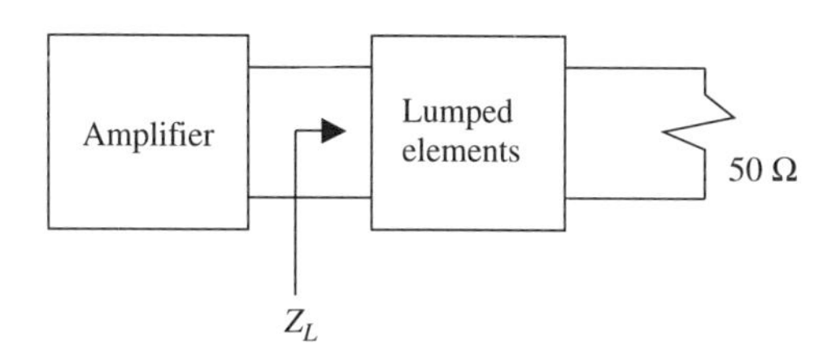

Figure P6.21

6.22. A 200-Ω load is to be matched with a 50-Ω line. Design (**a**) a resistive network and (**b**) a reactive network to match the load. (**c**) Compare the performance of the two designs.

7

IMPEDANCE TRANSFORMERS

In Chapter 6, several techniques were considered to match a given load impedance at a fixed frequency. These techniques included transmission line stubs as well as lumped elements. Note that lumped element circuits may not be practical at higher frequencies. Further, it may be necessary in certain cases to keep the reflection coefficient below a specified value over a given frequency band. In this chapter we present transmission line impedance transformers that can meet such requirements. The chapter begins with a single-section impedance transformer that provides perfect matching at a single frequency. A matching bandwidth can be increased at the cost of a higher reflection coefficient. This concept is used to design multisection transformers. The characteristic impedance of each section is controlled to obtain the desired passband response.

Multisection binomial transformers exhibit an almost flat reflection coefficient about the center frequency and increase gradually on either side. A wider bandwidth is achieved with an increased number of quarter-wave sections. Chebyshev transformers can provide even wider bandwidth with the same number of sections, but the reflection coefficient exhibits ripples in its passband. The chapter includes a procedure to design these multisection transformers as well as transmission line tapers. The chapter concludes with a brief discussion on the Bode–Fano constraints, which provide an insight into the trade-off between the bandwidth and the allowed reflection coefficient.

7.1 SINGLE-SECTION QUARTER-WAVE TRANSFORMERS

We considered a single-section quarter-wavelength transformer design problem in Example 3.5. In this section we present a detailed analysis of such circuits.

Radio-Frequency and Microwave Communication Circuits: Analysis and Design, Second Edition,
By Devendra K. Misra
ISBN 0-471-47873-3

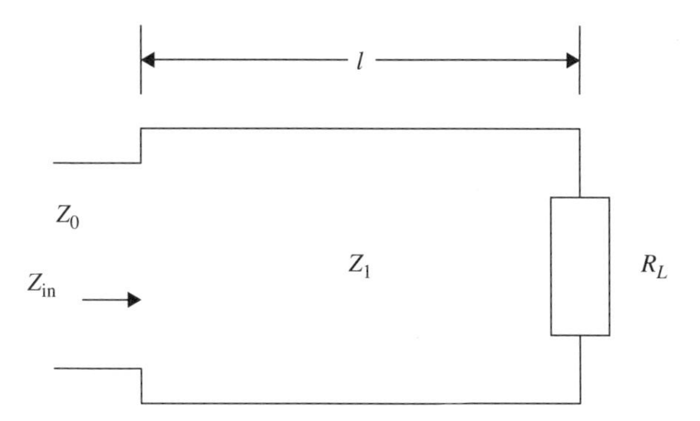

Figure 7.1 Single-section quarter-wave transformer.

Consider the load resistance R_L that is to be matched with a transmission line of characteristic impedance Z_0. Assume that a transmission line of length l and characteristic impedance Z_1 is connected between the two, as shown in Figure 7.1. Its input impedance Z_{in} is found as follows:

$$Z_{in} = Z_1 \frac{R_L + jZ_1 \tan \beta l}{Z_1 + jR_L \tan \beta l} \tag{7.1.1}$$

For $\beta l = 90^\circ$ (i.e., $l = \lambda/4$) and $Z_1 = \sqrt{Z_0 R_L}$, Z_{in} is equal to Z_0; hence, there is no reflected wave beyond this point toward the generator. However, it reappears at other frequencies when $\beta l \neq 90^\circ$. The corresponding reflection coefficient Γ_{in} can be determined as follows:

$$\begin{aligned}\Gamma_{in} &= \frac{Z_{in} - Z_0}{Z_{in} + Z_0} = \frac{Z_1[(R_L + jZ_1 \tan \beta l)/(Z_1 + jR_L \tan \beta l)] - Z_0}{Z_1[(R_L + jZ_1 \tan \beta l)/(Z_1 + jR_L \tan \beta l)] + Z_0} \\ &= \frac{R_L - Z_0}{R_L + Z_0 + j2\sqrt{Z_0 R_L} \tan \beta l} = \rho_{in} \exp(j\varphi)\end{aligned}$$

Therefore,

$$\begin{aligned}\rho_{in} &= \frac{R_L - Z_0}{[(R_L + Z_0)^2 + 4Z_0 R_L \tan^2 \beta l]^{1/2}} \\ &= \frac{1}{\{1 + [2\sqrt{Z_0 R_L}/(R_L - Z_0) \sec \beta l]^2\}^{1/2}}\end{aligned} \tag{7.1.2}$$

Variation in ρ_{in} with frequency is illustrated in Figure 7.2. For βl near 90°, it can be approximated as follows:

$$\rho_{in} \approx \frac{|R_L - Z_0|}{2\sqrt{Z_0 R_L} \tan \beta l} \approx \frac{|R_L - Z_0|}{2\sqrt{Z_0 R_L}} \cos \beta l \tag{7.1.3}$$

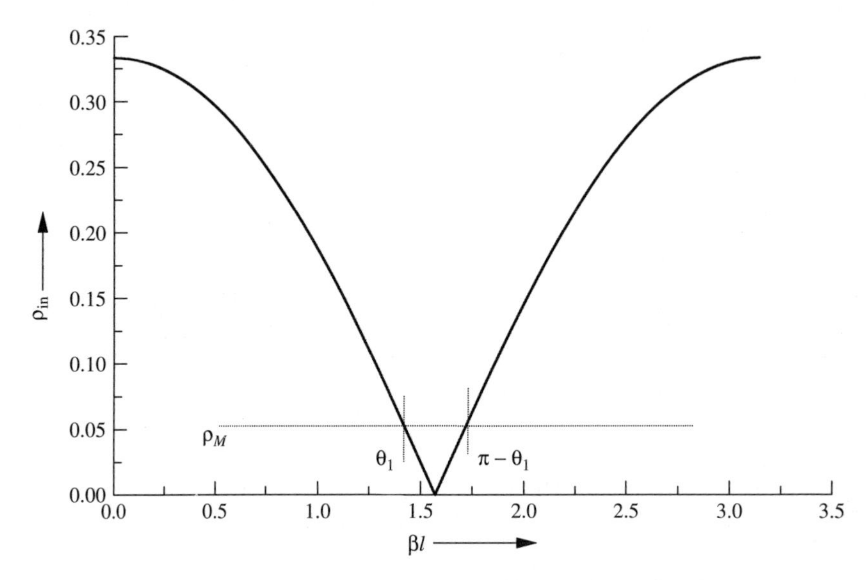

Figure 7.2 Reflection coefficient characteristics of a single-section impedance transformer used to match a 100-Ω load to a 50-Ω line.

If ρ_M is the maximum allowable reflection coefficient at the input, then

$$\cos\theta_1 = \left| \frac{2\rho_M\sqrt{Z_0 R_L}}{(R_L - Z_0)\sqrt{1-\rho_M^2}} \right| \qquad \theta_1 < \frac{\pi}{2} \tag{7.1.4}$$

In the case of a TEM wave propagating on the transmission line, $\beta l = (\pi/2)(f/f_0)$, where f_0 is the frequency at which $\beta l = \pi/2$. In this case, the bandwidth $f_2 - f_1 = \Delta f$ is given by

$$\Delta f = (f_2 - f_1) = 2(f_0 - f_1) = 2\left(f_0 - \frac{2f_0}{\pi}\theta_1\right) \tag{7.1.5}$$

and the fractional bandwidth is

$$\frac{\Delta f}{f_0} = 2 - \frac{4}{\pi}\cos^{-1}\left| \frac{2\rho_M\sqrt{Z_0 R_L}}{(R_L - Z_0)\sqrt{1-\rho_M^2}} \right| \tag{7.1.6}$$

Example 7.1 Design a single-section quarter-wave impedance transformer to match a 100-Ω load to a 50-Ω air-filled coaxial line at 900 MHz (Figure 7.3). Determine the range of frequencies over which the reflection coefficient remains below 0.05.

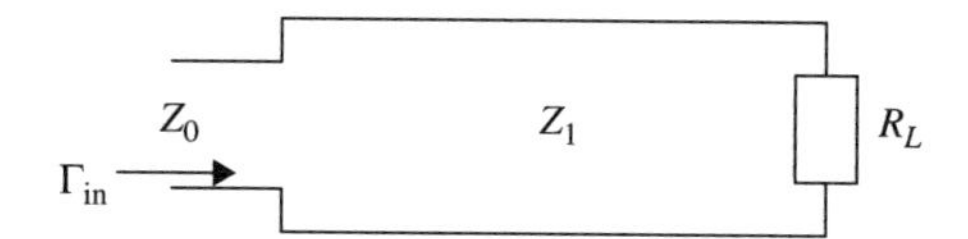

Figure 7.3 Setup for Example 7.1.

SOLUTION For $R_L = 100\,\Omega$ and $Z_0 = 50\,\Omega$,

$$Z_1 = \sqrt{100 \times 50} = 70.7106781\,\Omega$$

and

$$l = \frac{\lambda}{4} = \frac{3 \times 10^8}{4 \times 900 \times 10^6}\ \text{m} = 8.33\,\text{cm}$$

The magnitude of the reflection coefficient increases as βl changes from $\pi/2$ (i.e., the signal frequency changes from 900 MHz). If the maximum allowed ρ is $\rho_M = 0.05$ (VSWR = 1.1053), the fractional bandwidth is found to be

$$\frac{\Delta f}{f_0} = 2 - \frac{4}{\pi}\cos^{-1}\left|\frac{2\rho_M\sqrt{Z_0 R_L}}{(R_L - Z_0)\sqrt{1-\rho_M^2}}\right| = 0.180897$$

Therefore, 818.5964 MHz $\leq f \leq$ 981.4037 MHz or $1.4287 \leq \beta l \leq 1.7129$.

7.2 MULTISECTION QUARTER-WAVE TRANSFORMERS

Consider an N-section impedance transformer connected between a transmission line of characteristic impedance of Z_0 and load R_L, as shown in Figure 7.4. As indicated, the length of every section is the same, although their characteristic impedances are different. Impedance at the input of Nth section can be found as follows:

$$Z_{\text{in}}^N = Z_N \frac{\exp(j\beta l) + \Gamma_N \exp(-j\beta l)}{\exp(j\beta l) - \Gamma_N \exp(-j\beta l)} \tag{7.2.1}$$

where

$$\Gamma_N = \frac{R_L - Z_N}{R_L + Z_N} \tag{7.2.2}$$

The reflection coefficient seen by the $(N-1)$st section is

$$\Gamma'_{N-1} = \frac{Z_{\text{in}}^N - Z_{N-1}}{Z_{\text{in}}^N + Z_{N-1}} = \frac{Z_N(e^{j\beta l} + \Gamma_N e^{-j\beta l}) - Z_{N-1}(e^{j\beta l} - \Gamma_N e^{-j\beta l})}{Z_N(e^{j\beta l} + \Gamma_N e^{-j\beta l}) + Z_{N-1}(e^{j\beta l} - \Gamma_N e^{-j\beta l})}$$

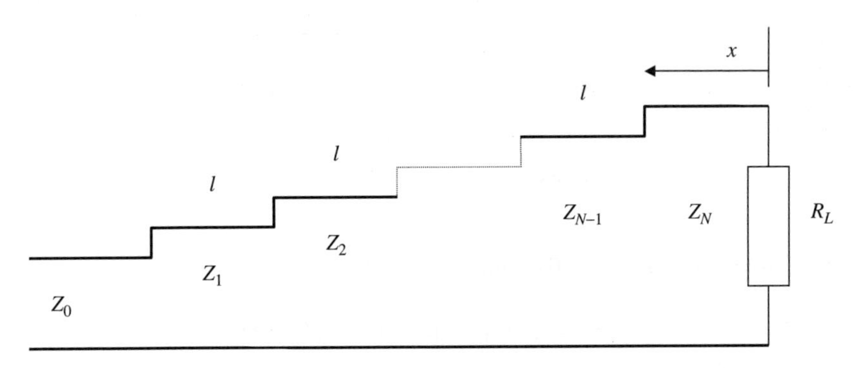

Figure 7.4 N-section impedance transformer.

or

$$\Gamma'_{N-1} = \frac{(Z_N - Z_{N-1})e^{j\beta l} + \Gamma_N(Z_N + Z_{N-1})e^{-j\beta l}}{(Z_N + Z_{N-1})e^{j\beta l} + \Gamma_N(Z_N - Z_{N-1})e^{-j\beta l}}$$

Therefore,

$$\Gamma'_{N-1} = \frac{\Gamma_{N-1} + \Gamma_N e^{-j2\beta l}}{1 + \Gamma_N \Gamma_{N-1} e^{-j2\beta l}} \tag{7.2.3}$$

where

$$\Gamma_{N-1} = \frac{Z_N - Z_{N-1}}{Z_N + Z_{N-1}} \tag{7.2.4}$$

If Z_N is close to R_L and Z_{N-1} is close to Z_N, then Γ_N and Γ_{N-1} are small quantities, and a first-order approximation can be assumed. Hence,

$$\Gamma'_{N-1} \approx \Gamma_{N-1} + \Gamma_N e^{-j2\beta l} \tag{7.2.5}$$

Similarly,

$$\Gamma'_{N-2} \approx \Gamma_{N-2} + \Gamma'_N e^{-j2\beta l} = \Gamma_{N-2} + \Gamma_{N-1} e^{-j2\beta l} + \Gamma_N e^{-j4\beta l}$$

Therefore, by induction, the reflection coefficient seen by the feeding line is

$$\Gamma \approx \Gamma_0 + \Gamma_1 e^{-j2\beta l} + \Gamma_2 e^{-j4\beta l} + \cdots + \Gamma_{\mathrm{N}-1} e^{-j2(N-1)\beta l} + \Gamma_{\mathrm{N}} e^{-j2N\beta l} \tag{7.2.6}$$

or

$$\Gamma = \sum_{n=0}^{N} \Gamma_n e^{-j2n\beta l} \tag{7.2.7}$$

where

$$\Gamma_n = \frac{Z_{n+1} - Z_n}{Z_{n+1} + Z_n} \tag{7.2.8}$$

Thus, we need a procedure to select Γ_n so that Γ is minimized over the desired frequency range. To this end, we recast equation (7.2.6) as follows:

$$\Gamma = \Gamma_0 + \Gamma_1 w + \Gamma_2 w^2 + \cdots + \Gamma_N w^N = \Gamma_N \prod_{n=1}^{N} (w - w_n) \tag{7.2.9}$$

where

$$\varphi = -2\beta l \tag{7.2.10}$$

and

$$w = e^{j\varphi} \tag{7.2.11}$$

Note that for $\beta l = 0$ (i.e., $\lambda \to \infty$), individual transformer sections in effect have no electrical length and the load R_L appears to be connected directly to the main line. Therefore,

$$\Gamma = \sum_{n=0}^{N} \Gamma_n = \frac{R_L - Z_0}{R_L + Z_0} \qquad (\because w = 1) \tag{7.2.12}$$

and only N of the $(N + 1)$-section reflection coefficients can be selected independently.

7.3 TRANSFORMER WITH UNIFORMLY DISTRIBUTED SECTION REFLECTION COEFFICIENTS

If all of the section reflection coefficients are equal, (7.2.9) can be simplified as follows:

$$\frac{\Gamma}{\Gamma_N} = 1 + w + w^2 + w^3 + \cdots + w^{N-1} + w^N = \frac{w^{N+1} - 1}{w - 1} \tag{7.3.1}$$

or

$$\frac{\Gamma}{\Gamma_N} = \frac{e^{j(N+1)\varphi} - 1}{e^{j\varphi} - 1} = e^{jN\varphi/2} \frac{\sin\{[(N+1)/2]\varphi\}}{\sin(\varphi/2)}$$

Hence,

$$|\Gamma| = \rho(\varphi) = \rho_N \left| \frac{\sin\{[(N+1)/2]\varphi\}}{\sin(\varphi/2)} \right| = (N+1)\rho_N \left| \frac{\sin\{[(N+1)/2]\varphi\}}{(N+1)\sin(\varphi/2)} \right| \tag{7.3.2}$$

and from (7.2.12),

$$\sum_{n=0}^{N} \Gamma_n = (N+1)\rho_N = \frac{R_L - Z_0}{R_L + Z_0} \tag{7.3.3}$$

Therefore, equation (7.3.2) can be written as follows:

$$\rho(\beta l) = \left|\frac{R_L - Z_0}{R_L + Z_0}\right| \cdot \left|\frac{\sin[(N+1)\beta l]}{(N+1)\sin\beta l}\right| \tag{7.3.4}$$

This can be viewed as an equation that describes magnitude ρ of the reflection coefficient as a function of frequency. As (7.3.4) indicates, a pattern of $\rho(\beta l)$ repeats periodically with an interval of π. It peaks at $n\pi$, where n is an integer including zero. Further, there are $N - 1$ minor lobes between two consecutive main peaks. The number of zeros between the two main peaks of $\rho(\beta l)$ is equal to the number of quarter-wave sections, N.

Consider that there are three quarter-wave sections connected between a 100-Ω load and a 50-Ω line. Its reflection coefficient characteristics can be found from (7.3.4), as illustrated in Figure 7.5. There are three zeros in it, one at $\beta l = \pi/2$ and the other two located symmetrically around this point. In other words, zeros occur at $\beta l = \pi/4$, $\pi/2$, and $3\pi/4$. When the number of quarter-wave sections is increased from three to six, the $\rho(\beta l)$ plot changes as illustrated in Figure 7.6.

For a six-section transformer, Figure 7.6 shows five minor lobes between two main peaks of $\rho(\beta l)$. One of these minor lobes has its maximum value (peak) at $\beta l = \pi/2$. Six zeros of this plot are located symmetrically: $\beta l = n\pi/7$, $n = 1, 2, \ldots, 6$. Thus, characteristics of $\rho(\beta l)$ can be summarized as follows:

- The pattern of $\rho(\beta l)$ repeats with an interval of π.
- There are N nulls and $N-1$ minor peaks in an interval.

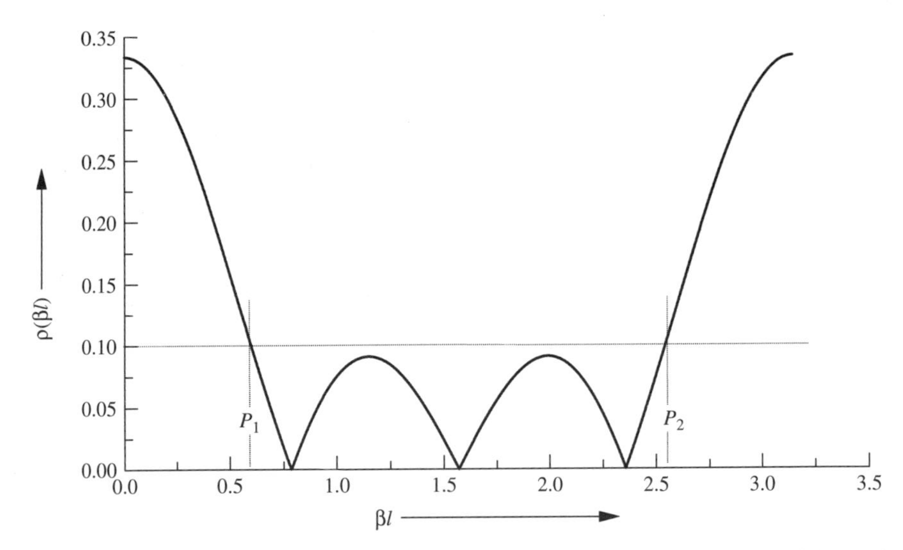

Figure 7.5 Reflection coefficient versus βl of a three-section transformer with equal section reflection coefficients for $R_L = 100\,\Omega$ and $Z_0 = 50\,\Omega$.

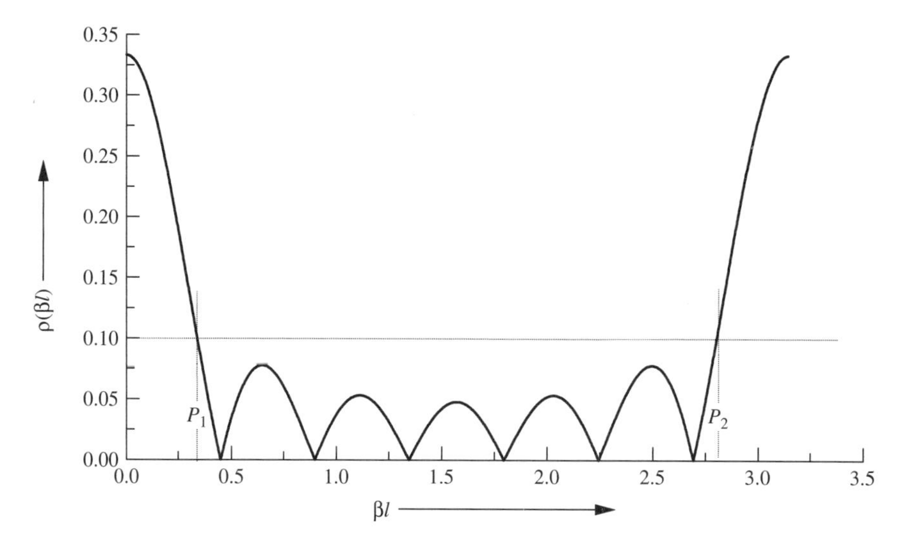

Figure 7.6 Reflection coefficient versus βl for a six-section transformer with equal section-reflection coefficients ($R_L = 100\,\Omega$ and $Z_0 = 50\,\Omega$).

- When N is odd, one of the nulls occurs at $\beta l = \pi/2$ (i.e., $l = \lambda/4$).
- If ρ_M is specified as an upper bound on ρ to define the frequency band, points P_1 and P_2 bound the acceptable range of βl. This range becomes larger as N increases.

Since

$$w^{N+1} - 1 = \prod_{n=0}^{N} (w - w_n) \tag{7.3.5}$$

where

$$w_n = e^{j[2\pi n/(N+1)]} \qquad n = 1, 2, \ldots, N \tag{7.3.6}$$

(7.3.1) may be written as follows:

$$\frac{\Gamma}{\Gamma_N} = \prod_{n=1}^{N} (w - w_n) = \prod_{n=1}^{N} (w - e^{j[2\pi n/(N+1)]}) \tag{7.3.7}$$

This equation is of the form of (7.2.9). It indicates that when section reflection coefficients are the same, roots are equispaced around the unit circle on the complex w-plane with the root at $w = 1$ deleted. This is illustrated in Figure 7.7 for $N = 3$. It follows that

$$\frac{\rho}{\rho_N} = \left| \prod_{n=1}^{N} (w - e^{j[2\pi n/(N+1)]}) \right| = \prod_{n=1}^{N} |(w - e^{j[2\pi n/(N+1)]})| \tag{7.3.8}$$

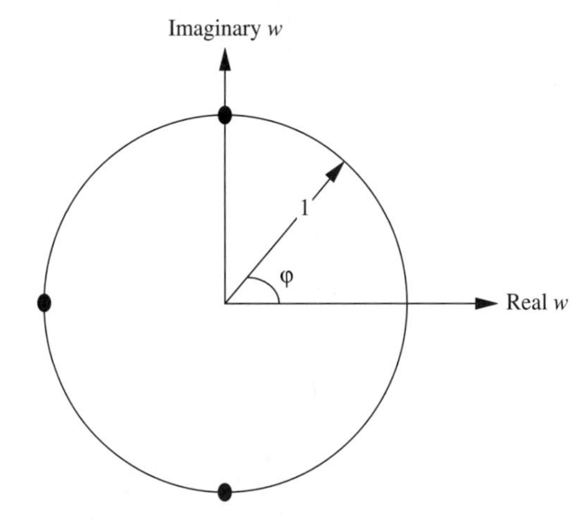

Figure 7.7 Location of zeros on a unit circle on the complex w-plane for $N = 3$.

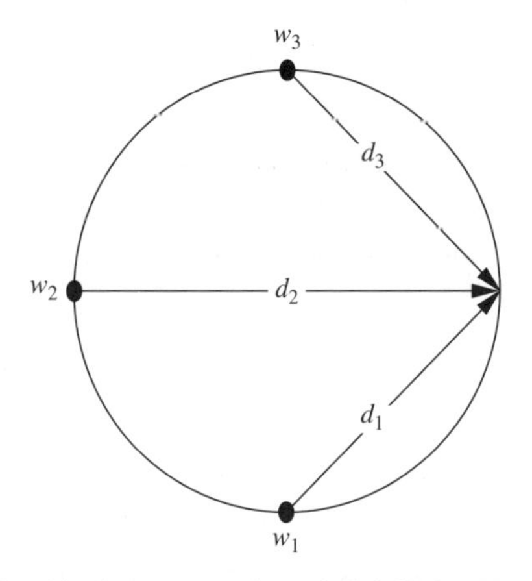

Figure 7.8 Graphical representation of (7.3.9) for $N = 3$ and $\varphi = 0$.

Since $w = e^{j\varphi}$ is constrained to lie on the unit circle, the distance between w and w_n is given by

$$d_n(\varphi) = |e^{j\varphi} - e^{j[2\pi n/(N+1)]}| \tag{7.3.9}$$

For the case of $N = 3$, $d_n(\varphi = 0)$ is illustrated in Figure 7.8. From (7.3.3), (7.3.8), and (7.3.9), we get

$$\rho(\varphi) = \frac{1}{N+1}\left|\frac{R_L - Z_0}{R_L + Z_0}\right| \prod_{n=1}^{N} d_n(\varphi) \tag{7.3.10}$$

Thus, as $\varphi = -2\beta l$ varies from 0 to 2π, $w = e^{j\varphi}$ makes one complete traverse of the unit circle, and distances $d_1, d_2, \ldots, d_N$ vary with φ. If w coincides with the root w_n, the distance d_n is zero. Consequently, the product of the distances is zero. Since the product of these distances is proportional to the reflection coefficient, $\rho(\varphi_n)$ goes to zero. It attains a local maximum whenever w is approximately halfway between successive roots.

Example 7.2 Design a four-section quarter-wavelength impedance transformer with uniform distribution of reflection coefficient to match a 100-Ω load to a 50-Ω air-filled coaxial line at 900 MHz. Determine the range of frequencies over which the reflection coefficient remains below 0.1. Compare this bandwidth with that obtained for a single-section impedance transformer.

SOLUTION From (7.3.3) with $N = 4$, we have

$$\Gamma_4 = \frac{1}{5}\frac{100 - 50}{100 + 50} = \frac{1}{15}$$

and from (7.2.2),

$$\frac{1}{15} = \frac{100 - Z_4}{100 + Z_4}$$

or

$$100 + Z_4 = 1500 - 15Z_4 \Rightarrow 16Z_4 = 1400 \Rightarrow Z_4 = 87.5\,\Omega$$

The characteristic impedance of other sections can be determined from (7.2.8) as follows:

$$\frac{1}{15} = \frac{Z_4 - Z_3}{Z_4 + Z_3} \Rightarrow 87.5 + Z_3 = 15 \times 87.5 - 15Z_3 \Rightarrow Z_3 = 76.5625\,\Omega$$

$$\frac{1}{15} = \frac{Z_3 - Z_2}{Z_3 + Z_2} \Rightarrow Z_2 = \frac{14 \times 76.5625}{16} = 66.9922\,\Omega \approx 67\,\Omega$$

and

$$\frac{1}{15} = \frac{Z_2 - Z_1}{Z_2 + Z_1} \Rightarrow Z_1 = \frac{14 \times 67}{16} = 58.625\,\Omega$$

The frequency range over which reflection coefficient remains below 0.1 is determined from (7.3.4) as follows:

$$0.1 = \left|\frac{1}{3} \times \frac{\sin 5\theta_m}{5\sin\theta_m}\right| \Rightarrow \left|\frac{\sin 5\theta_m}{\sin\theta_m}\right| = 1.5$$

where θ_m represents the value of βl at which magnitude of reflection coefficient is equal to 0.1. This is a transcendental equation that can be solved graphically. To that end, one needs to plot $|\sin 5\theta_m|$ and $1.5\,|\sin\theta_m|$ versus θ_m on the same graph and look for the intersection of two curves. Alternatively, a numerical

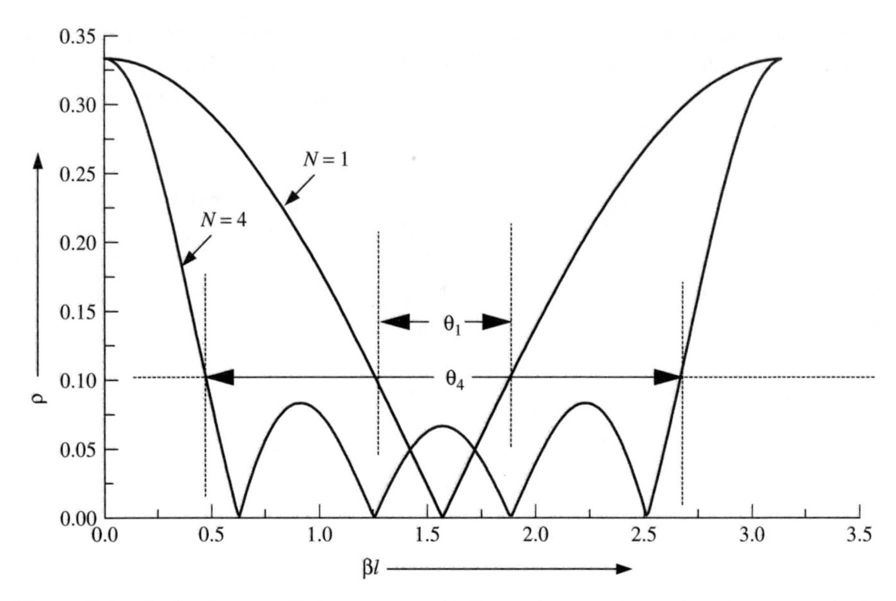

Figure 7.9 Reflection coefficient versus βl for a four-section ($N = 4$) and a single section ($N = 1$) of impedance transformer.

method can be employed to determine θ_m. Two solutions to this equation are found to be 0.476 and 2.665, respectively. Hence,

$$0.476 \le \beta l \le 2.665 \quad \text{or} \quad 272.7\,\text{MHz} \le f \le 1.5269\,\text{GHz}$$

Corresponding bandwidth with a single section can be evaluated as follows:

$$1.2841 \le \beta l \le 1.8575 \quad \text{or} \quad 735.74\,\text{MHz} \le f \le 1.0643\,\text{GHz}$$

Reflection coefficient characteristics for two cases are displayed in Figure 7.9. Clearly, it has a much wider bandwidth, with four sections in comparison with that of a single section.

7.4 BINOMIAL TRANSFORMERS

As shown in Figures 7.5 and 7.6, there are peaks and nulls in the passband of a multisection quarter-wavelength transformer with uniform section reflection coefficients. This characteristic can be traced to equispaced roots on the unit circle. One way to avoid this behavior is to place all the roots at a common point w equal to -1. With this setting, distances d_n are the same for all cases and $\prod_{n=1}^{N}(w - w_n)$ goes to zero only once. It occurs for φ equal to $-\pi$ (i.e., at $\beta l = \pi/2$). Thus, ρ is zero only for the frequency at which each section of the transformer is $\lambda/4$ long. With $w_n = -1$ for all n, equation (7.2.9) may be written as

$$\frac{\Gamma}{\Gamma_N} = \prod_{n=1}^{N} (w - w_n) = (w + 1)^N = \sum_{m=0}^{N} \frac{N!}{m!(N-m)!} w^m \tag{7.4.1}$$

The following binomial expansion is used in writing (7.4.1):

$$(1 + x)^m = \sum_{n=1}^{N} \frac{m!}{n!(m-n)!} x^n$$

A comparison of equation (7.4.1) with (7.2.7) indicates that

$$\frac{\Gamma_n}{\Gamma_N} = \frac{N!}{n!(N-n)!} \qquad n = 0, 1, 2, \ldots, N \tag{7.4.2}$$

and therefore the section reflection coefficients, normalized to Γ_N, are binomially distributed.

From equation (7.4.1),

$$\Gamma = \Gamma_N (w + 1)^N \Rightarrow \Gamma(\beta l) = \Gamma_N (e^{-j2\beta l} + 1)^N = \Gamma_N 2^N e^{-jN\beta l} (\cos \beta l)^N$$

or

$$\rho(\beta l) = \rho_N 2^N |\cos \beta l|^N \tag{7.4.3}$$

For $\beta l = 0$, load R_L is effectively connected to the input line. Therefore,

$$\rho(0) = \rho_N 2^N = \left| \frac{R_L - Z_0}{R_L + Z_0} \right| \tag{7.4.4}$$

and

$$\rho(\beta l) = \left| \frac{R_L - Z_0}{R_L + Z_0} \right| \times |\cos \beta l|^N \tag{7.4.5}$$

Reflection coefficient characteristics of multisection binomial transformers versus βl (in degrees) are illustrated in Figure 7.10. The reflection coefficient scale is normalized with the load reflection coefficient ρ_L. Unlike the preceding case of uniformly distributed section reflection coefficients, it shows a smooth characteristic without lobes.

Example 7.3 Design a four-section quarter-wavelength binomial impedance transformer to match a 100-Ω load to a 50-Ω air-filled coaxial line at 900 MHz. Determine the range of frequencies over which the reflection coefficient remains below 0.1. Compare this result with that obtained in Example 7.2.

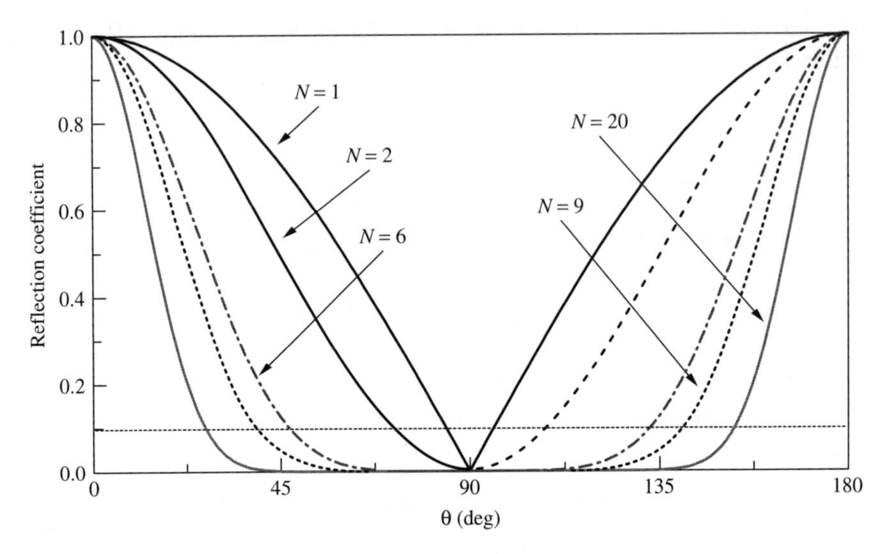

Figure 7.10 Reflection coefficient versus βl for a multisection binomial transformer.

SOLUTION With $N = 4$ and $n = 0$, 1, and 2 in (7.4.2), we get

$$\frac{\Gamma_0}{\Gamma_4} = \frac{4!}{0!4!} = \frac{\Gamma_4}{\Gamma_4} \Rightarrow \Gamma_0 = \Gamma_4$$

$$\frac{\Gamma_1}{\Gamma_4} = \frac{4!}{1!3!} = \frac{\Gamma_3}{\Gamma_4} \Rightarrow \Gamma_1 = \Gamma_3 = 4\Gamma_4$$

and

$$\frac{\Gamma_2}{\Gamma_4} = \frac{4!}{2!2!} = 6 \Rightarrow \Gamma_2 = 6\Gamma_4$$

From (7.4.4),

$$\rho(0) = \rho_N 2^N = \left|\frac{R_L - Z_0}{R_L + Z_0}\right| \Rightarrow \rho_4 = \frac{1}{2^4}\left|\frac{100 - 50}{100 + 50}\right| = \frac{1}{48} = 0.020833$$

and from (7.2.8),

$$\Gamma_n = \frac{Z_{n+1} - Z_n}{Z_{n+1} + Z_n} \Rightarrow \rho_n(Z_{n+1} + Z_n) = Z_{n+1} - Z_n$$

Therefore,

$$Z_n = \frac{1 - \rho_n}{1 + \rho_n} \times Z_{n+1} \tag{7.4.6}$$

Alternatively,

$$Z_{n+1} = \frac{1 + \rho_n}{1 - \rho_n} \times Z_n \tag{7.4.7}$$

Characteristic impedance of each section can be determined from (7.4.6), as follows:

$$Z_4 = \frac{1 - 1/48}{1 + 1/48} \times 100 = 95.9184\,\Omega$$

$$Z_3 = \frac{1 - 4/48}{1 + 4/48} \times 95.9184 = 81.1617\,\Omega$$

$$Z_2 = \frac{1 - 6/48}{1 + 6/48} \times 81.1617 = 63.1258\,\Omega$$

and

$$Z_1 = \frac{1 - 4/48}{1 + 4/48} \times 63.1258 = 53.4141\,\Omega$$

If we continue with this formula, we find that

$$Z_0 = \frac{1 - 1/48}{1 + 1/48} \times 53.4141 = 51.2339\,\Omega$$

This is different from the given value of 50 Ω. It happened because of the approximation involved in the formula. Error keeps building up if the characteristic impedances are determined proceeding in just one way. To minimize it, common practice is to determine the characteristic impedances up to halfway proceeding from the load side and then the remaining half from the input side. Thus, Z_1 and Z_2 should be determined from (7.4.7) as follows:

$$Z_1 = \frac{1 + 1/48}{1 - 1/48} \times 50 = 52.1277\,\Omega$$

and

$$Z_2 = \frac{1 + 4/48}{1 - 4/48} \times 50 = 61.6054\,\Omega$$

The frequency range over which the reflection coefficient remains below 0.1 can be determined from (7.4.5) as follows:

$$0.1 = \tfrac{1}{3}|\cos(\vartheta_M)|^4 \Rightarrow \vartheta_M = 0.7376$$

Therefore,

$$0.7376 \le \beta l \le 2.404 \quad \text{or} \quad 422.61\,\text{MHz} \le f \le 1.3774\,\text{GHz}$$

Clearly, it has a larger frequency bandwidth than that of a single section. However, it is less than the bandwidth obtained with uniformly distributed section reflection coefficients. Reflection coefficient as a function of βl is illustrated in Figure 7.11.

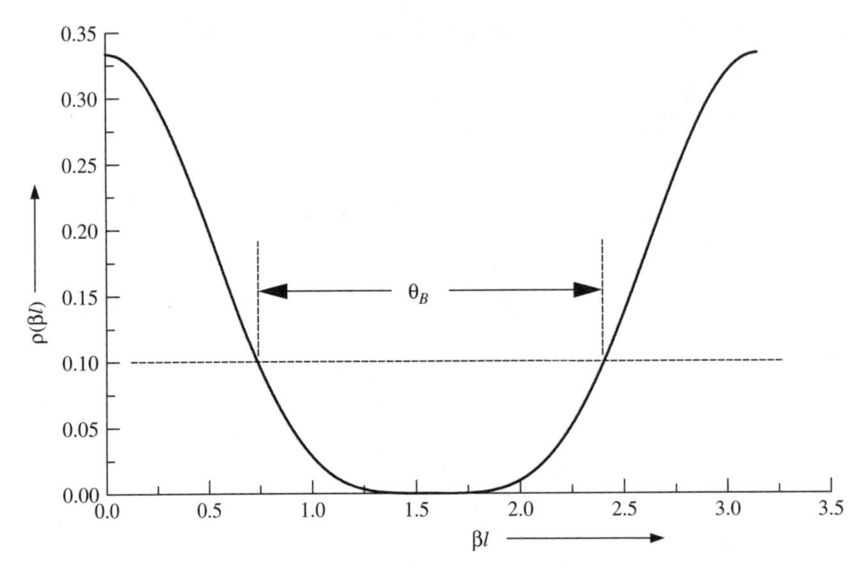

Figure 7.11 Reflection coefficient of a four-section binomial transformer versus βl.

7.5 CHEBYSHEV TRANSFORMERS

Consider again the case of a uniform three-section impedance transformer that is connected between a 50-Ω line and 100-Ω load. Distribution of zeros around the unit circle is shown in Figure 7.12 with solid points. Its frequency response is illustrated in Figure 7.13 as curve (a). Now, let us move two of these zeros to $\pm 120^\circ$ while keeping the remaining one fixed at 180°, as illustrated by unfilled

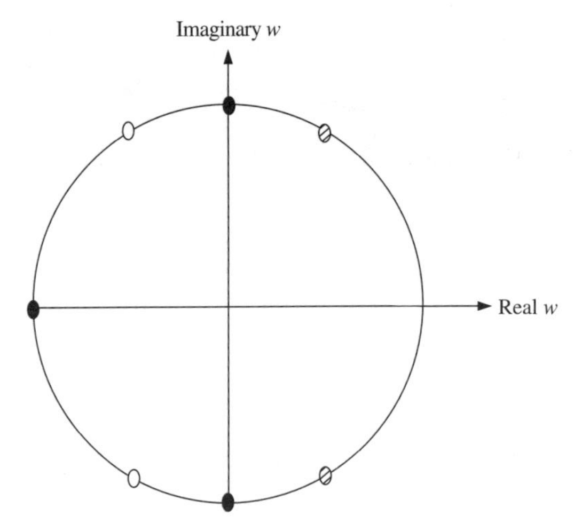

Figure 7.12 Distribution of zeros for a uniform three-section impedance transformer (solid), with two of those zeros moved to $\pm 120^\circ$ (unfilled) or to $\pm 60^\circ$ (hatched).

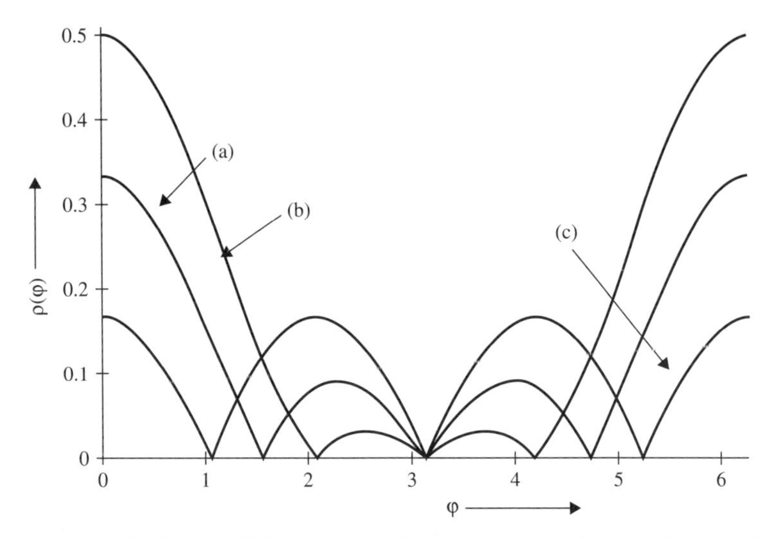

Figure 7.13 Reflection coefficient versus φ for (a) a uniform three-section transformer, (b) two zeros moved to ±120°, and (c) two zeros moved to ±60°.

points in Figure 7.12. With this change in the distribution of zeros on the complex w-plane, $\rho(\beta l)$ versus βl varies as shown by curve (b) in Figure 7.13. It shows relatively much lower peaks of intervening lobes, while the widths and heights of the two main lobes increase. On the other hand, if we move the two zeros to ±60°, the main peaks go down while intervening lobes rise. This is illustrated by hatched points in Figure 7.12 and by curve (c) in Figure 7.13. In this case, minor lobes increase while the main lobes are reduced in size. Note that zeros in this case are uniformly distributed around the unit circle, and therefore its $\rho(\beta l)$ characteristic has identical lobes.

Thus, the heights of intervening lobes decrease at the cost of the main lobe when zeros are moved closer together. On the other hand, moving the zeros farther apart raises the level of intervening lobes but reduces the main lobe. However, we need a systematic method to determine the location of each zero for a maximum permissible reflection coefficient, ρ_M, and the number of quarter-wave sections, N. An optimal distribution of zeros around the unit circle will keep the peaks of all passband lobes at the same height of ρ_M.

In order to have the magnitudes of all minor lobes in the passband equal, section reflection coefficients are determined by the characteristics of Chebyshev functions, named after the Russian mathematician who first studied them. These functions satisfy the following differential equation:

$$(1 - x^2)\frac{d^2T_m(x)}{dx^2} - x\frac{dT_m(x)}{dx} + m^2T_m(x) = 0 \qquad (7.5.1)$$

Chebyshev functions of degree m, represented by $T_m(x)$, are mth-degree polynomials that satisfy (7.5.1). The first four of these and a recurrence relation for higher-order Chebyshev polynomials are given as follows:

$$
\begin{aligned}
T_1(x) &= x \\
T_2(x) &= 2x^2 - 1 \\
T_3(x) &= 4x^3 - 3x \\
T_4(x) &= 8x^4 - 8x^2 + 1 \\
&\vdots \\
T_m(x) &= 2xT_{m-1}(x) - T_{m-2}(x)
\end{aligned}
$$

Alternatively,

$$
T_m(x) = \begin{cases} \cos(m\cos^{-1}x) & -1 \le x \le 1 \\ \cosh(m\cosh^{-1}x) & x \ge 1 \\ (-1)^m\cosh(m\cosh^{-1}|x|) & x \le -1 \end{cases} \tag{7.5.2}
$$

Note that for $x = \cos\theta$,

$$
T_m(\cos\theta) = \cos m\theta \tag{7.5.3}
$$

Figure 7.14 depicts Chebyshev polynomials of degrees 1 through 4. The following characteristics can be noted:

- The magnitudes of these polynomials oscillate between ± 1 for $-1 \le x \le 1$.
- For $|x| > 1$, $|T_m(x)|$ increases at a faster rate with x as m increases.

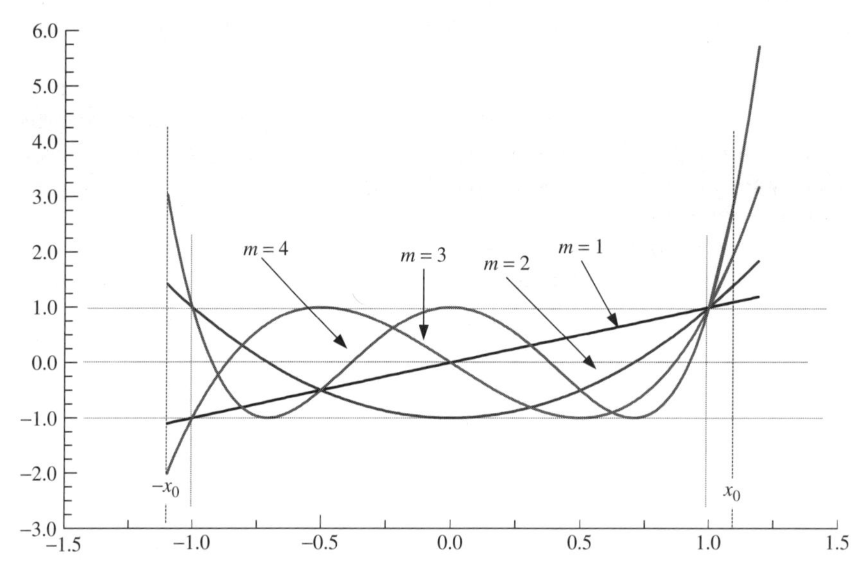

Figure 7.14 Chebyshev polynomials for m = 1, 2, 3, 4.

- The numbers of zeros are equal to the order of the polynomials. Zeros of an even-order polynomial are located symmetrically about the origin, with one of the minor lobes' peak at $x = 0$.
- Polynomials of odd orders have one zero at $x = 0$ while the remaining zeros are located symmetrically.

These characteristics of Chebyshev polynomials are utilized to design an impedance transformer that has ripples of equal magnitude in its passband. The number of quarter-wave sections determines the order of Chebyshev polynomials and the distribution of zeros on the complex w-plane is selected according to that. With x_0 properly selected, $|T_m(x)|$ corresponds precisely to $\rho(\beta l)$. This is done by linking βl to x of Chebyshev polynomial as follows:

$$x = x_0 \cos \beta l \tag{7.5.4}$$

Consider the design of a three-section equal-ripple impedance transformer. A Chebyshev polynomial of order 3 is appropriate for this case. Figure 7.15 illustrates $T_3(x)$ together with (7.5.4). It is to be noted that Chebyshev variable x and angle φ on the complex w-plane are related through (7.5.4) because φ is equal to $-2\beta l$. As φ varies from 0 to -2π, x changes as illustrated in Figure 7.15(b) and the corresponding $T_3(x)$ in Figure 7.15(a). Figure 7.16 shows $|T_3(\varphi)|$, which can represent the desired $\rho(\varphi)$ provided that

$$\frac{T_3(\varphi = 0)}{1} = \frac{T_3(x_0)}{1} = \frac{\left|\dfrac{(R_L - Z_0)}{(R_L + Z_0)}\right|}{\rho_M} \tag{7.5.5}$$

where ρ_M is the maximum allowed reflection coefficient in the passband.

For an m-section impedance transformer, (7.5.5) can be written as

$$T_m(x_0) = \left|\frac{R_L - Z_0}{R_L + Z_0}\right| \frac{1}{\rho_M} \tag{7.5.6}$$

Location x_0 can now be determined from (7.5.2) as follows:

$$x_0 = \cosh\left[\frac{1}{m} \times \cosh^{-1} T_m(x_0)\right] \tag{7.5.7}$$

With x_0 determined from (7.5.7), $|T_m(x)|$ represents the $\rho(\beta l)$ desired. Hence, zeros of the reflection coefficient are the same as those of $T_m(x)$. Since Chebyshev polynomials have zeros in the range $-1 < x < 1$, zeros of $\rho(\varphi)$ can be determined from (7.5.2). Therefore,

$$T_m(x_n) = 0 \Rightarrow \cos(m \times \cos^{-1} x_n) = 0 = \pm\cos\left[(2n - 1)\frac{\pi}{2}\right] \qquad n = 1, 2, \ldots$$

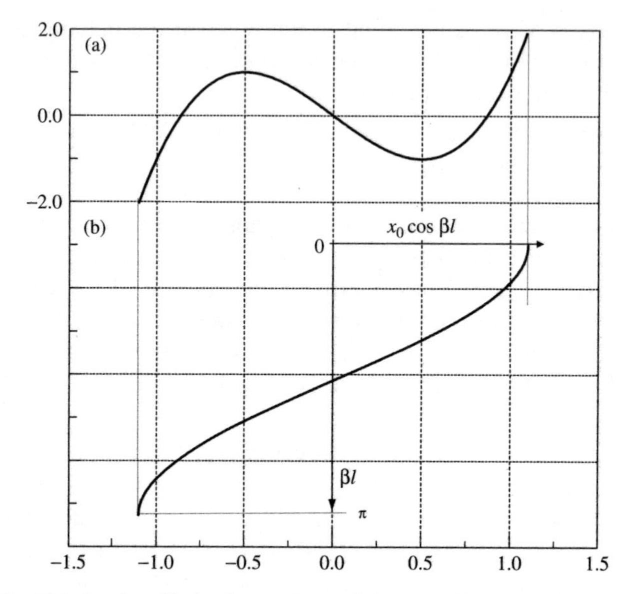

Figure 7.15 Third-order Chebyshev polynomial (a) and its variable, x versus βl (b).

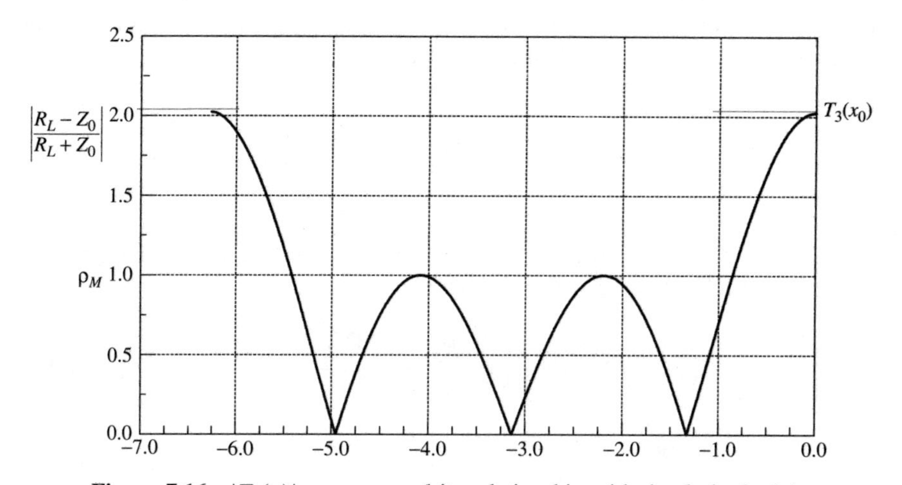

Figure 7.16 $|T_3(x)|$ versus φ and its relationship with the desired $\rho(\varphi)$.

where x_n is the location of the nth zero. Hence,

$$x_n = \pm \cos\left[(2n-1)\frac{\pi}{2m}\right] \tag{7.5.8}$$

and the corresponding φ_n can be evaluated from (7.5.4) as follows:

$$\varphi_n = 2\cos^{-1}\frac{x_n}{x_0} \tag{7.5.9}$$

The zeros of $\rho(\varphi)$, w_n, on the complex w-plane are now known because $w_n = e^{i\varphi_n}$. Equation (7.2.9) can be used to determine the section reflection coefficient, Γ_n. Z_n is, in turn, determined from equation (7.2.8). The bandwidth of the impedance transformer extends from $x = -1$ to $x = +1$. With $\beta l = \theta_M$ at $x = 1$, (7.5.4) gives

$$\cos\theta_M = \frac{1}{x_0}$$

Therefore, bandwidth of a Chebyshev transformer can be expressed as follows:

$$\cos^{-1}\frac{1}{x_0} < \beta l \leq \pi - \cos^{-1}\frac{1}{x_0} \tag{7.5.10}$$

Example 7.4 Reconsider the matching of a 100-Ω load to a 50-Ω line. Design an equal-ripple four-section quarter-wavelength impedance transformer and compare its bandwidth with those obtained earlier for $\rho_M = 0.1$.

SOLUTION From (7.5.6),

$$T_4(x_0) = \left|\frac{100-50}{100+50}\right| \times \frac{1}{0.1} = \frac{1}{0.3} = 3.3333$$

Therefore, x_0 can now be determined from (7.5.7) as follows:

$$x_0 = \cosh\left[\tfrac{1}{4}\cosh^{-1}(3.3333)\right]$$

In case inverse hyperbolic functions are not available on a calculator, the following procedure can be used to evaluate x_0. Assume that $y = \cosh^{-1}(3.3333) \Rightarrow 3.3333 = \cosh y = (e^y + e^{-y})/2$.

Hence,

$$e^y + e^{-y} = 2 \times 3.3333 = 6.6666$$

or

$$e^{2y} - 6.6666 \times e^y + 1 = 0 \Rightarrow e^y = \frac{6.6666 \pm \sqrt{(6.6666)^2 - 4}}{2} = 6.5131$$

and

$$0.1535y = \ln(6.5131) = 1.8738$$

Therefore,

$$\frac{1}{4}\cosh^{-1}(3.3333) = \frac{1.8738}{4} = 0.4684$$

and

$$x_0 = \cosh(0.4684) = \frac{e^{0.4684} + e^{-0.4684}}{2} = 1.1117$$

Note that the other solution of e^y produces $y = -1.874$ and $x_0 = 1.1117$.

From (7.5.8) and (7.5.9), we get

$$x_n = \pm 0.9239, \pm 0.3827$$

and

$$\varphi_n = -67.61^\circ, -139.71^\circ, -220.29^\circ, -292.39^\circ$$

Therefore,

$$w_1 = 0.381 - j0.925$$
$$w_2 = -0.763 - j0.647$$
$$w_3 = -0.763 + j0.647$$

and

$$w_4 = 0.381 + j0.925$$

Now, from equation (7.2.9) we get

$$\Gamma = \Gamma_4(1 + 0.764w + 0.837w^2 + 0.764w^3 + w^4)$$

Therefore,

$$\Gamma_0 = \Gamma_4, \Gamma_1 = \Gamma_3 = 0.764\Gamma_4 \quad \text{and} \quad \Gamma_2 = 0.837\Gamma_4$$

From equation (7.2.12), we have

$$\Gamma = \sum_{n=0}^{N} \Gamma_n = \frac{R_L - Z_0}{R_L + Z_0} \Rightarrow 4.365\Gamma_4 = \frac{R_L - Z_0}{R_L + Z_0}$$

$$\Gamma_4 = \frac{1}{4.365} \times \frac{100 - 50}{100 + 50} = 0.076$$

Hence,

$$\Gamma_0 = \Gamma_4 = 0.076, \quad \Gamma_1 = \Gamma_3 = 0.058, \quad \text{and} \quad \Gamma_2 = 0.064$$

Now, the characteristic impedance of each section can be determined from (7.2.8) as follows:

$$\Gamma_4 = \frac{Z_L - Z_4}{Z_L + Z_4} \Rightarrow Z_4 = \frac{1 - \Gamma_4}{1 + \Gamma_4} Z_L = 85.87\,\Omega$$

$$\Gamma_3 = \frac{Z_4 - Z_3}{Z_4 + Z_3} \Rightarrow Z_3 = \frac{1 - \Gamma_3}{1 + \Gamma_3} Z_4 = 76.46\,\Omega$$

$$\Gamma_2 = \frac{Z_3 - Z_2}{Z_3 + Z_2} \Rightarrow Z_2 = \frac{1 - \Gamma_2}{1 + \Gamma_2} Z_3 = 67.26\,\Omega$$

and

$$\Gamma_1 = \frac{Z_2 - Z_1}{Z_2 + Z_1} \Rightarrow Z_1 = \frac{1 - \Gamma_1}{1 + \Gamma_1} Z_2 = 59.89\ \Omega$$

To minimize the accumulating error in Z_n, it is advisable to calculate half of the impedance values from the load side and the other half from the input side. In other words, Z_1 and Z_2 should be determined as follows:

$$Z_1 = \frac{1 + \Gamma_0}{1 - \Gamma_0} Z_0 = 58.2251\ \Omega$$

and

$$Z_2 = \frac{1 + \Gamma_1}{1 - \Gamma_1} Z_1 = 65.3951\ \Omega$$

The bandwidth is determined from (7.5.10) as follows:

$$\cos \beta l = \frac{1}{x_0} = 0.899 \Rightarrow \beta l = 0.452$$

Therefore,

$$0.452 \leq \beta l \leq 2.6896$$

Thus, the bandwidth is greater than what was achieved from either a uniform or a binomial distribution of Γ_n coefficients.

7.6 EXACT FORMULATION AND DESIGN OF MULTISECTION IMPEDANCE TRANSFORMERS

The analysis and design presented so far is based on the assumption that the section reflection coefficients are small, as implied by (7.2.5). If this is not the case, an exact expression for the reflection coefficient must be used. Alternatively, there are design tables available in the literature (e.g., Matthaei et al., 1980) that can be used to synthesize an impedance transformer. In case of a two- or three-section Chebyshev transformer, the design procedure summarized below may be used as well.

Exact formulation of the multisection impedance transformer is conveniently developed via the *power loss ratio*, P_{LR}, defined as follows:

$$P_{LR} = \frac{\text{incident power}}{\text{power delivered to the load}}$$

If P_{in} represents the incident power and ρ_{in} is the input reflection coefficient, then

$$P_{\text{LR}} = \frac{P_{\text{in}}}{(1 - \rho_{\text{in}}^2) P_{\text{in}}} = \frac{1}{1 - \rho_{\text{in}}^2} \Rightarrow \rho_{\text{in}} = \sqrt{\frac{P_{\text{LR}} - 1}{P_{\text{LR}}}} \tag{7.6.1}$$

For any transformer, ρ_{in} can be determined from its input impedance Z_{in}. The power loss ratio P_{LR} can subsequently be evaluated from (7.6.1). This P_{LR} can be expressed in terms of $Q_{2N}(\cos\theta)$, an even polynomial of degree $2N$ in cos θ. Hence,

$$P_{LR} = 1 + Q_{2N}(\cos\theta) \tag{7.6.2}$$

Coefficients of $Q_{2N}(\cos\theta)$ are functions of various impedances Z_n. For an equal-ripple characteristic in the passband, a Chebyshev polynomial can be used to specify P_{LR} as follows:

$$P_{LR} = 1 + k^2 T_N^2(\sec\theta_M \cos\theta) \tag{7.6.3}$$

where k^2 is determined from the maximum value of P_{LR} in the passband. θ_M represents the value of βl that corresponds to the maximum allowed reflection coefficient ρ_M. Since $T_N^2(\sec\theta_M \cos\theta)$ has a maximum value of unity in the passband, the extreme value of P_{LR} will be $1 + k^2$. Various characteristic impedances are determined by equating (7.6.2) and (7.6.3). Further, (7.6.1) produces a relation between ρ_M and k^2, as follows:

$$\rho_M = \sqrt{\frac{k^2}{1+k^2}} \tag{7.6.4}$$

For the two-section impedance transformer shown in Figure 7.17, reflection coefficient characteristics as a function of θ are illustrated in Figure 7.18. The power loss ratio for this transformer is found to be

$$P_{LR} = 1 + \frac{(Z_L - Z_0)^2}{4Z_L Z_0} \frac{(\sec^2\theta_z \cos^2\theta - 1)^2}{\tan^4\theta_z} \tag{7.6.5}$$

where θ_z is the value of θ at the lower zero (i.e., $\theta_z < \pi/2$), as shown in Figure 7.18.

The maximum power loss ratio, $P_{LR\,max}$, is found to be

$$P_{LR\ max} = 1 + \frac{(Z_L - Z_0)^2}{4Z_L Z_0} \cot^4\theta_z \tag{7.6.6}$$

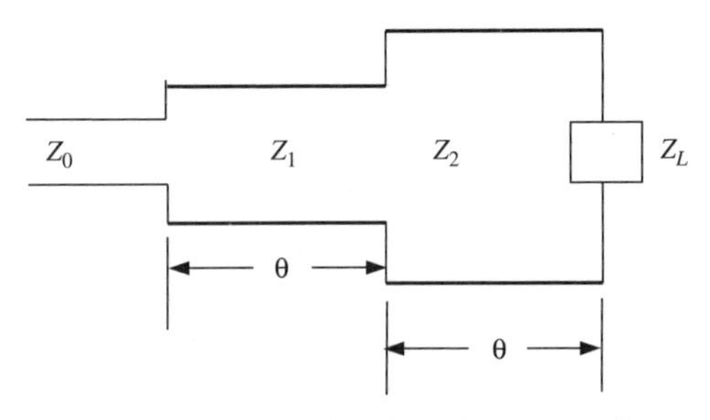

Figure 7.17 Two-section impedance transformer.

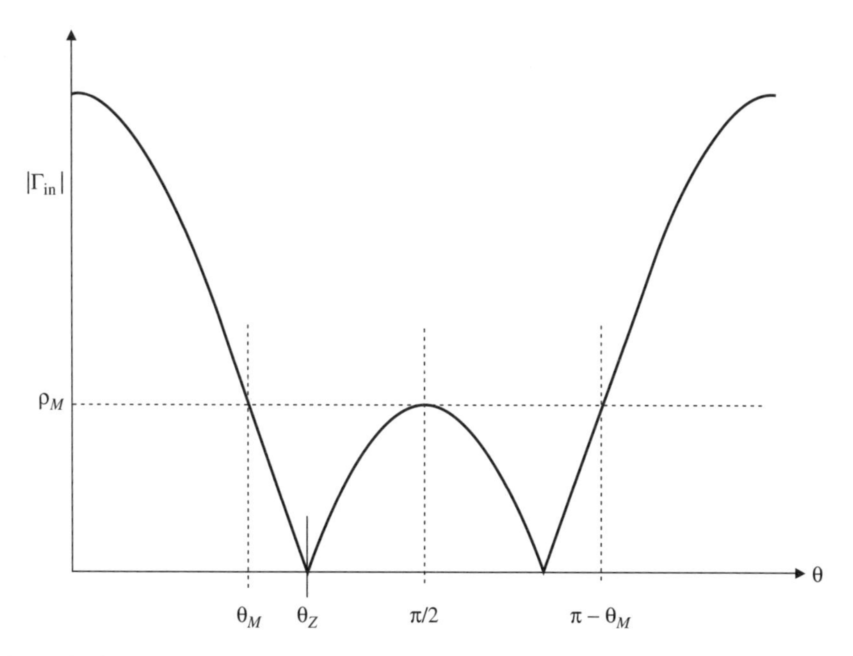

Figure 7.18 Reflection coefficient characteristics of a two-section Chebyshev transformer.

Hence,

$$k^2 = \frac{(Z_L - Z_0)^2}{4Z_L Z_0} \cot^4 \theta_z \tag{7.6.7}$$

Required values of characteristic impedances, Z_1 and Z_2, are determined from the equations

$$Z_1^2 = Z_0^2 \left[\frac{(Z_L - Z_0)^2}{4Z_0^2 \tan^4 \theta_z} + \frac{Z_L}{Z_0} \right]^{1/2} + \frac{(Z_L - Z_0) Z_0}{2 \tan^2 \theta_z} \tag{7.6.8}$$

and

$$Z_2 = \frac{Z_L}{Z_1} Z_0 \tag{7.6.9}$$

The passband edge, θ_M, is given by

$$\theta_M = \cos^{-1}(\sqrt{2} \cos \theta_z) \tag{7.6.10}$$

Hence, the bandwidth of this transformer is found as

$$\theta_M \le \theta = \beta l \le \pi - \theta_M \tag{7.6.11a}$$

For a TEM wave,

$$\frac{\Delta f}{f_0} = 2 - \frac{4\theta_M}{\pi} \tag{7.6.11b}$$

If the bandwidth is specified in a design problem along with Z_L and Z_0, then θ_M is known. Therefore, θ_z can be determined from (7.6.10). Impedances Z_1 and Z_2 are determined subsequently from (7.6.8) and (7.6.9), respectively. The corresponding maximum reflection coefficient ρ_M can easily be evaluated following k^2 from (7.6.7). On the other hand, if ρ_M is specified instead of θ_M, then k^2 is determined from (7.6.4). θ_z, in turn, is calculated from (7.6.7). Now, Z_1, Z_2, and θ_M can be determined from (7.6.8), (7.6.9), and (7.6.10), respectively.

In the limit $\theta_z \to \pi/2$, two zeros of ρ in Figure 7.18 come together to give a maximally flat transformer. In that case, (7.6.8) and (7.6.9) are simplified to give

$$Z_1 = Z_L^{1/4} \times Z_0^{3/4} \tag{7.6.12}$$

$$Z_2 = Z_L^{3/4} \times Z_0^{1/4} \tag{7.6.13}$$

and

$$\theta_M = \cos^{-1}(\cot\theta_z) \tag{7.6.14}$$

Example 7.5 Use the exact theory to design a two-section Chebyshev transformer to match a 500 Ω load to a 100-Ω line. The required fractional bandwidth is 0.6. What is the resulting value of ρ_M?

SOLUTION From (7.6.10) and (7.6.11),

$$\frac{\Delta f}{f_0} = 2 - \frac{4}{\pi}\cos^{-1}(\sqrt{2}\cos\theta_z) \Rightarrow \cos\theta_z = \frac{1}{\sqrt{2}}\cos\left(\frac{2 - \Delta f/f_0}{4/\pi}\right)$$

$$= \frac{1}{\sqrt{2}}\cos\left(\frac{2 - 0.6}{4/\pi}\right) = 0.321$$

Therefore, $\theta_z = 1.244$ rad and

$$\theta_M = \cos^{-1}(\sqrt{2}\cos\theta_z) = \cos^{-1}(\sqrt{2} \times 0.321) = 1.1$$

From (7.6.7),

$$k^2 = \frac{(Z_L - Z_0)^2}{4Z_L Z_0}\cot^4\theta_z = \frac{(\overline{Z}_L - 1)^2}{4\overline{Z}_L}\cot^4\theta_z = 0.0106$$

Therefore, from (7.6.4),

$$\rho_M = \sqrt{\frac{k^2}{1 + k^2}} = 0.1022$$

Now, from (7.6.8),

$$Z_1^2 = Z_0^2 \left[\frac{(Z_L - Z_0)^2}{4Z_0^2 \tan^4 \theta_z} + \frac{Z_L}{Z_0} \right]^{1/2} + \frac{(Z_L - Z_0)Z_0}{2\tan^2 \theta_z} \Rightarrow \overline{Z}_1^2$$

$$= \left[\frac{(\overline{Z}_L - 1)^2}{4\tan^4 \theta_z} + \overline{Z}_L \right]^{1/2} + \frac{\overline{Z}_L - 1}{2\tan^2 \theta_z} = 2.4776$$

Therefore,

$$\overline{Z}_1 = \sqrt{2.4776} = 1.57 \Rightarrow Z_1 = 157.41\ \Omega$$

and from (7.6.9),

$$Z_2 = \frac{Z_L}{Z_1} Z_0 = 317.65\ \Omega$$

Check: For $\theta = \pi/2$,

$$\overline{Z}_{\text{in}} = \frac{\overline{Z}_1^2}{\overline{Z}_2^2} \overline{Z}_L = \frac{1.5741^2}{3.1765^2}(5) = 1.2277$$

$$\Gamma = \frac{\overline{Z}_{\text{in}} - 1}{\overline{Z}_{\text{in}} + 1} = 0.1022 = \rho_m$$

Let us consider the design of a three-section impedance transformer. Figure 7.19 shows a three-section impedance transformer connected between the load Z_L and a transmission line of characteristic impedance Z_0. The reflection coefficient characteristic of such a Chebyshev transformer is illustrated in Figure 7.20. The power loss ratio, P_{LR}, of a three-section Chebyshev transformer is found to be

$$P_{\text{LR}} = 1 + \frac{(Z_L - Z_0)^2}{4Z_L Z_0} \frac{(\sec^2 \theta_z \cos^2 \theta - 1)^2 \cos^2 \theta}{\tan^4 \theta_z} \tag{7.6.15}$$

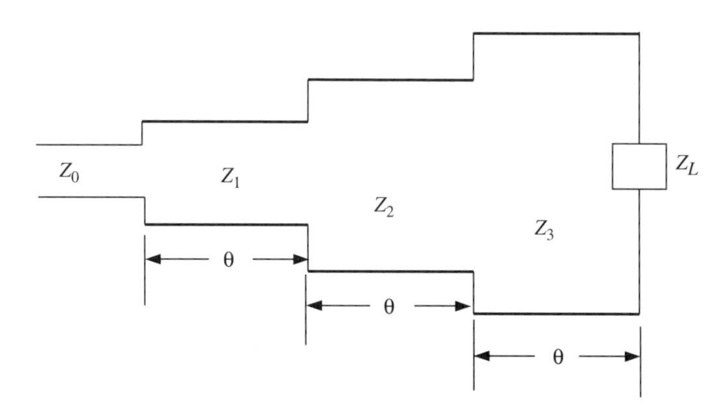

Figure 7.19 Three-section impedance transformer.

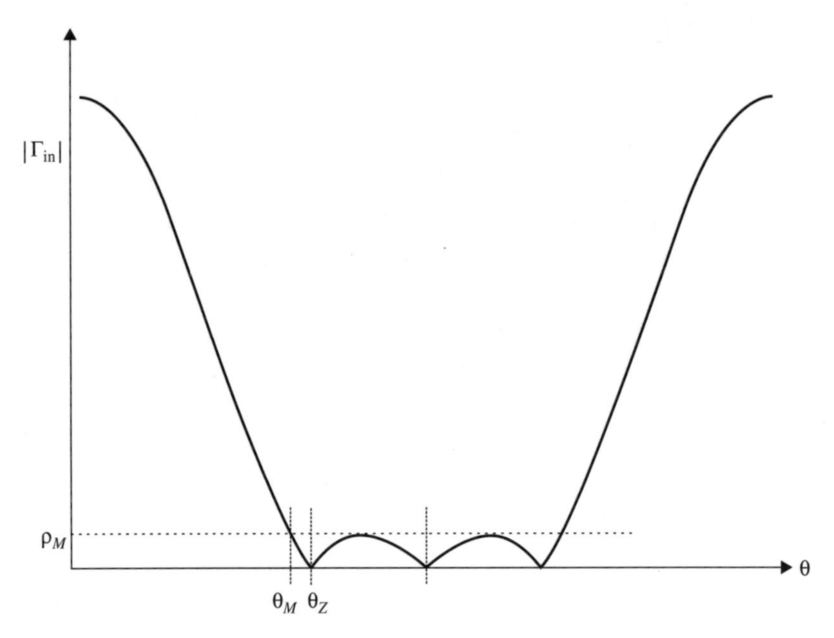

Figure 7.20 Reflection coefficient characteristics of a three-section Chebyshev transformer.

It attains the maximum allowed value at $\theta = \theta_M$, where

$$\theta_M = \cos^{-1}\left(\frac{2}{\sqrt{3}}\cos\theta_z\right) \tag{7.6.16}$$

Since the maximum power loss ratio must be equal to $1 + k^2$,

$$k^2 = \frac{(Z_L - Z_0)^2}{4Z_L Z_0}\left(\frac{2\cos\theta_z}{3\sqrt{3}\tan^2\theta_z}\right)^2 \tag{7.6.17}$$

Characteristic impedance Z_1 is obtained by solving the equation

$$\frac{Z_L - Z_0}{\tan^2\theta_z} = \frac{Z_1^2}{Z_0} + 2\left(\frac{Z_L}{Z_0}\right)^{1/2} Z_1 - \frac{Z_L Z_0^2}{Z_1^2} - 2\left(\frac{Z_L}{Z_0}\right)^{1/2} Z_1^{-1} Z_0^2 \tag{7.6.18}$$

The other two characteristic impedances, Z_2 and Z_3, are then determined as follows:

$$Z_2 = \sqrt{Z_L Z_0} \tag{7.6.19}$$

and

$$Z_3 = \frac{Z_L Z_0}{Z_1} \tag{7.6.20}$$

The range of the passband (bandwidth) is still given by (7.6.11) provided that θ_M is now computed from (7.6.16).

Example 7.6 Design a three-section Chebyshev transformer (exact theory) to match a 500-Ω load to a 100-Ω line. The required fractional bandwidth is 0.6. Compute ρ_M and compare it with that obtained in Example 7.5.

SOLUTION From (7.6.16) and (7.6.11),

$$\theta_M = \cos^{-1}\left(\frac{2}{\sqrt{3}}\cos\theta_z\right) \quad \text{and} \quad \frac{\Delta f}{f_0} = 0.6 \Rightarrow \frac{(\pi/2) - \theta_M}{\pi/2}$$

$$= 0.3 \Rightarrow \theta_M = \frac{\pi}{2}(1 - 0.3) = 1.1$$

Therefore,

$$\cos\theta_z = \frac{\sqrt{3}}{2}\cos\theta_M = 0.3932 \Rightarrow \theta_z = 1.17\,\text{rad}$$

Now, from (7.6.17),

$$\mathrm{k}^2 = \frac{(\mathrm{Z_L} - \mathrm{Z_0})^2}{4\mathrm{Z_L}\mathrm{Z_0}}\left(\frac{2\cos\theta_z}{3\sqrt{3}\tan^2\theta_z}\right)^2 = \frac{(\overline{\mathrm{Z}}_\mathrm{L} - 1)^2}{4\overline{\mathrm{Z}}_\mathrm{L}}\left(\frac{2\cos\theta_z}{3\sqrt{3}\tan^2\theta_z}\right)^2 = 6.125 \times 10^{-4}$$

and from (7.6.4),

$$\rho_M = \sqrt{\frac{k^2}{1 + k^2}} = 0.0247$$

which is approximately one-fourth of the previous (two-section transformer) case.

Z_1 is obtained by solving equation (7.6.18) as follows:

$$\frac{Z_L - Z_0}{\tan^2\theta_z} = \frac{Z_1^2}{Z_0} + 2\left(\frac{Z_L}{Z_0}\right)^{1/2} Z_1 - \frac{Z_L Z_0^2}{Z_1^2} - 2\left(\frac{Z_L}{Z_0}\right)^{1/2} Z_1^{-1} Z_0^2$$

or

$$\frac{\overline{Z}_L - 1}{\tan^2\theta_z} = \overline{Z}_1^2 + 2(\overline{Z}_L)^{1/2}\,\overline{Z}_1 - \frac{\overline{Z}_L}{\overline{Z}_1^2} - 2(\overline{Z}_L)^{1/2}\,\overline{Z}_1^{-1} = \frac{4}{\tan^2\theta_z} = 0.7314$$

Therefore,

$$\overline{Z}_1^2 + 2\sqrt{5}\,\overline{Z}_1 - \frac{5}{\overline{Z}_1^2} - \frac{2\sqrt{5}}{\overline{Z}_1} = 0.7314$$

or

$$\overline{Z}_1^4 + 2\sqrt{5}\,\overline{Z}_1^3 - 0.7314\,\overline{Z}_1^2 - 2\sqrt{5}\,\overline{Z}_1 - 5 = 0$$

Therefore,

$$\overline{Z}_1 = -4.4679, (-0.6392 - j0.6854), (-0.6392 + j0.6854), \text{ and } 1.2741$$

Thus, only one of these solutions can be physically realized because the others have a negative real part. After selecting $Z_1 = 127.42\ \Omega$, (7.6.19) and (7.6.20) give

$$Z_3 = \frac{Z_L Z_0}{Z_1} = 392.4277\ \Omega$$

and

$$Z_2 = \sqrt{Z_L Z_0} = 223.6068\ \Omega$$

If we use the approximate formulation, then following the procedure of Example 7.4, we find from (7.5.6) that

$$T_3(x_0) = \frac{500 - 100}{500 + 100} \times \frac{1}{0.02474} = 26.9458$$

Now, from (7.5.7),

$$x_0 = \cosh[\tfrac{1}{3}\cosh^{-1}(26.9458)] = 2.0208$$

and from (7.5.8),

$$x_n = \pm\cos\left[(2n-1)\frac{\pi}{6}\right] = \pm 0.866, 0$$

Using (7.5.9) we get

$$\varphi_n = 2.2558, 3.1416, \text{ and } 4.0274$$

Since $w_n = e^{j\varphi_n}$, we find that

$$w_1 = -0.6327 + j0.7744$$

$$w_2 = -0.6327 - j0.7744$$

$$w_3 = -1$$

and from (7.2.9),

$$\frac{\Gamma}{\Gamma_3} = (w - w_1)(w - w_2)(w - w_3) = w^3 + 2.2654w^2 + 2.2654w + 1$$

Hence,

$$\Gamma_0 = \Gamma_3$$

$$\Gamma_1 = \Gamma_2 = 2.2654\Gamma_3$$

Now, from (7.2.12),

$$\Gamma_0 + \Gamma_1 + \Gamma_2 + \Gamma_3 = 6.5307\Gamma_3 = \frac{500 - 100}{500 + 100}$$

Therefore,

$$\Gamma_3 = 0.1021 = \Gamma_0$$

$$\Gamma_1 = \Gamma_2 = 0.2313$$

Corresponding impedances are now determined from (7.2.8) as follows:

$$\left.\begin{aligned} \overline{Z}_1 &= \frac{1+\Gamma_0}{1-\Gamma_0} = 1.2274 \\ \overline{Z}_2 &= \frac{1+\Gamma_1}{1-\Gamma_1}\overline{Z}_1 = 2.4234 \end{aligned}\right\} \quad \text{calculated from the input side}$$

and

$$\overline{Z}_3 = \frac{1-\Gamma_3}{1+\Gamma_3}\overline{Z}_L = 4.0737 \qquad \text{calculated from the load side}$$

A comparison of these values with those obtained earlier using the exact formulation shows that the two sets are within 10% of deviation for this example. Note that for the fractional bandwidth of 0.6 with a single section, reflection coefficient ρ_M increases to 0.3762.

7.7 TAPERED TRANSMISSION LINES

Consider that the multisection impedance transformer of Figure 7.4 is replaced by a tapered transition of length L, as illustrated in Figure 7.21. The characteristic impedance of this transition is a continuous smooth function of distance, with its values at the two ends as Z_0 and R_L. An approximate theory of such a matching section can easily be developed on the basis of the analysis presented in Section 7.2.

The continuously tapered line can be modeled by a large number of incremental sections of length δz. One of these sections, connected at z, has a characteristic impedance of $Z + \delta Z$, and the one before it has a characteristic impedance of Z, as shown in Figure 7.21. These impedance values are conveniently normalized by Z_0 before obtaining the incremental reflection coefficient at distance z. Hence,

$$\delta\Gamma_0 = \frac{\overline{Z} + \delta\overline{Z} - \overline{Z}}{\overline{Z} + \delta\overline{Z} + \overline{Z}} \approx \frac{\delta\overline{Z}}{2\overline{Z}} \tag{7.7.1}$$

As $\delta z \to 0$, it can be written as

$$d\Gamma_0 = \frac{d\overline{Z}}{2\overline{Z}} = \frac{1}{2}\frac{d(\ln \overline{Z})}{dz}dz \tag{7.7.2}$$

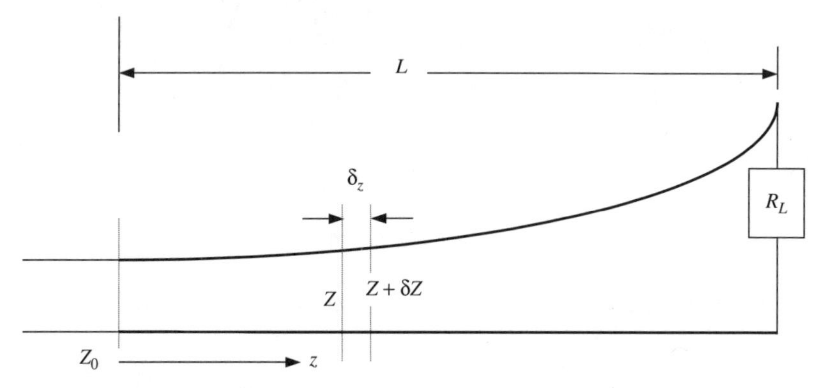

Figure 7.21 Tapered transition connected between load R_L and transmission line of characteristic impedance Z_0.

The corresponding incremental reflection coefficient $d\Gamma_{\text{in}}$ at the input can be written as follows:

$$d\Gamma_{\text{in}} \approx e^{-j2\beta z} d\Gamma_0 = e^{-j2\beta z} \left(\frac{1}{2}\right) \frac{d}{dz} (\ln \overline{Z})\, dz$$

The total reflection coefficient, Γ_{in}, at the input of the tapered section can be determined by summing up these incremental reflections with their appropriate phase angles. Hence,

$$\Gamma_{\text{in}} = \int_0^L d\Gamma_{\text{in}} = \frac{1}{2} \int_0^L e^{-j2\beta z} \frac{d}{dz} (\ln \overline{Z})\, dz \tag{7.7.3}$$

Therefore, Γ_{in} can be determined from (7.7.3) provided that $\overline{Z}(z)$ is given. However, the synthesis problem is a bit complex because in that case, $\overline{Z}(z)$ is to be determined for a specified Γ_{in}. Let us first consider a few examples of evaluating Γ_{in} for the given distributions of $d(\ln \overline{Z})/dz$.

Case 1 $\dfrac{d}{dz}(\ln \overline{Z})$ is constant over the entire length of the taper. Suppose that

$$\frac{d}{dz} \ln \overline{Z}(z) = C_1 \qquad 0 \le z \le L$$

where C_1 is a constant. On integrating this equation, we get

$$\ln \overline{Z}(z) = C_1 z + C_2$$

With $\overline{Z}(z = 0) = 1$ and $\overline{Z}(z = L) = \overline{R}_L$, constants C_1 and C_2 can be determined. Hence,

$$\ln \overline{Z} = \frac{z}{L} \ln \overline{R}_L \Rightarrow \overline{Z} = e^{(z/L) \ln \overline{R}_L} \tag{7.7.4}$$

Thus, the impedance is changing exponentially with distance. Therefore, this kind of taper is called an *exponential taper*. From (7.7.3), the total reflection coefficient is found as

$$\Gamma_{\text{in}} = \frac{1}{2}\int_0^L e^{-j2\beta z}\frac{d}{dz}\left(\frac{z}{L}\ln\overline{R}_L\right)\,dz = \frac{\ln\overline{R}_L}{2L}\int_0^L e^{-j2\beta z}\,dz = \frac{\ln\overline{R}_L}{2L}\left.\frac{e^{-j2\beta z}}{-j2\beta}\right|_0^L$$

or

$$\Gamma_{\text{in}} = \frac{\ln\overline{R}_L}{2L}\frac{e^{-j\beta L}-1}{-j2\beta} = \frac{\ln\overline{R}_L}{2}e^{-j\beta L}\frac{\sin\beta L}{\beta L} \Rightarrow \frac{2\rho_{\text{in}}}{\ln\overline{R}_L} = \frac{\sin\beta L}{\beta L} \tag{7.7.5}$$

where $\rho_{\text{in}} = |\Gamma_{\text{in}}|$. It is assumed in this evaluation that $\beta = 2\pi/\lambda$ is not changing with distance z. The right-hand side of (7.7.5) is displayed in Figure 7.22 as a function of βL. Since L is fixed for a given taper and β is related directly to signal frequency, this plot also represents the frequency response of an exponential taper.

Case 2 $\frac{d}{dz}(\ln\overline{Z})$ is a triangular function. If $\frac{d}{dz}(\ln\overline{Z})$ is a triangular function, defined as

$$\frac{d(\ln\overline{Z})}{dz} = \begin{cases} \dfrac{4z}{L^2}\ln\overline{R}_L & 0 \le z \le \dfrac{L}{2} \\ 4\left(\dfrac{1}{L}-\dfrac{z}{L^2}\right)\ln\overline{R}_L & \dfrac{L}{2} \le z \le L \end{cases} \tag{7.7.6}$$

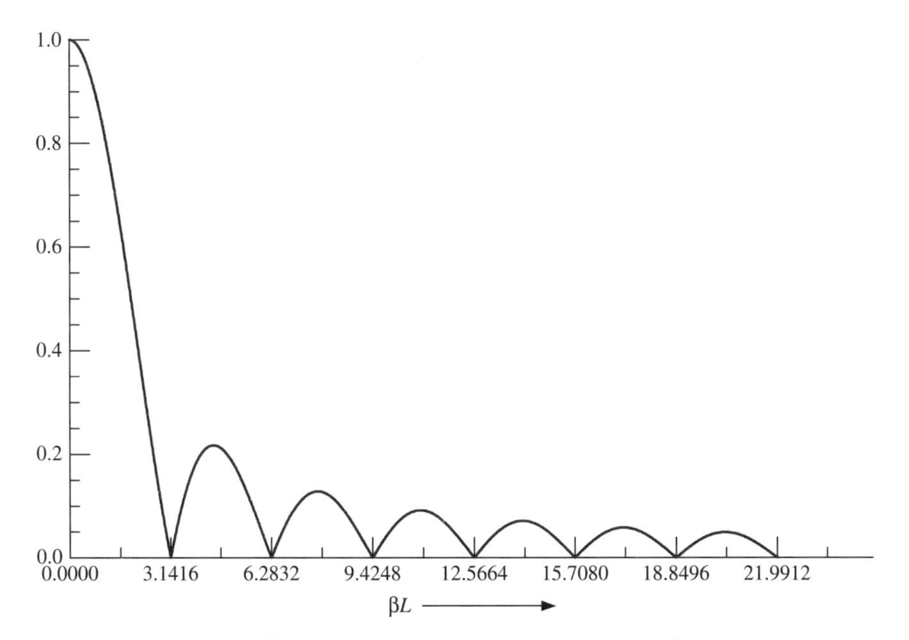

Figure 7.22 Frequency response of an exponential taper.

then

$$\ln \overline{Z} = \begin{cases} \int_0^{L/2} \frac{4z}{L^2} \ln \overline{R}_L \, dz & 0 \le z \le \frac{L}{2} \\ \int_{L/2}^{L} 4\left(\frac{1}{L} - \frac{z}{L^2}\right) \ln \overline{R}_L \, dz & \frac{L}{2} \le z \le L \end{cases}$$

or

$$\ln \overline{Z} = \begin{cases} \frac{2z^2}{L^2} \ln \overline{R}_L + c_1 & 0 \le z \le \frac{L}{2} \\ 4\left(\frac{z}{L} - \frac{z^2}{2L^2}\right) \ln \overline{R}_L + c_2 & \frac{L}{2} \le z \le L \end{cases}$$

Integration constants c_1 and c_2 are determined from the given conditions $Z(z = 0) = 1$ and $Z(z = L) = \overline{R}_L$. Hence,

$$\overline{Z} = \begin{cases} \exp\left\{\frac{2z^2}{L^2} \ln \overline{R}_L\right\} & 0 \le z \le \frac{L}{2} \\ \exp\left\{\left(\frac{4z}{L} - \frac{2z^2}{L^2} - 1\right) \ln \overline{R}_L\right\} & \frac{L}{2} \le z \le L \end{cases} \tag{7.7.7}$$

and therefore total reflection coefficient is found from (7.7.3) as

$$\Gamma_{\text{in}} = \frac{1}{2}\left[\int_0^{L/2} e^{-j2\beta z} \frac{4z}{L^2} \ln \overline{R}_L \, dz + \int_{L/2}^{L} e^{-j2\beta z} \frac{4}{L^2}(L - z) \ln \overline{R}_L \, dz\right]$$

or

$$\Gamma_{\text{in}} = \frac{2 \ln \overline{R}_L}{L^2}\left[\int_0^{L/2} z e^{-j2\beta z} \, dz + \int_{L/2}^{L} (L - z) e^{-j2\beta z} \, dz\right]$$

Since

$$\int z e^{-j2\beta z} \, dz = \frac{z e^{-j2\beta z}}{-j2\beta} - \frac{e^{-j2\beta z}}{(-j2\beta)^2}$$

and

$$\int e^{-j2\beta z} \, dz = \frac{e^{-j2\beta z}}{-j2\beta}$$

Γ_{in} can be found as follows:

$$\Gamma_{\text{in}} = \frac{1}{2} e^{-j\beta L} \ln \overline{R}_L \left[\frac{\sin(\beta L/2)}{\beta L/2}\right]^2 \tag{7.7.8}$$

Figure 7.23 shows normalized magnitude of Γ_{in} as a function of βL. As before, it also represents the frequency response of this taper of length L. When we compare it with that of Figure 7.22 for an exponential taper, we find that the

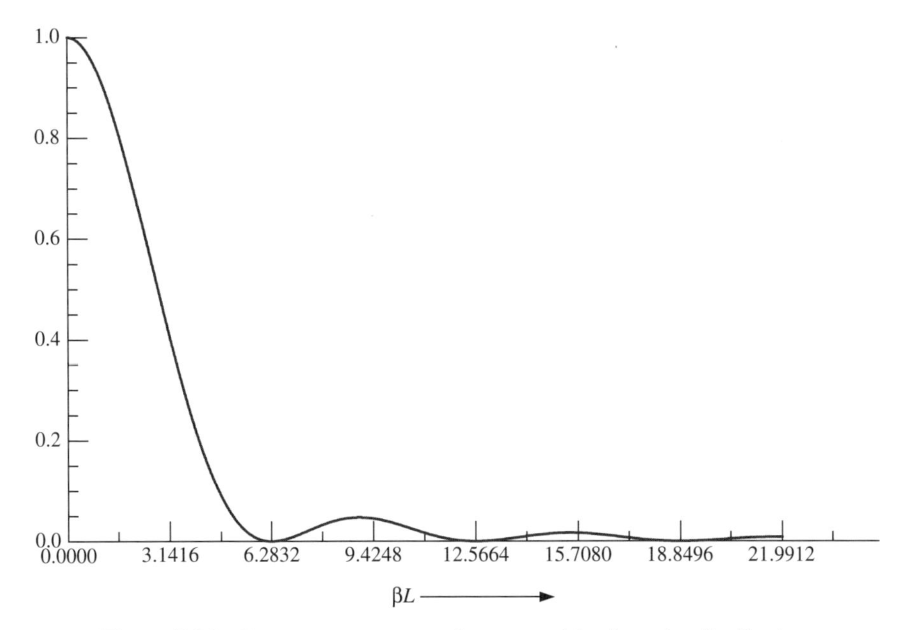

Figure 7.23 Frequency response of a taper with triangular distribution.

magnitude of minor lobes is reduced now, but the main lobe has become relatively wider. It affects the cutoff frequency of a given taper. For example, if the maximum allowed normalized reflection coefficient is 0.2, an exponential taper has a lower value of βL than that of a taper with triangular distribution. However, it is the other way around for a normalized reflection coefficient of 0.05. For a fixed length L, βL is proportional to frequency, and hence the lower value of βL has a lower cutoff frequency than in the other case.

Figure 7.24 depicts variation in normalized characteristic impedance along the length of two tapers. A coaxial line operating in the TEM mode can be used to design these tapers by varying the diameter of its inner conductor. Similarly, the narrow sidewall width of a TE_{10}-mode rectangular waveguide can be modified to get the desired taper while keeping its wide side constant.

Case 3 $\dfrac{d}{dz}(\ln \overline{Z})$ is a Gaussian distribution. Assume that

$$\frac{d}{dz}(\ln \overline{Z}) = K_1 e^{-\alpha(z-L/2)^2} \tag{7.7.9}$$

This distribution is centered around the midpoint of the tapered section. Its fall-off rate is governed by the coefficient α. Integrating it over distance z, we find that

$$\ln \overline{Z} = K_1 \int_0^z e^{-\alpha(z'-L/2)^2}\, dz' \tag{7.7.10}$$

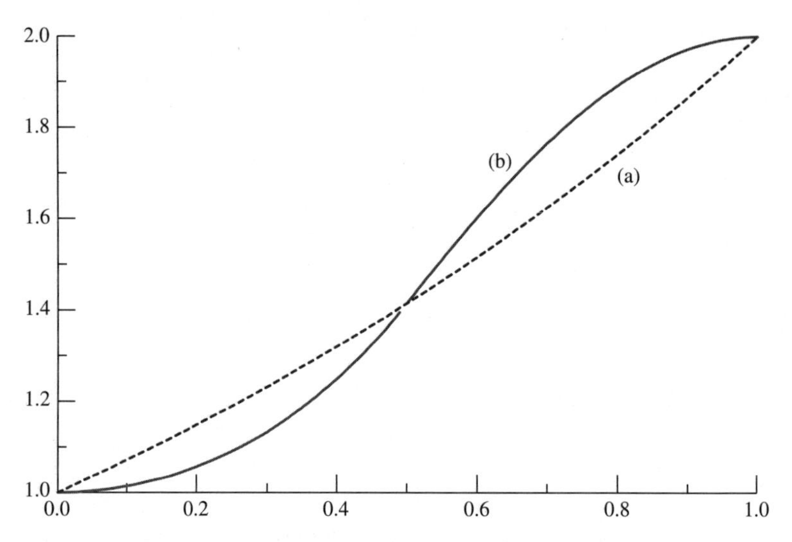

Figure 7.24 Distribution of normalized characteristic impedance along a tapered line terminated by $R_L = 2Z_0$ for (a) case 1 and (b) case 2.

Since $\overline{Z}(z = L) = \overline{R}_L$,

$$K_1 = \frac{\ln \overline{R}_L}{\displaystyle\int_0^L e^{-\alpha(z'-L/2)^2}\, dz'} \tag{7.7.11}$$

Therefore,

$$\ln \overline{Z} = \ln \overline{R}_L \frac{\displaystyle\int_0^z e^{-\alpha(z'-L/2)^2}\, dz'}{\displaystyle\int_0^L e^{-\alpha(z'-L/2)^2}\, dz'} \tag{7.7.12}$$

and

$$\Gamma_{\text{in}} = \frac{1}{2} K_1 \int_0^L e^{-j2\beta z} e^{-\alpha(z-L/2)^2}\, dz \tag{7.7.13}$$

With the following substitution of variable and associated limits of integration in (7.7.11) and (7.7.13),

$$z - \frac{L}{2} = x$$

$$dz = dx$$

$$z = 0 \rightarrow x = -\frac{L}{2}$$

and

$$z = L \rightarrow x = \frac{L}{2}$$

we get

$$\Gamma_{\text{in}} = \frac{1}{2}K_1 \int_{-L/2}^{L/2} e^{-j2\beta(x+L/2)} e^{-\alpha x^2}\, dx = \frac{1}{2}K_1 e^{-j\beta L} \int_{-L/2}^{L/2} e^{-j2\beta x} e^{-\alpha x^2}\, dx$$

or

$$\Gamma_{\text{in}} = \frac{1}{2}\ln \overline{R}_L\, e^{-j\beta l} \frac{\displaystyle\int_{-L/2}^{L/2} e^{-j2\beta x} e^{-\alpha x^2}\, dx}{\displaystyle\int_{-L/2}^{L/2} e^{-\alpha x^2}\, dx} = \frac{1}{2}\ln \overline{R}_L\, e^{-j\beta l} \frac{\displaystyle\int_{0}^{L/2} \cos 2\beta x\, e^{-\alpha x^2}\, dx}{\displaystyle\int_{0}^{L/2} e^{-\alpha x^2}\, dx} \tag{7.7.14}$$

It can be arranged after substituting $\sqrt{\alpha}\, x$ with y, and $\sqrt{\alpha}(L/2)$ with ξ as follows:

$$\Gamma_{\text{in}} = \frac{1}{2}\ln \overline{R}_L\, e^{-j\beta l} \frac{\displaystyle\int_{0}^{\xi} \cos[(\beta L/\xi)y] e^{-y^2}\, dy}{\displaystyle\int_{0}^{\xi} e^{-y^2}\, dy} \tag{7.7.15}$$

Figure 7.25 shows the normalized magnitude of the reflection coefficient versus βL for two different values of ξ. It indicates that the main lobe becomes wider, and the level of minor lobes goes down as ξ is increased.

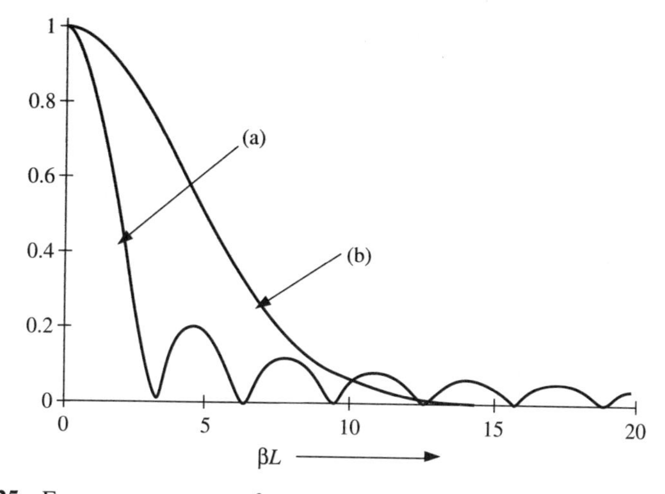

Figure 7.25 Frequency response of a tapered line with a normal distribution of normalized characteristic impedance for (a) $\xi = 0.3$ and (b) $\xi = 3$.

7.8 SYNTHESIS OF TRANSMISSION LINE TAPERS

The frequency response of a transmission line taper can be determined easily from (7.7.3). However, the inverse problem of determining its impedance variation that provides the desired frequency response needs further consideration. To this end, assume that $F(\alpha)$ is the Fourier transform of a function $f(x)$. In other words,

$$F(\alpha) = \int_{-\infty}^{\infty} f(x)e^{-j\alpha x}\,dx \tag{7.8.1}$$

and

$$f(x) = \frac{1}{2\pi}\int_{-\infty}^{\infty} F(\alpha)e^{j\alpha x}\,d\alpha \tag{7.8.2}$$

For convenience, we rewrite (7.7.3) as follows:

$$\Gamma_{\text{in}}(2\beta) = \int_{0}^{L}\left[\frac{1}{2}\frac{d}{dz}(\ln\overline{Z})\right]e^{-j2\beta z}\,dz$$

If $\frac{1}{2}\,d(\ln\overline{Z})/dz = 0$ for $-\infty \leq z < 0$ and for $z > L$, it may be written as

$$\Gamma_{\text{in}}(2\beta) = \int_{-\infty}^{\infty}\frac{1}{2}\frac{d}{dz}(\ln\overline{Z})e^{-j2\beta z}\,dz \tag{7.8.3}$$

A comparison of (7.8.3) with (7.8.1) indicates that $\Gamma_{\text{in}}(2\beta)$ represents the Fourier transform of $\frac{1}{2}\,d[\ln\overline{Z}(z)]/dz$. Hence, from (7.8.2),

$$\frac{1}{2}\frac{d(\ln\overline{Z})}{dz} = \frac{1}{2\pi}\int_{-\infty}^{\infty}\Gamma_{\text{in}}(2\beta)e^{j2\beta x}\,d(2\beta) \tag{7.8.4}$$

This equation can be used to design a taper that will have the desired reflection coefficient characteristics. However, only those $\Gamma_{\text{in}}(2\beta)$ that have an inverse Fourier transform limited to $0 \leq z \leq L$ (zero outside this range) can be realized. At this point, we can conveniently introduce the following normalized variables:

$$p = 2\pi\frac{z-(L/2)}{L} \tag{7.8.5}$$

and

$$u = \frac{\beta L}{\pi} = \frac{2L}{\lambda} \tag{7.8.6}$$

Therefore,

$$z = \frac{L}{2\pi}(p+\pi)$$

and

$$dz = \frac{L}{2\pi}dp$$

New integration limits are found as $-\pi \le p \le \pi$. Hence,

$$\Gamma_{\text{in}}(2\beta) = \frac{1}{2}\int_{-\pi}^{\pi} e^{-j2\beta(L/2+Lp/2\pi)}\, \frac{d}{dp}(\ln \overline{Z})\, \frac{dp}{dz}\frac{L}{2\pi}dp$$

or

$$\Gamma_{\text{in}}(2\beta) = \frac{1}{2}\, e^{-j\beta L} \int_{-\pi}^{\pi} e^{-jup}\frac{d}{dp}\,(\ln \overline{Z})\, dp \tag{7.8.7}$$

Now, if we define $g(p) = d(\ln \overline{Z})/dp$ and $F(u) = \int_{-\pi}^{\pi} e^{-jup} g(p)\, dp$, (7.8.7) can be expressed as follows:

$$\Gamma_{\text{in}}(2\beta) = \tfrac{1}{2}e^{-j\beta L} F(u)$$

or

$$\rho_{\text{in}}(2\beta) = \tfrac{1}{2}|F(u)|$$

Further,

$$F(u = 0) = \ln \overline{Z}_L \tag{7.8.8}$$

Thus, $F(u)$ and $g(p)$ form the Fourier transform pair. Therefore,

$$g(p) = \frac{1}{2\pi}\int_{-\infty}^{\infty} e^{jup}\, F(u)\, du = \begin{cases} \text{nonzero} & \text{for } |p| < \pi \\ 0 & \text{for } |p| > \pi \end{cases} \tag{7.8.9}$$

To satisfy the conditions embedded in writing (7.8.3), only those $F(u)$ that produce $g(p)$ as zero for $|p| > \pi$ can be synthesized. Suitable restrictions on $F(u)$ can be derived after expanding $g(p)$ in a complex Fourier series as follows:

$$g(p) = \begin{cases} \displaystyle\sum_{n=-\infty}^{\infty} a_n e^{jnp} & -\pi \le p \le \pi \\ 0 & |p| > \pi \end{cases} \tag{7.8.10}$$

Since $g(p)$ is a real function, constants $a_n = a_{-n}^*$. Therefore,

$$F(u) = \int_{-\pi}^{\pi} e^{-jup} \sum_{n=-\infty}^{\infty} a_n e^{jnp}\, dp = \sum_{n=-\infty}^{\infty} a_n \int_{-\pi}^{\pi} e^{-j(u-n)p}\, dp = \sum_{n=-\infty}^{\infty} a_n \left.\frac{e^{-j(u-n)p}}{-j(u-n)}\right|_{-\pi}^{\pi}$$

or

$$F(u) = \sum_{n=-\infty}^{\infty} a_n \frac{e^{-j(u-n)\pi} - e^{j(u-n)\pi}}{-j(u-n)} = 2\pi \sum_{n=-\infty}^{\infty} a_n \frac{\sin \pi(u-n)}{\pi(u-n)} \tag{7.8.11}$$

For $u = m$ (an integer),

$$\frac{\sin \pi(m-n)}{\pi(m-n)} = \begin{cases} 1 & \text{for } m = n \\ 0 & \text{for } m \ne n \end{cases}$$

Therefore,

$$F(n) = 2\pi a_n \Rightarrow a_n = \frac{F(n)}{2\pi}$$

and

$$F(u) = \sum_{n=-\infty}^{\infty} F(n) \frac{\sin \pi(u-n)}{\pi(u-n)} \tag{7.8.12}$$

This is the well-known *sampling theorem* used in communication theory. It states that $F(u)$ is uniquely reconstructed from a knowledge of sample values of $F(u)$ at $u = n$, where n is an integer (positive or negative) including zero (i.e., $u = n = 0, \pm 1, \pm 2, \pm 3, \ldots$).

To have greater flexibility in selecting $F(u)$, let us assume that $a_n = 0$ for all $|n| > N$. Therefore,

$$g(p) = \sum_{n=-N}^{N} a_n e^{jnp} \tag{7.8.13}$$

and from (7.8.11),

$$F(u) = 2\pi \sum_{n=-N}^{N} a_n(-1)^n \frac{\sin \pi u}{\pi(u-n)} = 2\pi \frac{\sin \pi u}{\pi u} \sum_{n=-N}^{N} a_n(-1)^n \frac{u}{u-n}$$

This series can be recognized as the partial-fraction expansion of a function, and therefore $F(u)$ can be expressed as follows:

$$F(u) = 2\pi \frac{\sin \pi u}{\pi u} \frac{Q(u)}{\prod_{n=1}^{N}(u^2 - n^2)} \tag{7.8.14}$$

where $Q(u)$ is an arbitrary polynomial of degree $2N$ in u subject to the restriction $Q(-u) = Q^*(u)$, so that $a_n = a_{-n}^*$. Further, it contains an arbitrary constant multiplier. In (7.8.14), $\sin \pi u$ has zeros at $u = \pm n$. However, the first $2N$ of these zeros are canceled by $(u^2 - n^2)$. These can be replaced by $2N$ new arbitrarily located zeros by proper choice of $Q(u)$.

Example 7.7 Design a transmission line taper to match a 100-Ω load to a 50-Ω line. The desired frequency response has a triple zero at $\beta L = \pm 2\pi$. Plot its frequency response and normalized characteristic impedance distribution.

SOLUTION The desired characteristic can be accomplished by moving zeros at ± 1 and ± 3 into the points $u = \pm 2$. Due to this triple zero, the reflection coefficient will remain small over a relatively wide range of βL around $\pm 2\pi$. Therefore,

$$Q(u) = C(u^2 - 4)^3$$

and

$$F(u) = 2\pi C \frac{\sin \pi u}{\pi u} \frac{(u^2 - 4)^3}{(u^2 - 1) \times (u^2 - 4) \times (u^2 - 9)}$$

Since $u = \beta L/\pi$, $F(u)$ versus u represents the frequency response of this taper. Further, it can be normalized with $F(0)$ as follows:

$$\left| \frac{F(u)}{F(0)} \right| = \frac{9}{16} \left(\frac{\sin \pi u}{\pi u} \right) \frac{(u^2 - 4)^3}{(u^2 - 1)(u^2 - 4)(u^2 - 9)}$$

A plot of this equation is illustrated in Figure 7.26. It shows that the normalized magnitude is very small around $u = 2$.

Since $F(0)$ is given by (7.8.8), multiplying constant C in $F(u)$ can be determined easily. Hence,

$$F(0) = 2\pi C \times 1 \times \frac{(-4)^3}{(-1)(-4)(-9)} = 2\pi C \times \frac{16}{9} = \ln \overline{Z}_L \Rightarrow C = \frac{9}{32\pi} \ln \overline{Z}_L$$

Further,

$$F(1) = \frac{2\pi C}{\pi} \left(\frac{\sin \pi u}{u^2 - 1} \right) \Bigg|_{u=1} \times \frac{(1-4)^3}{(1-4)(1-9)} = C\pi \frac{9}{8}$$

$$= \frac{9}{32\pi} \ln \overline{Z}_L \ \pi \frac{9}{8} = 0.3164 \ln \overline{Z}_L$$

$$F(2) = 0$$

$$F(3) = \frac{2\pi C}{3\pi} \left(\frac{\sin \pi u}{u^2 - 9} \right) \Bigg|_{u=3} \times \frac{(9-4)^3}{(9-1)(9-4)} = -\frac{25}{72} C\pi$$

$$= \left(-\frac{25}{72} \pi \right) \times \frac{9}{32\pi} \ln \overline{Z}_L = 0.09766 \ln \overline{Z}_L$$

and $F(n) = 0$ for $n > 3$.

Therefore,

$$a_0 = \frac{1}{2\pi} F(0) = \frac{1}{2\pi} \ln \overline{Z}_L$$

$$a_1 = a_{-1} = \frac{1}{2\pi} F(1) = \frac{0.3164}{2\pi} \ln \overline{Z}_L$$

$$a_2 = a_{-2} = 0$$

$$a_3 = a_{-3} = \frac{1}{2\pi} F(3) = -\frac{0.09766}{2\pi} \ln \overline{Z}_L$$

and

$$a_n = a_{-n} = 0 \qquad \text{for } n > 3$$

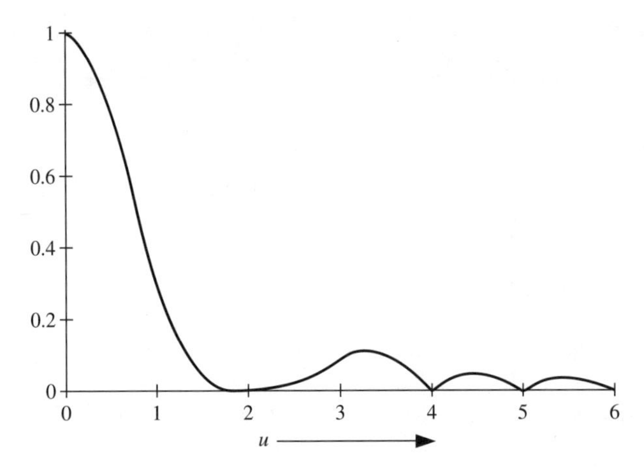

Figure 7.26 Normalized frequency response of the taper in Example 7.7.

Hence,

$$g(p) = \frac{d(\ln \overline{Z})}{dp} = \frac{\ln \overline{Z}_L}{2\pi}(a_0 + 2a_1 \cos p + 2a_3 \cos 3p)$$

$$= \frac{\ln \overline{Z}_L}{2\pi}(1 + 0.6328 \times \cos p - 0.1953 \times \cos 3p)$$

and

$$\ln \overline{Z} = \frac{\ln \overline{Z}_L}{2\pi}(p + 0.6328 \times \sin p - 0.0651 \times \sin 3p) + c_1$$

The integration constant c_1 can be evaluated as follows:

$$\ln \overline{Z} = 0 \text{ at } p = -\pi \quad \text{or} \quad \ln \overline{Z} = \ln \overline{Z}_L \text{ at } p = \pi$$

Therefore,

$$c_1 = 0.5 \ln \overline{Z}_L$$

and

$$\ln \overline{Z} = \frac{\ln \overline{Z}_L}{2\pi}(p + \pi + 0.6328 \sin p - 0.0651 \sin 3p)$$

where

$$p = \frac{2\pi}{L}\left(z - \frac{L}{2}\right)$$

Normalized impedance (Z) versus the length of the taper (z/L) is illustrated in Figure 7.27.

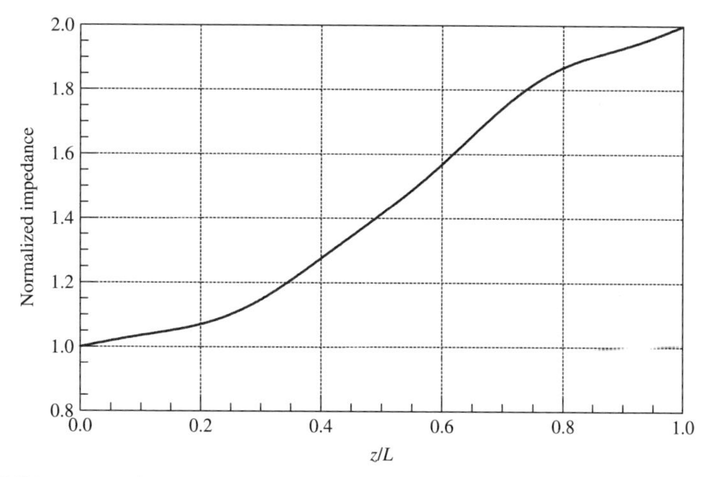

Figure 7.27 Normalized characteristic impedance variation along the length of the taper in Example 7.7.

Example 7.8 Design a taper to match a 100-Ω load to a 50-Ω transmission line. Its reflection coefficient has double zeros at $\beta L = \pm 2\pi$ and $\pm 3\pi$. Plot its frequency response and normalized characteristic impedance versus the normalized length of the taper.

SOLUTION This design can be achieved by moving the zeros at $u = \pm\, 1$ and $\pm\, 4$ into the points $\pm\, 2$ and $\pm\, 3$, respectively. Hence,

$$Q(u) = C(u^2 - 2^2)^2(u^2 - 3^2)^2 = C(u^2 - 4)^2(u^2 - 9)^2$$

and

$$F(u) = C\frac{\sin \pi u}{\pi u}\frac{(u^2 - 4)^2(u^2 - 9)^2}{(u^2 - 1)(u^2 - 4)(u^2 - 9)(u^2 - 16)}$$

Therefore, $F(0) = 2.25C$, $F(1) = 0.8C$, $F(2) = 0$, $F(3) = 0$, $F(4) = (7/40)C$, and $F(n) = 0$ for $n > 4$. From (7.8.8),

$$F(0) = \ln \overline{Z}_L \Rightarrow C = \frac{\ln \overline{Z}_L}{2.25}$$

and

$$g(p) = \frac{d(\ln \overline{Z})}{dp} = \sum_{n=-\infty}^{\infty} a_n e^{-jnp} \qquad a_n = \frac{F(n)}{2\pi}$$

or

$$\begin{aligned} g(p) &= \frac{2.25C}{2\pi} + \frac{0.8C}{2\pi}(e^{jp} + e^{-jp}) + \frac{7C}{(40)(2\pi)}(e^{j4p} + e^{-j4p}) \\ &= \frac{\ln \overline{Z}_L}{2\pi}(1 + 0.7111 \cos p + 0.1556 \cos 4p) \end{aligned}$$

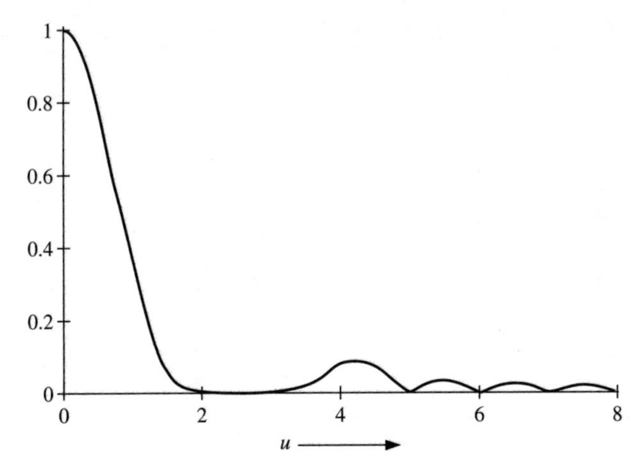

Figure 7.28 Frequency response of the taper in Example 7.8.

Therefore,

$$\ln \overline{Z} = \frac{\ln \overline{Z}_L}{2\pi}(p + 0.7111 \sin p + 0.0389 \sin 4p) + c_1$$

Since, $\ln \overline{Z}|_{p=-\pi} = 0$, $c_1 = \ln \overline{Z}_L/2$ and

$$\ln \overline{Z} = \frac{\ln \overline{Z}_L}{2\pi}(p + \pi + 0.7111 \sin p + 0.0389 \sin 4p)$$

where

$$p = \frac{2\pi}{L}\left(z - \frac{L}{2}\right)$$

Frequency response and normalized characteristic impedance distribution are shown in Figures 7.28 and 7.29, respectively.

Klopfenstein Taper

The preceding procedure indicates infinite possibilities of synthesizing a transmission line taper. A designer naturally looks for the best design. In other words, which design gives the shortest taper for a given ρ_M? The answer to this question is the Klopfenstein taper, which is derived from a stepped Chebyshev transformer. The design procedure is summarized here without its derivation.

The characteristic impedance variation of this taper versus distance z is given by

$$Z(z) = \sqrt{Z_0 R_L e^{\kappa}} \tag{7.8.15}$$

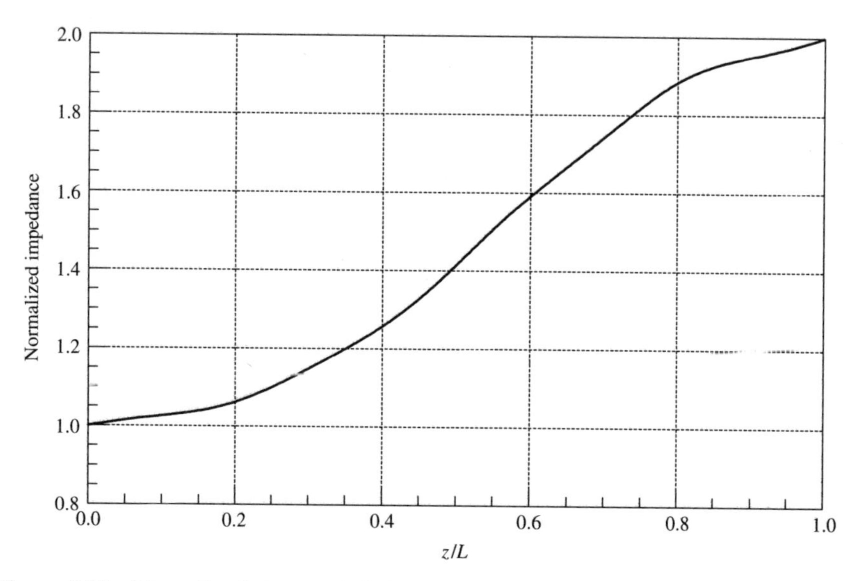

Figure 7.29 Normalized characteristic impedance variation of the taper in Example 7.8.

where

$$\kappa = \frac{A^2\Gamma_0}{\cosh A} f\left(\frac{2z}{L} - 1, A\right) \qquad 0 \le z \le L \tag{7.8.16}$$

$$f(\zeta, A) = \int_0^{\zeta} \frac{I_1(A\sqrt{1-y^2})}{A\sqrt{1-y^2}} dy \qquad |x| \le 1 \tag{7.8.17}$$

$$\Gamma_0 = \frac{R_L - Z_0}{R_L + Z_0} \tag{7.8.18}$$

Since the cutoff value of βL is equal to A, we have

$$A = \beta_0 L \tag{7.8.19}$$

For a maximum ripple ρ_M in its passband, it can be determined as

$$A = \cosh^{-1} \frac{|\Gamma_0|}{\rho_M} \tag{7.8.20}$$

$I_1(x)$ is the modified Bessel function of the first kind. $f(\zeta, A)$ needs to be evaluated numerically except for the following special cases:

$$f(0, A) = 0$$

$$f(x, 0) = \frac{x}{2}$$

$$f(1, A) = \frac{\cosh A - 1}{A^2}$$

The resulting input reflection coefficient as a function of βL (and hence, frequency) is given by

$$\Gamma_{\text{in}}(\beta L) = \Gamma_0 e^{-j\beta L} \frac{\cosh\sqrt{A^2 - (\beta L)^2}}{\cosh A} \tag{7.8.21}$$

Note that the hyperbolic cosine becomes the cosine function for an imaginary argument.

Example 7.9 Design a Klopfenstein taper to match a 100-Ω load to a 50-Ω transmission line. The maximum allowed reflection coefficient in its passband is 0.1. Plot the characteristic impedance variation along its normalized length and the input reflection coefficient versus βL.

SOLUTION From (7.8.18),

$$\Gamma_0 = \frac{R_L - Z_0}{R_L + Z_0} = \frac{100 - 50}{100 + 50} = \frac{1}{3}$$

TABLE 7.1 Characteristic Impedance of the Klopfenstein Taper (in Example 7.9) as a Function of Its Normalized Length

Z/L	Impedance (Ω)
0	56.000896
0.05	57.030043
0.1	58.169255
0.15	59.412337
0.2	60.757724
0.25	62.202518
0.3	63.74238
0.35	65.371433
0.4	67.082194
0.45	68.865533
0.5	70.710678
0.55	72.605261
0.6	74.535428
0.65	76.486009
0.7	78.440749
0.75	80.382597
0.8	82.294063
0.85	84.157604
0.9	85.95606
0.95	87.673089
1	89.29361

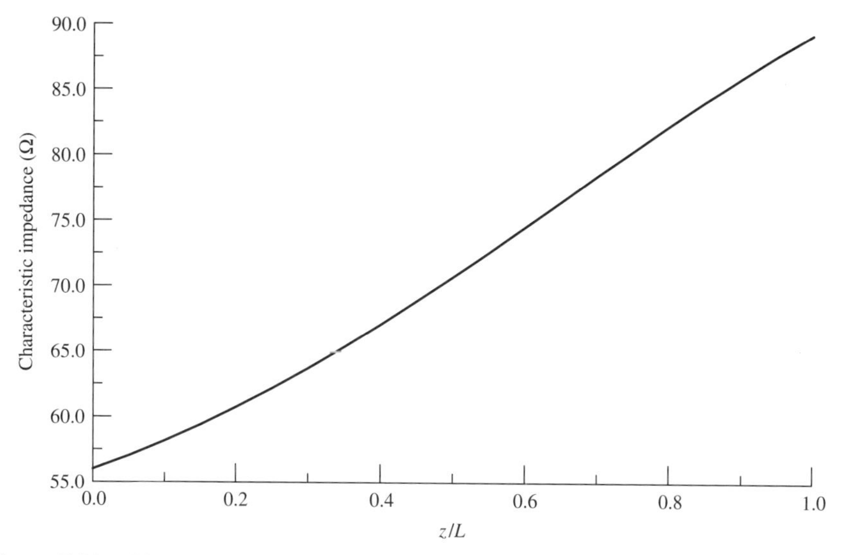

Figure 7.30 Characteristic impedance distribution along the normalized length of the Klopfenstein taper in Example 7.9.

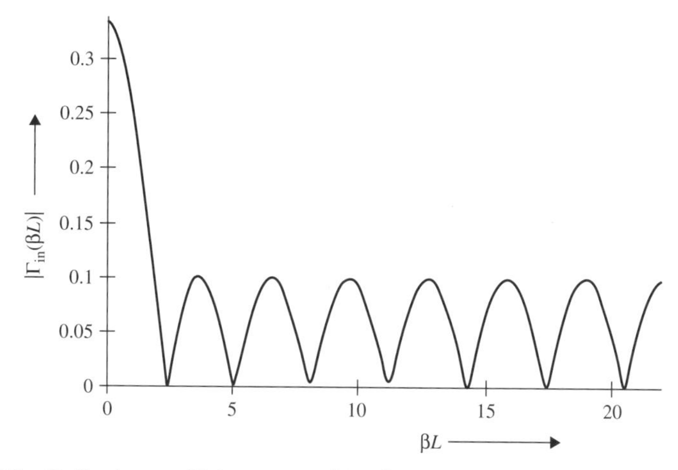

Figure 7.31 Reflection coefficient versus βL of the Klopfenstein taper in Example 7.9.

From (7.8.20),

$$A = \cosh^{-1} \frac{|\Gamma_0|}{\rho_M} = \cosh^{-1} \frac{\frac{1}{3}}{0.1} = 1.87382$$

Equation (7.8.15) is used to evaluate $Z(z)$ numerically. These results are shown in Table 7.1. These data are also displayed in Figure 7.30. The reflection coefficient of this taper is depicted in Figure 7.31.

7.9 BODE–FANO CONSTRAINTS FOR LOSSLESS MATCHING NETWORKS

Various matching circuits described in this and Chapter 6 indicate that zero reflection is possible only at selected discrete frequencies. Examples 6.6 and 6.7 show that the bandwidth over which the reflection coefficient remains below a specified value is increased with the number of components. A circuit designer would like to know whether a lossless passive network could be designed for perfect matching. Further, it will be helpful if associated constraints and design trade-offs are known. Bode–Fano constraints provide such means to the designer. Bode–Fano criteria provide optimum results under ideal conditions. These results may require approximation to implement the circuit. Table 7.2 summarizes these constraints for selected R–C and R–L loads. The detailed formulations are beyond the scope of this book.

As an example, if it is desired to design a reactive matching network at the output of a transistor amplifier, the situation is similar to that depicted in the top row of the table. Further, consider the case in which magnitude ρ of the reflection

TABLE 7.2 Bode–Fano Constraints to Match Certain Loads Using Passive Lossless Network

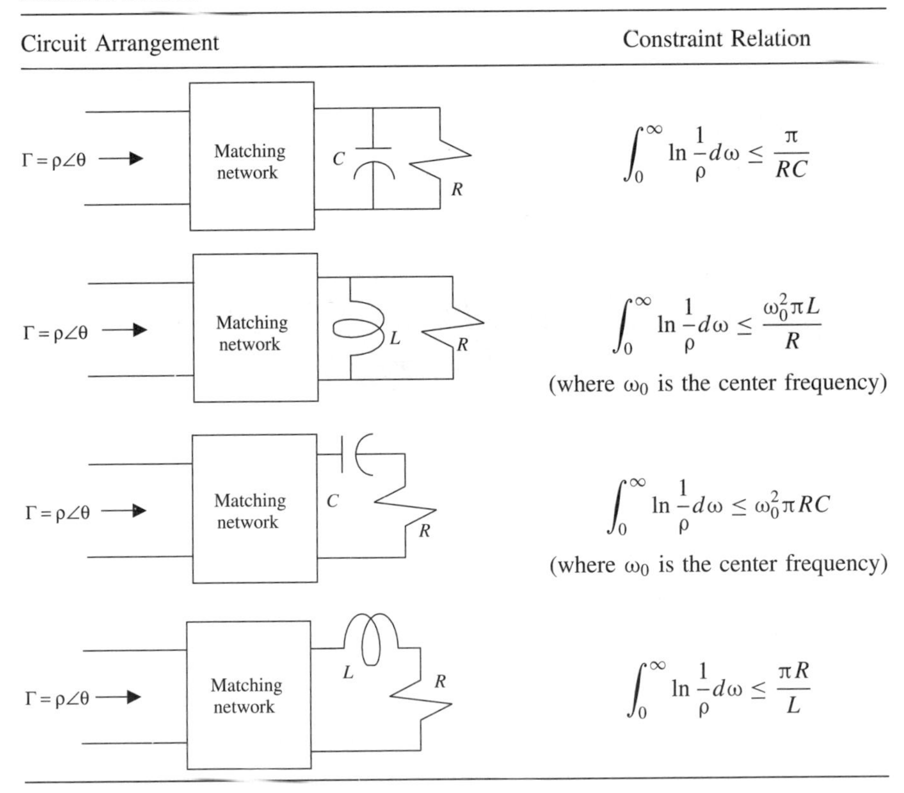

Circuit Arrangement	Constraint Relation
$\Gamma = \rho\angle\theta$ → Matching network, C, R	$\int_0^\infty \ln\frac{1}{\rho}\,d\omega \le \frac{\pi}{RC}$
$\Gamma = \rho\angle\theta$ → Matching network, L, R	$\int_0^\infty \ln\frac{1}{\rho}\,d\omega \le \frac{\omega_0^2\pi L}{R}$ (where ω_0 is the center frequency)
$\Gamma = \rho\angle\theta$ → Matching network, C, R	$\int_0^\infty \ln\frac{1}{\rho}\,d\omega \le \omega_0^2\pi RC$ (where ω_0 is the center frequency)
$\Gamma = \rho\angle\theta$ → Matching network, L, R	$\int_0^\infty \ln\frac{1}{\rho}\,d\omega \le \frac{\pi R}{L}$

coefficient remains constant at ρ_m ($\rho_m < 1$) over the frequency band from ω_1 to ω_2, whereas it stays at unity outside this band. The constraint for this circuit is found to be

$$\int_0^\infty \ln \frac{1}{\rho} d\omega = \ln \frac{1}{\rho_m} \int_{\omega_1}^{\omega_2} d\omega = (\omega_2 - \omega_1) \ln \frac{1}{\rho_m} \leq \frac{\pi}{RC} \tag{7.9.1}$$

This condition shows the trade-off associated with the bandwidth and the reflection coefficient. If R and C are given, the right-hand side of (7.9.1) is fixed. In this situation, a wider bandwidth ($\omega_2 - \omega_1$) is possible only at the cost of a higher reflection coefficient ρ_m. Further, ρ_m cannot go to zero unless the bandwidth $\omega_2 - \omega_1$ is zero. In other words, a perfect matching condition is achievable only at discrete frequencies.

SUGGESTED READING

Bahl, I., and P. Bhartia, *Microwave Solid State Circuit Design*. New York: Wiley, 1988.

Collin, R. E., *Foundations for Microwave Engineering*. New York: McGraw-Hill, 1992.

Elliott, R. S., *An Introduction to Guided Waves and Microwave Circuits*. Englewood Cliffs, NJ: Prentice Hall, 1993.

Fusco, V. F., *Microwave Circuits*. Englewood Cliffs, NJ: Prentice Hall, 1987.

Matthaei, G. L., L. Young, and E. M. T. Jones, *Microwave Filters, Impedance-Matching Networks, and Coupling Structures*. Dedham, MA: Artech House, 1980.

Pozar, D. M., *Microwave Engineering*. New York: Wiley, 1998.

PROBLEMS

7.1. The output impedance of a microwave receiver is 50 Ω. Find the required length and characteristic impedance of a transmission line that will match a 100-Ω load to the receiver. Signal frequency and phase velocity are 10 GHz and 2.4×10^8 m/s, respectively. What is the frequency band over which the reflection coefficient remains below 0.1?

7.2. A quarter-wave impedance-matching transformer is used to match a microwave source with internal impedance 10 Ω to a 50-Ω transmission line. If the source frequency is 3 GHz and the phase velocity on the line is equal to the speed of light in free space, find **(a)** the required length and characteristic impedance of the matching section and **(b)** the reflection coefficients at two ends of the matching section.

7.3. A 0.7-m-long transmission line short-circuited at one end has an input impedance of $-j68.8\,\Omega$. The signal frequency and phase velocity are 100 MHz and 2×10^8 m/s, respectively.

(a) What is the characteristic impedance of the transmission line?

(b) If a 200-Ω load is connected to the end of this line, what is the new input impedance?

(c) A quarter-wave matching transformer is to be used to match a 200-Ω load with the line. What will be the required length and characteristic impedance of the matching transformer?

7.4. Design a two-section binomial transformer to match a 100-Ω load to a 75-Ω line. What is the frequency band over which the reflection coefficient remains below 0.1?

7.5. Design a three-section binomial transformer to match a 75-Ω load to a 50-Ω transmission line. What is the fractional bandwidth for a VSWR less than 1.1?

7.6. Using the approximate theory, design a two-section Chebyshev transformer to match a 100-Ω load to a 75-Ω line. What is the frequency band over which the reflection coefficient remains below 0.1? Compare the results with those obtained in Problem 7.4.

7.7. Design a four-section Chebyshev transformer to match a network with input impedance 60 Ω to a transmission line of characteristic impedance 40 Ω. Find the bandwidth if the maximum VSWR allowed in its passband is 1.2.

7.8. Using the exact theory, design a two-section Chebyshev transformer to match a 10-Ω load to a 75-Ω line. Find the bandwidth over which the reflection coefficient remains below 0.05.

7.9. Using the exact theory, design a three-section Chebyshev transformer to match a 10-Ω load to a 75-Ω line. Find the bandwidth over which the reflection coefficient remains below 0.05. Compare the results with those obtained in Problem 7.8.

7.10. Design a transmission line taper to match a 75-Ω load to a 50-Ω line. The desired frequency response has a double zero at $\beta L = \pm 2\pi$. Plot its frequency response and normalized characteristic impedance distribution.

7.11. Design a transmission line taper to match a 40-Ω load to a 75-Ω line. Its reflection coefficient has triple zeros at $\beta L = \pm 2\pi$. Plot its frequency response and normalized characteristic impedance distribution versus the normalized length of the taper.

7.12. Design a Klopfenstein taper to match a 40-Ω load to a 75-Ω line. The maximum allowed reflection coefficient in its passband is 0.1. Plot the characteristic impedance variation along its normalized length and the input reflection coefficient versus βL. Compare the results with those obtained in Problem 7.11.

8

TWO-PORT NETWORKS

Electronic circuits are frequently needed for processing a given electrical signal to extract the desired information or characteristics. This includes boosting the strength of a weak signal or filtering out certain frequency bands, and so on. Most of these circuits can be modeled as a black box that contains a linear network comprising resistors, inductors, capacitors, and dependent sources. Thus, it may include electronic devices but not the independent sources. Further, it has four terminals, two for input and the other two for output of the signal. There may be a few more terminals to supply the bias voltage for electronic devices. However, these bias conditions are embedded in equivalent dependent sources. Hence, a large class of electronic circuits can be modeled as two-port networks. Parameters of the two-port completely describe its behavior in terms of voltage and current at each port. These parameters simplify the description of its operation when the two-port network is connected into a larger system.

Figure 8.1 shows a two-port network along with appropriate voltages and currents at its terminals. Sometimes, port 1 is called the input port and port 2, the output port. The upper terminal is customarily assumed to be positive with respect to the lower one on either side. Further, currents enter the positive terminals at each port. Since the linear network does not contain independent sources, the same currents leave respective negative terminals. There are several ways to characterize this network. Some of these parameters and relations among them are presented in this chapter, including impedance parameters, admittance parameters, hybrid

Radio-Frequency and Microwave Communication Circuits: Analysis and Design, Second Edition,
By Devendra K. Misra
ISBN 0-471-47873-3

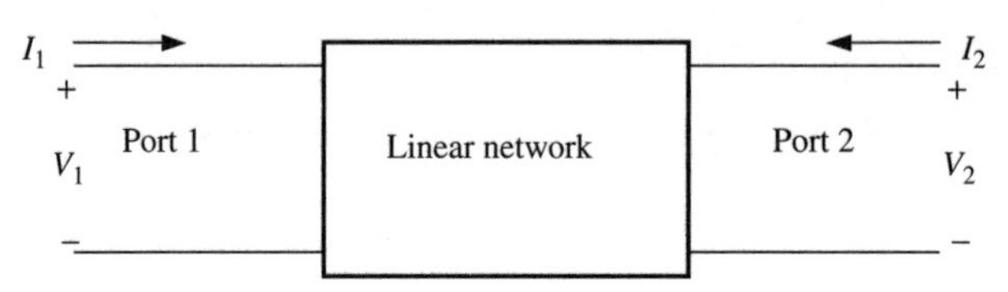

Figure 8.1 Two-port network.

parameters, and transmission parameters. Scattering parameters are introduced later in the chapter to characterize the high-frequency and microwave circuits.

8.1 IMPEDANCE PARAMETERS

Consider the two-port network shown in Figure 8.1. Since the network is linear, the superposition principle can be applied. Assuming that it contains no independent sources, voltage V_1 at port 1 can be expressed in terms of two currents as follows:

$$V_1 = Z_{11}I_1 + Z_{12}I_2 \tag{8.1.1}$$

Since V_1 is in volts and I_1 and I_2 are in amperes, parameters Z_{11} and Z_{12} must be in ohms. Therefore, these are called the *impedance parameters*. Similarly, we can write V_2 in terms of I_1 and I_2 as follows:

$$V_2 = Z_{21}I_1 + Z_{22}I_2 \tag{8.1.2}$$

Using the matrix representation, we can write

$$\begin{bmatrix} V_1 \\ V_2 \end{bmatrix} = \begin{bmatrix} Z_{11} & Z_{12} \\ Z_{21} & Z_{22} \end{bmatrix} \begin{bmatrix} I_1 \\ I_2 \end{bmatrix} \tag{8.1.3}$$

or

$$[V] = [Z][I] \tag{8.1.4}$$

where $[Z]$ is called the *impedance matrix* of the two-port network.

If port 2 of this network is left open, I_2 will be zero. In this condition, (8.1.1) and (8.1.2) give

$$Z_{11} = \left.\frac{V_1}{I_1}\right|_{I_2=0} \tag{8.1.5}$$

and

$$Z_{21} = \left.\frac{V_2}{I_1}\right|_{I_2=0} \tag{8.1.6}$$

Similarly, with a source connected at port 2 while port 1 is open circuit, we find that

$$Z_{12} = \left.\frac{V_1}{I_2}\right|_{I_1=0} \tag{8.1.7}$$

and

$$Z_{22} = \left.\frac{V_2}{I_2}\right|_{I_1=0} \tag{8.1.8}$$

Equations (8.1.5) through (8.1.8) define the impedance parameters of a two-port network.

Example 8.1 Find the impedance parameters for the two-port network shown in Figure 8.2.

SOLUTION If I_2 is zero, V_1 and V_2 can be found from Ohm's law as $6I_1$. Hence, from (8.1.5) and (8.1.6),

$$Z_{11} = \left.\frac{V_1}{I_1}\right|_{I_2=0} = \frac{6I_1}{I_1} = 6\,\Omega$$

and

$$Z_{21} = \left.\frac{V_2}{I_1}\right|_{I_2=0} = \frac{6I_1}{I_1} = 6\,\Omega$$

Similarly, when a source is connected at port 2 and port 1 has an open circuit, we find that

$$V_2 = V_1 = 6I_2$$

Hence, from (8.1.7) and (8.1.8),

$$Z_{12} = \left.\frac{V_1}{I_2}\right|_{I_1=0} = \frac{6I_2}{I_2} = 6\,\Omega$$

and

$$Z_{22} = \left.\frac{V_2}{I_2}\right|_{I_1=0} = \frac{6I_2}{I_2} = 6\,\Omega$$

Therefore,

$$\begin{bmatrix} Z_{11} & Z_{12} \\ Z_{21} & Z_{22} \end{bmatrix} = \begin{bmatrix} 6 & 6 \\ 6 & 6 \end{bmatrix}$$

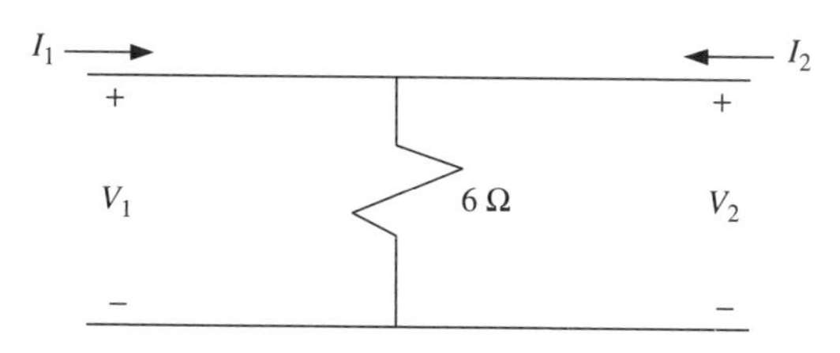

Figure 8.2 Two-port network for Example 8.1.

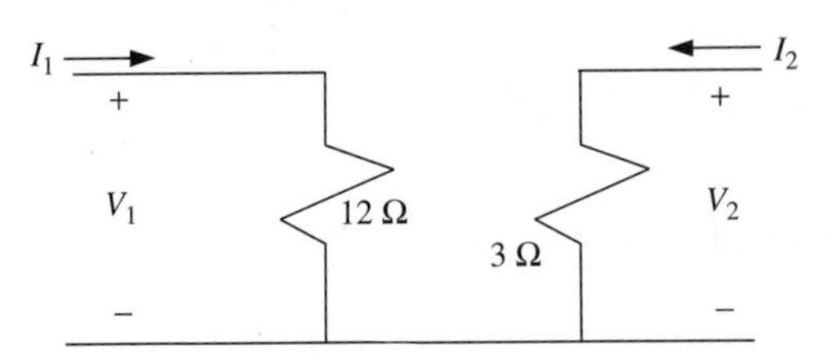

Figure 8.3 Two-port network for Example 8.2.

Example 8.2 Find the impedance parameters of the two-port network shown in Figure 8.3.

SOLUTION As before, assume that a source is connected at port 1 while port 2 is open. In this condition, $V_1 = 12I_1$ and $V_2 = 0$. Therefore,

$$Z_{11} = \left.\frac{V_1}{I_1}\right|_{I_2=0} = \frac{12I_1}{I_1} = 12\ \Omega$$

and

$$Z_{21} = \left.\frac{V_2}{I_1}\right|_{I_2=0} = 0$$

Similarly, when a source connected at port 2 and port 1 has an open circuit, we find that

$$V_2 = 3I_2 \quad \text{and} \quad V_1 = 0$$

Hence, from (8.1.7) and (8.1.8),

$$Z_{12} = \left.\frac{V_1}{I_2}\right|_{I_1=0} = 0$$

and

$$Z_{22} = \left.\frac{V_2}{I_2}\right|_{I_1=0} = \frac{3I_2}{I_2} = 3\ \Omega$$

Therefore,

$$\begin{bmatrix} Z_{11} & Z_{12} \\ Z_{21} & Z_{22} \end{bmatrix} = \begin{bmatrix} 12 & 0 \\ 0 & 3 \end{bmatrix}$$

Example 8.3 Find the impedance parameters for the two-port network shown in Figure 8.4.

SOLUTION Assuming that the source is connected at port 1 while port 2 is open, we find that

$$V_1 = (12 + 6)I_1 = 18I_1 \quad \text{and} \quad V_2 = 6I_1$$

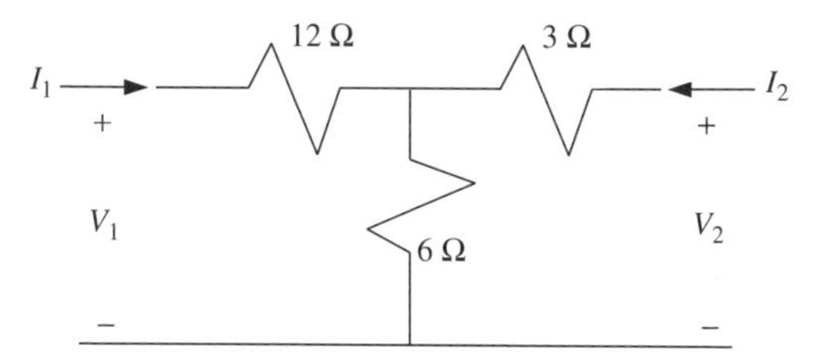

Figure 8.4 Two-port network for Example 8.3.

Note that there is no current flowing through a 3-Ω resistor because port 2 is open.

Therefore,

$$Z_{11} = \left.\frac{V_1}{I_1}\right|_{I_2=0} = \frac{18I_1}{I_1} = 18\ \Omega$$

and

$$Z_{21} = \left.\frac{V_2}{I_1}\right|_{I_2=0} = \frac{6I_1}{I_1} = 6\ \Omega$$

Similarly, with a source at port 2 and port 1 open circuit,

$$V_2 = (6+3)I_2 = 9I_2 \quad \text{and} \quad V_1 = 6I_2$$

This time, there is no current flowing through a 12-Ω resistor because port 1 is open. Hence, from (8.1.7) and (8.1.8),

$$Z_{12} = \left.\frac{V_1}{I_2}\right|_{I_1=0} = \frac{6I_2}{I_2} = 6\ \Omega$$

and

$$Z_{22} = \left.\frac{V_2}{I_2}\right|_{I_1=0} = \frac{9I_2}{I_2} = 9\ \Omega$$

Therefore,

$$\begin{bmatrix} Z_{11} & Z_{12} \\ Z_{21} & Z_{22} \end{bmatrix} = \begin{bmatrix} 18 & 6 \\ 6 & 9 \end{bmatrix}$$

An analysis of results obtained in Examples 8.1 to 8.3 indicates that Z_{12} and Z_{21} are equal for all three circuits. In fact, it is an inherent characteristic of these networks. It will hold for any reciprocal circuit. If a given circuit is symmetrical, Z_{11} will be equal to Z_{22} as well. Further, impedance parameters obtained in Example 8.3 are equal to the sum of the corresponding results found in Examples 8.1 and 8.2. This happens because if the circuits of these two examples are connected in series, we end up with the circuit of Example 8.3, shown in Figure 8.5.

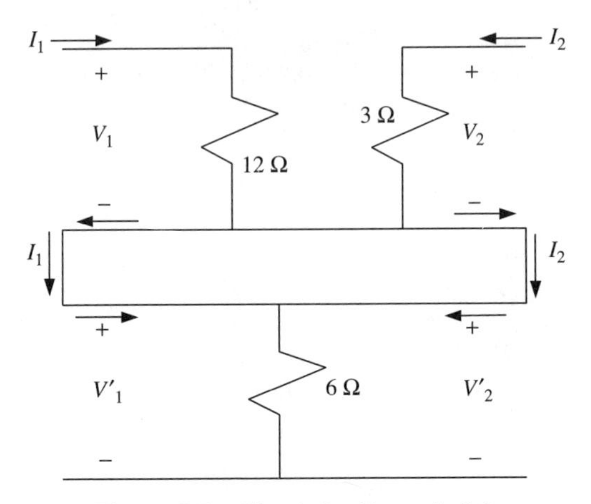

Figure 8.5 Circuit for Example 8.3.

Example 8.4 Find the impedance parameters for a transmission line network shown in Figure 8.6.

SOLUTION This circuit is symmetrical because interchanging port 1 and port 2 does not affect it. Therefore, Z_{22} must be equal to Z_{11}. Further, if current I at port 1 produces an open-circuit voltage V at port 2, current I injected at port 2 will produce V at port 1. Hence, it is a reciprocal circuit. Therefore, Z_{12} will be equal to Z_{21}. Assume that the source is connected at port 1 while the other port is open. If V_{in} is incident voltage at port 1, $V_{in}e^{-\gamma l}$ is the voltage at port 2. Since the reflection coefficient of an open circuit is $+1$, the reflected voltage at this port is equal to the incident voltage. Therefore, the reflected voltage reaching port 1 is $V_{in}e^{-2\gamma l}$. Hence,

$$V_1 = V_{in} + V_{in}e^{-2\gamma l}$$

$$V_2 = 2V_{in}e^{-\gamma l}$$

$$I_1 = \frac{V_{in}}{Z_0}(1 - e^{2\gamma l})$$

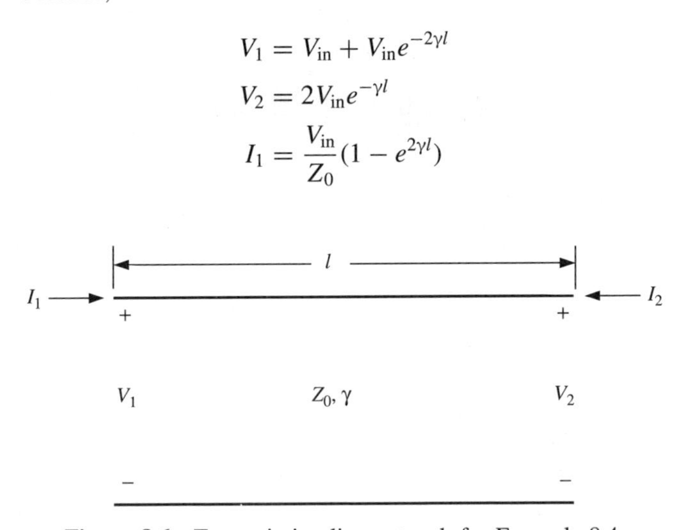

Figure 8.6 Transmission line network for Example 8.4.

and

$$I_2 = 0$$

Therefore,

$$Z_{11} = \left.\frac{V_1}{I_1}\right|_{I_2=0} = \frac{V_{\text{in}}(1 + e^{-2\gamma l})}{(V_{\text{in}}/Z_0)(1 - e^{-2\gamma l})} = Z_0 \frac{e^{+\gamma l} + e^{-\gamma l}}{e^{+\gamma l} - e^{-\gamma l}} = \frac{Z_0}{\tanh \gamma l} = Z_0 \coth \gamma l$$

and

$$Z_{21} = \left.\frac{V_2}{I_1}\right|_{I_2=0} = \frac{2V_{\text{in}} e^{-\gamma l}}{(V_{\text{in}}/Z_0)(1 - e^{-2\gamma l})} = Z_0 \frac{2}{e^{+\gamma l} - e^{-\gamma l}} = \frac{Z_0}{\sinh \gamma l}$$

For a lossless line, $\gamma = j\beta$, and therefore,

$$Z_{11} = \frac{Z_0}{j \tan \beta l} = -jZ_0 \cot \beta l$$

and

$$Z_{21} = \frac{Z_0}{j \sin \beta l} = -j \frac{Z_0}{\sin \beta l}$$

8.2 ADMITTANCE PARAMETERS

Consider again the two-port network shown in Figure 8.1. Since the network is linear, the superposition principle can be applied. Assuming that it contains no independent sources, current I_1 at port 1 can be expressed in terms of two voltages:

$$I_1 = Y_{11}V_1 + Y_{12}V_2 \tag{8.2.1}$$

Since I_1 is in amperes and V_1 and V_2 are in volts, parameters Y_{11} and Y_{12} must be in siemens. Therefore, these are called the *admittance parameters*. Similarly, we can write I_2 in terms of V_1 and V_2 as follows:

$$I_2 = Y_{21}V_1 + Y_{22}V_2 \tag{8.2.2}$$

Using the matrix representation, we can write

$$\begin{bmatrix} I_1 \\ I_2 \end{bmatrix} = \begin{bmatrix} Y_{11} & Y_{12} \\ Y_{21} & Y_{22} \end{bmatrix} \begin{bmatrix} V_1 \\ V_2 \end{bmatrix} \tag{8.2.3}$$

or

$$[I] = [Y][V] \tag{8.2.4}$$

where $[Y]$ is called the *admittance matrix* of the two-port network.

If port 2 of this network has a short circuit, V_2 will be zero. In this condition, (8.2.1) and (8.2.2) give

$$Y_{11} = \left.\frac{I_1}{V_1}\right|_{V_2=0} \tag{8.2.5}$$

and

$$Y_{21} = \left.\frac{I_2}{V_1}\right|_{V_2=0} \tag{8.2.6}$$

Similarly, with a source connected at port 2 and a short circuit at port 1,

$$Y_{12} = \left.\frac{I_1}{V_2}\right|_{V_1=0} \tag{8.2.7}$$

and

$$Y_{22} = \left.\frac{I_2}{V_2}\right|_{V_1=0} \tag{8.2.8}$$

Equations (8.2.5) through (8.2.8) define the admittance parameters of a two-port network.

Example 8.5 Find the admittance parameters of the circuit shown in Figure 8.7.

SOLUTION If V_2 is zero, I_1 is equal to $0.05\,V_1$ and I_2 is $-0.05\,V_1$. Hence, from (8.2.5) and (8.2.6),

$$Y_{11} = \left.\frac{I_1}{V_1}\right|_{V_2=0} = \frac{0.05\,V_1}{V_1} = 0.05\text{ S}$$

and

$$Y_{21} = \left.\frac{I_2}{V_1}\right|_{V_2=0} = \frac{-0.05\,V_1}{V_1} = -0.05\text{ S}$$

Similarly, with a source connected at port 2 and port 1 having a short circuit,

$$I_2 = -I_1 = 0.05\,V_2$$

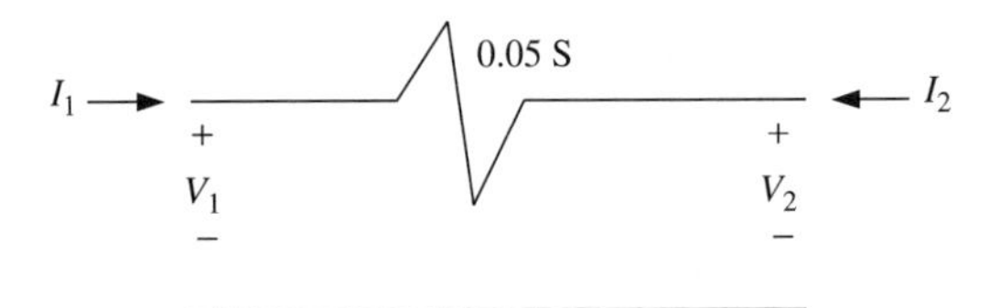

Figure 8.7 Circuit for Example 8.5.

Hence, from (8.2.7) and (8.2.8),

$$Y_{12} = \left.\frac{I_1}{V_2}\right|_{V_1=0} = \frac{-0.05\,V_2}{V_2} = -0.05\,\text{S}$$

and

$$Y_{22} = \left.\frac{I_2}{V_2}\right|_{V_1=0} = \frac{0.05\,V_2}{V_2} = 0.05\,\text{S}$$

Therefore,

$$\begin{bmatrix} Y_{11} & Y_{12} \\ Y_{21} & Y_{22} \end{bmatrix} = \begin{bmatrix} 0.05 & -0.05 \\ -0.05 & 0.05 \end{bmatrix}$$

Again we find that Y_{11} is equal to Y_{22} because this circuit is symmetrical. Similarly, Y_{12} is equal to Y_{21} because it is reciprocal.

Example 8.6 Find the admittance parameters for the two-port network shown in Figure 8.8.

SOLUTION Assuming that a source is connected at port 1 and port 2 has a short circuit, we find that

$$I_1 = \frac{0.1(0.2 + 0.025)}{0.1 + 0.2 + 0.025} V_1 = \frac{0.0225}{0.325} V_1 \qquad \text{A}$$

and if voltage across 0.2 S is V_N, then

$$V_N = \frac{I_1}{0.2 + 0.025} = \frac{0.0225}{0.225 \times 0.325} V_1 = \frac{V_1}{3.25} \qquad \text{V}$$

Therefore,

$$I_2 = -0.2\,V_N = -\frac{0.2}{3.25} V_1 \qquad \text{A}$$

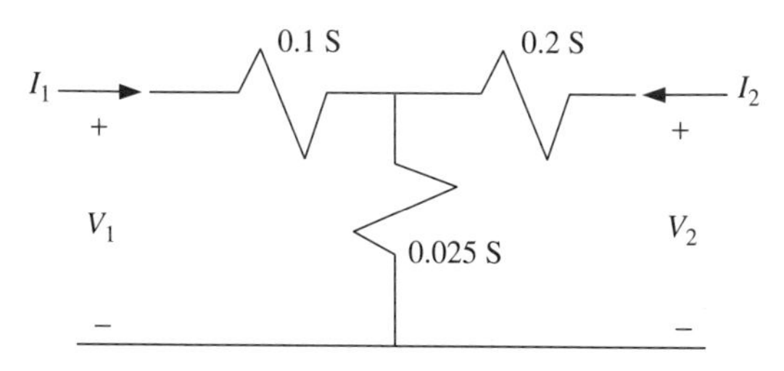

Figure 8.8 Two-port network for Example 8.6.

Hence, from (8.2.5) and (8.2.6),

$$Y_{11} = \left.\frac{I_1}{V_1}\right|_{V_2=0} = \frac{0.0225}{0.325} = 0.0692\,\text{S}$$

and

$$Y_{21} = \left.\frac{I_2}{V_1}\right|_{V_2=0} = -\frac{0.2}{3.25} = -0.0615\,\text{S}$$

Similarly, with a source at port 2 and port 1 having a short circuit,

$$I_2 = \frac{0.2(0.1 + 0.025)}{0.2 + 0.1 + 0.025} V_1 = \frac{0.025}{0.325} V_2 \qquad \text{A}$$

and if voltage across 0.1 S is V_M, then

$$V_M = \frac{I_2}{0.1 + 0.025} = \frac{0.025}{0.125 \times 0.325} V_2 = \frac{2\,V_2}{3.25} \qquad \text{V}$$

Therefore,

$$I_1 = -0.1\,V_M = -\frac{0.2}{3.25} V_2 \qquad \text{A}$$

Hence, from (8.2.7) and (8.2.8),

$$Y_{12} = \left.\frac{I_1}{V_2}\right|_{V_1=0} = -\frac{0.2}{3.25} = -0.0615\,\text{S}$$

and

$$Y_{22} = \left.\frac{I_2}{V_2}\right|_{V_1=0} = \frac{0.025}{0.325} = 0.0769\,\text{S}$$

Therefore,

$$\begin{bmatrix} Y_{11} & Y_{12} \\ Y_{21} & Y_{22} \end{bmatrix} = \begin{bmatrix} 0.0692 & -0.0615 \\ -0.0615 & 0.0769 \end{bmatrix}$$

As expected, $Y_{12} = Y_{21}$ but $Y_{11} \neq Y_{22}$. This is because the given circuit is reciprocal but is not symmetrical.

Example 8.7 Find the admittance parameters of the two-port network shown in Figure 8.9.

SOLUTION Assuming that a source is connected at port 1 and port 2 has a short circuit, we find that

$$I_1 = \left[0.05 + \frac{0.1(0.2 + 0.025)}{0.1 + 0.2 + 0.025}\right] V_1 = 0.1192\,V_1 \qquad \text{A}$$

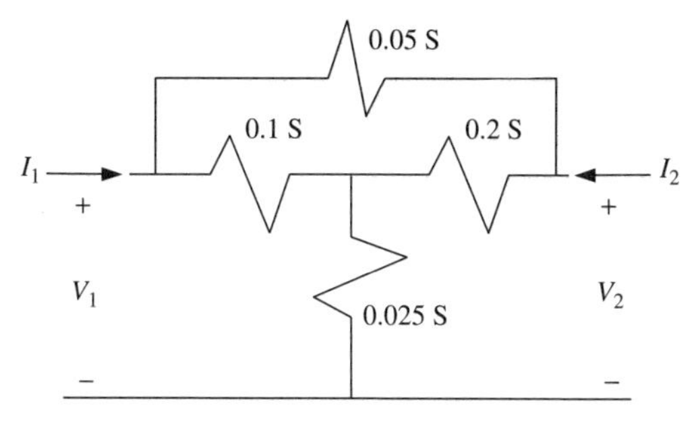

Figure 8.9 Two-port network for Example 8.7.

and if current through 0.05 S is I_N, then

$$I_N = \frac{0.05}{0.05 + [0.1(0.2 + 0.025)/(0.1 + 0.2 + 0.025)]} I_1 = 0.05\, V_1 \qquad \text{A}$$

Current through 0.1 S is $I_1 - I_N = 0.0692\, V_1$. Using the current-division rule, current I_M through 0.2 S is found as follows:

$$I_M = \frac{0.2}{0.2 + 0.025} 0.0692\, V_1 = 0.0615\, V_1 \qquad \text{A}$$

Hence, $I_2 = -(I_N + I_M) = -0.1115\, V_1$ A.

Now, from (8.2.5) and (8.2.6),

$$Y_{11} = \left.\frac{I_1}{V_1}\right|_{V_2=0} = 0.1192\,\text{S}$$

and

$$Y_{21} = \left.\frac{I_2}{V_1}\right|_{V_2=0} = -0.1115\,\text{S}$$

Similarly, with a source at port 2 and port 1 having a short circuit, current I_2 at port 2 is

$$I_2 = \left[0.05 + \frac{0.2(0.1 + 0.025)}{0.2 + 0.1 + 0.025}\right] V_2 = 0.1269\, V_2 \qquad \text{A}$$

and current I_N through 0.05 S can be found as follows:

$$I_\text{N} = \frac{0.05}{0.05 + [0.2(0.1 + 0.025)/(0.2 + 0.1 + 0.025)]} I_2 = 0.05\, V_2 \qquad \text{A}$$

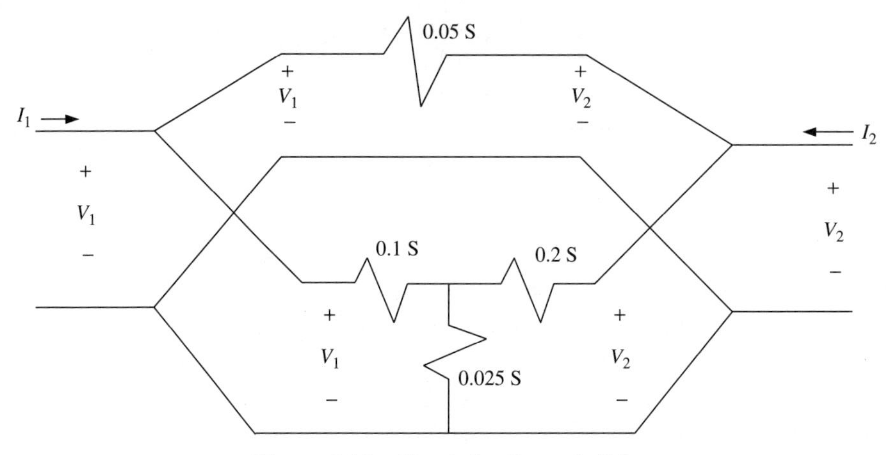

Figure 8.10 Circuit for Example 8.7.

Current through 0.2 S is $I_2 - I_N = 0.0769\, V_2$. Using the current-division rule one more time, the current I_M through 0.1 S is found as follows:

$$I_\mathrm{M} = \frac{0.1}{0.1 + 0.025} 0.0769\, V_2 = 0.0615 V_2 \qquad \mathrm{A}$$

Hence, $I_1 = -(I_N + I_M) = -0.1115\, V_2$ A. Therefore, from (8.2.7) and (8.2.8),

$$Y_{12} = \left. \frac{I_1}{V_2} \right|_{V_1=0} = -0.1115\,\mathrm{S}$$

and

$$Y_{22} = \left. \frac{I_2}{V_2} \right|_{V_1=0} = 0.1269\,\mathrm{S}$$

Therefore,

$$\begin{bmatrix} Y_{11} & Y_{12} \\ Y_{21} & Y_{22} \end{bmatrix} = \begin{bmatrix} 0.1192 & -0.1115 \\ -0.1115 & 0.1269 \end{bmatrix}$$

As expected, $Y_{12} = Y_{21}$ but $Y_{11} \neq Y_{22}$. This is because the given circuit is reciprocal but is not symmetrical. Further, we find that the admittance parameters obtained in Example 8.7 are equal to the sum of the corresponding impedance parameters of Examples 8.5 and 8.6. This is because when the circuits of these two examples are connected in parallel, we end up with the circuit of Example 8.7, shown in Figure 8.10.

Example 8.8 Find the admittance parameters of a transmission line of length l, shown in Figure 8.11.

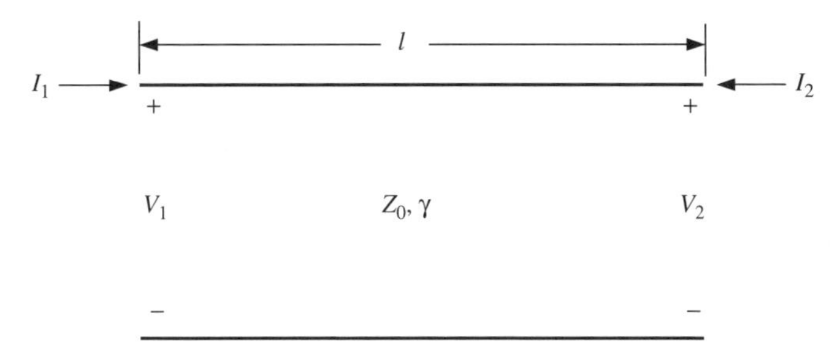

Figure 8.11 Setup for Example 8.8.

SOLUTION This circuit is symmetrical because interchanging port 1 and port 2 does not affect it. Therefore, Y_{22} must be equal Y_{11}. Further, if voltage V at port 1 produces a short-circuit current I at port 2, voltage V at port 2 will produce current I at port 1. Hence, it is a reciprocal circuit. Therefore, Y_{12} will be equal to Y_{21}. Assume that a source is connected at port 1 when the other port has a short circuit. If V_{in} is the incident voltage at port 1, it will appear as $V_{\text{in}}e^{-\gamma l}$ at port 2. Since the reflection coefficient of a short circuit is equal to -1, reflected voltage at this port is 180° out of phase with incident voltage. Therefore, the reflected voltage reaching at port 1 is $-V_{\text{in}}e^{-2\gamma l}$. Hence,

$$V_1 = V_{\text{in}} - V_{\text{in}}e^{-2\gamma l}$$

$$V_2 = 0$$

$$I_1 = \frac{V_{\text{in}}}{Z_0}(1 + e^{-2\gamma l})$$

and

$$I_2 = -\frac{2V_{\text{in}}}{Z_0}e^{-\gamma l}$$

Therefore,

$$Y_{11} = \left.\frac{I_1}{V_1}\right|_{V_2=0} = \frac{(V_{\text{in}}/Z_0)(1 + e^{-2\gamma l})}{V_{\text{in}}(1 - e^{-2\gamma l})} = \frac{e^{+\gamma l} + e^{-\gamma l}}{Z_0(e^{+\gamma l} - e^{-\gamma l})} = \frac{1}{Z_0 \tanh \gamma l}$$

and

$$Y_{21} = \left.\frac{I_2}{V_1}\right|_{V_2=0} = \frac{-(2V_{\text{in}}/Z_0)e^{-\gamma l}}{V_{\text{in}}(1 - e^{-2\gamma l})} = -\frac{2}{Z_0(e^{+\gamma l} - e^{-\gamma l})} = -\frac{1}{Z_0 \sinh \gamma l}$$

For a lossless line, $\gamma = j\beta$ and therefore

$$Y_{11} = \frac{1}{jZ_0 \tan \beta l}$$

and

$$Y_{21} = -\frac{1}{jZ_0 \sin \beta l} = j\frac{1}{Z_0 \sin \beta l}$$

8.3 HYBRID PARAMETERS

Reconsider the two-port network of Figure 8.1. Since the network is linear, the superposition principle can be applied. Assuming that it contains no independent sources, voltage V_1 at port 1 can be expressed in terms of current I_1 at port 1 and voltage V_2 at port 2:

$$V_1 = h_{11}I_1 + h_{12}V_2 \tag{8.3.1}$$

Similarly, we can write I_2 in terms of I_1 and V_2:

$$I_2 = h_{21}I_1 + h_{22}V_2 \tag{8.3.2}$$

Since V_1 and V_2 are in volts while I_1 and I_2 are in amperes, parameter h_{11} must be in ohms, h_{12} and h_{21} must be dimensionless, and h_{22} must be in siemens. Therefore, these are called the *hybrid parameters*.

Using the matrix representation, we can write

$$\begin{bmatrix} V_1 \\ I_2 \end{bmatrix} = \begin{bmatrix} h_{11} & h_{12} \\ h_{21} & h_{22} \end{bmatrix} \begin{bmatrix} I_1 \\ V_2 \end{bmatrix} \tag{8.3.3}$$

Hybrid parameters are especially important in transistor circuit analysis. These parameters are determined as follows. If port 2 has a short circuit, V_2 will be zero. In this condition, (8.3.1) and (8.3.2) give

$$h_{11} = \left.\frac{V_1}{I_1}\right|_{V_2=0} \tag{8.3.4}$$

and

$$h_{21} = \left.\frac{I_2}{I_1}\right|_{V_2=0} \tag{8.3.5}$$

Similarly, with a source connected at port 2 while port 1 is open,

$$h_{12} = \left.\frac{V_1}{V_2}\right|_{I_1=0} \tag{8.3.6}$$

and

$$h_{22} = \left.\frac{I_2}{V_2}\right|_{I_1=0} \tag{8.3.7}$$

Thus, parameters h_{11} and h_{21} represent the input impedance and the forward current gain, respectively, when a short circuit is at port 2. Similarly, h_{12} and h_{22} represent the reverse voltage gain and the output admittance, respectively, when port 1 has an open circuit. Because of this mix, these are called *hybrid parameters*. In transistor circuit analysis, these are generally denoted by h_i, h_f, h_r, and h_0, respectively.

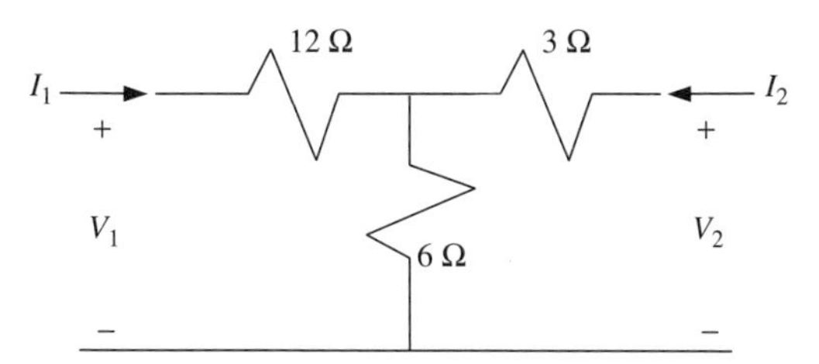

Figure 8.12 Two-port network for Example 8.9..

Example 8.9 Find the hybrid parameters of the two-port network shown in Figure 8.12.

SOLUTION With a short circuit at port 2,

$$V_1 = I_1\left(12 + \frac{6 \times 3}{6+3}\right) = 14I_1$$

and using the current-divider rule, we find that

$$I_2 = -\frac{6}{6+3}I_1 = -\frac{2}{3}I_1$$

Therefore, from (8.3.4) and (8.3.5),

$$h_{11} = \left.\frac{V_1}{I_1}\right|_{V_2=0} = 14\,\Omega$$

and

$$h_{21} = \left.\frac{I_2}{I_1}\right|_{V_2=0} = -\frac{2}{3}$$

Similarly, with a source connected at port 2 while port 1 has an open circuit,

$$V_2 = (3+6)I_2 = 9I_2$$

and

$$V_1 = 6I_2$$

because there is no current flowing through a 12-Ω resistor. Hence, from (8.3.6) and (8.3.7),

$$h_{12} = \left.\frac{V_1}{V_2}\right|_{I_1=0} = \frac{6I_2}{9I_2} = \frac{2}{3}$$

and

$$h_{22} = \left.\frac{I_2}{V_2}\right|_{I_1=0} = \frac{1}{9}\text{S}$$

Thus,

$$\begin{bmatrix} h_{11} & h_{12} \\ h_{21} & h_{22} \end{bmatrix} = \begin{bmatrix} 14\,\Omega & \frac{2}{3} \\ -\frac{2}{3} & \frac{1}{9}\text{ S} \end{bmatrix}$$

8.4 TRANSMISSION PARAMETERS

Reconsider the two-port network of Figure 8.1. Since the network is linear, the superposition principle can be applied. Assuming that it contains no independent sources, voltage V_1 and current I_1 at port 1 can be expressed in terms of current I_2 and voltage V_2 at port 2 as follows:

$$V_1 = AV_2 - BI_2 \tag{8.4.1}$$

Similarly, we can write I_1 in terms of I_2 and V_2 as follows:

$$I_1 = CV_2 - DI_2 \tag{8.4.2}$$

Since V_1 and V_2 are in volts while I_1 and I_2 are in amperes, parameters A and D must be dimensionless, B must be in ohms, and C must be in siemens.

Using the matrix representation, (8.4.2) can be written as follows:

$$\begin{bmatrix} V_1 \\ I_1 \end{bmatrix} = \begin{bmatrix} A & B \\ C & D \end{bmatrix} \begin{bmatrix} V_2 \\ -I_2 \end{bmatrix} \tag{8.4.3}$$

Transmission parameters (also known as *elements of chain matrix*) are especially important for analysis of circuits connected in cascade. These parameters are determined as follows. If port 2 has a short circuit, V_2 will be zero. Under this condition, (8.4.1) and (8.4.2) give

$$B = \left.\frac{V_1}{-I_2}\right|_{V_2=0} \tag{8.4.4}$$

and

$$D = \left.\frac{I_1}{-I_2}\right|_{V_2=0} \tag{8.4.5}$$

Similarly, with a source connected at the port 1 while port 2 is open, we find that

$$A = \left.\frac{V_1}{V_2}\right|_{I_2=0} \tag{8.4.6}$$

and

$$C = \left.\frac{I_1}{V_2}\right|_{I_2=0} \tag{8.4.7}$$

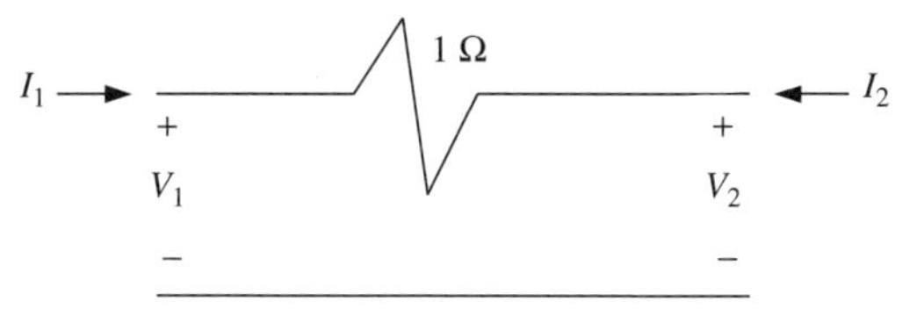

Figure 8.13 Network for Example 8.10.

Example 8.10 Determine the transmission parameters of the network shown in Figure 8.13.

SOLUTION With a source connected at port 1 while port 2 has a short circuit (so that V_2 is zero),

$$I_2 = -I_1 \quad \text{and} \quad V_1 = I_1 V$$

Therefore, from (8.4.4) and (8.4.5),

$$B = \left.\frac{V_1}{-I_2}\right|_{V_2=0} = 1\,\Omega$$

and

$$D = \left.\frac{I_1}{-I_2}\right|_{V_2=0} = 1$$

Similarly, with a source connected at port 1 while port 2 is open (so that I_2 is zero),

$$V_2 = V_1 \quad \text{and} \quad I_1 = 0$$

Now, from (8.4.6) and (8.4.7),

$$A = \left.\frac{V_1}{V_2}\right|_{I_2=0} = 1$$

and

$$C = \left.\frac{I_1}{V_2}\right|_{I_2=0} = 0$$

Hence, the transmission matrix of this network is

$$\begin{bmatrix} A & B \\ C & D \end{bmatrix} = \begin{bmatrix} 1 & 1 \\ 0 & 1 \end{bmatrix}$$

Example 8.11 Determine the transmission parameters of the network shown in Figure 8.14.

SOLUTION With a source connected at port 1 while port 2 has a short circuit (so that V_2 is zero),

$$I_2 = -I_1 \quad \text{and} \quad V_1 = 0\text{V}$$

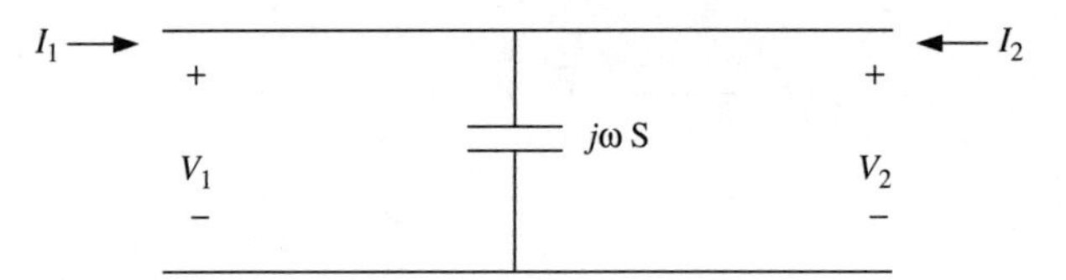

Figure 8.14 Network for Example 8.11.

Therefore, from (8.4.4) and (8.4.5),

$$B = \left.\frac{V_1}{-I_2}\right|_{V_2=0} = 0\,\Omega$$

and

$$D = \left.\frac{I_1}{-I_2}\right|_{V_2=0} = 1$$

Similarly, with a source connected at port 1 while port 2 is open (so that I_2 is zero),

$$V_2 = V_1 \quad \text{and} \quad I_1 = j\omega V_1 \qquad \text{A}$$

Now, from (8.4.6) and (8.4.7),

$$A = \left.\frac{V_1}{V_2}\right|_{I_2=0} = 1$$

and

$$C = \left.\frac{I_1}{V_2}\right|_{I_2=0} = j\omega \qquad \text{S}$$

Hence, transmission matrix of this network is

$$\begin{bmatrix} A & B \\ C & D \end{bmatrix} = \begin{bmatrix} 1 & 0 \\ j\omega & 1 \end{bmatrix}$$

Example 8.12 Determine the transmission parameters of the network shown in Figure 8.15.

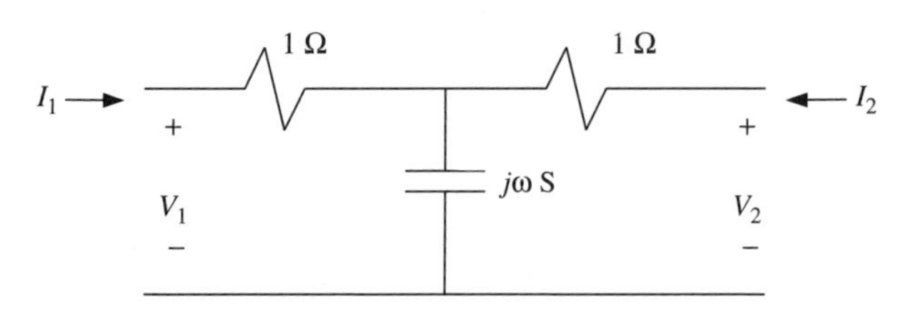

Figure 8.15 Network for Example 8.12.

SOLUTION With a source connected at port 1 while port 2 has a short circuit (so that V_2 is zero), we find that

$$V_1 = \left(1 + \frac{1}{1 + j\omega}\right) I_1 = \frac{2 + j\omega}{1 + j\omega} I_1$$

and

$$I_2 = -\frac{\frac{1}{j\omega}}{1/j\omega + 1} I_1 = -\frac{1}{1 + j\omega} I_1$$

Therefore, from (8.4.4) and (8.4.5),

$$B = \left.\frac{V_1}{-I_2}\right|_{V_2=0} = 2 + j\omega \qquad \Omega$$

and

$$D = \left.\frac{I_1}{-I_2}\right|_{V_2=0} = 1 + j\omega$$

Similarly, with a source connected at port 1 while port 2 is open (so that I_2 is zero),

$$V_1 = \left(1 + \frac{1}{j\omega}\right) I_1 = \frac{1 + j\omega}{j\omega} I_1$$

and

$$V_2 = \frac{1}{j\omega} I_1$$

Now, from (8.4.6) and (8.4.7),

$$A = \left.\frac{V_1}{V_2}\right|_{I_2=0} = 1 + j\omega$$

and

$$C = \left.\frac{I_1}{V_2}\right|_{I_2=0} = j\omega \qquad \text{S}$$

Hence,

$$\begin{bmatrix} A & B \\ C & D \end{bmatrix} = \begin{bmatrix} 1 + j\omega & 2 + j\omega \\ j\omega & 1 + j\omega \end{bmatrix}$$

Example 8.13 Find the transmission parameters of the transmission line shown in Fig. 8.16.

SOLUTION Assume that a source is connected at port 1 while the other port has a short circuit. If V_{in} is incident voltage at port 1, it will be $V_{in}e^{-\gamma l}$ at port 2.

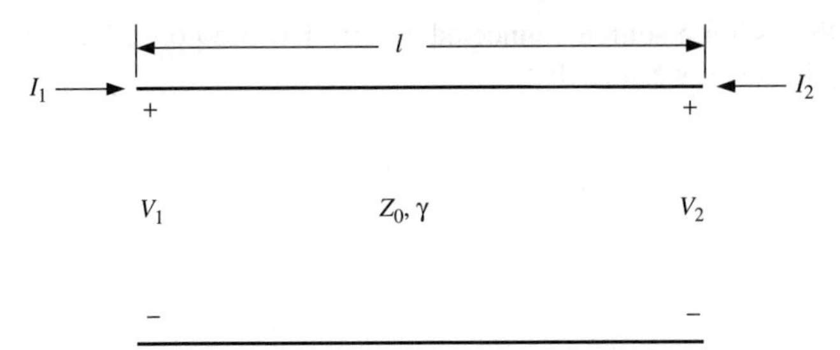

Figure 8.16 Transmission line for Example 8.13.

Since the reflection coefficient of the short circuit is -1, the reflected voltage at this port is 180° out of phase with the incident voltage. Therefore, the reflected voltage reaching at port 1 is $-V_{\text{in}}e^{-2\gamma l}$. Hence,

$$V_1 = V_{\text{in}} - V_{\text{in}}e^{-2\gamma l}$$

$$V_2 = 0$$

$$I_1 = \frac{V_{\text{in}}}{Z_0}(1 + e^{-2\gamma l})$$

and

$$I_2 = -\frac{2V_{\text{in}}}{Z_0}e^{-\gamma l}$$

Therefore, from (8.4.4) and (8.4.5),

$$B = \left.\frac{V_1}{-I_2}\right|_{V_2=0} = \frac{Z_0}{2e^{-\gamma l}}(1 - e^{-2\lambda l}) = Z_0\frac{e^{\gamma l} - e^{-\gamma l}}{2} \quad \Omega = Z_0 \sinh \gamma l \quad \Omega$$

and

$$D = \left.\frac{I_1}{-I_2}\right|_{V_2=0} = \frac{1 + e^{-2\lambda l}}{2e^{-\lambda l}} = \frac{e^{\gamma l} + e^{-\gamma l}}{2} = \cosh \gamma l$$

Now assume that port 2 has an open circuit while the source is still connected at port 1. If V_{in} is incident voltage at port 1, $V_{\text{in}}e^{-\gamma l}$ is at port 2. Since the reflection coefficient of an open circuit is $+1$, the reflected voltage at this port is equal to the incident voltage. Therefore, the reflected voltage reaching at port 1 is $V_{\text{in}}e^{-2\gamma l}$. Hence,

$$V_1 = V_{\text{in}} + V_{\text{in}}e^{-2\gamma l}$$

$$V_2 = 2V_{\text{in}}e^{-\gamma l}$$

$$I_1 = \frac{V_{\text{in}}}{Z_0}(1 - e^{-2\gamma l})$$

and

$$I_2 = 0$$

Now, from (8.4.6) and (8.4.7),

$$A = \left.\frac{V_1}{V_2}\right|_{I_2=0} = \frac{1 + e^{-2\gamma l}}{2e^{-\gamma l}} = \cosh \gamma l$$

and

$$C = \left.\frac{I_1}{V_2}\right|_{I_2=0} = \frac{1 - e^{-2\gamma l}}{2Z_0 e^{-\gamma l}} = \frac{1}{Z_0} \sinh \gamma l$$

Hence, the transmission matrix of a finite-length transmission line is

$$\begin{bmatrix} A & B \\ C & D \end{bmatrix} = \begin{bmatrix} \cosh \gamma l & Z_0 \sinh \gamma l \\ \frac{1}{Z_0} \sinh \gamma l & \cosh \gamma l \end{bmatrix}$$

For a lossless line, $\gamma = j\beta$, and therefore it simplifies to

$$\begin{bmatrix} A & B \\ C & D \end{bmatrix} = \begin{bmatrix} \cos \beta l & jZ_0 \sin \beta l \\ j\frac{1}{Z_0} \sin \beta l & \cos \beta l \end{bmatrix}$$

An analysis of results obtained in Examples 8.10 to 8.13 indicates that the following condition holds for all four circuits:

$$AD - BC = 1 \tag{8.4.8}$$

This is because these circuits are reciprocal. In other words, if a given circuit is known to be reciprocal, (8.4.8) must be satisfied. Further, we find that transmission parameter A is equal to D in all four cases. This always happens when a given circuit is reciprocal.

In Example 8.11, A and D are real, B is zero, and C is imaginary. For a lossless line in Example 8.13, A and D simplify to real numbers while C and D become purely imaginary. This characteristic of the transmission parameters is associated with any lossless circuit. A comparison of the circuits in Examples 8.10 to 8.12 reveals that the two-port network of Example 8.12 can be obtained by cascading that of Example 8.10 on the two sides of Example 8.11, as shown in Figure 8.17.

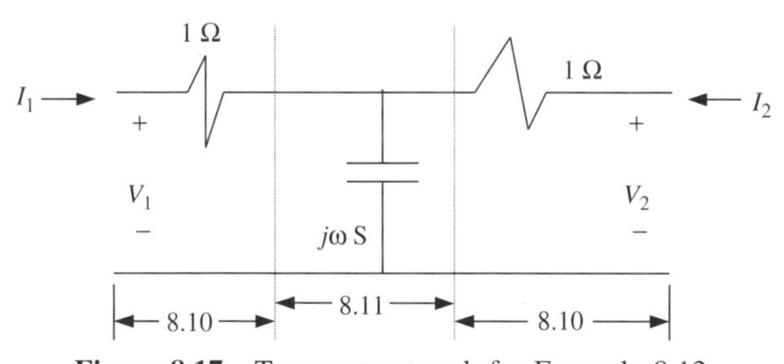

Figure 8.17 Two-port network for Example 8.12.

Therefore, the chain (or transmission) matrix for the network shown in Example 8.12 can be obtained after multiplying three chain matrices as follows:

$$\begin{bmatrix} 1 & 1 \\ 0 & 1 \end{bmatrix} \cdot \begin{bmatrix} 1 & 0 \\ j\omega & 1 \end{bmatrix} \cdot \begin{bmatrix} 1 & 1 \\ 0 & 1 \end{bmatrix} = \begin{bmatrix} 1 & 1 \\ 0 & 1 \end{bmatrix} \cdot \begin{bmatrix} 1 & 1 \\ j\omega & j\omega + 1 \end{bmatrix}$$

$$= \begin{bmatrix} 1 + j\omega & 2 + j\omega \\ j\omega & 1 + j\omega \end{bmatrix}$$

This shows that chain matrices are convenient in the analysis and design of networks connected in cascade.

8.5 CONVERSION OF IMPEDANCE, ADMITTANCE, CHAIN, AND HYBRID PARAMETERS

One type of network parameter can be converted into another via the respective defining equations. For example, the admittance parameters of a network can be found from its impedance parameters as follows. From (8.2.3) and (8.1.3), we find that

$$\begin{bmatrix} I_1 \\ I_2 \end{bmatrix} = \begin{bmatrix} Y_{11} & Y_{12} \\ Y_{21} & Y_{22} \end{bmatrix} \begin{bmatrix} V_1 \\ V_2 \end{bmatrix} = \begin{bmatrix} Z_{11} & Z_{12} \\ Z_{21} & Z_{22} \end{bmatrix}^{-1} \begin{bmatrix} V_1 \\ V_2 \end{bmatrix}$$

Hence,

$$\begin{bmatrix} Y_{11} & Y_{12} \\ Y_{21} & Y_{22} \end{bmatrix} = \begin{bmatrix} Z_{11} & Z_{12} \\ Z_{21} & Z_{22} \end{bmatrix}^{-1} = \frac{1}{Z_{11}Z_{22} - Z_{12}Z_{21}} \begin{bmatrix} Z_{22} & -Z_{12} \\ -Z_{21} & Z_{11} \end{bmatrix}$$

Similarly, (8.3.3) can be rearranged as follows:

$$\begin{bmatrix} I_1 \\ I_2 \end{bmatrix} = \begin{bmatrix} \frac{D}{B} & -\frac{AD - BC}{B} \\ -\frac{1}{B} & \frac{A}{B} \end{bmatrix} \begin{bmatrix} V_1 \\ V_2 \end{bmatrix}$$

Hence,

$$\begin{bmatrix} Y_{11} & Y_{12} \\ Y_{21} & Y_{22} \end{bmatrix} = \begin{bmatrix} \frac{D}{B} & -\frac{AD - BC}{B} \\ -\frac{1}{B} & \frac{A}{B} \end{bmatrix}$$

Relations between other parameters can be found following a similar procedure. These relations are given in Table 8.1.

8.6 SCATTERING PARAMETERS

As illustrated in preceding sections, Z-parameters are useful in analyzing series circuits while Y-parameters simplify the analysis of parallel (shunt) connected circuits. Similarly, transmission parameters are useful for chain or cascade circuits.

TABLE 8.1 Conversions Among the Impedance, Admittance, Chain, and Hybrid Parameters

$Z_{11} = \dfrac{Y_{22}}{Y_{11}Y_{22} - Y_{12}Y_{21}}$	$Z_{11} = \dfrac{A}{C}$	$Z_{11} = \dfrac{h_{11}h_{22} - h_{12}h_{21}}{h_{22}}$
$Z_{12} = \dfrac{-Y_{12}}{Y_{11}Y_{22} - Y_{12}Y_{21}}$	$Z_{12} = \dfrac{AD - BC}{C}$	$Z_{12} = \dfrac{h_{12}}{h_{22}}$
$Z_{21} = \dfrac{-Y_{21}}{Y_{11}Y_{22} - Y_{12}Y_{21}}$	$Z_{21} = \dfrac{1}{C}$	$Z_{21} = \dfrac{-h_{21}}{h_{22}}$
$Z_{22} = \dfrac{Y_{11}}{Y_{11}Y_{22} \quad Y_{12}Y_{21}}$	$Z_{22} = \dfrac{D}{C}$	$Z_{22} = \dfrac{1}{h_{22}}$
$Y_{11} = \dfrac{Z_{22}}{Z_{11}Z_{22} - Z_{12}Z_{21}}$	$Y_{11} = \dfrac{D}{B}$	$Y_{11} = \dfrac{1}{h_{11}}$
$Y_{12} = \dfrac{-Z_{12}}{Z_{11}Z_{22} - Z_{12}Z_{21}}$	$Y_{12} = \dfrac{-(AD - BC)}{B}$	$Y_{12} = \dfrac{-h_{12}}{h_{11}}$
$Y_{21} = \dfrac{-Z_{21}}{Z_{11}Z_{22} - Z_{12}Z_{21}}$	$Y_{21} = \dfrac{-1}{B}$	$Y_{21} = \dfrac{h_{21}}{h_{11}}$
$Y_{22} = \dfrac{Z_{11}}{Z_{11}Z_{22} - Z_{12}Z_{21}}$	$Y_{22} = \dfrac{A}{B}$	$Y_{22} = \dfrac{h_{11}h_{22} - h_{12}h_{21}}{h_{11}}$
$A = \dfrac{Z_{11}}{Z_{21}}$	$A = \dfrac{-Y_{22}}{Y_{21}}$	$A = \dfrac{-(h_{11}h_{22} - h_{12}h_{21})}{h_{21}}$
$B = \dfrac{Z_{11}Z_{22} - Z_{12}Z_{21}}{Z_{21}}$	$B = \dfrac{-1}{Y_{21}}$	$B = \dfrac{-h_{11}}{h_{21}}$
$C = \dfrac{1}{Z_{21}}$	$C = \dfrac{-(Y_{11}Y_{22} - Y_{12}Y_{21})}{Y_{21}}$	$C = \dfrac{-h_{22}}{h_{21}}$
$D = \dfrac{Z_{22}}{Z_{21}}$	$D = \dfrac{-Y_{11}}{Y_{21}}$	$D = \dfrac{-1}{h_{21}}$
$h_{11} = \dfrac{Z_{11}Z_{22} - Z_{12}Z_{21}}{Z_{22}}$	$h_{11} = \dfrac{1}{Y_{11}}$	$h_{11} = \dfrac{B}{D}$
$h_{12} = \dfrac{Z_{12}}{Z_{22}}$	$h_{12} = \dfrac{-Y_{12}}{Y_{11}}$	$h_{12} = \dfrac{AD - BC}{D}$
$h_{21} = \dfrac{-Z_{21}}{Z_{22}}$	$h_{21} = \dfrac{Y_{21}}{Y_{11}}$	$h_{21} = \dfrac{-1}{D}$
$h_{22} = \dfrac{1}{Z_{22}}$	$h_{22} = \dfrac{Y_{11}Y_{22} - Y_{12}Y_{21}}{Y_{11}}$	$h_{22} = \dfrac{C}{D}$

However, the characterization procedure of these parameters requires an open or short circuit at the other port. This extreme reflection makes it very difficult (and in certain cases, impossible) to determine the parameters of a network at radio and microwave frequencies. Therefore, a new representation based on traveling waves is defined. This is known as the *scattering matrix* of the network. Elements of this matrix are known as *scattering parameters.*

Figure 8.18 shows a network along with incident and reflected waves at its two ports. We adopt a convention of representing the incident wave by a_i and the reflected wave by b_i at the ith port. Hence, a_1 is an incident wave, while b_1 is a reflected wave at port 1. Similarly, a_2 and b_2 represent incident and reflected waves at port 2, respectively. Assume that a source is connected at port 1 that produces the incident wave a_1. A part of this wave is reflected back at the input (due to impedance mismatch), while the remaining signal is transmitted through the network. It may change in magnitude as well as in phase before emerging at port 2. Depending on the termination at this port, part of the signal is reflected back as input to port 2. Hence, reflected wave b_1 depends on incident signals a_1 and a_2 at the two ports. Similarly, emerging wave b_2 also depends on a_1 and a_2. Mathematically,

$$b_1 = S_{11}a_1 + S_{12}a_2 \tag{8.6.1}$$

$$b_2 = S_{21}a_1 + S_{22}a_2 \tag{8.6.2}$$

Using the matrix notation, we can write

$$\begin{bmatrix} b_1 \\ b_2 \end{bmatrix} = \begin{bmatrix} S_{11} & S_{12} \\ S_{21} & S_{22} \end{bmatrix} \begin{bmatrix} a_1 \\ a_2 \end{bmatrix} \tag{8.6.3}$$

or

$$[b] = [S][a] \tag{8.6.4}$$

where $[S]$ is called the *scattering matrix* of the two-port network; S_{ij} are known as the *scattering parameters* of this network, and a_i represents the incident wave at the ith port and b_i represents the reflected wave at the ith port.

If port 2 is matched terminated while a_1 is incident at port 1, a_2 is zero. In this condition, (8.6.1) and (8.6.2) give

$$S_{11} = \left.\frac{b_1}{a_1}\right|_{a_2=0} \tag{8.6.5}$$

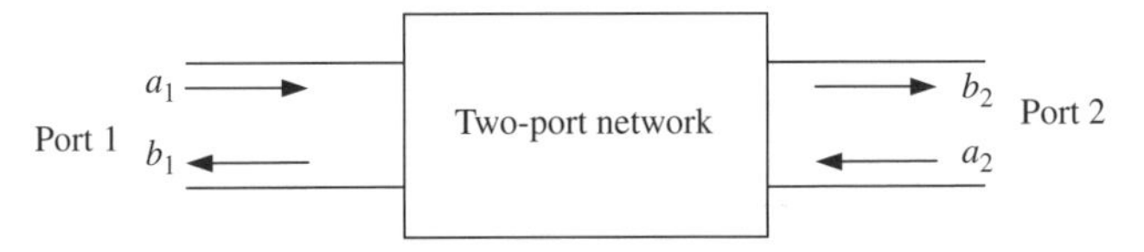

Figure 8.18 Two-port network with associated incident and reflected waves.

and

$$S_{21} = \left.\frac{b_2}{a_1}\right|_{a_2=0} \tag{8.6.6}$$

Similarly, with a source connected at port 2 while port 1 is terminated by a matched load, we find that

$$S_{12} = \left.\frac{b_1}{a_2}\right|_{a_1=0} \tag{8.6.7}$$

and

$$S_{22} = \left.\frac{b_2}{a_2}\right|_{a_1=0} \tag{8.6.8}$$

Hence, S_{ii} is the reflection coefficient Γ_i at the ith port when the other port is matched terminated. S_{ij} is the forward transmission coefficient of the jth port if i is greater than j, whereas it represents the reverse transmission coefficient if i is less than j with the other port terminated by a matched load.

We have not yet defined a_i and b_i in terms of voltage, current, or power. To that end, we write steady-state total voltage and current at the ith port as follows:

$$V_i = V_i^{\text{in}} + V_i^{\text{ref}} \tag{8.6.9}$$

and

$$I_i = \frac{1}{Z_{0i}}(V_i^{\text{in}} - V_i^{\text{ref}}) \tag{8.6.10}$$

where superscripts "in" and "ref" represent the incident and reflected voltages, respectively. Z_{0i} is the characteristic impedance at the ith port. Equations (8.6.9) and (8.6.10) can be solved to find incident and reflected voltages in terms of total voltage and current at the ith port. Hence,

$$V_i^{\text{in}} = \tfrac{1}{2}(V_i + Z_{0i}I_i) \tag{8.6.11}$$

and

$$V_i^{\text{ref}} = \tfrac{1}{2}(V_i - Z_{0i}I_i) \tag{8.6.12}$$

Assuming both of the ports to be lossless so that Z_{0i} is a real quantity, the average power incident at the ith port is

$$P_i^{\text{in}} = \frac{1}{2}\text{Re}\left[V_i^{\text{in}}(I_i^{\text{in}})^*\right] = \frac{1}{2}\text{Re}\left[V_i^{\text{in}}\left(\frac{V_i^{\text{in}}}{Z_{0i}}\right)^*\right] = \frac{1}{2Z_{0i}}|V_i^{\text{in}}|^2 \tag{8.6.13}$$

and average power reflected from the ith port is

$$P_i^{\text{ref}} = \frac{1}{2}\text{Re}\left[V_i^{\text{ref}}(I_i^{\text{ref}})^*\right] = \frac{1}{2}\text{Re}\left[V_i^{\text{ref}}\left(\frac{V_i^{\text{ref}}}{Z_{0i}}\right)^*\right] = \frac{1}{2Z_{0i}}|V_i^{\text{ref}}|^2 \tag{8.6.14}$$

The a_i and b_i are defined in such a way that the squares of their magnitudes represent the power flowing in respective directions. Hence,

$$a_i = \frac{V_i^{\text{in}}}{\sqrt{2Z_{0i}}} = \frac{1}{2}\left(\frac{V_i + Z_{0i}I_i}{\sqrt{2Z_{0i}}}\right) = \frac{1}{2\sqrt{2}}\left(\frac{V_i}{\sqrt{Z_{0i}}} + \sqrt{Z_{0i}}I_i\right) \quad (8.6.15)$$

and

$$b_i = \frac{V_i^{\text{ref}}}{\sqrt{2Z_{0i}}} = \frac{1}{2}\left(\frac{V_i - Z_{0i}I_i}{\sqrt{2Z_{0i}}}\right) = \frac{1}{2\sqrt{2}}\left(\frac{V_i}{\sqrt{Z_{0i}}} - \sqrt{Z_{0i}}I_i\right) \quad (8.6.16)$$

Therefore, units of a_i and b_i are

$$\sqrt{\text{watt}} = \frac{\text{volt}}{\sqrt{\text{ohm}}} = \text{ampere} \cdot \sqrt{\text{ohm}}$$

Power available from the source, P_{avs}, at port 1 is

$$P_{\text{avs}} = |a_1|^2$$

power reflected from port 1, P_{ref}, is

$$P_{\text{ref}} = |b_1|^2$$

and power delivered to the port (and hence to the network), P_d, is

$$P_d = P_{\text{avs}} - P_{\text{ref}} = |a_1|^2 - |b_1|^2$$

Consider the circuit arrangement shown in Figure 8.19. There is a voltage source V_{S1} connected at port 1 while port 2 is terminated by load impedance Z_L. The source impedance is Z_S. Various voltages, currents, and waves are as depicted at the two ports of this network. Further, it is assumed that the characteristic impedances at port 1 and port 2 are Z_{01} and Z_{02}, respectively. Input impedance Z_1 at port 1 of the network is defined as the impedance across its terminals when port 2 is terminated by load Z_L while source V_{S1} along with Z_S are disconnected. Similarly, output impedance Z_2 at port 2 of the network is defined as the impedance across its terminals with load Z_L disconnected and voltage source V_{S1} replaced by a short circuit. Hence, source impedance Z_S terminates

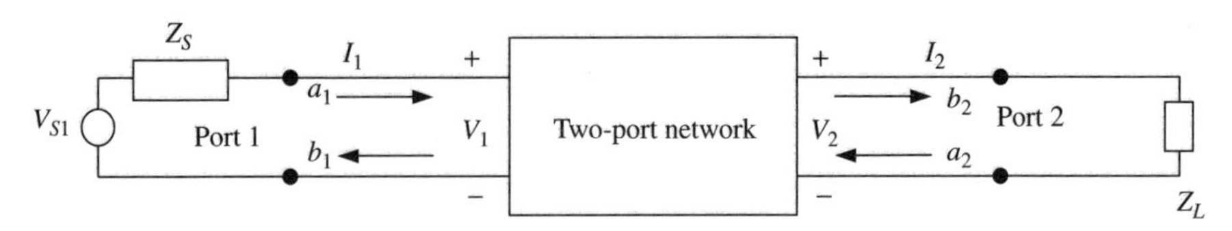

Figure 8.19 Two-port network with a voltage source connected at port 1 and port 2 is terminated.

port 1 of the network in this case. Input impedance Z_1 and output impedance Z_2 are responsible for input reflection coefficient Γ_1 and output reflection coefficient Γ_2, respectively. Hence, the ratio of b_1 to a_1 represents Γ_1 while that of b_2 to a_2 is Γ_2. For the two-port network, we can write

$$b_1 = S_{11}a_1 + S_{12}a_2 \tag{8.6.17}$$

and

$$b_2 = S_{21}a_1 + S_{22}a_2 \tag{8.6.18}$$

The load reflection coefficient Γ_L is

$$\Gamma_L = \frac{Z_L - Z_{02}}{Z_L + Z_{02}} = \frac{a_2}{b_2} \tag{8.6.19}$$

Note that b_2 leaves port 2, and therefore it is incident on the load Z_L. Similarly, the wave reflected back from the load enters port 2 as a_2. The source reflection coefficient Γ_S is found as

$$\Gamma_S = \frac{Z_S - Z_{01}}{Z_S + Z_{01}} = \frac{a_1}{b_1} \tag{8.6.20}$$

Since b_1 leaves port 1 of the network, it is the incident wave on Z_S, while a_1 is the reflected wave.

The input and output reflection coefficients are

$$\Gamma_1 = \frac{Z_1 - Z_{01}}{Z_1 + Z_{01}} = \frac{b_1}{a_1} \tag{8.6.21}$$

and

$$\Gamma_2 = \frac{Z_2 - Z_{02}}{Z_2 + Z_{02}} = \frac{b_2}{a_2} \tag{8.6.22}$$

Dividing (8.6.17) by a_1 and then using (8.6.21), we find that

$$\frac{b_1}{a_1} = \Gamma_1 = \frac{Z_1 - Z_{01}}{Z_1 + Z_{01}} = S_{11} + S_{12}\frac{a_2}{a_1} \tag{8.6.23}$$

Now, dividing (8.6.18) by a_2 and then combining with (8.6.19), we get

$$\frac{b_2}{a_2} = S_{22} + S_{21}\frac{a_1}{a_2} = \frac{1}{\Gamma_L} \Rightarrow \frac{a_1}{a_2} = \frac{1 - S_{22}\Gamma_L}{S_{21}\Gamma_L} \tag{8.6.24}$$

From (8.6.23) and (8.6.24),

$$\Gamma_1 = S_{11} + \frac{S_{12}S_{21}\Gamma_L}{1 - S_{22}\Gamma_L} \tag{8.6.25}$$

If a matched load is terminating port 2, then $\Gamma_L = 0$ and (7.6.25) simplifies to $\Gamma_1 = S_{11}$.

Similarly, from (8.6.18) and (8.6.22),

$$\frac{b_2}{a_2} = \Gamma_2 = \frac{Z_2 - Z_{02}}{Z_2 + Z_{02}} = S_{22} + S_{21}\frac{a_1}{a_2} \tag{8.6.26}$$

From (8.6.17) and (8.6.20), we have

$$\frac{b_1}{a_1} = S_{11} + S_{12}\frac{a_2}{a_1} = \frac{1}{\Gamma_S} \Rightarrow \frac{a_2}{a_1} = \frac{1 - S_{11}\Gamma_S}{S_{12}\Gamma_S} \tag{8.6.27}$$

Substituting (8.6.27) into (8.6.26), we get

$$\Gamma_2 = S_{22} + \frac{S_{21}S_{12}\Gamma_S}{1 - S_{11}\Gamma_S} \tag{8.6.28}$$

If Z_S is equal to Z_{01}, port 1 is matched and $\Gamma_S = 0$. Therefore, (8.6.28) simplifies to

$$\Gamma_2 = S_{22}$$

Hence, S_{11} and S_{22} can be found by evaluating the reflection coefficients at respective ports while the other port is matched terminated.

Let us determine the other two parameters, S_{21} and S_{12}, of the two-port network. Starting with (8.6.6) for S_{21}, we have

$$S_{21} = \left.\frac{b_2}{a_1}\right|_{a_2=0} \tag{8.6.6}$$

Now, a_2 is found from (8.6.15) with i as 2 and forcing it to zero, we get

$$a_2 = \frac{1}{2}\left(\frac{V_2 + Z_{02}I_2}{\sqrt{2Z_{02}}}\right) = 0 \Rightarrow V_2 = -Z_{02}I_2 \tag{8.6.29}$$

Substituting (8.6.29) in the expression for b_2 that is obtained from (8.6.16) with i as 2, we find that

$$b_2 = \frac{1}{2}\left(\frac{V_2 - Z_{02}I_2}{\sqrt{2Z_{02}}}\right) = -I_2\sqrt{\frac{Z_{02}}{2}} \tag{8.6.30}$$

An expression for a_1 is obtained from (8.6.15) with $i = 1$. It simplifies for Z_S equal to Z_{01} as follows:

$$a_1 = \frac{1}{2}\left(\frac{V_1 + Z_{01}I_1}{\sqrt{2Z_{01}}}\right) = \frac{V_{S1}}{2\sqrt{2Z_{01}}} \tag{8.6.31}$$

S_{21} is obtained by substituting (8.6.30) and (8.6.31) into (8.6.6) as follows:

$$S_{21} = \left.\frac{b_2}{a_1}\right|_{a_2=0} = \frac{-\sqrt{Z_{02}}I_2/\sqrt{2}}{V_{S1}/2\sqrt{2Z_{01}}} = \frac{2V_2}{V_{S1}}\sqrt{\frac{Z_{01}}{Z_{02}}} \tag{8.6.32}$$

Following a similar procedure, S_{12} may be found as

$$S_{12} = \left.\frac{b_1}{a_2}\right|_{a_1=0} = \frac{2V_1}{V_{S2}}\sqrt{\frac{Z_{02}}{Z_{01}}} \tag{8.6.33}$$

An analysis of S-parameters indicates that

$$|S_{11}|^2 = \left.\frac{|b_1|^2}{|a_1|^2}\right|_{a_2=0} = \frac{P_{\text{avs}} - P_d}{P_{\text{avs}}} \tag{8.6.34}$$

where P_{avs} is power available from the source and P_d is power delivered to port 1. These two powers will be equal if the source impedance is conjugate of Z_1; that is, the source is matched with port 1. Similarly, from (8.6.32),

$$|S_{21}|^2 = \frac{Z_{02}(I_2/\sqrt{2})^2}{(\frac{1}{4}Z_{01})(V_{S1}/\sqrt{2})^2} = \frac{Z_{02}(I_2/\sqrt{2})^2}{\frac{1}{2}[(1/2Z_{01})(V_{S1}/\sqrt{2})^2]} = \frac{P_{\text{AVN}}}{P_{\text{avs}}} \tag{8.6.35}$$

where P_{AVN} is power available at port 2 of the network. It will be equal to power delivered to a load that is matched to the port. This power ratio of (8.6.35) may be called the *transducer power gain.*

Following a similar procedure, it may be found that $|S_{22}|^2$ represents the ratio of power reflected from port 2 to power available from the source at port 2, while port 1 is terminated by a matched load Z_S and $|S_{12}|^2$ represents a reverse transducer power gain.

Shifting the Reference Planes

Consider the two-port network shown in Figure 8.20. Assume that a_i and b_i are incident and reflected waves, respectively, at unprimed reference planes of the ith

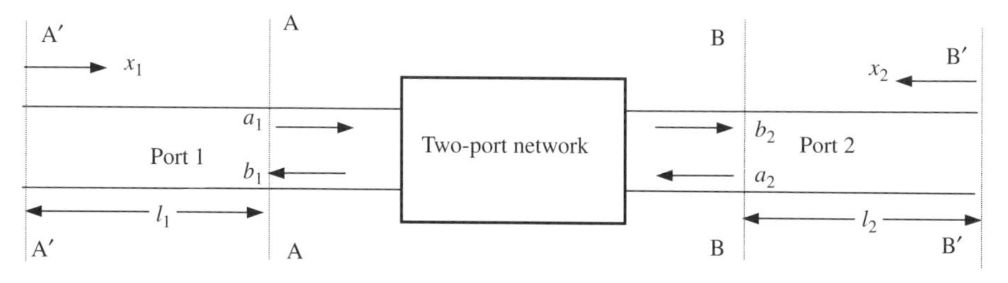

Figure 8.20 Two-port network with two reference planes on each side.

port. We use unprimed S-parameters for this case. Next, consider that plane A–A is shifted by a distance l_1 to A′–A′. At this plane, a_1' and b_1' represent inward- and outward-traveling waves, respectively. Similarly, a_2' and b_2' represent inward- and outward-traveling waves, respectively, at plane B′–B′. We denote the scattering parameters at primed planes by a prime on each as well. Hence,

$$\begin{bmatrix} b_1 \\ b_2 \end{bmatrix} = \begin{bmatrix} S_{11} & S_{12} \\ S_{21} & S_{22} \end{bmatrix} \begin{bmatrix} a_1 \\ a_2 \end{bmatrix} \tag{8.6.36}$$

and

$$\begin{bmatrix} b_1' \\ b_2' \end{bmatrix} = \begin{bmatrix} S_{11}' & S_{12}' \\ S_{21}' & S_{22}' \end{bmatrix} \begin{bmatrix} a_1' \\ a_2' \end{bmatrix} \tag{8.6.37}$$

Wave b_1 is delayed in phase by βl_1 as it travels from A to A′. This means that b_1 is ahead in phase with respect to b_1'. Hence,

$$b_1 = b_1' e^{j\beta l_1} \tag{8.6.38}$$

Wave a_1 comes from A′ to plane A. Therefore, it has a phase delay of βl_1 with respect to a_1'. Mathematically,

$$a_1 = a_1' e^{-j\beta l_1} \tag{8.6.39}$$

On the basis of similar considerations at port 2, we can write

$$b_2 = b_2' e^{j\beta l_2} \tag{8.6.40}$$

and

$$a_2 = a_2' e^{-j\beta l_2} \tag{8.6.41}$$

Substituting for b_1, a_1, b_2, and a_1 from (8.6.38) to (8.6.41) into (8.6.36), we get

$$\begin{bmatrix} b_1' e^{j\beta l_1} \\ b_2' e^{j\beta l_2} \end{bmatrix} = \begin{bmatrix} S_{11} & S_{12} \\ S_{21} & S_{22} \end{bmatrix} \begin{bmatrix} a_1' e^{-j\beta l_1} \\ a_2' e^{-j\beta l_2} \end{bmatrix} \tag{8.6.42}$$

We can rearrange this equation as follows:

$$\begin{bmatrix} b_1' \\ b_2' \end{bmatrix} = \begin{bmatrix} S_{11} e^{-j2\beta l_1} & S_{12} e^{-j\beta(l_1+l_2)} \\ S_{21} e^{-j\beta(l_1+l_2)} & S_{22} e^{-j2\beta l_2} \end{bmatrix} \begin{bmatrix} a_1' \\ a_2' \end{bmatrix} \tag{8.6.43}$$

Now, on comparing (8.6.43) with (8.6.37), we find that

$$\begin{bmatrix} S_{11}' & S_{12}' \\ S_{21}' & S_{22}' \end{bmatrix} = \begin{bmatrix} S_{11} e^{-j2\beta l_1} & S_{12} e^{-j\beta(l_1+l_2)} \\ S_{21} e^{-j\beta(l_1+l_2)} & S_{22} e^{-j2\beta l_2} \end{bmatrix} \tag{8.6.44}$$

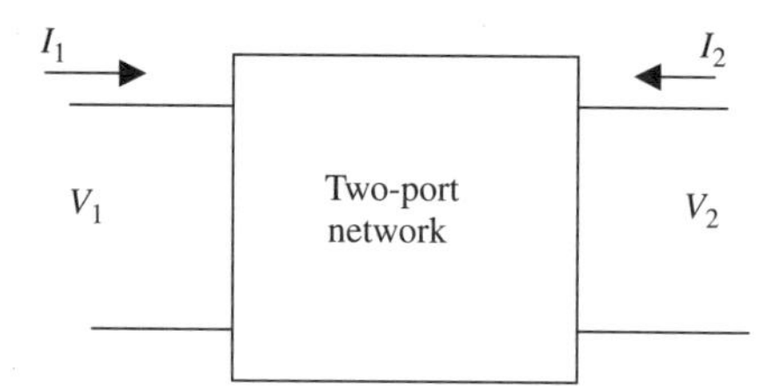

Figure 8.21 Network for Example 8.14.

Following a similar procedure, one can find the other relation as follows:

$$\begin{bmatrix} S_{11} & S_{12} \\ S_{21} & S_{22} \end{bmatrix} = \begin{bmatrix} S'_{11}e^{j2\beta l_1} & S'_{12}e^{j\beta(l_1+l_2)} \\ S'_{21}e^{j\beta(l_1+l_2)} & S'_{22}e^{j2\beta l_2} \end{bmatrix} \tag{8.6.45}$$

Example 8.14 The total voltages and currents at two ports of a network are found as follows: $V_1 = 10\ \angle 0°\text{V}$, $I_1 = 0.1\ \angle 40°\text{A}$, $V_2 = 12\ \angle 30°\text{V}$, and $I_2 = 0.15\ \angle 100°\text{A}$. Determine the incident and reflected voltages, assuming that the characteristic impedance is 50 Ω at both its ports (see Figure 8.21).

SOLUTION From (8.6.11) and (8.6.12) with $i = 1$, we find that

$$V_1^{\text{in}} = \tfrac{1}{2}(10\angle 0° + 50 \times 0.1\angle 40°) = 6.915 + j1.607\ \text{V}$$

and

$$V_1^{\text{ref}} = \tfrac{1}{2}(10\angle 0° - 50 \times 0.1\angle 40°) = 3.085 - j1.607\ \text{V}$$

Similarly, with $i = 2$, incident and reflected voltages at port 2 are found to be

$$V_2^{\text{in}} = \tfrac{1}{2}(12\angle 30° + 50 \times 0.15\angle 100°) = 4.545 + j6.695\ \text{V}$$

and

$$V_2^{\text{ref}} = \tfrac{1}{2}(12\angle 30° - 50 \times 0.15\angle 100°) = 5.845 - j0.691\ \text{V}$$

Example 8.15 Find the S-parameters of a series impedance Z connected between the two ports, as shown in Figure 8.22.

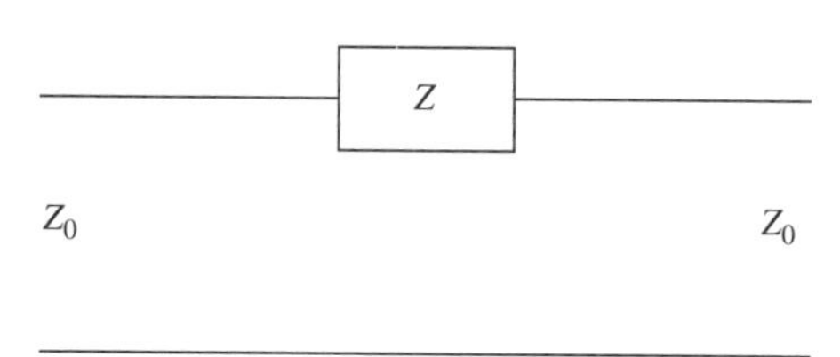

Figure 8.22 Setup for Example 8.15.

SOLUTION From (8.6.25), with $\Gamma_L = 0$ (i.e., port 2 is terminated by a matched load), we find that

$$S_{11} = \Gamma_1|_{a_2=0} = \frac{(Z + Z_0) - Z_0}{(Z + Z_0) + Z_0} = \frac{Z}{Z + 2Z_0}$$

Similarly, from (8.6.28), with $\Gamma_S = 0$ (i.e., port 1 is terminated by a matched load),

$$S_{22} = \Gamma_2|_{a_1=0} = \frac{(Z + Z_0) - Z_0}{(Z + Z_0) + Z_0} = \frac{Z}{Z + 2Z_0}$$

S_{21} and S_{12} are determined from (8.6.32) and (8.6.33), respectively. For evaluating S_{21}, we connect a voltage source V_{S1} at port 1 while port 2 is terminated by Z_0, as shown in Figure 8.23. The source impedance is Z_0.

Using the voltage-divider formula, we can write

$$V_2 = \frac{Z_0}{Z + 2Z_0} V_{S1}$$

Hence, from (8.6.32),

$$S_{21} = \frac{2V_2}{V_{S1}} = \frac{2Z_0}{Z + 2Z_0} = \frac{2Z_0 + Z - Z}{Z + 2Z_0} = 1 - \frac{Z}{Z + 2Z_0}$$

For evaluating S_{12}, we connect a voltage source V_{S2} at port 2 while port 1 is terminated by Z_0, as shown in Figure 8.24. The source impedance is Z_0. Using the voltage-divider rule again, we find that

$$V_1 = \frac{Z_0}{Z + 2Z_0} V_{S2}$$

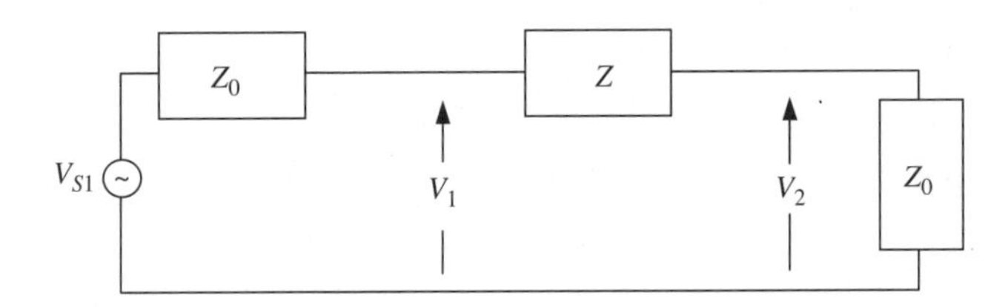

Figure 8.23 Setup to determine S_{21} for Example 8.15.

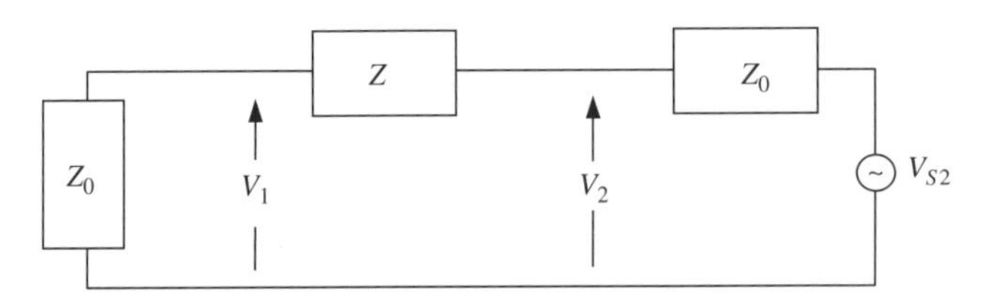

Figure 8.24 Setup to determine S_{12} for Example 8.15.

Now, from (8.6.33),

$$S_{12} = \frac{2V_1}{V_{S2}} = \frac{2Z_0}{Z + 2Z_0} = \frac{2Z_0 + Z - Z}{Z + 2Z_0} = 1 - \frac{Z}{Z + 2Z_0}$$

Therefore,

$$\begin{bmatrix} S_{11} & S_{12} \\ S_{21} & S_{22} \end{bmatrix} = \begin{bmatrix} \Gamma_1 & 1 - \Gamma_1 \\ 1 - \Gamma_1 & \Gamma_1 \end{bmatrix}$$

where

$$\Gamma_1 = \frac{Z}{Z + 2Z_0}$$

An analysis of the S-parameters indicates that S_{11} is equal to S_{22}. This is because the given two-port network is symmetrical. Further, its S_{12} is equal to S_{21} as well. This happens because this network is reciprocal.

Consider a special case where the series impedance Z is a purely reactive element, which means that Z is equal to jX. In that case, Γ_1 can be written as

$$\Gamma_1 = \frac{jX}{jX + 2Z_0}$$

Therefore,

$$\begin{aligned} |S_{11}|^2 + |S_{21}|^2 &= \left|\frac{jX}{jX + 2Z_0}\right|^2 + \left|1 - \frac{jX}{jX + 2Z_0}\right|^2 \\ &= \frac{X^2}{X^2 + (2Z_0)^2} + \frac{(2Z_0)^2}{X^2 + (2Z_0)^2} = 1 \end{aligned}$$

Similarly,

$$|S_{12}|^2 + |S_{22}|^2 = 1$$

On the other hand,

$$S_{11}S_{12}^* + S_{21}S_{22}^* = \frac{jX}{jX + 2Z_0} \times \frac{2Z_0}{-jX + 2Z_0} + \frac{2Z_0}{jX + 2Z_0} \times \frac{-jX}{-jX + 2Z_0} = 0$$

and

$$S_{12}S_{11}^* + S_{22}S_{21}^* = \frac{2Z_0}{jX + 2Z_0} \times \frac{-jX}{-jX + 2Z_0} + \frac{jX}{jX + 2Z_0} \times \frac{2Z_0}{-jX + 2Z_0} = 0$$

These characteristics of the scattering matrix can be summarized as follows:

$$\sum S_{ij}S_{ik}^* = \delta_{jk} = \begin{cases} 1 & \text{if } j = k \\ 0 & \text{otherwise} \end{cases}$$

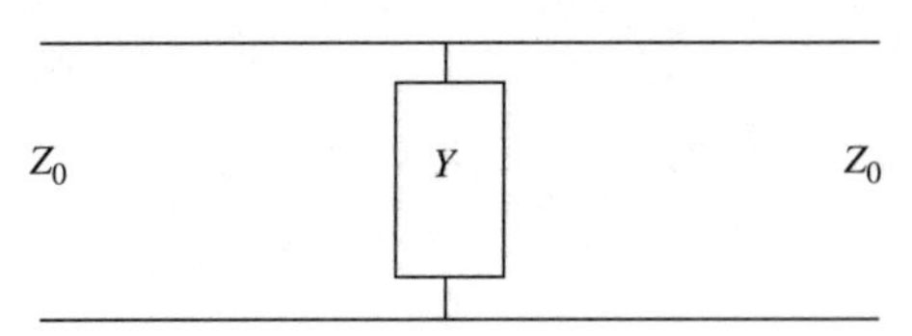

Figure 8.25 Network for Example 8.16.

When the elements of a matrix satisfy this condition, it is called a *unitary matrix*. Hence, the scattering matrix of a reactance circuit is unitary.

Example 8.16 Find the S-parameters of a shunt admittance Y connected between two ports, as shown in Figure 8.25.

SOLUTION From (8.6.25),

$$S_{11} = \Gamma_1|_{\Gamma_L=0} = \frac{Z_1 - Z_{01}}{Z_1 + Z_{01}} = \frac{1/Y_1 - 1/Y_0}{1/Y_1 + 1/Y_0}$$
$$= \frac{Y_0 - Y_1}{Y_0 + Y_1} = \frac{Y_0 - (Y + Y_0)}{Y_0 + (Y + Y_0)} = \frac{-Y}{2Y_0 + Y}$$

Similarly, from (8.6.28),

$$S_{22} = \Gamma_2|_{\Gamma_S=0} = \frac{Z_2 - Z_{02}}{Z_2 + Z_{02}} = \frac{1/Y_2 - 1/Y_0}{1/Y_2 + 1/Y_0}$$
$$= \frac{Y_0 - Y_2}{Y_0 + Y_2} = \frac{Y_0 - (Y + Y_0)}{Y_0 + (Y + Y_0)} = \frac{-Y}{2Y_0 + Y}$$

For evaluating S_{21}, voltage source V_{S1} is connected at port 1, and port 2 is terminated by a matched load, as shown in Figure 8.26.

Using the voltage-divider rule,

$$V_2 = \frac{1/(Y + Y_0)}{Z_0 + 1/(Y + Y_0)} V_{S1} = \frac{1}{Z_0(Y + Y_0) + 1} V_{S1} = \frac{Y_0}{Y + 2Y_0} V_{S1}$$

Now, from (8.6.32),

$$S_{21} = \left.\frac{2V_2}{V_{S1}}\right|_{\Gamma_L=0} = \frac{2Y_0}{Y + 2Y_0} = 1 - \frac{Y}{Y + 2Y_0} = 1 + \Gamma_1$$

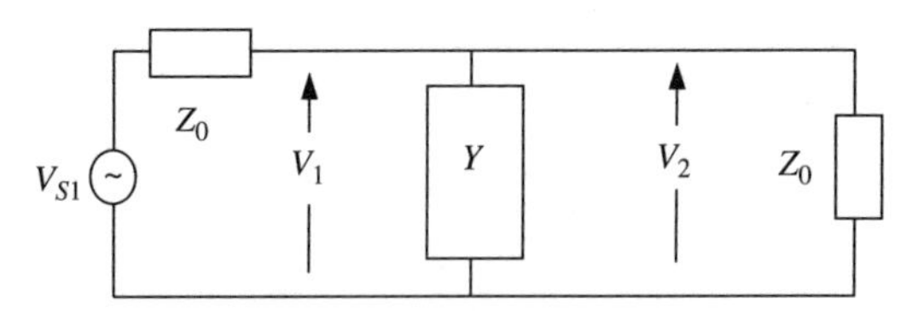

Figure 8.26 Setup to determine S_{21} for Example 8.16.

Similarly, when V_{S2} in series with Z_0 is connected at port 2 while port 1 is match-terminated, S_{12} can be found from (8.6.33) as follows:

$$S_{12} = \left.\frac{2V_1}{V_{S2}}\right|_{\Gamma_S=0} = \frac{2Y_0}{Y + 2Y_0} = 1 - \frac{Y}{Y + 2Y_0} = 1 + \Gamma_1$$

where

$$\Gamma_1 = -\frac{Y}{Y + 2Y_0}$$

As expected, $S_{11} = S_{22}$ and $S_{12} = S_{21}$, because this circuit is symmetrical as well as reciprocal. Further, for $Y = jB$, the network becomes lossless. In that case,

$$|S_{11}|^2 + |S_{21}|^2 = \frac{B^2}{B^2 + (2Y_0)^2} + \frac{(2Y_0)^2}{B^2 + (2Y_0)^2} = 1$$

and

$$S_{11}S_{12}^* + S_{21}S_{22}^* = \frac{-jB}{jB + 2Y_0} \times \frac{2Y_0}{-jB + 2Y_0} + \frac{2Y_0}{jB + 2Y_0} \times \frac{jB}{-jB + 2Y_0} = 0$$

Hence, with Y as jB, the scattering matrix is unitary. In fact, it can easily be proved that the scattering matrix of any lossless two-port network is unitary.

Example 8.17 An ideal transformer is designed to operate at 500 MHz. It has 1000 turns on its primary side and 100 turns on its secondary side. Assuming that it has a 50-Ω connector on each side, determine its S-parameters (see Figure 8.27).

SOLUTION For an ideal transformer,

$$\frac{V_1}{V_2} = \frac{-I_2}{I_1} = n \quad \text{and} \quad \frac{\mathrm{V}_1}{\mathrm{V}_2}\frac{(-I_2)}{I_1} = \frac{V_1/I_1}{V_2/(-I_2)} = \frac{Z_1}{Z_2} = n^2$$

Now, following the procedure used in preceding examples, we find that

$$S_{11} = \left.\frac{b_1}{a_1}\right|_{a_2=0} = \Gamma_1|_{a_2=0} = \frac{Z_0 n^2 - Z_0}{Z_0 n^2 + Z_0} = \frac{n^2 - 1}{n^2 + 1} = \frac{99}{101}$$

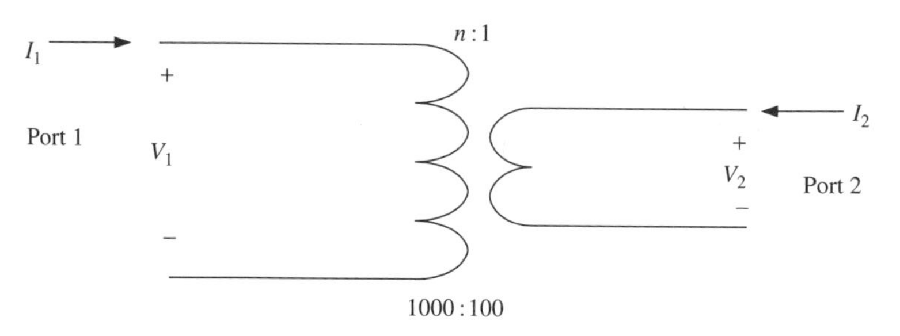

Figure 8.27 Transformer for Example 8.17.

and voltage V_1 on the primary side of the transformer due to a voltage source V_{S1} with its internal impedance Z_0 can be determined by the voltage-divider rule as follows:

$$V_1 = \frac{Z_0 n^2}{Z_0 n^2 + Z_0} V_{S1} \quad \Rightarrow \quad \frac{V_1}{V_{S1}} = \frac{n^2}{n^2+1} = \frac{nV_2}{V_{S1}}$$

Hence,

$$S_{21} = \left. \frac{2V_2}{V_{S1}} \right|_{a_2=0} = \frac{2n}{n^2+1} = \frac{20}{101}$$

Similarly, S_{22} and S_{12} are determined after connecting a voltage source V_{S2} with its internal impedance Z_0 at port 2. Port 1 is terminated by a matched load this time. Therefore,

$$S_{22} = \left. \frac{b_2}{a_2} \right|_{a_1=0} = \Gamma_2|_{a_1=0} = \frac{(Z_0/n^2) - Z_0}{(Z_0/n^2) + Z_0} = \frac{1-n^2}{1+n^2} = -\frac{99}{101}$$

and

$$V_2 = \frac{Z_0/n^2}{(Z_0/n^2) + Z_0} V_{S2} \Rightarrow \frac{V_2}{V_{S2}} = \frac{1/n^2}{(1/n^2)+1} = \frac{1}{1+n^2} = \frac{V_1/n}{V_{S2}}$$

Hence,

$$S_{12} = \left. \frac{2V_1}{V_{S2}} \right|_{a_1=0} = \frac{2n}{n^2+1} = \frac{20}{101}$$

Thus,

$$\begin{bmatrix} S_{11} & S_{12} \\ S_{21} & S_{22} \end{bmatrix} = \begin{bmatrix} \frac{99}{101} & \frac{20}{101} \\ \frac{20}{101} & -\frac{99}{101} \end{bmatrix}$$

This time, S_{11} is different from S_{22} because the network is not symmetrical. However, S_{12} is equal to S_{21} because it is reciprocal. Further, an ideal transformer is lossless. Hence, its scattering matrix must be unitary. Let us verify if that is the case.

$$|S_{11}|^2 + |S_{21}|^2 = \left(\frac{99}{101}\right)^2 + \left(\frac{20}{101}\right)^2 = \frac{10{,}201}{10{,}201} = 1$$

$$|S_{12}|^2 + |S_{22}|^2 = \left(\frac{20}{101}\right)^2 + \left(-\frac{99}{101}\right)^2 = \frac{10{,}201}{10{,}201} = 1$$

$$S_{11}S_{12}^* + S_{21}S_{22}^* = \frac{99}{101} \times \frac{20}{101} + \frac{20}{101} \times \left(-\frac{99}{101}\right) = 0$$

and

$$S_{12}S_{11}^* + S_{22}S_{21}^* = \frac{20}{101} \times \frac{99}{101} + \left(-\frac{99}{101}\right) \times \frac{20}{101} = 0$$

Hence, the scattering matrix is indeed unitary.

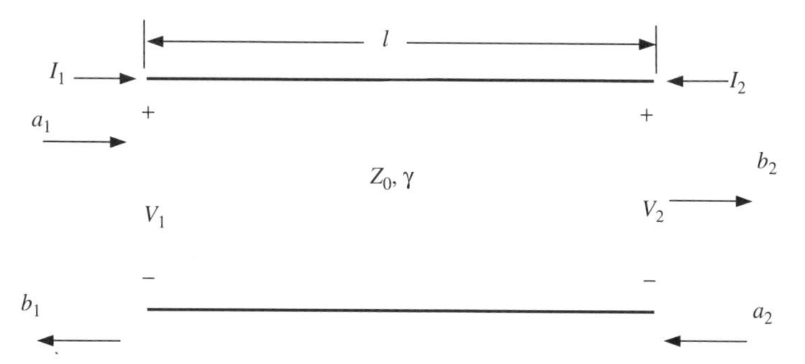

Figure 8.28 Transmission line network for Example 8.18.

Example 8.18 Find the scattering parameters of the transmission line network shown in Figure 8.28.

SOLUTION With port 2 match-terminated, wave a_1 entering port 1 emerges as $a_1 e^{-\gamma l}$ at port 2. This wave is absorbed by the termination, making a_2 zero. Further, b_1 is zero as well because Z_1 is equal to Z_0 in this case. Hence,

$$b_2 = a_1 e^{-\gamma l}$$

Similarly, when a source is connected at port 2 while port 1 is match-terminated, we find that a_1 and b_2 are zero and

$$b_1 = a_2 e^{-\gamma l}$$

Hence,

$$\begin{bmatrix} S_{11} & S_{12} \\ S_{21} & S_{22} \end{bmatrix} = \begin{bmatrix} 0 & e^{-\gamma l} \\ e^{-\gamma l} & 0 \end{bmatrix}$$

As expected, $S_{11} = S_{22}$ and $S_{12} = S_{21}$, because this circuit is symmetrical as well as reciprocal. Further, for $\gamma = j\beta$, the network becomes lossless. In that case,

$$\begin{bmatrix} S_{11} & S_{12} \\ S_{21} & S_{22} \end{bmatrix} = \begin{bmatrix} 0 & e^{-j\beta l} \\ e^{-j\beta l} & 0 \end{bmatrix}$$

Obviously, it is a unitary matrix now.

Example 8.19 Find the scattering parameters of the two-port network shown in Figure 8.29.

SOLUTION With port 2 matched-terminated by a 50-Ω load, impedance Z_1 at port 1 is

$$Z_1 = j50 + \frac{50(-j25)}{50 - j25} = j50 + 10 - j20 = 10 + j30\ \Omega$$

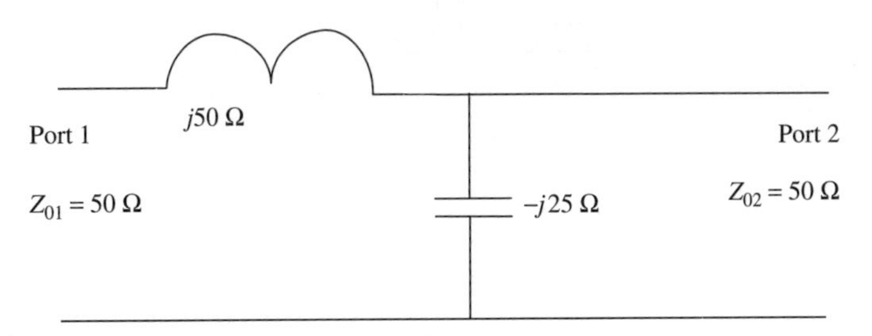

Figure 8.29 Two-port network for Example 8.19.

Therefore,

$$S_{11} = \Gamma_1|_{a_2=0} = \frac{Z_1 - Z_{01}}{Z_1 + Z_{01}} = \frac{10 + j30 - 50}{10 + j30 + 50} = 0.74536 \angle 116.565^\circ$$

Similarly, output impedance Z_2 at port 2 when port 1 is match-terminated by a 50-Ω load is

$$Z_2 = \frac{(50 + j50)(-j25)}{50 + j50 - j25} = 10 - j30\,\Omega$$

and from (8.6.32),

$$S_{22} = \Gamma_2|_{a_1=0} = \frac{Z_2 - Z_{02}}{Z_1 + Z_{02}} = \frac{10 - j30 - 50}{10 - j30 + 50} = 0.74536 \angle - 116.565^\circ$$

Now, let us connect a voltage source V_{S1} with its impedance at 50 Ω at port 1 while port 2 is matched, as shown in Figure 8.30. Using the voltage-divider rule, we find that

$$V_2 = \frac{10 - j20}{50 + j50 + 10 - j20} V_{S1}$$

Therefore,

$$S_{21} = \left.\frac{2V_2}{V_{S1}}\right|_{a_2=0} = 2\frac{10 - j20}{60 + j30} = 0.66666 \angle - 90^\circ$$

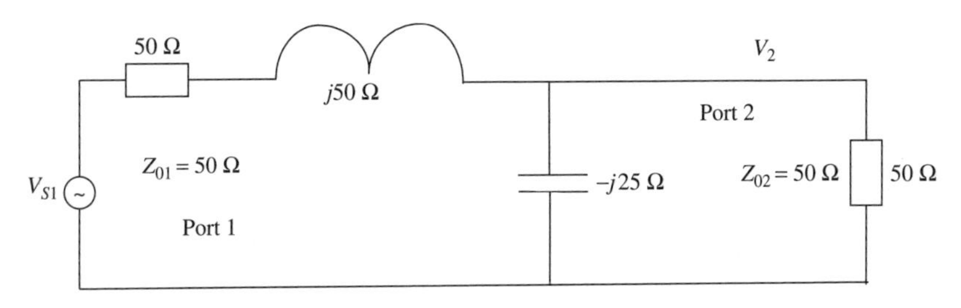

Figure 8.30 Setup to determine S_{21} for Example 8.19.

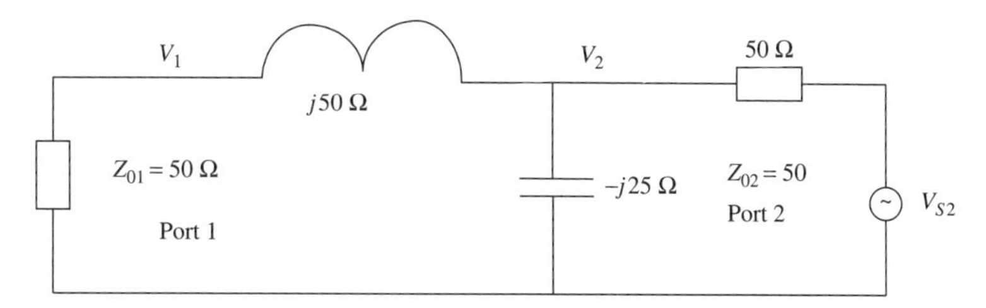

Figure 8.31 Setup to determine S_{12} for Example 8.19.

Now connect the voltage source at port 2 while port 1 is match-terminated as shown in Figure 8.31.

Using the voltage-divider rule again, we find that

$$V_2 = \frac{10 - j30}{50 + 10 - j30} V_{S2} = 0.4714\ \angle - 45^\circ\ \mathrm{V}$$

Using it one more time, V_1 is found as follows:

$$V_1 = \frac{50}{50 + j50} V_2 = \frac{1}{1 + j1} 0.4714\ \angle - 45^\circ = 0.33333\ \angle - 90^\circ\ \mathrm{V}$$

and now from (8.6.33), we get

$$S_{12} = \left.\frac{2V_1}{V_{S2}}\right|_{a_1=0} = 2 \times 0.33333\ \angle - 90^\circ = 0.66666\ \angle - 90^\circ$$

Hence,

$$\begin{bmatrix} S_{11} & S_{12} \\ S_{21} & S_{22} \end{bmatrix} = \begin{bmatrix} 0.74536\ \angle 116.565^\circ & 0.66666\ \angle - 90^\circ \\ 0.66666\ \angle - 90^\circ & 0.74536\ \angle - 116.565^\circ \end{bmatrix}$$

In this case, S_{11} is different from S_{22} because the network is not symmetrical. However, S_{12} is equal to S_{21} because the network is reciprocal. Further, this network is lossless. Hence, its scattering matrix must be unitary. It can be verified as follows:

$$\begin{aligned} |S_{11}|^2 + |S_{21}|^2 &= (0.74536)^2 + (0.66666)^2 = 0.99999 = 1 \\ |S_{12}|^2 + |S_{22}|^2 &= (0.66666)^2 + (0.74536)^2 = 0.99999 = 1 \\ S_{11}S_{12}^* + S_{21}S_{22}^* &= (0.74536\ \angle 116.565^\circ) \times (0.66666\ \angle 90^\circ) \\ &\quad + (0.66666\ \angle - 90^\circ) \times (0.74536\ \angle 116.565^\circ) = 0 \end{aligned}$$

and

$$
\begin{aligned}
S_{12}S_{11}^* + S_{22}S_{21}^* &= (0.66666 \angle -90^\circ) \times (0.74536 \angle -116.565^\circ) \\
&\quad + (0.74536 \angle -116.565^\circ) \times (0.66666 \angle 90^\circ) \\
&= 0
\end{aligned}
$$

Hence, this scattering matrix is indeed unitary.

Table 8.2 summarizes the characteristics of parameter matrices of reciprocal and symmetrical two-port networks. The properties of scattering parameters of lossless junctions are listed in Table 8.3.

TABLE 8.2 Properties of Parameters of Reciprocal and Symmetrical Two-Port Networks

Parameter Matrix	Properties
$\begin{bmatrix} Z_{11} & Z_{12} \\ Z_{21} & Z_{22} \end{bmatrix}$	$Z_{12} = Z_{21}$ $Z_{11} = Z_{22}$
$\begin{bmatrix} Y_{11} & Y_{12} \\ Y_{21} & Y_{22} \end{bmatrix}$	$Y_{12} = Y_{21}$ $Y_{11} = Y_{22}$
$\begin{bmatrix} A & B \\ C & D \end{bmatrix}$	$AD - BC = 1$ $A = D$
$\begin{bmatrix} S_{11} & S_{12} \\ S_{21} & S_{22} \end{bmatrix}$	$S_{12} = S_{21}$ $S_{11} = S_{22}$

TABLE 8.3 Properties of Scattering Matrix of Lossless Junctions

Properties	Explanation
Matrix [S] is symmetrical.	$[S]^t = [S]$, where $[S]^t$ is the transpose matrix of $[S]$. Consequently, $S_{ij} = S_{ji}$
Matrix [S] is unitary.	$[S]^a = [S^*]^t = [S]^{-1}$, where $[S]^a$ is the adjoint matrix of $[S]$, $[S^*]^t$ is the conjugate of $[S]^t$, and $[S]^{-1}$ is the inverse matrix of $[S]$. Consequently, $\sum_{i=1}^{N} S_{ij}S_{ik}^* = \delta_{jk} = \begin{cases} 1 & \text{for } j = k \\ 0 & \text{otherwise} \end{cases}$ Therefore, $\sum_{i=1}^{N} S_{ij}S_{ij}^* = \sum_{i=1}^{N} \lvert S_{ij}\rvert^2 = 1, \qquad j = 1, 2, 3, \ldots, N$

8.7 CONVERSION FROM IMPEDANCE, ADMITTANCE, CHAIN, AND HYBRID PARAMETERS TO SCATTERING PARAMETERS, OR VICE VERSA

From (8.1.3) we can write

$$\frac{V_i}{\sqrt{2Z_{0i}}} = \frac{1}{\sqrt{2Z_{0i}}}\sum_{k=1}^{2} Z_{ik}I_k = \sum_{k=1}^{2}\frac{Z_{ik}}{Z_{0i}}\sqrt{\frac{Z_{0i}}{2}}I_k$$

$$= \sum_{k=1}^{2}\overline{Z}_{ik}\sqrt{\frac{Z_{0i}}{2}}I_k \qquad \text{where } i = 1, 2$$

Therefore, from (8.6.15) and (8.6.16), we have

$$a_i = \frac{1}{2}\sum_{k=1}^{2}(\overline{Z}_{ik} + \delta_{ik})\sqrt{\frac{Z_{0i}}{2}}I_k \tag{8.7.1}$$

and

$$b_i = \frac{1}{2}\sum_{k=1}^{2}(\overline{Z}_{ik} - \delta_{ik})\sqrt{\frac{Z_{0i}}{2}}I_k \tag{8.7.2}$$

where

$$\delta_{ik} = \begin{cases} 1 & \text{for } i = k \\ 0 & \text{otherwise} \end{cases}$$

Equations (8.7.1) and (8.7.2) can be written in matrix form as follows:

$$[a] = \tfrac{1}{2}\{[\overline{Z}] + [U]\}[\overline{I}] \tag{8.7.3}$$

$$[b] = \tfrac{1}{2}\{[\overline{Z}] - [U]\}[\overline{I}] \tag{8.7.4}$$

where

$$\overline{I}_k = \sqrt{\frac{Z_{0i}}{2}}I_k$$

and $[U]$ is the unit matrix.

From (8.7.3) and (8.7.4), we have

$$[b] = \{[\overline{Z}] - [U]\}\{[\overline{Z}] + [U]\}^{-1}[a] \tag{8.7.5}$$

Hence,

$$[S] = \{[\overline{Z}] - [U]\}\{[\overline{Z}] + [U]\}^{-1} \tag{8.7.6}$$

TABLE 8.4 Conversion from (to) Impedance, Admittance, Chain, or Hybrid to (from) Scattering Parameters

$S_{11} = \dfrac{(Z_{11} - Z_{01}^*)(Z_{22} + Z_{02}) - Z_{12}Z_{21}}{(Z_{11} + Z_{01})(Z_{22} + Z_{02}) - Z_{12}Z_{21}}$	$Z_{11} = \dfrac{(Z_{01}^* + S_{11}Z_{01})(1 - S_{22}) + S_{12}S_{21}Z_{01}}{(1 - S_{11})(1 - S_{22}) - S_{12}S_{21}}$
$S_{12} = \dfrac{2Z_{12}\sqrt{R_{01}R_{02}}}{(Z_{11} + Z_{01})(Z_{22} + Z_{02}) - Z_{12}Z_{21}}$	$Z_{12} = \dfrac{2S_{12}\sqrt{R_{01}R_{02}}}{(1 - S_{11})(1 - S_{22}) - S_{12}S_{21}}$
$S_{21} = \dfrac{2Z_{21}\sqrt{R_{01}R_{02}}}{(Z_{11} + Z_{01})(Z_{22} + Z_{02}) - Z_{12}Z_{21}}$	$Z_{21} = \dfrac{2S_{21}\sqrt{R_{01}R_{02}}}{(1 - S_{11})(1 - S_{22}) - S_{12}S_{21}}$
$S_{22} = \dfrac{(Z_{11} + Z_{01})(Z_{22} - Z_{02}^*) - Z_{12}Z_{21}}{(Z_{11} + Z_{01})(Z_{22} + Z_{02}) - Z_{12}Z_{21}}$	$Z_{22} = \dfrac{(Z_{02}^* + S_{22}Z_{02})(1 - S_{11}) + S_{12}S_{21}Z_{02}}{(1 - S_{11})(1 - S_{22}) - S_{12}S_{21}}$
$S_{11} = \dfrac{(1 - Y_{11}Z_{01}^*)(1 + Y_{22}Z_{02}) + Y_{12}Y_{21}Z_{01}^*Z_{02}}{(1 + Y_{11}Z_{01})(1 + Y_{22}Z_{02}) - Y_{12}Y_{21}Z_{01}Z_{02}}$	$Y_{11} = \dfrac{(1 - S_{11})(Z_{02}^* + S_{22}Z_{02}) + S_{12}S_{21}Z_{02}}{(Z_{01}^* + S_{11}Z_{01})(Z_{02}^* + S_{22}Z_{02}) - S_{12}S_{21}Z_{01}Z_{02}}$
$S_{12} = \dfrac{-2Y_{12}\sqrt{R_{01}R_{02}}}{(1 + Y_{11}Z_{01})(1 + Y_{22}Z_{02}) - Y_{12}Y_{21}Z_{01}Z_{02}}$	$Y_{12} = \dfrac{-2S_{12}\sqrt{R_{01}R_{02}}}{(Z_{01}^* + S_{11}Z_{01})(Z_{02}^* + S_{22}Z_{02}) - S_{12}S_{21}Z_{01}Z_{02}}$
$S_{21} = \dfrac{-2Y_{21}\sqrt{R_{01}R_{02}}}{(1 + Y_{11}Z_{01})(1 + Y_{22}Z_{02}) - Y_{12}Y_{21}Z_{01}Z_{02}}$	$Y_{21} = \dfrac{-2S_{21}\sqrt{R_{01}R_{02}}}{(Z_{01}^* + S_{11}Z_{01})(Z_{02}^* + S_{22}Z_{02}) - S_{12}S_{21}Z_{01}Z_{02}}$
$S_{22} = \dfrac{(1 + Y_{11}Z_{01})(1 - Y_{22}Z_{02}^*) + Y_{12}Y_{21}Z_{01}Z_{02}^*}{(1 + Y_{11}Z_{01})(1 + Y_{22}Z_{02}) - Y_{12}Y_{21}Z_{01}Z_{02}}$	$Y_{22} = \dfrac{(Z_{01}^* + S_{11}Z_{01})(1 - S_{22}) + S_{12}S_{21}Z_{01}}{(Z_{01}^* + S_{11}Z_{01})(Z_{02}^* + S_{22}Z_{02}) - S_{12}S_{21}Z_{01}Z_{02}}$
$S_{11} = \dfrac{AZ_{02} + B - CZ_{01}^*Z_{02} - DZ_{01}^*}{AZ_{02} + B + CZ_{01}Z_{02} + DZ_{01}}$	$A = \dfrac{(Z_{01}^* + S_{11}Z_{01})(1 - S_{22}) + S_{12}S_{21}Z_{01}}{2S_{21}\sqrt{R_{01}R_{02}}}$
$S_{12} = \dfrac{2(AD - BC)(\sqrt{R_{01}R_{02}})}{AZ_{02} + B + CZ_{01}Z_{02} + DZ_{01}}$	$B = \dfrac{(Z_{01}^* + S_{11}Z_{01})(Z_{02}^* + S_{22}Z_{02}) - S_{12}S_{21}Z_{01}Z_{02}}{2S_{21}\sqrt{R_{01}R_{02}}}$
$S_{21} = \dfrac{2\sqrt{R_{01}R_{02}}}{AZ_{02} + B + CZ_{01}Z_{02} + DZ_{01}}$	$C = \dfrac{(1 - S_{11})(1 - S_{22}) - S_{12}S_{21}}{2S_{21}\sqrt{R_{01}R_{02}}}$
$S_{22} = \dfrac{-AZ_{02}^* + B - CZ_{01}Z_{02}^* + DZ_{01}}{AZ_{02} + B + CZ_{01}Z_{02} + DZ_{01}}$	$D = \dfrac{(1 - S_{11})(Z_{02}^* + S_{22}Z_{02}) + S_{12}S_{21}Z_{02}}{2S_{21}\sqrt{R_{01}R_{02}}}$
$S_{11} = \dfrac{(h_{11} - Z_{01}^*)(1 + h_{22}Z_{02}) - h_{12}h_{21}Z_{02}}{(Z_{01} + h_{11})(1 + h_{22}Z_{02}) - h_{12}h_{21}Z_{02}}$	$h_{11} = \dfrac{(Z_{01}^* + S_{11}Z_{01})(Z_{02}^* + S_{22}Z_{02}) + S_{12}S_{21}Z_{01}Z_{02}}{(1 - S_{11})(Z_{02}^* + S_{22}Z_{02}) + S_{12}S_{21}Z_{02}}$
$S_{12} = \dfrac{2h_{12}\sqrt{R_{01}R_{02}}}{(Z_{01} + h_{11})(1 + h_{22}Z_{02}) - h_{12}h_{21}Z_{02}}$	$h_{12} = \dfrac{2S_{12}\sqrt{R_{01}R_{02}}}{(1 - S_{11})(Z_{02}^* + S_{22}Z_{02}) + S_{12}S_{21}Z_{02}}$
$S_{21} = \dfrac{-2h_{21}\sqrt{R_{01}R_{02}}}{(Z_{01} + h_{11})(1 + h_{22}Z_{02}) - h_{12}h_{21}Z_{02}}$	$h_{21} = \dfrac{-2S_{21}\sqrt{R_{01}R_{02}}}{(1 - S_{11})(Z_{02}^* + S_{22}Z_{02}) + S_{12}S_{21}Z_{02}}$
$S_{22} = \dfrac{(Z_{01} + h_{11})(1 - h_{22}Z_{02}^*) + h_{12}h_{21}Z_{02}^*}{(Z_{01} + h_{11})(1 + h_{22}Z_{02}) - h_{12}h_{21}Z_{02}}$	$h_{22} = \dfrac{(1 - S_{11})(1 - S_{22}) - S_{12}S_{21}}{(1 - S_{11})(Z_{02}^* + S_{22}Z_{02}) + S_{12}S_{21}Z_{02}}$

or

$$[\overline{Z}] = \{[U] + [S]\}\{[U] - [S]\}^{-1} \tag{8.7.7}$$

Similarly, other conversion relations can be developed. Table 8.4 lists these equations when the two ports have different complex characteristic impedances. Characteristic impedance at port 1 is assumed Z_{01}, with its real part as R_{01}. Similarly, characteristic impedance at port 2 is Z_{02}, with its real part as R_{02}.

TABLE 8.5 Conversion Between Scattering and Chain Scattering Parameters

$S_{11} = \dfrac{T_{21}}{T_{11}}$	$T_{11} = \dfrac{1}{S_{21}}$
$S_{12} = \dfrac{T_{11}T_{22} - T_{12}T_{21}}{T_{11}}$	$T_{12} = \dfrac{-S_{22}}{S_{21}}$
$S_{21} = \dfrac{1}{T_{11}}$	$T_{21} = \dfrac{S_{11}}{S_{21}}$
$S_{22} = \dfrac{-T_{12}}{T_{11}}$	$T_{22} = \dfrac{-(S_{11}S_{22} - S_{12}S_{21})}{S_{21}}$

8.8 CHAIN SCATTERING PARAMETERS

Chain scattering parameters, also known as *scattering transfer parameters* or *T-parameters*, are useful for networks in cascade. These are defined on the basis of waves a_1 and b_1 at port 1 as dependent variables and waves a_2 and b_2 at port 2 as independent variables. Hence,

$$\begin{bmatrix} a_1 \\ b_1 \end{bmatrix} = \begin{bmatrix} T_{11} & T_{12} \\ T_{21} & T_{22} \end{bmatrix} \begin{bmatrix} b_2 \\ a_2 \end{bmatrix}$$

Conversion relations of chain scattering parameters with others can easily be developed. Conversions between *S*- and *T*-parameters are listed in Table 8.5.

SUGGESTED READING

Bahl, I., and P. Bhartia, *Microwave Solid State Circuit Design*. New York: Wiley, 1988.

Collin, R. E., *Foundations for Microwave Engineering*. New York: McGraw-Hill, 1992.

Elliott, R. S., *An Introduction to Guided Waves and Microwave Circuits*. Englewood Cliffs, NJ: Prentice Hall, 1993.

Fusco, V. F., *Microwave Circuits*. Englewood Cliffs, NJ: Prentice Hall, 1987.

Gonzalez, G., *Microwave Transistor Amplifiers*. Upper Saddle River, NJ: Prentice Hall, 1997.

Pozar, D. M., *Microwave Engineering*. New York: Wiley, 1998.

Ramo, S., J. R. Whinnery, and T. Van Duzer, *Fields and Waves in Communication Electronics*. New York: Wiley, 1994.

Rizzi, P. A., *Microwave Engineering*. Englewood Cliffs, NJ: Prentice Hall, 1988.

PROBLEMS

8.1. Determine the Z and Y matrixes for the two-port network shown in Figure P8.1. The signal frequency is 500 MHz.

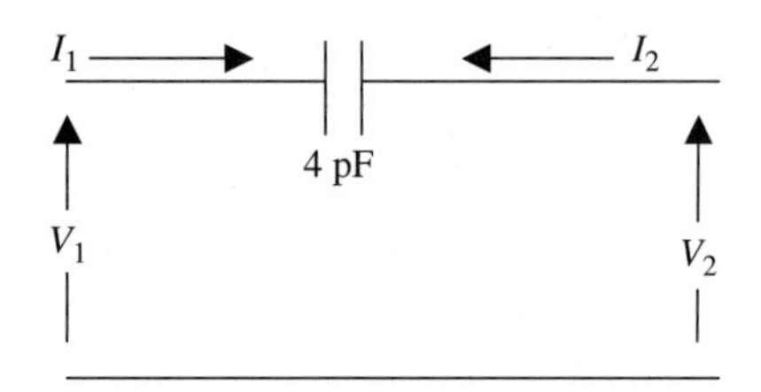

Figure P8.1

8.2. Determine the Z-parameters of the two-port network shown in Figure P8.2.

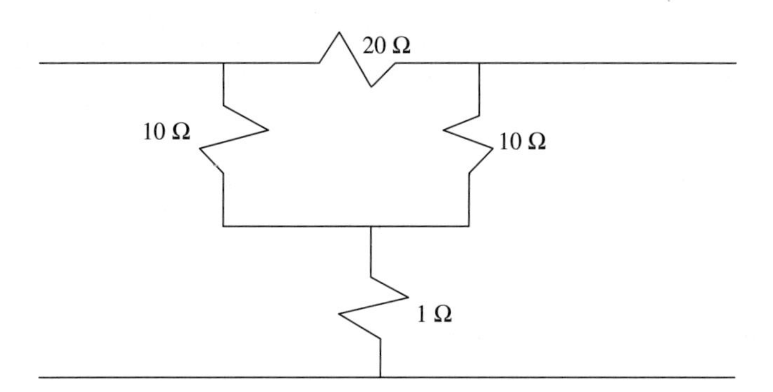

Figure P8.2

8.3. Determine the Z-parameters of the two-port network shown in Figure P8.3 operating at 800 MHz.

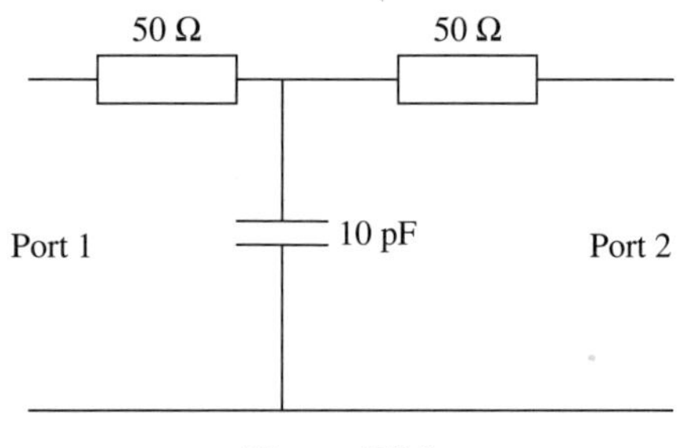

Figure P8.3

8.4. Determine the Z-matrix for the two-port network shown in Figure P8.4.

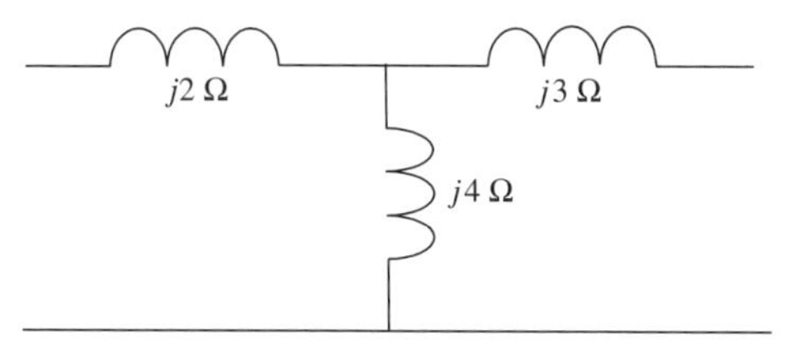

Figure P8.4

8.5. The two-port network shown in Figure P8.5 represents a high-frequency equivalent model of the transistor. If the transistor is operating at 10^8 rad/s, find its Z-parameters.

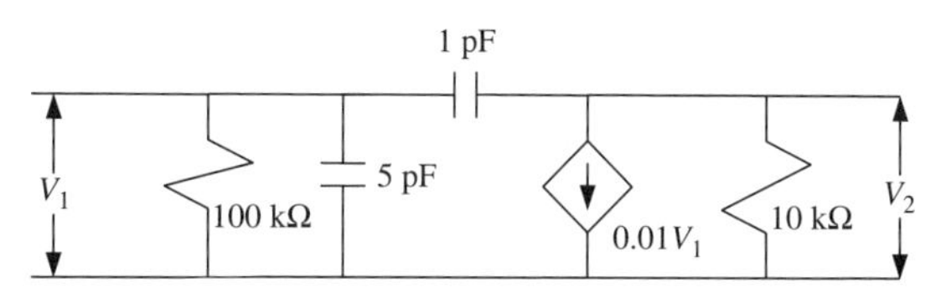

Figure P8.5

8.6. Determine the Y-parameters of the two-port network shown in Figure P8.6.

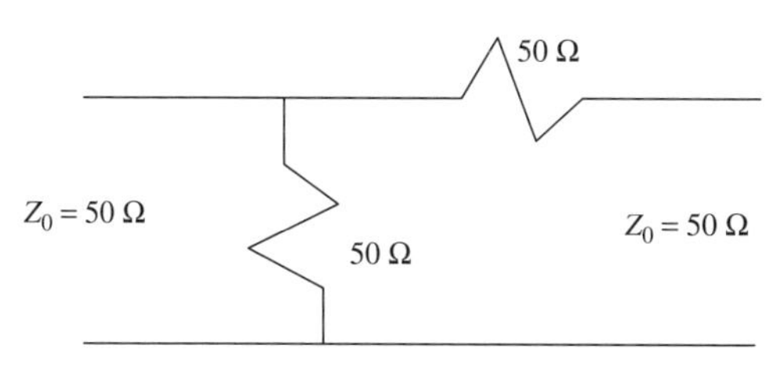

Figure P8.6

8.7. Find the Y-parameters of the two-port network shown in Figure P8.7.

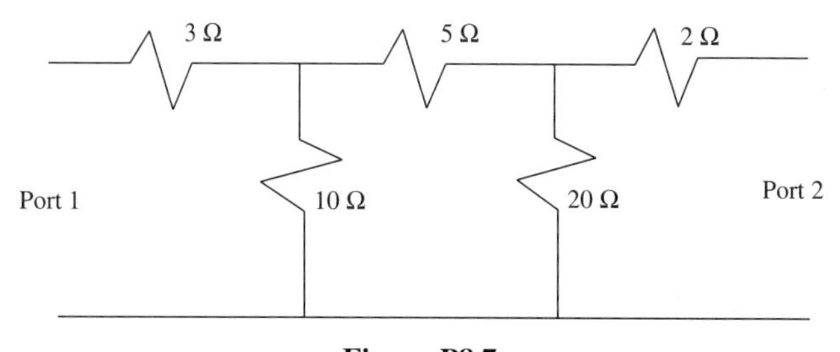

Figure P8.7

8.8. Measurements are performed with a two-port network as shown in Figure P8.8. The voltages and currents observed are tabulated in Table P8.8. However, there are a few blank entries in this table. Fill in those blanks and find the Y and Z parameters of this two-port network.

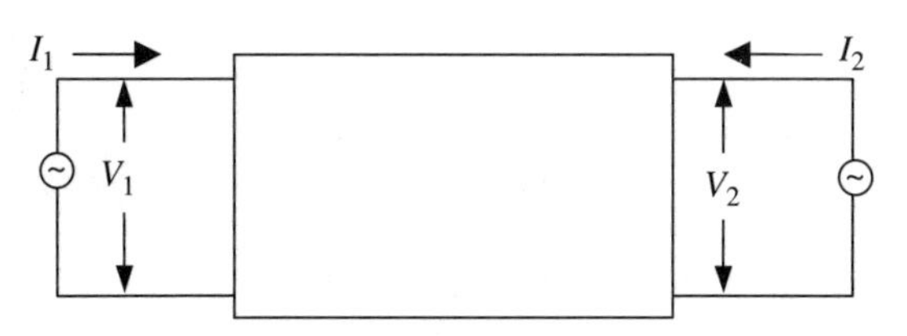

Figure P8.8

TABLE P8.8

Experiment	V_1 (V)	V_2 (V)	I_1 (A)	I_2 (A)
1	20	0	4	−8
2	50	100	−20	−5
3			5	0
4	100	50		
5			5	15

8.9. The Y-matrix of a two-port network is given as

$$[Y] = \begin{bmatrix} 10 & -5 \\ 50 & 20 \end{bmatrix} \text{ mS}$$

It is connected in a circuit as shown in Figure P8.9. Find the voltages V_1 and V_2.

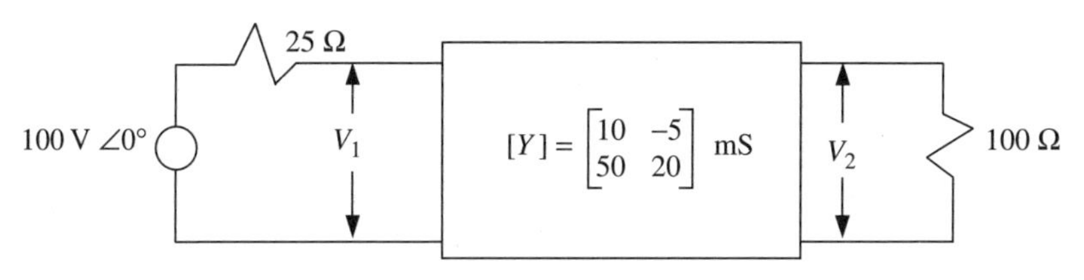

Figure P8.9

8.10. Determine the transmission and hybrid parameters for the networks shown in Figure P8.10.

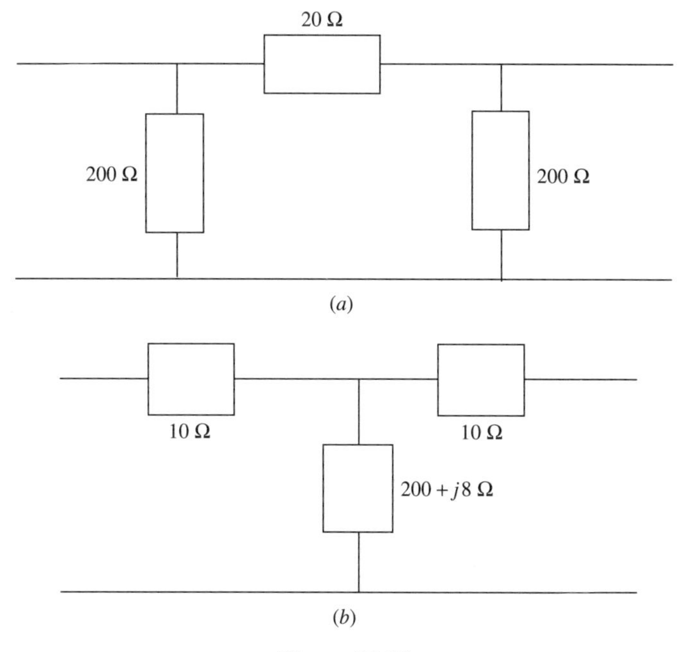

Figure P8.10

8.11. Determine the transmission parameters ($ABCD$) of the circuit shown in Figure P8.11 operating at $\omega = 10^6$ rad/s.

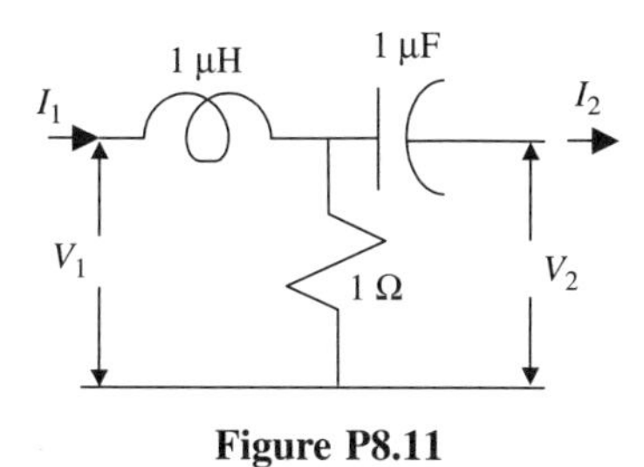

Figure P8.11

8.12. Find the transmission matrix of the two-port network shown in Figure P8.12.

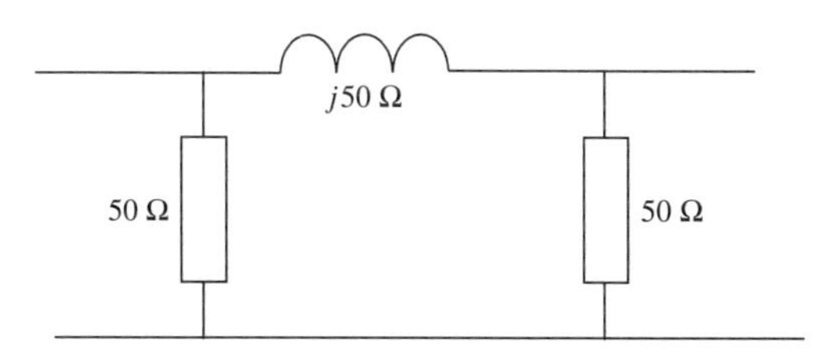

Figure P8.12

8.13. Find the transmission matrix (*ABCD* parameters) of the circuit shown in Figure P8.13 at 4 GHz.

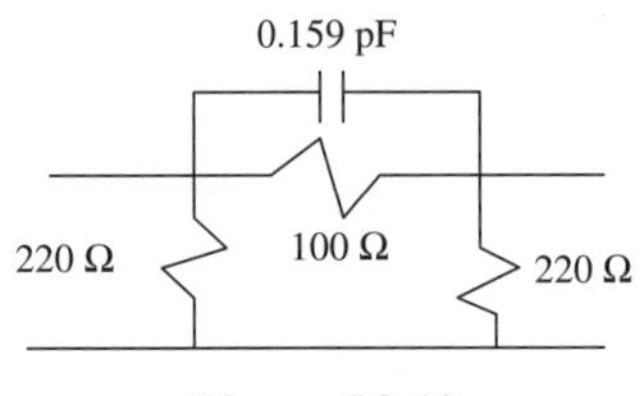

Figure P8.13

8.14. Find the transmission matrix of the two-port network shown in Figure P8.14.

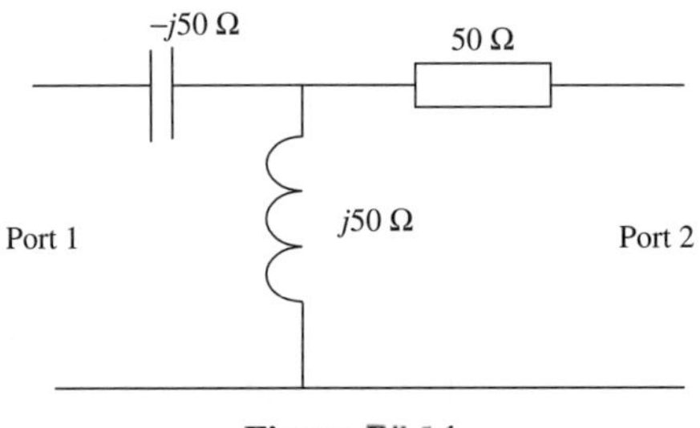

Figure P8.14

8.15. Determine the *S*-parameters for the network shown in Figure P8.15.

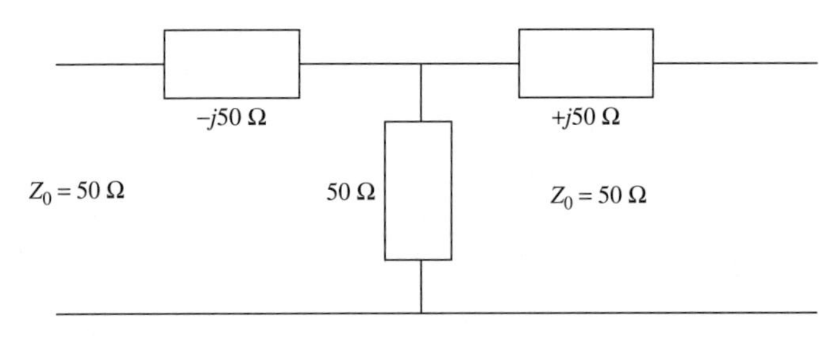

Figure P8.15

8.16. Calculate the *Z* and *S* parameters for the circuit shown in Figure P8.16 at 4 GHz. Both ports of the network have characteristic impedance of 50 Ω.

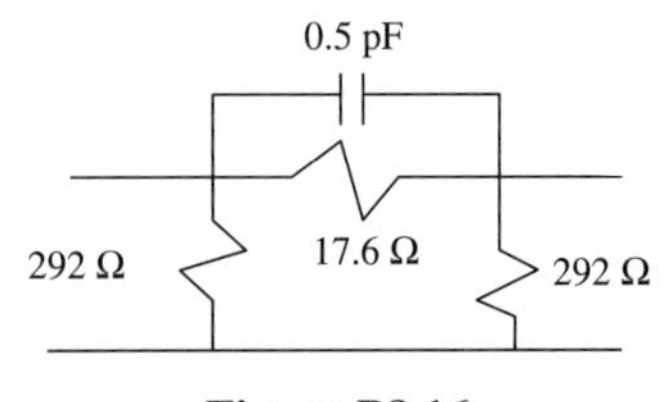

Figure P8.16

8.17. Determine the S-parameters of the two-port T-network shown in Figure P8.17.

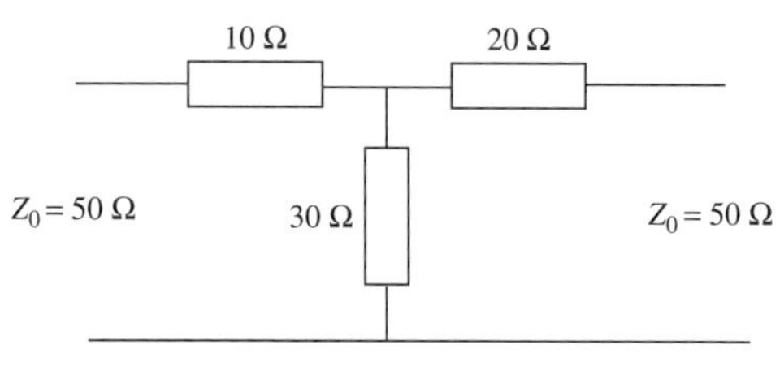

Figure P8.17

8.18. Determine the S-parameters of the two-port network shown in Figure P8.18.

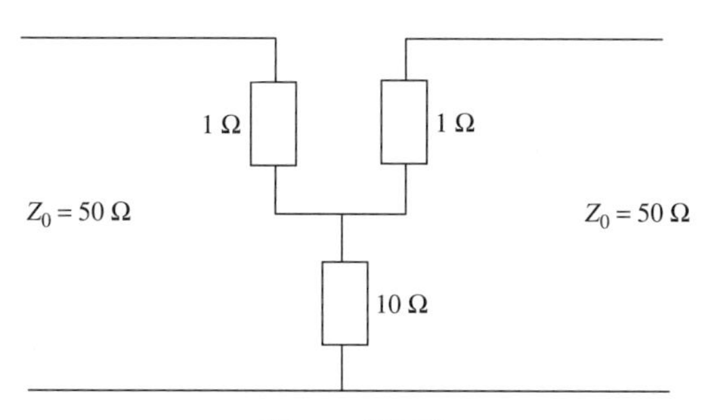

Figure P8.18

8.19. Determine the S-parameters of the two-port network shown in Figure P8.19.

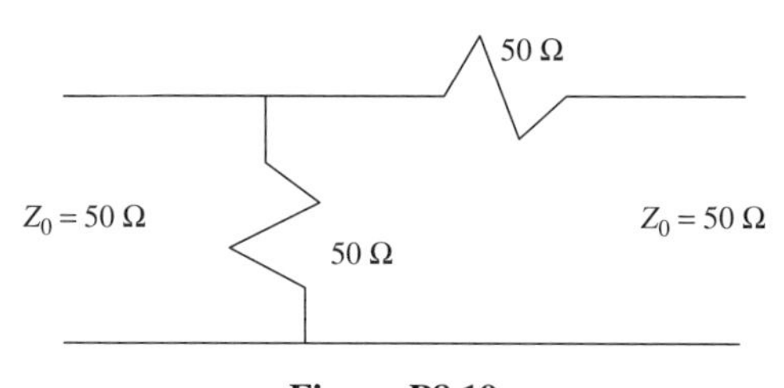

Figure P8.19

8.20. Determine the S-parameters of the two-port π-network shown in Figure P8.20.

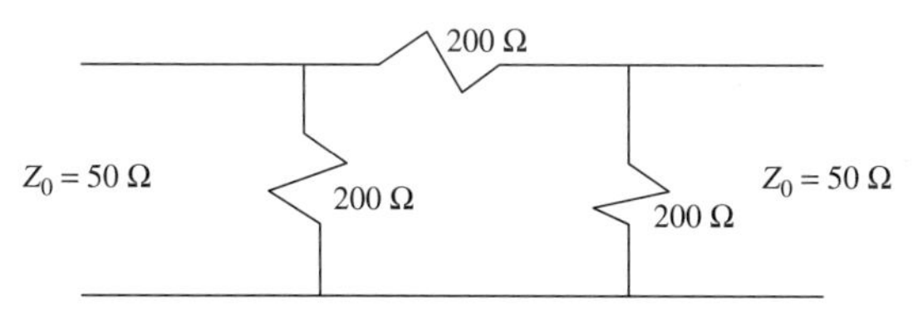

Figure P8.20

8.21. Scattering variables measured at a port are found as follows: $a = 8 - j5$ and $b = 2 + j5$. Normalizing impedance is $Z_0 = 50\,\Omega$. Calculate the corresponding voltage and current.

8.22. Scattering variables measured at a port are $a = 0.85\ \angle 45^\circ$ and $b = 0.25\ \angle 65^\circ$. The normalizing impedance is $Z_0 = 50\,\Omega$. Calculate the corresponding voltage and current.

8.23. Calculate the power flow into a port in each case for the following sets of scattering variables at the port:

(a) $a = 3 + j5, b = 0.2 - j0.1$

(b) $a = 40 - j30, b = 5 - j5$

8.24. Calculate the power flow into a port in each case for the following sets of scattering variables at the port:

(a) $a = 5\ \angle 60^\circ,\ b = 2\ \angle -45^\circ$

(b) $a = 50\ \angle 0^\circ,\ b = 10\ \angle 0^\circ$

8.25. Calculate the scattering variables a and b in each of the following cases:

(a) $V = 4 - j3$ V, $I = 0.3 + j0.4$ A

(b) $V = 10\ \angle -30^\circ$ V, $I = 0.7\ \angle 70^\circ$ A

Assume that characteristic impedance at the port is $100\,\Omega$.

8.26. Find the reflection coefficient for each of the following sets of scattering variables:

(a) $a = 0.3 + j0.4, b = 0.1 - 0.2$

(b) $a = -0.5 + j0.2, b = 0 - j0.1$

(c) $a = 0.5\ \angle -70^\circ,\ b = 0.3\ \angle 20^\circ$

(d) $a = 5\ \angle 0^\circ,\ b = 0.3\ \angle 90^\circ$

8.27. The scattering parameters of a two-port network are given as follows: $S_{11} = 0.687\ \angle 107^\circ$, $S_{21} = 1.72\ \angle -59^\circ$, $S_{12} = 0.114\ \angle -81^\circ$, and $S_{22} = 0.381\ \angle 153^\circ$. A source of 3 mV with an internal resistance of $50\,\Omega$ is connected at port 1. Assuming the characteristic impedance at its ports as $50\,\Omega$, determine the scattering variables a_1, b_1, a_2, and b_2 for the following load conditions: **(a)** $50\,\Omega$ and **(b)** $100\,\Omega$.

9

FILTER DESIGN

A circuit designer frequently requires filters to extract the desired frequency spectrum from a wide variety of electrical signals. If a circuit passes all signals from dc through a frequency ω_c but stops the rest of the spectrum, it is known as a *low-pass filter*. The frequency ω_c is called its *cutoff frequency*. Conversely, a *high-pass filter* stops all signals up to ω_c and passes those at higher frequencies. If a circuit passes only a finite frequency band that does not include zero (dc) and infinite frequency, it is called a *bandpass filter*. Similarly, a *bandstop filter* passes all signals except a finite band. Thus, bandpass and bandstop filters are specified by two cutoff frequencies to set the frequency band. If a filter is designed to block a single frequency, it is called a *notch filter*.

The ratio of the power delivered by a source to a load with and without a two-port network inserted in between is known as the *insertion loss* of that two-port. It is generally expressed in decibels. The fraction of the input power that is lost due to reflection at its input port is called the *return loss*. The ratio of the power delivered to a matched load to that supplied to it by a matched source is called the *attenuation* of that two-port network. Filters have been designed using active devices such as transistors and operational amplifiers, as well as with only passive devices (inductors and capacitors only). Therefore, these circuits may be classified as *active* or *passive filters*. Unlike passive filters, active filters can amplify the signal besides blocking the undesired frequencies. However, passive filters are economical and easy to design. Further, passive filters perform fairly well at higher frequencies. In this chapter we present the design procedure of these passive circuits.

Radio-Frequency and Microwave Communication Circuits: Analysis and Design, Second Edition,
By Devendra K. Misra
ISBN 0-471-47873-3

There are two methods available to synthesize passive filters. One of them is known as the *image parameter method* and the other as the *insertion-loss method*. The former provides a design that can pass or stop a certain frequency band, but its frequency response cannot be shaped. The insertion-loss method is more powerful in the sense that it provides a specified response of the filter. Both of these techniques are described in this chapter. The chapter concludes with a design overview of microwave filters.

9.1 IMAGE PARAMETER METHOD

Consider the two-port network shown in Figure 9.1. V_1 and V_2 represent voltages at its ports. Currents I_1 and I_2 are assumed as indicated in the figure. Note that I_1 is entering port 1 while I_2 is leaving port 2. Further, Z_{in} is the input impedance at port 1 when Z_{i2} terminates at port 2. Similarly, Z_0 is the output impedance with Z_{i1} connected at port 1. Z_{i1} and Z_{i2} are known as the *image impedance* of the network. Following the transmission parameter description of the two-port, we can write

$$V_1 - AV_2 + BI_2 \tag{9.1.1}$$

and

$$I_1 = CV_2 + DI_2 \tag{9.1.2}$$

Therefore, the impedance Z_{in} at its input can be found as

$$Z_{in} = \frac{V_1}{I_1} = \frac{AV_2 + BI_2}{CV_2 + DI_2} = \frac{AZ_{i2} + B}{CZ_{i2} + D} \tag{9.1.3}$$

Alternatively, (9.1.1) and (9.1.2) can be rearranged as follows after noting that $AD—BC$ must be unity for a reciprocal network. Hence,

$$V_2 = DV_1 - BI_1 \tag{9.1.4}$$

and

$$I_2 = -CV_1 + AI_1 \tag{9.1.5}$$

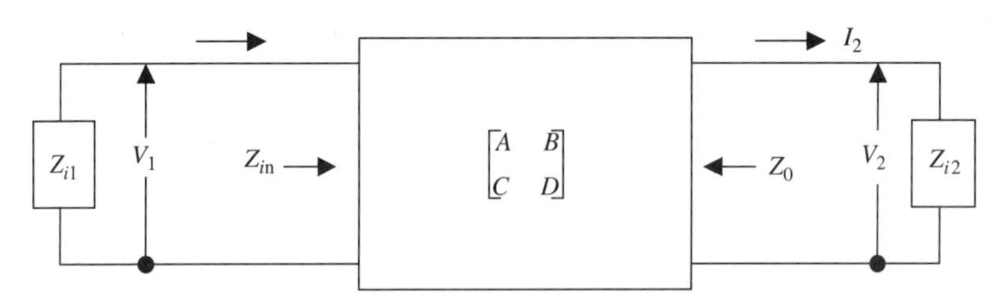

Figure 9.1 Two-port network with terminations.

Therefore, the output impedance Z_0 is found to be

$$Z_0 = -\frac{V_2}{I_2} = -\frac{DV_1 - BI_1}{-CV_1 + AI_1} = \frac{DZ_{i1} + B}{CZ_{i1} + A} \tag{9.1.6}$$

Note that

$$Z_{i1} = -\frac{V_1}{I_1}$$

For $Z_{\text{in}} = Z_{i1}$ and $Z_0 = Z_{i2}$, equations (9.1.3) and (9.1.6) give

$$Z_{i1}(CZ_{i2} + D) = AZ_{i2} + B \tag{9.1.7}$$

and

$$Z_{i2}(CZ_{i1} + A) = DZ_{i1} + B \tag{9.1.8}$$

Subtracting equation (9.1.8) from (9.1.7), we find that

$$Z_{i2} = \frac{D}{A} Z_{i1} \tag{9.1.9}$$

Now, substituting (9.1.9) into (9.1.7), we have

$$Z_{i1} = \sqrt{\frac{AB}{CD}} \tag{9.1.10}$$

Similarly, substituting (9.1.10) into (9.1.9) yields

$$Z_{i2} = \sqrt{\frac{BD}{AC}} \tag{9.1.11}$$

The transfer characteristics of the network can be formulated as follows. From (9.1.1),

$$\frac{V_1}{V_2} = A + B\frac{I_2}{V_2} = A + \frac{B}{Z_{i2}}$$

or

$$\frac{V_1}{V_2} = \sqrt{\frac{A}{D}}(\sqrt{AD} + \sqrt{BC})$$

For a reciprocal two-port network, $AD - BC$ is unity. Therefore,

$$\frac{V_2}{V_1} = \sqrt{\frac{D}{A}} \frac{AD - BC}{\sqrt{AD} + \sqrt{BC}} = \sqrt{\frac{D}{A}}(\sqrt{AD} - \sqrt{BC}) \tag{9.1.12}$$

Similarly, from (9.1.2),

$$\frac{I_2}{I_1} = \sqrt{\frac{A}{D}}(\sqrt{AD} - \sqrt{BC}) \tag{9.1.13}$$

Note that equation (9.1.12) is similar to (9.1.13) except that the multiplying coefficient in one is the reciprocal of the other. This coefficient may be interpreted as the transformer turns ratio. It is unity for symmetrical T and π networks. The propagation factor γ (equal to $\alpha + j\beta$, as usual) of the network can be defined as

$$e^{-\gamma} = \sqrt{AD} - \sqrt{BC}$$

Since $AD - BC = 1$, we find that

$$\cosh\gamma = \sqrt{AD} \tag{9.1.14}$$

The characteristic parameters of π and T networks are summarized in Table 9.1; the corresponding low-pass and high-pass circuits are given in Table 9.2.

TABLE 9.1 Parameters of T and π Networks

	π-Network	T-Network
	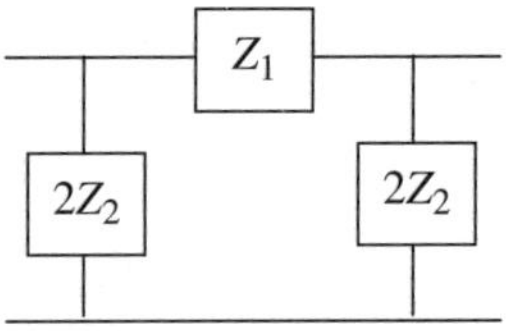	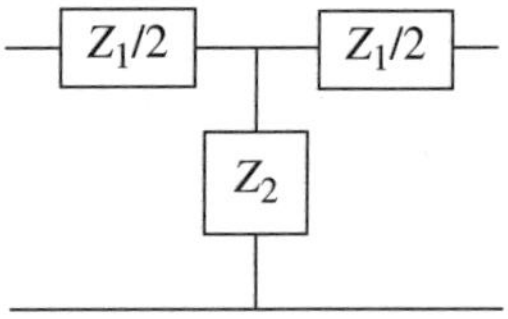
ABCD parameters	$A = 1 + \frac{Z_1}{2Z_2}$	$A = 1 + \frac{Z_1}{2Z_2}$
	$B = Z_1$	$B = Z_1 + \frac{Z_1^2}{4Z_2}$
	$C = \frac{1}{Z_2} + \frac{Z_1}{4Z_2^2}$	$C = \frac{1}{Z_2}$
	$D = 1 + \frac{Z_1}{2Z_2}$	$D = 1 + \frac{Z_1}{2Z_2}$
Image impedance	$Z_{i\pi} = \sqrt{\frac{Z_1 Z_2}{1 + Z_1/4Z_2}} = \frac{Z_1 Z_2}{Z_{iT}}$	$Z_{iT} = \sqrt{Z_1 Z_2\left(1 + \frac{Z_1}{4Z_2}\right)}$
Propagation constant, γ	$\cosh\gamma = 1 + \frac{Z_1}{2Z_2}$	$\cosh\gamma = 1 + \frac{Z_1}{2Z_2}$

TABLE 9.2 Constant-*k* Filter Sections

Filter Type	T-Section	π-Section
Low-pass	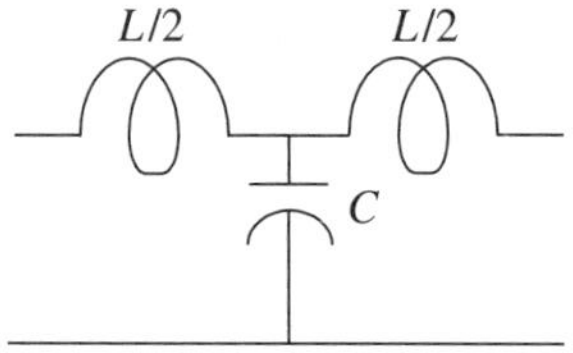 $L/2$ $L/2$ C	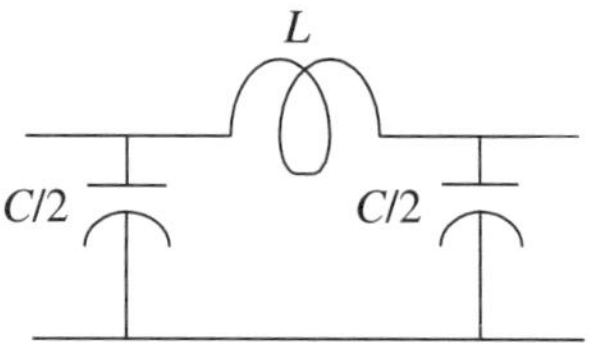 L $C/2$ $C/2$
High-pass	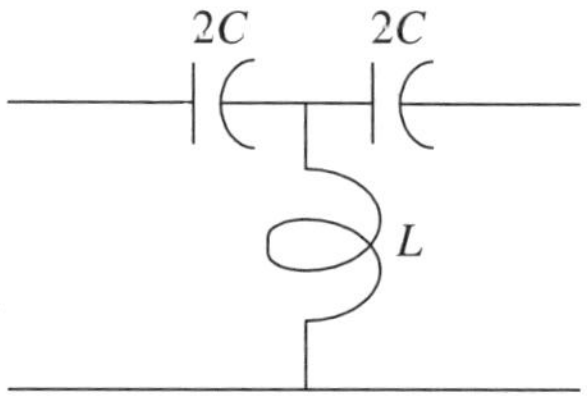 $2C$ $2C$ L	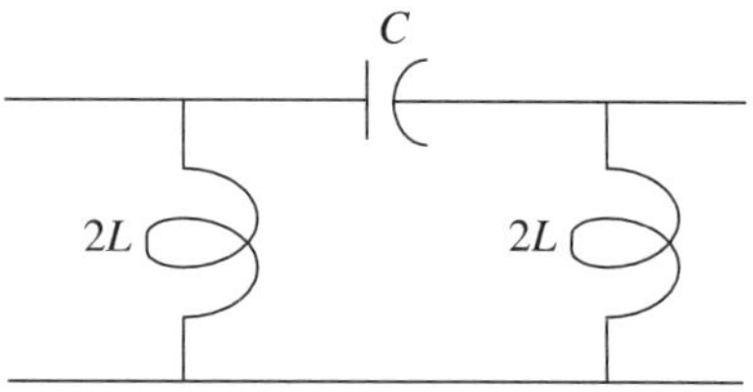 C $2L$ $2L$

For a low-pass T-section as illustrated in Table 9.2,

$$Z_1 = j\omega L \tag{9.1.15}$$

and

$$Z_2 = \frac{1}{j\omega C} \tag{9.1.16}$$

Therefore, its image impedance can be found from Table 9.1 as follows:

$$Z_{iT} = \sqrt{\frac{L}{C}\left(1 - \frac{\omega^2 LC}{4}\right)} \tag{9.1.17}$$

In the case of a dc signal, the second term inside the parentheses will be zero and the resulting image impedance is generally known as the *nominal impedance*, Z_0. Hence,

$$Z_0 = \sqrt{\frac{L}{C}}$$

Note that the image impedance goes to zero if $\omega^2 LC/4 = 1$. The frequency that satisfies this condition is known as the *cutoff frequency*, ω_c. Hence,

$$\omega_c = \frac{2}{\sqrt{LC}} \tag{9.1.18}$$

Similarly, for a high-pass T-section,

$$Z_1 = \frac{1}{j\omega C} \tag{9.1.19}$$

and

$$Z_2 = j\omega L \tag{9.1.20}$$

Therefore, its image impedance is found to be

$$Z_{i\mathrm{T}} = \sqrt{\frac{L}{C}\left(1 - \frac{1}{4\omega^2 LC}\right)} \tag{9.1.21}$$

The cutoff frequency of this circuit will be given as

$$\omega_c = \frac{1}{2\sqrt{LC}} \tag{9.1.22}$$

Example 9.1 Design a low-pass constant-k T-section that has a nominal impedance of 75 Ω and a cutoff frequency of 2 MHz. Plot its frequency response in the frequency band 100 kHz to 10 MHz.

SOLUTION Since the nominal impedance Z_0 must be 75 Ω,

$$Z_0 = 75 = \sqrt{\frac{L}{C}}$$

The cutoff frequency ω_c of a low-pass T-section is given by (9.1.18). Therefore,

$$\omega_c = 2\pi \times 2 \times 10^6 = \frac{2}{\sqrt{LC}}$$

These two equations can be solved for the inductance L and capacitance C, as follows:

$$L = 11.9366\,\mu\mathrm{H}$$

and

$$C = 2.122\,\mathrm{nF}$$

This circuit is illustrated in Figure 9.2. Note that inductance L calculated here is twice the value needed for a T-section. Propagation constant γ of this circuit is

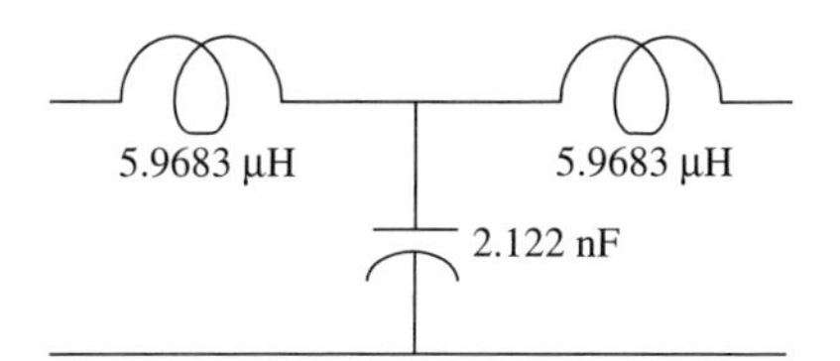

Figure 9.2 Low-pass constant-k T-section.

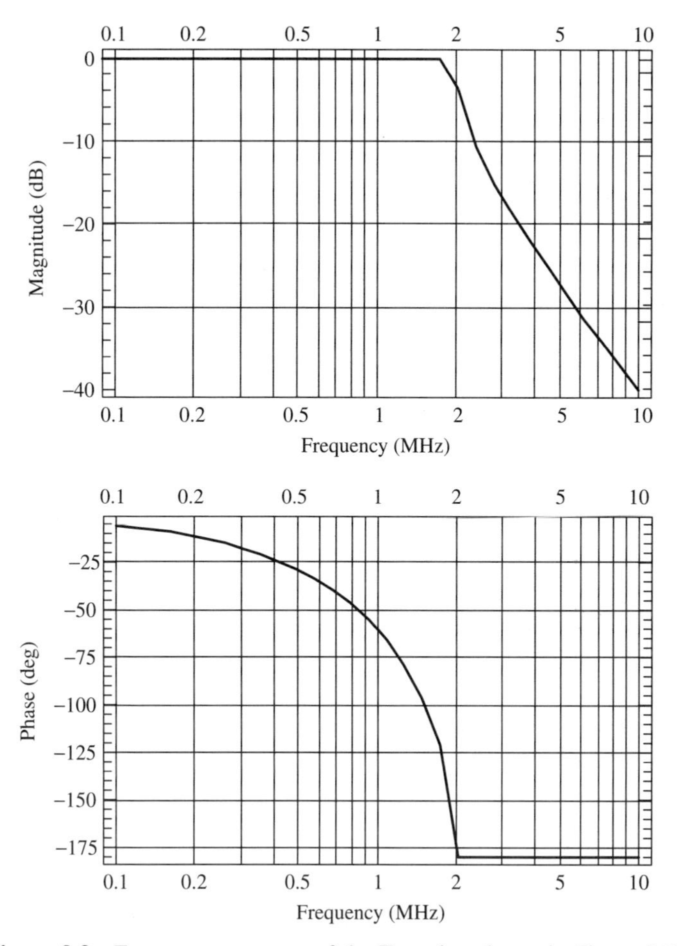

Figure 9.3 Frequency response of the T-section shown in Figure 9.2.

determined from the formula listed in Table 8.1. The transfer characteristics are then found as $e^{-\gamma}$.

The frequency response of the designed circuit is shown in Figure 9.3. The magnitude of the transfer function (ratio of the output to input voltages) remains constant at 0 dB for frequencies lower than 2 MHz. Therefore, the output magnitude will be equal to the input in this range. It falls by 3 dB if the signal frequency approaches 2 MHz. It falls continuously as the frequency increases further. Note that the phase angle of the transfer function remains constant at -180° beyond 2 MHz. However, it shows a nonlinear characteristic in the passband.

The image impedance Z_{iT} of this T-section can be found from (9.1.17). Its characteristics (magnitude and phase angle) with frequency are displayed in Figure 9.4. The magnitude of Z_{iT} reduces continuously as frequency is increased and becomes zero at the cutoff. The phase angle of Z_{iT} is zero in passband, and it changes to 90° in the stopband. Thus, image impedance is a variable resistance in the passband, whereas it switches to an inductive reactance in the stopband.

The frequency characteristics illustrated in Figures 9.3 and 9.4 are representative of any constant-*k* filter. There are two major drawbacks to this type of filter:

1. The signal attenuation rate after the cutoff point is not very sharp,
2. The image impedance is not constant with frequency. From a design point of view, it is important that it stay constant, at least in its passband.

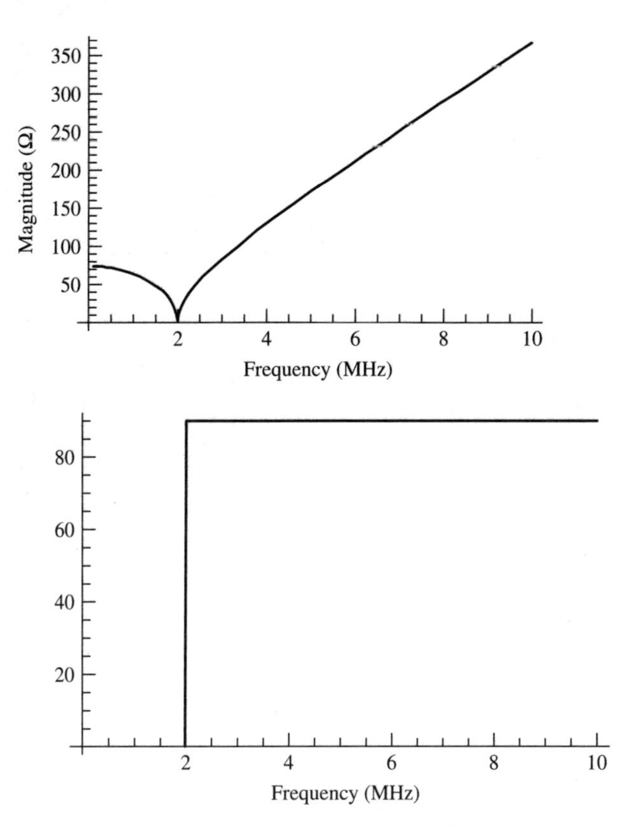

Figure 9.4 Image impedance of the constant-*k* filter of Figure 9.2 as a function of frequency.

These problems can be remedied using the techniques described in the following section.

m-Derived Filter Sections

Consider the two T-sections shown in Figure 9.5. The first network represents the constant-k filter considered in the preceding section, and the second is a new m-derived section. It is assumed that the two networks have the same image impedance. From Table 9.1 we can write

$$Z'_{i\mathrm{T}} = Z_{i\mathrm{T}} = \sqrt{Z'_1 Z'_2\left(1+\frac{Z'_1}{4Z'_2}\right)} = \sqrt{Z_1 Z_2\left(1+\frac{Z_1}{4Z_2}\right)}$$

It can be solved for Z'_2 as follows:

$$Z'_2 = \frac{1}{Z'_1}\left(Z_1 Z_2 + \frac{Z_1^2 - Z_1'^2}{4}\right) \tag{9.1.23}$$

Now assume that

$$Z'_1 = mZ_1 \tag{9.1.24}$$

Substituting (9.1.24) in to (9.1.23), we get

$$Z'_2 = \frac{Z_2}{m} + \frac{1-m^2}{4m}Z_1 \tag{9.1.25}$$

Thus, an m-derived section is designed from the values of components determined for the corresponding constant-k filter. The value of m is selected to sharpen the attenuation at cutoff or to control the image impedance characteristics in the passband.

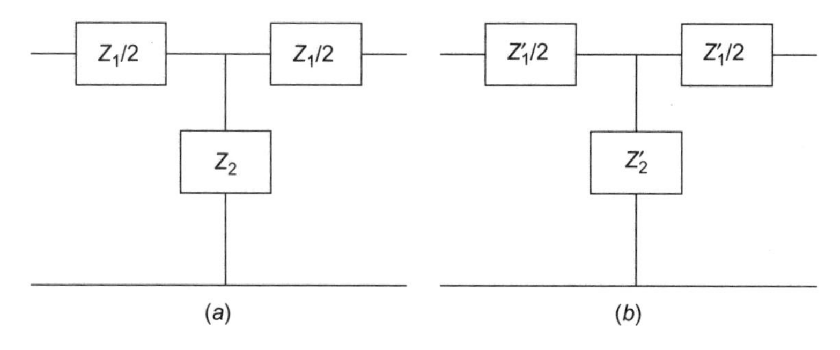

Figure 9.5 Constant-k T-section (a) and m-derived section (b).

For a low-pass filter, the m-derived section can be designed from the corresponding constant-k filter using (9.1.24) and (9.1.25) as follows:

$$Z_1' = j\omega mL \tag{9.1.26}$$

and

$$Z_2' = \frac{1-m^2}{4m} j\omega L + \frac{1}{j\omega mC} \tag{9.1.27}$$

Now we need to find its propagation constant γ and devise some way to control its attenuation around the cutoff. Expression for the propagation constant of a T-section is listed in Table 9.1. To find γ of this T-section, we first divide (9.1.26) by (9.1.27):

$$\frac{Z_1'}{Z_2'} = -\frac{\omega^2 m^2 LC}{1 - [(1-m^2)/4]\omega^2 LC} = -\frac{4\omega^2 m^2/\omega_c^2}{1-(1-m^2)\omega^2/\omega_c^2} \tag{9.1.28}$$

where

$$\omega_c = \frac{2}{\sqrt{LC}} \tag{9.1.29}$$

Using the formula listed in Table 9.1 and (9.1.28), the propagation constant γ is found as follows:

$$\cosh\gamma = 1 + \frac{Z_1'}{2Z_2'} = 1 - \frac{2\,(m\omega/\omega_c)^2}{1-(1-m^2)\,(\omega/\omega_c)^2}$$

or

$$\cosh\gamma = \frac{\omega_c^2 - \omega^2 - (m\omega)^2}{\omega_c^2 - (1-m^2)\omega^2} \tag{9.1.30}$$

Hence, the right-hand side of (9.1.30) is dependent on the frequency ω. It will go to infinity (and therefore, γ) if the following condition is satisfied:

$$\omega = \frac{\omega_c}{\sqrt{1-m^2}} = \omega_\infty \tag{9.1.31}$$

This condition can be used to sharpen the attenuation at cutoff. A small m will place ω_∞ close to ω. In other words, ω_∞ is selected a little higher than ω_c and the fractional value of m is then determined from (9.1.31). Z_1' and Z_2' of the m-derived section are subsequently found using (9.1.24) and (9.1.25), respectively.

Note that the image impedance of an m-derived T-section is the same as that of the corresponding constant-k network. However, it will be a function of m in the case of a π-network. Therefore, this characteristic can be utilized to design input and output networks of the filter so that the image impedance of the composite circuit stays almost constant in its passband. Further, an infinite cascade of T-sections can be considered as a π-network after splitting its shunt

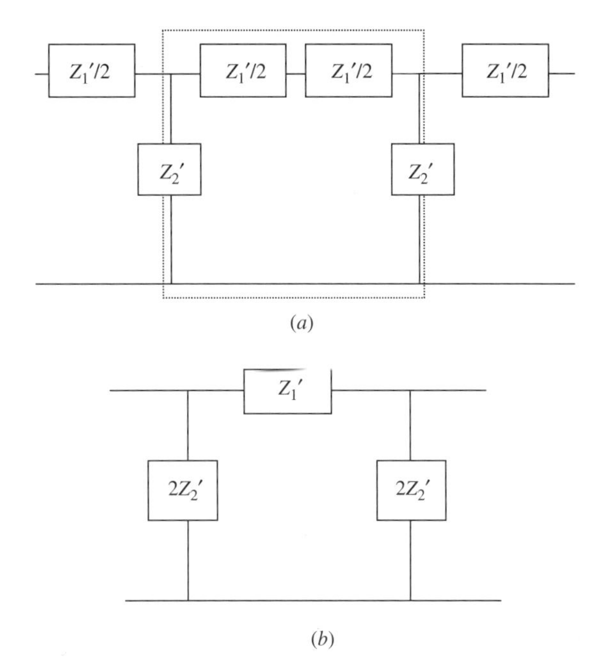

Figure 9.6 Deembedding of a π-network (*b*) from a cascaded T-network (*a*).

arm as illustrated in Figure 9.6. Note that the impedance Z_2' of the T-network is replaced by $2Z_2'$, while two halves of the series arms of the T-network give Z_1' of the π-network. Image impedance $Z_{i\pi}$ is found from Table 9.1 as follows:

$$Z_{i\pi} = \frac{Z_1' Z_2'}{Z_{iT}} = \frac{Z_1 Z_2 + [(1 - m^2)/4] Z_1^2}{Z_{iT}} \tag{9.1.32}$$

For the low-pass constant-k filter, we find from (9.1.15) to (9.1.18) that

$$Z_1 Z_2 = \frac{L}{C} = Z_0^2$$

$$Z_1^2 = -\omega^2 L^2 = \left(\frac{2Z_0\omega}{\omega_c}\right)^2$$

and,

$$Z_{iT} = Z_0\sqrt{1 - \left(\frac{\omega}{\omega_c}\right)^2}$$

Therefore,

$$Z_{i\pi} = \frac{1 - (1 - m^2)(\omega/\omega_c)^2}{\sqrt{1 - (\omega/\omega_c)^2}} Z_0$$

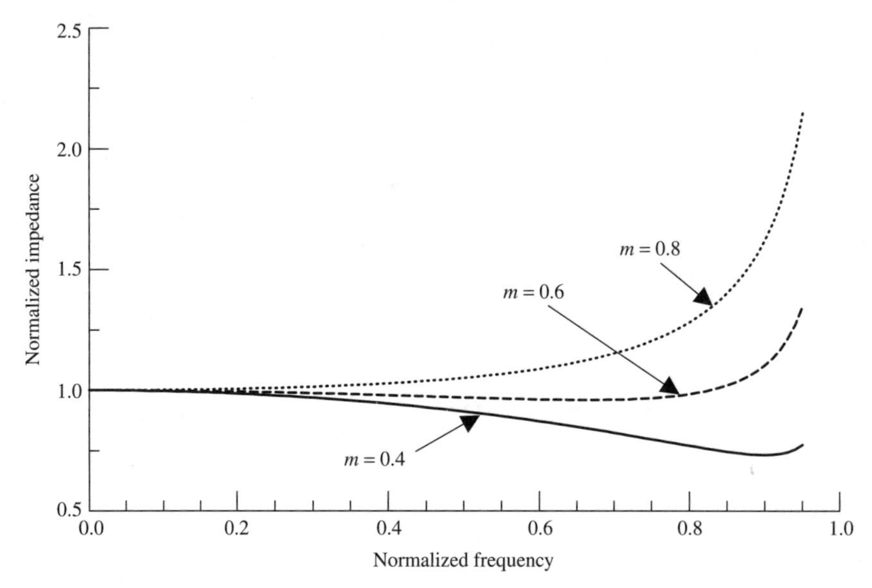

Figure 9.7 Normalized image impedance of π-network versus normalized frequency for three values of m.

or

$$\overline{Z}_{i\pi} = \frac{1-(1-m^2)\overline{\omega}^2}{\sqrt{(1-\overline{\omega}^2)}} \tag{9.1.33}$$

where $\overline{Z}_{i\pi} = Z_{i\pi}/Z_0$ may be called the *normalized image impedance* of a π-network and $\overline{\omega} = \omega/\omega_c$ the *normalized frequency*.

Equation (9.1.33) is displayed graphically in Figure 9.7. The normalized image impedance is close to unity for nearly 90% of the passband if m is kept around 0.6. Therefore, it can be used as pre- and poststages of the composite filter (after bisecting it) to control the input impedance. These formulas will be developed later in this section.

Example 9.2 Design an m-derived T-section low-pass filter with a cutoff frequency of 2 MHz and a nominal impedance of 75 Ω. Assume that $f_\infty = 2.05$ MHz. Plot the response of this filter in the frequency band of 100 kHz to 10 MHz.

SOLUTION From (9.1.31),

$$1 - m^2 = \left(\frac{f_c}{f_\infty}\right)^2$$

Therefore,

$$m = \sqrt{1-\left(\frac{f_c}{f_\infty}\right)^2} = \sqrt{1-\left(\frac{2}{2.05}\right)^2} = 0.2195$$

TABLE 9.3 Design Relations for Composite Filters

Low-Pass	High-Pass
Constant-k T-filter	*Constant-k T-filter*
L/2, L/2, C $Z_0 = \sqrt{L/C}$ $\omega_c = 2/\sqrt{LC}$ $L = 2Z_0/\omega_c$ $C = 2/Z_0\omega_c$	2C, 2C, L $Z_0 = \sqrt{L/C}$ $\omega_c = 1/2\sqrt{LC}$ $L = 0.5Z_0/\omega_c$ $C = 0.5/Z_0\omega_c$
m-derived T-Section	*m-derived T-Section*
(Values of L and C are the same as above)	(Values of L and C are the same as above)
$m = \sqrt{1 - \left(\frac{f_c}{f_\infty}\right)^2}$	$m = \sqrt{1 - \left(\frac{f_\infty}{f_c}\right)^2}$
mL/2, mL/2, $\frac{1-m^2}{4m}L$, mC	2C/m, 2C/m, L/m, $\frac{4m}{1-m^2}C$
Input and Output Matching Sections	Input and Output Matching Sections
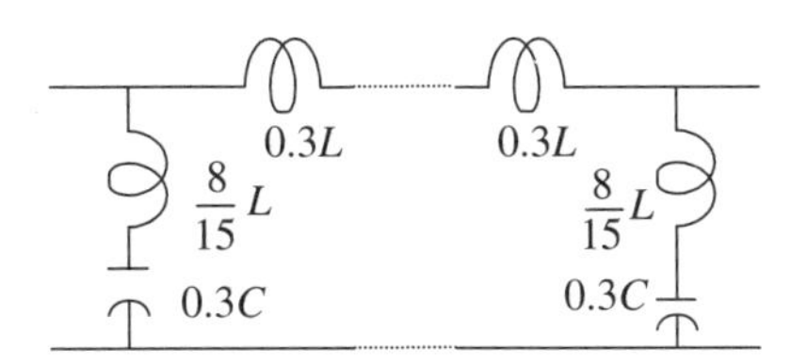	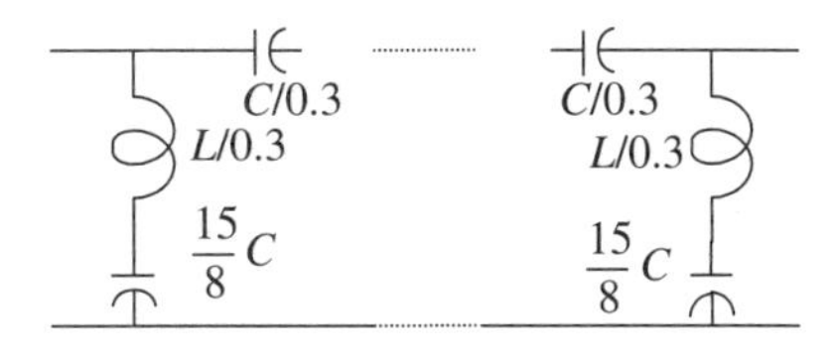

Using (9.1.24) and (9.1.25) or Table 9.3 and component values of the constant-k section obtained earlier in Example 9.1, the m-derived filter is designed as follows:

$$mL/2 = 0.2195 \times 5.9683 = 1.31\,\mu\text{H}$$

$$mC = 0.2195 \times 2.122 = 465.78\,\text{nF}$$

and

$$\frac{1 - m^2}{4m}L = 12.94\,\mu\text{H}$$

This filter circuit is illustrated in Figure 9.8.

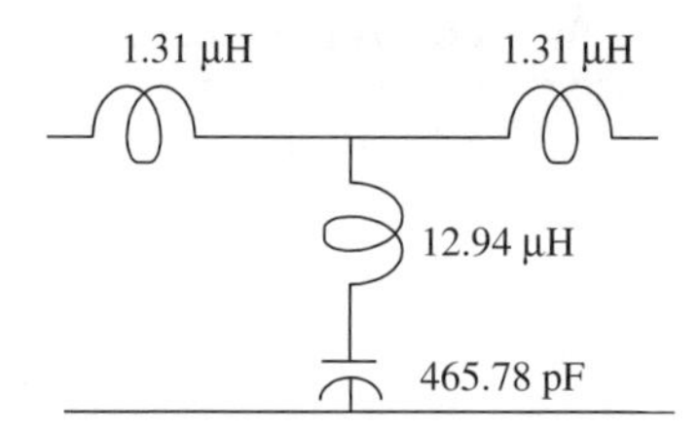

Figure 9.8 An m-derived T-section for Example 9.2.

As noted earlier, the image impedance of this circuit will be the same as that of the corresponding constant-k section. Hence, it will vary with frequency as illustrated in Figure 9.4. The propagation constant γ of this circuit is determined from the formula listed in Table 9.1. The transfer characteristics are then found as $e^{-\gamma}$. Its magnitude and phase characteristics versus frequency are illustrated in Figure 9.9. The transfer characteristics illustrated in Figure 9.9 indicate that the m-derived filter has a sharp change at a cutoff frequency of 2 MHz. However, the output signal rises to -4 dB in its stopband. On the other hand, the constant-k filter provides higher attenuation in its stopband. For example, the m-derived filter characteristic in Figure 9.9 shows only 4-dB attenuation at 6 MHz, whereas the corresponding constant-k T-section has an attenuation of more than 30 dB at this frequency, as depicted in Figure 9.3.

Composite Filters

As demonstrated through the preceding examples, the m-derived filter provides an infinitely sharp attenuation right at its cutoff. However, the attenuation in its stopband is unacceptably low. Contrary to this, the constant-k filter shows a higher attenuation in its stopband, although the change is unacceptably gradual at the transition from passband to stopband. One way to solve the problem is to cascade these two filters. Since image impedance stays the same in two cases, this cascading will not create a new impedance-matching problem. However, we still need to address the problem of image impedance variation with frequency at the input and output ports of the network.

As illustrated in Figure 9.7, the image impedance of the π-section with $m = 0.6$ remains almost constant over 90% of the passband. If this network is bisected to connect on either side of the cascaded constant-k and m-derived sections, it should provide the desired impedance characteristics. To verify these characteristics, let us consider a bisected π-section as shown in Figure 9.10. Its transmission parameters can be found easily following the procedure described in Chapter 8.

From (8.4.4) to (8.4.7), we find that

$$A = 1 + \frac{Z_1'}{4Z_2'} \tag{9.1.34}$$

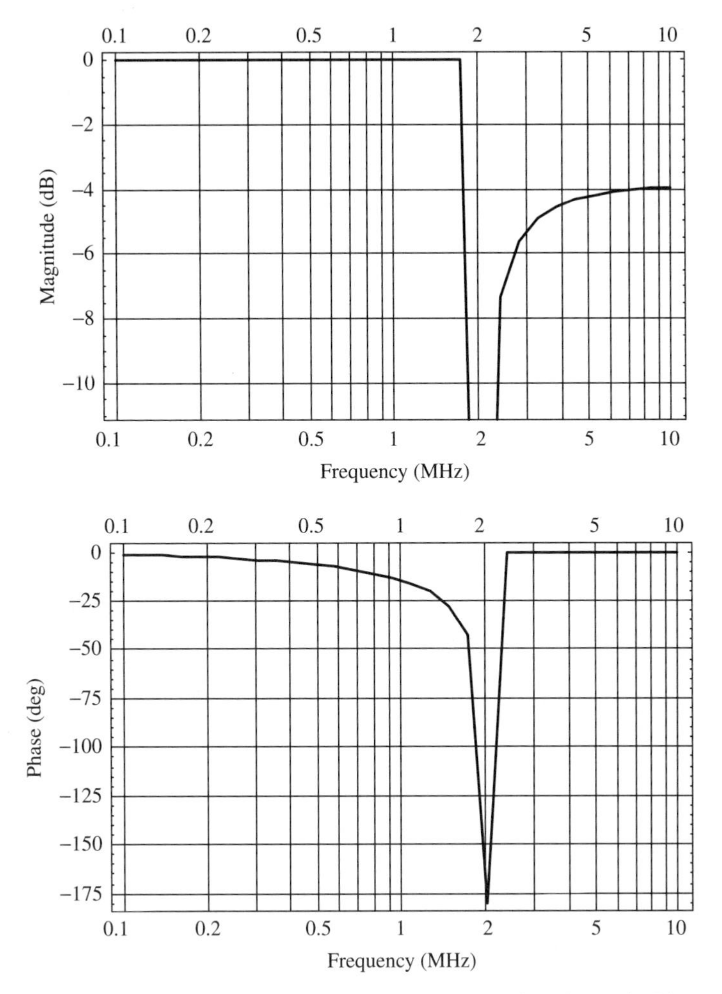

Figure 9.9 Frequency response of the m-derived T-section shown in Figure 9.8.

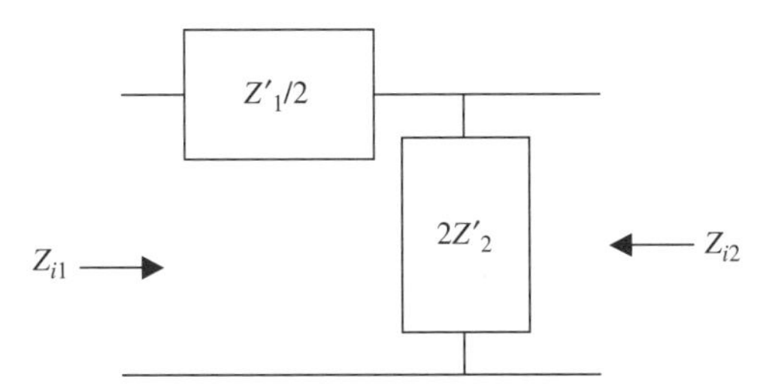

Figure 9.10 Right-hand side of the bisected π-section shown in Figure 9.6.

$$B = \frac{Z_1'}{2} \tag{9.1.35}$$

$$C = \frac{1}{2Z_2'} \tag{9.1.36}$$

and

$$D = 1 \tag{9.1.37}$$

Image impedance Z_{i1} and Z_{i2} can now be found from (9.1.10) and (9.1.11) as follows:

$$Z_{i1} = \sqrt{Z_1' Z_2' + \frac{Z_1'^2}{4}} = Z_{iT} \tag{9.1.38}$$

and

$$Z_{i2} = \sqrt{\frac{Z_1' Z_2'}{1 + Z_1'/4Z_2'}} = \frac{Z_1' Z_2'}{Z_{iT}} = Z_{i\pi} \tag{9.1.39}$$

Thus, the bisected π-section can be connected at the input and output ports of cascaded constant-k and m-derived sections to obtain a composite filter that solves the impedance problem as well. Relations for the design of these circuits are summarized in Table 9.3.

Example 9.3 Design a low-pass composite filter with a cutoff frequency of 2 MHz and an image impedance of 75 Ω in its passband. Assume that $f_\infty =$ 2.05 MHz. Plot its response in the frequency range 100 kHz to 10 MHz.

SOLUTION Constant-k and m-derived sections of this filter are designed in Examples 9.1 and 9.2, respectively. The filter's input and output matching sections can be designed as follows. With $m = 0.6$, we find that

$$\frac{mL}{2} = 3.581\,\mu\text{H}$$

$$\frac{mC}{2} = 636.6\,\text{pF}$$

and

$$\frac{1 - m^2}{2m} L = 6.366\,\mu\text{H}$$

This composite filter is depicted in Figure 9.11.

Figure 9.12 illustrates the frequency response of this composite filter. As it indicates, there is a fairly sharp change in output signal as the frequency changes from its passband to the stopband. At the same time, the output stays well below −40 dB in its stopband. Figure 9.13 shows variation in the image impedance

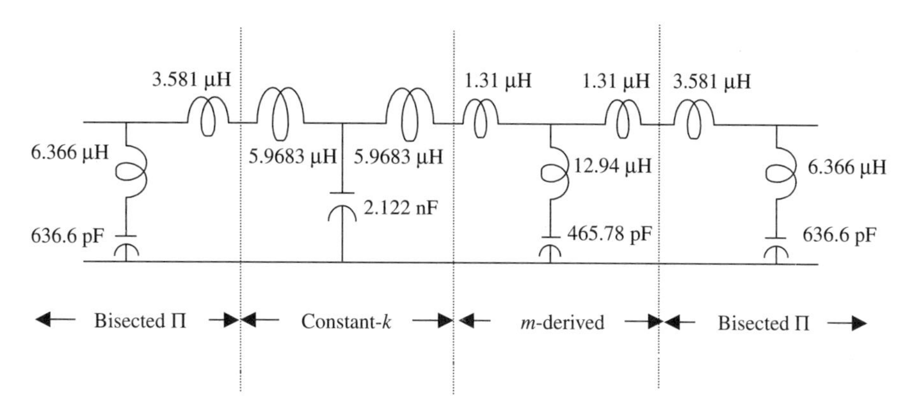

Figure 9.11 Composite filter of Example 9.3.

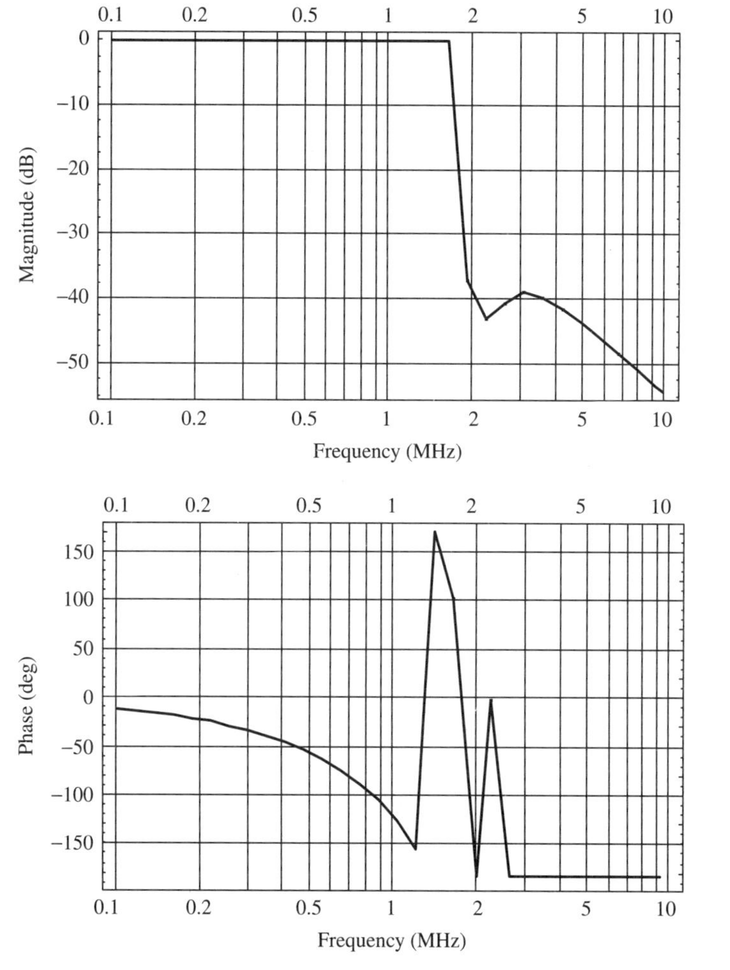

Figure 9.12 Frequency response of the composite filter of Figure 9.11.

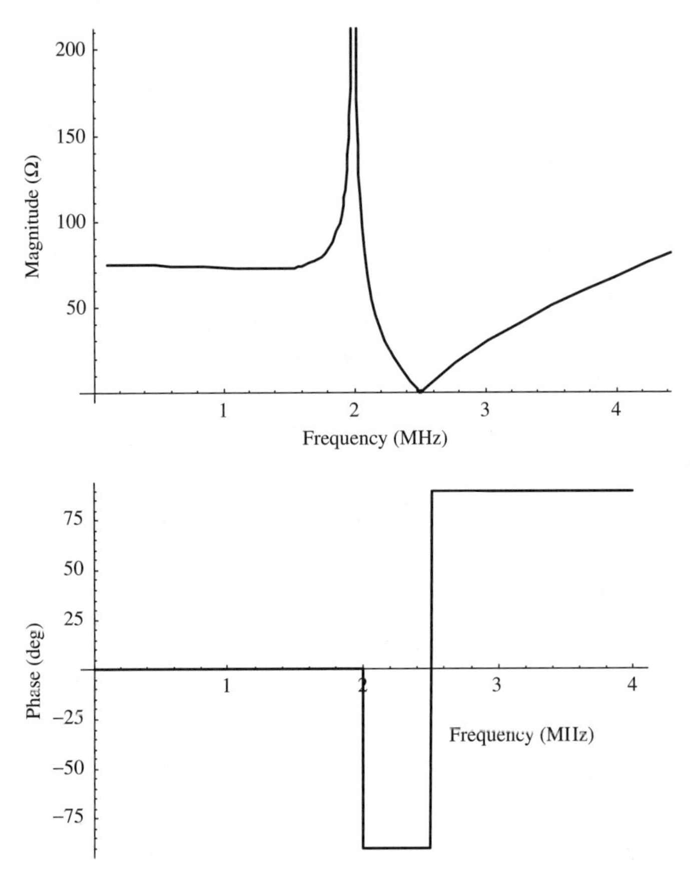

Figure 9.13 Image impedance of the composite filter of Figure 9.11 versus frequency.

of this filter as the signal frequency changes. This indicates that the image impedance stays at 75 Ω (pure real, because the phase angle is zero) over most of its passband.

Example 9.4 Design a high-pass composite filter with a nominal impedance of 75 Ω. It must pass all signals over 2 MHz. Assume that $f_\infty = 1.95$ MHz. Plot its characteristics in the frequency range 1 to 10 MHz.

SOLUTION From Table 9.3, we find the components of its constant-k section as follows:

$$L = \frac{75}{2 \times 2 \times \pi \times 2 \times 10^6} \text{H} = 2.984\,\mu\text{H}$$

and

$$C = \frac{1}{2 \times 2 \times \pi \times 2 \times 10^6 \times 75} \text{F} = 530.5\,\text{pF}$$

Similarly, the component values for its m-derived filter section are determined as follows:

$$m = \sqrt{1 - \left(\frac{f_\infty}{f_c}\right)^2} = \sqrt{1 - \left(\frac{1.95}{2}\right)^2} = 0.222$$

Hence,

$$\frac{2C}{m} = 4.775\,\text{nF}$$
$$\frac{L}{m} = 13.43\,\mu\text{H}$$

and

$$\frac{4m}{1-m^2}C = 0.496\,\text{nF}$$

The component values for the bisected π-section to be used at its input and output ports are found as

$$\frac{C}{0.3} = 1.768\,\text{nF}$$
$$\frac{L}{0.3} = 9.947\,\mu\text{H}$$

and

$$\frac{15}{8}C = 0.9947\,\text{nF}$$

The composite filter (after simplifying for the series capacitors in various sections) is illustrated in Figure 9.14.

The frequency response of this composite filter is depicted in Figure 9.15. It shows that the attenuation in its stopband stays below -40 dB, and the switching to passband is fairly sharp. As usual with this type of circuit, its phase characteristics may not be acceptable for certain applications because of inherent distortion.

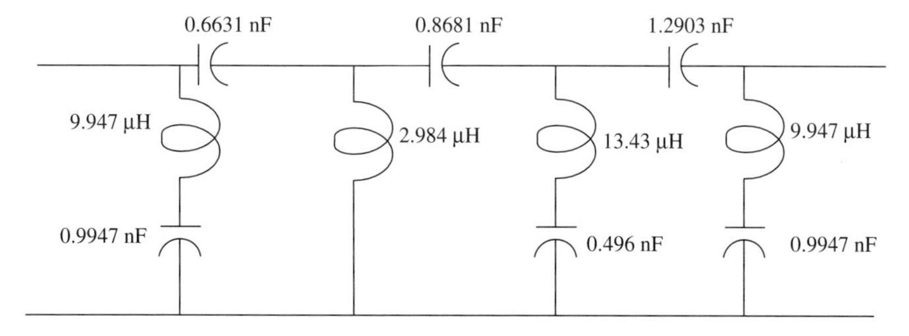

Figure 9.14 Composite high-pass filter with 2-MHz cutoff frequency.

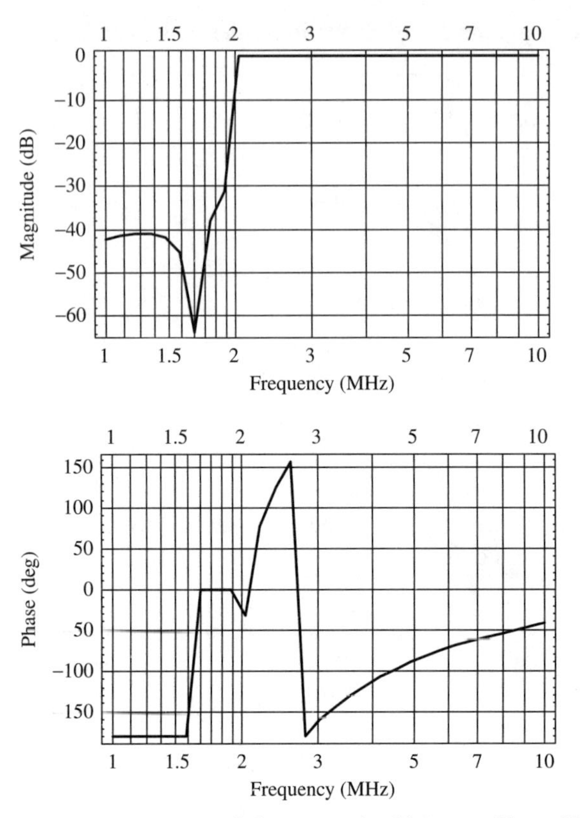

Figure 9.15 Frequency response of the composite high-pass filter of Figure 9.14.

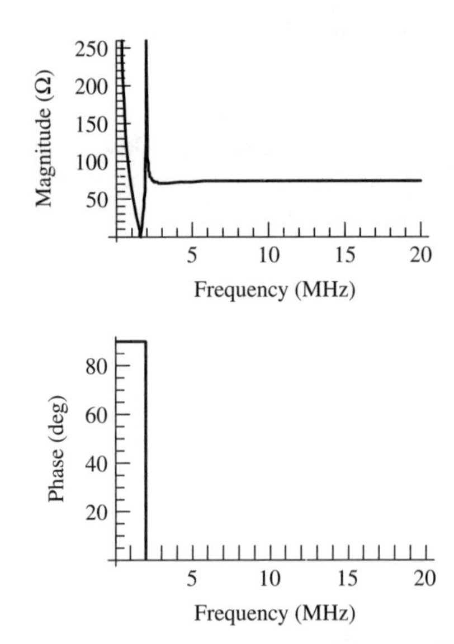

Figure 9.16 Image impedance of the composite high-pass filter of Figure 9.14 as a function of signal frequency.

As depicted in Figure 9.16, the image impedance of this composite filter is almost constant at 75 Ω. Note that its magnitude varies over a wide range in the stopband while its phase angle seems to remain constant at 90°. The phase angle of the image impedance is zero in the passband.

9.2 INSERTION-LOSS METHOD

The output of an ideal filter would be the same as its input in the passband, whereas it would be zero in the stopband. The phase response of this filter must be linear to avoid signal distortion. In reality, such circuits do not exist and a compromise is needed to design the filters. The image parameter method described in Section 9.1 provides a simple design procedure. However, the transfer characteristics of this circuit cannot be shaped as desired. On the other hand, the insertion-loss method provides ways to shape pass- and stopbands of the filter, although its design theory is much more complex.

The *power-loss ratio* of a two-port network is defined as the ratio of the power that is delivered to the load when it is connected directly at the generator to the power delivered when the network is inserted between the two. In other words,

$$P_{LR} = \frac{\text{power incident at port 1}}{\text{power delivered to the load connected at port 2}} = \frac{1}{1 - |\Gamma(\omega)|^2} \tag{9.2.1}$$

The power-loss ratio, P_{LR}, expressed in decibels, is generally known as the insertion loss of the network. It can be proved that $|\Gamma(\omega)|^2$ must be an even function of ω for a physically realizable network. Therefore, polynomials of ω^2 can represent it as follows.

$$|\Gamma(\omega)|^2 = \frac{f_1(\omega^2)}{f_1(\omega^2) + f_2(\omega^2)} \tag{9.2.2}$$

and

$$P_{LR} = 1 + \frac{f_1(\omega^2)}{f_2(\omega^2)} \tag{9.2.3}$$

where $f_1(\omega^2)$ and $f_2(\omega^2)$ are real polynomials in ω^2.

Alternatively, the magnitude of the voltage gain of the two-port network can be found as

$$|G(\omega)| = \frac{1}{\sqrt{P_{LR}}} = \frac{1}{\sqrt{1 + f_1(\omega^2)/f_2(\omega^2)}} \tag{9.2.4}$$

Hence, if P_{LR} is specified, $\Gamma(\omega)$ is fixed. Therefore, the insertion-loss method is similar to the impedance-matching methods discussed in Chapter 7.

Traditionally, the filter design begins with a lumped-element low-pass network that is synthesized by using normalized tables. It is subsequently scaled to the desired cutoff frequency and the impedance. Also, the low-pass prototype can

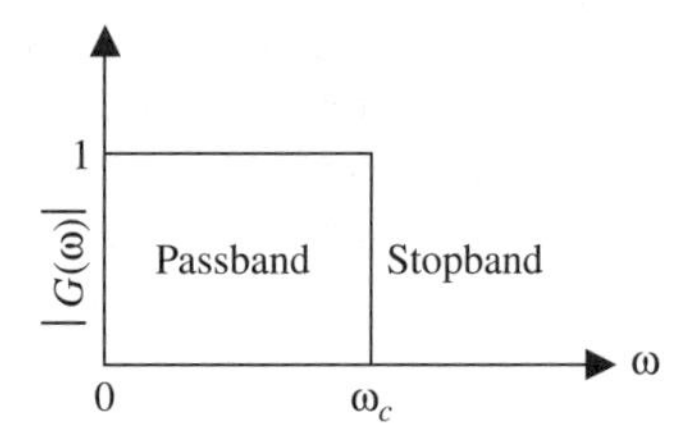

Figure 9.17 Characteristics of an ideal low-pass filter.

be transformed to obtain a high-pass, a bandpass, or a bandstop filter. These lumped-element filters are used as a starting point to design the transmission line filter. In this section we present the design procedure for two different types of lumped-element low-pass filters. It is followed by the transformation techniques used to design high-pass, bandpass, and bandstop filters.

Low-Pass Filters

As illustrated in Figure 9.17, an ideal low-pass filter will pass the signals below its cutoff frequency ω_c without attenuation while it will stop those with higher frequencies. Further, the transition from its passband to its stopband will be sharp. In reality, this type of filter cannot be designed. Several approximations to these characteristics are available that can be physically synthesized. Two of these are presented below.

Maximally Flat Filter

As its name suggests, this type of filter provides the flattest possible passband response. However, its transition from passband to stopband is gradual. This is also known as a binomial or Butterworth filter. The magnitude of its voltage gain (and hence, its frequency response) is given as follows:

$$|G(\omega)| = \frac{1}{\sqrt{1+\zeta\overline{\omega}^{2n}}} \qquad n = 1, 2, 3, \ldots \tag{9.2.5}$$

where ζ is a constant that controls the power-loss ratio at its band edge, n is the order, and $\overline{\omega}$ is the normalized frequency. For $|G(\omega)| = 0.7071$ (i.e., $-3\,\text{dB}$) at the band edge, ζ is unity.

Note that derivatives of $|G(\omega)|$ at normalized frequencies much below the band edge are close to zero. This guarantees maximum flatness in the passband. Further, (9.2.5) indicates that $|G(\omega)|$ will be a fractional number for $\overline{\omega}$ greater than zero. Therefore, it is generally known as the insertion loss, L, of the network when expressed in decibels. Hence,

$$L = -20\log_{10}\left[\frac{1}{\sqrt{1+\zeta\left(\omega/\omega_c\right)^{2n}}}\right] = 10\log_{10}\left[1+\zeta\left(\frac{\omega}{\omega_c}\right)^{2n}\right] \quad \text{dB} \tag{9.2.6}$$

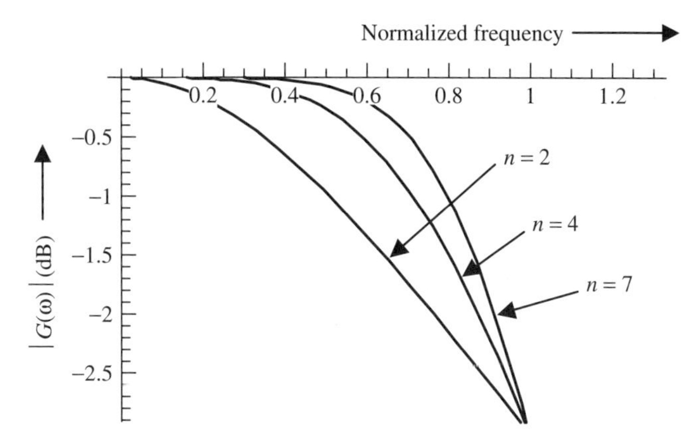

Figure 9.18 Characteristics of maximally flat low-pass filters.

or

$$\frac{L}{10} = \log_{10}\left[1 + \zeta\left(\frac{\omega}{\omega_c}\right)^{2n}\right]$$

or

$$\zeta\left(\frac{\omega}{\omega_c}\right)^{2n} = 10^{L/10} - 1 \tag{9.2.7}$$

Figure 9.18 illustrates the passband characteristics of this type of filter for three different values of n. The power-loss ratio at its band edge is assumed to be -3 dB, and therefore ζ is equal to unity. It shows that the passband becomes flatter with a sharper cutoff when its order n is increased. However, this relationship is not a linear one. This change in characteristics is significant for lower values of n.

Equation (9.2.7) can be used for determining the ζ and n values of a filter as follows. If the insertion loss at the band edge ($\omega = \omega_c$) is specified as $L = L_c$, (9.2.7) yields

$$\zeta = 10^{L_c/10} - 1 \tag{9.2.8}$$

Similarly, the required order n (and hence, number of elements required) of a filter can be determined for the L specified at a given stopband frequency. It is found as follows:

$$n = \frac{1}{2} \times \frac{\log_{10}(10^{L/10} - 1) - \log_{10}\zeta}{\log_{10}(\omega/\omega_c)} \tag{9.2.9}$$

Chebyshev Filter

A filter with a sharper cutoff can be realized at the cost of flatness in its passband. Chebyshev filters possess ripples in the passband but provide a sharp transition into the stopband. In this case, Chebyshev polynomials are used to represent the insertion loss. Mathematically,

$$|G(\omega)| = \frac{1}{\sqrt{1 + \zeta T_m^2(\overline{\omega})}} \qquad m = 1, 2, 3, \ldots \tag{9.2.10}$$

where ζ is a constant, $\overline{\omega}$ is normalized frequency, and $T_m(\omega)$ is a Chebyshev polynomial of the first kind and degree m, defined in (7.5.2). Figure 9.19 shows the frequency response of a typical Chebyshev filter for $m = 7$. It assumes that ripples up to -3 dB in its passband are acceptable. A comparison of this characteristic to that for a seventh-order Butterworth filter shown in Figure 9.18 indicates that it has a much sharper transition from passband to stopband. However, it is achieved at the cost of ripples in its passband.

As before, the insertion loss of a Chebyshev filter is found as follows:

$$L = -20\log_{10}\left[\frac{1}{\sqrt{1 + \zeta T_m^2(\omega/\omega_c)}}\right] = 10\log_{10}\left(1 + \zeta T_m^2 \frac{\omega}{\omega_c}\right)$$

or

$$L = \begin{cases} 10\log_{10}\left[1 + \zeta\cos^2\left(m\cos^{-1}\dfrac{\omega}{\omega_c}\right)\right] & 0 \le \omega \le \omega_c \\ 10\log_{10}\left[1 + \zeta\cosh^2\left(m\cosh^{-1}\dfrac{\omega}{\omega_c}\right)\right] & \omega_c < \omega \end{cases} \tag{9.2.11}$$

where

$$\zeta = 10^{0.1 \times G_r} - 1 \tag{9.2.12}$$

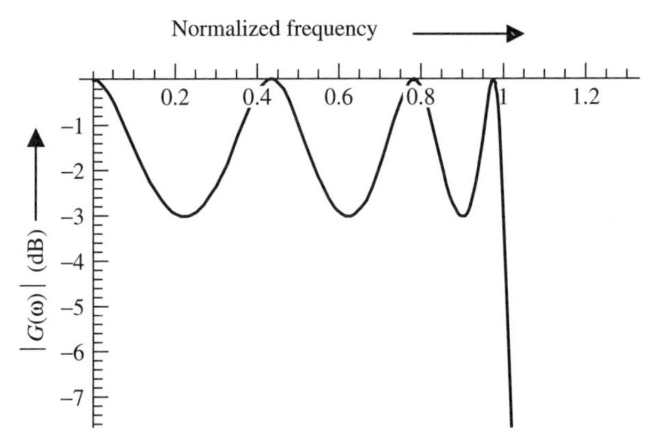

Figure 9.19 Characteristics of a low-pass Chebyshev filter for $m = 7$.

G_r is the ripple amplitude in decibels. The order m (and hence, number of elements) of a Chebyshev filter can be found from its characteristics as follows:

$$m = \frac{\cosh^{-1}\sqrt{(10^{0.1\times L} - 1)/(10^{0.1\times G_r} - 1)}}{\cosh^{-1}(\omega/\omega_c)} \tag{9.2.13}$$

where L is required insertion loss in decibels at a specified frequency ω.

Example 9.5 It is desired to design a maximally flat low-pass filter with at least 15 dB attenuation at $\omega = 1.3\,\omega_c$ and -3 dB at its band edge. How many elements will be required for this filter? If a Chebyshev filter is used with a 3-dB ripple in its passband, find the number of circuit elements.

SOLUTION From (9.2.8),

$$\zeta = 10^{L_c/10} - 1 = 10^{0.3} - 1 = 1$$

and from (9.2.9) we have

$$n = \frac{1}{2} \times \frac{\log_{10}(10^{L/10} - 1) - \log_{10}\zeta}{\log_{10}(\omega/\omega_c)} = 0.5 \times \frac{\log_{10}(10^{1.5} - 1)}{\log_{10}(1.3)} = 6.52$$

Therefore, seven elements will be needed for this maximally flat filter.

In the case of a Chebyshev filter, (9.2.13) gives

$$m = \frac{\cosh^{-1}\sqrt{(10^{0.1\times L} - 1)/(10^{0.1\times G_r} - 1)}}{\cosh^{-1}(\omega/\omega_c)} = \frac{\cosh^{-1}\sqrt{10^{1.5} - 1}}{\cosh^{-1}(1.3)} = 3.17$$

Hence, it will require only three elements. The characteristics of these two filters are illustrated in Figure 9.20.

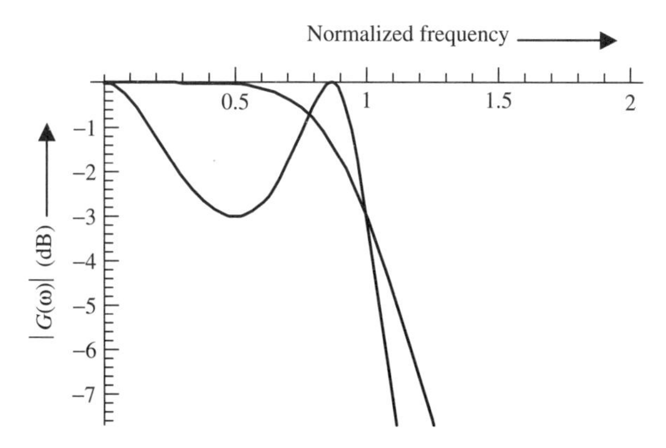

Figure 9.20 Characteristics of low-pass Butterworth and Chebyshev filters.

Low-Pass Filter Synthesis

The scope of this book does not include the theory of passive filter synthesis. Several excellent references are available in the literature for those who may be interested in this. The design procedure presented here is based on a doubly terminated low-pass ladder network, as shown in Figure 9.21. The g notation used in the figure signifies the roots of an nth-order transfer function that govern its characteristics. These represent the normalized reactance values of filter elements with a cutoff frequency $\omega_c = 1$. The source resistance is represented by g_0, and the load is g_{n+1}. The filter is made up of series inductors and shunt capacitors that are in the form of cascaded T-networks. Another possible configuration is the cascaded π-network that is obtained after replacing g_1 by a short circuit, connecting a capacitor across the load g_{n+1}, and renumbering filter elements 1 through n. Elements of this filter are determined from the n roots of the transfer function.

The transfer function is selected according to the pass- and stopband characteristics desired. Normalized values of elements are then found from the roots of that transfer function. These values are then adjusted according to the desired cutoff frequency and the source and load resistance. Design procedures for the maximally flat and the Chebyshev filters can be summarized as follows.

Assume that the cutoff frequency is given as follows:

$$\omega_c = 1 \tag{9.2.14}$$

Butterworth and Chebyshev filters can then be designed using the following formulas.

For a Butterworth filter,

$$g_0 = g_{n+1} = 1 \tag{9.2.15}$$

and

$$g_p = 2 \sin \frac{(2p-1)\pi}{2n} \qquad p = 1, 2, \cdots, n \tag{9.2.16}$$

Element values computed from (9.2.15) and (9.2.16) are given in Table 9.4 for n up to 7.

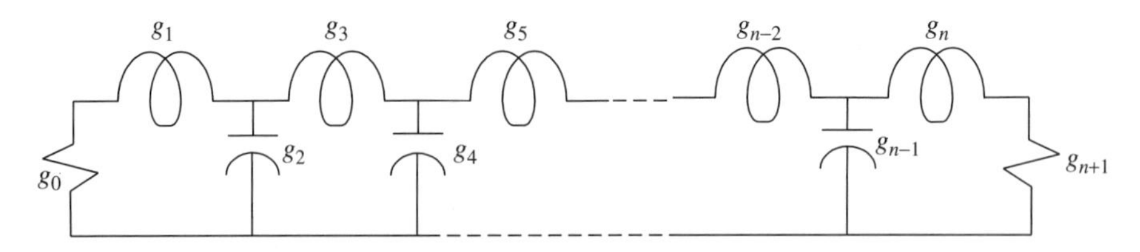

Figure 9.21 Low-pass ladder network prototype.

TABLE 9.4 Element Values for Low-Pass Binomial Filter Prototypes ($g_0 = 1$, $\omega_c = 1$)

n	g_1	g_2	g_3	g_4	g_5	g_6	g_7	g_8
1	2.0000	1.0000						
2	1.4142	1.4142	1					
3	1.0000	2.0000	1.0000	1.0000				
4	0.7654	1.8478	1.8478	0.7654	1.0000			
5	0.6180	1.6180	2.0000	1.6180	0.6180	1		
6	0.5176	1.4142	1.9319	1.9319	1.4142	0.5176	1	
7	0.4450	1.2470	1.8019	2.0000	1.8019	1.2470	0.445	1.0000

For a Chebyshev filter,

$$g_0 = 1 \tag{9.2.17}$$

$$g_{m+1} = \begin{cases} 1 & m \text{ is an odd number} \\ \coth \dfrac{\xi}{4} & m \text{ is an even number} \end{cases} \tag{9.2.18}$$

$$g_1 = \frac{2a_1}{\chi} \tag{9.2.19}$$

and

$$g_p = \frac{4a_{(p-1)}a_p}{b_{(p-1)}g_{(p-1)}} \qquad p = 2, 3, \cdots, m \tag{9.2.20}$$

where

$$\xi = \ln\left(\coth \frac{G_r}{17.37}\right) \tag{9.2.21}$$

$$\chi = \sinh \frac{\xi}{2m} \tag{9.2.22}$$

$$a_p = \sin \frac{(2p-1)\pi}{2m} \tag{9.2.23}$$

and

$$b_p = \chi^2 + \sin^2 \frac{p\pi}{m} \tag{9.2.24}$$

The element values for several low-pass Chebyshev filters that are computed from (9.2.17) to (9.2.24) are given in Tables 9.5 to 9.9.

TABLE 9.5 Element Values for Low-Pass Chebyshev Filter Prototypes ($g_0 = 1$, $\omega_c = 1$, and 0.1 dB ripple)

m	g_1	g_2	g_3	g_4	g_5	g_6	g_7	g_8
1	0.3053	1.000						
2	0.8431	0.6220	1.3554					
3	1.0316	1.1474	1.0316	1.0000				
4	1.1088	1.3062	1.7704	0.8181	1.3554			
5	1.1468	1.3712	1.9750	1.3712	1.1468	1.0000		
6	1.1681	1.4040	2.0562	1.5171	1.9029	0.8618	1.3554	
7	1.1812	1.4228	2.0967	1.5734	2.0967	1.4228	1.1812	1.0000

TABLE 9.6 Element Values for Low-Pass Chebyshev Filter Prototypes ($g_0 = 1$, $\omega_c = 1$, and 0.5 dB ripple)

m	g_1	g_2	g_3	g_4	g_5	g_6	g_7	g_8
1	0.6987	1.0000						
2	1.4029	0.7071	1.9841					
3	1.5963	1.0967	1.5963	1.0000				
4	1.6703	1.1926	2.3662	0.8419	1.9841			
5	1.7058	1.2296	2.5409	1.2296	1.7058	1.0000		
6	1.7254	1.2479	2.6064	1.3136	2.4759	0.8696	1.9841	
7	1.7373	1.2582	2.6383	1.3443	2.6383	1.2582	1.7373	1.0000

TABLE 9.7 Element Values for Low-Pass Chebyshev Filter Prototypes ($g_0 = 1$, $\omega_c = 1$, and 1.0 dB ripple)

m	g_1	g_2	g_3	g_4	g_5	g_6	g_7	g_8
1	1.0178	1.0000						
2	1.8220	0.6850	2.6599					
3	2.0237	0.9941	2.0237	1.0000				
4	2.0991	1.0644	2.8312	0.7892	2.6599			
5	2.1350	1.0911	3.0010	1.0911	2.1350	1.0000		
6	2.1547	1.1041	3.0635	1.1518	2.9368	0.8101	2.6599	
7	2.1666	1.1115	3.0937	1.1735	3.0937	1.1115	2.1666	1.0000

Scaling the Prototype to the Desired Cutoff Frequency and Load

- *Frequency scaling.* For scaling the frequency from 1 to ω_c, divide all normalized g values that represent capacitors or inductors by the desired cutoff frequency expressed in radians per second. Resistors are excluded from this operation.

TABLE 9.8 Element Values for Low-Pass Chebyshev Filter Prototypes ($g_0 = 1$, $\omega_c = 1$, and 2.0 dB ripple)

m	g_1	g_2	g_3	g_4	g_5	g_6	g_7	g_8
1	1.5297	1.0000						
2	2.4883	0.6075	4.0957					
3	2.7108	0.8326	2.7108	1.0000				
4	2.7926	0.8805	3.6064	0.6818	4.0957			
5	2.8311	0.8984	3.7829	0.8984	2.8311	1.0000		
6	2.8522	0.9071	3.8468	0.9392	3.7153	0.6964	4.0957	
7	2.8651	0.9120	3.8776	0.9536	3.8776	0.9120	2.8651	1.0000

TABLE 9.9 Element Values for Low-Pass Chebyshev Filter Prototypes ($g_0 = 1$, $\omega_c = 1$, and 3.0 dB ripple)

m	g_1	g_2	g_3	g_4	g_5	g_6	g_7	g_8
1	1.9954	1.0000						
2	3.1014	0.5339	5.8095					
3	3.3489	0.7117	3.3489	1.0000				
4	3.4391	0.7483	4.3473	0.5920	5.8095			
5	3.4815	0.7619	4.5378	0.7619	3.4815	1.0000		
6	3.5047	0.7685	4.6063	0.7929	4.4643	0.6033	5.8095	
7	3.5187	0.7722	4.6392	0.8038	4.6392	0.7722	3.5187	1.0000

- *Impedance scaling.* To scale g_0 and g_{n+1} to X Ω from unity, multiply all g values that represent resistors or inductors by X. On other hand, divide those g values representing capacitors by X.

Example 9.6 Design a Butterworth filter with a cutoff frequency of 10 MHz and an insertion loss of 30 dB at 40 MHz. It is to be used between a 50-Ω load and a generator with an internal resistance of 50 Ω.

SOLUTION From (9.2.9), we have

$$n = \frac{1}{2} \times \frac{\log_{10}(10^{30/10} - 1) - \log_{10}(10^{3/10} - 1)}{\log_{10}(40/10)}$$

$$= \frac{1}{2} \times \frac{\log_{10}(10^3 - 1) - \log_{10}(1.9953 - 1)}{\log_{10}(4)} \approx \frac{0.5 \times 3}{0.6} \approx 2.5$$

Since the number of elements must be an integer, selecting $n = 3$ will provide more than the attenuation specified at 40 MHz. From (9.2.15) and (9.2.16),

$$g_0 = g_4 = 1$$

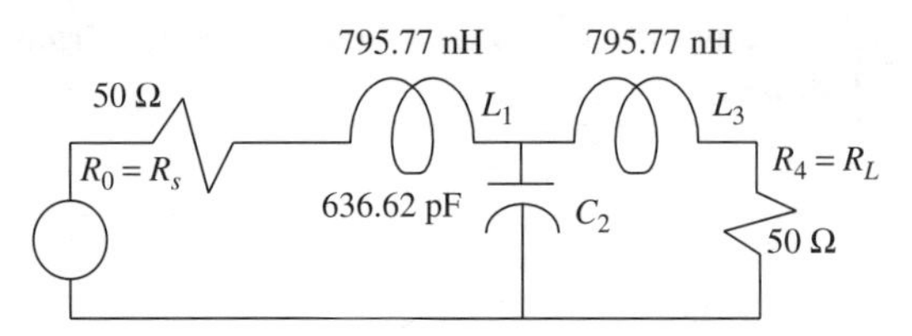

Figure 9.22 Maximally flat low-pass filter obtained in Example 9.6.

$$g_1 = 2\sin\frac{\pi}{6} = 1$$

$$g_2 = 2\sin\frac{\pi}{2} = 2$$

and

$$g_3 = 2\sin\frac{5\pi}{6} = 1$$

If we use the circuit arrangement illustrated in Figure 9.22, element values can be scaled to match the frequency and load resistance. Using the two rules, these values are found as follows:

$$L_1 = L_3 = 50 \times \frac{1}{2\pi \times 10^7}\text{H} = 795.77\,\text{nH}$$

and

$$C_2 = \frac{1}{50} \times \frac{2}{2\pi \times 10^7}\text{F} = 636.62\,\text{pF}$$

The frequency response of this filter is shown in Figure 9.23. It indicates a 6-dB insertion loss in the passband of the filter. This happens because the source resistance is considered a part of this circuit. In other words, this represents voltage across R_L with respect to source voltage, not to input of the circuit. Since source resistance is equal to the load, there is equal division of source voltage that results in -6 dB. Another 3-dB loss at the band edge shows a total of about 9 dB at 10 MHz. This characteristic shows that there is an insertion loss of over 40 dB at 40 MHz. It is clearly more than 30 dB from the band edge, as required by the example. Also, there is a nonlinear phase variation in its passband.

Alternatively, we can use a circuit topology as shown in Figure 9.24. In that case, component values are found by using the scaling rules:

$$L_2 = 50 \times \frac{1}{2\pi \times 10^7} \times 2 = 0.15915 \times 10^{-5}\,\text{H} = 1591.5\,\text{nH}$$

and

$$C_1 = C_3 = \frac{1}{50} \times \frac{1}{2\pi \times 10^7} \times 1 = \frac{10^{-9}}{\pi}\text{F} = 318.31\,\text{pF}$$

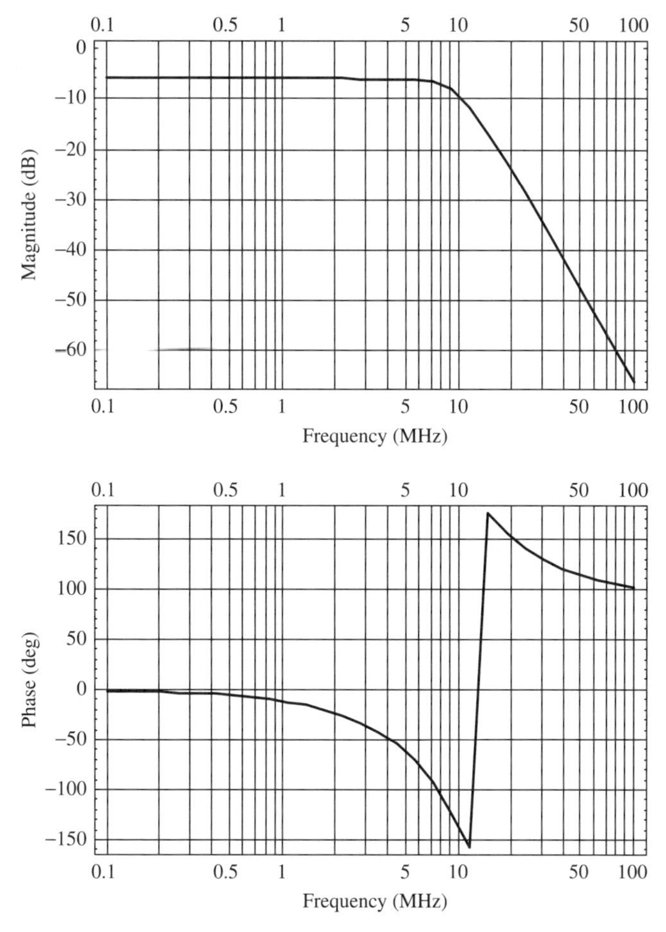

Figure 9.23 Characteristics of the low-pass filter shown in Figure 9.22.

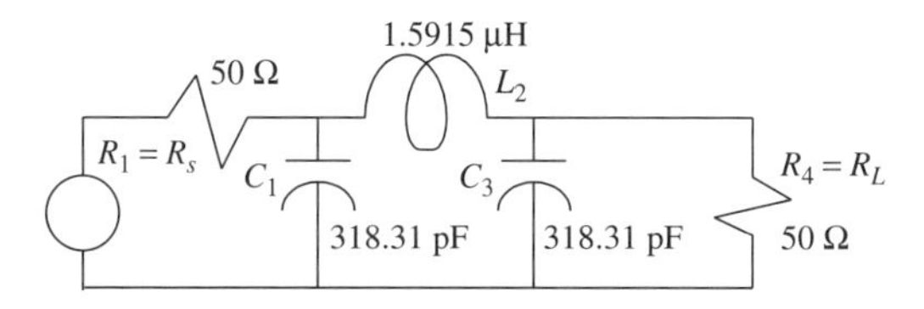

Figure 9.24 Alternative circuit for Example 9.6.

The frequency response of this circuit is illustrated in Figure 9.25. It is identical to that shown in Figure 9.23 for the earlier circuit.

Example 9.7 Design a low-pass Chebyshev filter that may have ripples no more than 0.01 dB in its passband. The filter must pass all frequencies up to 100 MHz

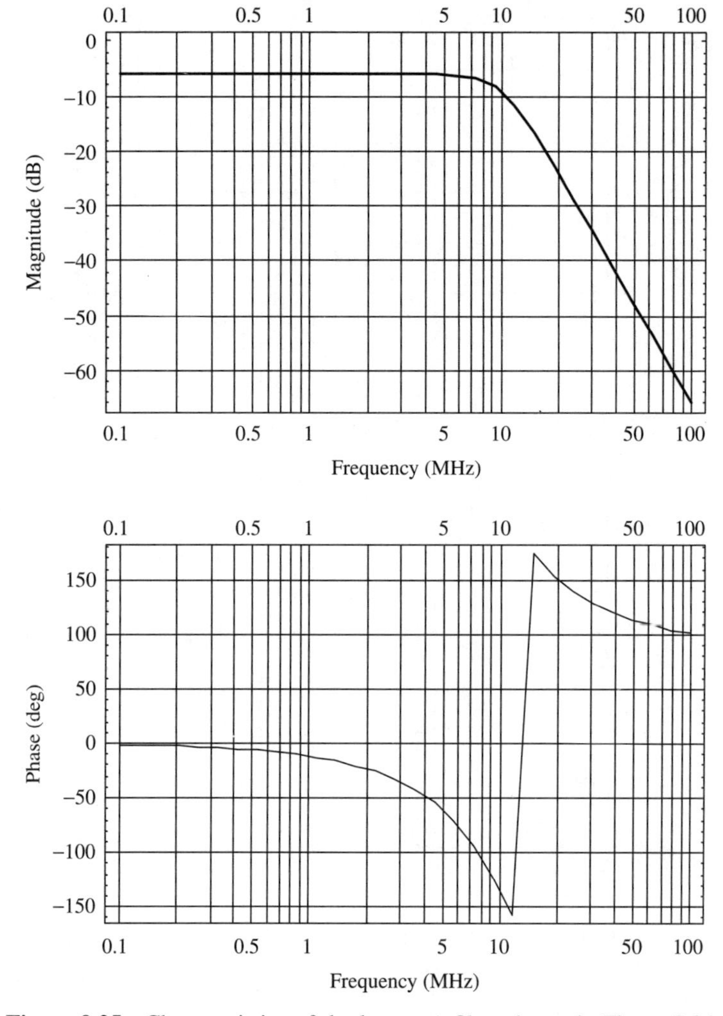

Figure 9.25 Characteristics of the low-pass filter shown in Figure 9.24.

and attenuate the signal at 400 MHz by at least 5 dB. The load and the source resistance are of 75 Ω each.

SOLUTION Since

$$G_r = 0.01\,\text{dB}$$

and

$$L\left(\frac{400}{100}\right) = 5\,\text{dB}$$

(9.2.13) gives

$$m = \frac{\cosh^{-1}\sqrt{(10^{0.5}-1)/(10^{0.001}-1)}}{\cosh^{-1}(4)} = 2$$

Since we want a symmetrical filter with 75 Ω on each side, we select $m = 3$ (an odd number). It will provide more than 5 dB of insertion loss at 400 MHz. The g values can now be determined from (9.2.17) to (9.2.24), as follows. From (9.2.23),

$$a_1 = \sin\frac{\pi}{6} = 0.5$$

$$a_2 = \sin\frac{\pi}{2} = 1$$

and

$$a_3 = \sin\frac{5\pi}{6} = 0.5$$

Next, from (9.2.21), (9.2.22), and (9.2.24), we get

$$\xi = \ln\left(\coth\frac{0.01}{17.37}\right) = 7.5$$

$$\chi = \sinh\frac{7.5}{6} = 1.6019$$

$$b_1 = 1.6019^2 + \sin^2\frac{\pi}{3} = 3.316$$

$$b_2 = 1.6019^2 + \sin^2\frac{2\pi}{3} = 3.316$$

and

$$b_3 = 1.6019^2 + \sin^2\frac{3\pi}{3} = 2.566$$

However, b_3 is not needed in further calculations.

Finally, from (9.2.17) and (9.2.20),

$$g_0 = g_4 = 1$$

$$g_1 = \frac{2 \times 0.5}{1.6019} = 0.62425$$

$$g_2 = \frac{4 \times 0.5 \times 1}{3.316 \times 0.62425} = 0.9662$$

and

$$g_3 = \frac{4 \times 1 \times 0.5}{3.316 \times 0.9662} = 0.62425$$

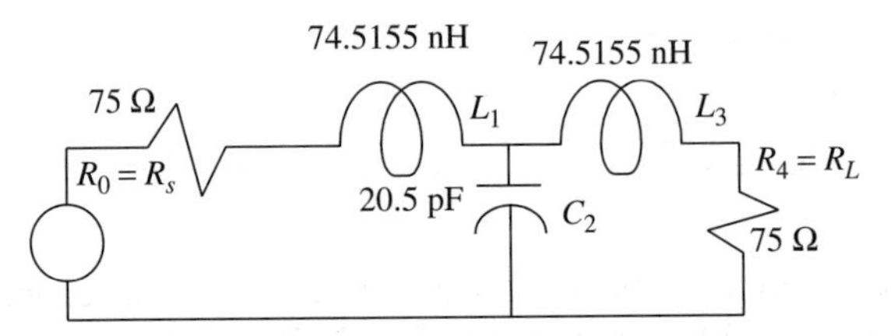

Figure 9.26 Low-pass Chebyshev filter circuit obtained in Example 9.7.

For the circuit topology of Figure 9.26, element values are found after applying the scaling rules as follows:

$$L_1 = L_3 = \frac{75 \times 0.62425}{2\pi \times 10^8}\,\text{H} = 74.5155\,\text{nH}$$

and

$$C_2 = \frac{1}{75} \times \frac{1}{2\pi \times 10^8} \times 0.9662\,\text{F} = 20.5\,\text{pF}$$

The frequency response of this filter is illustrated in Figure 9.27. As expected, there is over a 20-dB insertion loss at 400 MHz compared with its passband. Note that the magnitude of ripple allowed is so small that it does not show up in this figure. However, it is present there, as shown on an expanded scale in Figure 9.28.

Figure 9.28 shows the passband characteristics of this Chebyshev filter. Since the scale-of-magnitude plot is now expanded, the passband ripple is clearly visible. As expected, the ripple stays between −6.02 and −6.03 dB. Phase variation in this passband ranges from 0° to −70°. The band edge of this filter can be sharpened further either by using a higher-order filter or by allowing larger magnitudes of the ripple in its passband. The higher-order filter will require more elements because the two are directly related. The next example illustrates that a sharper transition between the bands can be obtained even with a three-element filter if ripple magnitudes up to 3 dB are acceptable.

Example 9.8 Reconsider Example 9.7 to design a low-pass filter that exhibits the Chebyshev response with 3-dB ripple in its passband, $m = 3$, and a cutoff frequency of 100 MHz. The filter must have 75 Ω at both its input and output ports.

SOLUTION From (8.2.23),

$$a_1 = \sin\frac{\pi}{6} = 0.5$$

$$a_2 = \sin\frac{\pi}{2} = 1$$

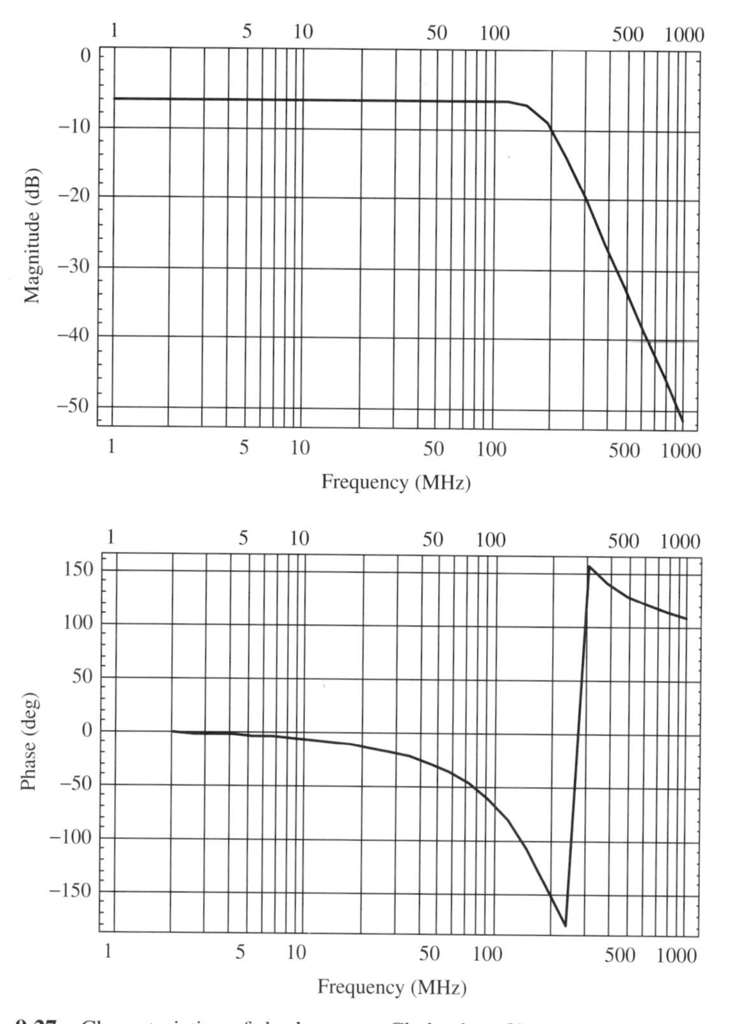

Figure 9.27 Characteristics of the low-pass Chebyshev filter shown in Figure 9.26.

and

$$a_3 = \sin \frac{5\pi}{6} = 0.5$$

Now, from (9.2.21), (9.2.22), and (9.2.24),

$$\xi = \ln\left(\coth \frac{3}{17.37}\right) = 1.7661$$

$$\chi = \sinh \frac{1.7661}{6} = 0.2986$$

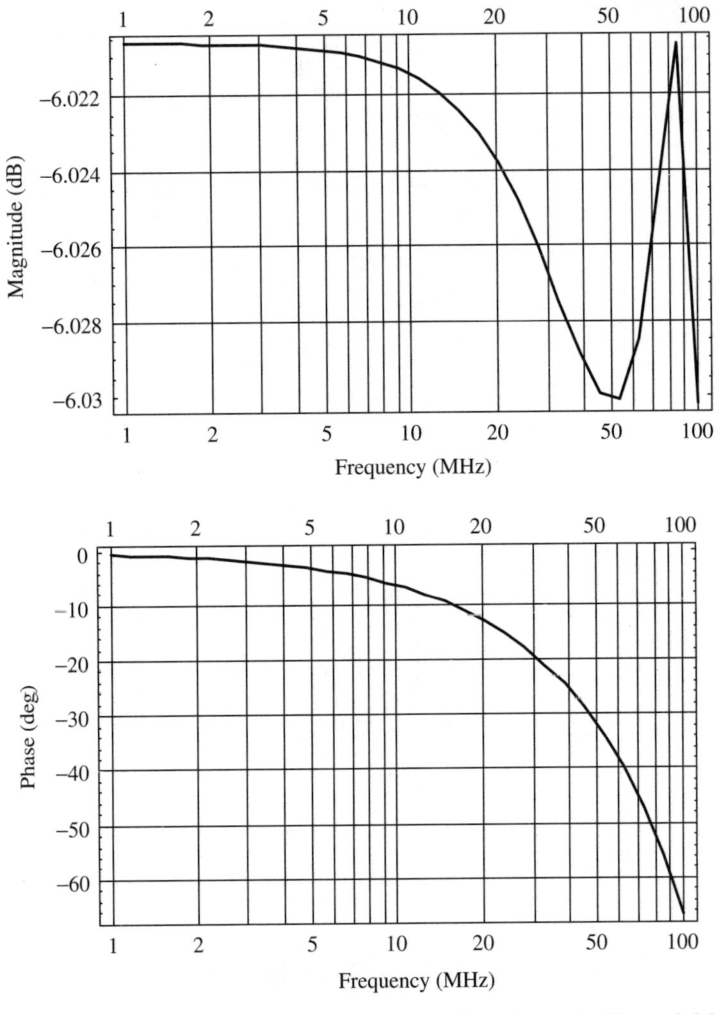

Figure 9.28 Passband characteristics of the filter shown in Figure 9.26.

$$b_1 = 0.2986^2 + \sin^2 \frac{\pi}{3} = 0.8392$$

$$b_2 = 0.2986^2 + \sin^2 \frac{2\pi}{3} = 0.8392$$

and

$$b_3 = 0.2986^2 + \sin^2 \frac{3\pi}{3} = 0.0892$$

However, b_3 is not needed.

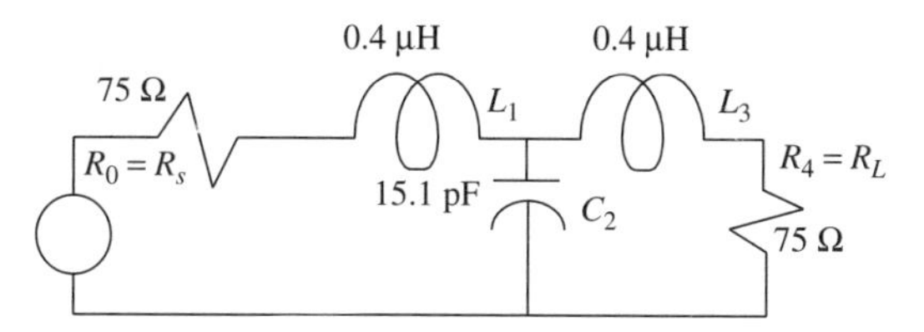

Figure 9.29 Low-pass Chebyshev filter circuit obtained in Example 9.8.

The g values for the filter can now be determined from (9.2.17) to (9.2.20), as follows:

$$g_0 = g_4 = 1$$

$$g_1 = \frac{2 \times 0.5}{0.2986} = 3.349$$

$$g_2 = \frac{4 \times 0.5 \times 1}{3.349 \times 0.8392} = 0.7116$$

and

$$g_3 = \frac{4 \times 1 \times 0.5}{0.7116 \times 0.8392} = 3.349$$

These g values are also listed in Table 9.9 for $m = 3$.

Elements of the circuit shown in Figure 9.29 are determined using the scaling rules, as follows (see Figure 9.30 for the characteristics):

$$L_1 = L_3 = \frac{75 \times 3.349}{2\pi \times 10^8} \text{H} = 0.4\,\mu\text{H}$$

and

$$C_2 = \frac{1}{75} \times \frac{1}{2\pi \times 10^8} \times 0.7116\,\text{F} = 15.1\,\text{pF}$$

High-Pass Filter

As mentioned earlier, a high-pass filter can be designed by transforming the low-pass prototype. This frequency transformation is illustrated in Figure 9.31. As illustrated, an ideal low-pass filter passes all signals up to the normalized frequency of unity with zero insertion loss, whereas it attenuates higher frequencies completely. On the other hand, a high-pass filter must pass all signals with frequencies higher than its cutoff frequency ω_c and stop the signals that have lower frequencies. Therefore, the following frequency transformation formula will transform a low-pass filter to a high-pass filter:

$$\overline{\omega} = -\frac{\omega_c}{\omega} \tag{9.2.25}$$

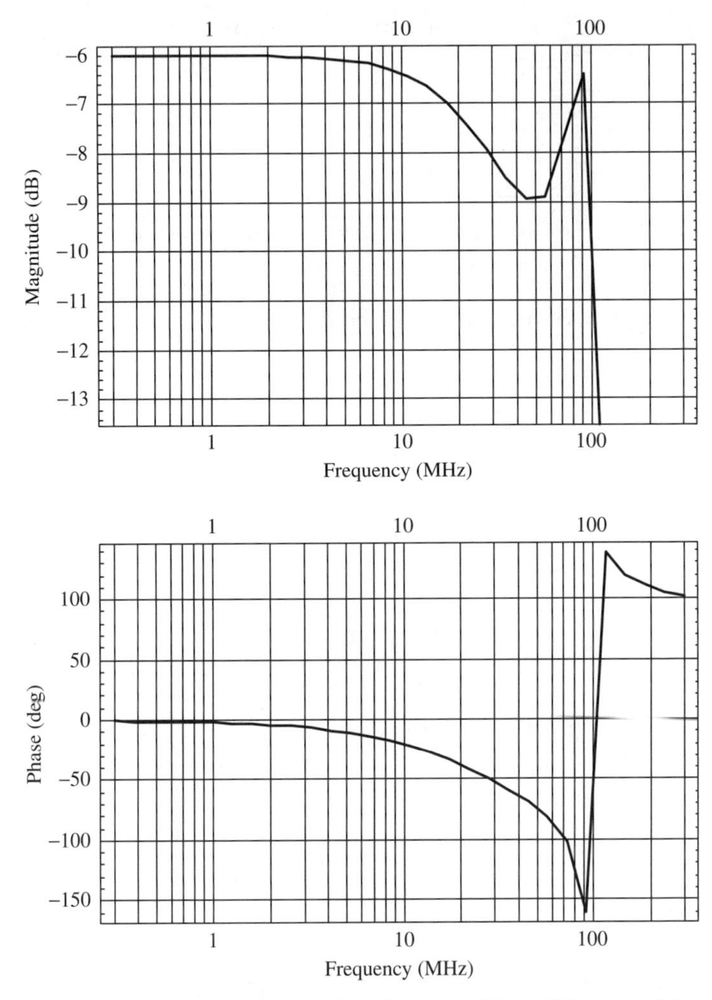

Figure 9.30 Characteristics of the low-pass filter of Example 9.8.

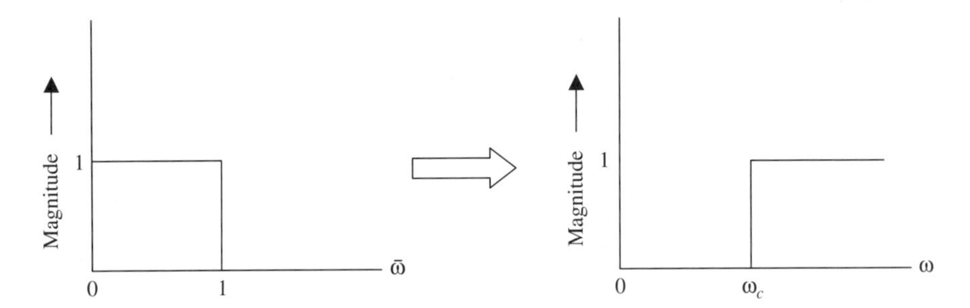

Figure 9.31 Transformation from a low-pass to a high-pass characteristic.

Thus, inductors and capacitors will change their places. Inductors will replace the shunt capacitors of the low-pass filter and capacitors will be connected in series, in place of inductors. These elements are determined as follows:

$$C_{\mathrm{HP}} = \frac{1}{\omega_c g_L} \tag{9.2.26}$$

and

$$L_{\mathrm{HP}} = \frac{1}{\omega_c g_C} \tag{9.2.27}$$

Capacitor C_{HP} and inductor L_{HP} can now be scaled as required by the load and source resistance.

Example 9.9 Design a high-pass Chebyshev filter with passband ripple magnitude less than 0.01 dB. It must pass all frequencies over 100 MHz and exhibit at least 5 dB of attenuation at 25 MHz. Assume that the load and source resistances are at 75 Ω each.

SOLUTION The low-pass filter designed in Example 9.7 provides the initial data for this high-pass filter. With $m = 3$, $g_L = 0.62425$, and $g_C = 0.9662$, we find from (9.2.26) and (9.2.27) that

$$C_{\mathrm{HP}} = \frac{1}{2\pi \times 10^8 \times 0.62425}\,\mathrm{F} = 2.5495\,\mathrm{nF}$$

and

$$L_{\mathrm{HP}} = \frac{1}{2\pi \times 10^8 \times 0.9662}\,\mathrm{H} = 1.6472\,\mathrm{nH}$$

Now, applying the resistance scaling, we get

$$C_1 = C_3 = \frac{2.5495}{75}\,\mathrm{nF} = 33.9933\,\mathrm{pF} = 34\,\mathrm{pF}$$

and

$$L_2 = 75 \times 1.6472 = 123.5\,\mathrm{nH}$$

The resulting high-pass Chebyshev filter is shown in Figure 9.32. Its frequency response is illustrated in Figure 9.33. As before, the source resistance is considered a part of the filter circuit and therefore its passband shows a 6-dB insertion loss.

Bandpass Filter

A bandpass filter can be designed by transforming the low-pass prototype as illustrated in Figure 9.34. Here, an ideal low-pass filter passes all signals up to

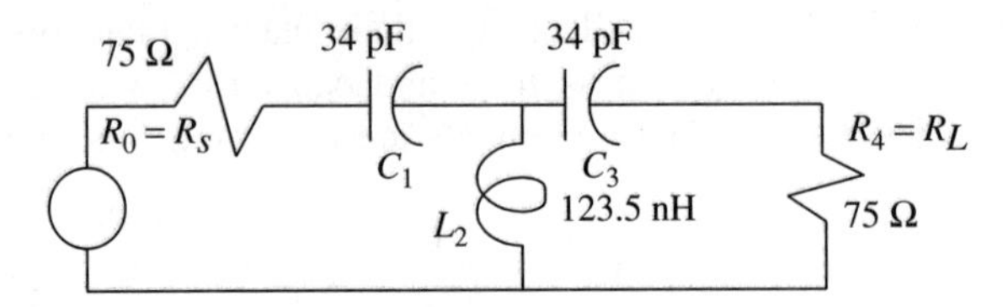

Figure 9.32 High-pass Chebyshev filter obtained for Example 9.8.

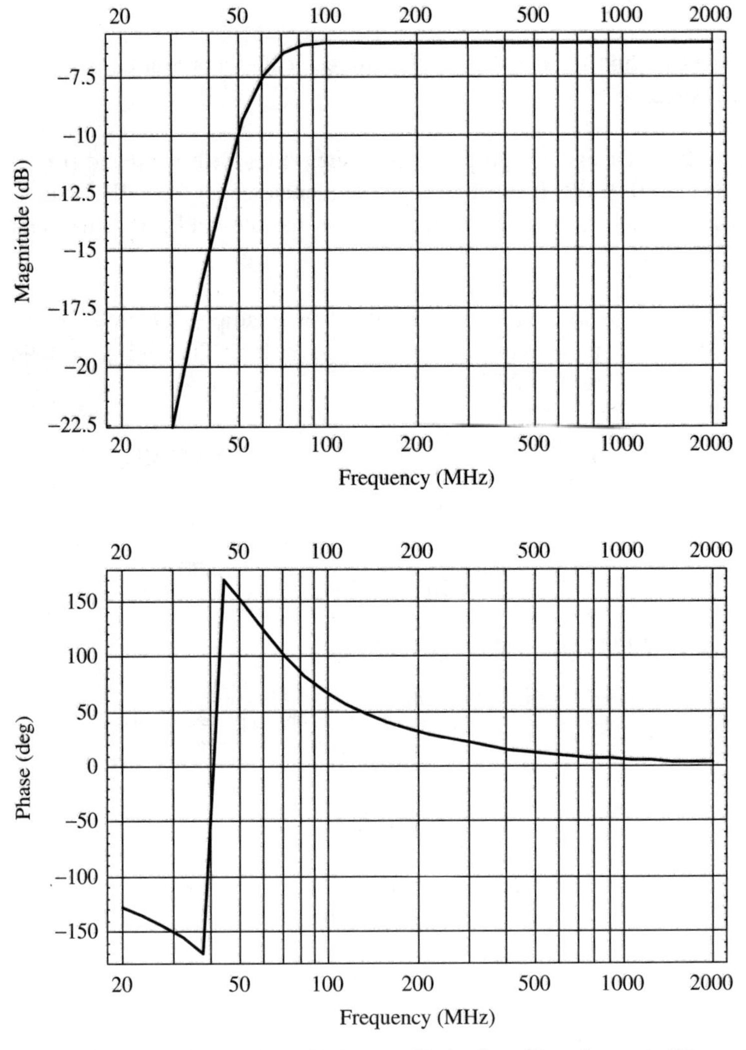

Figure 9.33 Characteristics of the high-pass Chebyshev filter shown in Figure 9.32.

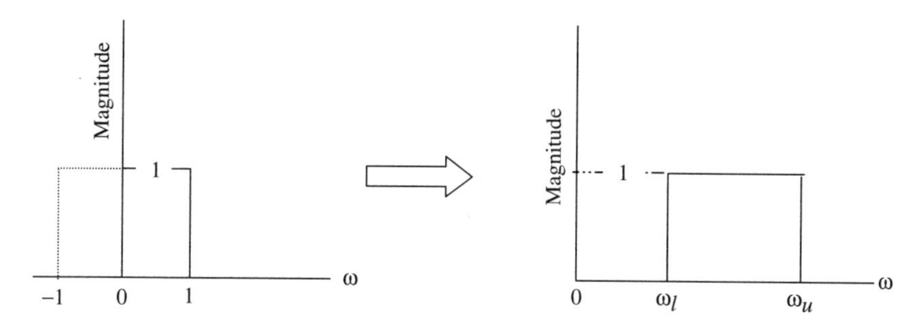

Figure 9.34 Transformation from a low-pass to a bandpass characteristic.

the normalized frequency of unity with zero insertion loss, whereas it attenuates higher frequencies completely. On the other hand, a bandpass filter must pass all signals with frequencies between ω_l and ω_u and stop the signals that are outside this frequency band. Hence, the following frequency relation transforms the response of a low-pass filter to a bandpass:

$$\overline{\omega} = \frac{1}{\omega_u - \omega_l} \frac{\omega^2 - \omega_0^2}{\omega} \tag{9.2.28}$$

where

$$\omega_0 = \sqrt{\omega_l \times \omega_u} \tag{9.2.29}$$

This transformation replaces the series inductor of low-pass prototype with an inductor L_{BP1} and a capacitor C_{BP1} that are connected in series. The component values are determined as follows:

$$C_{\text{BP1}} = \frac{\omega_u - \omega_l}{\omega_0^2 g_L} \tag{9.2.30}$$

and

$$L_{\text{BP1}} = \frac{g_L}{\omega_u - \omega_l} \tag{9.2.31}$$

Also, the capacitor C_{BP2} that is connected in parallel with an inductor L_{BP2} will replace the shunt capacitor of the low-pass prototype. These elements are found as follows:

$$L_{\text{BP2}} = \frac{\omega_u - \omega_l}{\omega_0^2 g_C} \tag{9.2.32}$$

and

$$C_{\text{BP2}} = \frac{g_C}{\omega_u - \omega_l} \tag{9.2.33}$$

These elements need to be scaled further as desired by the load and source resistance.

Example 9.10 Design a bandpass Chebyshev filter that exhibits no more than 0.01-dB ripples in its passband. It must pass signals in the frequency band 10 to 40 MHz with zero insertion loss. Assume that the load and source resistances are at 75 Ω each.

SOLUTION From (9.2.29), we have

$$f_0 = \sqrt{f_l f_u} = \sqrt{10^7 \times 40 \times 10^6} = 20 \times 10^6\,\text{Hz}$$

Now, using (9.2.30) to (9.2.33) with $g_L = 0.62425$ and $g_C = 0.9662$ from Example 9.7, we get

$$C_{\text{BP1}} = \frac{2\pi \times 10^6(40 - 10)}{(2\pi \times 20 \times 10^6)^2 \times 0.62424}\text{F} = 19.122\,\text{nF}$$

$$L_{\text{BP1}} = \frac{0.62424}{2\pi \times 30 \times 10^6}\text{H} = 3.3116\,\text{nH}$$

$$L_{\text{BP2}} = \frac{2\pi \times 10^6(40 \quad 10)}{(2\pi \times 20 \times 10^6)^2 \times 0.9662}\text{H} = 12.354\,\text{nH}$$

and

$$C_{\text{BP2}} = \frac{0.9662}{2\pi \times 30 \times 10^6}\text{F} = 5.1258\,\text{nF}$$

Next, values of elements are determined after applying the load and source resistance scaling. Hence,

$$C_1 = C_3 = \frac{19.122}{75}\text{nF} = 254.96\,\text{pF} \approx 255\,\text{pF}$$

$$L_1 = L_3 = 75 \times 3.3116\,\text{nH} = 0.2484\,\mu\text{H}$$

$$L_2 = 75 \times 12.354\,\text{nH} = 0.9266\,\mu\text{H}$$

and

$$C_2 = \frac{5.1258}{75}\text{nF} = 68.344\,\text{pF}$$

The resulting filter circuit is shown in Figure 9.35, and its frequency response is depicted in Figure 9.36. It illustrates that we indeed have transformed the low-pass prototype to a bandpass filter for the 10- to 40-MHz frequency band. As before, the source resistance is considered a part of the filter circuit, and therefore its passband shows a 6-dB insertion loss. Figure 9.37 shows its passband characteristics with expanded scales. It indicates that the ripple magnitude is limited to 0.01 dB, as desired.

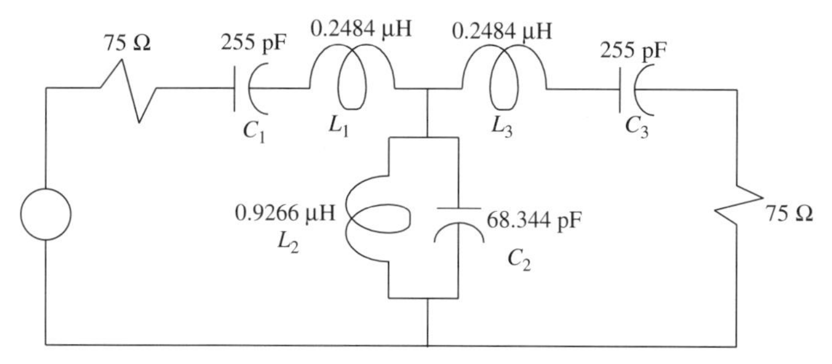

Figure 9.35 Bandpass filter circuit for Example 9.10.

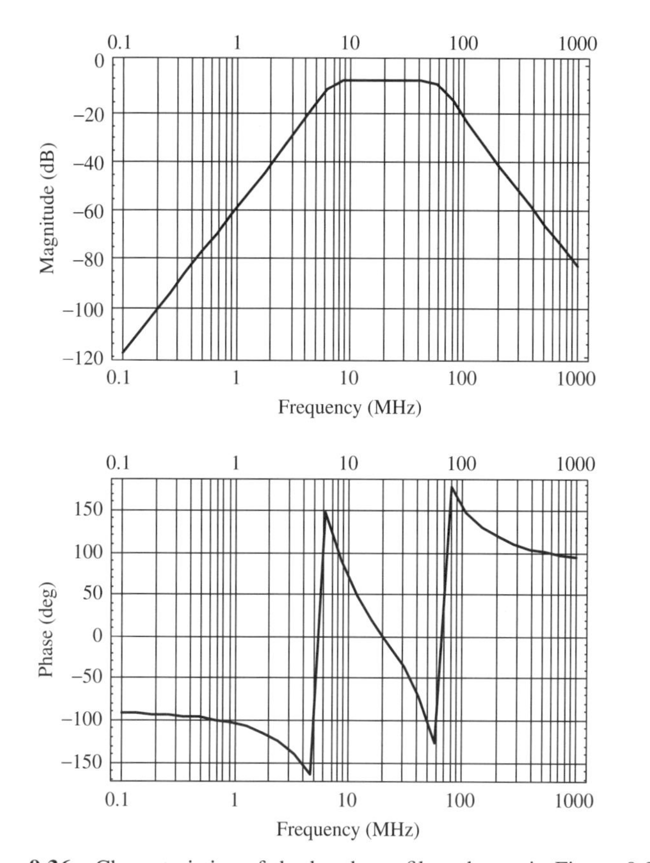

Figure 9.36 Characteristics of the bandpass filter shown in Figure 9.35.

Bandstop Filter

A bandstop filter can be realized by transforming the low-pass prototype as illustrated in Figure 9.38. Here, an ideal low-pass filter passes all signals up to the normalized frequency of unity with zero insertion loss, whereas it completely

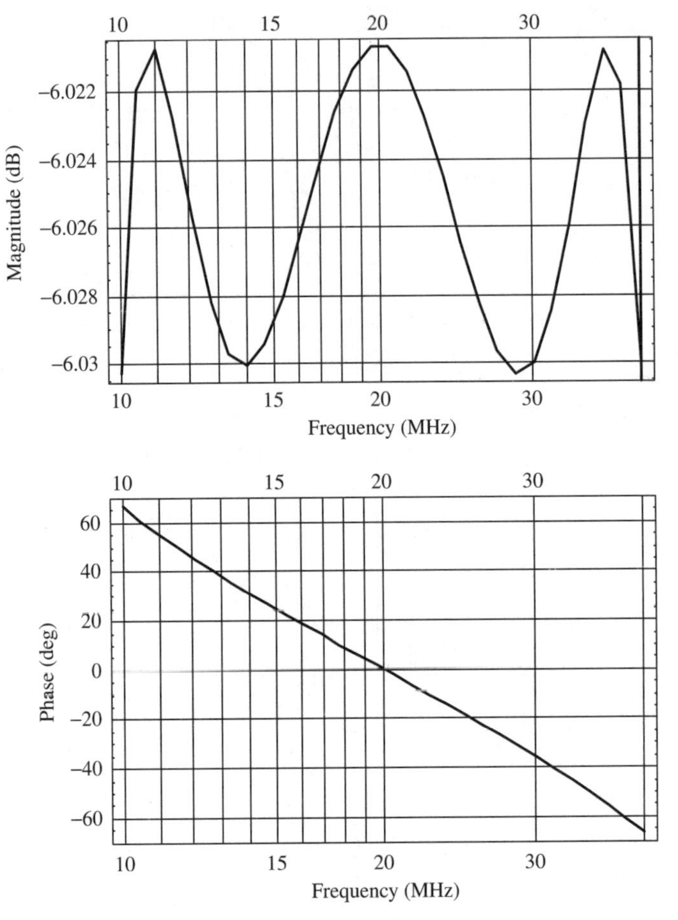

Figure 9.37 Passband characteristics of the bandpass filter shown in Figure 9.35.

attenuates higher frequencies. On the other hand, a bandstop filter must stop all signals with frequencies between ω_l and ω_u and pass the signals that are outside this frequency band. Hence, its characteristics are opposite to that of a bandpass filter considered earlier. The following frequency relation transforms the response of a low-pass filter to the bandstop:

$$\overline{\omega} = (\omega_u - \omega_l)\frac{\omega}{\omega^2 - \omega_0^2} \tag{9.2.34}$$

This transformation replaces the series inductor of low-pass prototype with an inductor L_{BS1} and a capacitor C_{BS1} that are connected in parallel. Component values are determined as follows:

$$L_{\text{BS1}} = \frac{(\omega_u - \omega_l)g_L}{\omega_0^2} \tag{9.2.35}$$

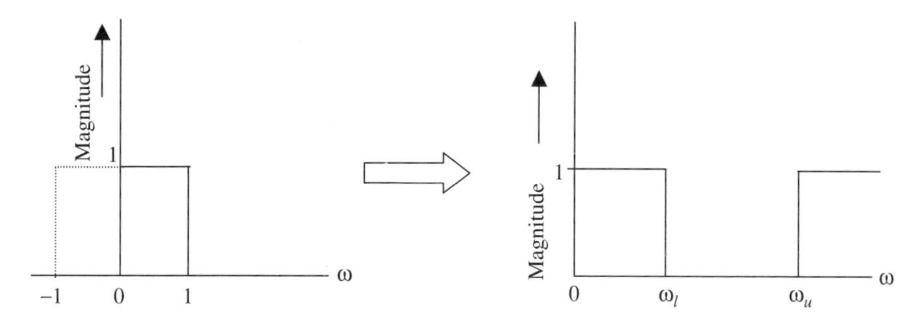

Figure 9.38 Transformation from a low-pass to a bandstop characteristic.

and

$$C_{BS1} = \frac{1}{(\omega_u - \omega_l)g_L} \tag{9.2.36}$$

Also, capacitor C_{BS2}, which is connected in series with an inductor L_{BS2}, will replace the shunt capacitor of low-pass prototype. These elements are found as follows:

$$L_{BS2} = \frac{1}{(\omega_u - \omega_l)g_C} \tag{9.2.37}$$

and

$$C_{BS2} = \frac{(\omega_u - \omega_l)g_C}{\omega_0^2} \tag{9.2.38}$$

These elements need to be further scaled as desired by the load and source resistance. Table 9.10 summarizes these transformations for the low-pass prototype filter.

Example 9.11 Design a maximally flat bandstop filter with $n = 3$. It must stop signals in the frequency range 10 to 40 MHz and pass the rest of the frequencies. Assume that the load and source resistances are at 75 Ω each.

SOLUTION From (9.2.29), we have

$$f_0 = \sqrt{f_l f_u} = \sqrt{10^7 \times 40 \times 10^6} = 20 \times 10^6 \text{ Hz}$$

Now, using (9.2.35) to (9.2.38) with $g_L = 1$ and $g_C = 2$ from Example 9.6, we get

$$L_{BS1} = \frac{2\pi \times 10^6(40 - 10)}{(2\pi \times 20 \times 10^6)^2} \times 1 \text{ H} = 11.94 \text{ nH}$$

$$C_{BS1} = \frac{1}{2\pi \times 10^6(40 - 10) \times 1} \text{F} = 5.305 \text{ nF}$$

TABLE 9.10 Filter Transformations

Filter	Circuit Elements	
Low-pass	g_L	g_C
High-pass	$\frac{1}{\omega_c g_L}$	$\frac{1}{\omega_c g_C}$
Bandpass	$\frac{\omega_u - \omega_l}{\omega_0^2 g_L}$ $\frac{g_L}{\omega_u - \omega_l}$	$\frac{\omega_u - \omega_l}{\omega_0^2 g_C}$ $\frac{g_C}{\omega_u - \omega_l}$
Bandstop	$\frac{(\omega_u - \omega_l)\, g_L}{\omega_0^2}$ $\frac{1}{(\omega_u - \omega_l)\, g_L}$	$\frac{1}{(\omega_u - \omega_l)\, g_C}$ $\frac{(\omega_u - \omega_l)\, g_C}{\omega_0^2}$

$$L_{\text{BS2}} = \frac{1}{2\pi \times 10^6 (40 - 10) \times 2} \text{H} = 2.653\,\text{nH}$$

and

$$C_{\text{BS2}} = \frac{2\pi \times 10^6 (40 - 10)}{(2\pi \times 20 \times 10^6)^2} \times 2\,\text{F} = 23.87\,\text{nF}$$

Values of required elements are determined next following the load and source resistance scaling. Hence,

$$C_1 = C_3 = \frac{5.305}{75}\text{nF} = 70.73\,\text{pF}$$

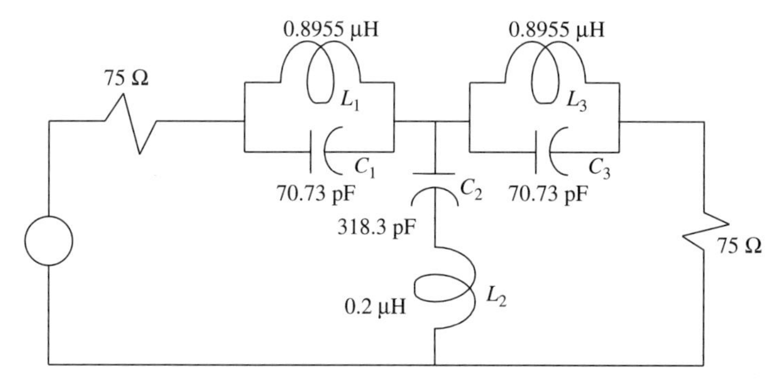

Figure 9.39 Bandstop circuit for Example 9.11.

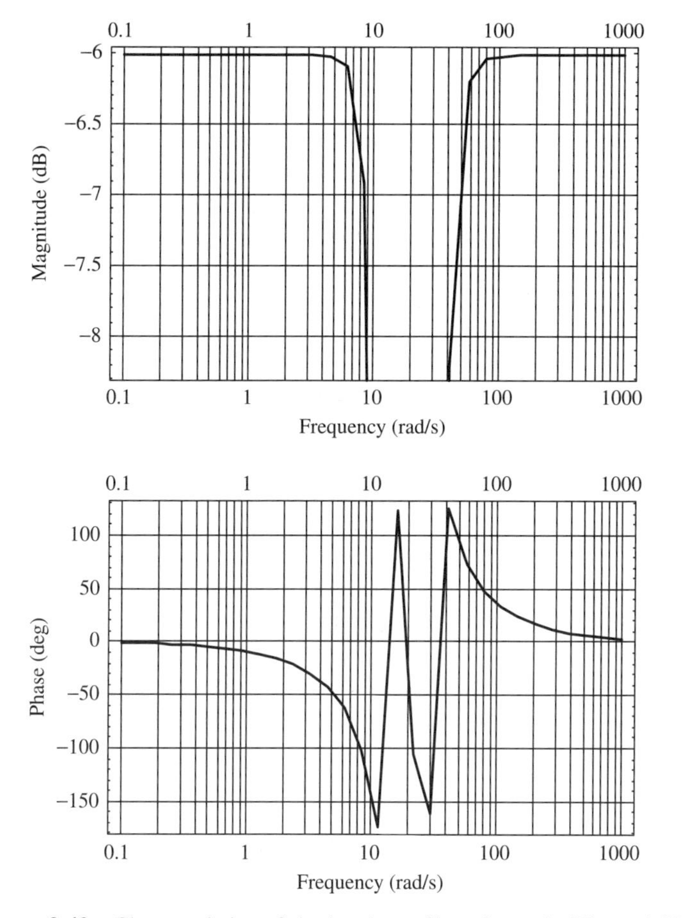

Figure 9.40 Characteristics of the bandstop filter shown in Figure 9.39.

$$L_1 = L_3 = 75 \times 11.94\,\text{nH} = 0.8955\,\mu\text{H}$$

$$L_2 = 75 \times 2.653\,\text{nH} = 0.1989\,\mu\text{H} \approx 0.2\,\mu\text{H}$$

and

$$C_2 = \frac{23.87}{75}\text{nF} = 318.3\,\text{pF}$$

The resulting filter circuit is shown in Figure 9.39. Its frequency response depicted in Figure 9.40 indicates that we have transformed the low-pass prototype to a bandstop filter for the 10- to 40-MHz frequency band. As before, the source resistance is considered a part of the filter circuit, and therefore its passband shows a 6-dB insertion loss.

9.3 MICROWAVE FILTERS

The filter circuits presented so far in this chapter use lumped elements. However, these may have practical limitations at microwave frequencies. When the signal wavelength is short, distances between the filter components need to be taken into account. Further, discrete components at such frequencies may cease to operate due to associated parasitic elements and need to be approximated with distributed components. As found in Chapter 3, transmission line stubs can be used in place of lumped elements. However, there may be certain practical problems in implementing the series stubs. This section begins with a technique to design a low-pass filter with only parallel connected lines of different characteristic impedance values. It is known as the *stepped impedance* (or high-Z low-Z) *filter*. Since this technique works mainly for low-pass filters, the procedure to transform series reactance to shunt that utilizes Kuroda's identities is summarized next. Redundant sections of the transmission line are used to separate filter elements, and therefore this procedure is known as *redundant filter synthesis*. Nonredundant circuit synthesis makes use of these sections to improve the filter response as well. Other methods of designing microwave filters include coupled transmission lines and resonant cavities. Impedance transformers, discussed in Chapter 7, are essentially bandpass filters with different impedance at the two-ports. Interested readers can find detailed design procedures of all these filters in the references at the end of this chapter.

Low-Pass Filter Design Using Stepped Impedance Distributed Elements

Consider a lossless transmission line of length d and characteristic impedance Z_0, as shown in Figure 9.41(a). If it is equivalent to the T-network shown in Figure 9.41(b), these two must have the same network parameters. The impedance matrix of the transmission line was determined in Example 8.4 as follows:

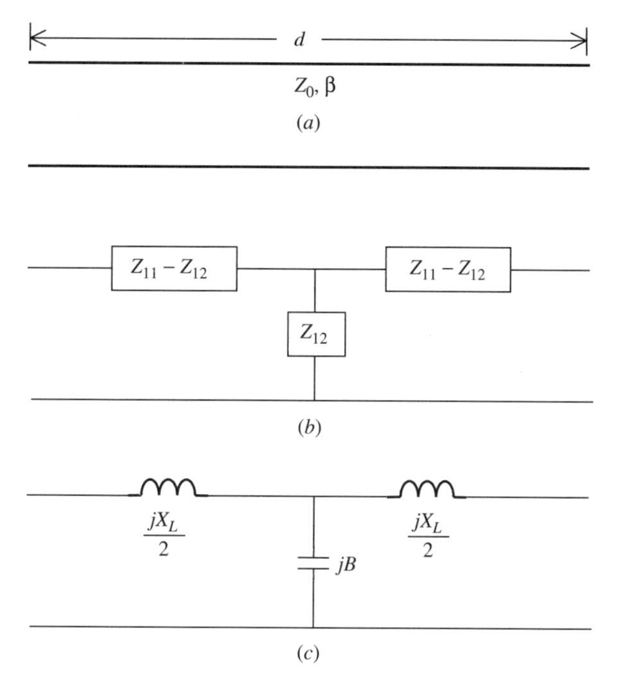

Figure 9.41 Lossless transmission line (*a*), equivalent symmetrical T-network (*b*), and (*c*) elements of T-networks.

$$[Z]_{\text{line}} = \begin{bmatrix} -jZ_0 \cot \beta d & -j\dfrac{Z_0}{\sin \beta d} \\ -j\dfrac{Z_0}{\sin \beta d} & -jZ_0 \cot \beta d \end{bmatrix} \tag{9.3.1}$$

The impedance matrix of the T-network, $[Z]_T$, may be found as

$$[Z]_{\text{T}} = \begin{bmatrix} Z_{11} & Z_{12} \\ Z_{12} & Z_{11} \end{bmatrix} \tag{9.3.2}$$

Therefore, if this T-network represents the transmission line, we can write

$$Z_{11} = -jZ_0 \cot \beta d \tag{9.3.3}$$

and

$$Z_{12} = -j\frac{Z_0}{\sin \beta d} \tag{9.3.4}$$

Hence, Z_{12} is a capacitive reactance. Circuit elements in the series arm of the T-network are found to be inductive reactance. This is found as follows:

$$Z_{11} - Z_{12} = -jZ_0 \frac{\cos \beta d - 1}{\sin \beta d} = jZ_0 \frac{\sin (\beta d/2)}{\cos (\beta d/2)} = jZ_0 \tan \frac{\beta d}{2} \tag{9.3.5}$$

Therefore, the series element of the T-network is an inductive reactance while its shunt element is capacitive, provided that $\beta d < \pi/2$. For $\beta d < \pi/4$, (9.3.4) and (9.3.5) give

$$B \approx \frac{\beta d}{Z_0} \tag{9.3.6}$$

and

$$X_L \approx Z_0 \beta d \tag{9.3.7}$$

Some special cases are as follows:

1. For Z_0 very large, (9.3.6) becomes negligible compared with (9.3.7). Therefore, the equivalent T-network (and hence the transmission line) represents an inductor that is given by

$$L = \frac{Z_0 \beta d}{\omega} \tag{9.3.8}$$

2. For Z_0 very small, (9.3.6) dominates over (9.3.7). Therefore, the transmission line is effectively representing a shunt capacitance in this case. It is given by

$$C = \frac{\beta d}{\omega Z_0} \tag{9.3.9}$$

Hence, high-impedance sections ($Z_0 = Z_h$) of the transmission line can replace the inductors, and low-impedance sections ($Z_0 = Z_m$) can replace the capacitors of a low-pass filter. In a design, these impedance values must be selected as far apart as possible and the section lengths are determined to satisfy (9.3.8) and (9.3.9). When combined with the scaling rules, electrical lengths of the inductive and capacitive sections are found as

$$(\beta d)_L = \frac{R_0 L}{Z_h} \quad \text{rad} \tag{9.3.10}$$

and

$$(\beta d)_C = \frac{Z_m C}{R_0} \quad \text{rad} \tag{8.3.11}$$

where L and C are normalized element values (g values) of the filter and R_0 is the filter impedance.

Example 9.12 Design a three-element maximally flat low-pass filter with its cutoff frequency as 1 GHz. It is to be used between a 50-Ω load and a generator with its internal impedance at 50 Ω. Assume that $Z_h = 150\ \Omega$ and $Z_m = 30\ \Omega$.

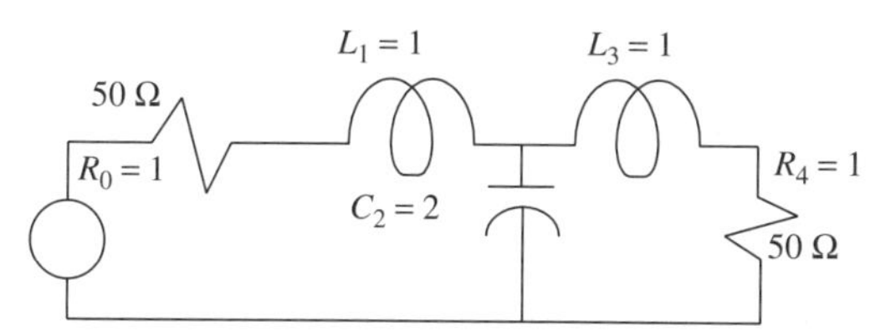

Figure 9.42 Maximally flat low-pass filter with normalized elements' values for Example 9.12.

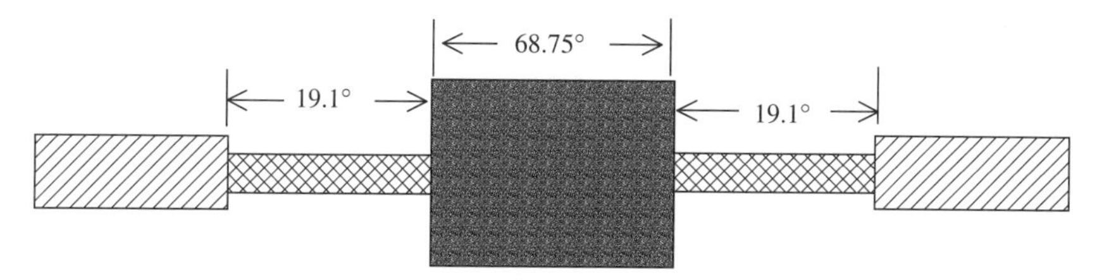

Figure 9.43 Maximally flat low-pass filter for Example 9.12.

SOLUTION From (9.2.15) and (9.2.16),

$$g_0 = g_4 = 1$$

$$g_1 = 2\sin\frac{\pi}{6} = 1$$

$$g_2 = 2\sin\frac{\pi}{2} = 2$$

and

$$g_3 = 2\sin\frac{5\pi}{6} = 1$$

If we use a circuit arrangement as illustrated in Figure 9.42, element values can be scaled for the frequency and the impedance. Using the scaling rules, these values are found as follows (see Figure 9.43):

$$\theta_1 = \theta_3 = (\beta d)_L = \frac{1 \times 50}{150} = 0.3333\,\text{rad} = 19.1^\circ$$

and

$$\theta_2 = (\beta d)_C = \frac{30 \times 2}{50} = 1.2\,\text{rad} = 68.75^\circ$$

Filter Synthesis Using the Redundant Elements

As mentioned earlier, transmission line sections can replace the lumped elements of a filter circuit. Richard's transformation provides the tools needed in such a design. Further, Kuroda's identities are used to transform series elements to a shunt configuration that facilitates the design.

Richard's Transformation

Richard proposed that open- and short-circuited lines could be synthesized like lumped elements through the following transformation:

$$\Omega = \tan \beta d = \tan \frac{\omega d}{v_p} \tag{9.3.12}$$

where v_p is the phase velocity of signal propagating on the line. Since the tangent function is periodic with a period of 2π, (9.3.12) is a periodic transformation. Substituting Ω in place of ω, the reactance of the inductor L and of the capacitor C may be written as follows:

$$jX_L = j\Omega L = jL \tan \beta d \tag{9.3.13}$$

and

$$jX_C = -j\frac{1}{\Omega C} = -j\frac{1}{C \tan \beta d} = -j\frac{1}{C} \cot \beta d \tag{9.3.14}$$

Comparison of (9.3.13) and (9.3.14) with the special cases considered in Section 3.2 indicates that the former represents a short-circuited line with its characteristic impedance as L, while the latter is an open-circuited with Z_0 as $1/C$. Filter impedance is assumed to be unity. To obtain the cutoff of a low-pass filter prototype at unity frequency, the following must hold:

$$\Omega = \tan \beta d = 1 \rightarrow \beta d = \frac{\pi}{4} \rightarrow d = \frac{\lambda}{8} \tag{9.3.15}$$

Note that the transformation holds only at the cutoff frequency ω_c (frequency corresponding to λ), and therefore the filter response will differ from that of its prototype. Further, the response repeats every $4\,\omega_c$. These transformations are illustrated in Figure 9.44. These stubs are known as the *commensurate lines* because of their equal lengths.

Kuroda's Identities

As mentioned earlier, Kuroda's identities facilitate the design of distributed element filters, providing the means to transform the series stubs into shunt, or vice versa, to separate the stubs physically, and to render the characteristic impedance realizable. Figure 9.45 illustrates four identities that are useful in such a transformation of networks. These networks employ the unit element (U.E.), which is basically a $\lambda/8$-long line at the cutoff frequency ω_c, with specified characteristic

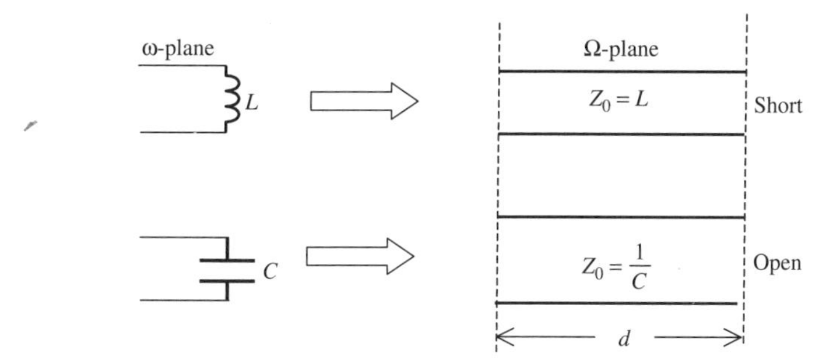

Figure 9.44 Distributed inductor and capacitor obtained from Richard's transformation.

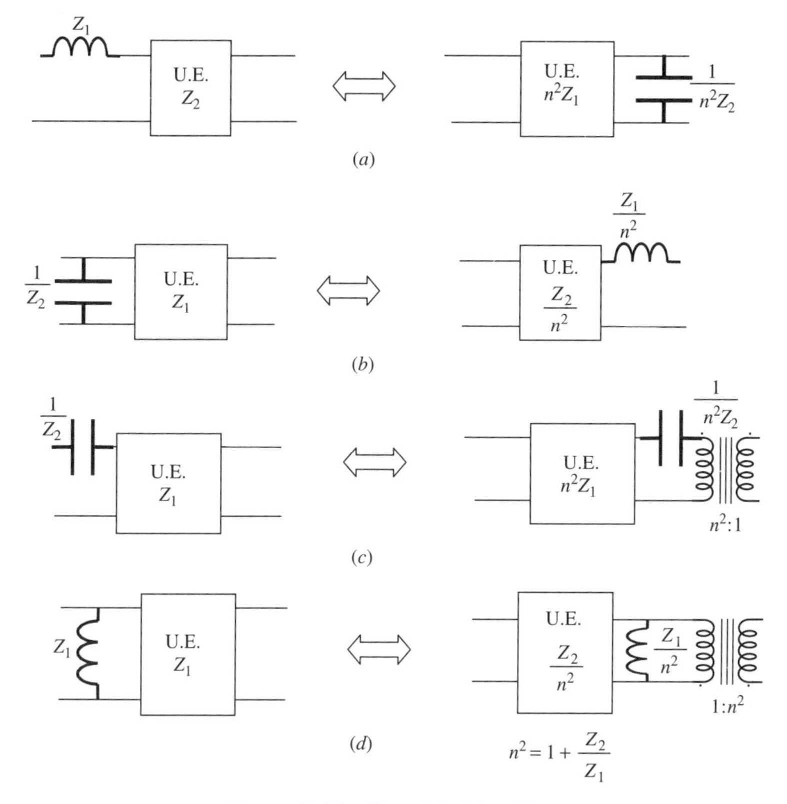

Figure 9.45 Kuroda's identities.

impedance. Inductors represent short-circuited stubs and capacitors open-circuited stubs. Obviously, its equivalent in lumped circuit theory does not exist.

The first identity is proved below. A similar procedure can be used to verify others. The circuits associated with the first identity can be redrawn as shown in Figure 9.46. A short-circuited stub of characteristic impedance Z_1 replaces the

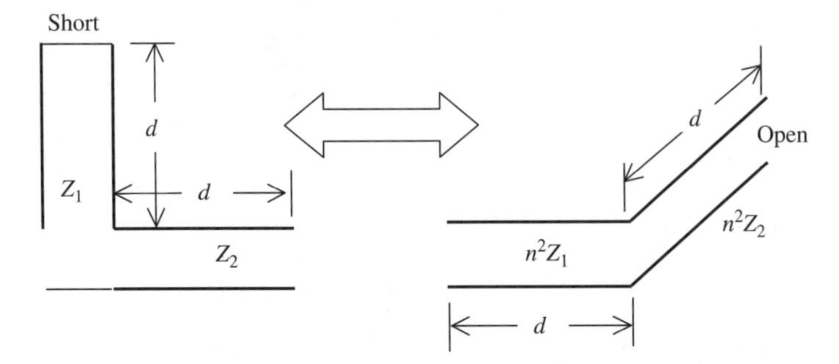

Figure 9.46 Illustration of Kuroda's first identity, shown in Figure 9.45.

series inductor that is followed by a unit element transmission line of characteristic impedance Z_2. Its transformation gives a unit element transmission line of characteristic impedance n^2Z_1 and an open-circuited stub of characteristic impedance n^2Z_2 in place of a shunt capacitor.

Following Example 8.10, the transmission matrix of a short-circuited series stub with input impedance Z_S can be found as follows. The impedance Z_S of the series stub is given as

$$Z_S = jZ_1 \tan \beta d = j\Omega Z_1 \tag{9.3.16}$$

$$\begin{bmatrix} A & B \\ C & D \end{bmatrix}_S = \begin{bmatrix} 1 & j\Omega Z_1 \\ 0 & 1 \end{bmatrix} \tag{9.3.17}$$

Following Example 8.13, the transmission matrix of the unit element is found to be

$$\begin{bmatrix} A & B \\ C & D \end{bmatrix}_{\text{U.E.}} = \begin{bmatrix} \cos \beta d & jZ_2 \sin \beta d \\ \dfrac{j \sin \beta d}{Z_2} & \cos \beta d \end{bmatrix} = \cos \beta d \begin{bmatrix} 1 & jZ_2 \tan \beta d \\ \dfrac{j \tan \beta d}{Z_2} & 1 \end{bmatrix}$$

or

$$\begin{bmatrix} A & B \\ C & D \end{bmatrix}_{\text{U.E.}} = \frac{1}{\sqrt{1+\Omega^2}} \begin{bmatrix} 1 & j\Omega Z_2 \\ \dfrac{j\Omega}{Z_2} & 1 \end{bmatrix} \tag{9.3.18}$$

Since these two elements are connected in cascade, the transmission matrix for the first circuit in Figure 9.46 is found from (9.3.17) and (9.3.18) as follows:

$$\begin{bmatrix} A & B \\ C & D \end{bmatrix}_a = \begin{bmatrix} A & B \\ C & D \end{bmatrix}_S \cdot \begin{bmatrix} A & B \\ C & D \end{bmatrix}_{\text{U.E.}} = \frac{1}{\sqrt{1+\Omega^2}} \begin{bmatrix} 1 - \dfrac{\Omega^2 Z_1}{Z_2} & j\Omega(Z_1 + Z_2) \\ \dfrac{j\Omega}{Z_2} & 1 \end{bmatrix} \tag{9.3.19}$$

Now for the second circuit shown in Figure 9.46, the transmission matrix for the unit element is

$$\begin{bmatrix} A & B \\ C & D \end{bmatrix}_{\text{U.E.}} = \cos\beta d \begin{bmatrix} 1 & jn^2 Z_1 \tan\beta d \\ \dfrac{j\tan\beta d}{n^2 Z_1} & 1 \end{bmatrix} = \frac{1}{\sqrt{1+\Omega^2}} \begin{bmatrix} 1 & j\Omega n^2 Z_1 \\ \dfrac{j\Omega}{n^2 Z_1} & 1 \end{bmatrix} \tag{9.3.20}$$

Since the impedance Z_P of the shunt stub is given as

$$Z_p = -jn^2 Z_2 \cot\beta d = -j\frac{n^2 Z_2}{\Omega} \tag{9.3.21}$$

its transmission matrix may be found, by following Example 8.11, as

$$\begin{bmatrix} A & B \\ C & D \end{bmatrix}_p = \begin{bmatrix} 1 & 0 \\ \dfrac{j\Omega}{n^2 Z_2} & 1 \end{bmatrix} \tag{9.3.22}$$

Hence, the transmission matrix of the circuit shown on right-hand side is found as

$$\begin{bmatrix} A & B \\ C & D \end{bmatrix}_b = \begin{bmatrix} A & B \\ C & D \end{bmatrix}_{\text{U.E.}} \cdot \begin{bmatrix} A & B \\ C & D \end{bmatrix}_p = \frac{1}{\sqrt{1+\Omega^2}} \begin{bmatrix} 1 - \dfrac{\Omega^2 Z_1}{Z_2} & j\Omega n^2 Z_1 \\ \dfrac{j\Omega}{n^2}\left(\dfrac{1}{Z_1} + \dfrac{1}{Z_2}\right) & 1 \end{bmatrix} \tag{9.3.23}$$

On substituting for n^2, it is easy to show that this matrix is equal to that found in (9.3.19).

Example 9.13 Design a three-element maximally flat low-pass filter with its cutoff frequency as 1 GHz. It is to be used between a 50-Ω load and a generator with its internal impedance at 50 Ω.

SOLUTION A step-by-step procedure to design this filter is as follows:

- *Step 1.*

As before, g values for the filter are found from (9.2.15) and (9.2.16) as

$$g_0 = g_4 = 1$$

$$g_1 = 2\sin\frac{\pi}{6} = 1$$

$$g_2 = 2\sin\frac{\pi}{2} = 2$$

and

$$g_3 = 2\sin\frac{5\pi}{6} = 1$$

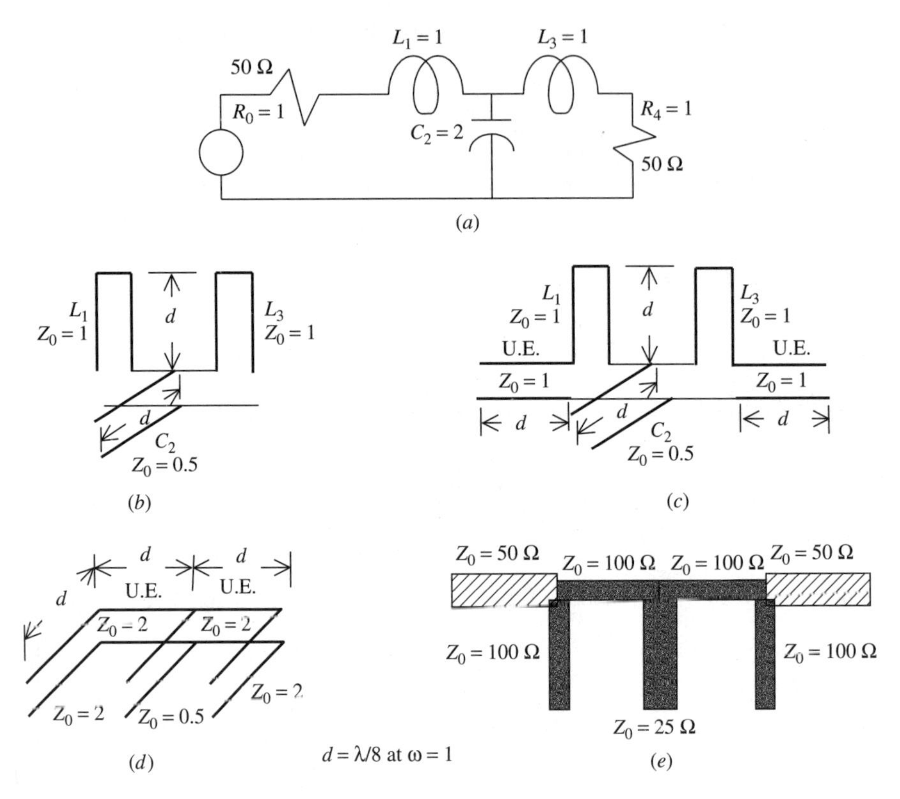

Figure 9.47 Maximally flat low-pass filter for Example 9.13.

The circuit is drawn with normalized values as shown in Figure 9.47(*a*).

- *Step 2*. Using Richard's transformation, the filter elements are replaced by stubs. Hence, short-circuited series stubs replace inductors L_1 and L_3, while an open-circuited shunt stub replaces the capacitor C_2. This is shown in Figure 9.47(*b*).
- *Step 3*. Since the circuit found in step 2 cannot be fabricated in its present form, unit elements are added on its two sides. This is shown in Figure 9.47(*c*).
- *Step 4*. Using Kuroda's identity shown in Figure 9.45(*a*), series stub and unit element combinations are transformed to an open-circuited shunt stub and the unit element. Since $Z_1 = Z_2 = 1$, equivalent capacitance C and the characteristic impedance of the corresponding stub are found to be $C = \frac{1}{2} = 0.5$ and $Z_0 = 2$. Also, the characteristic impedance of each unit element is found to be 2. This circuit is illustrated in Figure 9.47(*d*).
- *Step 5*. For impedance and frequency scaling, all normalized characteristic impedances are now multiplied by 50 Ω, and the stub lengths are selected as $\lambda/8$ at 1 GHz. This is shown in Figure 9.47(*e*).

SUGGESTED READING

Bahl, I., and P. Bhartia, *Microwave Solid State Circuit Design*. New York: Wiley, 1988.

Collin, R. E., *Foundations for Microwave Engineering*. New York: McGraw-Hill, 1992.

Davis, W. A., *Microwave Semiconductor Circuit Design*. New York: Van Nostrand Reinhold, 1984.

Elliott, R. S., *An Introduction to Guided Waves and Microwave Circuits*. Englewood Cliffs, NJ: Prentice Hall, 1993.

Fusco, V. F., *Microwave Circuits*. Englewood Cliffs, NJ: Prentice Hall, 1987.

Matthaei, G. L., L. Young, and E. M. T. Jones, *Microwave Filters, Impedance-Matching Networks, and Coupling Structures*. Dedham, MA: Artech House, 1980.

Pozar, D. M., *Microwave Engineering*. New York: Wiley, 1998.

PROBLEMS

9.1. Design a low-pass composite filter with nominal impedance of 50 Ω. It must pass all signals up to 10 MHz. Assume that $f_\infty = 10.05$ MHz. Plot its characteristics in the frequency range 1 to 100 MHz.

9.2. Design a high-pass T-section constant-k filter that has a nominal impedance of 50 Ω and a cutoff frequency of 10 MHz. Plot its frequency response in the frequency band 100 kHz to 100 MHz.

9.3. Design an m-derived T-section high-pass filter with a cutoff frequency of 10 MHz and nominal impedance of 50 Ω. Assume that $f_\infty = 9.95$ MHz. Plot the response of this filter in the frequency band 100 kHz to 100 MHz.

9.4. Design a high-pass composite filter with a cutoff frequency of 10 MHz and image impedance of 50 Ω in its passband. Assume that $f_\infty = 9.95$ MHz. Plot its response in the frequency range 100 kHz to 100 MHz.

9.5. It is desired to design a maximally flat low-pass filter with at least 25 dB of attenuation at $\omega = 2\omega_c$ and -1.5 dB at its band edge. How many elements will be required for this filter? If a Chebyshev filter is used with 1.5 dB of ripple in its passband, find the number of circuit elements.

9.6. Design a low-pass Butterworth filter with a cutoff frequency of 150 MHz and an insertion loss of 50 dB at 400 MHz. It is to be used between a 75-Ω load and a generator with its internal resistance as 75 Ω.

9.7. Design a low-pass Chebyshev filter that exhibits ripples no more than 2 dB in its passband. The filter must pass all frequencies up to 50 MHz and attenuate the signal at 100 MHz by at least 15 dB. The load and the source resistance are of 50 Ω each.

9.8. Reconsider Problem 9.7 to design a low-pass filter that exhibits the Chebyshev response with a 3-dB ripple in its passband. The filter must have 50 Ω at both its input and output ports.

9.9. Design a high-pass Chebyshev filter with a passband ripple magnitude below 3 dB. It must pass all frequencies over 50 MHz. A minimum of 15 dB of insertion loss is required at 25 MHz. Assume that the load and source resistances are of 75 Ω each.

9.10. Design a bandpass Chebyshev filter that exhibits no more than 2 dB of ripple in its passband. It must pass the signals that are in the frequency band 50 to 80 MHz with zero insertion loss. Assume that the load and source resistances are of 50 Ω each.

9.11. It is desired to design a maximally flat low-pass filter with at least 40 dB of attenuation at $f = 1.1$ GHz and -3 dB at $f = 1$ GHz. How many elements will be required for this filter? If a Chebyshev filter is used with 3 dB of ripple in its passband, find the number of circuit elements.

9.12. Design a Butterworth filter with a cutoff frequency of 960 MHz and an insertion loss of 15 dB at 1.5 GHz. It is to be used between a 50-Ω load and a generator with its internal resistance as 50 Ω.

9.13. Design a low-pass Chebyshev filter that exhibits ripples of no more than 1 dB in its passband. The filter must pass all frequencies up to 960 MHz and attenuate the signal at 1.5 GHz by at least 15 dB. The load and the source resistance are of 50 Ω each.

9.14. Reconsider Problem 9.13 to design a low-pass filter that exhibits the Chebyshev response with 2 dB of ripple in its passband. The filter must have 50 Ω at both its input and output ports.

9.15. Design a high-pass Chebyshev filter with a passband ripple magnitude below 0.5 dB. It must pass all frequencies over 500 MHz. A minimum of 20 dB of insertion loss is required at 200 MHz. Assume that the load and source resistances are of 50 Ω each.

9.16. Design a bandpass Chebyshev filter that exhibits no more than 0.1 dB of ripple in its passband. It must pass the signals that are in the frequency band 500 to 800 MHz with almost zero insertion loss and provide at least 15 dB of attenuation for signals below 300 MHz. Assume that the load and source resistances are of 75 Ω each.

9.17. Design a high-pass filter that exhibits no more than 2.5 dB of ripple in its passband. It must pass all frequencies over 100 MHz. A minimum of 40 dB of insertion loss is required at 25 MHz. Assume that the load and source resistances are of 50 Ω each.

9.18. Design a maximally flat bandstop filter with $n = 5$. It must stop the signals in the frequency range 50 to 80 MHz and pass the rest of the frequencies. Assume that the load and source resistances are of 50 Ω each.

9.19. Design a fifth-order low-pass Butterworth filter with a cutoff frequency of 5 GHz. It is to be used between a 75-Ω load and a generator with an internal resistance of 75 Ω.

9.20. Design a third-order low-pass Chebyshev filter that exhibits ripples of no more than 1 dB in its passband. The filter must pass all frequencies up to 1 GHz. The load and the source resistance are of 50 Ω each.

9.21. Design a fifth-order high-pass Chebyshev filter with passband ripple magnitude below 3 dB. It must pass all frequencies over 5 GHz. Assume that the load and source resistances are of 75 Ω each.

10

SIGNAL-FLOW GRAPHS AND THEIR APPLICATIONS

A *signal-flow graph* is a graphical means of portraying the relationship among the variables of a set of linear algebraic equations. S. J. Mason originally introduced it to represent the cause and effect of linear systems. Associated terms are defined in this chapter along with a procedure to draw a signal-flow graph for a given set of algebraic equations. Further, signal-flow graphs of microwave networks are obtained in terms of their S-parameters and associated reflection coefficients. The manipulation of signal-flow graphs is summarized to find the desired transfer functions. Finally, the relations for transducer power gain, available power gain, and operating power gain are formulated in this chapter.

Consider a linear network that has N input and output ports. It is described by a set of linear algebraic equations:

$$V_i = \sum_{j=1}^{N} Z_{ij} I_j \qquad i = 1, 2, \ldots, N$$

This says that the effect V_i at the ith port is a sum of gain times causes at its N ports. Hence, V_i represents the dependent variable (effect), and I_j are the independent variables (cause). *Nodes* or *junction points* of the signal-flow graph represent these variables. The nodes are connected by line segments called *branches* with an arrow on each directed toward the dependent node. The coefficient Z_{ij} is the gain of a branch that connects the ith dependent node with the

Radio-Frequency and Microwave Communication Circuits: Analysis and Design, Second Edition,
By Devendra K. Misra
ISBN 0-471-47873-3

jth independent node. A signal can be transmitted through a branch only in the direction of the arrow.

The basic properties of a signal-flow graph can be summarized as follows:

- A signal-flow graph can be used only when the system is linear.
- A set of algebraic equations must be in the form of effects as functions of causes before its signal-flow graph can be drawn.
- A node is used to represent each variable. Normally, these are arranged from left to right, following a succession of inputs (causes) and outputs (effects) of the network.
- Nodes are connected together by branches with an arrow directed toward the dependent node.
- Signals travel along the branches only in the direction of the arrows.
- Signal I_k traveling along a branch that connects nodes V_i and I_k is multiplied by the branch gain Z_{ik}. The dependent node (effect) V_i is equal to the sum of the branch gain times the corresponding independent nodes (causes).

Example 10.1 The output b_1 of a system is caused by two inputs a_1 and a_2 as represented by the equation

$$b_1 = S_{11}a_1 + S_{12}a_2$$

Find its signal-flow graph.

SOLUTION There are two independent variables and one dependent variable in the equation. Locate these three nodes and connect nodes a_1 and a_2 with b_1, as shown in Figure 10.1. The arrows on two branches are directed toward effect b_1. Coefficients of input (cause) are shown as the gain of that branch.

Example 10.2 Output b_2 of a system is caused by two inputs a_1 and a_2 as represented by the equation

$$b_2 = S_{21}a_1 + S_{22}a_2$$

Find its signal-flow graph.

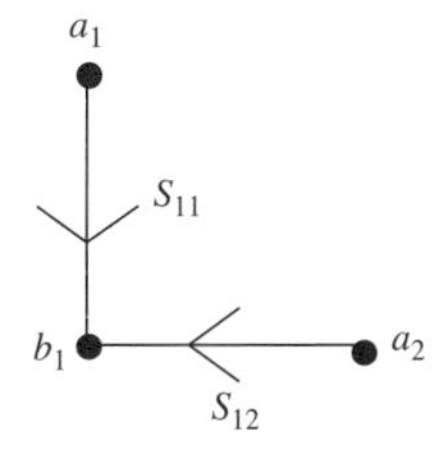

Figure 10.1 Signal-flow graph representation of Example 10.1.

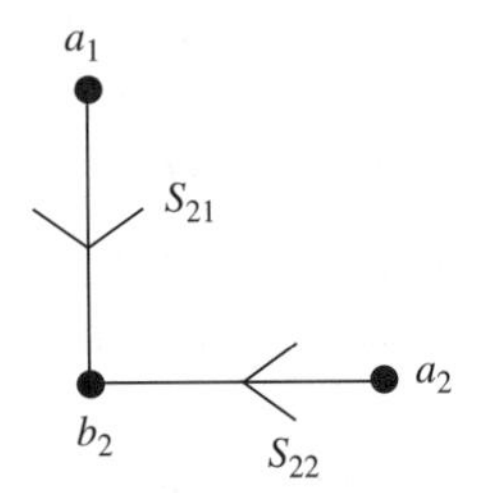

Figure 10.2 Signal-flow graph representation of Example 10.2.

SOLUTION There are again two independent variables and one dependent variable in the equation. Locate these three nodes and connect node a_1 with b_2 and a_2 with b_2, as shown in Figure 10.2. The arrows on two branches are directed toward effect b_2. Coefficients of input (cause) are shown as the gain of that branch.

Example 10.3 The input–output characteristics of a two-port network are given by the set of linear algebraic equations

$$b_1 = S_{11}a_1 + S_{12}a_2$$
$$b_2 = S_{21}a_1 + S_{22}a_2$$

Find its signal-flow graph.

SOLUTION There are two independent variables a_1 and a_2, and two dependent variables b_1 and b_2 in this set of equations. Locate the four nodes and then connect nodes a_1 and a_2 with b_1. Similarly, connect a_1 and a_2 with b_2, as shown in Figure 10.3. Arrows on the branches are directed toward effects b_1 and b_2. The coefficients of each input (cause) are shown as the gain of that branch.

Example 10.4 The following set of linear algebraic equations represents the input–output relations of a multiport network. Find the corresponding signal-flow graph.

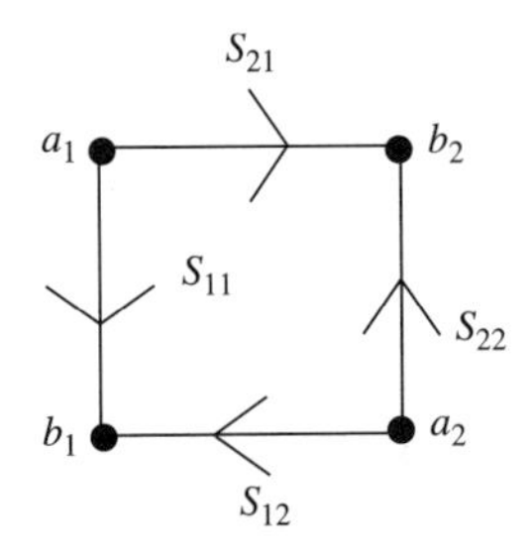

Figure 10.3 Signal-flow graph representation of Example 10.3.

SOLUTION

$$X_1 = R_1 + \frac{1}{2+s}X_2$$

$$X_2 = -4X_1 + R_2 - 7Y_1 - \frac{1}{s+4}X_2$$

$$Y_1 = \frac{s}{s^2+3}X_2$$

$$Y_2 = 10X_1 - sY_1$$

In the first equation, R_1 and X_2 represent the causes and X_1 is the effect. Hence, the signal-flow graph representing this equation can be drawn as illustrated in Figure 10.4.

Now consider the second equation. X_1 is the independent variable in it. Further, X_2 appears as cause as well as effect. This means that a branch must start and finish at the X_2 node. Hence, when this equation is combined with the first, the signal-flow graph will look as illustrated in Figure 10.5. Next, we add to it the signal-flow graph of the third equation. It has Y_1 as the effect and X_2 as the cause. It is depicted in Figure 10.6. Finally, the last equation has Y_2 as a dependent variable, and X_1 and Y_1 are two independent variables. A complete signal-flow graph representation is obtained after superimposing it as shown in Figure 10.7.

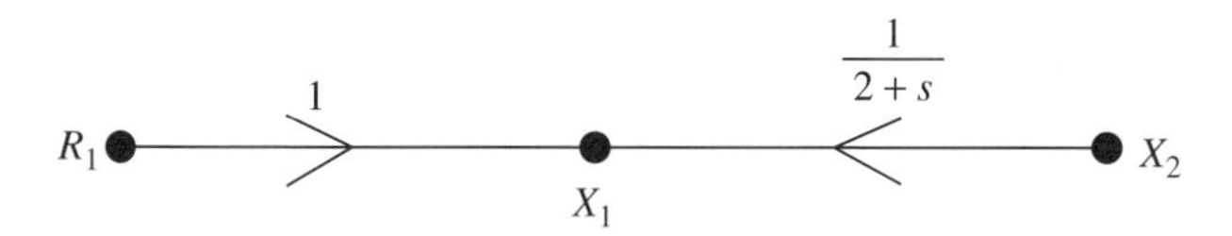

Figure 10.4 Signal-flow graph representation of the first equation of Example 10.4.

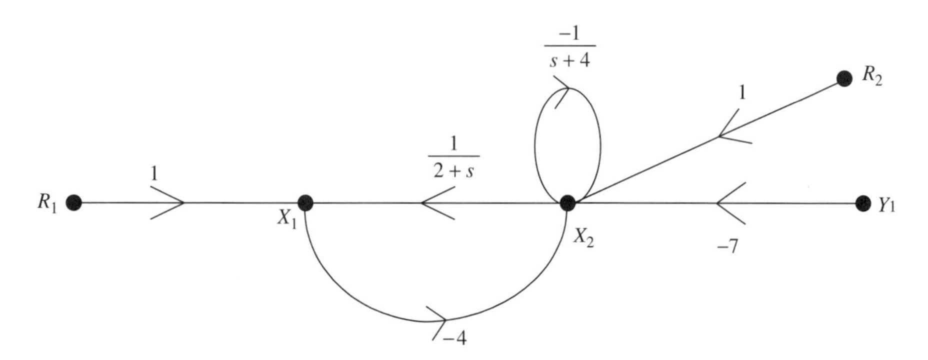

Figure 10.5 Signal-flow graph representation of the first two equations of Example 10.4.

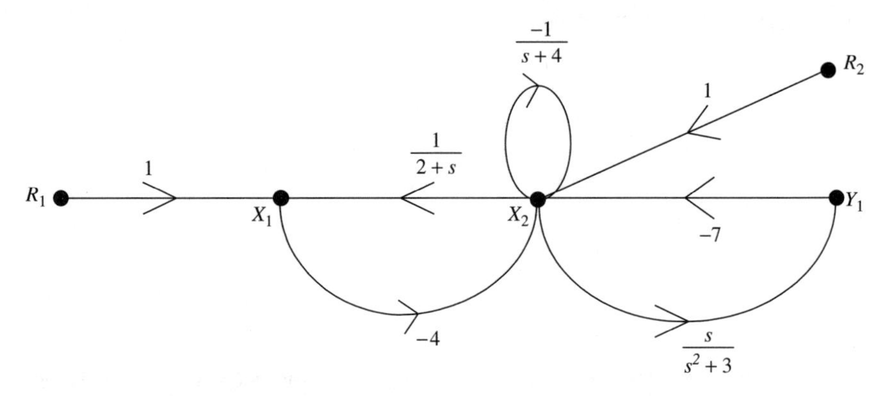

Figure 10.6 Signal-flow graph representation of the first three equations of Example 10.4.

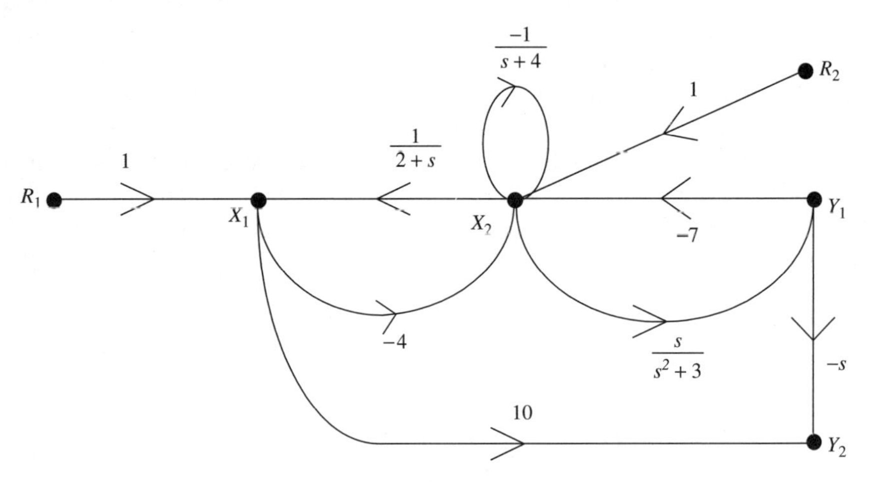

Figure 10.7 Complete signal-flow graph representation of Example 10.4.

10.1 DEFINITIONS AND MANIPULATION OF SIGNAL-FLOW GRAPHS

Before we proceed with manipulation of signal-flow graphs, it will be useful to define a few remaining terms.

Input and Output Nodes A node that has only outgoing branches is defined as an *input node* or *source*. Similarly, an *output node* or *sink* has only incoming branches. For example, R_1, R_2, and Y_1 are the input nodes in the signal-flow graph shown in Figure 10.8. This corresponds to the first two equations of Example 10.4. There is no output node (exclude the dotted branches) in it because X_1 and X_2 have both outgoing as well as incoming branches. Nodes X_1 and X_2 in Figure 10.8 can be made the output nodes by adding an outgoing branch of

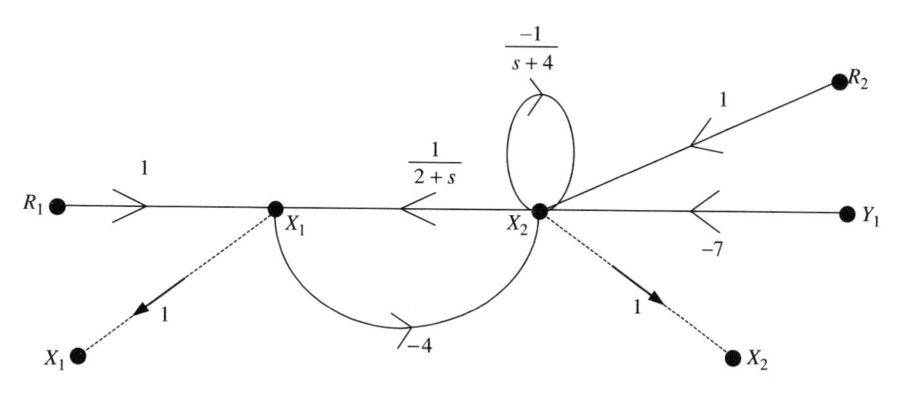

Figure 10.8 Signal-flow graph with R_1, R_2, and Y_1 as input nodes.

unity gain to each node. This is illustrated in Figure 10.8 by dotted branches. It is equivalent to adding $X_1 = X_1$ and $X_2 = X_2$ in the original set of equations. Thus, any nonoutput node can be made an output node in this way. However, this procedure cannot be used to convert these nodes to input nodes because that changes the equations. If an incoming branch of unity gain is added to node X_1, the corresponding equation is modified as follows:

$$X_1 = X_1 + R_1 + \frac{1}{2+s} X_2$$

However, X_1 can be made an input node by rearranging it as follows. The corresponding signal-flow graph is illustrated in Figure 10.9. It may be noted that R_1 is now an output node:

$$R_1 = X_1 - \frac{1}{2+s} X_2$$

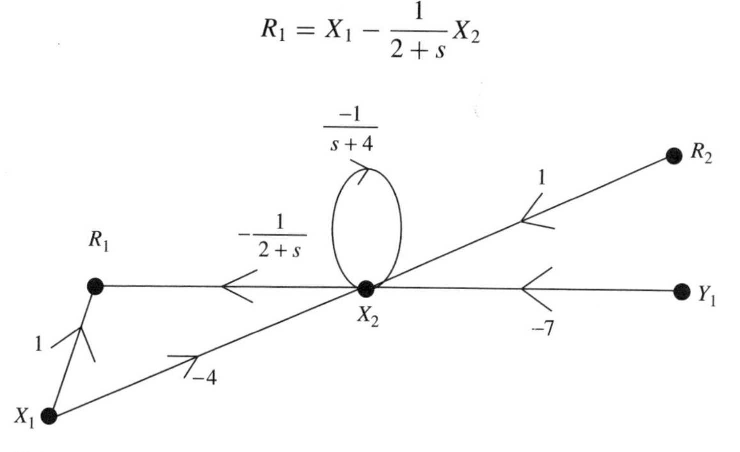

Figure 10.9 Signal-flow graph with R_1 as an output, and X_1, R_2, and Y_1 as the input nodes.

Path A continuous succession of branches traversed in the same direction is called the *path*. It is known as a *forward path* if it starts at an input node and ends at an output node without hitting a node more than once. The product of branch gains along a path is defined as the *path gain*. For example, there are two forward paths between nodes X_1 and R_1 in Figure 10.9. One of these forward paths is just one branch connecting the two nodes with a path gain of 1. The other forward path is X_1 to X_2 to R_1. Its path gain is $4/(2+s)$.

Loop A loop is a path that originates and ends at the same node without encountering other nodes more than once along its traverse. When a branch originates and terminates at the same node, it is called a *self-loop*. The path gain of a loop is defined as the *loop gain*.

Once the signal-flow graph is drawn, the ratio of an output to an input node (while other inputs, if there are more than one, are assumed to be zero) can be obtained by using *rules of reduction*. Alternatively, *Mason's rule* may be used. However, the latter rule is prone to error if the signal-flow graph is too complex. The reduction rules are generally recommended for such cases and are given as follows:

Rule 1 When there is only one incoming and one outgoing branch at a node (i.e., two branches are connected in series), it can be replaced by a direct branch with branch gain equal to the product of the two. This is illustrated in Figure 10.10.

Rule 2 Two or more parallel paths connecting two nodes can be merged into a single path with a gain that is equal to the sum of the original path gains, as depicted in Figure 10.11.

Rule 3 A self-loop of gain G at a node can be eliminated by multiplying its input branches by $1/(1-G)$. This is shown graphically in Figure 10.12.

Rule 4 A node that has one output and two or more input branches can be split in such a way that each node has just one input and one output branch, as shown in Figure 10.13.

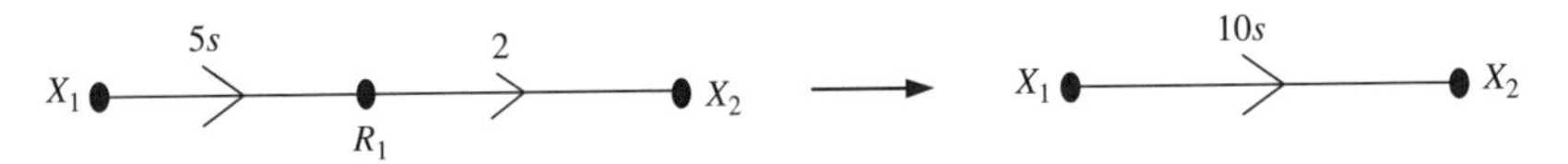

Figure 10.10 Graphical illustration of rule 1.

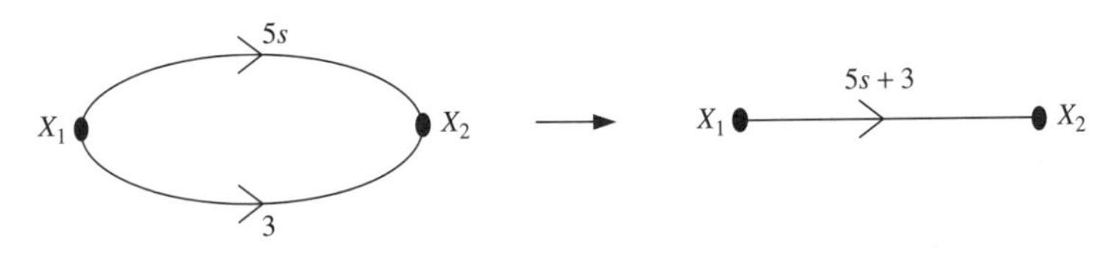

Figure 10.11 Graphical illustration of rule 2.

Figure 10.12 Graphical illustration of rule 3.

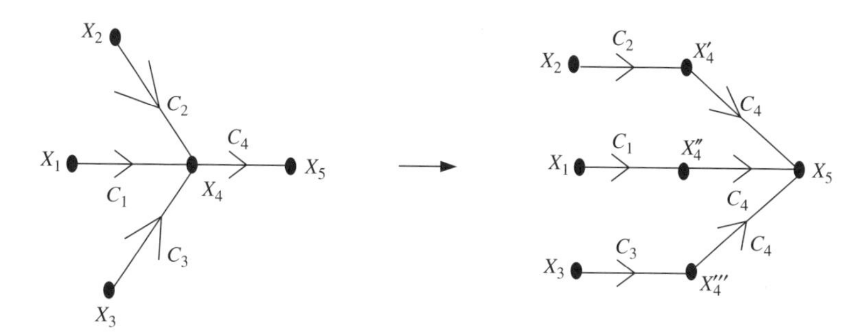

Figure 10.13 Graphical illustration of rule 4.

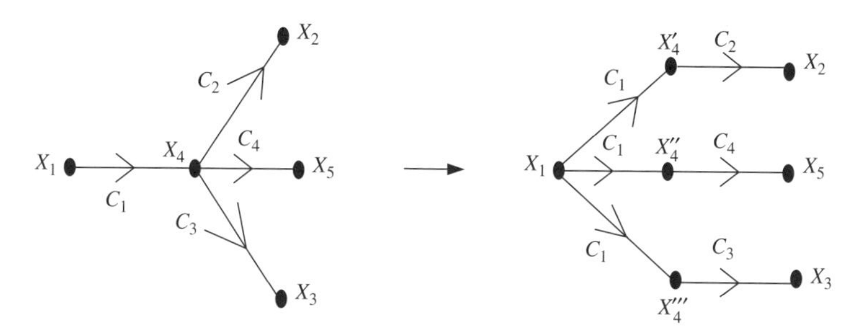

Figure 10.14 Graphical illustration of rule 5.

Rule 5 This is similar to rule 4. A node that has one input and two or more output branches can be split in such a way that each node has just one input and one output branch. This is shown in Figure 10.14.

Mason's Gain Rule Ratio T of the effect (output) to that of the cause (input) can be found using Mason's rule as follows:

$$T(s) = \frac{P_1\Delta_1 + P_2\Delta_2 + P_3\Delta_3 + \cdots}{\Delta} \tag{10.1.1}$$

where, P_i is the gain of the ith forward path,

$$\Delta = 1 - \sum L(1) + \sum L(2) - \sum L(3) + \cdots \tag{10.1.2}$$

$$\Delta_1 = 1 - \sum L(1)^{(1)} + \sum L(2)^{(1)} - \sum L(3)^{(1)} + \cdots \tag{10.1.3}$$

$$\Delta_2 = 1 - \sum L(1)^{(2)} + \sum L(2)^{(2)} - \sum L(3)^{(2)} + \cdots \tag{10.1.4}$$

$$\Delta_3 = 1 - \sum L(1)^{(3)} + \cdots \tag{10.1.5}$$

$$\vdots$$

$\Sigma L(1)$ stands for the sum of all first-order loop gains, $\Sigma L(2)$ is the sum of all second-order loop gains, and so on. $\Sigma L(1)^{(1)}$ denotes the sum of those first-order loop gains that do not touch path P_1 at any node, $\Sigma L(2)^{(1)}$ denotes the sum of those second-order loop gains that do not touch the path of P_1 at any point, $\Sigma L(1)^{(2)}$ denotes the sum of those first-order loops that do not touch path P_2 at any point, and so on. First-order loop gain was defined earlier. *Second-order loop gain* is the product of two first-order loops that do not touch at any point. Similarly, *third-order loop gain* is the product of three first-order loops that do not touch at any point.

Example 10.5 A signal-flow graph of a two-port network is given in Figure 10.15. Using Mason's rule, find its transfer function *Y/R*.

SOLUTION There are three forward paths from node R to node Y. Corresponding path gains are found as follows:

$$P_1 = 1 \times 1 \times \frac{1}{s+1} \times 1 \frac{s}{s+2} \times 3 \times 1 = \frac{3s}{(s+1)(s+2)}$$

$$P_2 = 1 \times 6 \times 1 = 6$$

and

$$P_3 = 1 \times 1 \times \frac{1}{s+1} \times (-4) \times 1 = -\frac{4}{s+1}$$

Next, it has two loops. The loop gains are

$$L_1 = -\frac{3}{s+1}$$

and

$$L_2 = -\frac{5s}{s+2}$$

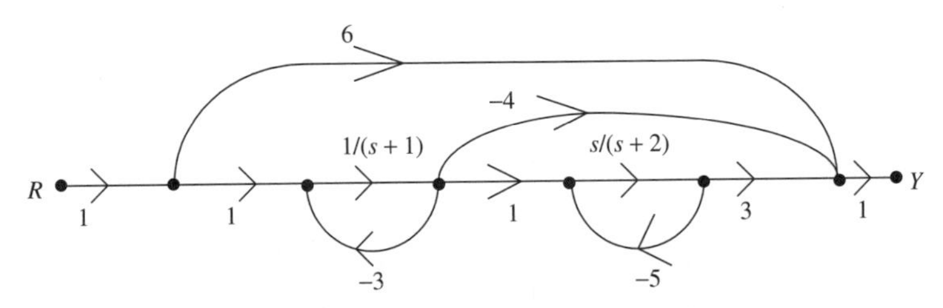

Figure 10.15 Signal-flow graph of Example 10.5.

Using Mason's rule, we find that

$$\frac{Y}{R} = \frac{P_1 + P_2(1 - L_1 - L_2 + L_1 L_2) + P_3(1 - L_2)}{1 - L_1 - L_2 + L_1 L_2}$$

10.2 SIGNAL-FLOW GRAPH REPRESENTATION OF A VOLTAGE SOURCE

Consider an ideal voltage source $E_S \angle\, 0°$ in series with source impedance Z_S, as shown in Figure 10.16. It is a single-port network with terminal voltage and current V_S and I_S, respectively. It is to be noted that the direction of current flow is assumed as entering the port, consistent with that of the two-port networks considered earlier. Further, the incident and reflected waves at this port are assumed to be a_S and b_S, respectively. The characteristic impedance at the port is assumed to be Z_0.

Using the usual circuit analysis procedure, the total terminal voltage V_S can be found as follows:

$$V_S = V_S^{\text{in}} + V_S^{\text{ref}} = E_S + Z_S I_S = E_S + Z_S(I_S^{\text{in}} + I_S^{\text{ref}}) = E_S + Z_S \frac{V_S^{\text{in}} - V_S^{\text{ref}}}{Z_0} \tag{10.2.1}$$

where superscripts "in" and "ref" on V_S and I_S are used to indicate the corresponding incident and reflected quantities. This equation can be rearranged as follows:

$$\left(1 + \frac{Z_S}{Z_0}\right) V_S^{\text{ref}} = E_S - \left(1 - \frac{Z_S}{Z_0}\right) V_S^{\text{in}}$$

or

$$V_S^{\text{ref}} = \frac{Z_0}{Z_0 + Z_S} E_S - \frac{Z_0 - Z_S}{Z_0 + Z_S} V_S^{\text{in}}$$

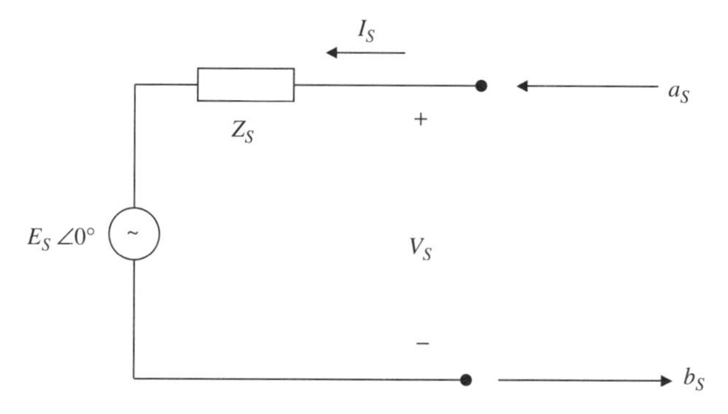

Figure 10.16 Incident and reflected waves at the output port of a voltage source.

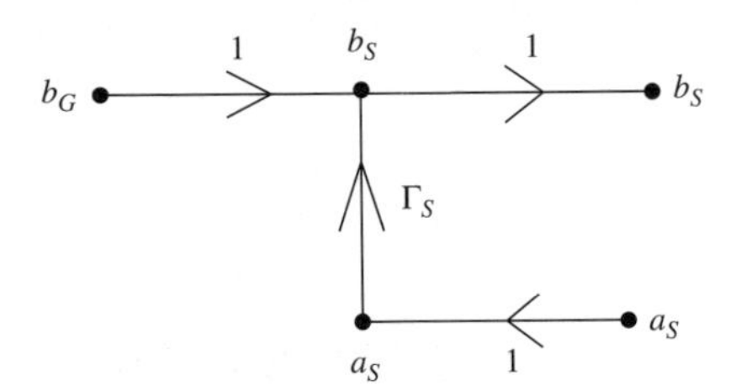

Figure 10.17 Signal-flow graph representation for a voltage source.

Dividing it by $\sqrt{2Z_0}$, we find that

$$\frac{V_S^{\text{ref}}}{\sqrt{2Z_0}} = \frac{Z_0}{Z_0 + Z_S}\frac{E_S}{\sqrt{2Z_0}} - \frac{Z_0 - Z_S}{Z_0 + Z_S}\frac{V_S^{\text{in}}}{\sqrt{2Z_0}}$$

Using (8.6.15) and (8.6.16), this can be written as follows:

$$b_S = b_G + \Gamma_S a_S \tag{10.2.2}$$

where

$$b_S = \frac{V_S^{\text{ref}}}{\sqrt{2Z_0}} \tag{10.2.3}$$

$$b_G = \frac{\sqrt{Z_0}E_S}{\sqrt{2}(Z_0 + Z_S)} \tag{10.2.4}$$

$$a_S = \frac{V_S^{\text{in}}}{\sqrt{2Z_0}} \tag{10.2.5}$$

and

$$\Gamma_S = \frac{Z_S - Z_0}{Z_S + Z_0} \tag{10.2.6}$$

From (10.2.2), the signal flow graph for a voltage source can be drawn as shown in Figure 10.17.

10.3 SIGNAL-FLOW GRAPH REPRESENTATION OF A PASSIVE SINGLE-PORT DEVICE

Consider load impedance Z_L as shown in Figure 10.18. It is a single-port device with port voltage and current V_L and I_L, respectively. The incident and reflected waves at the port are assumed to be a_L and b_L, respectively. Further, characteristic impedance at the port is Z_0. Following the usual circuit analysis rules, we find that

$$V_L = V_L^{\text{in}} + V_L^{\text{ref}} = Z_L I_L = Z_L(I_L^{\text{in}} + I_L^{\text{ref}}) = Z_L\frac{V_L^{\text{in}} - V_L^{\text{ref}}}{Z_0}$$

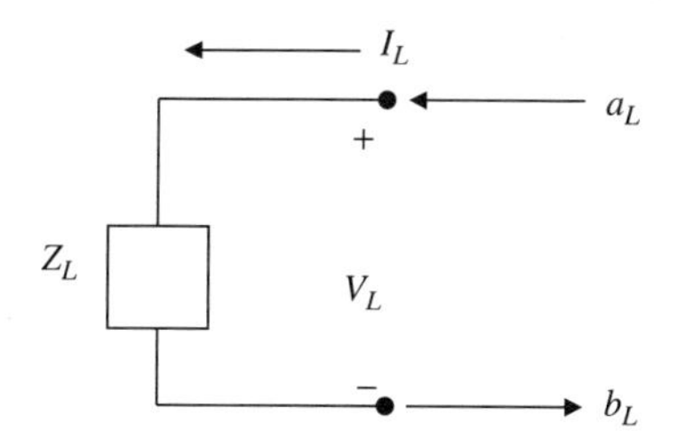

Figure 10.18 Passive one-port circuit.

or

$$\left(1 + \frac{Z_L}{Z_0}\right) V_L^{\text{ref}} = \left(\frac{Z_L}{Z_0} - 1\right) V_L^{\text{in}} \tag{10.3.1}$$

or

$$V_L^{\text{ref}} = \frac{Z_L - Z_0}{Z_L + Z_0} V_L^{\text{in}} = \Gamma_L V_L^{\text{in}} \tag{10.3.2}$$

After dividing by $\sqrt{2Z_0}$, we use (8.6.15) and (8.6.16) to find that

$$b_L = \Gamma_L a_L \tag{10.3.3}$$

where

$$b_L = \frac{V_L^{\text{ref}}}{\sqrt{2Z_0}} \tag{10.3.4}$$

$$a_L = \frac{V_L^{\text{in}}}{\sqrt{2Z_0}} \tag{10.3.5}$$

and

$$\Gamma_L = \frac{Z_L - Z_0}{Z_L + Z_0} \tag{10.3.6}$$

A signal-flow graph can be drawn on the basis of (10.3.3), as shown in Figure 10.19.

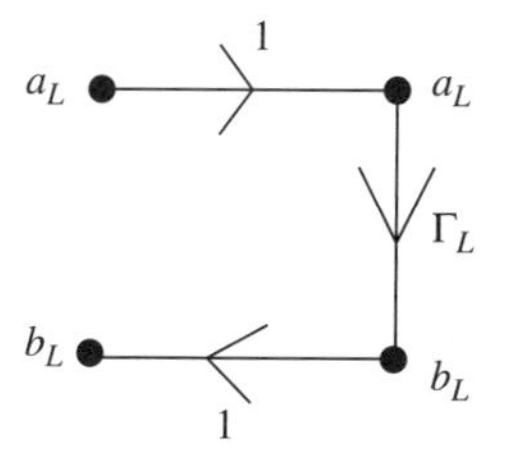

Figure 10.19 Signal-flow graph of a one-port passive device.

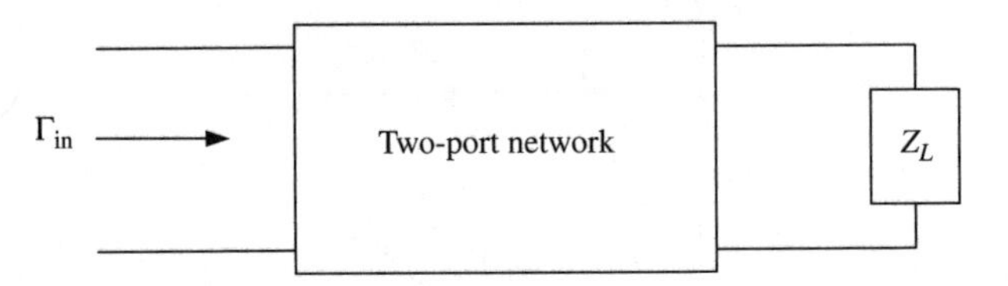

Figure 10.20 Two-port network with termination.

Example 10.6 Impedance Z_L terminates port 2 (the output) of a two-port network as shown in Figure 10.20. Draw the signal-flow graph and determine the reflection coefficient at its input port using Mason's rule.

SOLUTION As shown in Figure 10.21, combining the signal-flow graphs of a passive load and that of a two-port network obtained in Example 10.3, we can get the representation for this network. Its input reflection coefficient is given by the ratio of b_1 to a_1. For Mason's rule we find that there are two forward paths, $a_1 \to b_1$ and $a_1 \to b_2 \to a_L \to b_L \to a_2 \to b_1$. There is one loop, $b_2 \to a_L \to b_L \to a_2 \to b_2$, which does not touch the former path. The corresponding path and loop gains are

$$P_1 = S_{11}$$

$$P_2 = S_{21} \times 1 \times \Gamma_L \times 1 \times S_{12} = \Gamma_L S_{21} S_{12}$$

and

$$L_1 = 1 \times \Gamma_L \times 1 \times S_{22} = \Gamma_L S_{22}$$

Hence,

$$\Gamma_{in} = \frac{b_1}{a_1} = \frac{P_1(1 - L_1) + P_2}{1 - L_1} = \frac{S_{11}(1 - \Gamma_L S_{22}) + \Gamma_L S_{21} S_{12}}{1 - \Gamma_L S_{22}}$$

$$= S_{11} + \frac{\Gamma_L S_{21} S_{12}}{1 - \Gamma_L S_{22}}$$

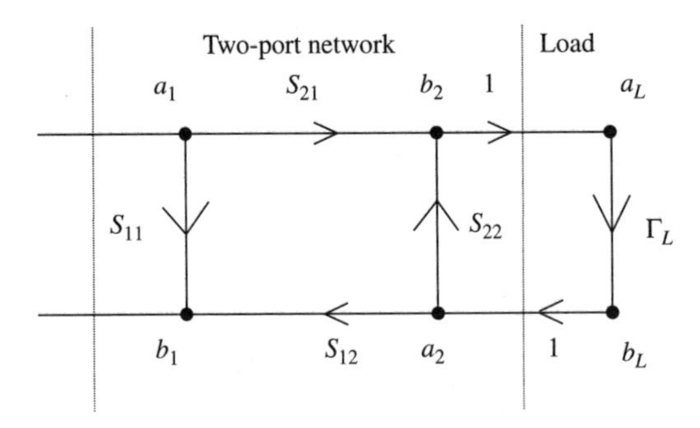

Figure 10.21 Signal-flow graph representation of the network shown in Figure 10.20.

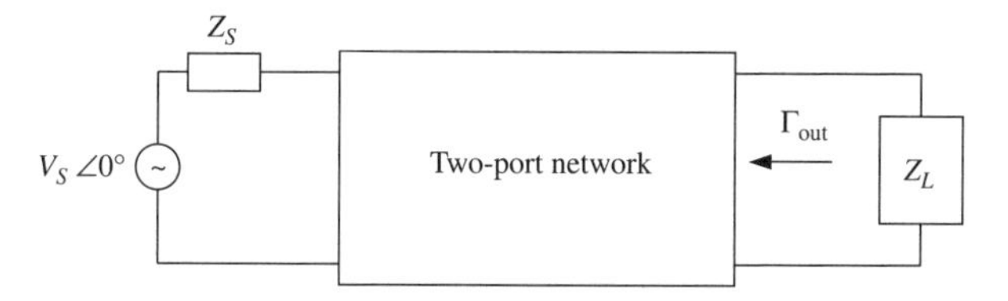

Figure 10.22 Two-port network with the source and termination.

Example 10.7 A voltage source is connected at the input port of a two-port network and the load impedance Z_L terminates its output, as shown in Figure 10.22. Draw its signal-flow graph and find the output reflection coefficient Γ_{out}.

SOLUTION A signal-flow graph of this circuit can be drawn by combining the results of Example 10.6 with those of a voltage source representation obtained in Section 10.2. This is illustrated in Figure 10.23. The output reflection coefficient is defined as the ratio of b_2 to a_2 with load disconnected, and the source impedance Z_S terminates port 1 (the input). Hence, there are two forward paths from a_2 to b_2, $a_2 \to b_2$, and $a_2 \to b_1 \to a_S \to b_S \to a_1 \to b_2$. There is only one loop (because the load is disconnected), $b_1 \to a_S \to b_S \to a_1 \to b_1$. The path and loop gains are found as follows:

$$P_1 = S_{22}$$

$$P_2 = S_{12} \times 1 \times \Gamma_S \times 1 \times S_{21} = \Gamma_S S_{12} S_{21}$$

and

$$L_1 = 1 \times \Gamma_S \times 1 \times S_{11} = \Gamma_S S_{11}$$

Therefore, from Mason's rule, we have

$$\Gamma_{out} = \frac{b_2}{a_2} = \frac{P_1(1 - L_1) + P_2}{1 - L_1} = \frac{S_{22}(1 - S_{11}\Gamma_S) + S_{21}S_{12}\Gamma_S}{1 - S_{11}\Gamma_S}$$

$$= S_{22} + \frac{S_{21}S_{12}\Gamma_S}{1 - S_{11}\Gamma_S}$$

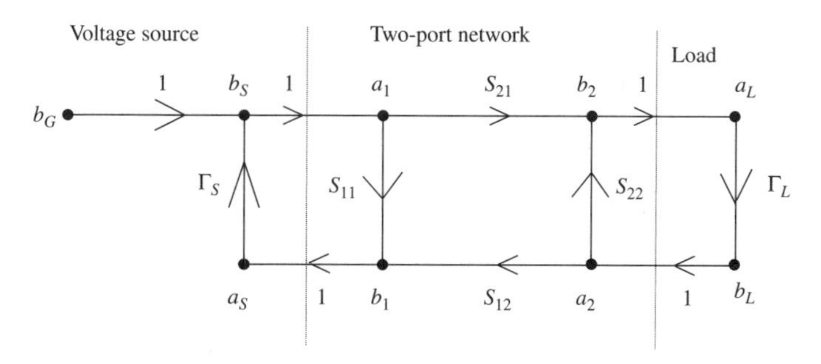

Figure 10.23 Signal-flow graph representation of the network shown in Figure 10.22.

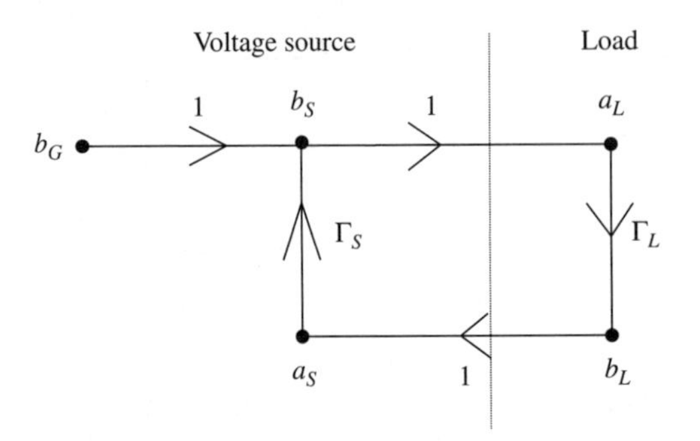

Figure 10.24 Signal-flow graph for Example 10.8.

Example 10.8 The signal-flow graph shown in Figure 10.24 represents a voltage source that is terminated by a passive load. Analyze the power transfer characteristics of this circuit and establish the conditions for maximum power transfer.

SOLUTION Using the defining equations of a_i and b_i, we can develop the following relations:

$$\text{power output of the source} = |b_S|^2$$

$$\text{power reflected back into the source} = |a_S|^2$$

Hence, power delivered by the source, P_d, is

$$P_d = |b_S|^2 - |a_S|^2$$

Similarly, we can write the following relations for power at the load:

$$\text{power incident on the load} = |a_L|^2$$

$$\text{power reflected from the load} = |b_L|^2 = |\Gamma_L|^2|a_L|^2$$

where Γ_L is the load reflection coefficient. Hence, power absorbed by the load, P_L, is given by

$$P_L = |a_L|^2 - |b_L|^2 = |a_L|^2(1 - |\Gamma_L|^2) = |b_S|^2(1 - |\Gamma_L|^2)$$

Since

$$b_S = b_G + \Gamma_S a_S = b_G + \Gamma_S b_L = b_G + \Gamma_S \Gamma_L a_L = b_G + \Gamma_S \Gamma_L b_S$$

we find that

$$b_S = \frac{b_G}{1 - \Gamma_S \Gamma_L}$$

and

$$a_S = \frac{1}{\Gamma_S}(b_S - b_G) = \frac{1}{\Gamma_S}\left(\frac{b_G}{1 - \Gamma_S\Gamma_L} - b_G\right) = \frac{\Gamma_L b_G}{1 - \Gamma_S\Gamma_L}$$

Hence,

$$P_d = |b_S|^2 - |a_S|^2 = \left|\frac{b_G}{1 - \Gamma_S\Gamma_L}\right|^2 - \left|\frac{\Gamma_L b_G}{1 - \Gamma_S\Gamma_L}\right|^2$$

$$= \left|\frac{b_G}{1 - \Gamma_S\Gamma_L}\right|^2 (1 - |\Gamma_L|^2) = |b_S|^2(1 - |\Gamma_L|^2) = P_L$$

This simply says that the power delivered by the source is equal to the power absorbed by the load. We can use this equation to establish the load condition under which the source delivers maximum power. For that, we consider ways to maximize.

$$P_L = \left|\frac{b_G}{1 - \Gamma_S\Gamma_L}\right|^2 (1 - |\Gamma_L|^2)$$

For P_L to be a maximum, its denominator $1 - \Gamma_S\Gamma_L$ is minimized. Let us analyze this term more carefully using the graphical method. It says that there is a phasor $\Gamma_S\Gamma_L$ at a distance of unity from the origin that rotates with a change in Γ_L (Γ_S is assumed constant), as illustrated in Figure 10.25. Hence, the denominator has an extreme value whenever $\Gamma_S\Gamma_L$ is a pure real number. If this number is positive real, the denominator is minimum. On the other hand, the denominator is maximized when $\Gamma_S\Gamma_L$ is a negative real number.

For $\Gamma_S\Gamma_L$ to be a positive real number, the load reflection coefficient Γ_L must be the complex conjugate of the source reflection coefficient Γ_S. In other words,

$$\Gamma_L = \Gamma_S^* \Rightarrow \frac{Z_L - Z_0}{Z_L + Z_0} = \left(\frac{Z_S - Z_0}{Z_S + Z_0}\right)^*$$

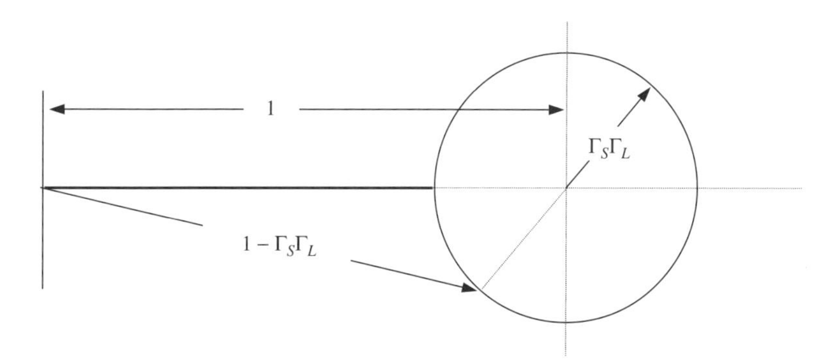

Figure 10.25 Graphical representation of $(1 - \Gamma_S\Gamma_L)$.

For a real Z_0 it reduces to

$$Z_L = Z_S^*$$

The value of this maximum power is

$$P_L = \frac{|b_G|^2}{(1 - |\Gamma_S|^2)} = P_{\text{avs}}$$

where P_{avs} is the maximum power available from the source. It is a well-known maximum power transfer theorem of electrical circuit theory.

Example 10.9 Draw the signal-flow graph of the three-port network shown in Figure 10.26. Find the transfer characteristics b_3/b_S using the graph.

SOLUTION For a three-port network, the scattering equations are given as follows:

$$b_1 = S_{11}a_1 + S_{12}a_2 + S_{13}a_3$$
$$b_2 = S_{21}a_1 + S_{22}a_2 + S_{23}a_3$$
$$b_3 = S_{31}a_1 + S_{32}a_2 + S_{33}a_3$$

For the voltage source connected at port 1, we have

$$b_S = b_G + \Gamma_S a_S$$

Since the wave coming out from the source is incident on port 1, b_S is equal to a_1. Therefore, we can write

$$a_1 = b_G + \Gamma_S b_1$$

For the load Z_L connected at port 2,

$$b_L = \Gamma_L a_L$$

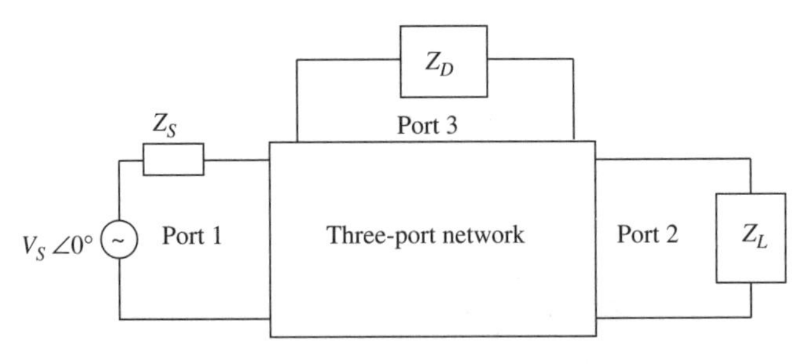

Figure 10.26 Three-port network of Example 10.9.

Again, the wave emerging from port 2 of the network is incident on the load. Similarly, the wave reflected back from the load is incident on port 2 of the network. Hence,

$$b_2 = a_L$$

and

$$b_L = a_2$$

Therefore, the relation above at the load Z_L can be modified as follows:

$$a_2 = \Gamma_L b_2$$

Following a similar procedure for the load Z_D connected at port 3, we find that

$$b_D = \Gamma_D a_D \Rightarrow a_3 = \Gamma_D b_3$$

Using these equations, a signal-flow graph can be drawn for this circuit as shown in Figure 10.27.

For b_3/b_s, there are two forward paths and eight loops in this flow graph. Path gains P_i and loop gains L_i are found as follows:

$$P_1 = S_{31}$$

$$P_2 = S_{21}S_{32}\Gamma_L$$

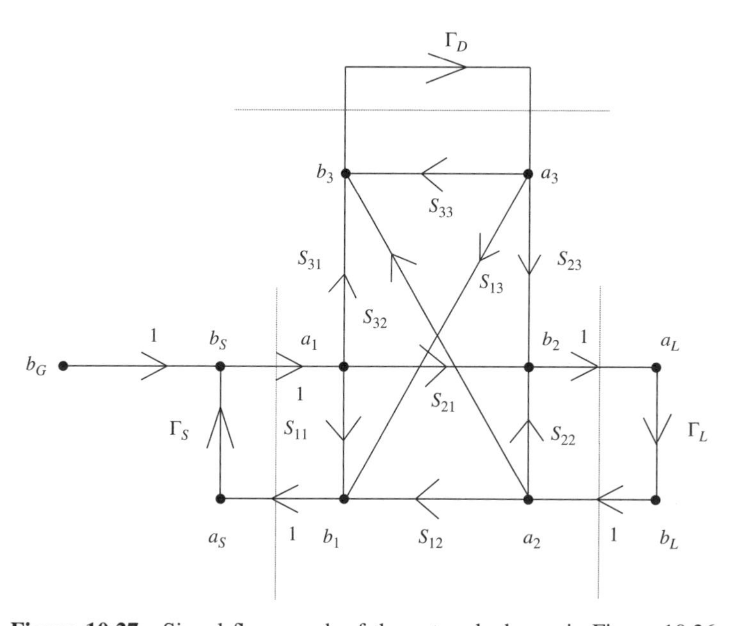

Figure 10.27 Signal-flow graph of the network shown in Figure 10.26.

$$L_1 = S_{11}\Gamma_S$$
$$L_2 = S_{22}\Gamma_L$$
$$L_3 = S_{33}\Gamma_D$$
$$L_4 = S_{21}S_{12}\Gamma_L\Gamma_S$$
$$L_5 = S_{31}S_{23}S_{12}\Gamma_D\Gamma_L\Gamma_S$$
$$L_6 = S_{13}S_{21}S_{32}\Gamma_D\Gamma_L\Gamma_S$$
$$L_7 = S_{23}S_{32}\Gamma_D\Gamma_L$$

and

$$L_8 = S_{13}S_{31}\Gamma_D\Gamma_S$$

Therefore, various terms of Mason's rule can be evaluated as follows:

$$\Delta_1 = 1 - S_{22}\Gamma_L$$
$$\Delta_2 = 1$$
$$\Sigma L(1) = L_1 + L_2 + L_3 + L_4 + L_5 + L_6 + L_7 + L_8$$
$$\Sigma L(2) = L_1L_2 + L_1L_3 + L_2L_3 + L_3L_4 + L_1L_7 + L_2L_8$$
$$\Sigma L(3) = L_1L_2L_3$$

Using Mason's rule we can find that

$$\frac{b_3}{b_S} = \frac{P_1\Delta_1 + P_2\Delta_2}{1 - \sum L(1) + \sum L(2) - \sum L(3)}$$

Note that this signal-flow graph is too complex, and therefore one can easily miss a few loops. In cases like this, it is prudent to simplify the signal-flow graph using the five rules mentioned earlier.

10.4 POWER GAIN EQUATIONS

We have seen in the preceding section that maximum power is transferred when load reflection coefficient Γ_L is the conjugate of the source reflection coefficient Γ_S. It can be generalized for any port of the network. In the case of an amplifier, maximum power will be applied to its input port if its input reflection coefficient Γ_{in} is the complex conjugate of the source reflection coefficient Γ_S. If this is not the case, the input will be less than the maximum power available from the source. Similarly, maximum amplified power will be transferred to the load only if load reflection coefficient Γ_L is the complex conjugate of output reflection coefficient Γ_{out}. Part of the power available at the output of the amplifier will be reflected back if there is a mismatch. Therefore, the power gain of an amplifier can be defined at least three different ways as follows:

$$\text{transducer power gain, } G_T = \frac{P_L}{P_{\text{avs}}} = \frac{\text{power delivered to the load}}{\text{power available from the source}}$$

$$\text{operating power gain, } G_P = \frac{P_L}{P_{\text{in}}} = \frac{\text{power delivered to the load}}{\text{power input to the network}}$$

and

$$\text{available power gain, } G_A = \frac{P_{\text{AVN}}}{P_{\text{avs}}} = \frac{\text{power available from the network}}{\text{power available from the source}}$$

where P_{avs} is the power available from the source, P_{in} the power input to the network, P_{AVN} the power available from the network, and P_L the power delivered to the load. Therefore, the transducer power gain will be equal to that of operating power gain if the input reflection coefficient Γ_{in} is the complex conjugate of the source reflection coefficient Γ_S. That is,

$$P_{\text{avs}} = P_{\text{in}}|_{\Gamma_{\text{in}}=\Gamma_S^*}$$

Similarly, the available power gain will be equal to the transducer power gain if the following condition holds at the output port:

$$P_{\text{AVN}} = P_L|_{\Gamma_L=\Gamma_{\text{out}}^*}$$

In this section we formulate these power gain relations in terms of scattering parameters and reflection coefficients of the circuit. To facilitate the formulation, we reproduce in Figure 10.28 the signal-flow graph obtained in Example 10.7.

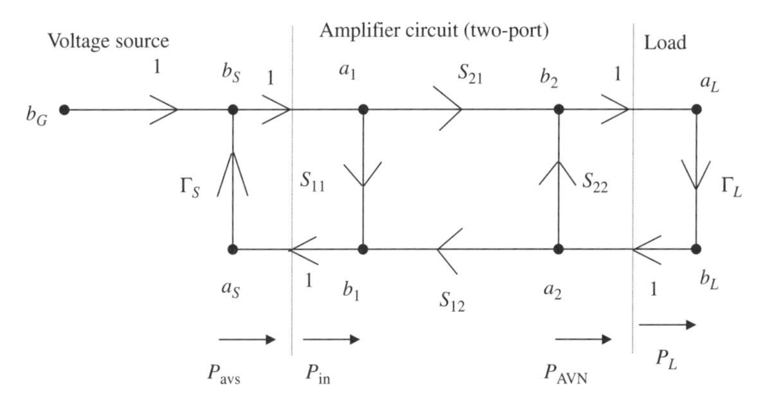

Figure 10.28 Signal-flow graph for an amplifier circuit with voltage source and terminating load.

Transducer Power Gain

To develop the relation for transducer power gain, we first determine the power delivered to load P_L and the power available from the source P_{avs}:

$$P_L = |a_L|^2 - |b_L|^2 = |a_L|^2(1 - |\Gamma_L|^2) = |b_2|^2(1 - |\Gamma_L|^2) \tag{10.4.1}$$

We derived an expression for P_{avs} in Example 10.7. That is rewritten here:

$$P_{avs} = |b_S|^2 - |a_S|^2 = \frac{|b_G|^2}{1 - |\Gamma_S|^2} \tag{10.4.2}$$

Therefore, the transducer power gain is

$$G_T = \frac{|b_2|^2}{|b_G|^2}(1 - |\Gamma_L|^2)(1 - |\Gamma_S|^2) \tag{10.4.3}$$

We now need to find an expression for b_2/b_G. Mason's rule can be used to formulate it. There is only one forward path in this case. However, there are three loops. The path gain P_1 and loop gains L_i are

$$P_1 = S_{21}$$

$$L_1 = S_{11}\Gamma_S$$

$$L_2 = S_{22}\Gamma_L$$

and

$$L_3 = S_{12}S_{21}\Gamma_S\Gamma_L$$

Therefore,

$$\begin{aligned}\frac{b_2}{b_G} &= \frac{P_1}{1 - (L_1 + L_2 + L_3) + L_1L_2} \\ &= \frac{S_{21}}{1 - (S_{11}\Gamma_S + S_{22}\Gamma_L + S_{21}\Gamma_L S_{12}\Gamma_S) + S_{11}\Gamma_S S_{22}\Gamma_L}\end{aligned} \tag{10.4.4}$$

Substituting (10.4.4) into (10.4.3), we get

$$G_T = \frac{|S_{21}|^2(1 - |\Gamma_L|^2)(1 - |\Gamma_S|^2)}{|1 - (S_{11}\Gamma_S + S_{22}\Gamma_L + S_{21}\Gamma_L S_{12}\Gamma_S) + S_{11}\Gamma_S S_{22}\Gamma_L|^2} \tag{10.4.5}$$

It can be simplified as follows:

$$G_T = \frac{|S_{21}|^2(1 - |\Gamma_L|^2)(1 - |\Gamma_S|^2)}{|(1 - S_{11}\Gamma_S)(1 - S_{22}\Gamma_L) - S_{21}S_{12}\Gamma_L\Gamma_S|^2}$$

or

$$G_T = \frac{|S_{21}|^2(1-|\Gamma_L|^2)(1-|\Gamma_S|^2)}{|(1-S_{22}\Gamma_L)[1-S_{11}\Gamma_S - S_{21}S_{12}\Gamma_L\Gamma_S/(1-S_{22}\Gamma_L)]|^2}$$

or

$$G_T = \frac{|S_{21}|^2(1-|\Gamma_L|^2)(1-|\Gamma_S|^2)}{|(1-S_{22}\Gamma_L)(1-\Gamma_{\text{in}}\Gamma_S)|^2}$$

$$= \frac{1-|\Gamma_S|^2}{|1-\Gamma_{\text{in}}\Gamma_S|^2}|S_{21}|^2\frac{1-|\Gamma_L|^2}{|1-S_{22}\Gamma_L|^2} \qquad (10.4.6)$$

where

$$\Gamma_{\text{in}} = S_{11} + \frac{S_{21}S_{12}\Gamma_L}{1-S_{22}\Gamma_L} \qquad (10.4.7)$$

This is the input reflection coefficient formulated in Example 10.6.
Alternatively, (10.4.5) can be written as

$$G_T = \frac{|S_{21}|^2(1-|\Gamma_L|^2)(1-|\Gamma_S|^2)}{|(1-S_{11}\Gamma_S)(1-\Gamma_{\text{out}}\Gamma_L)|^2} = \frac{1-|\Gamma_S|^2}{|1-S_{11}\Gamma_S|^2}|S_{21}|^2\frac{1-|\Gamma_L|^2}{|1-\Gamma_{\text{out}}\Gamma_L|^2} \qquad (10.4.8)$$

where

$$\Gamma_{\text{out}} = S_{22} + \frac{S_{21}S_{12}\Gamma_S}{1-S_{11}\Gamma_S} \qquad (10.4.9)$$

This is the output reflection coefficient formulated in Example 10.7.

Special Case Consider the case when the two-port network is unilateral. This means that there is an output signal at port 2 when a source is connected at port 1. However, a source (cause) connected at port 2 does not show its response (effect) at port 1. In terms of scattering parameters of the two-port, S_{12} is equal to zero in this case, and therefore, the input and output reflection coefficients in (10.4.7) and (10.4.9) simplify as follows:

$$\Gamma_{\text{in}} = S_{11}$$

and

$$\Gamma_{\text{out}} = S_{22}$$

The transducer power gain G_T as given by (10.4.6) or (10.4.8) simplifies to

$$G_T|_{S_{12}=0} = G_{TU} = \frac{1-|\Gamma_S|^2}{|1-S_{11}\Gamma_S|^2}|S_{21}|^2\frac{1-|\Gamma_L|^2}{|1-S_{22}\Gamma_L|^2} = G_S G_0 G_L \qquad (10.4.10)$$

where G_{TU} may be called the *unilateral transducer power gain*, while

$$G_S = \frac{1-|\Gamma_S|^2}{|1-S_{11}\Gamma_S|^2} \qquad (10.4.11)$$

$$G_0 = |S_{21}|^2 \qquad (10.4.12)$$

and

$$G_L = \frac{1 - |\Gamma_L|^2}{|1 - S_{22}\Gamma_L|^2} \tag{10.4.13}$$

Equation (10.4.10) indicates that maximum G_{TU} is obtained when the input and output ports are conjugate matched. Hence,

$$G_{TU_{\max}} = G_{S_{\max}} G_0 G_{L_{\max}} \tag{10.4.14}$$

where

$$G_{S_{\max}} = G_S|_{\Gamma_S = S_{11}^*} \tag{10.4.15}$$

and

$$G_{L_{\max}} = G_L|_{\Gamma_L = S_{22}^*} \tag{10.4.16}$$

Operating Power Gain

To find the operating power gain, we require the power delivered to the load and power input to the circuit. We evaluated the power delivered to the load in (10.4.1). The power input to the circuit is found as follows:

$$P_{\text{in}} = |a_1|^2 - |b_1|^2 = |a_1|^2(1 - |\Gamma_{\text{in}}|^2)$$

Hence,

$$G_P = \frac{P_L}{P_{\text{in}}} = \frac{|b_2|^2(1 - |\Gamma_L|^2)}{|a_1|^2(1 - |\Gamma_{\text{in}}|^2)} \tag{10.4.17}$$

Now we evaluate b_2/a_1 using Mason's rule. With the source disconnected, we find that there is only one forward path and one loop with gains S_{21} and $S_{22}\Gamma_L$, respectively. Hence,

$$\frac{b_2}{a_1} = \frac{S_{21}}{1 - S_{22}\Gamma_L} \tag{10.4.18}$$

Substituting (10.4.18) into (10.4.17), we find that

$$G_P = \frac{1}{1 - |\Gamma_{\text{in}}|^2} |S_{21}|^2 \frac{1 - |\Gamma_L|^2}{|1 - S_{22}\Gamma_L|^2} \tag{10.4.19}$$

Available Power Gain

Expressions for maximum powers available from the network and the source are needed to determine the available power gain. Maximum power available from the network P_{AVN} is delivered to the load if the load reflection coefficient is a complex conjugate of the output. Using (10.4.1), we find that

$$P_{\text{AVN}} = P_L|_{\Gamma_L = \Gamma_{\text{out}}^*} = |b_2|^2(1 - |\Gamma_L|^2) = |b_2|^2(1 - |\Gamma_{\text{out}}|^2) \tag{10.4.20}$$

The power available from the source is given by (10.4.2). Hence,

$$G_A = \frac{P_{\text{AVN}}}{P_{\text{avs}}} = \frac{|b_2|^2}{|b_G|^2}(1 - |\Gamma_{\text{out}}|^2)(1 - |\Gamma_S|^2) \tag{10.4.21}$$

Next, using Mason's rule in the signal-flow graph of Figure 10.28, we find that

$$\frac{b_2}{b_G} = \frac{S_{21}}{1 - (S_{11}\Gamma_S + S_{22}\Gamma_L + S_{21}\Gamma_L S_{12}\Gamma_S) + S_{11}\Gamma_S S_{22}\Gamma_L}$$
$$= \frac{S_{21}}{(1 - S_{11}\Gamma_S)(1 - \Gamma_{\text{out}}\Gamma_L)}$$

and

$$\left.\frac{b_2}{b_G}\right|_{\Gamma_L=\Gamma_{\text{out}}^*} = \frac{S_{21}}{(1 - S_{11}\Gamma_S)(1 - |\Gamma_{\text{out}}|^2)} \tag{10.4.22}$$

Now, substituting (10.4.22) into (10.4.21), we get

$$G_A = \frac{1 - |\Gamma_S|^2}{|1 - S_{11}\Gamma_S|^2}\,|S_{21}|^2\,\frac{1}{1 - |\Gamma_{\text{out}}|^2} \tag{10.4.23}$$

Example 10.10 The scattering parameters of a microwave amplifier are found at 800 MHz as follows ($Z_0 = 50\,\Omega$): $S_{11} = 0.45\ \angle 150^\circ$, $S_{12} = 0.01\ \angle -10^\circ$, $S_{21} = 2.05\ \angle 10^\circ$, and $S_{22} = 0.4\ \angle -150^\circ$. The source and load impedances are 20 and 30 Ω, respectively. Determine the transducer power gain, operating power gain, and available power gain.

SOLUTION First we determine the source reflection coefficient Γ_S and the load reflection coefficient Γ_L as follows:

$$\Gamma_S = \frac{Z_S - Z_0}{Z_S + Z_0} = \frac{20 - 50}{20 + 50} = -\frac{30}{70} = -0.429$$

and

$$\Gamma_L = \frac{Z_L - Z_0}{Z_L + Z_0} = \frac{30 - 50}{30 + 50} = -\frac{20}{80} = -0.25$$

Now the input reflection coefficient Γ_{in} and the output reflection coefficient Γ_{out} can be calculated from (10.4.7) and (10.4.9), respectively:

$$\Gamma_{\text{in}} = S_{11} + \frac{S_{21}S_{12}\Gamma_L}{1 - S_{22}\Gamma_L} = 0.45\ \angle 150^\circ + \frac{(0.01\ \angle -10^\circ)(2.05\ \angle 10^\circ)(-0.25)}{1 - (0.4\ \angle -150^\circ)(-0.25)}$$

or

$$\Gamma_{\text{in}} = -0.3953 + j0.2253 = 0.455\ \angle 150.32^\circ$$

and

$$\Gamma_{\text{out}} = S_{22} + \frac{S_{21}S_{12}\Gamma_S}{1 - S_{11}\Gamma_S} = 0.4\ \angle -150^\circ + \frac{(0.01\ \angle -10^\circ)(2.05\ \angle 10^\circ)(-0.429)}{1 - (0.45\ \angle 150^\circ)(-0.429)}$$

or

$$\Gamma_{\text{out}} = -0.3568 - j0.1988 = 0.4084\ \angle -150.87^\circ$$

The transducer power gain G_T can now be calculated from (10.4.6) or (10.4.8) as follows:

$$G_T = \frac{|S_{21}|^2(1 - |\Gamma_L|^2)(1 - |\Gamma_S|^2)}{|(1 - S_{22}\Gamma_L)(1 - \Gamma_{\text{in}}\Gamma_S)|^2} = \frac{|S_{21}|^2(1 - |\Gamma_L|^2)(1 - |\Gamma_S|^2)}{|(1 - S_{11}\Gamma_S)(1 - \Gamma_{\text{out}}\Gamma_L)|^2}$$

Hence,

$$G_T = \frac{1 - (0.429)^2}{|1 - (0.45\ \angle 150^\circ)(-0.429)|^2}(2.05)^2 \frac{1 - (0.25^2)}{|1 - (0.4084\ \angle -150.87^\circ)(-0.25)|^2}$$
$$= 5.4872$$

Similarly, from (10.4.19),

$$G_P = \frac{1}{1 - |\Gamma_{\text{in}}|^2}\ |S_{21}|^2\ \frac{1 - |\Gamma_L|^2}{|1 - S_{22}\Gamma_L|^2}$$

Therefore,

$$G_P = \frac{1}{1 - (0.455^2)}(2.05^2)\frac{1 - (0.25^2)}{|1 - (0.4\ \angle -150^\circ)(-0.25)|^2} = 5.9374$$

The available power gain is calculated from (10.4.23) as follows:

$$G_A = \frac{1 - |\Gamma_S|^2}{|1 - S_{11}\Gamma_S|^2}\ |S_{21}|^2\ \frac{1}{1 - |\Gamma_{\text{out}}|^2}$$

Hence,

$$G_A = \frac{1 - (0.429)^2}{|1 - (0.45\ \angle 150^\circ)(-0.429)|^2}(2.05^2)\frac{1}{1 - (0.4084)^2} = 5.8552$$

The three power gains can be expressed in decibels as follows:

$$G_T(\text{dB}) = 10\log(5.4872) \approx 7.4\,\text{dB}$$

$$G_P(\text{dB}) = 10\log(5.9374) \approx 7.7\,\text{dB}$$

and

$$G_A(\text{dB}) = 10\log(5.8552) \approx 7.7\,\text{dB}$$

Example 10.11 The scattering parameters of a microwave amplifier are found at 2 GHz as follows ($Z_0 = 50\ \Omega$): $S_{11} = 0.97\ \angle - 43^\circ$, $S_{12} = 0.0$, $S_{21} = 3.39\ \angle 140^\circ$, and $S_{22} = 0.63\ \angle - 32^\circ$. The source and load reflection coefficients are $0.97\ \angle 43^\circ$ and $0.63\ \angle 32^\circ$, respectively. Determine the transducer power gain, operating power gain, and available power gain.

SOLUTION Since S_{12} is zero, this amplifier is unilateral. Therefore, its input and output reflection coefficients from (10.4.7) and (10.4.9) may be found as follows:

$$\text{input reflection coefficient } \Gamma_{in} = S_{11} = 0.97\ \angle - 43^\circ$$

$$\text{output reflection coefficient } \Gamma_{out} = S_{22} = 0.63\ \angle - 32^\circ$$

On comparing the input reflection coefficient with that of the source, we find that one is the complex conjugate of the other. Hence, the source is matched with the input port, and therefore the power available from the source is equal to the power input. Similarly, we find that the load reflection coefficient is the conjugate of the output reflection coefficient. Hence, power delivered to the load will be equal to power available from the network at port 2. As may be verified from (10.4.10), (10.4.19), and (10.4.23), these conditions make all three power gains equal. From (10.4.10),

$$G_{TU} = \frac{1 - |\Gamma_S|^2}{|1 - S_{11}\Gamma_S|^2} |S_{21}|^2 \frac{1 - |\Gamma_L|^2}{|1 - S_{22}\Gamma_L|^2} = \frac{1 - (0.97)^2}{|1 - (0.97)^2|^2} (3.39)^2$$
$$\times \frac{1 - (0.63)^2}{|1 - (0.63)^2|^2} = 322.42$$

From (10.4.19),

$$G_P = \frac{1}{1 - |\Gamma_{in}|^2} |S_{21}|^2 \frac{1 - |\Gamma_L|^2}{|1 - S_{22}\Gamma_L|^2} = \frac{1}{1 - (0.97)^2} (3.39)^2 \frac{1 - (0.63)^2}{|1 - (0.63)^2|^2}$$
$$= 322.42$$

From (10.4.23),

$$G_A = \frac{1 - |\Gamma_S|^2}{|1 - S_{11}\Gamma_S|} |S_{21}|^2 \frac{1}{1 - |\Gamma_{out}|^2} = \frac{1 - (0.97)^2}{|1 - (0.97)^2|^2} (3.39)^2 \frac{1}{1 - (0.63)^2}$$
$$= 322.42$$

SUGGESTED READING

Collin, R. E., *Foundations for Microwave Engineering*. New York: McGraw-Hill, 1992.

Davis, W. A., *Microwave Semiconductor Circuit Design*. New York: Van Nostrand Reinhold, 1984.

Fusco, V. F., *Microwave Circuits*. Englewood Cliffs, NJ: Prentice Hall, 1987.

Gonzalez, G., *Microwave Transistor Amplifiers*. Upper Saddle River, NJ: Prentice Hall, 1997.

Pozar, D. M., *Microwave Engineering*. New York: Wiley, 1998.

Wolff, E. A., and R. Kaul, *Microwave Engineering and Systems Applications*. New York: Wiley, 1988.

PROBLEMS

10.1. Use Mason's rule to find $Y(s)/R(s)$ for the signal-flow graph shown in Figure P10.1.

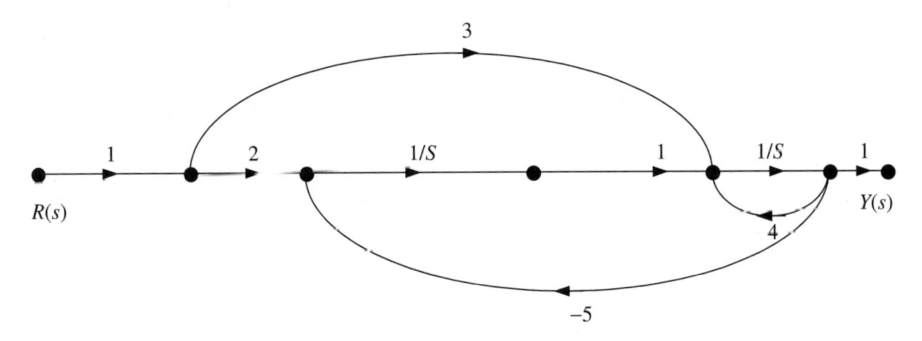

Figure P10.1

10.2. Use Mason's gain rule to find $Y_5(s)/Y_1(s)$ and $Y_2(s)/Y_1(s)$ for the system shown in Figure P10.2.

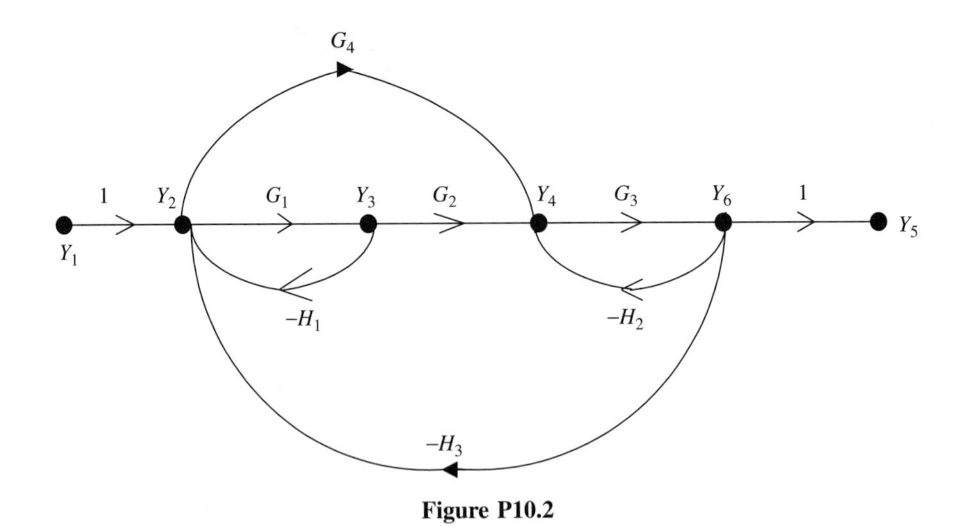

Figure P10.2

10.3. Use Mason's gain rule to find $Y_2(s)/R_3(s)$ for the system shown in Figure P10.3.

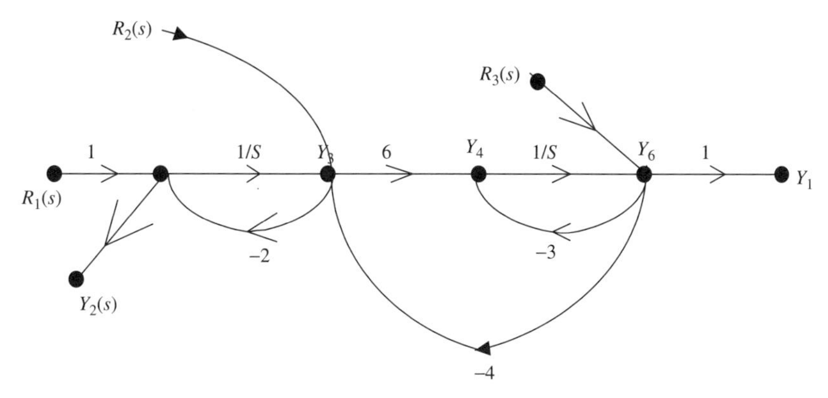

Figure P10.3

10.4. Use Mason's gain rule to find $Y(s)/R(s)$ for the signal-flow graph shown in Figure P10.4.

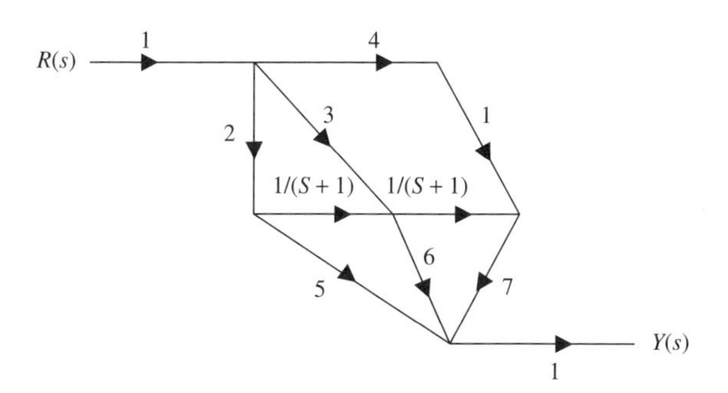

Figure P10.4

10.5. Draw the signal-flow graph for the circuit shown in Figure P10.5. Use Mason's gain rule to determine the power ratios P_2/P_1 and P_3/P_1.

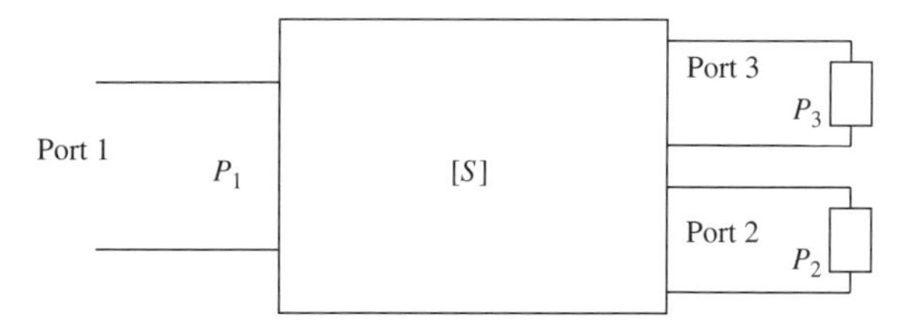

Figure P10.5

$$[S] = \begin{bmatrix} 0 & S_{12} & 0 \\ S_{21} & 0 & S_{23} \\ 0 & S_{32} & 0 \end{bmatrix}$$

Γ_{in} is the reflection coefficient at port 1, and Γ_2 and Γ_3 are the reflection coefficients at ports 2 and 3, respectively.

10.6. A voltage source has an internal impedance of $100\,\Omega$, and an electromotive force of 3 V. Using $Z_0 = 50\,\Omega$, find the reflection coefficient and scattering variables a and b for the source.

10.7. A 10-mV voltage source has an internal impedance of $100 + j50\,\Omega$. Assuming that the characteristic impedance Z_0 is $50\,\Omega$, find the reflection coefficient and scattering variables a and b for the source. Draw the signal-flow graph.

10.8. A voltage source of $1\ \angle 0^\circ$ V has an internal resistance of $80\,\Omega$. It is connected to a load of $30 + j40\,\Omega$. Assuming that the characteristic impedance is $50\,\Omega$, find the scattering variables and draw the signal-flow graph.

10.9. The scattering parameters of a two-port network are $S_{11} = 0.26 - j0.16$, $S_{12} = S_{21} = 0.42$, and $S_{22} = 0.36 - j0.57$.

(a) Determine the input reflection coefficient and transducer power loss for $Z_S = Z_L = Z_0$.

(b) Do the same for $Z_S = Z_0$ and $Z_L = 3Z_0$.

10.10. Two two-port networks are connected in cascade. The scattering parameters of the first two-port network are $S_{11} = 0.28$, $S_{12} = S_{21} = 0.56$, and $S_{22} = 0.28 - j0.56$. For the second network, $S_{11} = S_{22} = -j0.54$ and $S_{12} = S_{21} = 0.84$. Determine the overall scattering and transmission parameters.

10.11. Draw the flow graph representation for the two cascaded circuits of Problem 10.10. Using Mason's rule, determine the following: (**a**) the input reflection coefficient when the second network is match-terminated and (**b**) the forward transmission coefficient of the overall network.

10.12. A GaAs MESFET has the following S-parameters measured at 8 GHz with a 50-Ω reference: $S_{11} = 0.26\ \angle -55^\circ$, $S_{12} = 0.08\ \angle 80^\circ$, $S_{21} = 2.14\ \angle 65^\circ$, and $S_{22} = 0.82\ \angle -30^\circ$. For $Z_L = 40 + j160\,\Omega$ and $\Gamma_S = 0.31\ \angle -76.87^\circ$, determine the operating power gain of the circuit.

10.13. The S-parameters of a certain microwave transistor measured at 3 GHz with a 50-Ω reference are $S_{11} = 0.406\ \angle -100^\circ$, $S_{12} = 0$, $S_{21} = 5.0\ \angle 50^\circ$, and $S_{22} = 0.9\ \angle -60^\circ$. Calculate the maximum unilateral power gain.

10.14. **(a)** Find the value of the source impedance that results in maximum power delivered to the load in the circuit shown in Figure P10.14. Evaluate the maximum power delivered to the load.

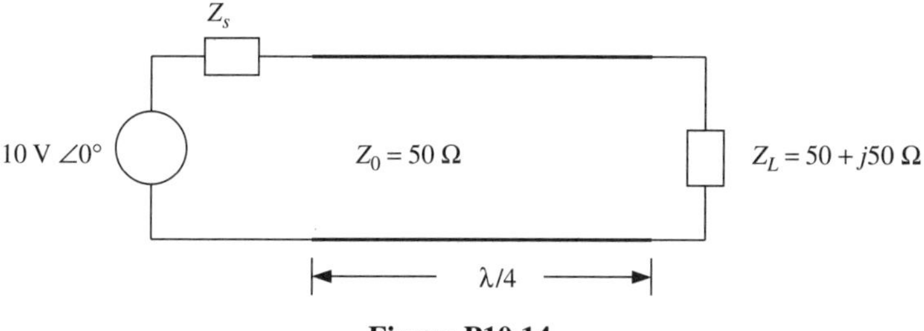

Figure P10.14

(b) Using the value of Z_s from part (a), find the Thévenin equivalent circuit at the load end and evaluate the power delivered to the load.

11

TRANSISTOR AMPLIFIER DESIGN

Amplifiers are among the basic building blocks of an electronic system. Although vacuum-tube devices are still used in high-power microwave circuits, transistors—silicon bipolar junction devices, GaAs MESFETs, heterojunction bipolar transistors (HBTs), and high-electron mobility transistors (HEMTs)—are common in many radio-frequency and microwave designs. The chapter begins with the stability considerations for a two-port network and the formulation of relevant conditions in terms of its scattering parameters. Expressions for input and output stability circles are presented next to facilitate the design of amplifier circuits. Design procedures for various small-signal single-stage amplifiers are discussed for unilateral as well as bilateral transistors. Noise figure considerations in amplifier design are discussed in the following section. An overview of broadband amplifiers is included. Small-signal equivalent circuits and biasing mechanisms for various transistors are also summarized in subsequent sections.

11.1 STABILITY CONSIDERATIONS

Consider a two-port network that is terminated by load Z_L as shown in Figure 11.1. A voltage source V_S with internal impedance Z_S is connected at its input port. Reflection coefficients at its input and output ports are Γ_{in} and Γ_{out}, respectively. The source reflection coefficient is Γ_S and the load reflection coefficient is Γ_L. Expressions for input and output reflection coefficients were formulated in Examples 10.6 and 10.7. For this two-port to be unconditionally stable at a given frequency, the

Radio-Frequency and Microwave Communication Circuits: Analysis and Design, Second Edition, By Devendra K. Misra
ISBN 0-471-47873-3

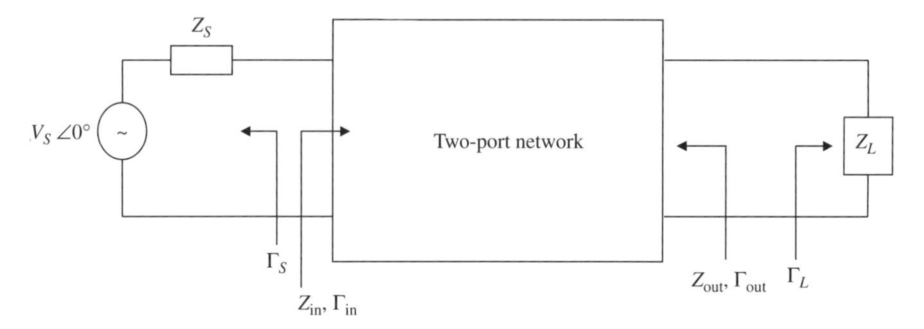

Figure 11.1 Two-port network with voltage source at its input and a load terminating the output port.

following inequalities must hold:

$$|\Gamma_S| < 1 \tag{11.1.1}$$

$$|\Gamma_L| < 1 \tag{11.1.2}$$

$$|\Gamma_{in}| = \left| S_{11} + \frac{S_{21}S_{12}\Gamma_L}{1 - S_{22}\Gamma_L} \right| < 1 \tag{11.1.3}$$

and

$$|\Gamma_{out}| = \left| S_{22} + \frac{S_{21}S_{12}\Gamma_S}{1 - S_{11}\Gamma_S} \right| < 1 \tag{11.1.4}$$

Condition (11.1.3) can be rearranged as follows:

$$\left| S_{11} + \frac{S_{22}S_{21}S_{12}\Gamma_L + (S_{12}S_{21} - S_{12}S_{21})}{S_{22}(1 - S_{22}\Gamma_L)} \right| < 1$$

or

$$\left| \frac{1}{S_{22}} \left(\Delta + \frac{S_{21}S_{12}}{1 - S_{22}\Gamma_L} \right) \right| < 1 \tag{11.1.5}$$

where

$$\Delta = S_{11}S_{22} - S_{12}S_{21} \tag{11.1.6}$$

Since

$$1 - S_{22}\Gamma_L \underset{\Gamma_L = 1}{\longrightarrow} 1 - S_{22} = 1 - |S_{22}| \exp(j\theta) = A \tag{11.1.7}$$

This traces a circle on the complex plane as θ varies from zero to 2π. It is illustrated in Figure 11.2. Further, $1/[1 - |S_{22}| \exp(j\theta)]$ represents a circle of radius r with its center located at d, where

$$r = \frac{1}{2} \left(\frac{1}{1 - |S_{22}|} - \frac{1}{1 + |S_{22}|} \right) = \frac{|S_{22}|}{1 - |S_{22}|^2} \tag{11.1.8}$$

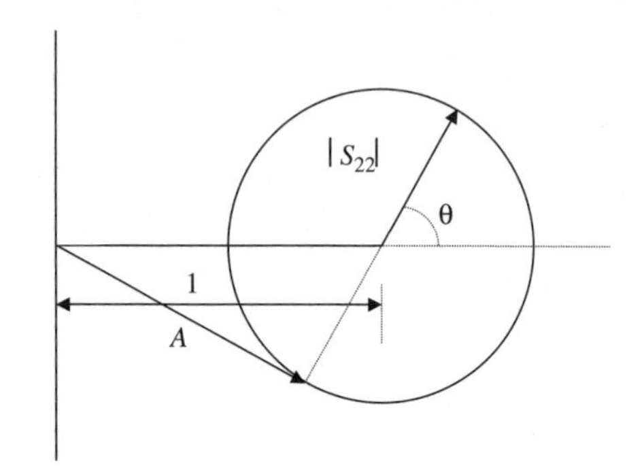

Figure 11.2 Graphical representation of (11.1.7).

and

$$d = \frac{1}{2}\left(\frac{1}{1+|S_{22}|} + \frac{1}{1-|S_{22}|}\right) = \frac{1}{1-|S_{22}|^2} \tag{11.1.9}$$

Hence, for $|\Gamma_L| \le 1$ and $\angle\Gamma_L = \theta$, condition (11.1.5) may be written as

$$\left|\frac{1}{|S_{22}|}\left(\Delta + S_{21}S_{12}\left\{\frac{1+|S_{22}|\exp(j\theta)}{1-|S_{22}|^2}\right\}\right)\right| < 1$$

or

$$\frac{1}{|S_{22}|}\left|\Delta + \frac{S_{12}S_{21}}{1-|S_{22}|^2} + \frac{S_{12}S_{21}|S_{22}|\exp(j\theta)}{1-|S_{22}|^2}\right| < 1 \tag{11.1.10}$$

Now, using the Minkowski inequality,

$$\left(\sum_{k=1}^{n}|a_k+b_k|^p\right)^{1/p} \le \left(\sum_{k=1}^{n}|a_k|^p\right)^{1/p} + \left(\sum_{k=1}^{n}|b_k|^p\right)^{1/p} \tag{11.1.11}$$

we find that (11.1.10) is satisfied if

$$\frac{1}{|S_{22}|}\left|\Delta + \frac{S_{12}S_{21}}{1-|S_{22}|^2}\right| + \frac{|S_{12}S_{21}|}{1-|S_{22}|^2} < 1$$

or

$$\frac{1}{|S_{22}|}\left|\Delta + \frac{S_{12}S_{21}}{1-|S_{22}|^2}\right| < 1 - \frac{|S_{12}S_{21}|}{1-|S_{22}|^2} \tag{11.1.12}$$

Since the left-hand side of (11.1.12) is always a positive number, this inequality will be satisfied if the following is true:

$$1 - |S_{22}|^2 - |S_{12}S_{21}| > 0 \tag{11.1.13}$$

Similarly, stability condition (11.1.4) will be satisfied if

$$1 - |S_{11}|^2 - |S_{12}S_{21}| > 0 \tag{11.1.14}$$

Adding (11.1.13) and (11.1.14), we get

$$2 - |S_{11}|^2 - |S_{22}|^2 - 2|S_{12}S_{21}| > 0$$

or

$$1 - \tfrac{1}{2}(|S_{11}|^2 + |S_{22}|^2) > |S_{12}S_{21}| \tag{11.1.15}$$

From (11.1.6) and (11.1.15), we have

$$|\Delta| < |S_{11}S_{22}| + |S_{12}S_{21}| < |S_{11}S_{22}| + 1 - \tfrac{1}{2}(|S_{11}|^2 + |S_{22}|^2)$$

or

$$|\Delta| < 1 - \tfrac{1}{2}(|S_{11}| - |S_{22}|)^2 \Rightarrow |\Delta| < 1 \tag{11.1.16}$$

Multiplying (11.1.13) and (11.1.14), we get

$$(1 - |S_{22}|^2 - |S_{12}S_{21}|)(1 - |S_{11}|^2 - |S_{12}S_{21}|) > 0$$

or

$$1 - |S_{11}|^2 - |S_{22}|^2 - 2|S_{12}S_{21}| + \zeta > 0 \tag{11.1.17}$$

where

$$\zeta = |S_{11}|^2|S_{22}|^2 + |S_{12}S_{21}|^2 + |S_{12}S_{21}|(|S_{11}|^2 + |S_{22}|^2)$$

From the self-evident identity,

$$|S_{12}S_{21}|(|S_{11}| - |S_{22}|)^2 \geq 0$$

it can be proved that

$$\zeta \leq |\Delta|^2$$

Therefore, (11.1.17) can be written as:

$$1 - |S_{11}|^2 - |S_{22}|^2 - 2|S_{12}S_{21}| + |\Delta|^2 > 0$$

or

$$1 - |S_{11}|^2 - |S_{22}|^2 + |\Delta|^2 > 2|S_{12}S_{21}|$$

or

$$\frac{1 - |S_{11}|^2 - |S_{22}|^2 + |\Delta|^2}{2|S_{12}S_{21}|} > 1$$

Therefore,

$$k = \frac{1 - |S_{11}|^2 - |S_{22}|^2 + |\Delta|^2}{2|S_{12}S_{21}|} > 1 \tag{11.1.18}$$

If the S-parameters of a transistor satisfy conditions (11.1.16) and (11.1.18), it is stable for any passive load and generator impedance. In other words, this transistor is *unconditionally stable*. On the other hand, it may be *conditionally stable* (stable for limited values of load or source impedance) if one or both of these conditions are violated. It means that the transistor can provide stable operation for a restricted range of Γ_S and Γ_L. A simple procedure to find these stable regions is to test inequalities (11.1.3) and (11.1.4) for particular load and source impedances. An alternative graphical approach is to find the circles of instability for load and generator reflection coefficients on a Smith chart. The latter approach is presented below.

From the expression of input reflection coefficient (10.4.7), we find that

$$\Gamma_{in} = S_{11} + \frac{S_{21}S_{12}\Gamma_L}{1 - S_{22}\Gamma_L} \Rightarrow \Gamma_{in}(1 - S_{22}\Gamma_L) = S_{11}(1 - S_{22}\Gamma_L) + S_{21}S_{12}\Gamma_L$$

or

$$\Gamma_{in} = S_{11} - \Gamma_L(S_{11}S_{22} - S_{12}S_{21} - \Gamma_{in}S_{22}) \Rightarrow \Gamma_L = \frac{S_{11} - \Gamma_{in}}{\Delta - \Gamma_{in}S_{22}}$$

or

$$\Gamma_L = \frac{S_{11} - \Gamma_{in}}{\Delta - \Gamma_{in}S_{22}}\frac{S_{22}}{S_{22}} = \frac{1}{S_{22}}\frac{S_{11}S_{22} - \Gamma_{in}S_{22} - S_{12}S_{21} + S_{12}S_{21}}{\Delta - \Gamma_{in}S_{22}}$$

or

$$\Gamma_L = \frac{1}{S_{22}}\left(1 + \frac{S_{12}S_{21}}{\Delta - \Gamma_{in}S_{22}}\right) = \frac{1}{\Delta S_{22}}\left(\Delta + \frac{S_{12}S_{21}}{1 - \Gamma_{in}\,\Delta^{-1}S_{22}}\right) \quad (11.1.19)$$

As before, $1 - \Gamma_{in}S_{22}\Delta^{-1}$ represents a circle on the complex plane. It is centered at 1 with radius $|\Gamma_{in}S_{22}\Delta^{-1}|$; the reciprocal of this expression is another circle with center at

$$\frac{1}{2}\left(\frac{1}{1 + |\Delta^{-1}S_{22}|} + \frac{1}{1 - |\Delta^{-1}S_{22}|}\right) = \frac{1}{1 - |\Delta^{-1}S_{22}|^2}$$

and radius

$$\frac{1}{2}\left(\frac{1}{1 - |\Delta^{-1}S_{22}|} - \frac{1}{1 + |\Delta^{-1}S_{22}|}\right) = \frac{|\Delta^{-1}S_{22}|}{1 - |\Delta^{-1}S_{22}|^2}$$

Since $|\Gamma_{in}| < 1$, the region of stability will include all points on the Smith chart outside this circle. From (11.1.19), the center of the load impedance circle, C_L, is

$$C_L = \frac{1}{\Delta S_{22}}\left(\Delta + \frac{S_{12}S_{21}}{1 - |\Delta^{-1}S_{22}|^2}\right) = \frac{1}{\Delta S_{22}}\left(\Delta + \frac{S_{12}S_{21}|\Delta|^2}{|\Delta|^2 - |S_{22}|^2}\right)$$

or

$$C_L = \frac{1}{S_{22}}\left(1 + \frac{S_{12}S_{21}\Delta^*}{|\Delta|^2 - |S_{22}|^2}\right) = \frac{1}{S_{22}}\frac{|\Delta|^2 - |S_{22}|^2 + S_{12}S_{21}\Delta^*}{|\Delta|^2 - |S_{22}|^2}$$

or

$$C_L = \frac{1}{S_{22}} \frac{\Delta^*(\Delta + S_{12}S_{21}) - |S_{22}|^2}{|\Delta|^2 - |S_{22}|^2} = \frac{1}{S_{22}} \frac{\Delta^* S_{11} S_{22} - |S_{22}|^2}{|\Delta|^2 - |S_{22}|^2}$$

Therefore,

$$C_L = \frac{\Delta^* S_{11} - S_{22}^*}{|\Delta|^2 - |S_{22}|^2} = \frac{(S_{22} - \Delta S_{11}^*)^*}{|S_{22}|^2 - |\Delta|^2} \tag{11.1.20}$$

Its radius, r_L, is given by

$$r_L = \frac{1}{|\Delta S_{22}|} \left| \frac{S_{12}S_{21}|\Delta^{-1} S_{22}|}{1 - |\Delta^{-1} S_{22}|^2} \right| = \left| \frac{S_{12}S_{21}}{|\Delta|^2 - |S_{22}|^2} \right| \tag{11.1.21}$$

As explained following (11.1.19), this circle represents the locus of points over which the input reflection coefficient Γ_{in} is equal to unity. On one side of this circle, the input reflection coefficient is less than unity (stable region); on its other side it exceeds 1 (unstable region). When load reflection coefficient Γ_L is zero (i.e., a matched termination is used), Γ_{in} is equal to S_{11}. Hence, the center of the Smith chart (reflection coefficient equal to zero) represents a stable point if $|S_{11}|$ is less than unity. On the other hand, it represents unstable impedance for $|S_{11}|$ greater than unity. If $\Gamma_L = 0$ is located outside the stability circle and is found stable, all outside points are stable. Similarly, if $\Gamma_L = 0$ is inside the stability circle and is found stable, all enclosed points are stable. If $\Gamma_L = 0$ is unstable, all points on that side of the stability circle are unstable.

Similarly, the locus of Γ_S can be derived from (11.1.4), with its center C_S and its radius r_S given as

$$C_S = \frac{\Delta^* S_{22} - S_{11}^*}{|\Delta|^2 - |S_{11}|^2} = \frac{(S_{11} - \Delta S_{22}^*)^*}{|S_{11}|^2 - |\Delta|^2} \tag{11.1.22}$$

and

$$r_S = \frac{1}{|\Delta S_{11}|} \left| \frac{S_{12}S_{21}|\Delta^{-1} S_{11}|}{1 - |\Delta^{-1} S_{11}|^2} \right| = \left| \frac{S_{12}S_{21}}{|\Delta|^2 - |S_{11}|^2} \right| \tag{11.1.23}$$

This circle represents the locus of points over which output reflection coefficient Γ_{out} is equal to unity. On one side of this circle, the output reflection coefficient is less than unity (stable region), whereas on its other side it exceeds 1 (unstable region). When the source reflection coefficient Γ_S is zero, Γ_{out} is equal to S_{22}. Hence, the center of the Smith chart (reflection coefficient equal to zero) represents a stable point if $|S_{22}|$ is less than unity. On the other hand, it represents an unstable impedance point for $|S_{22}|$ greater than unity. If $\Gamma_S = 0$ is located outside the stability circle and is found stable, all outside source-impedance points are stable. Similarly, if $\Gamma_S = 0$ is inside the stability circle and is found stable, all enclosed points are stable. If $\Gamma_S = 0$ is unstable, all points on that side of the stability circle are unstable.

Example 11.1 The S-parameters of a properly biased transistor are found at 2 GHz as follows (50-Ω reference impedance): $S_{11} = 0.894\ \angle{-60.6^\circ}$, $S_{12} = 0.02\ \angle 62.4^\circ$, $S_{21} = 3.122\ \angle 123.6^\circ$, and $S_{22} = 0.781\ \angle{-27.6^\circ}$. Determine its stability and plot the stability circles if the transistor is potentially unstable (see Figure 11.3).

SOLUTION From (11.1.16) and (11.1.18), we get

$$|\Delta| = |S_{11}S_{22} - S_{12}S_{21}| = 0.6964$$

and

$$k = \frac{1 + |\Delta|^2 - |S_{11}|^2 - |S_{22}|^2}{2|S_{12}S_{21}|} = 0.6071$$

Since one of the conditions for stability failed above, this transistor is potentially unstable.

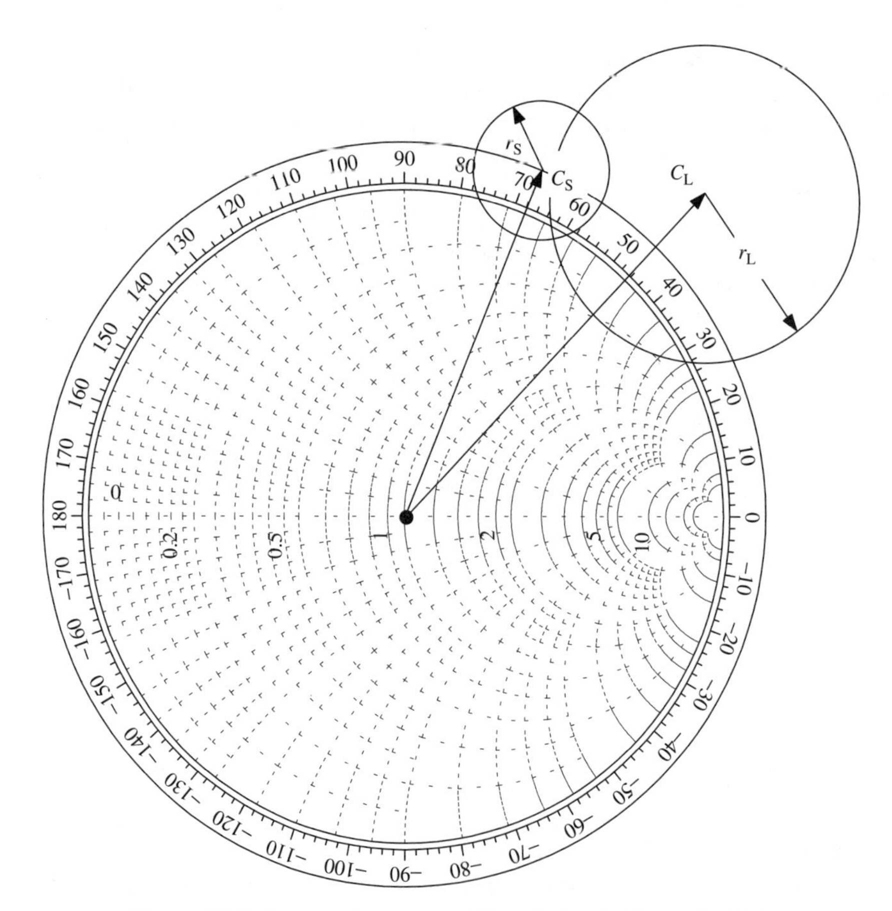

Figure 11.3 Input and output stability circles for Example 11.1.

Using (11.1.20) to (11.1.23), we can determine the stability circles as follows. For the output stability circle $C_L = 1.36\ \angle 46.7^\circ$ and $r_L = 0.5$. Since $|S_{11}|$ is 0.894, $\Gamma_L = 0$ represents a stable load point. Further, this point is located outside the stability circle, and therefore all points outside this circle are stable.

For the input stability circle $C_S = 1.13\ \angle 68.5^\circ$ and $r_S = 0.2$. Since $|S_{22}|$ is 0.781, $\Gamma_S = 0$ represents a stable source impedance point. Further, this point is located outside the stability circle, and therefore all points outside this circle are stable.

For the output stability circle, draw a radial line at 46.7°. With the radius of the Smith chart as unity, locate the center at 1.36 on this line. It can be done as follows. Measure the radius of the Smith chart using a ruler scale. Supposing that it is d millimeters, the location of the stability circle is then at $1.36d$ millimeters away on this radial line. Similarly, the radius of the stability circle is $0.5d$ millimeters. The load impedance must lie outside this circle for the circuit to be stable. Following a similar procedure, the input stability circle is drawn with its radius as $0.2d$ millimeters and center at $1.13d$ millimeters on the radial line at 68.5°. To have a stable design, the source impedance must lie outside this circle.

11.2 AMPLIFIER DESIGN FOR MAXIMUM GAIN

In this section we first consider the design of an amplifier that uses a unilateral transistor ($S_{12} = 0$) and has maximum possible gain. A design procedure using a bilateral transistor ($S_{12} \neq 0$) is developed next that requires simultaneous conjugate matching at its two ports.

Unilateral Case

When S_{12} is zero, the input reflection coefficient Γ_{in} reduces to S_{11} and the output reflection coefficient Γ_{out} simplifies to S_{22}. To obtain maximum gain, the source and load reflection coefficients must be equal to S_{11}^* and S_{22}^*, respectively. Further, the stability conditions for a unilateral transistor simplify to $|S_{11}| < 1$ and $|S_{22}| < 1$.

Example 11.2 The S-parameters of a properly biased BJT are found at 1 GHz as follows (with $Z_0 = 50\,\Omega$): $S_{11} = 0.60\ \angle -155^\circ$, $S_{22} = 0.48\ \angle -20^\circ$, $S_{12} = 0$, and $S_{21} = 6\ \angle 180^\circ$. Determine the maximum gain possible with this transistor and design an RF circuit that can provide this gain.

SOLUTION

1. Stability check:

$$k = \frac{1 - |S_{11}|^2 - |S_{22}|^2 + |\Delta|^2}{2|S_{12}S_{21}|} = \infty$$

because $S_{12} = 0$ and

$$|\Delta| = |S_{11}S_{22} - S_{12}S_{21}| = |S_{11}S_{22}| = 0.2909$$

Since both of the conditions are satisfied, the transistor is unconditionally stable.

2. The maximum possible power gain of the transistor is found as

$$G_{\mathrm{TU}} = \frac{1 - |\Gamma_s|^2}{|1 - S_{11}\Gamma_s|^2}|S_{21}|^2 \frac{1 - |\Gamma_L|^2}{|1 - S_{22}\Gamma_L|^2}$$

and

$$G_{TU_{\max}} = \frac{1 - |S_{11}^*|^2}{|1 - |S_{11}|^2|^2}|S_{21}|^2 \frac{1 - |S_{22}^*|^2}{|1 - |S_{22}|^2|^2}$$

$$= \frac{1}{1 - 0.606^2}\left(6^2\right)\frac{1}{1 - 0.48^2} = 73.9257$$

or

$$G_{\mathrm{TU}_{\max}} = 10\log_{10}(73.9257)\ \mathrm{dB} = 18.688\ \mathrm{dB}$$

3. For the maximum unilateral power gain,

$$\Gamma_S = S_{11}^* = 0.606\ \angle 155^\circ \quad \text{and} \quad \Gamma_L = S_{22}^* = 0.48\ \angle 20^\circ$$

The component values are determined from the Smith chart as illustrated in Figure 11.4.

Corresponding to the reflection coefficient's magnitude 0.606, the VSWR is 4.08. Hence, point D in Figure 11.4 represents the source impedance Z_S. Similarly, point B can be identified as the load impedance Z_L. Alternatively, the normalized impedance can be computed from the reflection coefficient. Then, points D and B can be located on the Smith chart. Now, we need to transform the zero reflection coefficient (normalized impedance of 1) to Γ_S (point D) at the input and to Γ_L (point B) at the output. One way to achieve this is to move first on the unity conductance circle from the center of the chart to point C and then on the constant-resistance circle to reach point D. Thus, it requires a shunt capacitor and then a series inductor to match at the input. Similarly, we can follow the unity-resistance circle from 1 to point A and then the constant-conductance circle to reach point B. Thus, a capacitor is connected in series with the load and then an inductor in shunt to match its output side. Actual values of the components are determined as follows.

The normalized susceptance at point C is estimated as $j1.7$. Hence, the shunt capacitor on the source side must provide a susceptance of $1.7/50 = 0.034$ S.

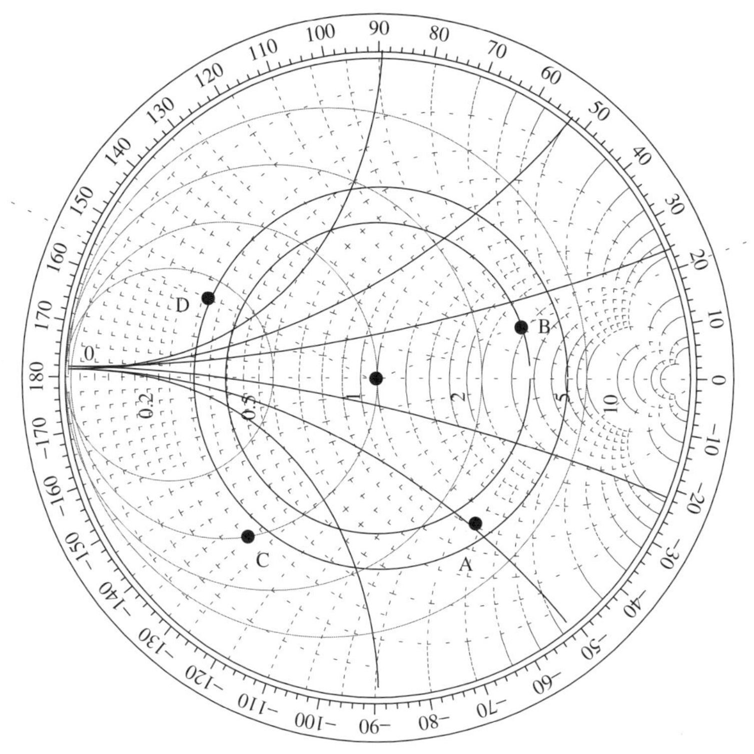

Figure 11.4 Smith chart illustrating the design of Example 11.2.

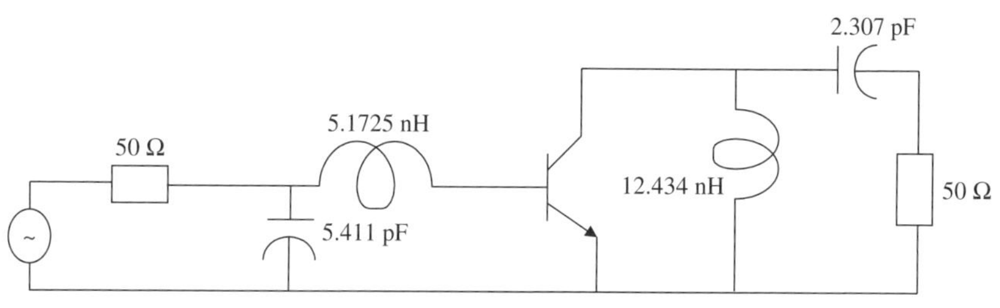

Figure 11.5 RF circuit designed for Example 11.2.

Since the signal frequency is 1 GHz, this capacitance must be 5.411 pF. Now the change in normalized reactance from point C to point D is determined as $j0.2 - (-j0.45) = j0.65$. The positive sign indicates that it is an inductor of reactance $0.65 \times 50 = 32.5\,\Omega$. The corresponding inductance is found as 5.1725 nH. Similarly, the normalized reactance at point A is estimated as $-j1.38$. Hence, the load requires a series capacitor of 2.307 pF. For transforming the susceptance of point A to that of point B, we need an inductance of $(-0.16 - 0.48)/50 = -0.0128$ S. It is found to be a shunt inductor of 12.434 nH. This circuit is illustrated in Figure 11.5.

Bilateral Case (Simultaneous Conjugate Matching)

To obtain the maximum possible gain, a bilateral transistor must be matched at both of its ports simultaneously, as shown in Figure 11.6. When its output port is properly terminated, the input side of the transistor is matched with the source such that $\Gamma_{in} = \Gamma_S^*$. Similarly, the output port of transistor is matched with the load when its input is matched terminated. Mathematically,

$$\Gamma_{in} = \Gamma_S^*$$

and

$$\Gamma_{out} = \Gamma_L^*$$

Therefore,

$$\Gamma_S^* = S_{11} + \frac{S_{12}S_{21}\Gamma_L}{1 - S_{22}\Gamma_L} \tag{11.2.1}$$

and

$$\Gamma_L^* = S_{22} + \frac{S_{12}S_{21}\Gamma_S}{1 - S_{11}\Gamma_S} \tag{11.2.2}$$

From (11.2.1),

$$\Gamma_S = S_{11}^* + \frac{S_{12}^* S_{21}^*}{(1/\Gamma_L^*) - S_{22}^*} \tag{11.2.3}$$

and from (11.2.2),

$$\Gamma_L^* = \frac{S_{22} - (S_{11}S_{22} - S_{12}S_{21})\Gamma_S}{1 - S_{11}\Gamma_S} = \frac{S_{22} - \Gamma_S \Delta}{1 - S_{11}\Gamma_S} \tag{11.2.4}$$

Substituting (11.2.4) in (11.2.3), we have

$$\Gamma_S = S_{11}^* + \frac{S_{12}^* S_{21}^*}{[(1 - S_{11}\Gamma_S)/(S_{22} - \Gamma_S \Delta)] - S_{22}^*}$$

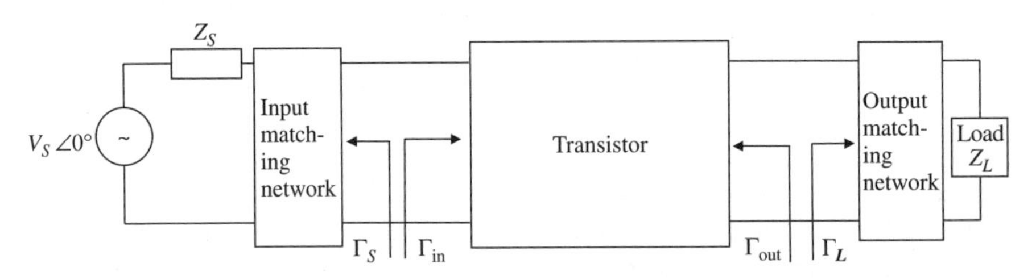

Figure 11.6 Bilateral transistor with input- and output-matching networks.

or

$$\Gamma_S = S_{11}^* + \frac{S_{12}^* S_{21}^* (S_{22} - \Gamma_S \Delta)}{1 - S_{11}\Gamma_S - |S_{22}|^2 + \Gamma_S S_{22}^* \Delta} = S_{11}^* + \frac{S_{12}^* S_{21}^* (S_{22} - \Gamma_S \Delta)}{1 - |S_{22}|^2 - (S_{11} - S_{22}^* \Delta)\Gamma_S}$$

or

$$(1 - |S_{22}|^2)\Gamma_S - (S_{11} - S_{22}^*\Delta)\Gamma_S^2 = S_{11}^*(1 - |S_{22}|^2) - S_{11}^*(S_{11} - S_{22}^*\Delta)\Gamma_S + S_{12}^* S_{21}^*(S_{22} - \Gamma_S \Delta)$$

or

$$(S_{11} - S_{22}^*\Delta)\Gamma_S^2 + (|\Delta|^2 - |S_{11}|^2 + |S_{22}|^2 - 1)\Gamma_S + (S_{11}^* - S_{22}\Delta^*) = 0$$

This is a quadratic equation in Γ_S and its solution can be found as follows:

$$\Gamma_S = \frac{B_1 \pm \sqrt{B_1^2 - 4|C_1|^2}}{2C_1} = \Gamma_{MS} \tag{11.2.5}$$

where

$$B_1 = 1 + |S_{11}|^2 - |S_{22}|^2 - |\Delta|^2$$

and

$$C_1 = S_{11} - S_{22}^*\Delta$$

Similarly, a quadratic equation for Γ_L can be formulated. Solutions to that equation are found to be

$$\Gamma_{ML} = \frac{B_2 \pm \sqrt{B_2^2 - 4|C_2|^2}}{2C_2} \tag{11.2.6}$$

where

$$B_2 = 1 + |S_{22}|^2 - |S_{11}|^2 - |\Delta|^2$$

and

$$C_2 = S_{22} - S_{11}^*\Delta$$

If $|B_1/2C_1| > 1$ and $B_1 > 0$ in equation (11.2.5), the solution with a minus sign produces $|\Gamma_{MS}| < 1$ and that with a positive sign produces $|\Gamma_{MS}| > 1$. On other hand, if $|B_1/2C_1| > 1$ with $B_1 < 0$ in this equation, the solution with a plus sign produces $|\Gamma_{MS}| < 1$ and the solution with a minus sign produces $|\Gamma_{MS}| > 1$. Similar considerations apply to equation (11.2.6) as well.

Observation $|B_i/2C_i| > 1$ implies that $|k| > 1$.

Proof:

$$\left|\frac{B_i}{2C_i}\right| > 1 \Rightarrow \left|\frac{1 + |S_{11}|^2 - |S_{22}|^2 - |\Delta|^2}{2(S_{11} - S_{22}^*\Delta)}\right| > 1$$

or

$$|1 + |S_{11}|^2 - |S_{22}|^2 - |\Delta|^2| > 2|(S_{11} - S_{22}^*\Delta)|$$

Squaring on both sides of this inequality, we get

$$|1 + |S_{11}|^2 - |S_{22}|^2 - |\Delta|^2|^2 > 4|(S_{11} - S_{22}^*\Delta)|^2$$

and

$$\begin{aligned}|S_{11} - S_{22}^*\Delta|^2 &= (S_{11} - S_{22}^*\Delta)(S_{11}^* - S_{22}\Delta^*)\\ &= |S_{12}S_{21}|^2 + (1 - |S_{22}|^2)(|S_{11}|^2 - |\Delta|^2)\end{aligned}$$

Therefore,

$$|1 + |S_{11}|^2 - |S_{22}|^2 - |\Delta|^2|^2 > 4|S_{12}S_{21}|^2 + 4(1 - |S_{22}|^2)(|S_{11}|^2 - |\Delta|^2)$$

or

$$|1 + |S_{11}|^2 - |S_{22}|^2 - |\Delta|^2|^2 - 4(1 - |S_{22}|^2)(|S_{11}|^2 - |\Delta|^2) > 4|S_{12}S_{21}|^2$$

or

$$(1 - |S_{22}|^2 - |S_{11}|^2 + |\Delta|^2)^2 > 4|S_{12}S_{21}|^2$$

or

$$|1 - |S_{22}|^2 - |S_{11}|^2 + |\Delta|^2| > 2|S_{12}S_{21}|$$

Therefore,

$$\frac{|1 - |S_{22}|^2 - |S_{11}|^2 + |\Delta|^2|}{2|S_{12}S_{21}|} > 1 \Rightarrow k > 1$$

Also, it can be proved that $|\Delta| < 1$ implies that $B_1 > 0$ and $B_2 > 0$. Therefore, minus signs must be used in equations (11.2.5) and (11.2.6).

Since

$$G_{T\,\max} = \frac{(1 - |\Gamma_{MS}|^2)|S_{21}|^2(1 - |\Gamma_{ML}|^2)}{|(1 - S_{11}\Gamma_{MS})(1 - S_{22}\Gamma_{ML}) - S_{12}S_{21}\Gamma_{ML}\Gamma_{MS}|^2}$$

substituting equations (11.2.5), (11.2.6), and k from (11.1.18) gives

$$G_{T\,\max} = \frac{|S_{21}|}{|S_{12}|}(k - \sqrt{k^2 - 1}) \tag{11.2.7}$$

Maximum stable gain is defined as the value of $G_{T\,\max}$ when $k = 1$. Therefore,

$$G_{MSG} = \left|\frac{S_{21}}{S_{12}}\right| \tag{11.2.8}$$

Example 11.3 The S-parameters of a properly biased GaAs FET HFET-1101 are measured using a 50-Ω network analyzer at 6 GHz as follows: $S_{11} = 0.614\ \angle -167.4^\circ$, $S_{21} = 2.187\ \angle 32.4^\circ$, $S_{12} = 0.046\ \angle 65^\circ$, and $S_{22} = 0.716\ \angle -83^\circ$. Design an amplifier using this transistor for a maximum possible gain.

SOLUTION First we need to test for stability using (11.1.16) and (11.1.18).

$$|\Delta| = |S_{11}S_{22} - S_{12}S_{21}| = 0.3419$$

and

$$k = \frac{1 - |S_{11}|^2 - |S_{22}|^2 + |\Delta|^2}{2|S_{12}S_{21}|} = 1.1296$$

Since both of the conditions are satisfied, the transistor is unconditionally stable. For the maximum possible gain, we need to use simultaneous conjugate matching. Therefore, the source and load reflection coefficients are determined from (11.2.5) and (11.2.6) as follows:

$$\Gamma_{MS} = \frac{B_1 - \sqrt{B_1^2 - 4|C_1|^2}}{2C_1} = 0.8673\ \angle 169.76^\circ$$

and

$$\Gamma_{ML} = \frac{B_2 - \sqrt{B_2^2 - 4|C_2|^2}}{2C_2} = 0.9011\ \angle 84.48^\circ$$

There are many ways to synthesize these circuits. The design of one of the circuits is discussed here. As illustrated in Figure 11.7, normalized impedance points A and C are identified corresponding to Γ_{MS} and Γ_{ML}, respectively. The respective normalized admittance points B and D are located next. One way to transform the normalized admittance of unity to that of point B is to add a shunt susceptance (normalized) of about $j2.7$ at point E. This can be achieved with a capacitor or a stub of 0.194λ, as shown in Figure 11.8. Now, for moving from point E to point B, we require a transmission line of about 0.065λ. Similarly, an open-circuited shunt stub of 0.297λ will transform the unity value to $1 - j3.4$ at point F, and a transmission line length of 0.093λ will transform it to the desired value at point D.

Note that both of the shunt stubs are asymmetrical about the main line. A symmetrical stub is preferable. It can be achieved via two shunt-connected stubs of susceptance $j1.35$ at the input end and of $-j1.7$ at the output. This circuit

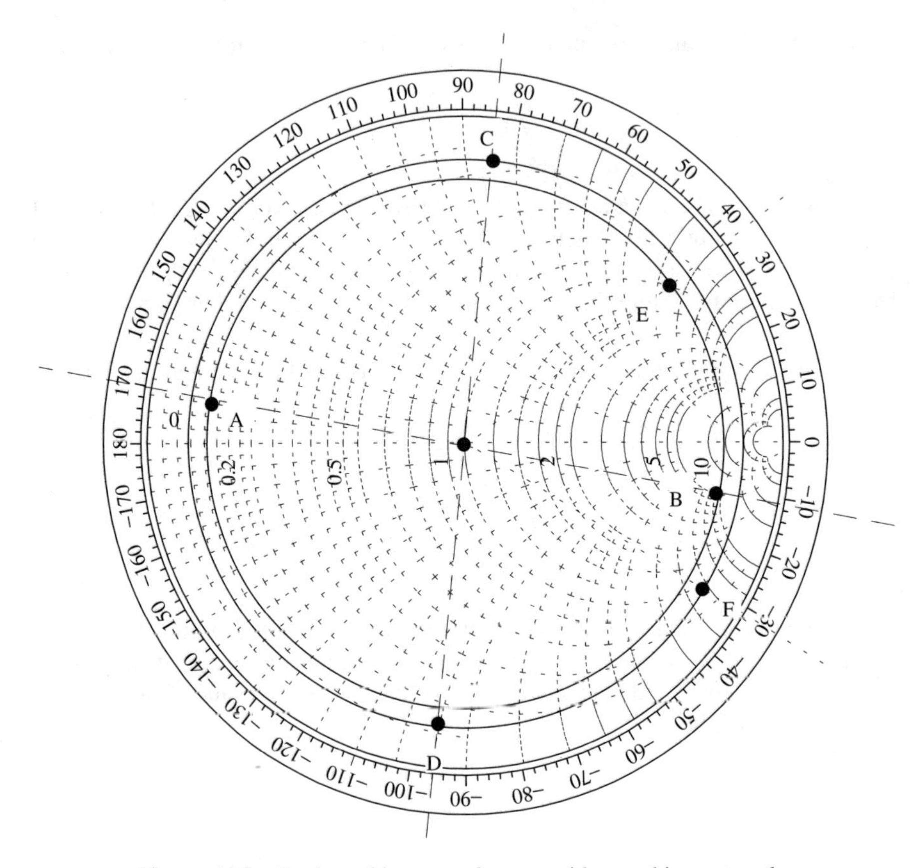

Figure 11.7 Design of input- and output-side matching networks.

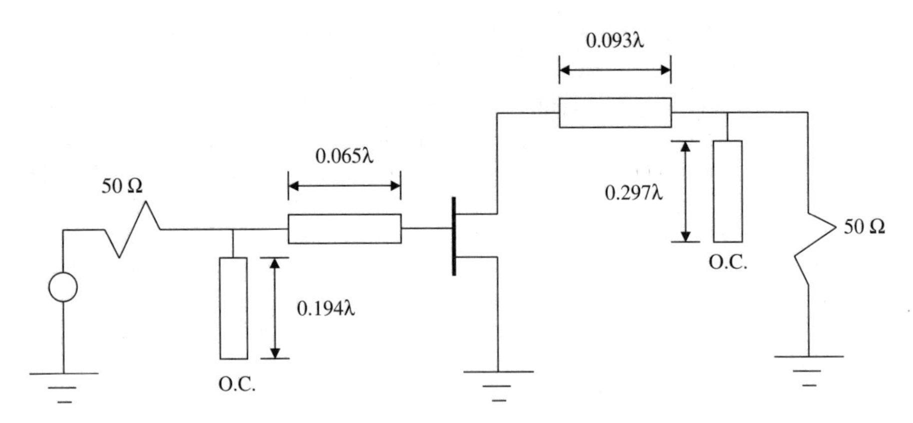

Figure 11.8 RF part of the amplifier circuit for Example 11.3.

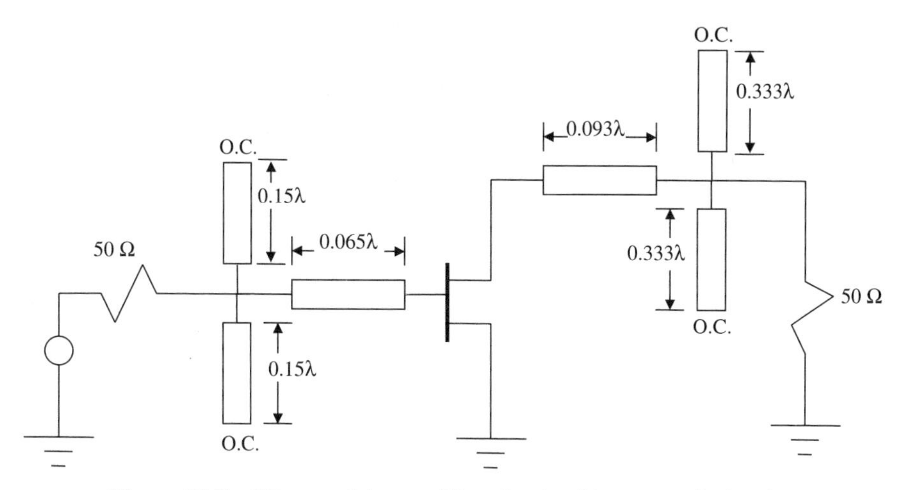

Figure 11.9 RF part of the amplifier circuit with symmetrical stubs.

is illustrated in Figure 11.9. The value of the gain is evaluated from (11.2.7) as follows:

$$G_{T\max} = \frac{|S_{21}|}{|S_{12}|}(k - \sqrt{k^2 - 1}) = 28.728 \Rightarrow 10\log(28.728) = 14.58\,\mathrm{dB}$$

Unilateral Figure of Merit In preceding examples we found that the design with a bilateral transistor is a bit complex compared with a unilateral case. The procedure for the bilateral transistor becomes even more cumbersome when the specified gain is less than its maximum possible value. It can be simplified by assuming that a bilateral transistor is unilateral. The unilateral figure of merit provides an estimate of the error associated with this assumption. It can be formulated from (10.4.8) and (10.4.10) as follows:

$$\frac{G_T}{G_{TU}} = \frac{|1 - S_{22}\Gamma_L|^2}{|1 - \Gamma_{\text{out}}\Gamma_L|^2} = \left|\frac{1 - S_{22}\Gamma_L}{1 - S_{22}\Gamma_L - S_{12}S_{21}\Gamma_S\Gamma_L/(1 - S_{11}\Gamma_S)}\right|^2$$

$$\frac{|1 - S_{22}\Gamma_L|^2}{|1 - \Gamma_{\text{out}}\Gamma_L|^2} = \left|\frac{1}{1 - S_{12}S_{21}\Gamma_S\Gamma_L/(1 - S_{22}\Gamma_L)(1 - S_{11}\Gamma_S)}\right|^2$$

or

$$\frac{G_T}{G_{TU}} = \left|\frac{1}{1 - X}\right|^2$$

where

$$X = \frac{S_{12}S_{21}\Gamma_S\Gamma_L}{(1 - S_{22}\Gamma_L)(1 - S_{11}\Gamma_S)}$$

Therefore, the bounds of this gain ratio are given by

$$\frac{1}{(1+|X|)^2} < \frac{G_T}{G_{TU}} < \frac{1}{(1-|X|)^2}$$

When $\Gamma_S = S_{11}^*$ and $\Gamma_L = S_{22}^*$, G_{TU} has a maximum value. In this case, the maximum error introduced by using G_{TU} in place of G_T ranges as follows:

$$\frac{1}{(1+U)^2} < \frac{G_T}{G_{TU}} < \frac{1}{(1-U)^2} \tag{11.2.9}$$

where

$$U = \frac{|S_{12}||S_{21}||S_{11}||S_{22}|}{(1-|S_{11}|^2)(1-|S_{22}|^2)} \tag{11.2.10}$$

The parameter U is known as the *unilateral figure of merit.*

Example 11.4 The scattering parameters of two transistors are as given below. Compare the unilateral figures of merit of the two. For transistor A, $S_{11} = 0.45\ \angle 150^\circ$, $S_{12} = 0.01\ \angle -10^\circ$, $S_{21} = 2.05\ \angle 10^\circ$, and $S_{22} = 0.4\ \angle -150^\circ$. For transistor B, $S_{11} = 0.641\ \angle -171.3^\circ$, $S_{12} = 0.057\ \angle 16.3^\circ$, $S_{21} = 2.058\ \angle 28.5^\circ$, and $S_{22} = 0.572\ \angle -95.7^\circ$.

SOLUTION From (11.2.10) we find that

$$U = \frac{0.01 \times 2.05 \times 0.45 \times 0.4}{(1-0.45^2)(1-0.4^2)} = 0.00551 = U_{\mathrm{A}}$$

for transistor A. Similarly, for transistor B,

$$U = \frac{0.057 \times 2.058 \times 0.641 \times 0.572}{(1-0.641^2)(1-0.572^2)} = 0.1085 = U_{\mathrm{B}}$$

Hence, the error bounds for these two transistors can be determined from (11.2.9) as follows. For transistor A,

$$0.9891 < \frac{G_T}{G_{TU}} < 1.0055$$

and for transistor B,

$$0.8138 < \frac{G_T}{G_{TU}} < 1.2582$$

Alternatively, these two results can be expressed in decibels as follows.

$$-0.0476\,\mathrm{dB} < \frac{G_T}{G_{TU}} < 0.0238\,\mathrm{dB}$$

and

$$-0.8948\,\text{dB} < \frac{G_T}{G_{TU}} < 0.9976\,\text{dB}$$

Conclusion: If $S_{12} = 0$ can be assumed for a transistor without introducing significant error, the design procedure will be much simpler than that in the bilateral case.

11.3 CONSTANT-GAIN CIRCLES

In Section 11.2 we considered the design of amplifiers for maximum possible gains. Now, let us consider the design procedure for other amplifier circuits. We split it again into two cases, unilateral and bilateral transistors.

Unilateral Case

We consider two different cases regarding unilateral transistors. In one case it is assumed that the transistor is unconditionally stable because $|S_{11}|$ and $|S_{22}|$ are less than unity. In the other case, one or both of these parameters may be greater than unity. Thus, it makes $|\Delta|$ greater than 1. From (10.4.10),

$$G_{TU} = \frac{1 - |\Gamma_S|^2}{|1 - S_{11}\Gamma_S|^2}|S_{21}|^2 \frac{1 - |\Gamma_L|^2}{|1 - S_{22}\Gamma_L|^2} = G_S G_0 G_L$$

Expressions of G_S and G_L in this equation are similar in appearance. Therefore, we can express them by the following general form:

$$G_i = \frac{1 - |\Gamma_i|^2}{|1 - S_{ii}\Gamma_i|^2} \qquad \begin{cases} i = S, & ii = 11 \\ i = L, & ii = 22 \end{cases} \tag{11.3.1}$$

We consider next two different cases of unilateral transistors. In one, the transistor is unconditionally stable, and in the other case it is potentially unstable. If the unilateral transistor is unconditionally stable, $|S_{ii}| < 1$. Therefore, maximum G_i in (11.3.1) will be given as

$$G_{i\,\max} = \frac{1}{1 - |S_{ii}|^2} \tag{11.3.2}$$

Impedances that produce $G_{i\,\max}(\Gamma_i = S_{ii}^*)$ are called *optimum terminations.* Therefore,

$$0 \le G_i \le G_{i\,\max}$$

Values of Γ_i that produce a constant gain G_i lie in a circle on the Smith chart. These circles are called *constant-gain circles.*

We define the normalized gain factor g_i as follows:

$$g_i = \frac{G_i}{G_{i\,\max}} = G_i(1 - |S_{ii}|^2) \tag{11.3.3}$$

Hence,

$$0 \le g_i \le 1$$

From (11.3.1) and (11.3.2), we can write

$$g_i = \frac{1 - |\Gamma_i|^2}{|1 - S_{ii}\Gamma_i|^2}(1 - |S_{ii}|^2) \Rightarrow g_i|1 - S_{ii}\Gamma_i|^2 = (1 - |\Gamma_i|^2)(1 - |S_{ii}|^2)$$

or

$$g_i(1 - S_{ii}\Gamma_i)(1 - S_{ii}^*\Gamma_i^*) = 1 - |S_{ii}|^2 - |\Gamma_i|^2 + |\Gamma_i|^2|S_{ii}|^2$$

or

$$g_i(1 - S_{ii}\Gamma_i - S_{ii}^*\Gamma_i^* + |S_{ii}|^2|\Gamma_i|^2) = 1 - |S_{ii}|^2 - |\Gamma_i|^2(1 - |S_{ii}|^2)$$

or

$$(g_i|S_{ii}|^2 + 1 - |S_{ii}|^2)|\Gamma_i|^2 - g_i(S_{ii}\Gamma_i + S_{ii}^*\Gamma_i^*) = 1 - g_i - |S_{ii}|^2$$

or

$$\Gamma_i\Gamma_i^* - g_i\frac{S_{ii}\Gamma_i + S_{ii}^*\Gamma_i^*}{1 - (1 - g_i)|S_{ii}|^2} + \frac{g_i^2|S_{ii}|^2}{[1 - (1 - g_i)|S_{ii}|^2]^2} = \frac{1 - g_i - |S_{ii}|^2}{1 - (1 - g_i)|S_{ii}|^2} + \frac{g_i^2|S_{ii}|^2}{[1 - (1 - g_i)|S_{ii}|^2]^2}$$

Therefore,

$$\left|\Gamma_i - \frac{g_i S_{ii}^*}{1 - (1 - g_i)|S_{ii}|^2}\right|^2 = \frac{(1 - g_i - |S_{ii}|^2)[1 - (1 - g_i)|S_{ii}|^2] + |S_{ii}|^2 g_i^2}{[1 - (1 - g_i)|S_{ii}|^2]^2} \tag{11.3.4}$$

which is the equation of a circle, with its center d_i and radius R_i given as follows:

$$d_i = \frac{g_i S_{ii}^*}{1 - (1 - g_i)|S_{ii}|^2} \tag{11.3.5}$$

and

$$R_i = \frac{(1 - |S_{ii}|^2)\sqrt{1 - g_i}}{1 - (1 - g_i)|S_{ii}|^2} \tag{11.3.6}$$

Example 11.5 The S-parameters of a MESFET are given in Table 11.1 ($Z_0 = 50\,\Omega$). Plot the constant-gain circles at 4 GHz for $G_L = 0$ and 1 dB and $G_S = 2$ and 3 dB. Using these plots, design an amplifier for a gain of 11 dB. Calculate

TABLE 11.1 S-Parameters of a MESFET

f(GHz)	S_{11}	S_{21}	S_{12}	S_{22}
3	$0.8\angle -90^\circ$	$2.8\angle 100^\circ$	0	$0.66\angle -50^\circ$
4	$0.75\angle -120^\circ$	$2.5\angle 80^\circ$	0	$0.6\angle -70^\circ$
5	$0.71\angle -140^\circ$	$2.3\angle 60^\circ$	0	$0.68\angle -85^\circ$

and plot its transducer power gain and input return loss in the frequency band 3 to 5 GHz.

SOLUTION Since S_{12} is zero, this transistor is unilateral. Hence,

$$k = \infty \text{ and } |\Delta| < 1 \qquad (\text{because } |S_{11}| < 1 \text{ and } |S_{22}| < 1)$$

Therefore, it is unconditionally stable. From (10.4.14) to (10.4.16) we have

$$G_{S\max} = \frac{1}{1 - |S_{11}|^2} = \frac{1}{1 - 0.75^2} = 2.28857 = 3.59\,\text{dB}$$

$$G_{L\max} = \frac{1}{1 - |S_{22}|^2} = \frac{1}{1 - 0.6^2} = 1.5625 = 1.92\,\text{dB}$$

and

$$G_0 = |S_{21}|^2 = 2.5^2 = 6.25 = 7.96\,\text{dB}$$

Therefore,

$$G_{TU\max} = 3.59 + 1.92 + 7.96 = 13.47\,\text{dB}$$

Thus, the maximum possible gain is 2.47 dB higher than the desired value of 11 dB. Obviously, this transistor can be used for the present design.

The constant-gain circles can be determined from (11.3.5) and (11.3.6). These results are shown in Table 11.2. As illustrated in Figure 11.10, the gain circles are drawn from these data. Since G_0 is found as 8 dB (approximately), the remaining 3 dB needs to be obtained through G_S and G_L. If we select G_S as 3 dB, G_L must be 0 dB. Alternatively, we can use G_S and G_L as 2 and 1 dB, respectively, to obtain a transducer power gain of $7.96 + 2 + 1 \approx 11$ dB.

Let us select point A on a 2-dB G_S circle and design the input-side network. The corresponding admittance is found at point B, and therefore a normalized

TABLE 11.2 Results for Constant-Gain Circles

$G_S = 3\,\text{dB} \approx 2$	$g_S = 0.875$	$d_S = 0.706\angle 120^\circ$	$R_S = 0.166$
$G_S = 2\,\text{dB} = 1.58$	$g_S = 0.691$	$d_S = 0.627\angle 120^\circ$	$R_S = 0.294$
$G_L = 1\,\text{dB} = 1.26$	$g_L = 0.8064$	$d_L = 0.52\angle 70^\circ$	$R_L = 0.303$
$G_L = 0\,\text{dB} = 1$	$g_L = 0.64$	$d_L = 0.44\angle 70^\circ$	$R_L = 0.44$

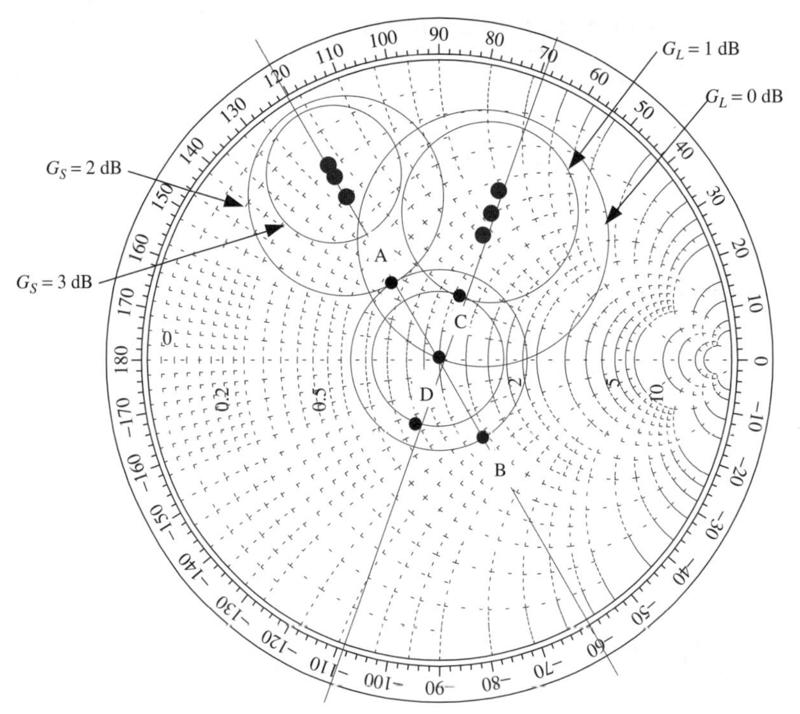

Figure 11.10 Constant-gain circles and the network design for Example 11.4.

capacitive susceptance of $j0.62$ is needed in parallel with the source admittance to reach the input VSWR circle. An open-ended 0.09λ-long shunt stub can be used for this. The normalized admittance is now $1 + j0.62$. This admittance can be transformed to that of point B by a 0.183λ-long section of transmission line. Similarly, point C can be used to obtain $G_L = 1\,\text{dB}$. A normalized reactance of $j0.48$ in series with a 50-Ω load can be used to synthesize this impedance. Alternatively, the corresponding admittance point D is identified. Hence, a shunt susceptance of $-j0.35$ (an open-circuit stub of 0.431λ), and then a transmission line length of 0.044λ can provide the desired admittance. This circuit is illustrated in Figure 11.11.

The return loss is found by expressing $|\Gamma_{\text{in}}|$ in decibels. Since

$$\Gamma_{\text{in}} = S_{11} + \frac{S_{12}S_{21}\Gamma_L}{1 - S_{22}\Gamma_L}$$

and $S_{12} = 0$, $\Gamma_{\text{in}} = S_{11}$; therefore, $|\Gamma_{\text{in}}\,(3\,\text{GHz})| = 0.8$, $|\Gamma_{\text{in}}\,(4\,\text{GHz})| = 0.75$, and $|\Gamma_{\text{in}}\,(5\,\text{GHz})\,| = 0.71$:

$$\text{return loss at } 3\,\text{GHz} = 20\log_{10}(0.8) = -1.94\,\text{dB}$$

$$\text{return loss at } 4\,\text{GHz} = 20\log_{10}(0.75) = -2.5\,\text{dB}$$

$$\text{return loss at } 5\,\text{GHz} = 20\log_{10}(0.71) = -2.97\,\text{dB}$$

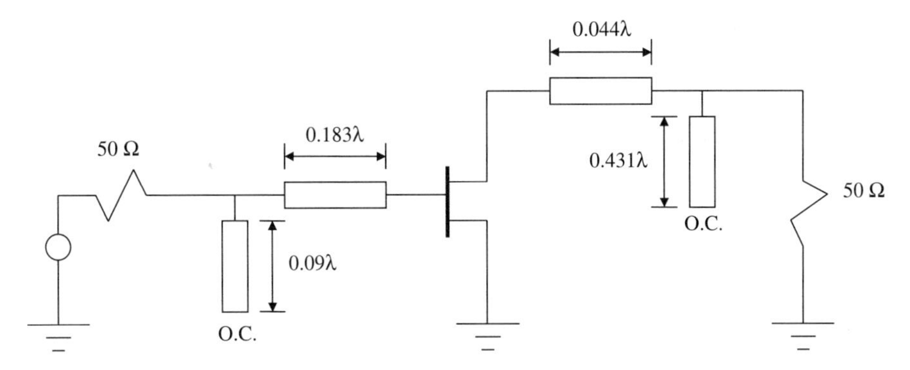

Figure 11.11 RF circuit designed for Example 11.4.

The transducer power gain at 4 GHz is 11 dB (because we designed the circuit for this gain). However, it will be different at other frequencies. We can evaluate it from (10.4.16) as follows:

$$G_{TU} = \frac{1 - |\Gamma_S|^2}{|1 - S_{11}\Gamma_S|^2}|S_{21}|^2\frac{1 - |\Gamma_L|^2}{|1 - S_{22}\Gamma_L|^2}$$

Note that Γ_S, Γ_L, and the S-parameters of the transistor are frequency dependent. Therefore, we need to determine reflection coefficients at other frequencies before using the formula above. For a circuit designed with reactive discrete components, the new reactances can be evaluated easily. The corresponding reflection coefficients can, in turn, be determined using the appropriate formula. However, we used transmission lines in our design. Electrical lengths of these lines will be different at other frequencies. We can calculate new electrical lengths by replacing λ as follows:

$$\lambda \to \frac{f_{\text{new}}}{f_{\text{design}}}\lambda_{\text{new}}$$

At 3 GHz, original lengths must be multiplied by $\frac{3}{4} = 0.75$ to adjust for the change in frequency. Similarly, it must be multiplied by $\frac{5}{4} = 1.25$ for 5 GHz. The new reflection coefficients can be determined using the Smith chart. The results are summarized below.

Γ_S Calculations:

Lengths at 4 GHz	3 GHz: Lengths and Γ_S		5 GHz: Lengths and Γ_S	
0.09	0.068	0.24 ∠158°	0.113	0.41 ∠81°
0.183	0.137		0.229	

Γ_L Calculations:

Lengths at 4 GHz	3 GHz: Lengths and Γ_L		5 GHz: Lengths and Γ_L	
0.044	0.033	$0.72\ \angle 109^\circ$	0.055	$0.15\ \angle -151^\circ$
0.431	0.323		0.539	

Therefore,

$$G_{TU}(3\,\text{GHz}) = \frac{1 - 0.24^2}{|1 - (0.24\ \angle 158^\circ)(0.8\ \angle -90^\circ)|^2} \times 2.8^2 \times \frac{1 - 0.72^2}{|1 - (0.72\ \angle 109^\circ)(0.66\ \angle -50^\circ)|^2}$$

or

$$G_{TU}(3\,\text{GHz}) = 5.4117 = 10\log(5.4117) = 7.33\,\text{dB}$$

Similarly,

$$G_{TU}(5\,\text{GHz}) = \frac{1 - 0.41^2}{|1 - (0.71\ \angle -140^\circ)(0.41\ \angle 81^\circ)|^2} \times 2.3^2 \times \frac{1 - 0.15^2}{|1 - (0.58\ \angle -85^\circ)(0.15\ \angle -141^\circ)|^2}$$

or

$$G_{TU}(5\,\text{GHz}) = \frac{4.4008}{0.7849 \times 1.1284} = 4.9688 = 6.96\,\text{dB}$$

These return-loss and gain characteristics are displayed in Figure 11.12.

If a unilateral transistor is potentially unstable, $|S_{ii}| > 1$. For $|S_{ii}| > 1$, the real part of the corresponding impedance will be negative. Further, G_i in (11.3.1) will be infinite for $\Gamma_i = 1/S_{ii}$. In other words, the total loop resistance on the input side (for $i = S$) or on the output side (for $i = L$) is zero. This is the characteristic of an oscillator. Hence, this circuit can oscillate. We can still use the same Smith chart to determine the corresponding impedance provided that the magnitude of reflection coefficient is assumed as $1/|S_{ii}|$ instead of $|S_{ii}|$, while its phase angle is the same as that of S_{ii}, and the resistance scale is interpreted as negative. The reactance scale of the Smith chart is not affected.

It can be shown that the location and radii of constant-gain circles are still given by (11.3.5) and (11.3.6). The centers of these circles are located along a radial line passing through $1/S_{ii}$ on the Smith chart. To prevent oscillations, Γ_i must be selected such that the loop resistance is a positive number.

Example 11.6 A GaAs FET is biased at $V_{ds} = 5$ V and $I_{ds} = 10$ mA. A 50-Ω ANA is used to determine its S-parameters at 1 GHz. These are found as follows: $S_{11} = 2.27\ \angle -120^\circ$, $S_{21} = 4\ \angle 50^\circ$, $S_{12} = 0$, and $S_{22} = 0.6\ \angle -80^\circ$.

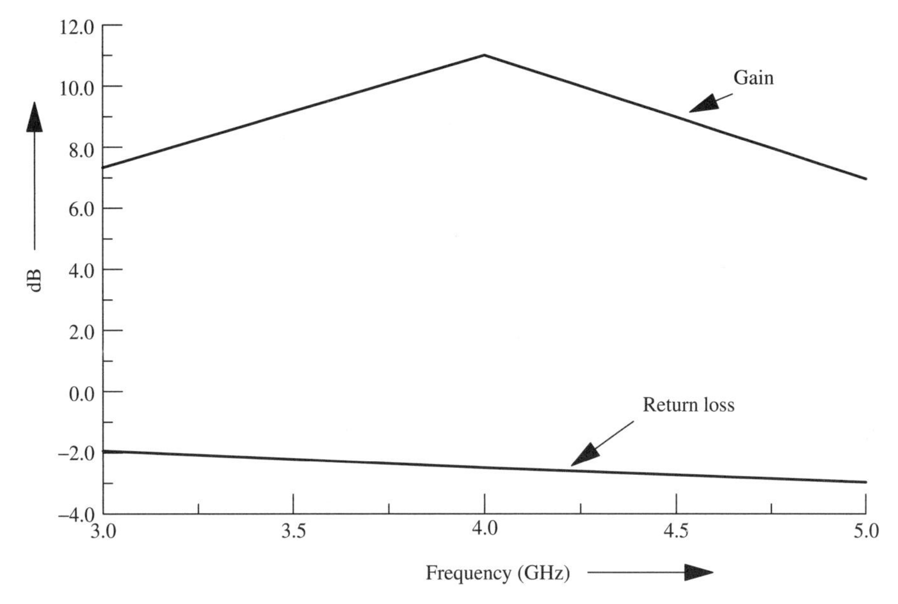

Figure 11.12 Gain and return loss versus frequency.

(**a**) Use a Smith chart to determine its input impedance and indicate on it the source impedance region(s) where the circuit is unstable.

(**b**) Plot the constant-gain circles for $G_S = 3\,\text{dB}$ and $G_S = 5\,\text{dB}$ on the same Smith chart.

(**c**) Find a source impedance that provides $G_S = 3\,\text{dB}$ with maximum possible degree of stability. Also, determine the load impedance that gives maximum G_L. Design the input and output networks.

(**d**) Find the gain (in decibels) of your amplifier circuit.

SOLUTION

(**a**) First we locate $1/2.27 = 0.4405$ at $\angle -120^\circ$ on the Smith chart. It is depicted as point P in Figure 11.13. This point gives the corresponding impedance if the resistance scale is interpreted as negative. Thus, the normalized input impedance is about $-0.49 - j0.46$. Hence,

$$Z_{\text{in}} = 50(-0.49 - j0.46) = -24.5 - j23\,\Omega$$

Therefore, if we use a source that has impedance with its real part less than $24.5\,\Omega$, the loop resistance on the input side will stay negative. That means that one has to select a source impedance that lies inside the resistive circle of 0.49. Outside this circle is unstable.

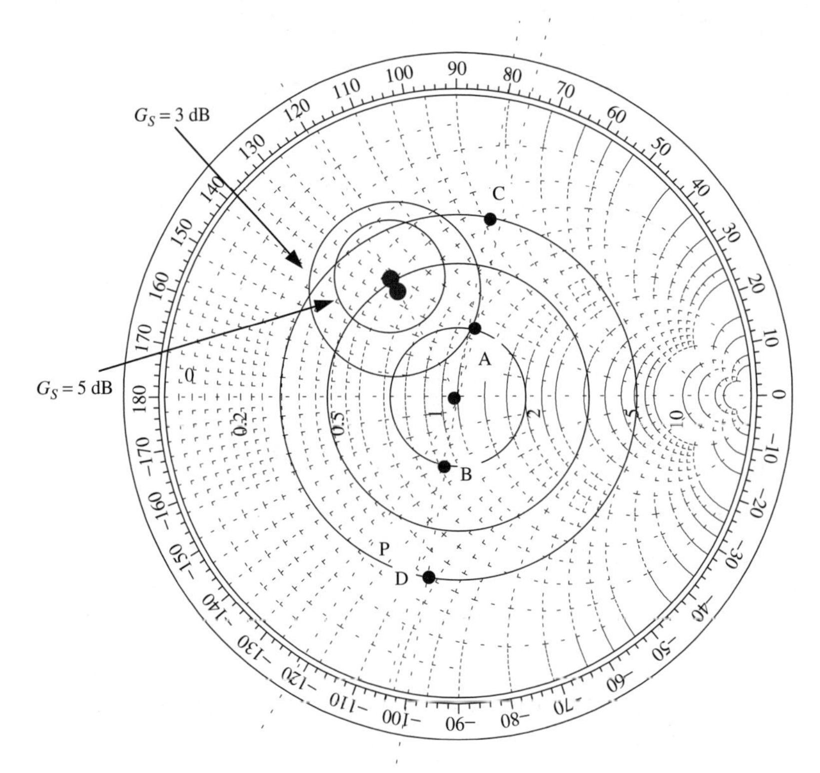

Figure 11.13 Amplifier design for Example 11.6.

(**b**) From (11.3.5) and (11.3.6),

$$d_S = \frac{g_S S_{11}^*}{1 - (1 - g_S)|S_{11}|^2} \quad \text{and} \quad R_S = \frac{(1 - |S_{11}|^2)\sqrt{1 - g_S}}{1 - (1 - g_S)|S_{11}|^2}$$

where $g_S = G_S(1 - |S_{11}|^2)$; the locations and radii of constant-G_S circles are found as tabulated here:

$$G_S = 5\,\text{dB} \Rightarrow 10^{0.5} = 3.1623 \qquad G_S = 3\,\text{dB} \Rightarrow 10^{0.3} = 2$$

$$g_S = 3.1623(1 - 2.27^2) = -13.1327 \qquad g_S = 2(1 - 2.27^2) = -8.3058$$

$$R_S = \frac{(1 - 2.27^2)\sqrt{1 + 13.1327}}{1 - (1 + 13.1327)2.27^2} = 0.2174 \qquad R_S = \frac{(1 - 2.27^2)\sqrt{1 + 8.3058}}{1 - (1 + 8.3058)2.27^2} = 0.2698$$

$$d_S = \frac{-13.1327(2.27\ \angle 120^\circ)}{1 - (1 + 13.1327)2.27^2} = 0.45\ \angle 120^\circ \qquad d_S = \frac{-8.3058(2.27\ \angle 120^\circ)}{1 - (1 + 8.3058)2.27^2} = 0.4015\ \angle 120^\circ$$

These constant-G_S circles are shown on the Smith chart in Figure 11.13.

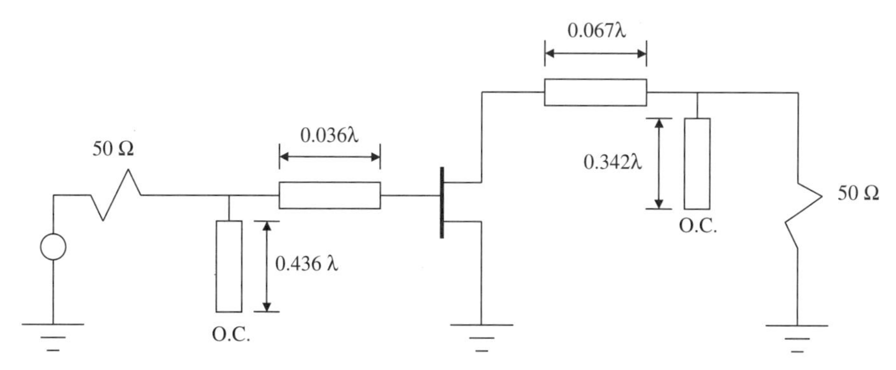

Figure 11.14 RF circuit for the amplifier of Example 11.6.

(c) To obtain $G_S = 3\,\text{dB}$ with maximum degree of stability, we select point A for the source impedance because it has a maximum possible real part. Therefore,

$$\overline{Z}_S \approx 1 + j0.5 \quad \text{or} \quad \Gamma_S = 0.24\ \angle 76^\circ$$

With $Z_S = 50 + j25\ \Omega$, the loop resistance in the input side is $50 - 24.5 = 25.5\,\Omega$. It is a positive value, and therefore the input circuit will be stable. For maximum G_L we select $\Gamma_L = S_{22}^* = 0.6\ \angle 80^\circ$. It is depicted as point C in Figure 11.13. The corresponding impedance Z_L is found to be

$$Z_L = 50(0.55 + j1.03) = 27.5 + j51.5\,\Omega$$

Now, the input and output circuits can be designed for these values. One such circuit is shown in Figure 11.14.

(d) As designed, $G_S = 3\,\text{dB}$. From (10.4.16),

$$G_{L\max} = \frac{1}{1 - |S_{22}|^2} = \frac{1}{1 - 0.6^2} = 1.5625 = 1.9382\,\text{dB}$$

and

$$G_0 = |S_{21}|^2 = 4^2 = 16 = 12.0412\,\text{dB}$$

Therefore,

$$G_{TU} = 3 + 1.9382 + 12.0412 = 16.9794\,\text{dB}$$

Bilateral Case

If a microwave transistor cannot be assumed to be unilateral, the design procedure becomes tedious for a less than maximum possible transducer power gain. In this case, the operating or available power gain approach is preferred because of its simplicity. The design equations for these circuits are developed in this section.

Unconditionally Stable Case The operating power gain of an amplifier is given by (10.4.19). For convenience, it is reproduced here:

$$G_P = \frac{(1 - |\Gamma_L|^2)|S_{21}|^2}{(|1 - S_{22}\Gamma_L|^2)(1 - |\Gamma_{in}|^2)} \qquad (10.4.19)$$

and the input reflection coefficient is

$$\Gamma_{in} = S_{11} + \frac{S_{12}S_{21}\Gamma_L}{1 - S_{22}\Gamma_L} = \frac{S_{11} - \Gamma_L\Delta}{1 - S_{22}\Gamma_L}$$

Therefore,

$$G_P = \frac{(1 - |\Gamma_L|^2)|S_{21}|^2}{(|1 - S_{22}\Gamma_L|^2) - (|S_{11} - \Gamma_L\Delta|^2)} = |S_{21}|^2 g_p \qquad (11.3.7)$$

where

$$g_p = \frac{1 - |\Gamma_L|^2}{(|1 - S_{22}\Gamma_L|^2) - (|S_{11} - \Gamma_L\Delta|^2)} \qquad (11.3.8)$$

The equation for g_p can be simplified and rearranged as follows:

$$\left|\Gamma_L - \frac{g_p C_2^*}{1 + g_p(|S_{22}|^2 - |\Delta|^2)}\right|^2 = \frac{[1 - g_p(1 - |S_{11}|^2)][1 + g_p(|S_{22}|^2 - |\Delta|^2) + g_p^2|C_2|^2]}{[1 + g_p(|S_{22}|^2 - |\Delta|^2)]^2} \qquad (11.3.9)$$

where

$$C_2 = S_{22} - S_{11}^*\Delta \qquad (11.3.10)$$

Equation (11.3.9) can be simplified further as follows:

$$\left|\Gamma_L - \frac{g_p C_2^*}{1 + g_p(|S_{22}|^2 - |\Delta|^2)}\right|^2 = \frac{1 - 2k|S_{12}S_{21}|g_p + |S_{12}S_{21}|^2 g_p^2}{[1 + g_p(|S_{22}|^2 - |\Delta|^2)]^2}$$

This represents a circle with its center c_p and radius R_p given as

$$c_p = \frac{g_p C_2^*}{1 + g_p(|S_{22}|^2 - |\Delta|^2)} \qquad (11.3.11)$$

and

$$R_p = \frac{\sqrt{(1 - 2k|S_{12}S_{21}|g_p + |S_{12}S_{21}|^2 g_p^2)}}{1 + g_p(|S_{22}|^2 - |\Delta|^2)} \qquad (11.3.12)$$

For $R_p = 0$, (11.3.12) can be solved for g_p, which represents its maximum value. It is given as

$$g_p = g_{p\max} = \frac{1}{|S_{12}S_{21}|}(k - \sqrt{k^2 - 1})$$

and from (11.3.7) we have

$$G_{p\max} = \frac{|S_{21}|}{|S_{12}|}(k - \sqrt{k^2 - 1}) \qquad (11.3.13)$$

A comparison of (11.3.13) with (11.2.7) indicates that the maximum operating power gain is equal to that of the maximum transducer power gain.

Following a similar procedure, expressions for center c_a and radius R_a of the constant available power gain circle can be formulated. These relations are

$$c_a = \frac{g_a C_1^*}{1 + g_a(|S_{11}|^2 - |\Delta|^2)} \qquad (11.3.14)$$

and

$$R_a = \frac{\sqrt{(1 - 2k|S_{12}S_{21}|g_a + |S_{12}S_{21}|^2\ g_a^2)}}{1 + g_a(|S_{11}|^2 - |\Delta|^2)} \qquad (11.3.15)$$

where

$$C_1 = S_{11} - S_{22}^*\Delta \qquad (11.3.16)$$

and

$$g_a = \frac{1 - |\Gamma_S|^2}{(|1 - S_{11}\Gamma_S|^2) - (|S_{22} - \Gamma_S\Delta|^2)} \qquad (11.3.17)$$

Example 11.7 A GaAs FET is biased at $V_{ds} = 4\,\text{V}$ and $I_{ds} = 0.5\ I_{dss}$. Its S-parameters are given at 6 GHz as follows: $S_{11} = 0.641\ \angle{-171.3^\circ}$, $S_{12} = 0.057\ \angle 16.3^\circ$, $S_{21} = 2.058\ \angle 28.5^\circ$, and $S_{22} = 0.572\ \angle{-95.7^\circ}$. Using this transistor, design an amplifier that provides an operating power gain of 9 dB.

SOLUTION Since $k = 1.5037$ and $|\Delta| = 0.3014$, this transistor is unconditionally stable. Further, $G_{p\max}$ is found to be 11.38 dB. Hence, it can be used to get a gain of 9 dB. The corresponding circle data are found from (11.3.11) and (11.3.12) as follows: $c_p = 0.5083\angle 103.94^\circ$ and $R_p = 0.4309$. This circle is drawn on a Smith chart (Figure 11.15) and the load reflection coefficient is selected as $0.36\ \angle 50^\circ$. The corresponding input reflection coefficient is calculated as $0.63\ \angle{-175.6^\circ}$. Hence, Γ_S must be equal to $0.63\ \angle 175.6^\circ$ (i.e., conjugate of input reflection coefficient). These load- and source-impedance points are depicted in Figure 11.15 as C and A, respectively. The corresponding admittance points are identified as D and B on this chart. The load-side network is designed by moving from point O to point F and then to point D. It is achieved by adding an open-circuited shunt stub of length 0.394λ and then a transmission line of 0.083λ. For the source side we can follow the path O–E–B, and

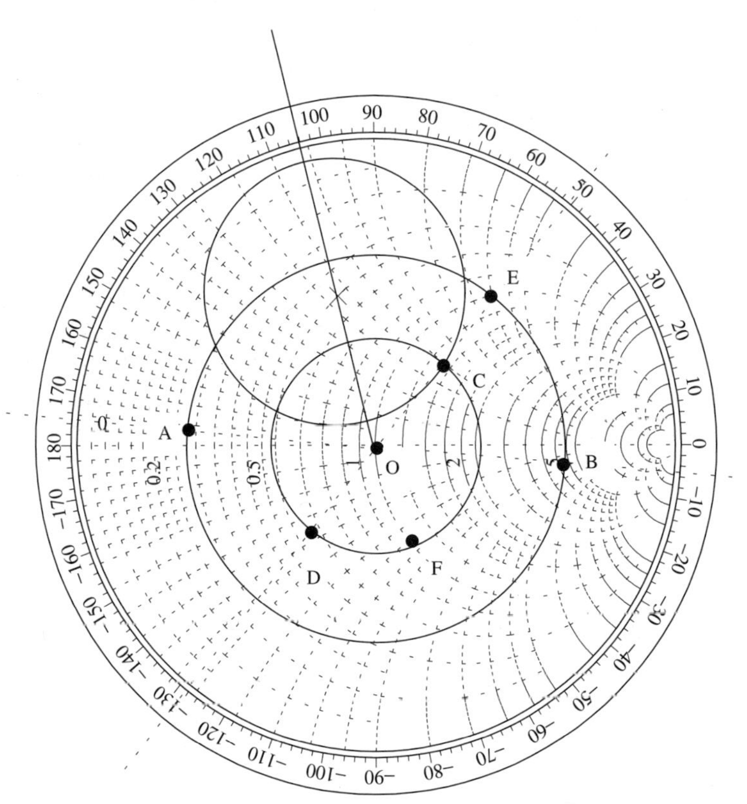

Figure 11.15 Matching network design for Example 11.7.

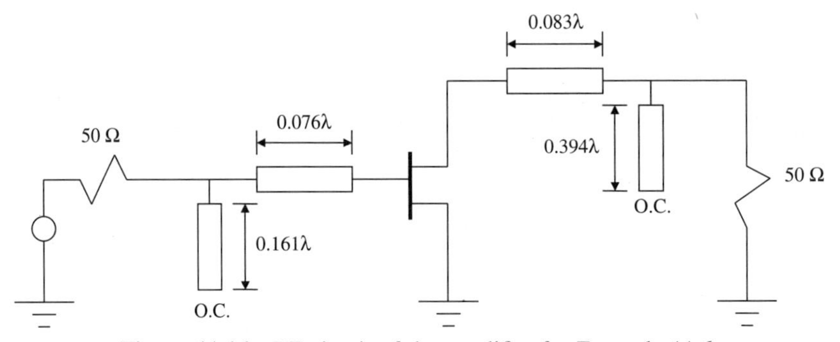

Figure 11.16 RF circuit of the amplifier for Example 11.6.

therefore an open-circuited shunt stub of 0.161λ followed by a 0.076λ-long transmission line can provide the desired admittance. The designed circuit is shown in Figure 11.16.

Potentially Unstable Case Operating power gain circles for a bilateral potentially unstable transistor can still be found from (11.3.9) and (11.3.10). However, the load impedance point is selected such that it is in the stable region. Further, the conjugate of its input reflection coefficient must be a stable point because it represents the source reflection coefficient. Similarly, the available power gain circles can be drawn on a Smith chart using (11.3.12) and (11.3.13) along with the stability circles of the transistor. The source impedance point is selected such that it is in the stable region and provides the desired gain. Further, conjugate of the corresponding output reflection coefficient must lie in the stable region as well because it represents the load.

Example 11.8 A GaAs FET is biased at $I_{ds} = 5\,\text{mA}$ and $V_{ds} = 5\,\text{V}$ to measure its scattering parameters at 8 GHz. Using a 50-Ω system, these data are as follows: $S_{11} = 0.5\ \angle{-180^\circ}$, $S_{12} = 0.08\ \angle 30^\circ$, $S_{21} = 2.5\ \angle 70^\circ$, and $S_{22} = 0.8\ \angle{-100^\circ}$. Design an amplifier using this transistor for an operating power gain of 10 dB.

SOLUTION From (11.1.16) and (11.1.18) we find that

$$|\Delta| = 0.2228$$

and

$$k = 0.3991$$

Since k is less than 1, this transistor is potentially unstable. The input and output stability circles are determined from (11.1.20)–(11.1.23) as follows:

For the input stability circle:

$$C_S = 1.6713\ \angle 170.59^\circ$$

and

$$r_S = 0.9983$$

Since $|S_{22}|$ is 0.8, $\Gamma_S = 0$ represents a stable source impedance point. Further, this point is located outside the stability circle, and therefore all points outside this circle are stable.

For the output stability circle,

$$C_L = 1.1769\ \angle 97.17^\circ$$

and

$$r_L = 0.3388$$

Since $|S_{11}|$ is 0.5, $\Gamma_L = 0$ represents a stable load impedance point. Further, this point is located outside the stability circle, and therefore, all points outside this circle are stable. The data for a 10-dB gain circle are found from (11.3.11) and (11.3.12) as

$$C_p = 0.5717\ \angle 97.2^\circ \quad \text{and} \quad R_p = 0.4733$$

These circles are drawn on the Smith chart shown in Figure 11.17. Since this transistor is potentially unstable for certain load and input impedances, we need to avoid those regions. If we select $\Gamma_L = 0.1\ \angle 97^\circ$ (point C in Figure 11.17),

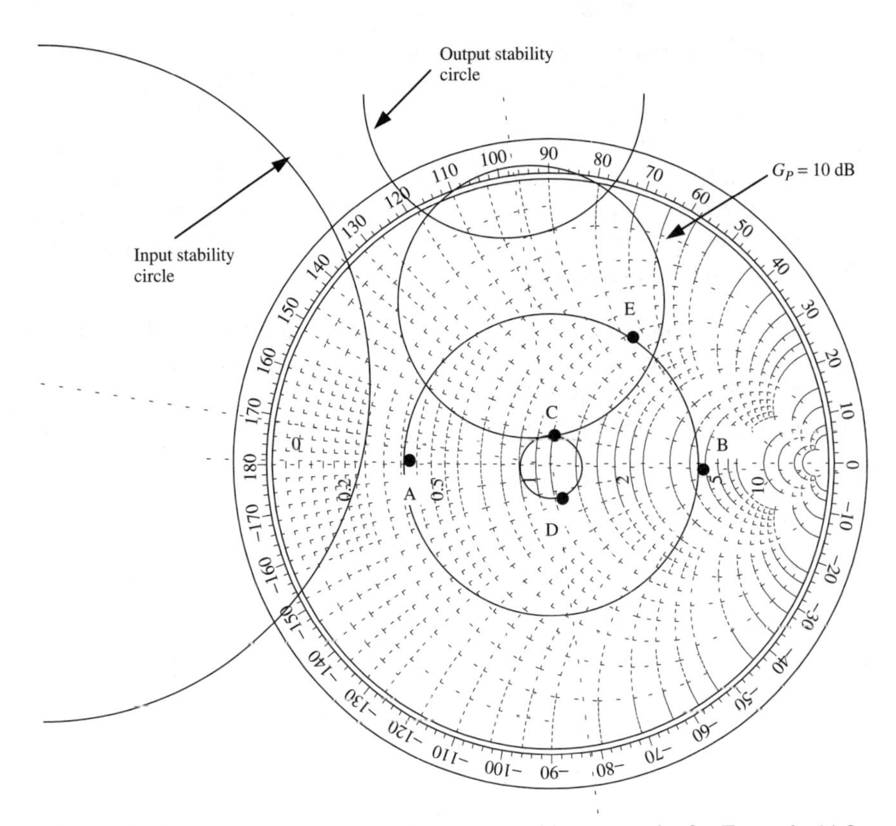

Figure 11.17 Design the input- and output-matching networks for Example 11.8.

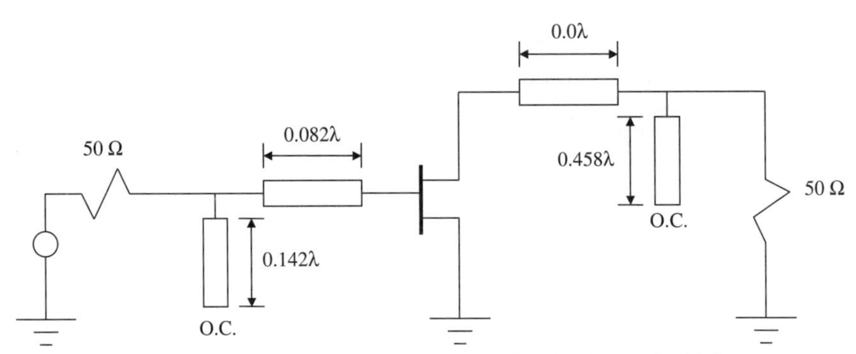

Figure 11.18 RF circuit of the amplifier for Example 11.8.

the corresponding input reflection coefficient is calculated as 0.5209 $\angle -179.3^\circ$. Therefore, Γ_S is 0.5209 $\angle$ 179.3°. We find that this point lies in the stable region. One of the possible designs is illustrated in Figure 11.18.

Example 11.9 Design an amplifier for maximum possible transducer power gain, $G_{T\,\max}$, using a BJT that is biased at $V_{CE} = 10\,\text{V}$, $I_C = 4\,\text{mA}$. Its

S-parameters are found as follows using a 50-Ω system at 750 MHz: $S_{11} = 0.277\ \angle -59^\circ$, $S_{12} = 0.078\ \angle 93^\circ$, $S_{21} = 1.92\ \angle 64^\circ$, and $S_{22} = 0.848\ \angle -31^\circ$.

SOLUTION From (11.1.16) and (11.1.18), $k = 1.0325$ and $|\Delta| = 0.3242$. Hence, the transistor is unconditionally stable. For a maximum possible transducer power gain, we need to look for a simultaneous conjugate matching. From (11.2.5) and (11.2.6), we find that

$$\Gamma_{MS} = 0.7298\ \angle 135.44^\circ \quad \text{and} \quad \Gamma_{ML} = 0.9511\ \angle 33.85^\circ$$

The corresponding normalized impedances are identified in Figure 11.19 as $0.1818 + j0.398$ (point A) and $0.2937 + j3.2619$ (point C), respectively. Points B and D represent respective normalized admittances on this Smith chart. For the load side, move from 1 to point E on the unity conductance circle and then from E to D on the constant-VSWR circle. Hence, an open-circuited shunt stub of 0.278λ followed by a 0.176λ-long transmission line is needed for the load side. Similarly, a shunt stub of 0.315λ and a 0.004λ-long section of the transmission line transform the normalized admittance of 1 to point F and then to B on the source side. The final circuit is illustrated in Figure 11.20.

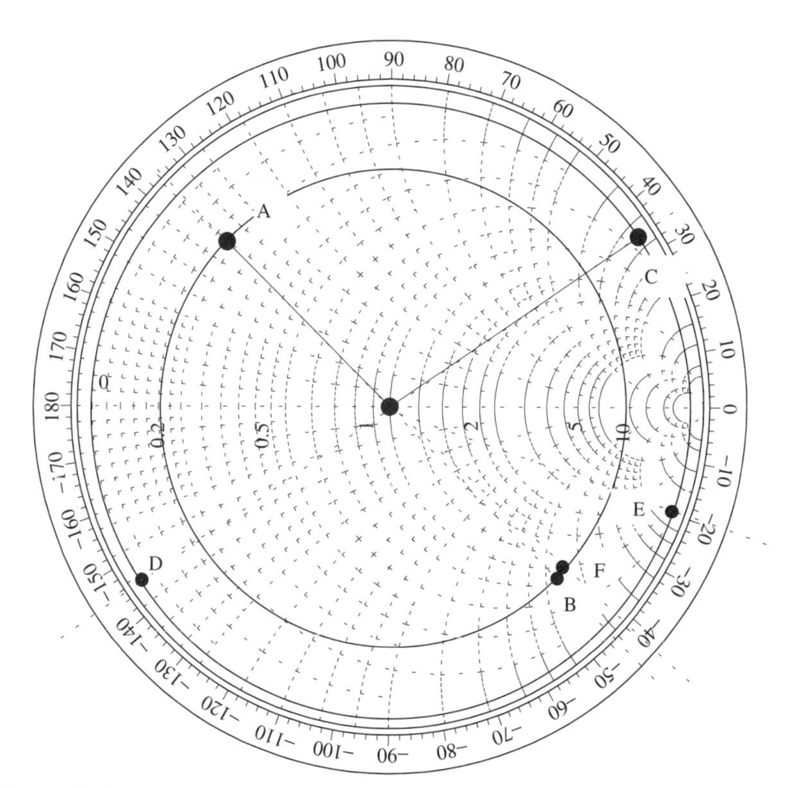

Figure 11.19 Design of input- and output-matching networks for Example 11.9.

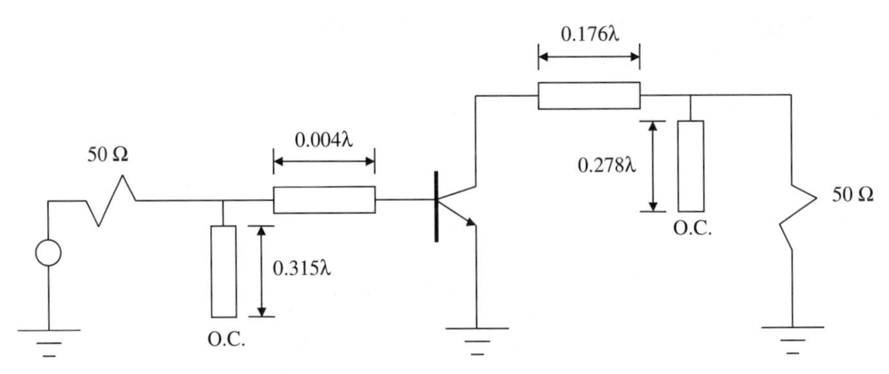

Figure 11.20 RF circuit of the amplifier for Example 11.9.

Example 11.10 Using the BJT of Example 11.9, design an amplifier for $G_p =$ 10 dB at 750 MHz. Also, determine the reflection coefficients required to obtain maximum operating power gain and show that they are identical to those obtained in Example 11.9 for the simultaneous conjugate match.

SOLUTION From Example 11.9 we know that the transistor is unconditionally stable. For an operating power gain of 10 dB, the location and radius of the constant-gain circle are determined from (11.3.11) and (11.3.12) as follows: $c_p = 0.7812\ \angle 33.85^\circ$ and $R_p = 0.2142$. This gain circle is drawn on a Smith chart as shown in Figure 11.21. For an operating power gain of 10 dB, the load impedance must be selected on this circle. If we select a normalized impedance of $1.8 + j1.6$ (point C), the load reflection coefficient is $0.56\ \angle 34^\circ$. The corresponding input reflection coefficient is found at $0.3455\ \angle -56.45^\circ$. Therefore, the source reflection coefficient Γ_S must be equal to $0.3455\ \angle 56.45^\circ$. Further, for a maximum operating power gain (i.e., 12.8074 dB), the gain circle converges to a point that is located at $0.9511\ \angle 33.85^\circ$. Hence, this point represents the load reflection coefficient that is needed to obtain maximum operating power gain. Further, the corresponding input reflection coefficient is found to be $0.7298\ \angle -135.44^\circ$. Therefore, $\Gamma_S = 0.7298\ \angle 135.44^\circ$. A comparison of these results with those obtained in Example 11.9 indicates that the results are identical in both cases.

One of the possible RF circuits for obtaining an operating power gain of 10 dB can be designed as follows. After locating the load and source impedance points C and A, respectively, constant-VSWR circles are drawn. The corresponding normalized admittance points D and B are then found. A normalized shunt susceptance of $-j1.3$ will transform the 50-Ω load impedance (normalized admittance of 1) to $1 - j1.3$ (point E); then a 0.124λ-long transmission line will move it to point D. An open-circuited 0.356λ-long stub can be used to obtain a $-j1.3$ susceptance.

For the design of an input matching network, a normalized shunt susceptance of $-j0.75$ may be used to transform a 50-Ω source impedance (normalized

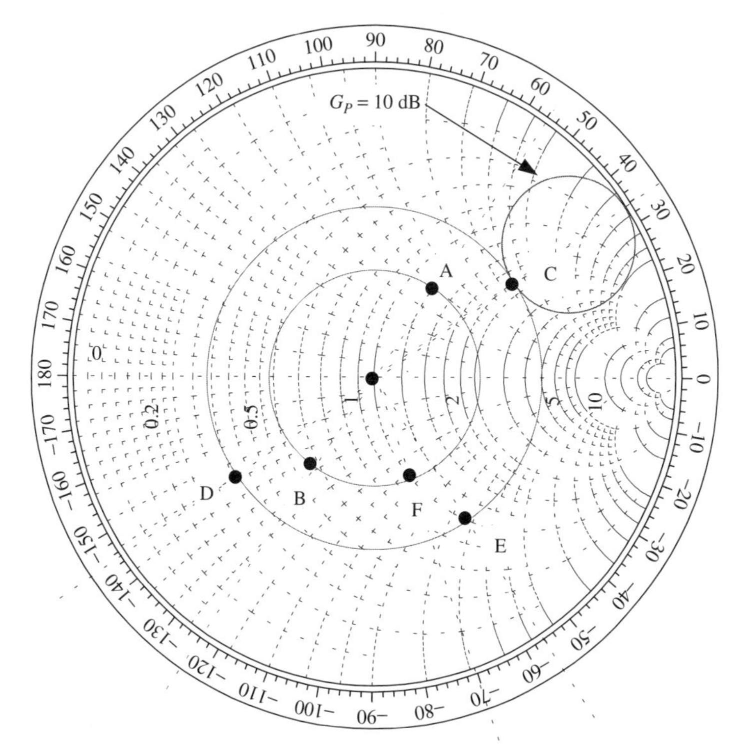

Figure 11.21 Design of input- and output-matching networks for Example 11.10.

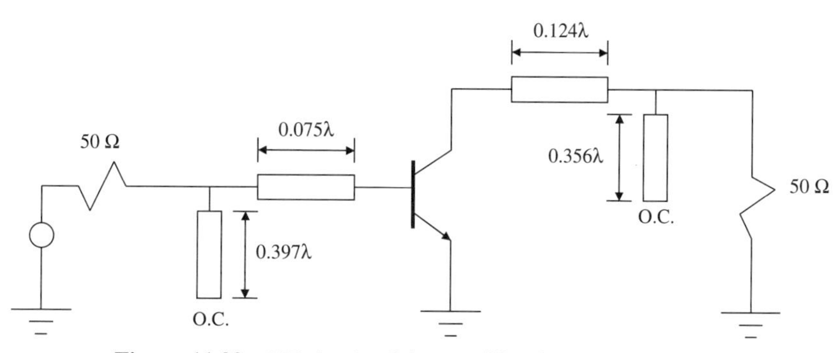

Figure 11.22 RF circuit of the amplifier for Example 11.10.

admittance of 1) to $1 - j0.75$ (point F). A 0.075λ-long transmission line is then connected to transform this admittance to that of point B. A 0.397λ-long open-circuited stub can be used to obtain the desired susceptance of $-j0.75$. The final circuit is illustrated in Figure 11.22.

Example 11.11 A suitably biased GaAs FET has the following S-parameters measured at 2 GHz with a 50-Ω system: $S_{11} = 0.7\ \angle{-65^\circ}$, $S_{12} = 0.03\ \angle 60^\circ$,

$S_{21} = 3.2\ \angle 110^\circ$, and $S_{22} = 0.8\ \angle -30^\circ$. Determine the stability and design an amplifier for $G_p = 10$ dB.

SOLUTION From (11.1.16) and (11.1.18), $|\Delta| = 0.5764$ and $k = 1.053$. Hence, the transistor is unconditionally stable.

The circle parameters for an operating power gain of 10 dB are found from (11.3.11) and (11.3.12) as $c_p = 0.3061\ \angle 39.45^\circ$ and $R_p = 0.6926$. This circle is drawn on a Smith chart as shown in Figure 11.23. If we select a normalized load of $0.45 + j0$ (point C), the corresponding load reflection coefficient will be $0.38\ \angle 180^\circ$. This gives an input reflection coefficient of $0.714\ \angle -63^\circ$. This means that the source reflection coefficient Γ_S must be $0.714\ \angle 63^\circ$. The corresponding normalized impedance may be found as $0.569 + j1.4769$ (point A). These data are used to draw the VSWR circles, as shown in the figure. Matching networks can be designed to synthesize the normalized admittances of points D (for the load side) and B (for the source side). A normalized shunt susceptance of $j0.8$ moves 50 Ω (center of the chart) to point E and then a 0.094λ-long transmission line can transform it to the desired value at point D. The shunt susceptance of $j0.8$ can be obtained via a capacitor of 1.27 pF. Similarly, a shunt inductor of

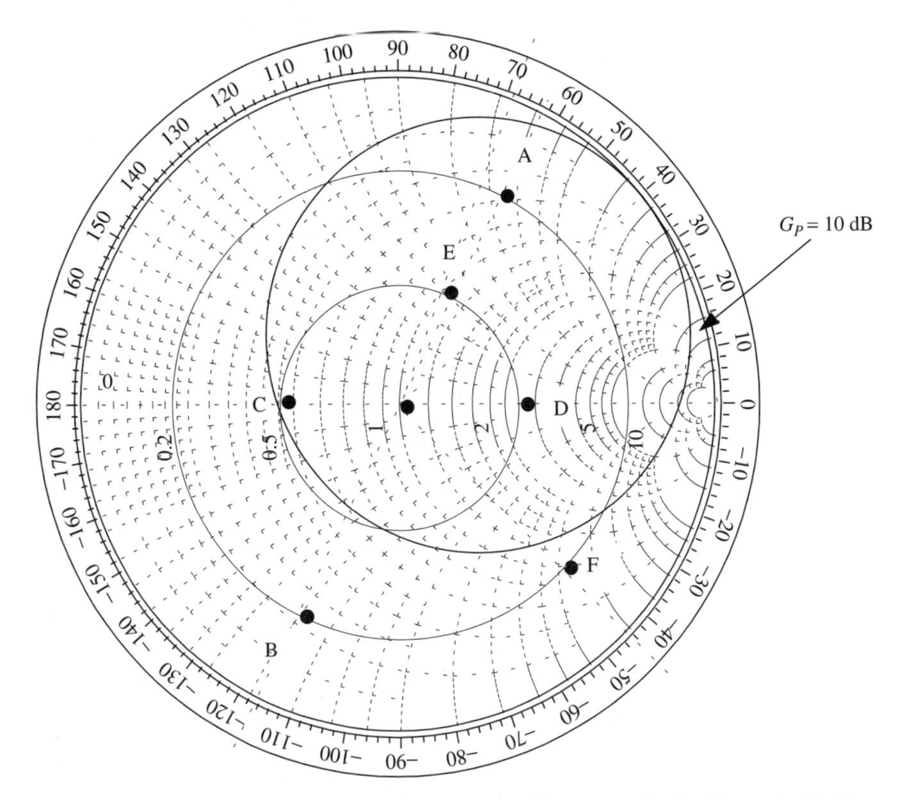

Figure 11.23 Design of input- and output-matching networks for Example 11.11.

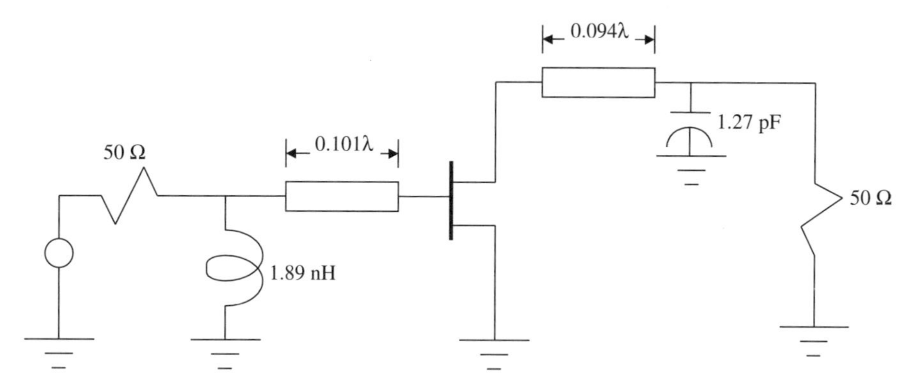

Figure 11.24 Circuit of the amplifier for Example 11.11.

1.89 nH adds a normalized susceptance of $-j2.1$ at the source side (point F), and then a 0.101λ-long transmission line can transform that to the desired value of $0.22 - j0.6$ (point B). The final circuit is illustrated in Figure 11.24.

11.4 CONSTANT NOISE FIGURE CIRCLES

Noise characteristics of the two-port networks are considered in Chapter 2. We found that the noise factor of an amplifier can be expressed via (2.5.21) as follows:

$$F = F_{\min} + \frac{R_n}{G_S}|Y_S - Y_{\text{opt}}|^2 \tag{11.4.1}$$

where $F_{\min}$ = minimum noise factor of the transistor (attained for $Y_S = Y_{\text{opt}}$)
R_n = equivalent noise resistance of the transistor
G_S = real part of the source admittance
$Y_S = G_S + jB_S = \dfrac{1}{Z_0}\dfrac{1-\Gamma_S}{1+\Gamma_S}$ = source admittance presented to the transistor
Γ_S = source reflection coefficient seen by the transistor
$Y_{\text{opt}} = \dfrac{1}{Z_0}\dfrac{1-\Gamma_{\text{opt}}}{1+\Gamma_{\text{opt}}}$ = optimum source admittance that results in minimum noise figure
Γ_{opt} = optimum source reflection coefficient that results in minimum noise figure

Since

$$|Y_S - Y_{\text{opt}}|^2 = \left|\frac{1}{Z_0}\frac{1-\Gamma_S}{1+\Gamma_S} - \frac{1}{Z_0}\frac{1-\Gamma_{\text{opt}}}{1+\Gamma_{\text{opt}}}\right|^2 = \frac{4}{|Z_0|^2}\frac{|\Gamma_S - \Gamma_{\text{opt}}|^2}{|1+\Gamma_S|^2 \cdot |1+\Gamma_{\text{opt}}|^2} \tag{11.4.2}$$

and

$$G_S = \mathrm{Re}(Y_S) = \frac{1}{2Z_0}\left(\frac{1-\Gamma_S}{1+\Gamma_S} + \frac{1-\Gamma_S^*}{1+\Gamma_S^*}\right) = \frac{1-|\Gamma_S|^2}{Z_0|1+\Gamma_S|^2} \tag{11.4.3}$$

(11.4.1) can be written as

$$F = F_{\min} + \frac{4R_n}{Z_0}\frac{|\Gamma_S - \Gamma_{\text{opt}}|^2}{(1-|\Gamma_S|^2)|1+\Gamma_{\text{opt}}|^2}$$

or

$$\frac{|\Gamma_S - \Gamma_{\text{opt}}|^2}{1-|\Gamma_S|^2} = \frac{F - F_{\min}}{4(R_n/Z_0)}|1+\Gamma_{\text{opt}}|^2 = N \tag{11.4.4}$$

where N is called the *noise figure parameter.*

From (11.4.4), we have

$$|\Gamma_S - \Gamma_{\text{opt}}|^2 = N(1-|\Gamma_S|^2)$$

or

$$(\Gamma_S - \Gamma_{\text{opt}})(\Gamma_S^* - \Gamma_{\text{opt}}^*) = \Gamma_S\Gamma_S^* - (\Gamma_S^*\Gamma_{\text{opt}} + \Gamma_S\Gamma_{\text{opt}}^*) + \Gamma_{\text{opt}}\Gamma_{\text{opt}}^* = N - N|\Gamma_S|^2$$

or

$$\Gamma_S\Gamma_S^* - \frac{\Gamma_S^*\Gamma_{\text{opt}} + \Gamma_S\Gamma_{\text{opt}}^*}{N+1} + \frac{|\Gamma_{\text{opt}}|^2}{(N+1)^2} = \frac{N - \Gamma_{\text{opt}}\Gamma_{\text{opt}}^*}{N+1} + \frac{|\Gamma_{\text{opt}}|^2}{(N+1)^2}$$

or

$$\left|\Gamma_S - \frac{\Gamma_{\text{opt}}}{N+1}\right|^2 = \frac{N(N+1) - N|\Gamma_{\text{opt}}|^2}{(N+1)^2}$$

Hence,

$$\left|\Gamma_S - \frac{\Gamma_{\text{opt}}}{N+1}\right| = \frac{\sqrt{N(N+1) - N|\Gamma_{\text{opt}}|^2}}{N+1} \tag{11.4.5}$$

Equation (11.4.5) represents a circle on the complex Γ_S-plane. Its center C_{NF} and radius R_{NF} are given by

$$C_{\text{NF}} = \frac{\Gamma_{\text{opt}}}{N+1} \tag{11.4.6}$$

and

$$R_{\text{NF}} = \frac{\sqrt{N(N+1) - N|\Gamma_{\text{opt}}|^2}}{N+1} \tag{11.4.7}$$

Using (11.4.6) and (11.4.7), constant noise–figure circles can be drawn on a Smith chart. The source reflection coefficient Γ_S is selected inside this circle to keep the noise figure of the amplifier within a specified noise factor F.

Example 11.12 A bipolar junction transistor is biased at $V_{CE} = 4\,\text{V}$ and $I_{CE} = 30\,\text{mA}$. Its scattering and noise parameters are measured at 1 GHz with a 50-Ω system as follows: $S_{11} = 0.707\ \angle{-155^\circ}$, $S_{21} = 5.0\ \angle 180^\circ$, $S_{12} = 0$, $S_{22} = 0.51\ \angle{-20^\circ}$, $F_{\min} = 3\,\text{dB}$, $R_n = 4\,\Omega$, and $\Gamma_{\text{opt}} = 0.45\ \angle 180^\circ$.

Design the input- and output-matching networks for this transistor so that it produces a power gain of 16 dB and a noise figure of less than 3.5 dB.

SOLUTION From (11.1.16) and (11.1.18), we find that $k = \infty$ and $|\Delta| = 0.3606$. Therefore, the transistor is unconditionally stable. From (10.4.14) to (10.4.16),

$$G_{S\,\max} = 3\,\text{dB}$$

$$G_{L\,\max} = 1.31\,\text{dB}$$

and

$$G_0 = 13.98\,\text{dB}$$

Hence,

$$G_{TU\,\max} = 3 + 1.31 + 13.98 = 18.29\,\text{dB}$$

Therefore, this transistor can be used for a power gain of 16 dB. Since G_0 is approximately 14 dB, the remaining 2 dB can be set via G_S and G_L. If we design an input-matching network that provides G_S as 1.22 dB, we need an output-matching network for a G_L of 0.78 dB. The G_S and G_L circles are found from (11.3.5) and (11.3.6) as follows:

$$d_S = 0.5634\ \angle 155^\circ$$

$$R_S = 0.3496$$

$$d_L = 0.4655\ \angle 20^\circ$$

and

$$R_L = 0.2581$$

The 3.5-dB noise figure circle is found from (11.4.6) and (11.4.7) as follows:

$$C_{\text{NF}} = 0.3658\ \angle 180^\circ$$

and

$$R_{\text{NF}} = 0.3953$$

These gain and noise circles are drawn on a Smith chart as shown in Figure 11.25. If we select point B on the 1.22-dB G_S circle as Γ_S, the noise figure will stay

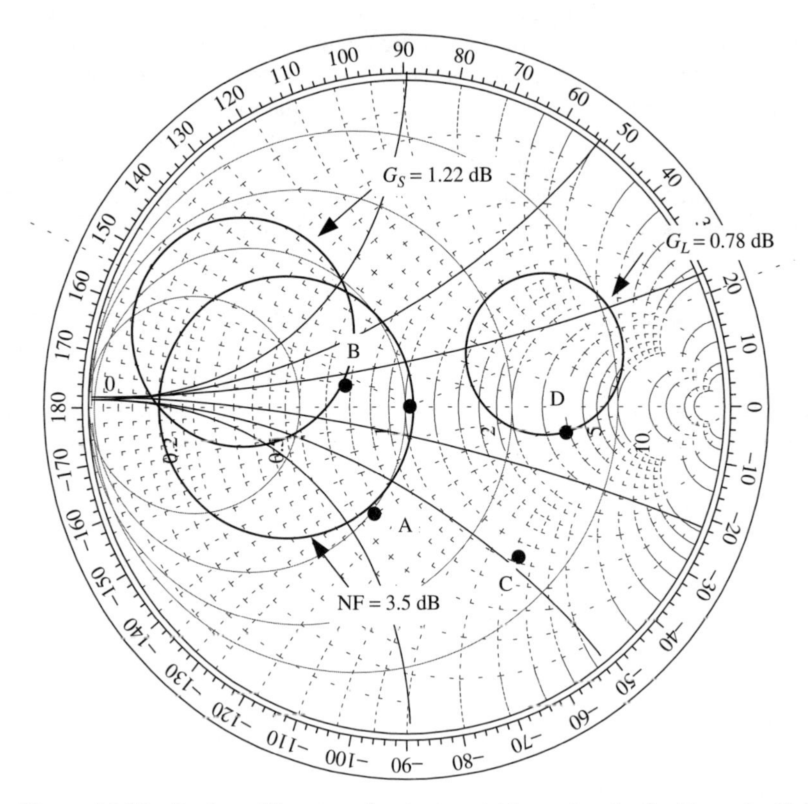

Figure 11.25 Design of input- and output-matching networks for Example 11.12.

within 3.5 dB. The load-side matching network is designed so that it provides G_L as 0.78 dB. We select point D for Γ_L. The two matching networks are designed below using discrete components.

For the design of a source-side network, we can move on the unity-conductance circle from 1 to point A. It requires a shunt capacitor for a normalized susceptance of $+j0.75$. A capacitor of 2.39 pF can provide this susceptance at 1 GHz. The next step is to reach point B from A by moving on the constant-resistance circle of 0.65, which requires an inductive reactance of normalized value $+j0.56$ and hence an inductor of 4.46 nH. Similarly, the load-side network can be designed following the unity-resistance circle from 1 to point C. It requires a capacitor of 2.27 pF in series with a 50-Ω output. Next, move from point C to D on the conductance circle of 0.35. Hence, it requires an inductor of 19.89 nH. The complete RF circuit is shown in Figure 11.26.

Example 11.13 A GaAs FET has the following scattering and noise parameters at 4 GHz, measured with a 50-Ω system: $S_{11} = 0.6\ \angle -60^\circ$, $S_{21} = 1.9\ \angle 81^\circ$, $S_{12} = 0.05\ \angle 26^\circ$, $S_{22} = 0.5\ \angle -60^\circ$, $F_{\min} = 1.6\,\text{dB}$, $R_n = 20\,\Omega$, and $\Gamma_{\text{opt}} = 0.62\ \angle 100^\circ$.

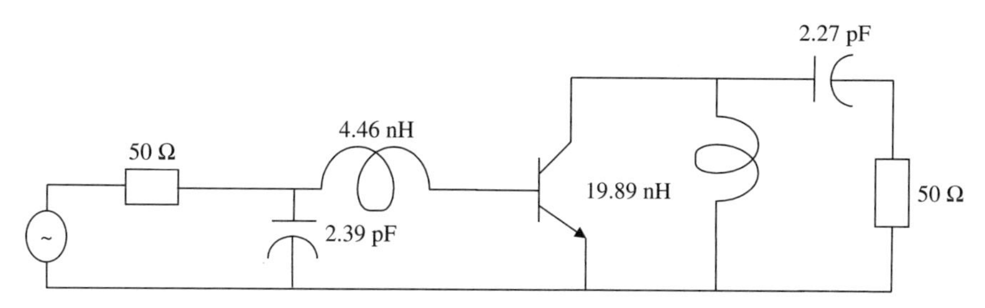

Figure 11.26 RF circuit design for Example 11.12.

(**a**) Assuming the FET to be unilateral, design an amplifier for a maximum possible gain and a noise figure no more than 2.0 dB. Estimate the error introduced in G_T due to this assumption.

(**b**) Redesign the amplifier in part (a), with FET being bilateral.

SOLUTION From (11.1.16) and (11.1.18), $|\Delta| = 0.3713$ and $k = 2.778$. Hence, the transistor is unconditionally stable.

(**a**) The unilateral figure of merit is found from (11.2.10) as 0.0594. Therefore,

$$-0.501\,\text{dB} < \frac{G_T}{G_{TU}} < 0.5938\,\text{dB}$$

This means that the maximum error in the gain of the amplifier due to this assumption will be on the order of ± 0.5 dB.

Using (10.4.14) to (10.4.16), we find that

$$G_{S\,\max} = 1.9382\,\text{dB}$$

$$G_{L\,\max} = 1.25\,\text{dB}$$

and

$$G_0 = 5.5751\,\text{dB}$$

Therefore,

$$G_{TU\,\max} = 1.9382 + 1.25 + 5.5751 = 8.7627\,\text{dB}$$

Since we are looking for a maximum possible gain with a noise figure less than 2 dB, we should use load impedance that provides a conjugate match. Note that this selection does not influence the noise figure. Thus, we select $\Gamma_L = S_{22}^* = 0.5\ \angle 60^\circ$. It is depicted by point C on the Smith chart in Figure 11.27. For the design of the source side, we first determine the relevant gain and noise circles.

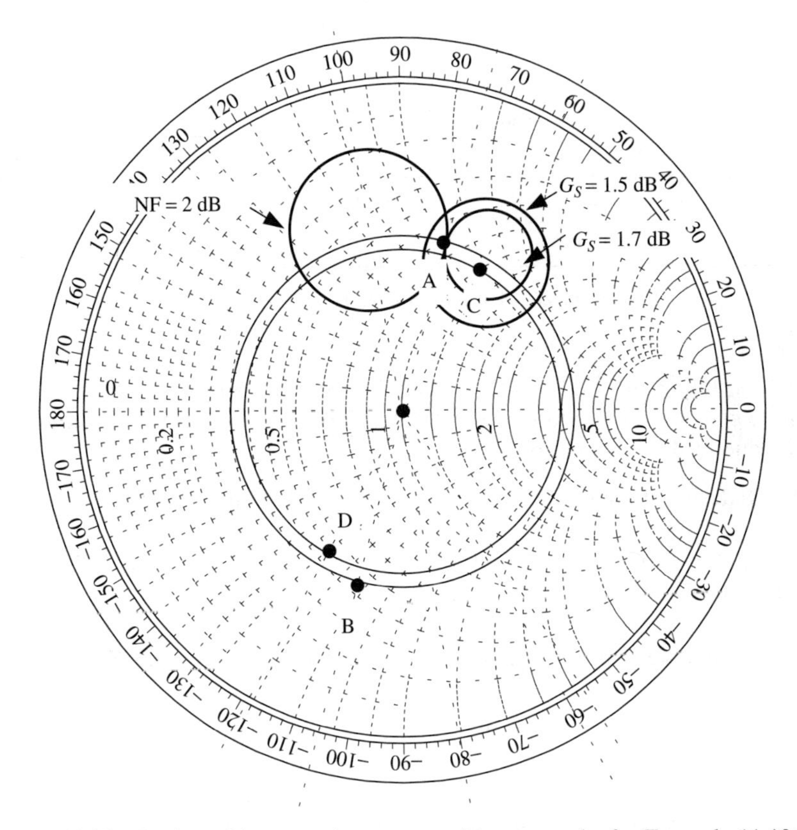

Figure 11.27 Design of input- and output-matching networks for Example 11.13(a).

From (11.3.5) and (11.3.6), the constant-gain circles for $G_S = 1.5\,\text{dB}$ and 1.7 dB are determined as follows. For $G_S = 1.5\,\text{dB}$,

$$d_S = 0.5618\ \angle 60^\circ,\ R_S = 0.2054$$

For $G_S = 1.7\,\text{dB}$,

$$d_S = 0.5791\ \angle 60^\circ,\ R_S = 0.1507$$

The 2-dB noise figure circle is found from (10.4.6) and (10.4.7) as follows:

$$C_{\text{NF}} = 0.5627\ \angle 100^\circ$$

and

$$R_{\text{NF}} = 0.2454$$

These circles are drawn on a Smith chart as shown in Figure 11.27. We find that the 2-dB noise figure circle almost touches the $G_S = 1.7$-dB circle at a point.

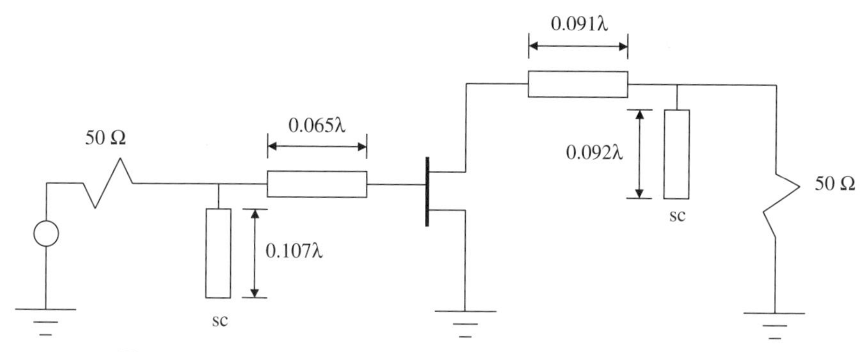

Figure 11.28 RF circuit of the amplifier for Example 11.13(a).

Further, the 1.5-dB gain circle has a larger radius. It suggests that a gain larger than 1.7 dB will have a higher noise figure. As illustrated, the optimum Γ_S for this problem is point A. A circuit designed with short-circuited shunt stubs and transmission line lengths is shown in Figure 11.28.

(**b**) When the transistor is bilateral, we need to use the operating power gain or available power gain approach for the design. The operating power gain approach provides information about the load impedance. Source impedance is determined subsequently as a complex conjugate of the input impedance. If this value of the source impedance does not satisfy the noise figure specification, one needs to select a different load impedance. Thus, several iterations may be needed before the noise and gain requirements are satisfied. On the other hand, the available power gain approach provides the source impedance. The load impedance is then determined as a complex conjugate of the output impedance. Hence, a suitable source impedance can be determined that simultaneously satisfies both the gain and the noise characteristics. Hence, the available power gain approach does not require iterative calculations, and therefore it is preferable in the present case.

From (11.3.14) and (11.3.15), the 8-dB gain circle is found as follows: $c_a = 0.5274\ \angle 64.76^\circ$ and $R_a = 0.2339$. Further, the maximum possible available power gain is found to be 8.5 dB. It represents a point at $0.5722\ \angle 64.76^\circ$. The noise circle stays the same as in part (a). These circles are drawn on a Smith chart as shown in Figure 11.29. If we select Γ_S as $0.535\ \angle 90^\circ$, the noise figure stays well below 2 dB. The corresponding output reflection coefficient is found to be $0.5041\ \angle -67.83^\circ$. Hence, the load reflection coefficient must be $0.5041\ \angle 67.83^\circ$. Now, the input and output matching networks can easily be designed. The transducer power gain of this circuit is found from (10.4.8) to be 8 dB.

A comparison of the gain obtained in part (b) with that in part (a) shows that the two differ by 0.76 dB. However, the uncertainty predicted in part (a) is about ± 0.5 dB. The 8.2-dB gain circle shown in Figure 11.29 indicates that the gain in part (b) can be increased to a little over 8.2 dB. That will bring the two values within the range predicted.

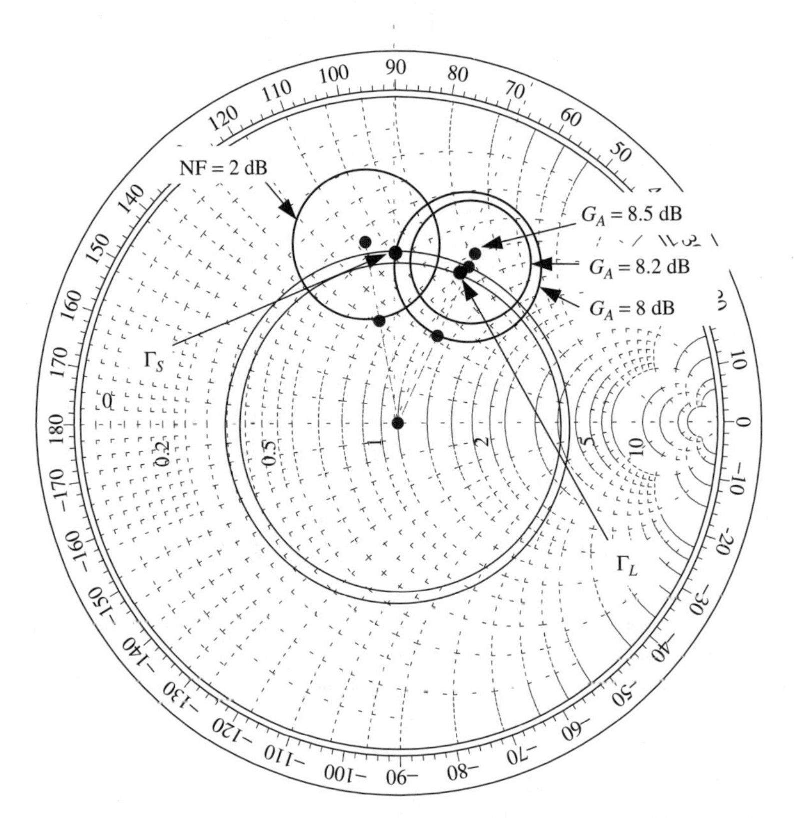

Figure 11.29 Available power gain circles and RF circuit design for Example 11.13(b).

Example 11.14 An Avantek low-noise silicon bipolar transistor, AT-41410, is biased at $V_{CE} = 8\,\text{V}$ and $I_C = 10\,\text{mA}$. Its scattering and noise parameters are measured at 1 GHz with a 50-Ω system. The results are found as follows: $S_{11} = 0.6\,\angle{-163^\circ}$, $S_{21} = 7.12\,\angle 86^\circ$, $S_{12} = 0.039\,\angle 35^\circ$, $S_{22} = 0.5\,\angle{-38^\circ}$, $F_{\min} = 1.3\,\text{dB}$, $R_n = 8\,\Omega$, and $\Gamma_{\text{opt}} = 0.06\,\angle 49^\circ$. Design the input and output matching networks of an amplifier that provides a power gain of 16 dB with a noise figure below 2.5 dB.

SOLUTION From (11.1.16) and (11.1.18), $|\Delta| = 0.1892$ and $k = 0.7667$. Hence, the transistor is potentially unstable. The input stability circle is determined from (11.1.22) and (11.1.23) as follows: $C_S = 1.7456\,\angle 171.69^\circ$ and $r_S = 0.8566$. Since $|C_S|$ is larger than r_S, this circle does not enclose the center of a Smith chart ($\Gamma_S = 0$). Further, the output reflection coefficient Γ_{out} is equal to S_{22} for $\Gamma_S = 0$. Hence, $|\Gamma_{\text{out}}|$ is 0.5 at this point and provides a stable circuit with this transistor. Therefore, the stability circle encloses the unstable region (i.e., the inside is unstable).

Similarly, the output stability circle is found from (11.1.20) and (11.1.21) as follows: $C_L = 2.1608\,\angle 50.8^\circ$ and $r_L = 1.2965$. Again, the center of a Smith chart ($\Gamma_L = 0$) is outside this circle because $|C_L|$ is larger than r_L. Since the input reflection coefficient Γ_{in} is equal to S_{11} at this point, $|\Gamma_{\text{in}}|$ is 0.6. Hence, it represents the load impedance that will provide a stable operation for this transistor. Therefore, all impedances on a Smith chart that are outside the stability circle represent the stable region, and those enclosed by the stability circle are unstable.

Since the transistor is bilateral and we want a noise figure below 2.5 dB, we use the available power gain approach. The 16-dB gain circle is found from (11.3.14) and (11.3.15) as follows: $c_a = 0.3542\ \angle 171.69^\circ$ and $R_a = 0.6731$. The 2.5-dB noise figure circle is found from (11.4.6) and (11.4.7) as follows: $C_{\text{NF}} = 0.0346\ \angle 40^\circ$ and $R_{\text{NF}} = 0.6502$. These circles are drawn on a Smith chart illustrated in Figure 11.30. If we select Γ_S at point B ($\Gamma_S = 0.13\ \angle 0^\circ$), Γ_{out} is found to be $0.469\ \angle{-36.45^\circ}$. Hence, the load reflection coefficient Γ_L must be equal to $0.469\ \angle 36.45^\circ$ (point D).

For an input-side network, we can move from point E to A on the unity-resistance circle by adding a normalized reactance of about $-j0.55$ (i.e., $-j27.5\,\Omega$) in series with $50\,\Omega$. It requires a capacitor of 5.7874 pF at 1 GHz. We then move on the conductance circle of 0.77 to reach point B from A. It means that we need a normalized susceptance of $-j0.43$ (i.e., $-j0.0086$ S). Hence, an inductor of 18.5064 nH connected in shunt will suffice.

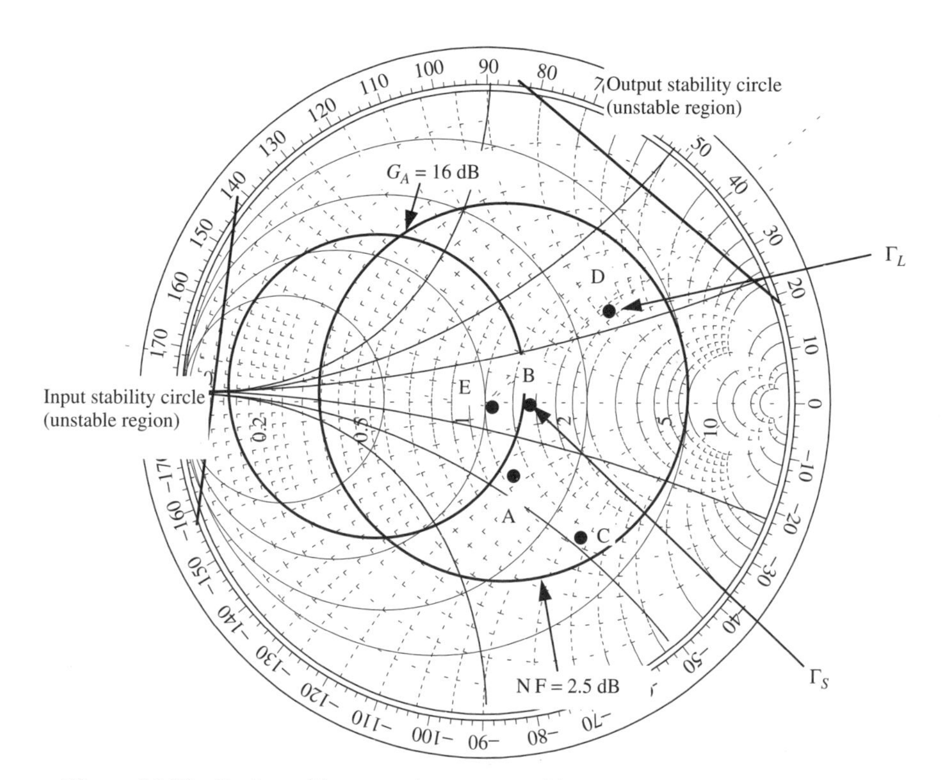

Figure 11.30 Design of input- and output-matching networks for Example 11.14.

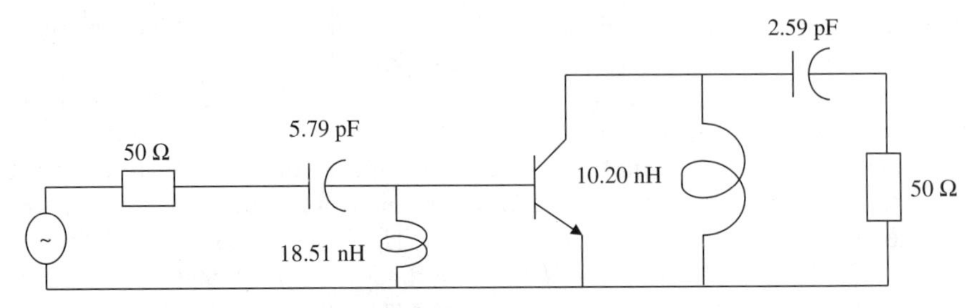

Figure 11.31 RF circuit for Example 11.14.

Similarly, an output-matching network can be designed following the path from E to C on the unity-resistance circle and then from C to D along the conductance circle of 0.4. It requires a series reactance of about $-j1.23$ (i.e., $-j61.5\,\Omega$) and then a shunt susceptance of approximately $-j0.78$ (i.e., $-j0.0156\,\text{S}$). Hence, a capacitor of 2.59 pF in series with the 50-Ω load and then a shunt inductor of 10.20 nH are needed, as illustrated in Figure 11.31.

11.5 BROADBAND AMPLIFIERS

Signals carrying information generally possess a finite bandwidth, and therefore the electronics employed to process such signals need to have characteristics constant over that bandwidth. The amplifier designs presented in preceding sections are valid at only a single frequency. This is primarily because the design of matching networks is frequency sensitive and governed by the Bode–Fano constraints discussed in Chapter 7. Some of the techniques used to broaden the bandwidth of amplifiers are summarized below.

Resistive Matching

As discussed in Section 6.3, the resistive-matching networks are independent of frequency and hence can be used to design broadband amplifiers. The upper limit will be determined from the frequencies when the resistances cease to work due to associated parasitic elements. Further, the noise figure of such amplifiers may be unacceptable.

Compensating Networks

Since the reactance of a capacitor is inversely related with frequency, it will have a higher value at lower frequencies than will the higher end of the band. On the other hand, the inductive reactance exhibits a direct relation with frequency, providing low reactance at the lower end of the band. Hence, these components can be used to devise compensating networks that are effective at the two edges of the frequency band. However, the design may be too complex and can adversely affect matching at the input and the output.

Negative Feedback

Negative feedback can widen the bandwidth of amplifier and improve the matching at its input and output. However, the gain of the amplifier is reduced and it may adversely affect the noise figure unless another relatively low-noise amplifier is added in the forward path before the original amplifier. Figure 11.32 illustrates a simplified feedback arrangement of the amplifier. If V_i and V_o are voltages at its input and output, respectively, A_v is the voltage gain of the amplifier, and β is the transfer function of the feedback network, we can write

$$V_o = (V_i - \beta V_o)A_V \tag{11.5.1}$$

Rearranging this equation, we find that

$$\frac{V_o}{V_i} = G_V = \frac{A_V}{1 + A_V\beta} \tag{11.5.2}$$

Since the voltage gain characteristic of the amplifier at higher frequencies can be expressed as

$$A_V = \frac{A_0}{1 + j(f/f_c)} \tag{11.5.3}$$

G_V is found to be

$$G_V = \frac{A_0/(1 + A_0\beta)}{1 + j[f/f_c(1 + A_0\beta)]} \tag{11.5.4}$$

where A_0 is the midband voltage gain, f_c the corner frequency (3-dB frequency), and f the signal frequency in hertz. From (11.5.4), the new corner frequency, f_c', is found as

$$f_c' = f_c(1 + A_0\beta) \tag{11.5.5}$$

Hence, the cutoff frequency of the overall gain is increased by a factor of $1 + A_0\beta$. However, the new midband voltage gain, given below, is reduced by the same factor. Thus, the gain–bandwidth product of the circuit remains the same:

$$A_0' = \frac{A_0}{1 + A_0\beta} \tag{11.5.6}$$

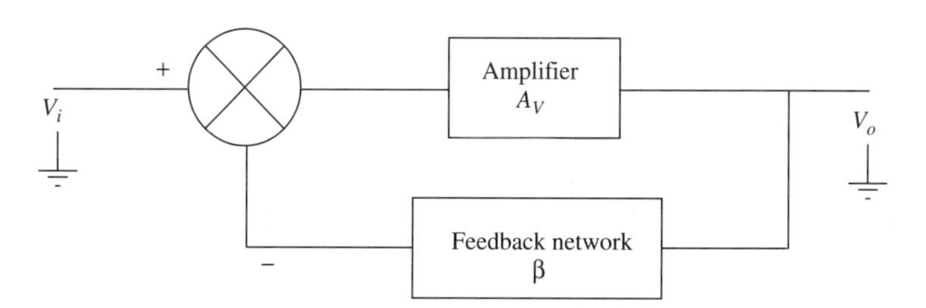

Figure 11.32 Simplified feedback system for an amplifier.

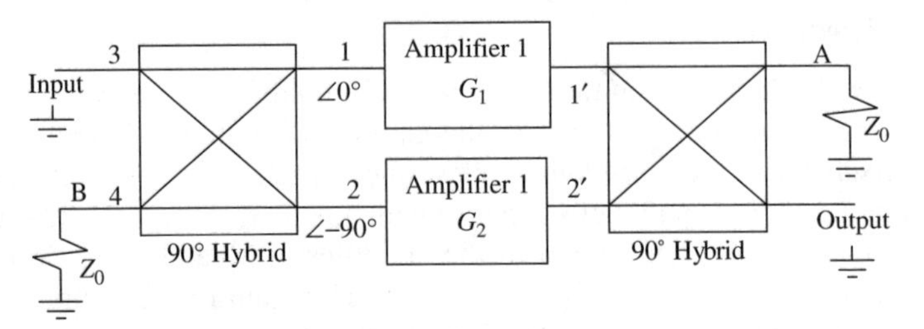

Figure 11.33 Block diagram of a balanced amplifier.

Balanced Circuits

Another way to improve the matching over wider bandwidth is illustrated in Figure 11.33. It employs two amplifiers in conjunction with two 90°-hybrid junctions. An ideal 90°-hybrid splits its input power equally in the forward direction while there is no power coupling to its fourth port. Further, the branched-out signal lags behind the direct path by 90°. Hence, the power input at port 3 of the first hybrid junction appears at port 1 and port 2 but is not coupled at all to port 4. Similarly, the signal entering port 1 divides equally between port 4 ($\angle - 90°$) and port 3 ($\angle 0°$).

Now assume that the two amplifiers in Figure 11.33 have identical input impedance. Signals fed to these amplifiers have the same magnitudes but a phase difference of 90°. Therefore, the signals reflected back would have identical magnitudes but maintain the phase difference. The reflected signals enter ports 1 and 2 and split with equal powers at ports 3 and 4. Thus, two signals appearing back at port 3 are equal in magnitude but 180° out of phase with respect to each other and therefore cancel out. On the other hand, signals appearing at port 4 are in phase (since both went through the 90° delay). However, port 4 is matched terminated and this power is dissipated. Signals amplified by the two amplifiers are fed to a second hybrid that is matched terminated at port A. Signals entering at its ports 1′ and 2′ appear at the output (only 50% power of each channel) while canceling out at port A. Thus, the overall gain of an ideal balanced amplifier is equal to that of an amplifier connected in one of its channels.

Traveling Wave Amplifiers

Traveling wave amplifiers (also known as *distributed amplifiers*) use discrete transistors in a distributed manner, as illustrated in Figure 11.34. In this technique, lumped inductors form artificial transmission lines in conjunction with the input and output capacitance of the transistors. The input signal travels through the gate line that is terminated by impedance Z_G at the end. An amplified signal is available via the drain line that is terminated by impedance Z_D at its other end. This arrangement provides the possibility of increasing the gain–bandwidth product of the amplifier.

The gate of each FET taps off the input signal traveling along the gate line and transfers it to the drain line through its transconductance. The remaining input

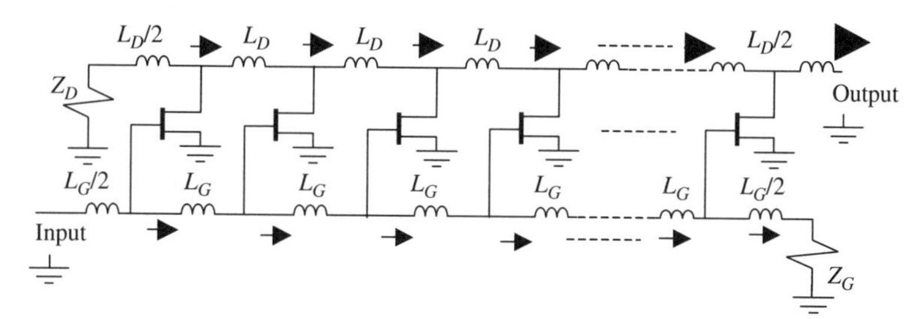

Figure 11.34 Schematic circuit of a traveling wave amplifier.

signal is dissipated in terminating impedance Z_G. The drain line parameters are selected such that the amplified signals available from each transistor are added in a forward traveling wave and an amplified signal is available at the output. This happens when the phase velocities on the drain and gate lines are same. Any signal propagating on the drain line in the opposite direction is dissipated in Z_D.

11.6 SMALL-SIGNAL EQUIVALENT-CIRCUIT MODELS OF TRANSISTORS

Bipolar Junction Transistor

Figure 11.35 illustrates a small-signal equivalent model of the BJT. In this circuit, r_b is the resistance between its base terminal and the apparent base. It is created because of low doping of the base. $C_{b'e'}$ represents diffusion capacitance C_D and junction capacitance C_{je} connected in parallel; $r_{b'e'}$ is the resistance at the emitter–base junction and $r_{b'c'}$ is the resistance at the base–collector junction. $C_{b'c'}$ is capacitance between base and collector; r_c represents the collector resistance because the collector is lightly doped; r_o is output resistance representing the finite slope of its $I_c - V_{CE}$ characteristics. Ideally, r_o will be infinite.

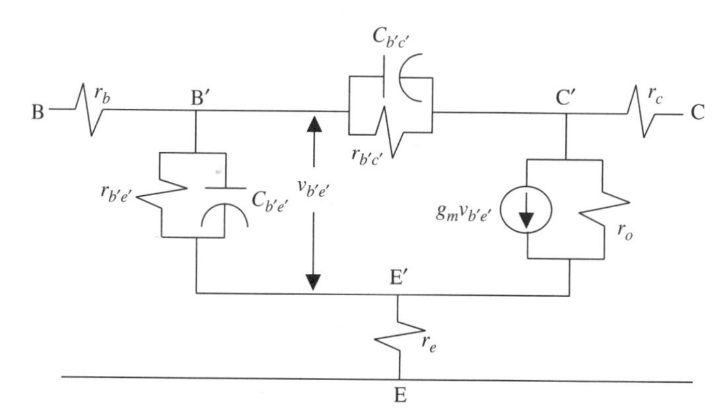

Figure 11.35 Small-signal equivalent circuit of a BJT.

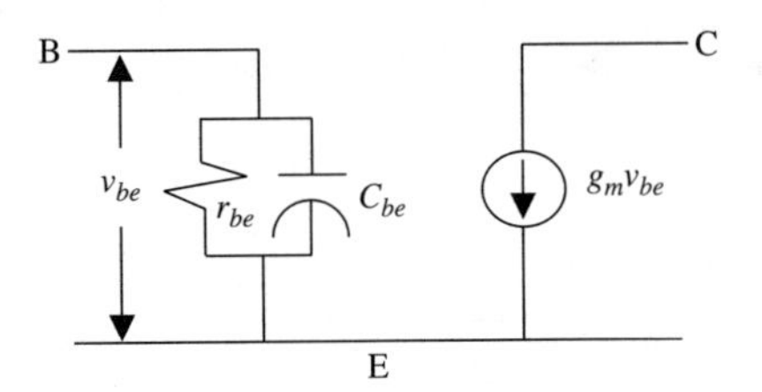

Figure 11.36 Simplified small-signal equivalent circuit of a BJT.

The resistance r_e represents emitter resistance, and g_m is the transconductance. Various parameters are related as follows:

$$\beta = h_{FE} = \frac{I_C}{I_B} \tag{11.6.1}$$

$$\alpha = \frac{I_C}{I_E} = \frac{\beta}{1+\beta} \tag{11.6.2}$$

$$g_m = \left.\frac{\delta i_C}{\delta v_{b'e'}}\right|_{V_{CE}} = \frac{q_e I_C}{kT} \tag{11.6.3}$$

$$r_{b'e'} = \left[\left.\frac{\delta i_b}{\delta v_{EB}}\right|_{V_{BC}}\right]^{-1} = \frac{\beta}{g_m} \tag{11.6.4}$$

$$r_{b'c'} = \left[\left.\frac{\delta i_b}{\delta v_{EC}}\right|_{V_{EB}}\right]^{-1} \approx \infty \tag{11.6.5}$$

$$C_{b'c'} = \frac{C_{b'c'o}}{(1 - v_{CB}/V_{\text{BIP}})^m} \tag{11.6.6}$$

$$C_{je} = \frac{C_{je0}}{[1 - (v_{EB}/V_{\text{BIP}})]^m} \tag{11.6.7}$$

and

$$C_D = \frac{q_e I_E}{kT}\tau_e \tag{11.6.8}$$

I_C, I_B, and I_E are the collector, base, and emitter currents, respectively, T is temperature in kelvin, q_e is electronic charge, k is the Boltzmann constant, $C_{b'c'o}$ is zero-biased capacitance, V_{BIP} is the built-in potential of the base–collector junction, m is equal to $\frac{1}{2}$ for uniform doping and $\frac{1}{3}$ for graded doping, and τ_e is the base transit time. A simplified small-signal equivalent of the BJT is shown in Figure 11.36.

Metal-Oxide-Semiconductor Field-Effect Transistor

The small-signal equivalent circuit of a MOSFET is illustrated in Figure 11.37. It is assumed that both the source and the substrate are grounded. R_S and R_D are

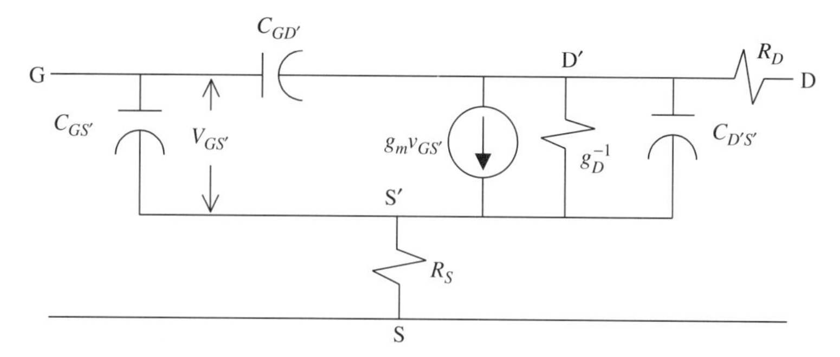

Figure 11.37 Small-signal equivalent circuit of a MOSFET.

the series resistance at the source and drain, respectively. The oxide overlay gives rise to capacitance that degrades the high-frequency performance. It includes the source–gate and drain–gate capacitances. $C_{GS'}$ and $C_{GD'}$ represent these as well as respective parasitic capacitances. $C_{D'S'}$ includes the capacitance between drain and substrate. The drain conductance g_D and transconductance g_m are defined as follows:

$$g_D = \left. \frac{\partial I_D}{\partial V_D} \right|_{V_G = \text{constant}} \tag{11.6.9}$$

and

$$g_m = \left. \frac{\partial I_D}{\partial V_G} \right|_{V_D = \text{constant}} \tag{11.6.10}$$

Metal-Semiconductor Field-Effect Transistors

The small-signal equivalent circuit of a MESFET is illustrated in Figure 11.38. R_G, R_S, and R_D represent the extrinsic parasitic resistances that are in series with the gate, the source, and the drain, respectively. $C_{D'S'}$ is the drain–source

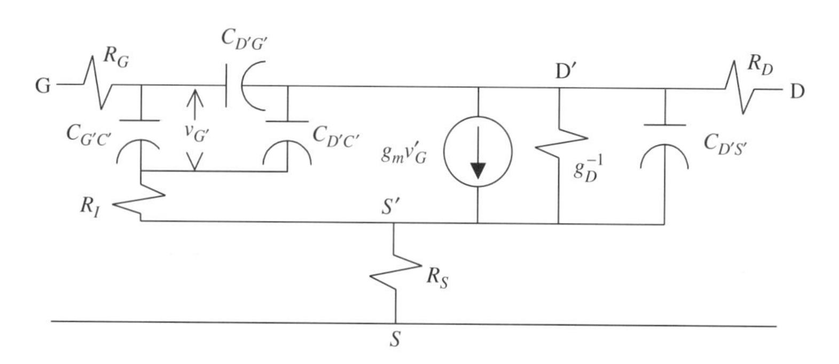

Figure 11.38 Small-signal equivalent circuit of a MESFET.

capacitance. Transconductance g_m and drain conductance g_D are as defined earlier. R_I is the channel resistance that is responsible for a finite charging time to capacitor $C_{G'C'}$ between the gate and the channel. $C_{D'C'}$ represents the drain-to-channel capacitance.

11.7 DC BIAS CIRCUITS FOR TRANSISTORS

Transistor circuits require dc bias that provides the desired quiescent point. Further, it should hold the operation stable over a range of temperatures. Resistive circuits used at lower frequencies can be employed in the RF range as well. However, sometimes these circuits may not work satisfactorily at higher frequencies. For example, a resistance in parallel with a bypass capacitor is frequently used at the emitter to provide stable operation at lower frequencies. This circuit may not work at microwave frequencies because it can produce oscillation. Further, the resistance in an amplifier circuit can degrade the noise figure. Active bias networks provide certain advantages over the resistive circuits.

Figure 11.39 illustrates a resistive bias network with voltage feedback. Inductors L_C and L_B are used to block RF from going toward R_C and R_B. At the same time, dc bias passes through them to respective terminals of BJT without loss (assuming that the inductors have zero resistance). If V_{BE} represents base–emitter voltage of the transistor,

$$V_{BE} + R_B I_B + R_C I_E = V_{CC} \tag{11.7.1}$$

Since

$$I_B = I_E - I_C = (1 - \alpha) I_E = \frac{I_E}{1 + \beta} \tag{11.7.2}$$

(11.7.1) can be written as

$$I_E = \frac{V_{CC} - V_{BE}}{R_C + R_B/(1 + \beta)} \tag{11.7.3}$$

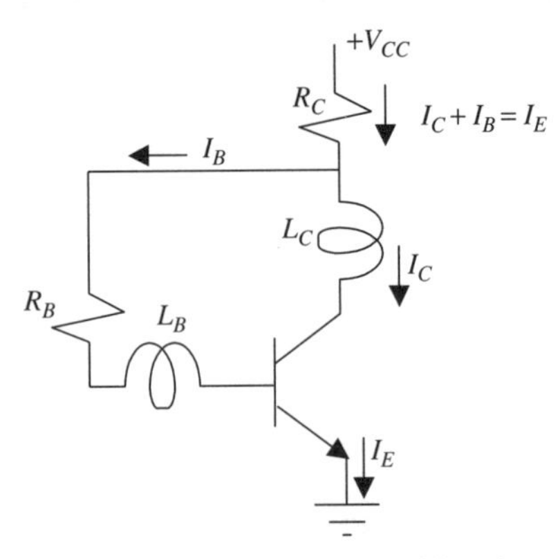

Figure 11.39 Transistor bias circuit with voltage feedback.

For I_E to be less sensitive to changes in V_{BE} and β, the following conditions must be satisfied:

$$V_{CC} \gg V_{BE} \tag{11.7.4}$$

and

$$R_C \gg \frac{R_B}{1+\beta} \tag{11.7.5}$$

Hence, it is advisable to use high V_{CC} and R_C. However, there are practical limitations. Another option is to use a small R_B, but it will limit V_{CE} and therefore the swing in output. It is up to the circuit designer to weigh these options of the trade-off and to pick the components accordingly.

An alternative biasing network that uses a bypassed resistor at the emitter is illustrated in Figure 11.40. As before, inductors L_C, L_{B1}, and L_{B2} are used to block the RF. On the other hand, capacitor C_E is used to bypass RF. Figure 11.41 shows the Thévenin equivalent of this circuit for the dc condition where

$$V_{\text{Th}} = \frac{R_2}{R_1 + R_2} V_{CC} \tag{11.7.6}$$

and

$$R_{\text{Th}} = \frac{R_1 R_2}{R_1 + R_2} \tag{11.7.7}$$

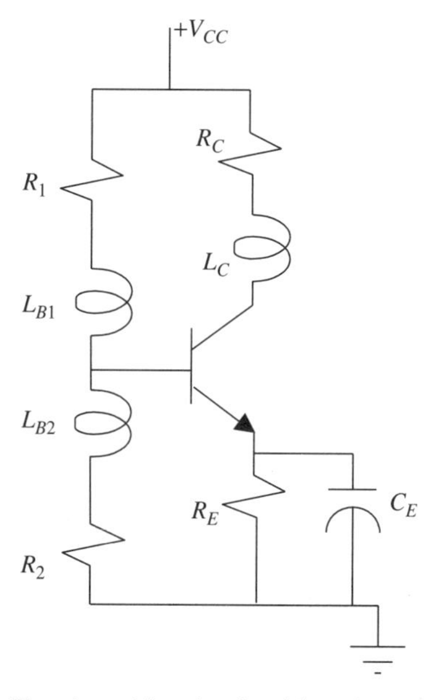

Figure 11.40 Transistor bias circuit with emitter bypassed resistor.

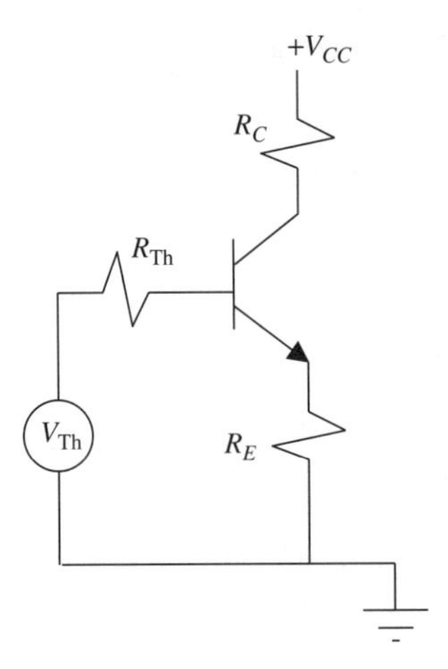

Figure 11.41 Equivalent of the circuit shown in Figure 11.40.

From the equivalent circuit shown in Figure 11.41,

$$V_{Th} = R_{Th} I_B + V_{BE} + R_E I_E = \left(\frac{R_{Th}}{1+\beta} + R_E \right) I_E + V_{BE} \tag{11.7.8}$$

Hence,

$$I_E = \frac{V_{Th} - V_{BE}}{R_E + R_{Th}/(1+\beta)} \tag{11.7.9}$$

Therefore, the following conditions must be met for stable I_E:

$$V_{Th} \gg V_{BE} \tag{11.7.10}$$

and

$$R_E \gg \frac{R_{Th}}{1+\beta} \tag{11.7.11}$$

Note that for a fixed V_{CC}, V_{Th} cannot be increased arbitrarily because it reduces V_{CB} and hence V_{CE}. Smaller V_{CE} means limited output swing. Further, large R_E will reduce V_{CE} as well. On the other hand, smaller R_{Th} will mean low input impedance of the circuit. It is up to the circuit designer to work within these contradictory requirements. As a rule of thumb, V_{Th} is selected no more than 15 to 20% of V_{CC}.

For a low-noise and low-power design, the drain current of the MESFET is selected around $0.15 I_{DSS}$ (I_{DSS} is the drain current in the saturation region with $V_{GS} = 0$), with V_{DS} just enough to keep it in saturation. A higher drain current point (generally around $0.95 I_{DSS}$) is selected for low-noise, high-power design. Higher V_{DS} is used for high-efficiency and high-power applications. Figure 11.42 shows a bias circuit that requires a dual positive power supply. V_S is applied to

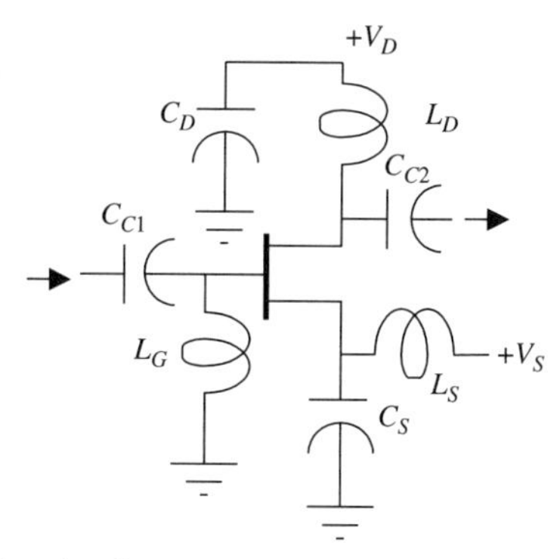

Figure 11.42 Bias circuit employing two unipolar (positive) dc sources.

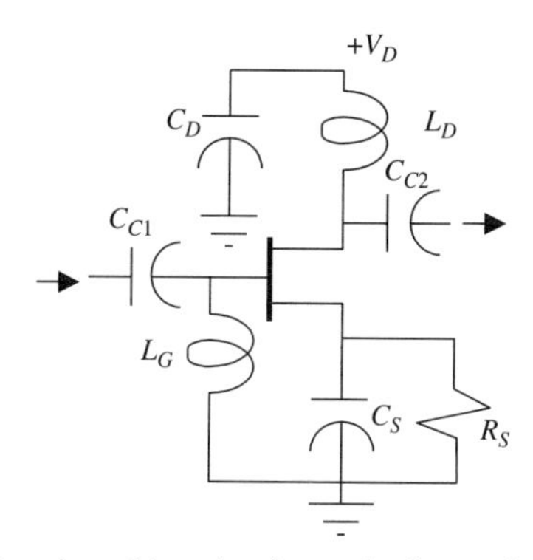

Figure 11.43 Transistor bias circuit employing a single power supply.

the source terminal of the MESFET via inductor L_S. Capacitor C_S grounds the source at RF while L_S blocks it from the dc supply. The inductor L_G grounds the gate for dc while stopping the RF. Similarly, bias voltage V_D is applied to the drain terminal via inductor L_D. However, it stops the RF from going toward the bias supply that is further ensured by capacitor C_D. This circuit is recommended for low-noise, high-gain, high-power, and high-efficiency applications.

Another bias circuit that requires only a single power supply is illustrated in Figure 11.43. The voltage drop across R_S is applied to the gate via inductor L_G. This resistance can also be used to adjust the gain of this circuit. As mentioned earlier, it will be generating thermal noise as well. Note that a quarter-wavelength-long line with a short circuit at one end can replace the RF-blocking inductors. The short circuit is generally achieved via a capacitor connected across the dc power supply.

SUGGESTED READING

Bahl, I. J., and P. Bhartia, *Microwave Solid State Circuit Design*. New York: Wiley, 1988.
Collin, R. E., *Foundations for Microwave Engineering*. New York: McGraw-Hill, 1992.

Davis, W. A., *Microwave Semiconductor Circuit Design*. New York: Van Nostrand Reinhold, 1984.

Gonzalez, G., *Microwave Transistor Amplifiers*. Upper Saddle River, NJ: Prentice Hall, 1997.

Pozar, D. M., *Microwave Engineering*. New York: Wiley, 1998.

Vendelin, G. D., A. Pavio, and U. L. Rhode, *Microwave Circuit Design Using Linear and Non-linear Techniques*. New York: Wiley, 1990.

Wolff, E. A., and R. Kaul, *Microwave Engineering and Systems Applications*. New York: Wiley, 1988.

PROBLEMS

11.1. The S-parameters of a certain microwave transistor are measured at 3 GHz with a 50-Ω reference resistance and found as $S_{11} = 0.505\ \angle{-150^\circ}$, $S_{12} = 0$, $S_{21} = 7\ \angle 180^\circ$, and $S_{22} = 0.45\ \angle - 20^\circ$. Calculate the maximum unilateral power gain and design an amplifier using this transistor for maximum power gain.

11.2. An NE868495-4 GaAs power FET has the following S-parameters measured at $V_{ds} = 9\,\text{V}$ and $I_{ds} = 600\,\text{mA}$ at 5 GHz with a 50-Ω line: $S_{11} = 0.45\ \angle 163^\circ$, $S_{12} = 0.04\ \angle 40^\circ$, $S_{21} = 2.55\ \angle{-106^\circ}$, and $S_{22} = 0.46\ \angle{-65^\circ}$. Use this transistor to design an amplifier for the maximum power gain at 5 GHz.

11.3. Design an amplifier for maximum possible gain at 600 MHz using a M/ACOM MA4T64435 bipolar junction transistor in the common-emitter configuration. Its S-parameters, measured at $V_{CE} = 8\,\text{V}$ and $I_C = 10\,\text{mA}$ ($Z_0 = 50\,\Omega$), are as follows: $S_{11} = 0.514\ \angle{-101^\circ}$, $S_{21} = 9.562\ \angle 112.2^\circ$, $S_{12} = 0.062\ \angle 49^\circ$, and $S_{22} = 0.544\ \angle{-53^\circ}$. Calculate the gain in decibels for your circuit.

11.4. Design an amplifier for a maximum possible transducer power gain at 2 GHz, using an Avantek ATF-45171 GaAs FET with following common-source S-parameters measured at $V_{DS} = 9\,\text{V}$ and $I_{DS} = 250\,\text{mA}$ ($Z_0 = 50\,\Omega$): $S_{11} = 0.83\ \angle{-137^\circ}$, $S_{21} = 3.45\ \angle 83^\circ$, $S_{12} = 0.048\ \angle 19^\circ$, and $S_{22} = 0.26\ \angle{-91^\circ}$. Calculate the gain in decibels for your circuit.

11.5. An NE720 GaAs MESFET biased at $V_{ds} = 4\,\text{V}$ and $I_{ds} = 30\,\text{mA}$ has the following S-parameters at 4 GHz: $S_{11} = 0.70\ \angle{-127^\circ}$, $S_{12} = 0$, $S_{21} = 3.13\ \angle 76^\circ$, and $S_{22} = 0.47\ \angle{-30^\circ}$. Plot (**a**) the input constant-gain circles for 3, 2, 1, 0, −1, and −2 dB and (**b**) the output constant-gain circles for the same decibel levels.

11.6. An NE41137 GaAs FET has the following S-parameters measured at $V_{ds} = 5\,\text{V}$ and $I_{ds} = 10\,\text{mA}$ at 3 GHz with a 50-Ω resistance: $S_{11} = 0.38\ \angle{-169^\circ}$, $S_{12} = 0$, $S_{21} = 1.33\ \angle{-39^\circ}$, and $S_{22} = 0.95\ \angle{-66^\circ}$. Design an amplifier using this transistor for a power gain of 6 dB.

11.7. An Avantek ATF-13036 GaAs FET in common-source configuration has the following S-parameters measured at $V_{DS} = 2.5\,\text{V}$ and $I_{DS} = 20\,\text{mA}$ at 6 GHz with a 50-Ω line: $S_{11} = 0.55\ \angle{-137^\circ}$, $S_{21} = 3.75\ \angle 52^\circ$, $S_{12} = 0.112\ \angle 14^\circ$, and $S_{22} = 0.3\ \angle{-80^\circ}$. Design an amplifier using this transistor for an available power gain of 10 dB.

11.8. An NE868898-7 GaAs FET has the following S-parameters measured at $V_{ds} = 9\,\text{V}$ and $I_{ds} = 1.2\,\text{A}$ at 7 GHz: $S_{11} = 0.42\ \angle 155^\circ$, $S_{12} = 0.13\ \angle{-10^\circ}$, $S_{21} = 2.16\ \angle 25^\circ$, and $S_{22} = 0.51\ \angle{-70^\circ}$.

(a) Draw the input and output stability circles if the transistor is potentially unstable.

(b) Find the maximum operating power gain.

(c) Plot the power gain circles for 8 and 6 dB.

11.9. The DXL 3503A (chip) Ku-band medium-power GaAs FET has the following S-parameters measured at $V_{ds} = 6\,\text{V}$ and $I_{ds} = 0.5I_{dss}$ at 18 GHz with a 50-Ω line: $S_{11} = 0.64\ \angle{-160^\circ}$, $S_{12} = 0.08\ \angle 127^\circ$, $S_{21} = 0.81\ \angle 23^\circ$, and $S_{22} = 0.77\ \angle{-78^\circ}$. This transistor is to be used for a high-gain amplifier. Design the input- and output-matching networks using balanced stubs for a maximum possible power gain at 18 GHz.

11.10. A M/ACOM BJT MA4T64433 in common-emitter configuration has the following S-parameters measured at $V_{CE} = 8\,\text{V}$ and $I_C = 25\,\text{mA}$ at 1 GHz with a 50-Ω line: $S_{11} = 0.406\ \angle{-155^\circ}$, $S_{21} = 6.432\ \angle 89.6^\circ$, $S_{12} = 0.064\ \angle 53.1^\circ$, and $S_{22} = 0.336\ \angle{-57.1^\circ}$. Design an amplifier using this transistor for an available power gain of 10 dB.

11.11. The S-parameters of a BJT at 2 GHz in a 50-Ω system are $S_{11} = 1.5\ \angle{-100^\circ}$, $S_{21} = 5.0\ \angle 50^\circ$, $S_{12} = 0.0\ \angle 0^\circ$, and $S_{22} = 0.9\ \angle{-60^\circ}$.

(a) Calculate the input impedance and the optimum output termination.

(b) Determine the unstable region on the Smith chart and construct constant-gain circles for $G_s = 3\,\text{dB}$ and $G_s = 4\,\text{dB}$.

(c) Design the input-matching network for $G_s = 4\,\text{dB}$ with the greatest degree of stability.

(d) Determine G_{TU} in decibels for your design.

11.12. An Avantek low-noise silicon bipolar transistor, AT-41470, has the following S-parameters measured at $V_{CE} = 8\,\text{V}$ and $I_C = 25\,\text{mA}$ at 100 MHz with a 50-Ω line: $S_{11} = 0.64\ \angle{-62^\circ}$, $S_{21} = 42.11\ \angle 147^\circ$, $S_{12} = 0.009\ \angle 75^\circ$, $S_{22} = 0.85\ \angle{-19^\circ}$, $F_{\text{min}} = 1.2\,\text{dB}$, $\Gamma_{\text{opt}} = 0.12\ \angle 5^\circ$, and $R_n = 8.5\,\Omega$. Design input and output matching networks of the amplifier for $G_T = 35\,\text{dB}$ and a low-noise figure of 2 dB. For design purposes, assume that the device is unilateral, and calculate the maximum error in G_T resulting from this assumption.

11.13. An Avantek low-noise GaAs FET, ATF-13100, has the following S-parameters measured at $V_{DS} = 2.5\,\text{V}$ and $I_{DS} = 20\,\text{mA}$ at 4 GHz with a

50-Ω line: $S_{11} = 0.8\ \angle{-58^\circ}$, $S_{21} = 4.54\ \angle 126^\circ$, $S_{12} = 0.085\ \angle 59^\circ$, $S_{22} = 0.49\ \angle{-33^\circ}$, $F_{\min} = 0.5\,\text{dB}$, $\Gamma_{\text{opt}} = 0.60\ \angle 30^\circ$, and $R_n = 16\,\Omega$. Design input- and output-matching networks of the amplifier for $G_T = 12\,\text{dB}$ and a low-noise figure of 2 dB. For design purposes, assume that the device is unilateral, and calculate the maximum error in G_T resulting from this assumption.

11.14. An HP HXTR-6101 microwave transistor has the following parameters measured at 4 GHz with a reference resistance of 50 Ω: minimum noise figure = 2.5 dB, $\Gamma_{\text{opt}} = 0.475\ \angle 155^\circ$, and $R_n = 3.5\,\Omega$. Plot the noise figure circles for given values of F at 2.5, 3.0, 3.5, 4.0, and 5.0 dB.

11.15. A transistor has the following parameters: $S_{11} = 0.5\ \angle 160^\circ$, $S_{12} = 0.06\ \angle 50^\circ$, $S_{21} = 3.6\ \angle 60^\circ$, and $S_{22} = 0.5\ \angle{-45^\circ}$, $\Gamma_{\text{opt}} = 0.4\ \angle 145^\circ$, $R_n = 0.4\,\Omega$, and $F_{\min} = 1.6\,\text{dB}$. Design an amplifier with the best possible noise figure for a power gain of 10 dB. Also, the VSWR on either side should be less than 2.

11.16. A DXL 1503A-P70 X-band low-noise GaAs FET is biased at $V_{ds} = 3.5\,\text{V}$ and $I_{ds} = 12\,\text{mA}$. Its S-parameters are measured at 12 GHz with a 50 -Ω line as follows: $S_{11} = 0.48\ \angle{-130^\circ}$, $S_{12} = 0.01\ \angle 92^\circ$, $S_{21} = 2.22\ \angle 75^\circ$, and $S_{22} = 0.52\ \angle{-65^\circ}$, $\Gamma_{\text{opt}} = 0.45\ \angle 175^\circ$, $R_n = 3.5\Omega$, and $F_{\min} = 2.5\,\text{dB}$. Design input- and output-matching networks using discrete components for a maximum amplifier gain of 9 dB and a low-noise figure of 3 dB.

12

OSCILLATOR DESIGN

Oscillator circuits are used for generating the periodic signals that are needed in various applications. These circuits convert a part of dc power into periodic output and do not require a periodic signal as input. The chapter begins with the basic principle of sinusoidal oscillator circuits. Subsequently, several transistor circuits are analyzed to establish their design procedures. Ceramic resonant circuits are frequently used to generate reference signals, and voltage-controlled oscillators are important in modern frequency synthesizer design using the phase-locked loop. Fundamentals of these circuits are discussed in this chapter. Diode oscillators used at microwave frequencies are also summarized. The chapter ends with a description of microwave transistor circuits using S-parameters.

12.1 FEEDBACK AND BASIC CONCEPTS

Solid-state oscillators use a diode or a transistor in conjunction with the passive circuit to produce sinusoidal steady-state signals. Transients or electrical noise triggers oscillations initially. A properly designed circuit sustains these oscillations subsequently. This process requires a nonlinear active device. In addition, since the device is producing RF power, it must have a negative resistance.

The basic principle of an oscillator circuit can be explained via a linear feedback system as illustrated in Figure 12.1. Assume that a part of output Y is fed back to the system along with an input signal X. As indicated, the transfer

Radio-Frequency and Microwave Communication Circuits: Analysis and Design, Second Edition,
By Devendra K. Misra
ISBN 0-471-47873-3

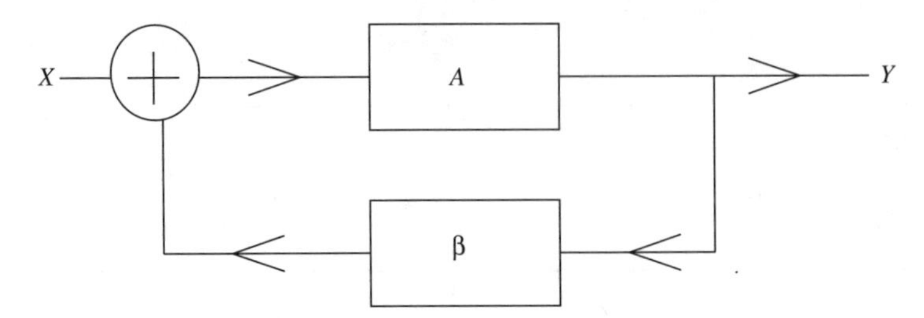

Figure 12.1 Simple feedback system.

function of the forward-connected subsystem is A, while the feedback path has a subsystem with its transfer function as β. Therefore,

$$Y = A(X + \beta Y)$$

The closed-loop gain T (generally called the *transfer function*) of this system is found from this equation as

$$T = \frac{Y}{X} = \frac{A}{1 - A\beta} \tag{12.1.1}$$

Product $A\beta$ is known as the *loop gain*. It is a product of the transfer functions of individual units in the loop. Numerator A is called the *forward path gain* because it represents the gain of a signal traveling from input to output.

For a loop gain of unity, T becomes infinite. Hence, the circuit has an output signal Y without an input signal X and the system oscillates. The condition $A\beta = 1$ is known as the *Barkhausen criterion*. Note that if the signal $A\beta$ is subtracted from X before it is fed to A, the denominator of (12.1.1) changes to $1 + A\beta$. In this case, the system oscillates for $A\beta = -1$. This is known as the *Nyquist criterion*. Since the output of an amplifier is generally 180° out of phase with its input, it may be a more appropriate description for that case.

Example 12.1 In the circuit shown in Figure 12.2, $R_1 = 10\,\text{k}\Omega$, $R_2 = 1\,\Omega$, $C = 0.002\,\mu\text{F}$, and $L = 1\,\mu\text{H}$. If the circuit is oscillating, find the voltage gain A of the forward amplifier and the frequency of oscillation.

SOLUTION For this circuit,

$$\beta = \frac{j\omega L}{R_1 - \omega^2 R_2 CL + j\omega(L + R_1 R_2 C)}$$

Therefore, $A\beta = 1$ only if

$$\omega = \sqrt{\frac{R_1}{R_2 LC}} = 2236.07 \times 10^6\ \text{rad/s} \rightarrow f = 355.88\ \text{MHz}$$

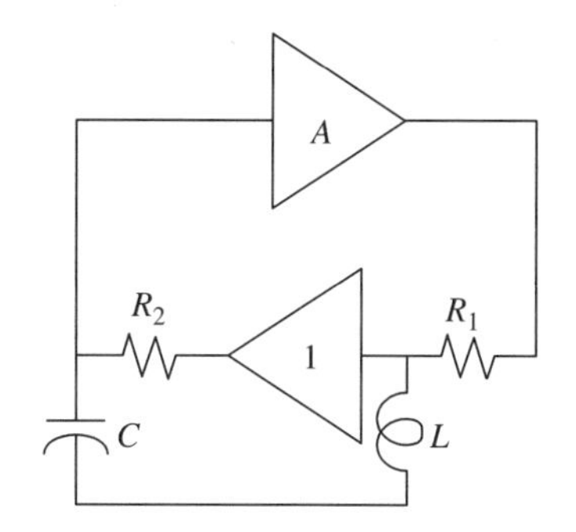

Figure 12.2 Circuit for Example 12.1.

and

$$A = 1 + R_1 R_2 \frac{C}{L} = 21$$

Generalized Oscillator Circuit

Consider the transistor circuit illustrated in Figure 12.3. Device T in this circuit may be a bipolar transistor or a FET. If it is a BJT, terminals 1, 2, and 4 represent the base, emitter, and collector, respectively. On the other hand, these may be the gate, source, and drain terminals if it is a FET. Its small-signal equivalent circuit is shown in Figure 12.4. The boxed part of this figure represents the transistor's equivalent, with g_m being its transconductance, and Y_i and Y_o its input and output admittances, respectively. Application of Kirchhoff's current law at nodes 1, 2,

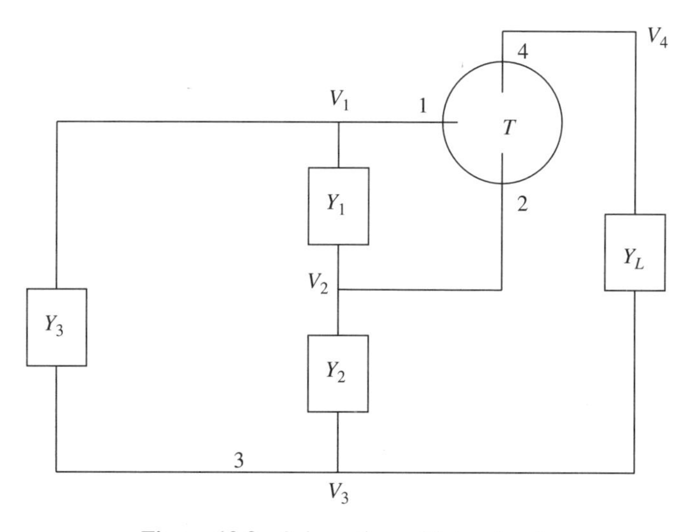

Figure 12.3 Schematic oscillator circuit.

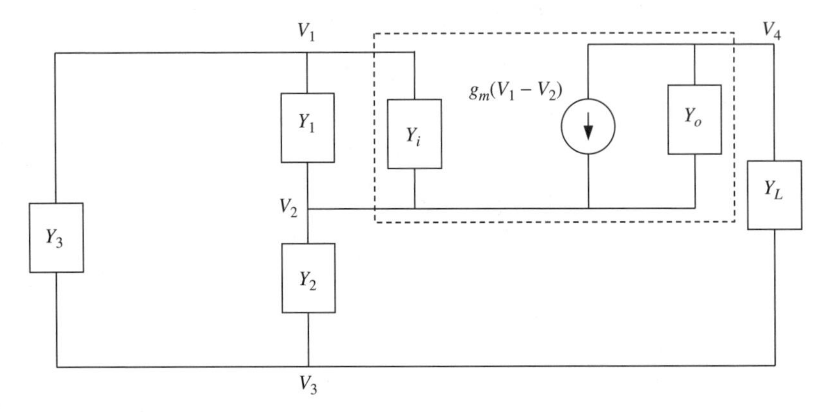

Figure 12.4 Electrical equivalent of the schematic oscillator circuit.

3, and 4 gives

$$Y_3(V_1 - V_3) + Y_1(V_1 - V_2) + Y_i(V_1 - V_2) = 0 \tag{12.1.2}$$

$$-Y_1(V_1 - V_2) - Y_2(V_3 - V_2) - Y_i(V_1 - V_2) - g_m(V_1 - V_2) - Y_o(V_4 - V_2) = 0 \tag{12.1.3}$$

$$-Y_3(V_1 - V_3) - Y_2(V_2 - V_3) - Y_L(V_4 - V_3) = 0 \tag{12.1.4}$$

and

$$g_m(V_1 - V_2) + Y_o(V_4 - V_2) + Y_L(V_4 - V_3) = 0 \tag{12.1.5}$$

Simplifying (12.1.2) to (12.1.5), we have

$$(Y_1 + Y_3 + Y_i)V_1 - (Y_1 + Y_i)V_2 - Y_3V_3 = 0 \tag{12.1.6}$$

$$-(Y_1 + Y_i + g_m)V_1 + (Y_1 + Y_2 + Y_i + g_m + Y_o)V_2 - Y_2V_3 - Y_oV_4 = 0 \tag{12.1.7}$$

$$-Y_3V_1 - Y_2V_2 + (Y_2 + Y_3 + Y_L)V_3 - Y_LV_4 = 0 \tag{12.1.8}$$

and

$$g_mV_1 - (g_m + Y_o)V_2 - Y_LV_3 + (Y_o + Y_L)V_4 = 0 \tag{12.1.9}$$

These equations can be written in matrix form as

$$\begin{bmatrix} Y_1 + Y_3 + Y_i & -(Y_1 + Y_i) & -Y_3 & 0 \\ -(Y_1 + Y_i + g_m) & Y_1 + Y_2 + Y_i + g_m + Y_o & -Y_2 & -Y_o \\ -Y_3 & -Y_2 & Y_2 + Y_3 + Y_L & -Y_L \\ g_m & -(g_m + Y_o) & -Y_L & Y_o + Y_L \end{bmatrix} \times \begin{bmatrix} V_1 \\ V_2 \\ V_3 \\ V_4 \end{bmatrix} = 0 \tag{12.1.10}$$

For a nontrivial solution to this system of equations, the determinant of the coefficient matrix must be zero. It sets constraints on the nature of circuit components that will be explained later.

Equation (12.1.10) represents the most general formulation. It can be simplified for specific circuits as follows:

1. If a node is connected to the ground, that column and row are removed from (12.1.10). For example, if node 1 is grounded, the first row as well as the first column will be removed from (12.1.10).
2. If two nodes are connected, the corresponding columns and rows of the coefficient matrix are added together. For example, if nodes 3 and 4 are connected, rows 3 and 4 as well as columns 3 and 4 are replaced by their sums as follows:

$$\begin{bmatrix} Y_1 + Y_3 + Y_i & -(Y_1 + Y_i) & -Y_3 \\ -(Y_1 + Y_i + g_m) & Y_1 + Y_2 + Y_i + g_m + Y_o & -(Y_2 + Y_o) \\ -Y_3 + g_m & -(g_m + Y_2 + Y_o) & Y_2 + Y_3 + Y_o \end{bmatrix} \begin{bmatrix} V_1 \\ V_2 \\ V_3 \end{bmatrix} = 0 \tag{12.1.11}$$

For a common-emitter BJT (or a common-source FET) circuit, $V_2 = 0$, and therefore, row 2 and column 2 are removed from (12.1.11). Hence, it simplifies further as follows:

$$\begin{vmatrix} Y_1 + Y_3 + Y_i & -Y_3 \\ -Y_3 + g_m & Y_2 + Y_3 + Y_o \end{vmatrix} = 0 \tag{12.1.12}$$

Therefore,

$$Y_1Y_2 + Y_1Y_3 + Y_1Y_o + Y_2Y_3 + Y_oY_3 + Y_2Y_i + Y_3Y_i + Y_iY_o + g_mY_3 = 0 \tag{12.1.13}$$

If $Y_i = G_i$ (pure real), $Y_o = G_o$ (pure real), and the other three admittances (Y_1, Y_2, and Y_3) are purely susceptive, separating its real and imaginary parts, we get

$$B_1B_2 + B_2B_3 + B_1B_3 = G_iG_o \tag{12.1.14}$$

and

$$B_1G_o + B_2G_i + B_3(g_m + G_i + G_o) = 0 \tag{12.1.15}$$

Note that the junction capacitance at the input and output of the transistor can be included in B_1 and B_2. Similarly, the losses associated with reactive components as well as the load can be included with G_i and G_o. Thus, these equations represent a fairly general situation. In most cases, either the input or output conductance may be assumed to be close to zero. Therefore, equation (12.1.14) is satisfied only when at least one susceptance is different from the other two (i.e., if one is capacitive, the other two must be inductive, or vice versa). In case of a common-emitter BJT circuit, the output impedance is very high and therefore G_o is approximately zero. In this case, equation (12.1.15) reduces to

$$\left(\frac{g_m}{G_i}+1\right)B_3+B_2=0=B_2+(1+\beta)B_3$$

or

$$X_3=-(1+\beta)X_2 \tag{12.1.16}$$

where $\beta = g_m/G_i$ represents the small-signal current gain of a common-emitter circuit. This equation indicates that if X_2 is an inductor, X_3 is a capacitor, or vice versa. Further, equation (12.1.14) may be written as

$$X_1+X_2+X_3=0 \tag{12.1.17}$$

For a common-source FET circuit, G_i is almost zero. Therefore, equation (12.1.15) simplifies to

$$X_1=-\frac{G_o}{G_o+g_m}X_3 \tag{12.1.18}$$

This requires that if X_1 is an inductor, X_3 is a capacitor, or vice versa. Substitution of X_3 from (12.1.18) into (12.1.17) gives

$$X_2=\frac{g_m}{G_o}X_1 \tag{12.1.19}$$

Hence, X_1 and X_2 must be of the same kind (inductive or capacitive), whereas X_3 different from the two. If X_1 and X_2 are inductive, X_3 must be a capacitive reactance. This type of oscillator circuit is called a *Hartley oscillator*. On the other hand, X_3 is an inductor if capacitors are used for X_1 and X_2. This circuit is called a *Colpitts oscillator*. Figure 12.5 illustrates the RF sections of these two circuits (excluding the transistor's biasing network). A BJT Hartley oscillator with its bias arrangement is shown in Figure 12.6.

Resonant frequency of the Hartley oscillator is obtained via (12.1.17) as follows:

$$\omega L_1+\omega L_2-\frac{1}{\omega C_3}=0$$

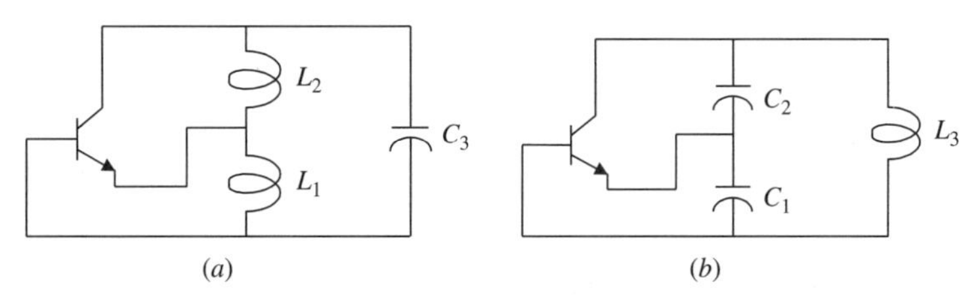

Figure 12.5 Simplified circuits of (*a*) Hartley and (*b*) Colpitts oscillators.

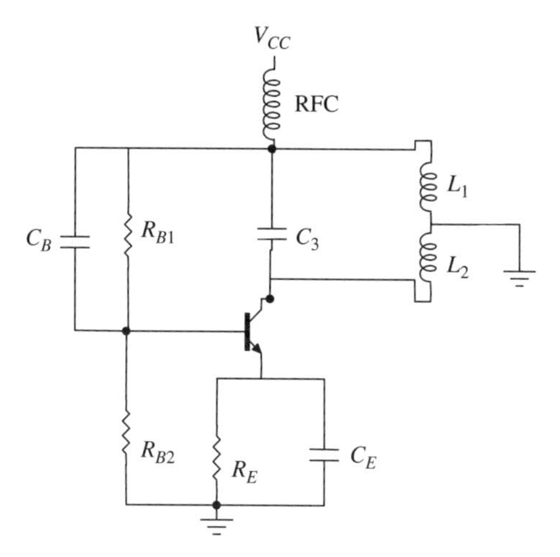

Figure 12.6 Biased BJT Hartley oscillator circuit.

or

$$\omega^2 = \frac{1}{C_3(L_1 + L_2)} \tag{12.1.20}$$

Note that this relation assumes that there is no mutual coupling between L_1 and L_2. If the coupling factor is κ (see Section 5.3), the correction term is $2\kappa(L_1L_2)^{0.5}$ that should be added to $L_1 + L_2$.

Similarly, the resonant frequency of a Colpitts oscillator is found from (12.1.17) as follows:

$$-\frac{1}{\omega C_1} - \frac{1}{\omega C_2} + \omega L_3 = 0$$

or

$$\omega^2 = \frac{C_1 + C_2}{C_1C_2L_3} \tag{12.1.21}$$

Resistors R_{B1}, R_{B2}, and R_E in Figure 12.6 are determined according to the bias point selected for a transistor. Capacitors C_B and C_E must bypass the RF, and therefore these should have relatively high values. C_E is selected such that its reactance at the design frequency is negligible in comparison with R_E. Similarly, the parallel combination of R_{B1} and R_{B2} must be infinitely large compared with the reactance of C_B. The RF choke (RFC) offers an infinitely large reactance at the RF while it passes dc with almost zero resistance. Thus, it blocks the ac signal from reaching the dc supply. Since capacitors C_B and C_E have almost zero reactance at RF, the node that connects L_1 and C_3 is connected electrically to the base of BJT. Also, the grounded junction of L_1 and L_2 is effectively connected to the emitter. Hence, the circuit depicted in Figure 12.6 is essentially the same for the RF as that shown in Figure 12.5(a).

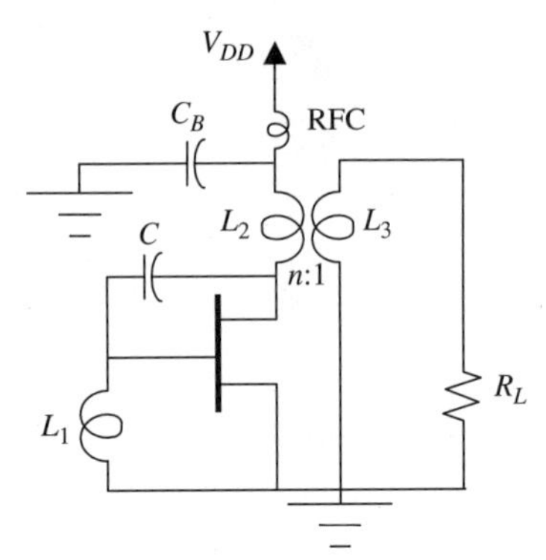

Figure 12.7 Hartley oscillator for Example 12.2.

Capacitor C_3 and total inductance $L_1 + L_2$ are determined such that (12.1.17) is satisfied at the desired frequency of oscillations. C_3 and L_2 satisfy (12.1.16) as well when the oscillator circuit operates. A Colpitts oscillator will require capacitors C_1 and C_2 in place of inductors and an inductor L_3 that replaces the capacitor C_3.

Example 12.2 The Hartley circuit shown in Figure 12.7 is oscillating at 150 MHz. If the transconductance g_m of the FET is 4.5 mS, the load resistance R_L is 50 Ω, and there is no coupling between L_1 and L_2 whereas L_2 and L_3 are tightly coupled, find the values of the circuit components.

SOLUTION RFC is used in this circuit to block, and the capacitor C_B to bypass, the RF signal. Thus, the unknown values are only L_1, L_2, L_3, and C. The load resistance R_L contributes to G_o via the coupling of L_2 and L_3 (see Example 8.17). If $L_1 = L_2 = 1$ nH, then from (12.1.19), $G_o = g_m$. Therefore,

$$n = \sqrt{\frac{1}{G_o R_L}} = \sqrt{\frac{1}{4.5 \times 10^{-3} \times 50}} = 2.1082$$

For a tightly coupled case, (5.3.10) gives

$$L_3 = \frac{L_2}{n^2} = \frac{1}{2.1082^2}\,\text{nH} = 0.225\,\text{nH}$$

The capacitor C is found from (12.1.20) as follows:

$$C = \frac{1}{(L_1 + L_2)\omega^2} = \frac{1}{2 \times 10^{-9} \times (2\pi \times 150 \times 10^6)^2}\,\text{F} = 562.9\,\text{pF}$$

Example 12.3 The Colpitts oscillator shown in Figure 12.8 is oscillating at 200 MHz. The transconductance g_m of the FET is 4.5 mS, $R_D = 50\,\Omega$, $R_G =$

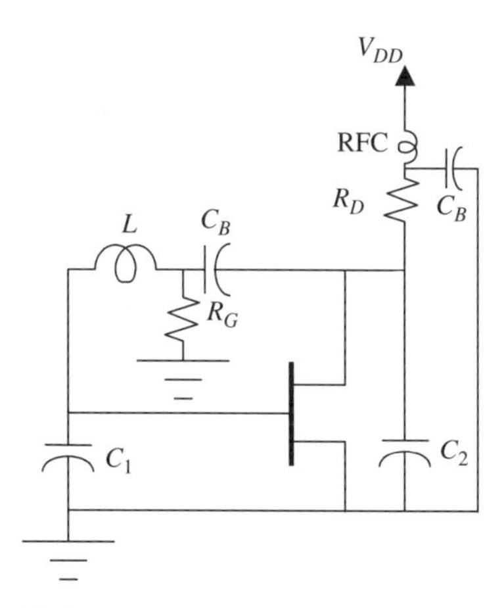

Figure 12.8 Colpitts oscillator for Example 12.3.

200 kΩ, and the two coupling capacitors C_B are large, to provide almost a short circuit for the signal. Find the values of the remaining components.

SOLUTION Capacitors C_B are large in value such that an RF signal is passed through almost unchanged. The resistor R_G provides a dc path to the gate. Resistors R_G and R_D appear in parallel at the output. Therefore,

$$R_o = \frac{1}{G_o} \approx \frac{R_G R_D}{R_G + R_D} = \frac{200 \times 10^3 \times 50}{200 \times 10^3 + 50} = 49.9875\ \Omega$$

From (12.1.19),

$$\frac{C_1}{C_2} = \frac{g_m}{G_o} = 4.5 \times 10^{-3} \times 49.9875 \to C_1 = 0.225 C_2$$

and from (12.1.21),

$$\frac{C_1 + C_2}{C_1 C_2} = \frac{1}{C_2} + \frac{1}{C_1} = L\omega^2$$

Assuming that $L = 1$ nH, C_2 is found from the two equations above as follows:

$$C_2 = \frac{49}{9 \times 10^{-9} \times (2\pi \times 200 \times 10^6)^2} \text{F} = 3.4477\ \text{nF}$$

Therefore, $C_1 = 77.7$ pF. A BJT-based Colpitts oscillator is shown in Figure 12.9. Resistors R_{B1}, R_{B2}, and R_E are determined from the usual procedure of biasing a transistor. Reactance of the capacitor C_{B1} must be negligible compared with parallel resistances R_{B1} and R_{B2}. Similarly, the reactance of C_{B2} must be negligible compared with that of the inductor L_3. The purpose of capacitor C_{B2} is

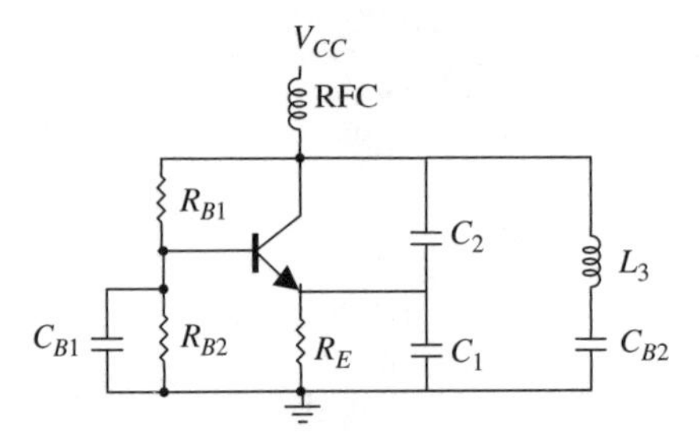

Figure 12.9 Biased BJT Colpitts oscillator circuit.

to protect the dc supply from short-circuiting via L_3 and RFC. Since capacitors C_{B1} and C_{B2} have negligible reactance, the ac equivalent of this circuit is the same as that shown in Figure 12.5(*b*). C_1, C_2, and L_3 are determined from the resonance condition (12.1.21). Also, (12.1.16) holds at the resonance.

As described in the preceding paragraph, capacitor C_{B2} provides almost a short circuit in the desired frequency range, and the inductor L_3 is selected such that (12.1.21) is satisfied. An alternative design procedure that provides better stability of the frequency is as follows. L_3 is selected larger than needed to satisfy (12.1.21), and then C_{B2} is determined to bring it down to the desired value at resonance. This type of circuit is called a *Clapp oscillator*. A FET-based Clapp oscillator circuit is shown in Figure 12.10. It is very similar to the Colpitts design and operation except for the selection of C_{B2}, which is connected in series with the inductor. At the design frequency, the series inductor–capacitor combination provides the same inductive reactance as that of the Colpitts circuit. However, if there is a drift in frequency, the reactance of this combination changes rapidly. This can be explained further with the help of Figure 12.11.

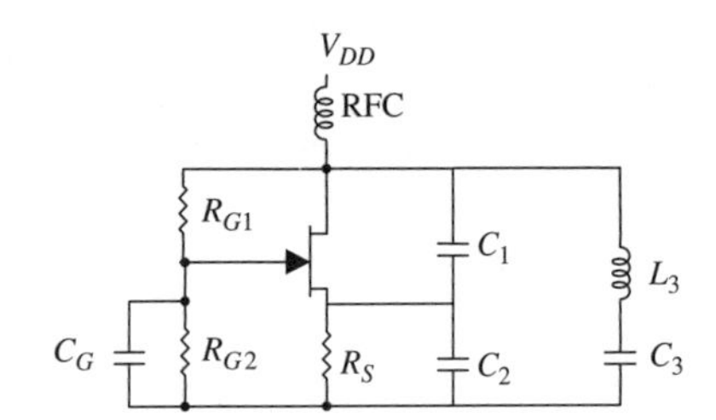

Figure 12.10 FET-based Clapp oscillator circuit.

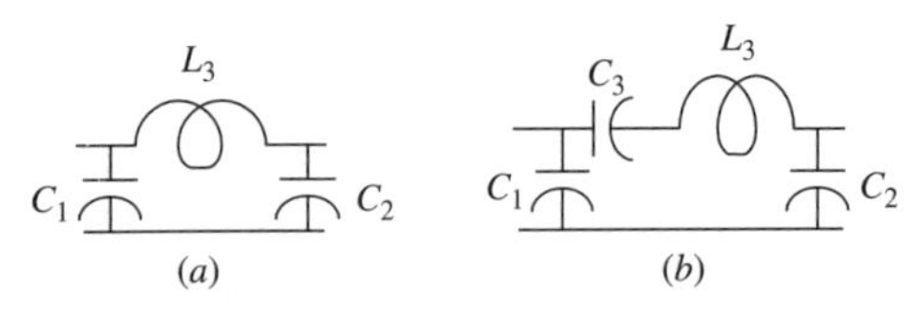

Figure 12.11 Resonant circuits for Colpitts (*a*) and Clapp (*b*) oscillators.

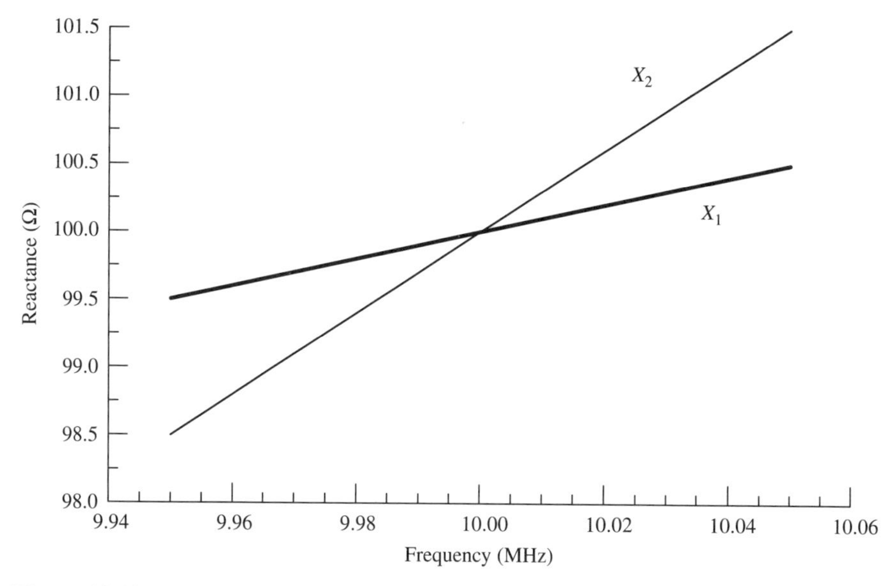

Figure 12.12 Reactance of inductive branch versus frequency for Colpitts (X_1) and Clapp (X_2) circuits.

Figure 12.11 illustrates the resonant circuits of Colpitts and Clapp oscillators. An obvious difference between the two circuits is the capacitor C_3, which is connected in series with L_3. Note that unlike C_3, the blocking capacitor C_{B2} shown in Figure 12.9 does not affect RF operation. Reactance X_1 of the series branch in the Colpitts circuit is ωL_3, whereas it is $X_2 = \omega L_3 - 1/\omega C_3$ in the case of the Clapp oscillator. If inductor L_3 in the former case is selected as 1.59 μH and the circuit is resonating at 10 MHz, the change in its reactance around resonance is as shown in Figure 12.12. The series branch of the Clapp circuit has the same inductive reactance at the resonance if $L_3 = 3.18$ μH and $C_3 = 159$ pF. However, the rate of change of reactance with frequency is now higher compared with X_1. This characteristic helps in reducing the drift in oscillation frequency.

Another Interpretation of the Oscillator Circuit

Ideal inductors and capacitors store electrical energy in the form of magnetic and electric fields, respectively. If such a capacitor with initial charge is connected across an ideal inductor, it discharges through that. Since there is no loss in this system, the inductor recharges the capacitor back and the process repeats. However, real inductors and capacitors are far from being ideal. Energy losses in the inductor and the capacitor can be represented by a resistance r_1 in this loop. Oscillations die out because of these losses. As shown in Figure 12.13, if a negative resistance $-r_1$ can be introduced in the loop, the effective resistance becomes zero. In other words, if a circuit can be devised to compensate for the

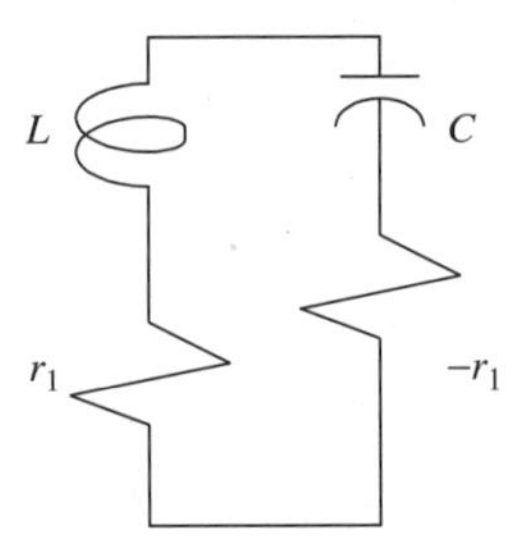

Figure 12.13 Ideal oscillator circuit.

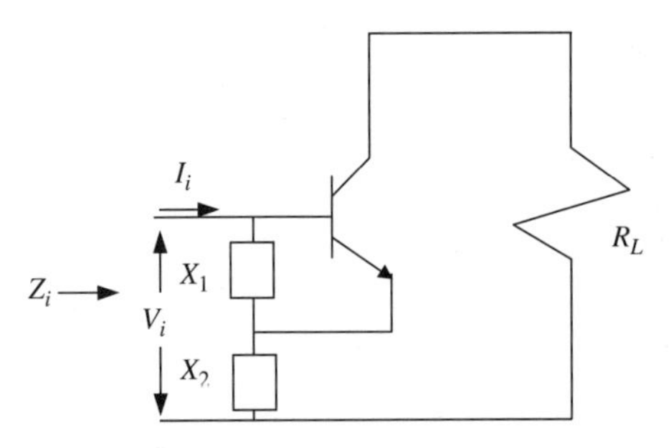

Figure 12.14 BJT circuit to obtain negative resistance.

losses, oscillations can be sustained. This can be done using an active circuit, as illustrated in Figure 12.14.

Consider the transistor circuit shown in Figure 12.14. X_1 and X_2 are arbitrary reactance, and for simplicity, the dc bias circuit is not shown in this circuit. For analysis, a small-signal equivalent circuit can be drawn as shown in Figure 12.15. Using Kirchhoff's voltage law, we can write

$$V_i = I_i(X_1 + X_2) - I_b(X_1 - \beta X_2) \tag{12.1.22}$$

and

$$0 = I_i(X_1) - I_b(X_1 + r_\pi) \tag{12.1.23}$$

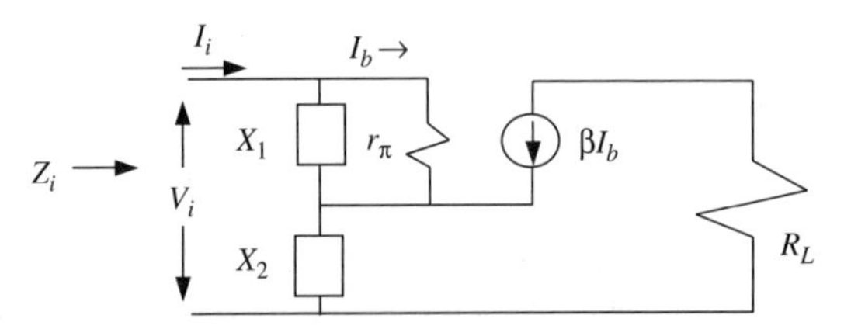

Figure 12.15 Small-signal equivalent of Figure 12.14 with the output impedance of the BJT neglected.

Equation (12.1.23) can be rearranged as

$$I_b = \frac{X_1}{X_1 + r_\pi} I_i \tag{12.1.24}$$

Substituting (12.1.24) into (12.1.22), we find that

$$V_i = I_i \left(X_1 + X_2 - \frac{X_1 - \beta X_2}{X_1 + r_\pi} X_1 \right) \tag{12.1.25}$$

The impedance Z_i across its input terminal can now be determined as follows:

$$Z_i = \frac{V_i}{I_i} = \left(X_1 + X_2 - \frac{X_1 - \beta X_2}{X_1 + r_\pi} X_1 \right) = \frac{(1 + \beta) X_1 X_2 + (X_1 + X_2) r_\pi}{X_1 + r_\pi} \tag{12.1.26}$$

For $X_1 \ll r_\pi$, the following approximation can be made:

$$Z_i \approx \frac{(1 + \beta) X_1 X_2}{r_\pi} + X_1 + X_2 \tag{12.1.27}$$

For X_1 and X_2 to be capacitive, it simplifies to

$$Z_i \approx -\frac{1 + \beta}{r_\pi} \frac{1}{\omega^2 C_1 C_2} - j \frac{C_1 + C_2}{\omega C_1 C_2} \tag{12.1.28}$$

Since β, r_π, and g_m of a BJT are related as

$$\frac{1 + \beta}{r_\pi} \approx g_m$$

(12.1.28) can be further simplified. Hence,

$$Z_i \approx -\frac{g_m}{\omega^2 C_1 C_2} - \frac{j}{\omega} \frac{C_1 + C_2}{C_1 C_2} \tag{12.1.29}$$

Therefore, if this circuit is used to replace capacitor C of Figure 12.13 and the following condition is satisfied, the oscillations can be sustained:

$$r_1 = \frac{g_m}{\omega^2 C_1 C_2} \tag{12.1.30}$$

The frequency of these oscillations is given as

$$\omega = \frac{1}{\sqrt{L C_1 C_2 / (C_1 + C_2)}} \tag{12.1.31}$$

A comparison of this equation with (12.1.21) indicates that it is basically the Colpitts oscillator.

On the other hand, if X_1 and X_2 are inductive, (12.1.27) gives the relation

$$Z_i \approx -g_m\omega^2 L_1 L_2 + j\omega(L_1 + L_2) \tag{12.1.32}$$

Now, if this circuit replaces inductor L of Figure 12.13 and the condition

$$r_1 = g_{\rm m}\omega^2 L_1 L_2 \tag{12.1.33}$$

is satisfied, sustained oscillations are possible. The frequency of these oscillations is

$$\omega = \frac{1}{\sqrt{C(L_1 + L_2)}} \tag{12.1.34}$$

This is identical to (12.1.20), the Hartley oscillator frequency.

12.2 CRYSTAL OSCILLATORS

Quartz and ceramic crystals are used in oscillator circuits for additional stability of frequency. They provide a fairly high Q value (on the order of 100,000) that shows a small drift with temperature (on the order of 0.001% per °C). A simplified electrical equivalent circuit of a crystal is illustrated in Figure 12.16. As this equivalent circuit indicates, the crystal exhibits both series and parallel resonant modes. For example, the terminal impedance of a crystal with typical values of $C_P = 29\,\text{pF}$, $L_S = 58\,\text{mH}$, $C_S = 0.054\,\text{pF}$, and $R_S = 15\,\Omega$ exhibits a distinct minimum and maximum with frequency, as shown in Figure 12.17. Its primary characteristics may be summarized as follows:

- The magnitude of the terminal impedance decreases up to around 2.844 MHz while its phase angle remains constant at −90°. Hence, it is effectively a capacitor in this frequency range.
- The magnitude of the impedance dips around 2.844 MHz and its phase angle goes through a sharp change from −90° to 90°. It has series resonance at this frequency.

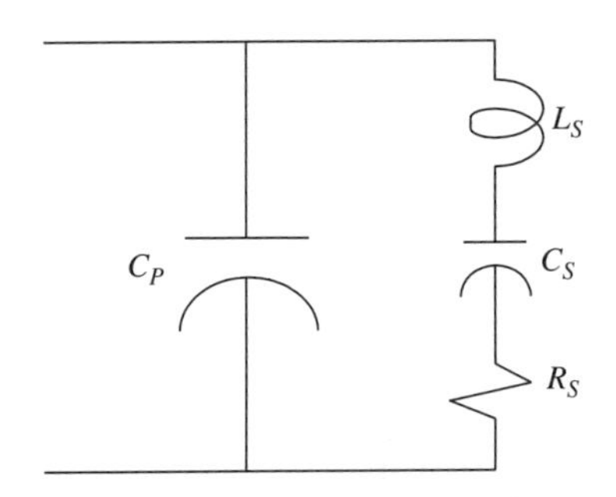

Figure 12.16 Equivalent circuit of a crystal.

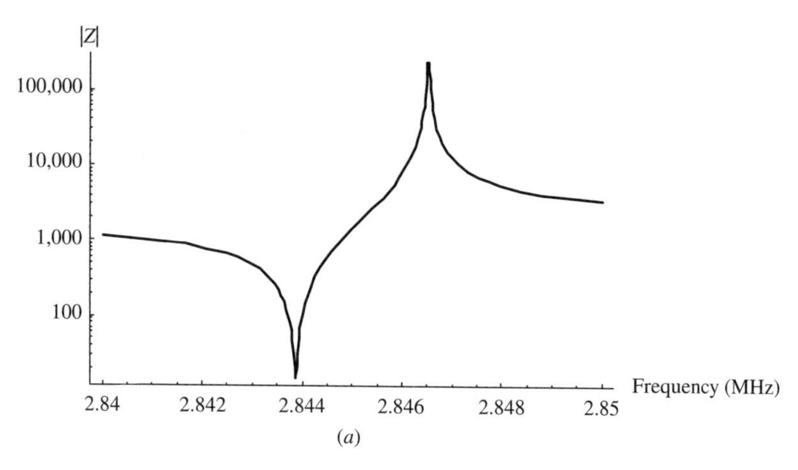

(*a*)

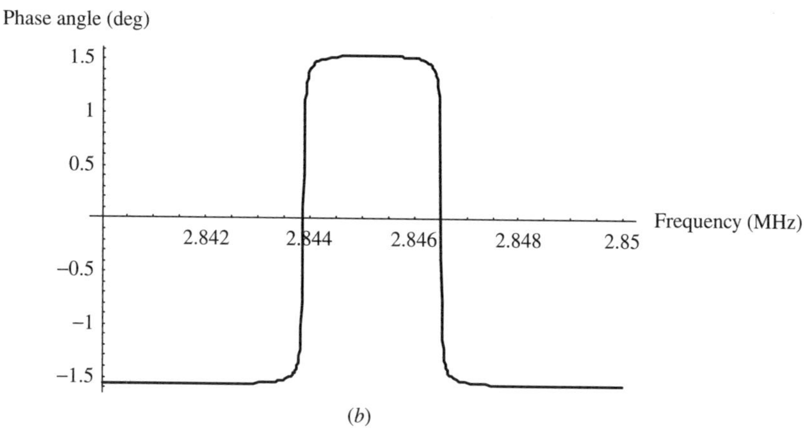

(*b*)

Figure 12.17 Magnitude (*a*) and phase angle (*b*) of the terminal impedance as a function of frequency.

- The magnitude of the impedance has a maximum around 2.8465 MHz, where its phase angle changes back to $-90°$ from $90°$. It exhibits a parallel resonance around this frequency.
- The phase angle of the impedance remains constant at $90°$ in the frequency range 2.844 to 2.8465 MHz while its magnitude increases. Hence, it is effectively an inductor.
- Beyond 2.8465 MHz, the phase angle stays at $-90°$ while its magnitude goes down with frequency. Therefore, it is changed back to a capacitor.

Series resonant frequency ω_S and parallel resonant frequency ω_P of the crystal can be found from its equivalent circuit. These are given as follows:

$$\omega_S = \frac{1}{\sqrt{L_S C_S}} \tag{12.2.1}$$

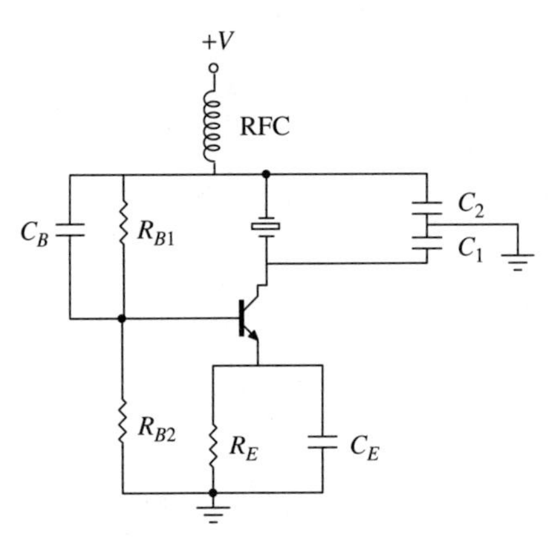

Figure 12.18 Pierce oscillator circuit.

and

$$\omega_P = \omega_S \sqrt{1 + \frac{C_S}{C_P}} \tag{12.2.2}$$

Hence, the frequency range $\Delta\omega$ over which the crystal behaves as an inductor can be determined as follows:

$$\omega_P - \omega_S = \Delta\omega = \omega_S \left[\sqrt{\left(1 + \frac{C_S}{C_P}\right)} - 1 \right] \approx \frac{C_S}{2C_P}\omega_S \tag{12.2.3}$$

$\Delta\omega$ is known as the *pulling figure* of the crystal. Typically, ω_P is less than 1% higher than ω_S. For an oscillator design, a crystal is selected such that the frequency of oscillation falls between ω_S and ω_P. Therefore, the crystal operates basically as an inductor in the oscillator circuit. A BJT oscillator circuit using the crystal is shown in Figure 12.18. It is known as a *Pierce oscillator*. A comparison of its RF equivalent circuit with that shown in Figure 12.5(*b*) indicates that the Pierce circuit is similar to the Colpitts oscillator, with inductor L_3 replaced by the crystal. As mentioned earlier, the crystal provides very stable frequency of oscillation over a wide range of temperature. The main drawback of a crystal oscillator circuit is that its tuning range is relatively small. It is achieved by adding a capacitor in parallel with the crystal. In this way, the parallel resonant frequency ω_P can be decreased up to the series resonant frequency ω_S.

12.3 ELECTRONIC TUNING OF OSCILLATORS

In most of the circuits considered so far, the capacitance of the tuned circuit can be varied to change the frequency of oscillation. It can be done electronically by using a varactor diode and controlling its bias voltage. There are two basic types of varactors: abrupt and hyperabrupt junctions. *Abrupt junction diodes* provide very high Q values and also operate over a very wide tuning voltage range (typically, 0 to 60 V). These diodes provide an excellent phase noise performance because of their high Q value.

Hyperabrupt-type diodes exhibit a quadratic characteristic of the capacitance with applied voltage. Therefore, these varactors provide a much more linear tuning characteristic than does the abrupt type. These diodes are preferred for tuning over a wide frequency band. An octave tuning range can be covered in less than 20 V. The main disadvantage of these diodes is that they have a much lower Q value, and therefore the phase noise is higher than that obtained from the abrupt junction diodes.

The capacitance of a varactor diode is related to its bias voltage as follows:

$$C = \frac{A}{(V_R + V_B)^n} \tag{12.3.1}$$

A is a constant, V_R the applied reverse bias voltage, and V_B the built-in potential, 0.7 V for silicon diodes and 1.2 V for GaAs diodes. For the following analysis, we can write

$$C = \frac{A}{V^n} \tag{12.3.2}$$

In this equation, A represents capacitance of the diode when V is 1 V. Also, n is a number between 0.3 and 0.6 but can be as high as 2 for a hyperabrupt junction. The resonant circuit of a typical voltage-controlled oscillator (VCO) has a parallel tuned circuit consisting of inductor L, fixed capacitor C_f, and the varactor diode with capacitance C. Therefore, its frequency of oscillation can be written as

$$\omega = \frac{1}{\sqrt{L(C_f + C)}} = \frac{1}{\sqrt{L\left(C_f + A/V^n\right)}} \tag{12.3.3}$$

Let ω_0 be the angular frequency of an unmodulated carrier and V_0 and C_0 be the corresponding values of V and C. Then

$$\omega_0^2 = \frac{1}{L(C_f + C_0)} = \frac{1}{L\left(C_f + A/V_0^n\right)} \tag{12.3.4}$$

Further, the carrier frequency deviates from ω_0 by $\delta\omega$ for a voltage change of δV. Therefore,

$$(\omega_0 + \delta\omega)^2 = \frac{1}{L[C_f + A/(V_0 + \delta V)^n]} \Rightarrow (\omega_0 + \delta\omega)^{-2}$$
$$= L[C_f + A(V_0 + \delta V)^{-n}] \tag{12.3.5}$$

Dividing (12.3.5) by (12.3.4), we have

$$\left(\frac{\omega_0 + \delta\omega}{\omega_0}\right)^2 = \frac{C_f + C_0}{C_f + A(V_0 + \delta V)^{-n}}$$
$$= \frac{C_f + C_0}{C_f + AV_0^{-n}\,(1 + \delta V/V_0)^{-n}}$$

or

$$\left(1 + \frac{\delta\omega}{\omega_0}\right)^2 = \frac{C_f + C_0}{C_f + C_0\,(1 + \delta V/V_0)^{-n}} \Rightarrow \left(1 + \frac{\delta\omega}{\omega_0}\right)^{-2}$$
$$= \frac{C_f + C_0\,(1 + \delta V/V_0)^{-n}}{C_f + C_0}$$

or

$$1 - 2\frac{\delta\omega}{\omega_0} \approx \frac{C_f + C_0[1 - n(\delta V/V_0)]}{C_f + C_0} = 1 - n\frac{\delta V}{V_0} \times \frac{C_0}{C_f + C_0}$$

Hence,

$$\frac{\delta\omega}{\delta V} = \frac{n\omega_0}{2V_0}\frac{C_0}{C_f + C_0} = K_1 \tag{12.3.6}$$

K_1 is called the *tuning sensitivity* of the oscillator. It is expressed in radians per second per volt.

Example 12.4 Figure 12.19 shows a voltage-controlled oscillator circuit. Assume that $R_s = 1.8\,k\Omega$, $C_1 = 150\,\text{pF}$, $C_2 = 72\,\text{pF}$, $L = 32\,\text{nH}$, and the capacitance of the varactor diode changes from 3.5 pF to 32 pF as its bias voltage V_{control} is varied. Further, two capacitors C_B are large enough to short-circuit the RF signal. Determine the frequency range of this oscillator.

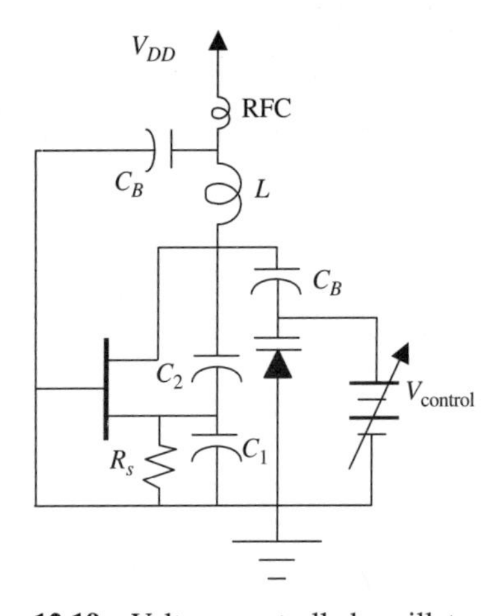

Figure 12.19 Voltage-controlled oscillator circuit.

SOLUTION From (12.1.31),

$$f = \frac{1}{2\pi\sqrt{LC}} \qquad \text{Hz}$$

where

$$C = C_{\text{var}} + \frac{C_1 C_2}{C_1 + C_2}$$

For the varactor capacitance $C_{\text{var}} = 3.5\,\text{pF}$, C is calculated as 52.15 pF and the corresponding frequency is found to be 123.2 MHz. When C_{var} is 32 pF, C is found as 80.65 pF that gives f as 99.07 MHz. Therefore, the frequency range of this oscillator is 99.07 to 123.2 MHz.

12.4 PHASE-LOCKED LOOP

A *phase-locked loop* (PLL) is a feedback system that is used to lock the output frequency and phase to the frequency and phase of a reference signal at its input. The reference waveform can be of many different types, including sinusoidal and digital. PLLs have been used for various applications, including filtering, frequency synthesis, motor speed control, frequency modulation, demodulation, and signal detection. The basic PLL consists of a voltage-controlled oscillator (VCO), a phase detector (PD), and a filter. In its most general form, the PLL may also contain a mixer and a frequency divider, as shown in Figure 12.20. In the steady state, the output frequency is expressed as follows:

$$f_o = f_m \pm N f_r \tag{12.4.1}$$

Hence, the output frequency can be controlled by varying N, f_r, or f_m.

It is helpful to consider the PLL in terms of phase rather than frequency. This is done by replacing f_o, f_m, and f_r with θ_o, θ_m, and θ_r, respectively. Further, the transfer characteristics of each building block need to be formulated before the PLL can be analyzed.

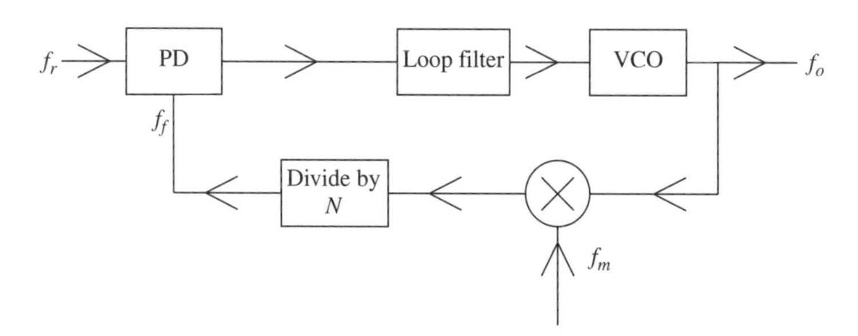

Figure 12.20 Block diagram of a PLL system.

Phase Detector

With the loop in lock, the output of the phase detector is a direct voltage V_e that is a function of the phase difference $\theta_d = \theta_r - \theta_f$. If input frequency f_r is equal to f_f, V_e must be zero. In commonly used analog phase detectors, V_e is a sinusoidal, triangular, or sawtooth function of θ_d. It is equal to zero when θ_d is equal to $\pi/2$ for the sinusoidal and triangular types, and π for the sawtooth type. Therefore, it is convenient to plot V_e versus a shifted angle θ_e as shown in Figures 12.21 to 12.23 for a direct comparison of these three types of detectors. Hence, the transfer characteristic of a sinusoidal-type phase detector can be expressed as follows:

$$V_e = A \sin \theta_e \qquad -\frac{\pi}{2} \leq \theta_e \leq \frac{\pi}{2} \tag{12.4.2}$$

This can be approximated around $\theta_e \approx 0$ by the expression

$$V_e \approx A\theta_e$$

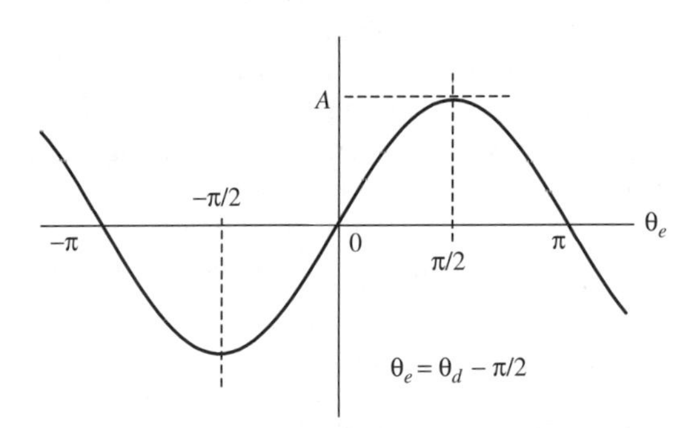

Figure 12.21 Sinusoidal output of the PD.

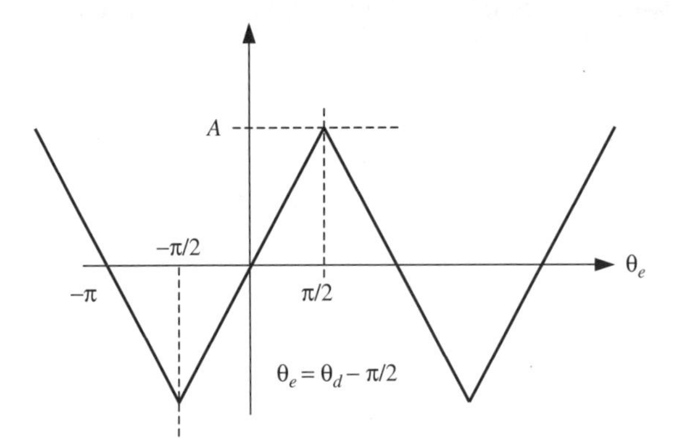

Figure 12.22 Triangular wave.

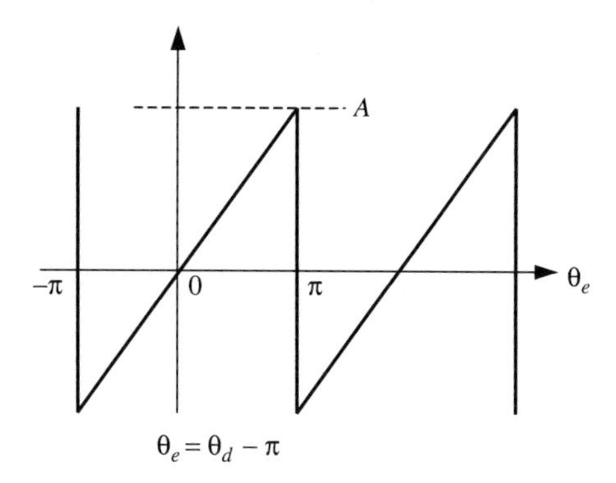

Figure 12.23 Sawtooth wave.

In the case of a triangular output of the phase detector, the transfer characteristic can be expressed as

$$V_e = \frac{2A}{\pi}\theta_e \qquad -\frac{\pi}{2} \le \theta_e \le \frac{\pi}{2} \tag{12.4.3}$$

From the transfer characteristic of the sawtooth-type phase detector illustrated in Figure 12.23, we can write

$$V_e = \frac{A}{\pi}\theta_e \qquad -\pi \le \theta_e \le \pi \tag{12.4.4}$$

Since V_e is zero in steady state, the gain factor K_d (volts per radians) in all three cases is

$$\frac{V_e}{\theta_e} = K_d \tag{12.4.5}$$

Voltage-Controlled Oscillator

As described earlier, a varactor diode is generally used in the resonant circuit of an oscillator. Its bias voltage is controlled to change the frequency of oscillation. Therefore, the transfer characteristic of an ideal voltage-controlled oscillator (VCO) has a linear relation, as depicted in Figure 12.24. Hence, the output frequency of a VCO can be expressed as follows:

$$f_o = f_S + k_o V_d \qquad \text{Hz}$$

or

$$\omega_0 = \omega_S + K_o V_d \qquad \text{rad/s}$$

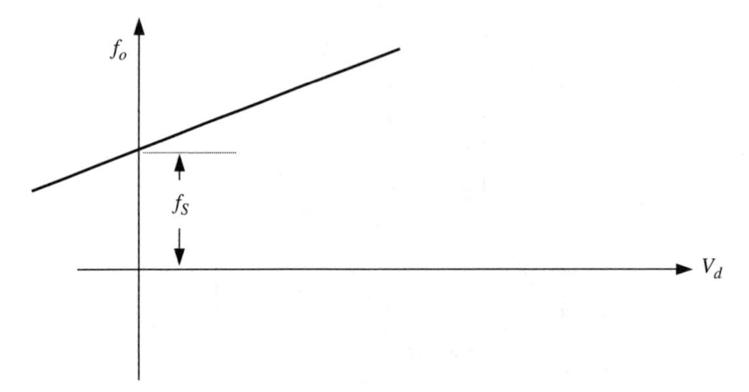

Figure 12.24 Characteristic of a voltage-controlled oscillator.

or

$$\omega_0 = \omega_S + \delta\omega \qquad \text{rad/s}$$

$$\theta(t) = \int_0^t \omega_0 dt = \omega_S t + \int_0^t \delta\omega dt = \theta_S + \theta_0(t)$$

Therefore,

$$\theta_0(t) = \int_0^t \delta\omega dt \Rightarrow \frac{d\theta_0(t)}{dt} = \delta\omega = K_o V_d \tag{12.4.6}$$

In the s-domain,

$$s\theta_0(s) = K_o V_d(s) \Rightarrow \frac{\theta_0(s)}{V_d(s)} = \frac{K_o}{s}$$

Hence, VCO acts as an integrator.

Loop Filters

A low-pass filter is connected right after the phase detector to suppress its output harmonics. Generally, it is a simple first-order filter. Sometimes higher-order filters are also employed to suppress additional ac components. The transfer characteristics of selected loop filters are summarized below.

Lead–Lag Filter. A typical lead–lag filter uses two resistors and a capacitor, as illustrated in Figure 12.25. Its transfer function, $F(s)$, can be found as follows:

$$\frac{V_o}{V_i} = \frac{R_2 + 1/sC_2}{R_1 + R_2 + 1/sC_2} = \frac{1 + sR_2C_2}{1 + s(R_1C_2 + R_2C_2)}$$

Hence,

$$F(s) = \frac{1 + \tau_2 s}{1 + \tau_1 s} \tag{12.4.7}$$

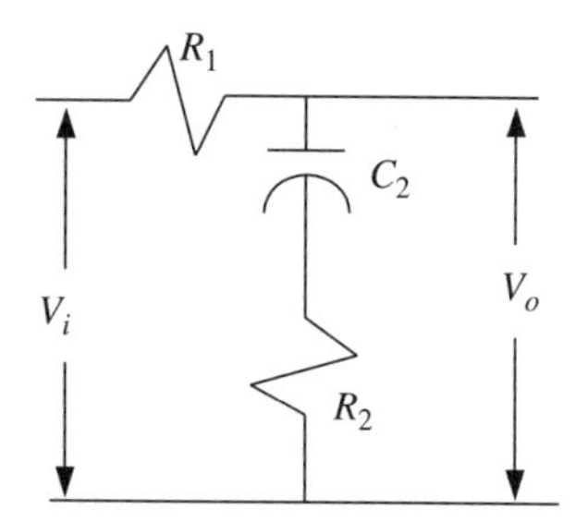

Figure 12.25 Lead–lag filter.

where

$$\tau_2 = R_2C_2 \tag{12.4.8}$$

and

$$\tau_1 = (R_1 + R_2)C_2 \tag{12.4.9}$$

Typical magnitude and phase characteristics versus frequency (the Bode plot) of a lead–lag filter are illustrated in Figure 12.26. The time constants τ_1 and τ_2 used to draw these characteristics were 0.1 and 0.01 s, respectively. Note that the changes in these characteristics occur at $1/\tau_1$ and $1/\tau_2$.

Integrator and Lead Filter. This type of filter generally requires an op amp along with two resistors and a capacitor. As illustrated in Figure 12.27, the feedback path of the op amp uses a capacitor in series with resistance. Assuming that the op amp is ideal and is being used in inverting configuration, the transfer characteristics of this filter can be found as follows:

$$\frac{V_o}{V_i} = -\frac{R_2 + 1/sC_2}{R_1} = -\frac{1 + sR_2C_2}{sR_1C_2}$$

$$F(s) = \frac{1 + s\tau_2}{s\tau_1} \tag{12.4.10}$$

where

$$\tau_1 = R_1C_2 \tag{12.4.11}$$

and

$$\tau_2 = R_2C_2 \tag{12.4.12}$$

Typical frequency response characteristics (the Bode plot) of an integrator and lead filter are illustrated in Figure 12.28. Time constants τ_1 and τ_2 used for this illustration are 0.1 and 0.01 s, respectively.

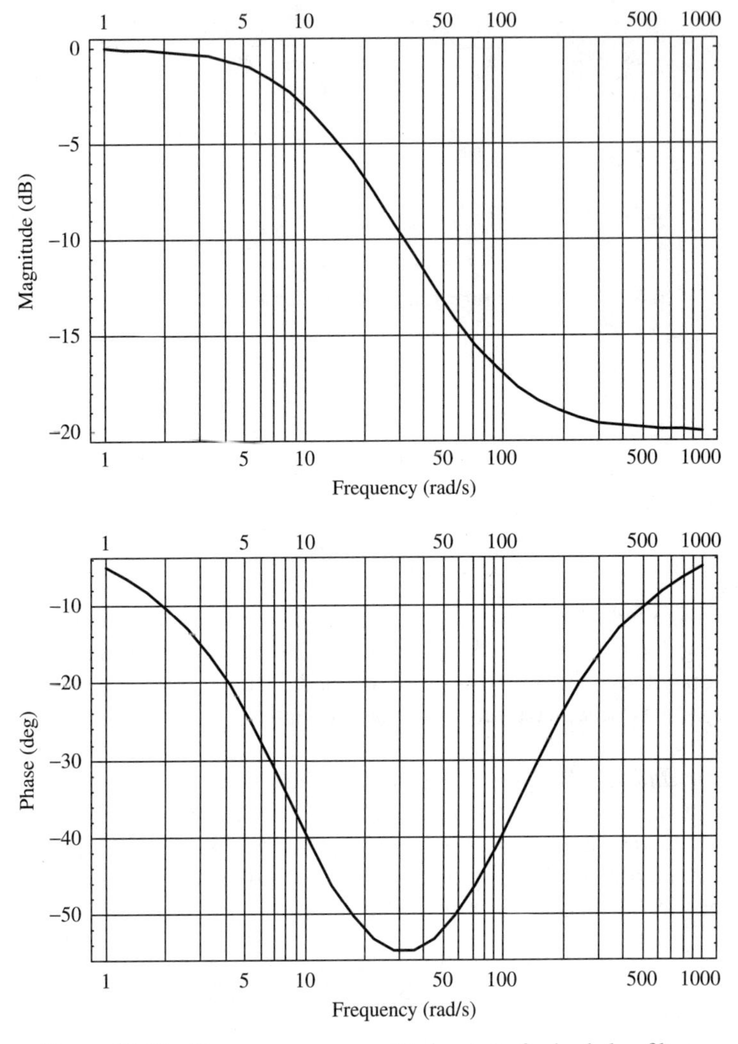

Figure 12.26 Frequency response (Bode plot) of a lead–lag filter.

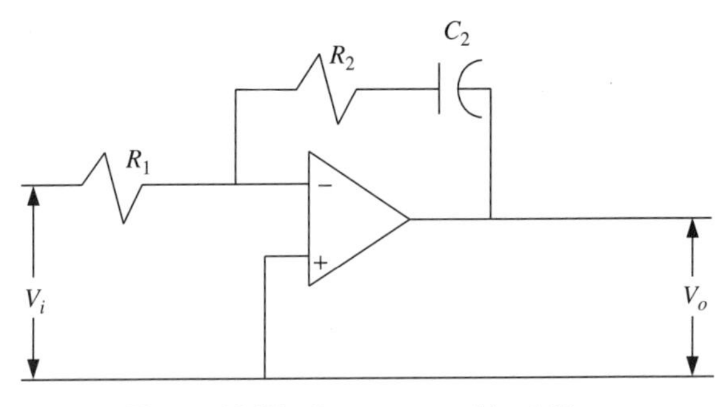

Figure 12.27 Integrator and lead filter.

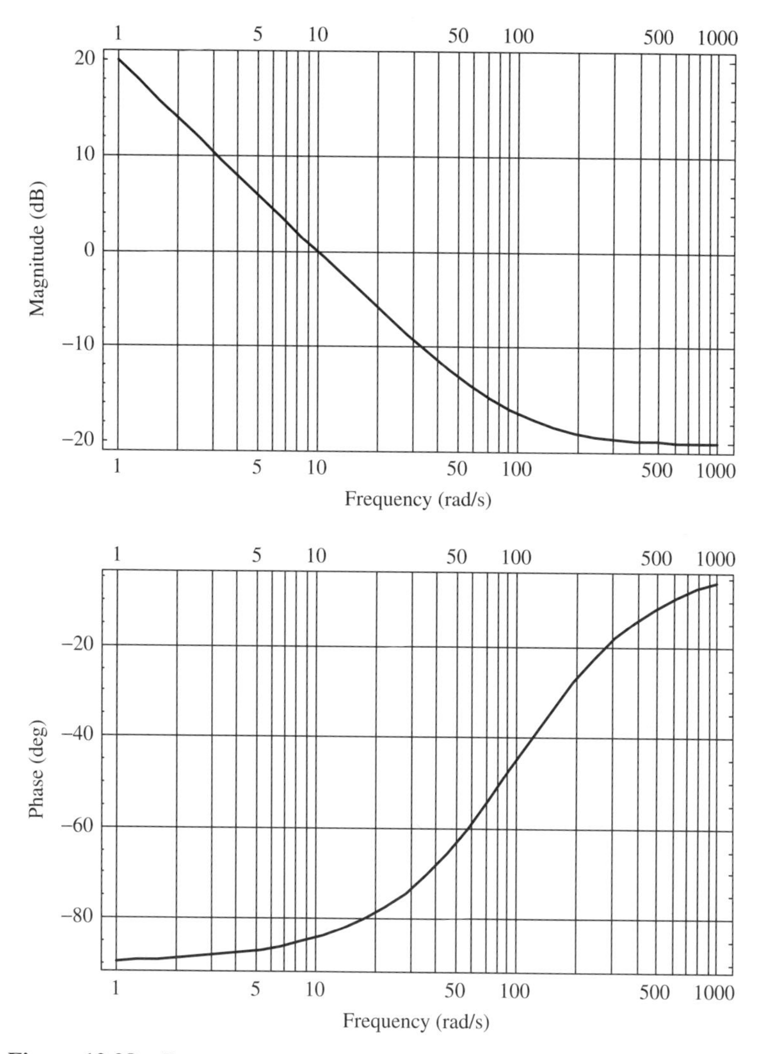

Figure 12.28 Frequency response (Bode plot) of integrator and lead filter.

Note again that there are significant changes occurring in these characteristics at frequencies that are equal to $1/\tau_1$ and $1/\tau_2$. For frequencies less than 10 rad/s, the phase angle is almost constant at -90°, whereas the magnitude changes at a rate of 20 dB/decade. It represents the characteristics of an integrator. The phase angle becomes zero for frequencies greater than 100 rad/s. The magnitude becomes constant at -20 dB as well.

Integrator and Lead–Lag Filter. Another active filter that is used in the loop is shown in Figure 12.29. It employs two capacitors in the feedback loop of the op amp. Assuming again that the op amp is ideal and is connected in inverting

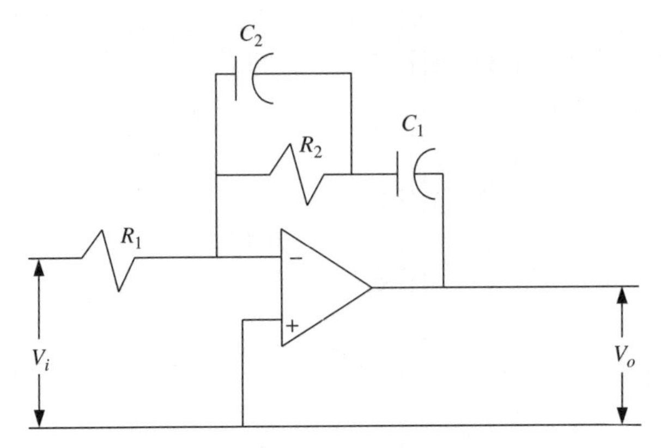

Figure 12.29 Integrator and lead–lag filter.

configuration, the transfer function of this filter can be found as follows:

$$\frac{V_o}{V_i} = -\frac{1}{R_1}\left(\frac{R_2/sC_2}{R_2 + 1/sC_2} + \frac{1}{sC_1}\right) = -\frac{1}{R_1}\left(\frac{R_2}{1 + sR_2C_2} + \frac{1}{sC_1}\right)$$

$$= -\frac{R_2/R_1}{1 + sR_2C_2} - \frac{1}{sR_1C_1}$$

or

$$\frac{V_o}{V_i} = -\frac{sR_2C_1 + 1 + sR_2C_2}{sR_1C_1(1 + sR_2C_2)}$$

This can be rearranged as follows:

$$F(s) = -\frac{1}{s\tau_1}\frac{1 + s\tau_2}{1 + s\tau_3} \tag{12.4.13}$$

where

$$\tau_1 = C_1R_1 \tag{12.4.14}$$

$$\tau_2 = R_2(C_1 + C_2) \tag{12.4.15}$$

and

$$\tau_3 = R_2C_2 \tag{12.4.16}$$

The frequency response of a typical integrator and lead–lag filter is shown in Figure 12.30. Time constants τ_1, τ_2, and τ_3 are assumed to be 1, 0.1, and 0.01 s, respectively. The corresponding frequencies are 1, 10, and 100 rad/s. At frequencies below 1 rad/s, the magnitude of the transfer function decreases at a rate of 20 dB/decade while its phase angle stays at $-90°$. Similar characteristics may be

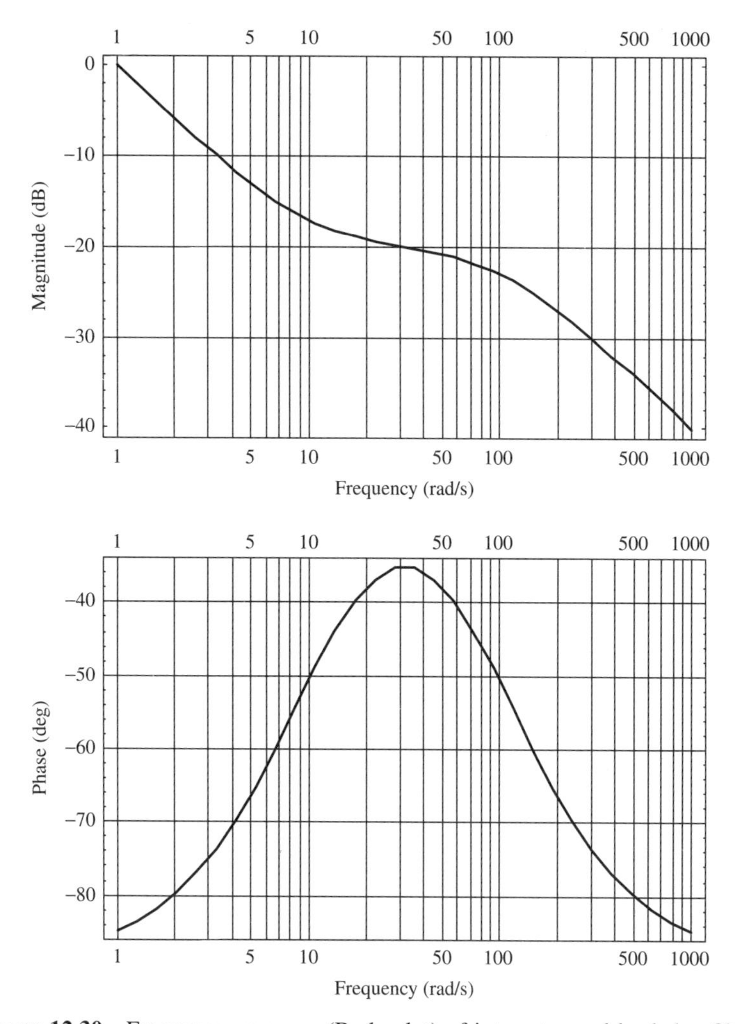

Figure 12.30 Frequency response (Bode plot) of integrator and lead–lag filter.

observed above 100 rad/s. It is a typical integrator characteristic. A significant change in the frequency response may be observed at 10 rad/s as well.

An equivalent block diagram of the PLL can be drawn as shown in Figure 12.31. Output θ_o of the VCO can be controlled by voltage V, reference signal θ_r, or modulator input θ_m. To understand the working of the PLL when it is nearly locked, the transfer function for each case can be formulated with the help of the figure.

A general formula for the transfer function $T(s)$ of a single-feedback-loop system can be written with the help of Figure 12.1:

$$T(s) = \frac{\text{forward gain}}{1 + \text{loop gain}} \tag{12.4.17}$$

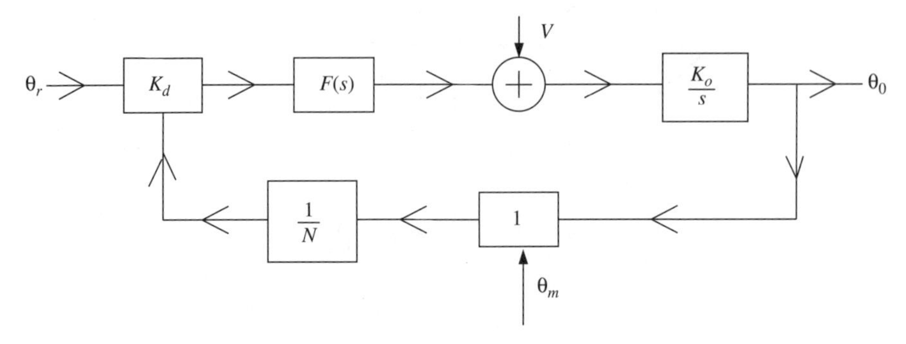

Figure 12.31 Equivalent block diagram of a PLL.

Since there are three different ways to control the phase of the VCO, the transfer function for each case can be found via (12.4.17), as follows:

1. When phase θ_m of the mixer input controls the VCO output θ_o, its transfer function $T_1(s)$ is found to be

$$T_1(s) = \frac{\theta_o(s)}{\theta_m(s)} = \frac{(1/N)K_d F(s)(K_o/s)}{1 + (1/N)K_d F(s)(K_o/s)1} = \frac{KF(s)}{s + KF(s)} \tag{12.4.18}$$

where

$$K = \frac{K_d K_o}{N} \tag{12.4.19}$$

2. When phase θ_r of the reference input controls the output θ_o of the VCO, its transfer function $T_2(s)$ will be

$$T_2(s) = \frac{\theta_o(s)}{\theta_r(s)} = \frac{K_d F(s) K_o/s}{1 + K[F(s)/s]} = \frac{NKF(s)}{s + KF(s)} \tag{12.4.20}$$

3. Similarly, for the case of input voltage V controlling θ_o, its transfer function $T_3(s)$ is found to be

$$T_3(s) = \frac{\theta_o(s)}{V} = \frac{K_o/s}{1 + K[F(s)/s]} = \frac{K_o}{s + KF(s)} \tag{12.4.21}$$

It is obvious from (12.4.18) to (12.4.21) that the transfer function $F(s)$ of the filter significantly affects the working of a PLL. Various types of loop filters have been used to obtain the desired PLL characteristics, including the three loop filters described earlier: the lead–lag filter, an integrator in combination with a lead network, and an integrator combined with a lead–lag network. The number of poles of the loop gain $[KF(s)/s]$ at the origin determines the type of phase-locked loop. Hence, a phase-locked loop with a passive lead–lag filter (without a charge pump) is of type 1 because its loop gain has only one pole at the origin.

Phase-detector output is a voltage that drives the loop filter. There is a significant improvement in performance if a current source (or sink) drives the loop filter. There is a double pole at the origin if either of the remaining two filters is used (an integrator in combination with a lead or a lead–lag network). Therefore, it is a type 2 phase-locked loop now because the integrator adds another pole at the origin. Further, the highest degree of the characteristic polynomial equation [$1 + KF(s)/s = 0$] is called the *order* of that phase-locked loop. Hence, if a PLL is employing the passive lead–lag filter of Figure 12.25, it is a second-order type 1 system. On the other hand, it will be second-order type 2 with the integrator and lead filter of Figure 12.27, and third-order type 2 with the integrator and lead–lag filter shown in Figure 12.29.

Equations (12.4.18) to (12.4.21) are used to analyze the stability of a PLL and to develop the relevant design procedures. As an example, consider the case where a lead–lag filter is being used and the frequency divider is excluded from the loop (i.e., $N = 1$). The transfer function $T_2(s)$ for this PLL can be obtained by substituting (12.4.7) into (12.4.20) as follows:

$$T_2(s) = \frac{K_v(1 + \tau_2 s)}{s(1 + \tau_1 s) + K_v(1 + \tau_2 s)}$$

where $K_v = K_d K_o$ is the dc gain of the loop.

Generally, T_2(s) is rearranged in the following form for further analysis of the PLL.

$$T_2(s) = \frac{\omega_n^2 + (2\zeta\omega_n - \omega_n^2/K_v)s}{s^2 + 2\zeta\omega_n s + \omega_n^2}$$

Coefficients ω_n and ζ are called the *natural frequency* and the *damping factor*, respectively. These are defined as

$$\omega_n = \sqrt{\frac{K_v}{\tau_1}}$$

and

$$\zeta = \frac{\omega_n}{2}\left(\tau_2 + \frac{1}{K_v}\right)$$

These coefficients must be selected such that the PLL is stable and performs according to the specifications. See Table 12.1 for a summary of PLL components.

Phase-Locked-Loop Terminology

Hold-in Range. This is also called the *lock range*, the *tracking range*, or the *synchronization range*. If reference frequency f_r is changed slowly from the free-running frequency f_f, VCO frequency f_o tracks f_r until the phase error θ_e approaches $\pm\pi/2$ for sinusoidal and triangular phase detectors or $\pm\pi$ for a

TABLE 12.1 Summary of PLL Components

Element	Transfer Function	Remarks
VCO	$\frac{K_o}{s}$	K_o is the slope of the oscillator frequency to the voltage characteristic. It is expressed in radians per second per volt.
Mixer	1	The change of output phase equals the change of input phase.
Divider	$\frac{1}{N}$	N is the division ratio.
Phase detector	K_d	K_d is the slope of the phase detector voltage to the phase characteristic in volts per radian.
Filter	$F(s)$	The transfer function depends on the type of filter used.

sawtooth-type phase detector. It is assumed here that the VCO is capable of sufficient frequency deviation and that the loop filter and amplifier (if any) are not overdriven.

From the transfer function of a sinusoidal phase detector as given by (12.4.2), we find that

$$\sin\theta_e = \frac{V_e}{K_d}$$

The output voltage V_e of the phase detector is applied to the input of the voltage-controlled oscillator. Therefore, using (12.4.6) with $V_d = V_e$,

$$\sin\theta_e = \frac{V_e}{K_d} = \frac{\delta\omega}{K_o K_d} \tag{12.4.22}$$

Since θ_e ranges between -90° and $+90^\circ$, the hold-in range, $\delta\omega_H$, for this case is found to be

$$\delta\omega_H = \pm K_d K_o \tag{12.4.23}$$

Hence, if there is no frequency divider in the loop (i.e., $N = 1$), the hold-in range is equal to dc loop gain.

Similarly, the hold-in range with a triangular or sawtooth phase detector with linear characteristics can be determined via (12.4.3) to (12.4.5). It may be found as

$$\delta\omega_H = \pm K_d K_o \theta_{e\,\max} \tag{12.4.24}$$

Note that $\theta_{e\,\max}$ is $\pi/2$ for triangular and π for sawtooth phase detectors.

Example 12.5 Free-running frequency and the gain factor of a voltage-controlled oscillator are given as 100 MHz and 100 kHz/V, respectively. It is being used in a PLL along with a sinusoidal phase detector with a maximum output of 2 V at

$\theta_e = \pi/2$ radians. If there are no amplifiers or frequency divider in the loop, what is its hold-in range?

SOLUTION From (12.4.2),

$$K_d = \frac{V_e}{\sin \theta_e} = 2$$

Similarly, from the voltage-controlled oscillator characteristics,

$$K_o = \frac{\delta\omega}{V_d} = 2\pi \times 100 \times 1000 \,\text{rad/s} \cdot \text{V}$$

Hence,

$$\delta\omega_H = \pm 2 \times 2\pi \times 10^5 \,\text{rad/s}$$

or

$$\delta f_H = \pm 200 \,\text{kHz}$$

Example 12.6 If the phase detector in Example 12.5 is replaced by a triangular type with $V_{e\,\max} = A = 2\,\text{V}$ at $\theta_{e\,\max} = \pi/2$ radians, find the new holding range.

SOLUTION From (12.4.5),

$$K_d = \frac{2}{\pi/2} = \frac{4}{\pi}$$

Since the voltage-controlled oscillator is still the same, $K_o = 2\pi \times 10^5$ rad/s ·V. Hence, the new hold-in range will be given as follows:

$$\delta\omega_H = \frac{4}{\pi} \times 2\pi \times 10^5 \times \frac{\pi}{2} = 4\pi \times 10^5 \,\text{rad/s}$$

or

$$\delta f_H = 200 \,\text{kHz}.$$

Acquisition of Lock. Suppose that a signal with frequency f_r is applied to the loop as shown in Figure 12.20. If this frequency is different from the feedback signal frequency f_f, the loop tries to capture or acquire the signal. Two frequencies are equalized if the initial difference is not too large. In general, maximum radian frequency difference $\pm(\omega_r - \omega_f)$ must be smaller than the hold-in range. Acquisition of lock is a nonlinear process whose description is beyond the scope of this chapter. Some general conclusions based on a commonly used lead–lag filter are summarized below.

Lock-in Range. If the frequency difference $|\omega_r - \omega_f|$ is less than a 3-dB bandwidth of closed-loop transfer function $T_2(s)$, the loop locks up without slipping cycles. The maximum lock-in range, $\delta\omega_L$, may be found by the formula

$$\delta\omega_L = \pm\frac{K\tau_2}{\tau_1} \tag{12.4.25}$$

K is the dc loop gain, $\tau_1 = (R_1 + R_2)C_2$, and $\tau_2 = R_2C_2$, as defined earlier. The lock-in range can be expressed in terms of natural frequency ω_n and damping factor ζ of closed-loop transfer function $T_2(s)$ with a lead–lag filter. It is given as follows:

$$\delta\omega_L \approx \pm 2\zeta\omega_n \tag{12.4.26}$$

where

$$\omega_n = \sqrt{\frac{K}{\tau_1}} \tag{12.4.27}$$

and

$$\zeta = \frac{\omega_n}{2}\left(\tau_2 + \frac{1}{K}\right) \tag{12.4.28}$$

Pull-in Range. Suppose that frequency difference $|\omega_r - \omega_f|$ is initially outside the lock-in range but is inside the pull-in range. In this case, the difference frequency signal at the output of the phase detector is nonlinear and contains a dc component that gradually shifts the VCO frequency $f_o(f_f = f_o/N)$ toward f_r until lock-in occurs. For high-gain loops, an approximate formula for the pull-in range is

$$\delta\omega_P \approx \pm\sqrt{2(2\zeta\omega_n K - \omega_n^2)} \tag{12.4.29}$$

It is desirable for the loop to have a large bandwidth that helps in capture initially. However, it increases the transmission of noise. These contradictory conditions may be satisfied in two different ways:

1. Use a larger bandwidth initially but reduce it after the lock is acquired.
2. Use a small bandwidth but sweep the frequency of the voltage-controlled oscillator until the lock is acquired.

Example 12.7 In the PLL shown in Figure 12.32, peak amplitudes of the reference input and VCO output are both 0.75 V. An analog multiplier is being used as a phase detector that has a 2-V output when its reference input, V_{in}, and the oscillator signal, V_{osc}, are both 2 V. The free-running frequency of the VCO is 10 MHz, which reduces to zero when its control voltage V_{cntl} is reduced by 1 V.

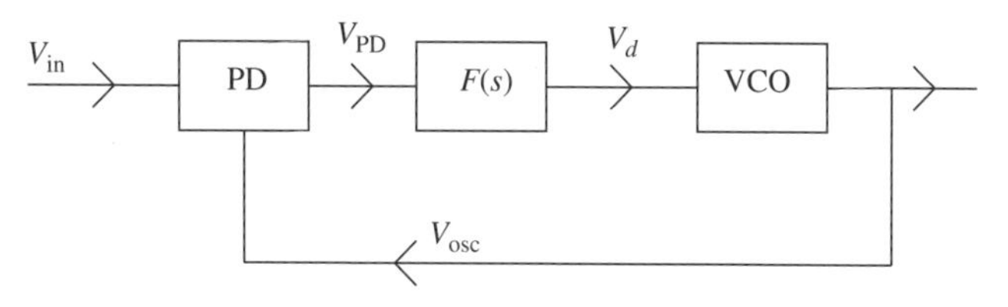

Figure 12.32 PPL for Example 12.7.

Find the phase difference between the input and VCO output when it is in lock with an input frequency of (**a**) 11 MHz and (**b**) 9 MHz.

SOLUTION Since the output of an analog multiplier is proportional to the product of its two inputs, we can write

$$V_{PD} = K_m V_{in} V_{osc}$$

Hence, the proportionality constant K_m is

$$K_m = \frac{2}{2 \times 2} = 0.5\,\text{V}^{-1}$$

For

$$V_{in} = E_{in} \sin \omega t$$

and

$$V_{osc} = E_{osc} \cos(\omega t - \phi_d)$$

$$V_{PD} = K_m E_{in} E_{osc} \sin \omega t \cos(\omega t - \phi_d)$$

$$= 0.5 K_m E_{in} E_{osc} [\sin \phi_d + \sin(2\,\omega t - \phi_d)]$$

Since the output of the phase detector is passed through a low-pass filter before it is applied to the voltage-controlled oscillator, the controlling voltage V_d may be written as

$$V_d = 0.5 K_m E_{in} E_{osc} \sin \phi_d \approx 0.5 K_m E_{in} E_{osc} \phi_d$$

Hence, the transfer function K_d is

$$K_d = \frac{V_d}{\phi_d} = 0.5 K_m E_{in} E_{osc} = 0.5 \times 0.5 \times 0.75^2 = 0.1406\,\text{V/rad}$$

Transfer function K_o of the voltage-controlled oscillator can be found from (12.4.6) as

$$K_o = \frac{\delta\omega}{V_d} = \frac{2\pi \times 10^7}{1} = 2\pi \times 10^7\,\text{rad/s} \cdot \text{V}$$

Since the output frequency of the voltage-controlled oscillator is given as

$$\omega_0 = K_o V_d + \omega_s = \omega_{\text{in}}$$

we can write

$$V_d = \frac{\omega_{\text{in}} - \omega_s}{K_o}$$

Hence,

$$\phi_d = \frac{V_d}{K_d} = \frac{\omega_{\text{in}} - \omega_s}{K_o K_d}$$

(**a**) The input signal frequency in this case is 11 MHz while the free-running frequency of the voltage-controlled oscillator is 10 MHz. Hence,

$$\phi_d = \frac{2\pi(11\,\text{MHz} - 10\,\text{MHz})}{K_o K_d} = 0.7111\,\text{rad} = 40.74^\circ$$

Since there is a constant phase shift of 90° that is always present, actual phase difference between the input signal and the output of VCO will be 90° − 40.74° = 49.26°.

(**b**) Since the input signal frequency in this case is 9 MHz, we find that

$$\phi_d = \frac{2\pi(9\,\text{MHz} - 10\,\text{MHz})}{K_o K_d} = -0.7111\,\text{rad} = -40.74^\circ$$

As before, because of a constant phase shift of 90°, the actual phase difference between the input signal and the output of the VCO will be 90° + 40.74° = 130.74°.

Further Analysis of the PLL

Consider the simplified block diagram of the PLL shown in Figure 12.33. It consists of only a phase detector, a low-pass filter, and a voltage-controlled oscillator. Assume that the reference and the output signals are v_r and v_o, respectively. Hence,

$$v_r(t) = A_r \sin[\omega t + \theta_r(t)] \tag{12.4.30}$$

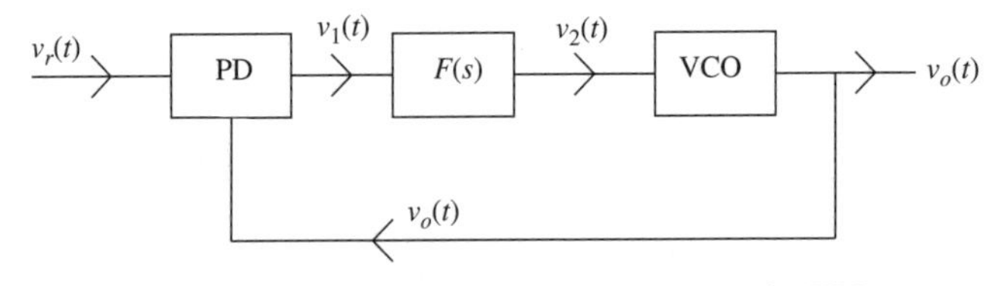

Figure 12.33 Simplified block diagram of a PLL.

and

$$v_o(t) = A_o \cos[\omega t + \theta_o(t)] \tag{12.4.31}$$

Note that output frequency of the VCO varies with $v_2(t)$; therefore,

$$\theta_o(t) = K_o \int^t v_2(\tau) d\tau \tag{12.4.32}$$

K_o is in radians per second per volt, as defined earlier. The output of the phase detector is

$$v_1(t) = K_m v_r(t) v_o(t) = K_m A_r A_o \sin[\omega t + \theta_r(t)] \cos[\omega t + \theta_o(t)]$$

or

$$v_1(t) = \frac{K_m A_r A_o}{2} \{\sin[\theta_r(t) - \theta_o(t)] + \sin[2\,\omega_o t + \theta_r(t) + \theta_o(t)]\} \tag{12.4.33}$$

The low-pass filter stops the sum frequency component (the second term). Hence,

$$v_2(t) = K_d \sin[\theta_e(t)] \otimes f(t) \tag{12.4.34}$$

where $\otimes$ represents the convolution of two terms, $\theta_e(t)$ is equal to $\theta_r(t) - \theta_o(t)$, and K_d is equal to $0.5 K_m A_r A_o$.

The overall equation describing the operation of the PLL may be found as follows:

$$\frac{d\theta_e(t)}{dt} = \frac{d\theta_r(t)}{dt} - \frac{d\theta_o(t)}{dt} = \frac{d\theta_r(t)}{dt} - \frac{d}{dt}\left\{K_d K_o \int^t \sin[\theta_e(\tau)] \otimes f(\tau) d\tau\right\}$$

or

$$\frac{d\theta_e(t)}{dt} = \frac{d\theta_r(t)}{dt} - K_d K_o \sin[\theta_e(t)] \otimes f(t) \tag{12.4.35}$$

For $\sin[\theta_e(t)] \approx \theta_e(t)$, (12.4.35) can be approximated as

$$\frac{d\theta_e(t)}{dt} \approx \frac{d\theta_r(t)}{dt} - K_d K_o \theta_e(t) \otimes f(t)$$

Taking the Laplace transform of this equation with initial conditions as zero, it can be transformed into the frequency domain. Hence,

$$s\theta_e(s) = s\theta_r(s) - K_d K_o \theta_e(s) F(s)$$

or

$$[s + K_d K_o F(s)][\theta_r(s) - \theta_o(s)] = s\theta_r(s)$$

or

$$K_d K_o F(s) \theta_r(s) = [s + K_d K_o F(s)]\theta_o(s)$$

Hence, the closed-loop transfer function $T(s)$ is

$$T(s) = \frac{\theta_o(s)}{\theta_r(s)} = \frac{K_d K_o F(s)}{s + K_d K_o F(s)} \tag{12.4.36}$$

This is identical to (12.4.20) if the frequency division is assumed as unity.

Example 12.8 Analyze the PLL system shown in Figure 12.34. By controlling the frequency divider, its output frequency can be changed from 2 MHz to 3 MHz in steps of 100 kHz. Transfer function of the phase detector, K_d, is 0.5 V/rad. Free-running frequency and a gain factor of VCO are 2.5 MHz and 10^7 rad/s·V, respectively. A voltage gain of the amplifier is 10. Design a passive lead–lag low-pass filter that can be used in the loop.

SOLUTION With f_m as zero (i.e., without the mixer), the output frequency is found from (12.4.1) as

$$f_o = N f_r$$

Since it has a resolution of 100 kHz, its reference frequency must be the same. Next, the range of N is determined from the equation above and the desired frequency band. For f_o to be 2 MHz, N must be equal to 20. Similarly, it is 30 for an output frequency of 3 MHz. Hence,

$$20 \le N \le 30$$

From (12.4.6),

$$V_d = \frac{\omega_o - \omega_f}{K_o} = \frac{2\pi}{10^7}(2 \times 10^6 - 2.5 \times 10^6) = -\frac{\pi}{10} \quad \text{V}$$

Going back around the loop, input to the amplifier (or output of the phase detector) can be found as

$$V_e = \frac{V_d}{K_a} = -\frac{\pi}{100} \quad \text{V}$$

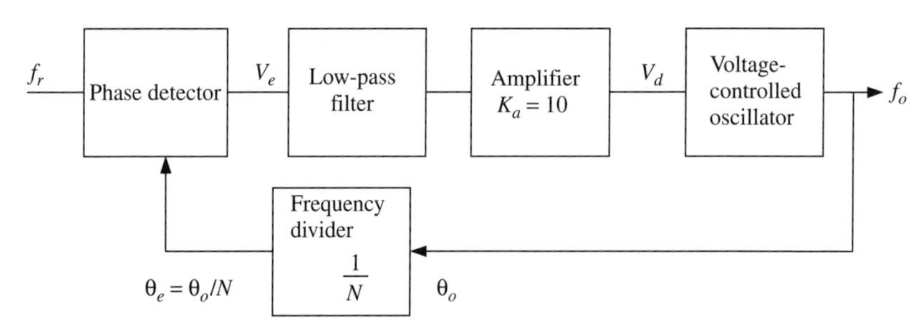

Figure 12.34 PPL for Example 12.8.

and the output of the frequency divider may be found from (12.4.5) as follows:

$$\theta_e = \frac{V_e}{K_d} = -\frac{\pi}{50} \qquad \text{rad}$$

Therefore, θ_o at the output frequency of 2 MHz is given as follows:

$$\theta_o = -\frac{2}{5}\pi = -1.2566\,\text{rad}$$

The dc loop gain K can be found via (12.4.19) as

$$K = \frac{K_d K_a K_o}{N} = \frac{0.5 \times 10 \times 10^7}{20} = 2.5 \times 10^6$$

The pull-in range $\delta\omega_p$ is given by (12.4.29) assuming that $K \gg \omega_n$.
For $\omega_n = 10^4$ rad/s and $\zeta = 0.8$, the pull-in range is

$$\delta\omega \approx 2.8249 \times 10^5\,\text{rad/s}$$

or

$$\delta f_p = 4.5 \times 10^4\,\text{Hz} = 45\,\text{kHz}$$

A lag–lead filter can be designed as follows. From (12.4.27),

$$\tau_1 = \frac{K}{\omega_n^2} = \frac{2.5 \times 10^6}{10^8} = 0.025\,\text{s} = (R_1 + R_2)C_2$$

and from (12.4.28),

$$\zeta = \frac{\omega_n}{2}\left(\tau_2 + \frac{1}{K}\right) \Rightarrow \tau_2 = \frac{2\zeta}{\omega_n} - \frac{1}{K} = \frac{2 \times 0.8}{10^4} - \frac{1}{2.5 \times 10^6}$$
$$= 1.596 \times 10^{-4}\text{s} = R_2 C_2$$

For $C_2 = 0.5\ \mu\text{F}$,

$$R_2 = \frac{1.596 \times 10^{-4}}{0.5 \times 10^{-6}} = 319.2\,\Omega$$

and

$$R_1 = \frac{\tau_1}{C_2} - R_2 = \frac{0.025}{0.5 \times 10^{-6}} - 319.2 = 50{,}000 - 319.2$$
$$= 49{,}680.8\,\Omega \approx 50\,\text{k}\Omega$$

12.5 FREQUENCY SYNTHESIZERS

The frequency of a reference oscillator can be multiplied and its harmonics can be generated with the help of a nonlinear device. The desired frequency component can then be filtered out. This type of frequency generation is called *direct synthesis*. However, it has several limitations, including constraints on the filter design. An alternative approach uses a PLL (analog or digital) to synthesize the desired signal. This is known as *indirect synthesis*. Another technique, known as *direct digital synthesis* (DDS), uses a digital computer and a digital-to-analog converter for generating the desired signal. In this section we consider indirect frequency synthesis. Mixers and frequency multipliers are described in Chapter 13.

Example 12.8 illustrated the basic principle of an indirect frequency synthesizer. A simplified PLL frequency synthesizer is illustrated in Figure 12.35. As discussed in Section 11.4, the feedback signal frequency f_d is equal to the reference frequency f_r if the loop is in the locked state. Since the frequency divider divides the output frequency by N, the output frequency f_o is equal to Nf_r. If N is an integer, the frequency step is equal to reference frequency f_r. For a fine resolution, the reference frequency should be very low. On the other hand, lower f_r increases the switching time. As a rule of thumb, the product of switching time t_s and the reference frequency f_r is equal to 25. Therefore, the switching time is 25 s for the reference frequency of 1 Hz. Since a shorter switching time is desired, it limits the frequency resolution of this system.

As illustrated in Figure 12.36, one way to improve the frequency resolution while keeping the switching time short (the reference frequency high) is to add

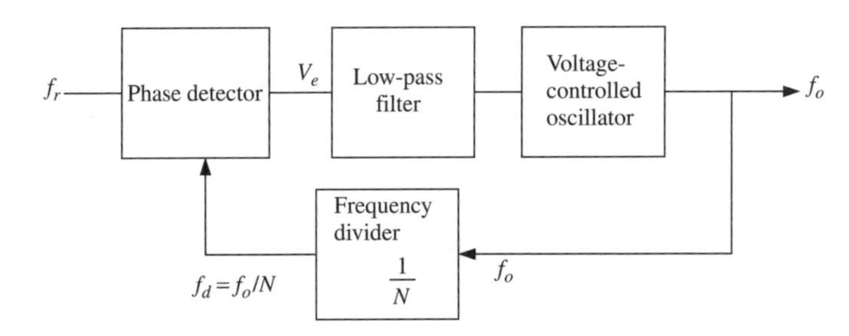

Figure 12.35 PLL frequency synthesizer.

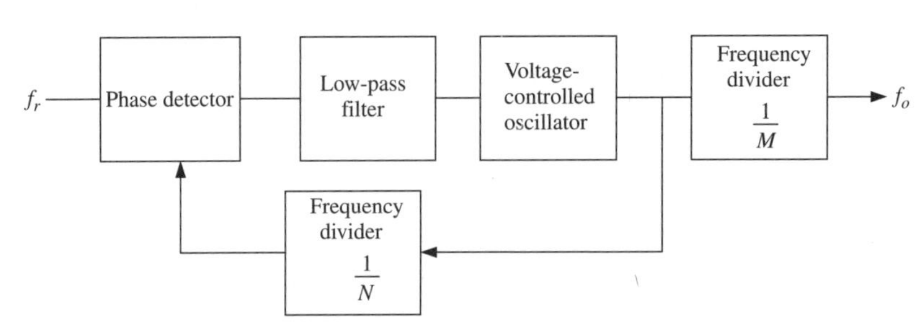

Figure 12.36 PLL followed by a frequency divider.

another frequency divider at its output. Hence, the output frequency in this system will be given as

$$f_o = N\frac{f_r}{M} \tag{12.5.1}$$

Thus, the frequency resolution in this system is f_r/M, which shows an improvement over the previous arrangement. However, a close analysis of this system indicates that the signal frequency in its phase-locked-loop section may become excessively large. This is explained further via the following example.

Example 12.9 Design a PLL frequency synthesizer that covers the frequency range from 100 to 110 MHz with a resolution of 1 kHz.

SOLUTION With the single loop of Figure 12.35, the reference frequency must be equal to the resolution required. Hence,

$$f_r = 1\,\text{kHz}$$

Since the output frequency $f_o = Nf_r$, the range of N is

$$10^5 \leq N \leq 11 \times 10^4$$

As explained earlier, the switching time in this case is on the order of 25 ms. If this switching time is not acceptable and we want to keep it below 250 μs, an arrangement similar to that illustrated in Figure 12.36 may be used. A reference frequency of 100 kHz provides the switching time of 250 μs. In this case, setting M of the output frequency can be determined as

$$M = \frac{f_r}{\text{frequency resolution}} = \frac{100\,\text{kHz}}{1\,\text{kHz}} = 100$$

To cover the desired frequency range, the voltage-controlled oscillator must generate frequencies from 10 to 11 GHz. This is because the frequency divider divides output of the voltage-controlled oscillator by M before the signal appearing at the output.

The range of N can be determined after dividing the frequency of the voltage-controlled oscillator by the reference frequency. Hence,

$$10^5 \leq N \leq 11 \times 10^4$$

Thus, this may not be a good technique to get better frequency resolution. However, this concept is extended in practice where multiple loops are used to obtain fine resolution. It is illustrated through the following example.

Example 12.10 Design a PLL frequency synthesizer for the frequency range 1 to 50 MHz with increments of 1 kHz.

SOLUTION Note that if a single-loop synthesizer is used in this case, its reference frequency should be of 1 kHz. That makes its switching time on the order of 0.025 s and the setting of frequency-divider ranges as

$$1 \times 10^3 \leq N \leq 50 \times 10^3$$

An alternative design that uses three phase-locked loops is illustrated in Figure 12.37. Loops A and B use a higher-reference frequency (say, 100 kHz), while the output of A (f_A) works as the reference for loop C after it is divided by M. The following relation holds when the loops are in the locked state:

$$f_A = f_o - f_B$$

Therefore, the output frequency f_o is equal to the sum of other two, f_A and f_B.

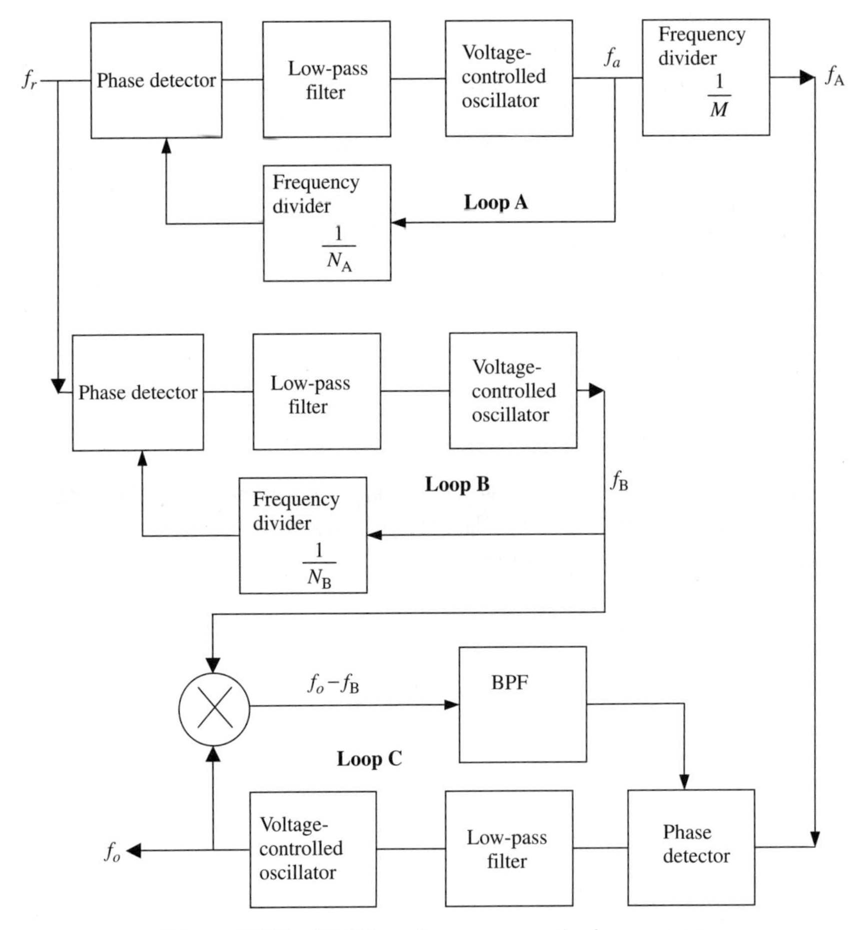

Figure 12.37 Multiloop frequency synthesizer system.

If a mixer is used to combine f_A and f_B directly, its sum and difference frequency components may be too close to separate the two via a bandpass filter. However, the present technique of using a phase-locked loop C for this purpose accomplishes the desired separation. Since the reference frequency is 100 kHz, the output frequency of loop A, f_a, can be varied in steps of 100 kHz, and 1 kHz resolution in the output can be achieved with M as 100. Hence,

$$f_a = 10^5 N_A$$

and

$$f_A = \frac{f_a}{100} = 10^3 \times N_A$$

Thus, f_A varies in 1-kHz steps. Loop A can generate increments of 1 and 10 kHz in the output frequency. Similarly, loop B produces 0.1- and 1-MHz steps in the output frequency. Reference frequency f_A to loop C should be selected in such a way that it keeps its switching time short. If, for example, f_A is equal to 1 kHz, the switching time of loop C is on the order of 0.025 s. Since it is the longest time among the three loops, it sets the overall response time of the synthesizer. To reduce the response time of loop C, f_A may be increased to 500 kHz so that

$$500\,\text{kHz} \le f_A \le 599\,\text{kHz}$$

Therefore, the output frequency of loop A (i.e., before dividing by 100) ranges as follows:

$$5 \times 10^7\,\text{Hz} \le f_a \le 5.99 \times 10^7\,\text{Hz}$$

The range of the frequency divider setting, N_A, can be determined easily as f_a/f_r. Hence, $500 \le N_A \le 599$. Since $f_B = f_o - f_A$, f_B is reduced by 500 kHz,

$$(1 - 0.5)\text{MHz} \le f_B \le (50 - 0.5)\text{MHz}$$

or

$$0.5\,\text{MHz} \le f_B \le 49.5\,\text{MHz}$$

Hence, the frequency divider setting, N_B, in loop B ranges as follows: $5 \le N_B \le 495$. Loops A and B use the same reference frequency and therefore have the same response time, on the order of 250 μs. On the other hand, loop C has a response time on the order of 50 μs. Thus, the overall response time of the frequency synthesizer is approximately 250 μs. At the same time, we achieved a frequency resolution of 1 kHz. This idea can be extended to reduce the frequency resolution even below 1 kHz while keeping the overall response time short by using additional loops.

12.6 ONE-PORT NEGATIVE RESISTANCE OSCILLATORS

Properly biased Gunn and IMPATT (impact ionization avalanche transit time) diodes exhibit negative resistance characteristics that can be utilized in conjunction with an external resonant circuit to design a solid-state microwave oscillator. The Gunn diode is a transferred-electron device that uses a bulk semiconductor (usually, GaAs or InP) as opposed to a *p-n* junction. Its dc-to-RF conversion efficiency is generally less than 10%.

The IMPATT diode uses a reverse-biased *p-n* junction to generate microwave power. The diode material is usually silicon or GaAs, and it is operated with a relatively high voltage, on the order of 70 to 100 V, to achieve a reverse-biased avalanche breakdown current. When coupled with a high-Q resonator and biased at an appropriate operating point, the diode exhibits a negative resistance effect and sustained oscillations are obtained. Its output power is much higher (on the order of tens or hundreds of watts) than that produced by a Gunn diode. On the other hand, Gunn diodes produce relatively low FM noise.

Figure 12.38 shows the RF equivalent circuits for a one-port negative resistance oscillator where Z_G is the input impedance (Y_G is the corresponding input admittance) of the active device (i.e., a biased diode). Figure 12.38(*b*) is an equivalent parallel arrangement of the circuit shown in Figure 12.38(*a*). In general, the impedance Z_G (or admittance Y_G) is current (or voltage) and frequency dependent.

Applying Kirchhoff's loop equation in the circuit shown in Figure 12.38(*a*), we get

$$(Z_G + Z_L)I = 0$$

If oscillations are occurring, current I in the loop is nonzero. In such a situation, the following conditions must be satisfied:

$$R_G + R_L = 0 \tag{12.6.1}$$

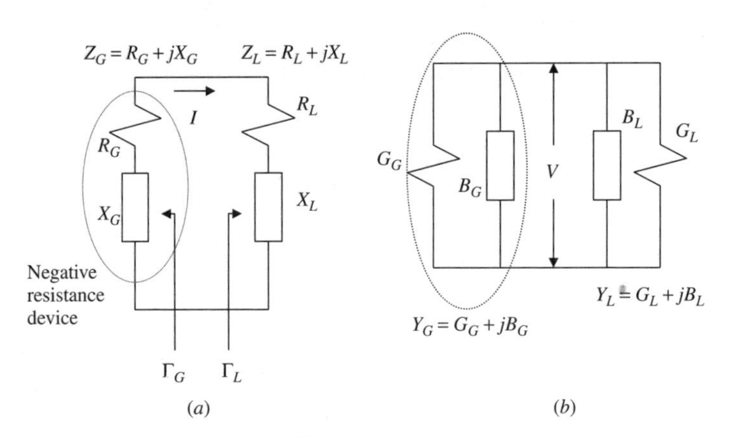

Figure 12.38 Series (*a*) and parallel (*b*) equivalent circuits for one-port negative resistance microwave oscillators.

and

$$X_G + X_L = 0 \qquad (12.6.2)$$

Similarly, applying Kirchhoff's node equation in the circuit shown in Figure 12.38 (*b*), we get

$$(Y_G + Y_L)V = 0$$

Again, voltage V across the circuit is not equal to zero if oscillations are occurring. Therefore, this condition is satisfied only if $Y_G + Y_L = 0$. Hence,

$$G_G + G_L = 0 \qquad (12.6.3)$$

and

$$B_G + B_L = 0 \qquad (12.6.4)$$

Since R_L (or G_L) has a positive value that is greater than zero, the conditions above require a negative R_G (or $-G_G$). Negative resistance implies a source because a positive resistance dissipates the energy.

Note that the impedance Z_G is nonlinearly related with the current (or voltage) and the frequency. Conditions (12.6.1) to (12.6.4) represent the steady-state behavior of the circuit. Initially, it is necessary for the overall circuit to be unstable at a certain frequency (i.e., the sum of R_G and R_L is a negative number). Any transient excitation or noise then causes the oscillation to build up at a frequency ω. As the current (or voltage) increases, R_G (or G_G) becomes less negative until it reaches a value such that (12.6.1) to (12.6.4) are satisfied.

For the startup of oscillation, negative resistance of the active device in a series circuit must exceed the load resistance by about 20% (i.e., $R_G \approx -1.2R_L$). Similarly, G_G is approximately $-1.2G_L$ to start the oscillations in the parallel circuit. For the stability of an oscillator, high-Q resonant circuits such as cavities and dielectric resonators are used.

Example 12.11 A 6-GHz oscillator is to be designed using a negative-resistance diode with its reflection coefficient Γ_G as $1.25 \angle 40^\circ$ (measured with $Z_0 = 50\,\Omega$ as reference). Determine its load impedance.

SOLUTION From

$$\Gamma_G = \frac{\overline{Z}_G - 1}{\overline{Z}_G + 1}$$

$$\overline{Z}_G = \frac{1 + \Gamma_G}{1 - \Gamma_G} = \frac{1 + 1.25 \angle 40^\circ}{1 - 1.25 \angle 40^\circ} = \frac{1.9576 + j0.8035}{0.0424 - j0.8035} = \frac{2.1161 \angle 22.32^\circ}{0.8046 \angle -86.98^\circ}$$

or

$$\overline{Z}_G = 2.63 \angle 109.3^\circ = -0.8693 + j2.4822$$

If a given reflection coefficient has magnitude greater than unity, we can still use a Smith chart to find the corresponding impedance provided that we follow the procedure described in Section 11.3. Thus, the impedance point can be identified on a Smith chart, as shown in Figure 12.39. The corresponding value is found to be the same as obtained above through the formula. Hence,

$$Z_G = R_G + jX_G = -43.5 + j124.11\ \Omega$$

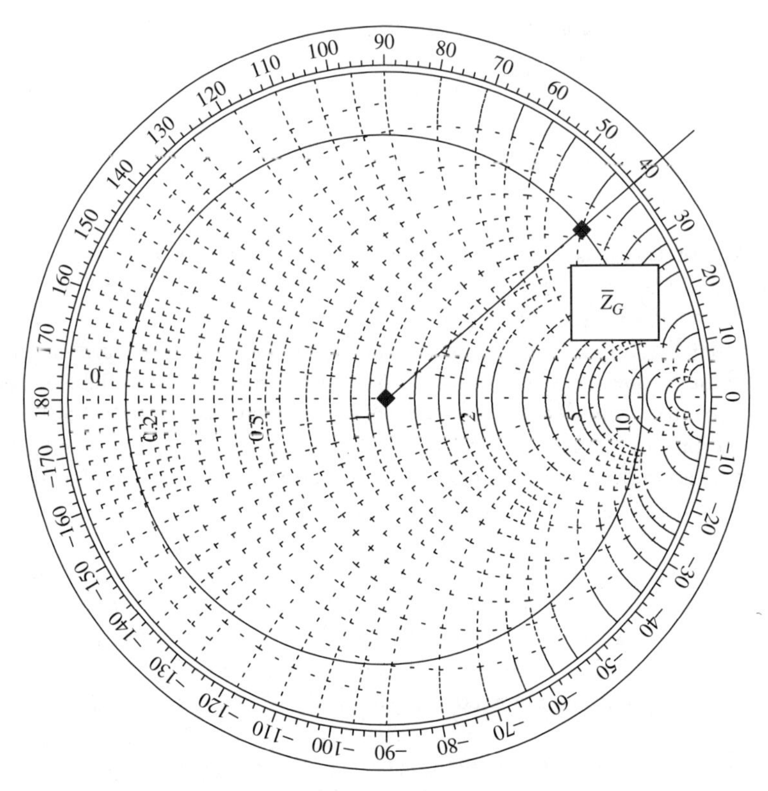

Figure 12.39 Graphical method of determining the input impedance of the diode.

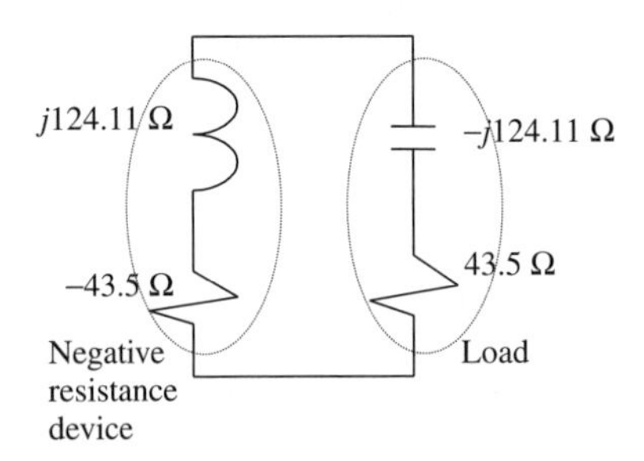

Figure 12.40 Equivalent circuit of Example 12.11.

Therefore, the load impedance is given as

$$Z_L = 43.5 - j124.11\,\Omega$$

The corresponding equivalent circuit is illustrated in Figure 12.40.

12.7 MICROWAVE TRANSISTOR OSCILLATORS

Several oscillator circuits using transistors were considered in Section 12.1 along with the basic concepts. In this section we present a design procedure that is based on the S-parameters of the transistor. The RF circuit arrangement of a microwave transistor oscillator is shown in Figure 12.41. A field-effect transistor shown as the device in this figure is in common-source configuration, although any other terminal can be used in its place, provided that the corresponding S-parameters are known. Further, a BJT can also be used in its place. In other words, the design procedure described here is general and applies to any transistor circuit configuration as long as its S-parameters are known.

Unlike the amplifier circuit, the transistor for an oscillator design must be unstable. Therefore, the opposite of conditions (11.1.16) and (11.1.18) must hold in this case. If the stability parameter k is not less than unity (or $|\Delta|$ is not larger than unity), an external positive feedback can be used to make the device unstable. Equations (12.6.1) to (12.6.4) can be recast as follows:

$$\Gamma_{\text{in}}\Gamma_G = 1 \tag{12.7.1}$$

or

$$\Gamma_{\text{out}}\Gamma_L = 1 \tag{12.7.2}$$

These two conditions state that passive terminations Z_G and Z_L must be added for resonating the input and output ports of the active device at the frequency

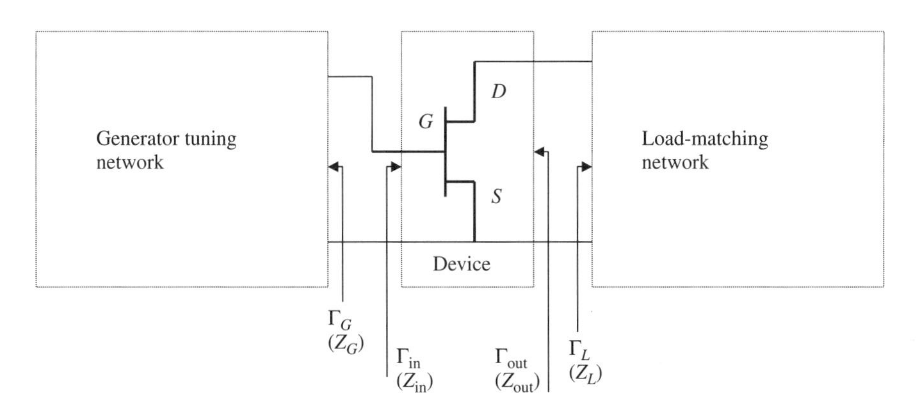

Figure 12.41 Microwave transistor oscillator circuit.

of oscillation. These two conditions are complementary to each other [i.e., if condition (12.7.1) is satisfied, (12.7.2) is also satisfied, and vice versa]. It can be proved as follows.

From the expression for input reflection coefficient obtained in Example 10.6, we can write

$$\Gamma_{\text{in}} = S_{11} + \frac{S_{12}S_{21}\Gamma_L}{1 - S_{22}\Gamma_L} = \frac{S_{11} - \Delta\Gamma_L}{1 - S_{22}\Gamma_L} \Rightarrow \frac{1}{\Gamma_{\text{in}}} = \frac{1 - S_{22}\Gamma_L}{S_{11} - \Delta\Gamma_L} \tag{12.7.3}$$

where $\Delta = S_{11}S_{22} - S_{12}S_{21}$, as defined in Chapter 11. If (12.7.1) is satisfied, then

$$\Gamma_G = \frac{1}{\Gamma_{\text{in}}} = \frac{1 - S_{22}\Gamma_L}{S_{11} - \Delta\Gamma_L} \Rightarrow S_{11}\Gamma_G - \Delta\Gamma_L\Gamma_G = 1 - S_{22}\Gamma_L \Rightarrow$$

$$\Gamma_L = \frac{1 - S_{11}\Gamma_G}{S_{22} - \Delta\Gamma_G} \tag{12.7.4}$$

and the output reflection obtained in Example 10.7 can be rearranged as

$$\Gamma_{\text{out}} = S_{22} + \frac{S_{12}S_{21}\Gamma_G}{1 - S_{11}\Gamma_G} = \frac{S_{22} - \Delta\Gamma_G}{1 - S_{11}\Gamma_G} \tag{12.7.5}$$

From (12.7.4) and (12.7.5),

$$\Gamma_{\text{out}}\Gamma_L = 1$$

Similarly, it can be proved that if condition (12.7.2) is satisfied, (12.7.1) is also satisfied.

After a transistor configuration is selected, the output stability circle can be drawn on the Γ_L plane, and Γ_L is selected to produce a large value of negative resistance at the input of the transistor. Then, Γ_G (and hence Z_G) can be chosen to match the input impedance Z_{in}. If small-signal S-parameters are used in such design, a fairly large negative value of R_{in} must be selected initially so that the sum of R_G and R_{in} is a negative number. As oscillations build up, R_{in} becomes less and less negative. If its initial value is less negative, the oscillation may cease when the power buildup cause R_{in} to increase to the point where $R_G + R_{\text{in}}$ is greater than zero. In practice, a value of $R_G = -R_{\text{in}}/3$ is typically used. To resonate the circuit, the reactive part of Z_G is selected as $-X_{\text{in}}$.

Example 12.12 Using a GaAs FET in the common-gate configuration, design an oscillator at 4 GHz. To provide positive feedback, a 4-nH inductor is connected in series with its gate. The S-parameters of the resulting circuit are ($Z_0 = 50\,\Omega$) $S_{11} = 2.18\angle -35^\circ$, $S_{21} = 2.75\angle 96^\circ$, $S_{12} = 1.26\angle 18^\circ$, and $S_{22} = 0.52\angle 155^\circ$.

SOLUTION From (11.1.16) and (11.1.18),

$$\Delta = S_{11}S_{22} - S_{21}S_{12} = 2.34\ \angle -68.9^\circ$$

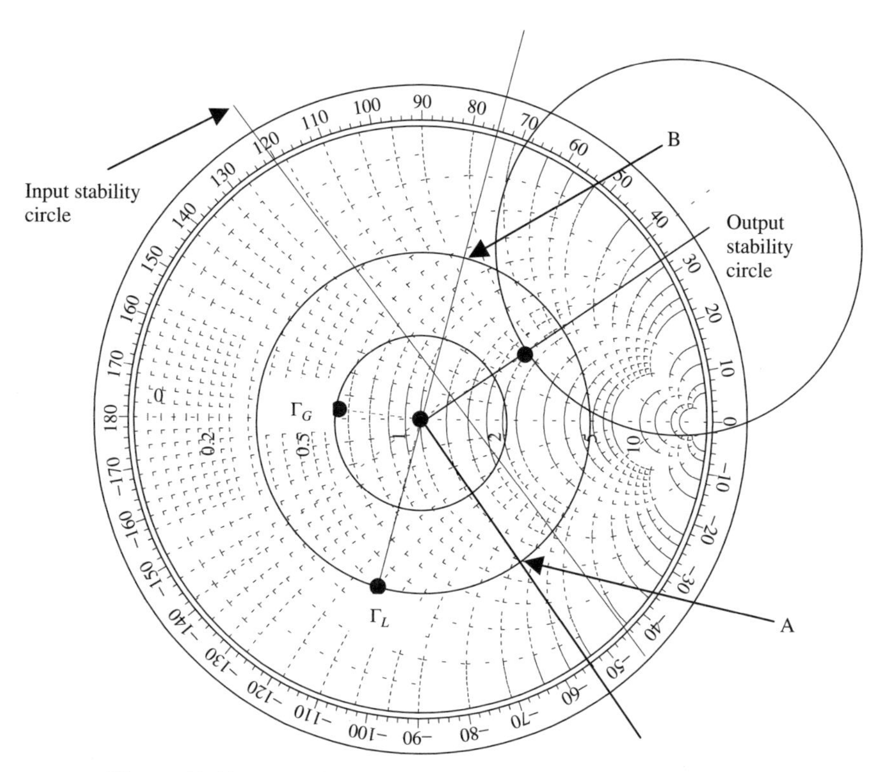

Figure 12.42 Input and output stability circles for Example 12.12.

and

$$k = \frac{1 - |S_{11}|^2 - |S_{22}|^2 + |\Delta|^2}{2|S_{12}S_{21}|} = 0.21$$

Therefore, the transistor and inductor combination is potentially unstable, and it can be used for an oscillator design (see Figure 12.42).

From (11.1.22) and (11.1.23), the input stability circle is determined as follows:

$$\text{center } C_s = \frac{(S_{11} - \Delta S_{22}^*)^*}{|S_{11}|^2 - |\Delta|^2} = 4.67 \angle -141.8^\circ$$

and

$$\text{radius } r_s = \left| \frac{S_{12}S_{21}}{|S_{11}|^2 - |\Delta|^2} \right| = 4.77$$

Further, it may be found that the inside of this circle is stable.

Similarly, the output stability circle is determined from (11.1.20) and (11.1.21) as follows:

$$\text{center } C_L = \frac{(S_{22} - \Delta S_{11}^*)^*}{|S_{22}|^2 - |\Delta|^2} = 1.08 \angle 33.1^\circ$$

and

$$\text{radius } r_L = \left| \frac{S_{12}S_{21}}{|S_{22}|^2 - |\Delta|^2} \right| = 0.67$$

Again, the inside of this circle is determined as the stable region.

We want to select Z_L in the unstable region for which ρ_{in} is large. Thus, after several trials, we selected $\Gamma_L = 0.59 \angle{-104^\circ}$. A single-stub network can be designed for this Γ_L. A 0.096λ-long shunt stub with a short circuit at its other end transforms 50 Ω to the admittance at point A, and then a 0.321λ-long line takes it to point B (the desired admittance).

Next, we calculate Γ_{in} for this Γ_L as follows:

$$\Gamma_{\text{in}} = S_{11} + \frac{S_{12}S_{21}\Gamma_L}{1 - S_{22}\Gamma_L} = 3.9644 \angle - 2.4252^\circ$$

Therefore,

$$Z_{\text{in}} = 50 \times \frac{1 + \Gamma_{\text{in}}}{1 - \Gamma_{\text{in}}} = 50(-1.6733 - j0.03815) = -83.67 - j1.91\,\Omega$$

$$Z_G = \frac{83.67}{3} + j1.91 \approx 27.9 + j1.91\,\Omega \Rightarrow \Gamma_G = \frac{Z_G - 50}{Z_G + 1} = 0.2847\angle 173.6^\circ$$

$$\text{VSWR} = (1 + 0.2847)/(1 - 0.2847) = 1.796$$

Hence, Z_G can be synthesized via a 27.9-Ω resistance in series with a 0.01λ-long line. The oscillator circuit designed is illustrated in Figure 12.43.

Example 12.13 Design a microwave oscillator at 2.75 GHz using a BJT in its common-base configuration. The S-parameters of the transistor are $S_{11} = 0.9\angle 150^\circ$, $S_{21} = 1.7\angle{-80^\circ}$, $S_{12} = 0.07\angle 120^\circ$, and $S_{22} = 1.08\angle{-56^\circ}$.

SOLUTION Note that this transistor is potentially unstable ($k = -0.6447$). This unstable region on a Smith chart can be increased further using an external feedback. A 1.45-nH inductor in its base terminal, shown in Figure 12.44, is found to provide optimum response. Before we proceed with the design, we

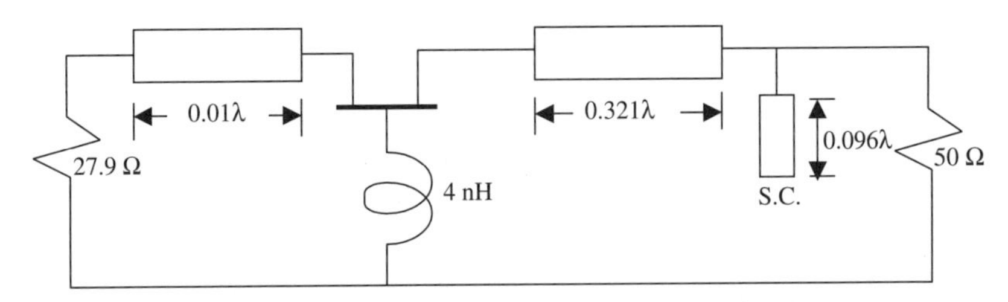

Figure 12.43 RF circuit of the GaAs FET oscillator for Example 12.12.

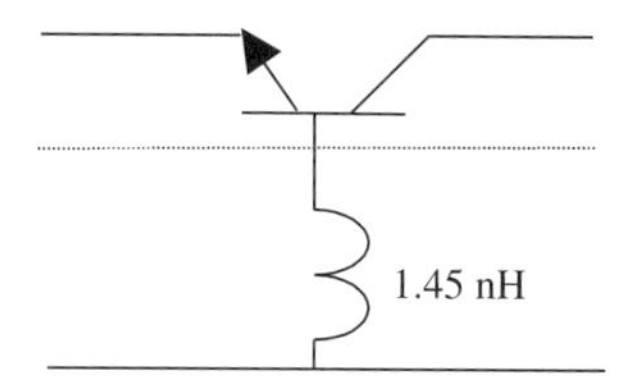

Figure 12.44 Transistor setup for Example 12.13.

need to determine the S-parameters of this transistor–inductor combination. Since these are connected in series, its impedance parameters can be found easily by adding impedance matrices of the two. The impedance matrix of the transistor can be determined from the corresponding S-parameters and the formulas tabulated in Table 8.4. The calculation proceeds as follows:

1. *Determination of the normalized Z-parameters of the BJT*

$$\Delta_1 = (1 - S_{11})(1 - S_{22}) - S_{12}S_{21} = 1.0164 + j1.3386 = 1.6807\angle 52.79^\circ$$

$$\overline{Z}_{11} = \frac{(1 + S_{11})(1 - S_{22}) + S_{21}S_{12}}{\Delta_1} = 0.3005\angle 63.58^\circ = 0.1337 + j0.2691$$

$$\overline{Z}_{12} = \frac{2S_{12}}{\Delta_1} = 0.0833\angle 67.21^\circ = 0.0323 + j0.0768$$

$$\overline{Z}_{21} = \frac{2S_{21}}{\Delta_1} = 2.023\angle - 132.79^\circ = -1.3743 - j1.4846$$

and

$$\overline{Z}_{22} = \frac{(1 - S_{11})(1 + S_{22}) + S_{21}S_{12}}{\Delta_1} = 2.0154\angle - 94.15^\circ = -0.1459 - j2.0101$$

2. *Z- matrix of the inductor section*. The impedance of the inductor,

$$Z_{\text{inductor}} = j2\pi \times 2.75 \times 10^9 \times 1.45 \times 10^{-9}\,\Omega = j25.0542\,\Omega$$

or

$$\overline{Z}_{\text{inductor}} = \frac{j25.0542}{50} = j0.5011$$

Therefore, the Z-matrix of the inductor section will be given as

$$[\overline{Z}_{\text{inductor}}] = \begin{bmatrix} j0.5011 & j0.5011 \\ j0.5011 & j0.5011 \end{bmatrix}$$

3. *Z-impedance of the BJT and inductor combination*

$$\begin{bmatrix} \overline{Z}'_{11} & \overline{Z}'_{12} \\ \overline{Z}'_{21} & \overline{Z}'_{22} \end{bmatrix} = \begin{bmatrix} \overline{Z}_{11} & \overline{Z}_{12} \\ \overline{Z}_{21} & \overline{Z}_{22} \end{bmatrix} + [\overline{Z}_{\text{inductor}}]$$

$$= \begin{bmatrix} 0.1337 + j0.7702 & 0.0323 + j0.5779 \\ -1.3743 - j0.9067 & -0.1459 - j1.509 \end{bmatrix}$$

4. *Determination of S-matrix of the BJT and inductor combination*. The S-parameters of this circuit can be determined easily from its Z-parameters using the formulas given in Table 8.4. Various steps of this calculation follow:

$$\Delta = (\overline{Z}_{11} + 1)(\overline{Z}_{22} + 1) - \overline{Z}'_{12}\overline{Z}'_{21} = 1.651 - j0.2294 = 1.6669\angle - 7.91^\circ$$

$$S'_{12} = \frac{2\overline{Z}'_{12}}{\Delta} = 0.6945\angle 94.71^\circ$$

$$S'_{21} = \frac{2\overline{Z}'_{21}}{\Delta} = 1.9755\angle - 138.68^\circ$$

$$S'_{11} = \frac{(\overline{Z}'_{11} - 1)(\overline{Z}'_{22} + 1) - \overline{Z}'_{12}\overline{Z}'_{21}}{\Delta} = 1.6733\angle 99.1^\circ$$

and

$$S'_{22} = \frac{(\overline{Z}'_{11} + 1)(\overline{Z}'_{22} - 1) - \overline{Z}'_{12}\overline{Z}'_{21}}{\Delta} = 1.13\angle - 101.3^\circ$$

Now from (11.1.16) and (11.1.18),

$$\Delta = S_{11}S_{22} - S_{21}S_{12} = 1.2601\angle 44.29^\circ$$

and

$$k = \frac{1 - |S_{11}|^2 - |S_{22}|^2 + |\Delta|^2}{2|S_{12}S_{21}|} = -0.5426$$

Therefore, the transistor and inductor combination is potentially unstable, and it can be used for an oscillator design. It may be interesting to know that without the feedback inductor (just for the transistor), Δ and k are $0.9072\ \angle\ 100.09^\circ$ and -0.6447, respectively. From (11.1.22) and (11.1.23), the input stability circle is determined as follows:

$$\text{center } C_s = \frac{(S_{11} - \Delta S^*_{22})^*}{|S_{11}|^2 - |\Delta|^2} = 1.0262\angle - 42.96^\circ$$

and

$$\text{radius } r_s = \left| \frac{S_{12}S_{21}}{|S_{11}|^2 - |\Delta|^2} \right| = 1.132$$

Further, it may be found that the outside of this circle is stable.

Similarly, the output stability circle of the circuit is determined from (11.1.20) and (11.1.21) as follows:

$$\text{center } C_L = \frac{(S_{22} - \Delta S_{11}^*)^*}{|S_{22}|^2 - |\Delta|^2} = 5.0241\angle 23.18^\circ$$

and

$$\text{radius } r_L = \left| \frac{S_{12}S_{21}}{|S_{22}|^2 - |\Delta|^2} \right| = 4.4107$$

This time, the inside of this circle is found to be stable.

There are two output stability circles illustrated in Fig. 12.45. Circle 1 encloses the unstable region for the BJT alone (i.e., the feedback inductor is not used). This

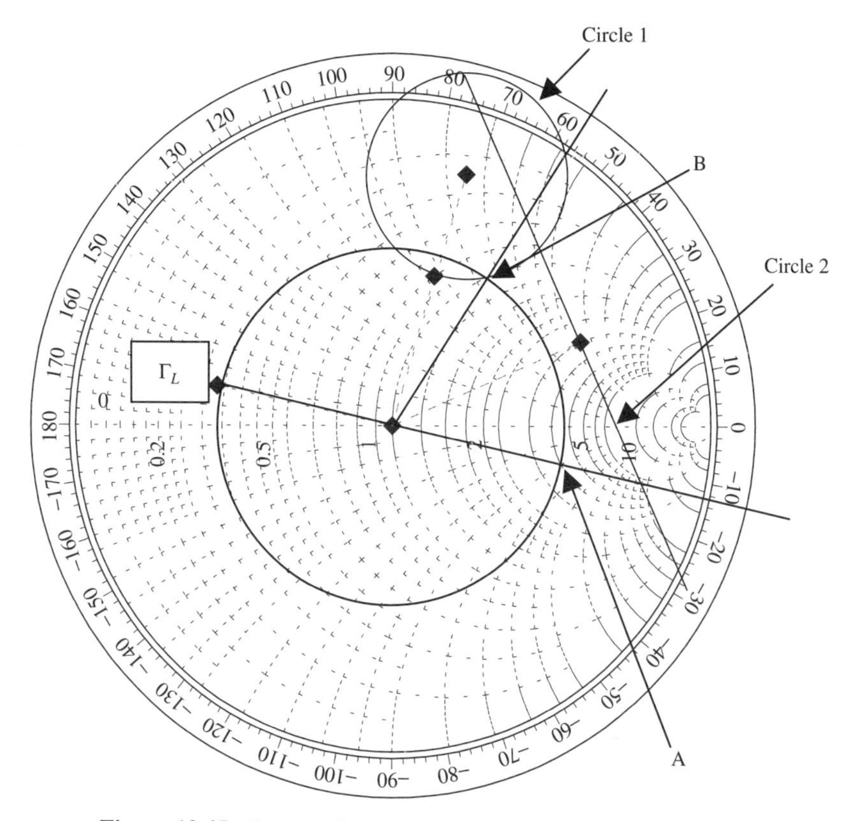

Figure 12.45 Input and output stability circles for Example 12.13.

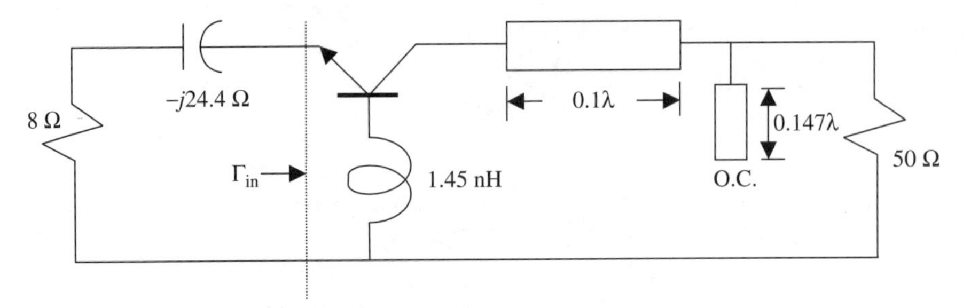

Figure 12.46 RF circuit of the BJT oscillator for Example 12.13.

means that the load impedance of the oscillator circle must be selected inside this circle. On the other hand, the range of impedance increases significantly with the feedback inductor. It is illustrated by circle 2, which encloses the stable region.

We want to select Z_L in the unstable region for which ρ_{in} is large. Thus, after several trials, we selected $\Gamma_L = 0.5689\angle 167.8^\circ$. A single-stub network can be designed for this Γ_L. A shunt stub of 0.147λ with an open circuit at its other end transforms 50 Ω to point B, and then a 0.1λ-long line provides the desired admittance (point A).

Next, we calculate Γ_{in} for this Γ_L as follows (see Figure 12.46):

$$\Gamma_{in} = S_{11} + \frac{S_{12}S_{21}\Gamma_L}{1 - S_{22}\Gamma_L} = 2.1726\angle 118.8375^\circ$$

Therefore,

$$Z_{in} = 50 \times \frac{1 + \Gamma_{in}}{1 - \Gamma_{in}} = -23.8 + j24.35\ \Omega$$

and

$$Z_G = \frac{23.8}{3} - j24.35 \approx 8 - j24.4\ \Omega$$

Three-Port *S*-Parameter Description of the Transistor

Transistor parameters are generally specified as a two-port device while keeping one of its terminals as a reference point (common emitter, common source, etc.). However, it may be advantageous in certain design instances to use a different terminal as a common electrode. Further, it may be prudent sometimes to add an external feedback that enhances the desired characteristics of a device. A three-port description of the transistor facilitates such design procedures. This section describes a computation method to determine three-port *S*-parameters of the transistor (or any other three-terminal circuit) from its two-port description. These parameters are used subsequently to design the feedback network for the device as well as for changing its common terminal.

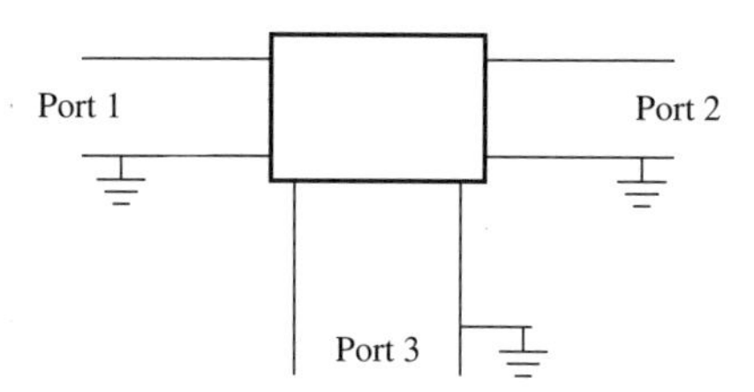

Figure 12.47 Three-port network.

Consider a three-port network as illustrated in Figure 12.47. Incident and reflected waves at the three ports are related through its S-parameters as follows.

$$\begin{bmatrix} b'_1 \\ b'_2 \\ b'_3 \end{bmatrix} = \begin{bmatrix} S'_{11} & S'_{12} & S'_{13} \\ S'_{21} & S'_{22} & S'_{23} \\ S'_{31} & S'_{32} & S'_{33} \end{bmatrix} \begin{bmatrix} a'_1 \\ a'_2 \\ a'_3 \end{bmatrix} \tag{12.7.6}$$

It can be shown that these S-parameters satisfy the following two conditions, and therefore all nine elements of the scattering matrix are not independent:

$$\sum_{i=1}^{3} S'_{ij} = 1 \qquad j = 1, 2, 3 \tag{12.7.7}$$

and

$$\sum_{j=1}^{3} S'_{ij} = 1 \qquad i = 1, 2, 3 \tag{12.7.8}$$

Condition (12.7.7) can be proved when there is negligible stray capacitance at each port. In that case, the algebraic sum of the currents entering these ports must be zero. Hence,

$$\sum_{n=1}^{3} i_n = 0$$

Using (8.6.15) and (8.6.16), we can write

$$\sum_{i=1}^{3} (a'_i - b'_i) = 0$$

or

$$\sum_{i=1}^{3} \left(a'_i - \sum_{j=1}^{3} S'_{ij} a'_j \right) = 0 \tag{12.7.9}$$

If we select $a'_2 = a'_3 = 0$, (12.7.9) simplifies to

$$a'_1 - \sum S'_{i1} a'_1 = 0$$

or

$$\sum_{i=1}^{3} S'_{i1} = 0 \tag{12.7.10}$$

Similarly, for $a'_1 = a'_3 = 0$,

$$\sum_{i=1}^{3} S'_{i2} = 0 \tag{12.7.11}$$

and for $a'_1 = a'_2 = 0$,

$$\sum_{i=1}^{3} S'_{i3} = 0 \tag{12.7.12}$$

Hence, condition (12.7.7) is verified.

The condition (12.7.8) can be verified as follows. When every port has the same voltage there will be no input current provided that the stray capacitance from each terminal to ground is zero. Since $i_i = a_i - b_i$, b'_i must be equal to a'_i. Consequently, $b_1 = b_2 = b_3 = a'$ for $a_1 = a_2 = a_3 = a'$. Hence,

$$b'_i = \sum_{j=1}^{3} S'_{ij} a'_j = a' \sum_{j=1}^{3} S'_{ij} = a' \qquad i = 1, 2, 3$$

or

$$\sum_{j=1}^{3} S'_{ij} = 1 \qquad i = 1, 2, 3$$

That proves condition (12.7.8).

Using (8.6.15) and (8.6.16), we find that the normalized voltage at port 3 is $V_3 = a'_3 + b'_3$. Let us now define new voltage waves at ports 1 and 2 as follows:

$$a_1 = a'_1 - \frac{V_3}{2} = a'_1 - \frac{a'_3 + b'_3}{2} \tag{12.7.13}$$

$$b_1 = b'_1 - \frac{V_3}{2} = b'_1 - \frac{a'_3 + b'_3}{2} \tag{12.7.14}$$

$$a_2 = a'_2 - \frac{V_3}{2} = a'_2 - \frac{a'_3 + b'_3}{2} \tag{12.7.15}$$

and

$$b_2 = b'_2 - \frac{V_3}{2} = b'_2 - \frac{a'_3 + b'_3}{2} \tag{12.7.16}$$

The new total voltages at ports 1 and 2 are now

$$V_1 = a_1 + b_1 = a_1' + b_1' - V_3 \tag{12.7.17}$$

and

$$V_2 = a_2 + b_2 = a_2' + b_2' - V_3 \tag{12.7.18}$$

Thus, these voltages are referred to the port 3 voltage.

Note that these definitions of new incident and reflected waves do not change the port currents; that is,

$$b_1 - a_1 = b_1' - a_1' \tag{12.7.19}$$

and

$$b_2 - a_2 = b_2' - a_2' \tag{12.7.20}$$

The two-port scattering matrix that relates b_1 and b_2 to a_1 and a_2 will now make the port 3 terminal common. Hence,

$$b_1 = S_{11}a_1 + S_{12}a_2 \tag{12.7.21a}$$

$$b_2 = S_{21}a_1 + S_{22}a_2 \tag{12.7.21b}$$

Combining (12.7.13) to (12.7.21), we have

$$b_1' - \frac{a_3' + b_3'}{2} = S_{11}\left(a_1' - \frac{a_3' + b_3'}{2}\right) + S_{12}\left(a_2' - \frac{a_3' + b_3'}{2}\right) \tag{12.7.22a}$$

and

$$b_2' - \frac{a_3' + b_3'}{2} = S_{21}\left(a_1' - \frac{a_3' + b_3'}{2}\right) + S_{22}\left(a_2' - \frac{a_3' + b_3'}{2}\right) \tag{12.7.22b}$$

From Kirchhoff's current law, the sum of all currents flowing into the three terminals must be zero. Therefore, we can write

$$a_1' - b_1' + a_2' - b_2' + a_3' - b_3' = 0 \tag{12.7.23}$$

Rearranging (12.7.22) and (12.7.23), we get

$$b_1' = S_{11}a_1' + S_{12}a_2' + (1 - S_{11} - S_{12})\frac{a_3' + b_3'}{2} \tag{12.7.24a}$$

$$b_2' = S_{21}a_1' + S_{22}a_2' + (1 - S_{21} - S_{22})\frac{a_3' + b_3'}{2} \tag{12.7.24b}$$

and

$$b'_3 = a'_1 + a'_2 + a'_3 - b'_1 - b'_2$$
$$= a'_1 + a'_2 + a'_3 - S_{11}a'_1 - S_{12}a'_2 - (1 - S_{11} - S_{12})\frac{a'_3 + b'_3}{2}$$
$$- S_{21}a'_1 - S_{22}a'_2 - (1 - S_{21} - S_{22})\frac{a'_3 + b'_3}{2}$$

or

$$b'_3 = \frac{2\Delta_{12}}{4 - \Delta}a'_1 + \frac{2\Delta_{21}}{4 - \Delta}a'_2 + \frac{\Delta}{4 - \Delta}a'_3 \tag{12.7.25}$$

where

$$\Delta_{12} = 1 - S_{11} - S_{21} \tag{12.7.26a}$$
$$\Delta_{21} = 1 - S_{12} - S_{22} \tag{12.7.26b}$$
$$\Delta = S_{11} + S_{22} + S_{12} + S_{21} = 2 - \Delta_{12} - \Delta_{21} = 2 - \Delta_{11} - \Delta_{22} \tag{12.7.26c}$$
$$\Delta_{11} = 1 - S_{11} - S_{12} \tag{12.7.26.d}$$
$$\Delta_{22} = 1 - S_{21} - S_{22} \tag{12.7.26.e}$$

Now from (12.7.24) and (12.7.25), we find that

$$b'_1 = \left(S_{11} + \frac{\Delta_{11}\Delta_{12}}{4 - \Delta}\right)a'_1 + \left(S_{12} + \frac{\Delta_{11}\Delta_{21}}{4 - \Delta}\right)a'_2 + \frac{2\Delta_{11}}{4 - \Delta}a'_3 \tag{12.7.27a}$$
$$b'_2 = \left(S_{21} + \frac{\Delta_{22}\Delta_{12}}{4 - \Delta}\right)a'_1 + \left(S_{22} + \frac{\Delta_{22}\Delta_{21}}{4 - \Delta}\right)a'_2 + \frac{2\Delta_{22}}{4 - \Delta}a'_3 \tag{12.7.27b}$$

Equations (12.7.25) and (12.7.27) can be expressed in matrix form as

$$\begin{bmatrix} b'_1 \\ b'_2 \\ b'_3 \end{bmatrix} = \begin{bmatrix} S_{11} + \dfrac{\Delta_{11}\Delta_{12}}{4 - \Delta} & S_{12} + \dfrac{\Delta_{11}\Delta_{21}}{4 - \Delta} & \dfrac{2\Delta_{11}}{4 - \Delta} \\ S_{21} + \dfrac{\Delta_{22}\Delta_{12}}{4 - \Delta} & S_{22} + \dfrac{\Delta_{22}\Delta_{21}}{4 - \Delta} & \dfrac{2\Delta_{22}}{4 - \Delta} \\ \dfrac{2\Delta_{12}}{4 - \Delta} & \dfrac{2\Delta_{21}}{4 - \Delta} & \dfrac{\Delta}{4 - \Delta} \end{bmatrix} \begin{bmatrix} a'_1 \\ a'_2 \\ a'_3 \end{bmatrix} \tag{12.7.28}$$

Therefore, if two-port S-parameters of a three-terminal device are given with terminal 3 common, the corresponding three-port scattering matrix can be found from (12.7.28). If we are given the scattering matrix of a transistor in its common-emitter configuration, it can be converted to a three-port device (four-terminal device) where the fourth terminal is a common ground point and its nine scattering parameters can be found via (12.7.28).

Feedback Network Design Consideration

If a load impedance Z is connected between terminal 3 and the ground, then $b'_3 = \Gamma a'_3$, where

$$\Gamma = \frac{Z - Z_0}{Z + Z_0}$$

Therefore,

$$b'_3 = \Gamma a'_3 = S'_{31}a'_1 + S'_{32}a'_2 + S'_{33}\Gamma a'_3$$

or

$$a'_3 = -\frac{S'_{31}}{S'_{33} - \Gamma^{-1}}a'_1 - \frac{S'_{32}}{S'_{33} - \Gamma^{-1}}a'_2 \tag{12.7.29}$$

Scattering parameters of the resulting two-port network can be found from (12.7.6) and (12.7.29) as follows:

$$b'_1 = \left(S'_{11} - \frac{S'_{13}S'_{31}}{S'_{33} - \Gamma^{-1}}\right)a'_1 + \left(S'_{12} - \frac{S'_{13}S'_{32}}{S'_{33} - \Gamma^{-1}}\right)a'_2 \tag{12.7.30}$$

and

$$b'_2 = \left(S'_{21} - \frac{S'_{23}S'_{31}}{S'_{33} - \Gamma^{-1}}\right)a'_1 + \left(S'_{22} - \frac{S'_{23}S'_{32}}{S'_{33} - \Gamma^{-1}}\right)a'_2 \tag{12.7.31}$$

These two equations can be expressed in matrix form as follows and the corresponding new S_{ijn} can be identified:

$$\begin{bmatrix} b'_1 \\ b'_2 \end{bmatrix} = \begin{bmatrix} \left(S'_{11} - \dfrac{S'_{13}S'_{31}}{S'_{33} - \Gamma^{-1}}\right) & \left(S'_{12} - \dfrac{S'_{13}S'_{32}}{S'_{33} - \Gamma^{-1}}\right) \\ \left(S'_{21} - \dfrac{S'_{23}S'_{31}}{S'_{33} - \Gamma^{-1}}\right) & \left(S'_{22} - \dfrac{S'_{23}S'_{32}}{S'_{33} - \Gamma^{-1}}\right) \end{bmatrix} \begin{bmatrix} a'_1 \\ a'_2 \end{bmatrix}$$
$$= \begin{bmatrix} S_{11\mathrm{n}} & S_{12\mathrm{n}} \\ S_{21\mathrm{n}} & S_{22\mathrm{n}} \end{bmatrix} \begin{bmatrix} a'_1 \\ a'_2 \end{bmatrix} \tag{12.7.32}$$

If port 3 is grounded (i.e., $V_3 = 0$), Γ in (12.7.32) is -1.

A similar procedure may be adopted for the other ports. In other words, if port 1 is terminated by impedance Z, $b_1 = \Gamma a_1$. The S-parameters of this new two-port network that is formed by ports 2 and 3 of the original three-port network can be determined from (12.7.6). Suppose that the common-source S-parameters of a FET are given, whereas the design procedure requires its common-gate S-parameters; then the computation proceeds as follows. First, we determine its three-port S-parameters using (12.7.28). The source terminal forms the third port of the FET while its gate and drain terminals remain as ports 1 and 2 of the three-port network. Next, we switch back to a two-port network by short-circuiting port 1. If a feedback network is needed, that impedance is used to terminate this port.

As (12.7.32) indicates, the S-parameters of the new two-port network can be modified to a certain extent by adjusting the terminating impedance Z. Note that its reflection coefficient Γ will satisfy the equations

$$\Gamma = \frac{S_{11n} - S'_{11}}{S_{11n}S'_{33} - \Lambda_1} \tag{12.7.33}$$

and

$$\Gamma = \frac{S_{22n} - S'_{22}}{S_{22n}S'_{33} - \Lambda_2} \tag{12.7.34}$$

where

$$\Lambda_1 = S'_{11}S'_{33} - S'_{31}S'_{13} \tag{12.7.35}$$

and

$$\Lambda_2 = S'_{22}S'_{33} - S'_{32}S'_{23} \tag{12.7.36}$$

In the case of a reactive termination at port 3, $|\Gamma| = 1$. It maps a circle in the S_{11n}-plane, the center C_1 and radius R_1 of which are given as

$$C_1 = \frac{S'_{11} - \Lambda_1 S'^{*}_{33}}{1 - |S'_{33}|^2} \tag{12.7.37}$$

and

$$R_1 = \frac{|S'_{31}S'_{13}|}{|1 - |S'_{33}|^2|} \tag{12.7.38}$$

Similarly, center C_2 and radius R_2 of the circle for S_{22n} are found as follows:

$$C_2 = \frac{S'_{22} - \Lambda_2 S'^{*}_{33}}{1 - |S'_{33}|^2} \tag{12.7.39}$$

and

$$R_2 = \frac{|S'_{32}S'_{23}|}{|1 - |S'_{33}|^2|} \tag{12.7.40}$$

These circles are drawn on a Smith chart and S_{11n} and S_{22n} are optimized for the desired feedback. In the case of an oscillator design, a positive feedback is desired. In other words, the inverse of the conditions (11.1.16) and (11.1.18) must hold so that the transistor becomes potentially unstable. Alternatively, the μ-factor can be determined from the following equation, and its magnitude should be kept to a small value that is less than unity. (Les Besser, *Applied Microwave and Wireless*, pp. 44–55, Spring 1995). The transistor is unconditionally stable for magnitudes greater than unity.

$$\mu = \frac{1 - |S_{22n}|^2}{|S_{11n} - \Delta_n S_{22n}^*| + |S_{21n} S_{12n}|} \tag{12.7.41}$$

where

$$\Delta_n = S_{11n} S_{22n} - S_{12n} S_{21n} \tag{12.7.42}$$

Example 12.14 Reconsider the BJT of Example 12.13. Find its three-port S-parameter description. Use the inductive termination at its base and verify whether that provides optimum feedback for the oscillator design.

SOLUTION From (12.7.28), the three-port S-parameters are found as follows:

$$\begin{bmatrix} S'_{11} & S'_{12} & S'_{13} \\ S'_{21} & S'_{22} & S'_{23} \\ S'_{31} & S'_{32} & S'_{33} \end{bmatrix} = \begin{bmatrix} 0.396\ \angle 84.4^\circ & 0.393\ \angle 29.3^\circ & 0.852\ \angle -43.5^\circ \\ 0.582\ \angle -79.1^\circ & 0.534\ \angle -54.7^\circ & 1.163\ \angle 60^\circ \\ 0.87\ \angle 11.8^\circ & 0.425\ \angle 34.9^\circ & 0.466\ \angle -115.4^\circ \end{bmatrix}$$

The reactance of a 1.45-nH inductor is $j25.0542\,\Omega$ and its reflection coefficient Γ can be found easily for $Z_0 = 50\,\Omega$. The S-parameters of the resulting two-port

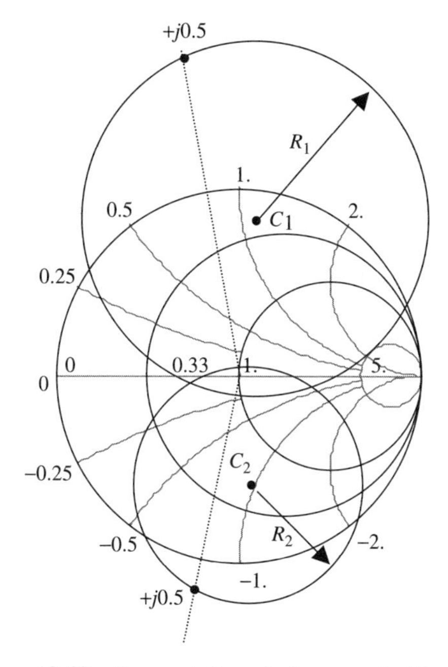

Figure 12.48 S_{11n} and S_{22n} circles on a Smith chart.

network are found from (12.7.32) as

$$\begin{bmatrix} S_{11n} & S_{12n} \\ S_{21n} & S_{22n} \end{bmatrix} = \begin{bmatrix} 1.722 \angle 100.1^\circ & 0.714 \angle 94.8^\circ \\ 2.083 \angle -136.4^\circ & 1.163 \angle -102.5^\circ \end{bmatrix}$$

These results are very close to those obtained in Example 12.13. Minor deviations are attributed to the roundoff errors involved.

To verify the feedback provided by the inductor, we first plot S_{11n} and S_{22n} circles on a Smith chart. To that end, the information of these circles is found from (12.7.37) to (12.7.40). The results are $C_1 = 0.837 \angle 84^\circ$, $R_1 = 0.947$, $C_2 = 0.587 \angle -84.6^\circ$, and $R_2 = 0.631$. These two circles are illustrated in Figure 12.48. It may easily be verified that the points indicated on these circles correspond to S_{11n} and S_{22n} of the two-port network obtained with an inductive termination of $j0.5$ at the base terminal. Further, the magnitude of the μ-factor of the transistor (without feedback inductor) in common-base configuration is 0.674. It means that the transistor is potentially unstable. When the inductor is added to its base, the magnitude of the μ-factor decreases to 0.126. Hence, the inductor enhances the positive feedback further to facilitate the oscillator design.

SUGGESTED READING

Bahl, I. J., and Bhartia, P., *Microwave Solid State Circuit Design*. New York: Wiley, 1988.

Collin, R. E., *Foundations for Microwave Engineering*. New York: McGraw-Hill, 1992.

Davis, W. A., *Microwave Semiconductor Circuit Design*. New York: Van Nostrand Reinhold, 1984.

Gonzalez, G., *Microwave Transistor Amplifiers*. Upper Saddle River, NJ: Prentice Hall, 1997.

Larson, L. E. (ed.), *RF and Microwave Circuit Design for Wireless Communications*. Boston: Artech House, 1996.

Pozar, D. M., *Microwave Engineering*. New York: Wiley, 1998.

Rohde, U. L., *Microwave and Wireless Synthesizers*. New York: Wiley, 1997.

Smith, J. R., *Modern Communication Circuits*. New York: McGraw-Hill, 1998.

Vendelin, G. D., Pavio, A., and Rhode, U. L., *Microwave Circuit Design Using Linear and Non-linear Techniques*. New York: Wiley, 1990.

PROBLEMS

12.1. In Figure P12.1, the gain of the amplifier is constant with frequency [i.e., $A_v(j\omega) = A_0$] and

$$\beta(j\omega) = \frac{10^{-6}}{1 - j(\omega - 2\pi \times 10^9)}$$

Use the Barkhausen criteria to determine the frequency of oscillation and the required value of A_o.

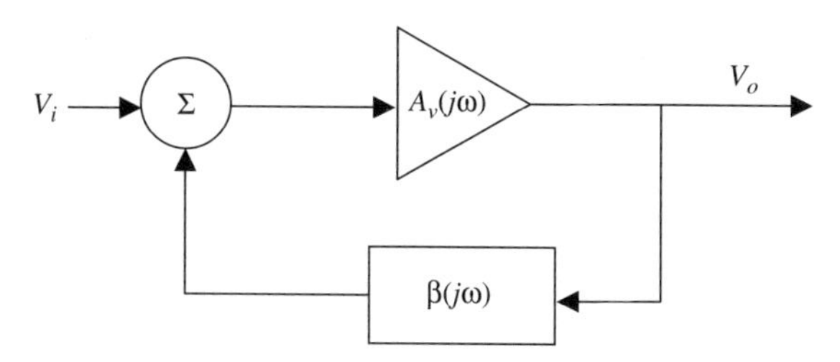

Figure P12.1

12.2. In Figure P12.2, the gain of the amplifier $[A_v(j\omega)]$ and the feedback factor $\beta(j\omega)$ are given as follows:

$$A_v(j\omega) = -\frac{8.45}{1 + j(8\,\omega \times 10^{-10})}$$

and

$$\beta(j\omega) = \frac{1}{(1 + j5\,\omega \times 10^{-10})^2}$$

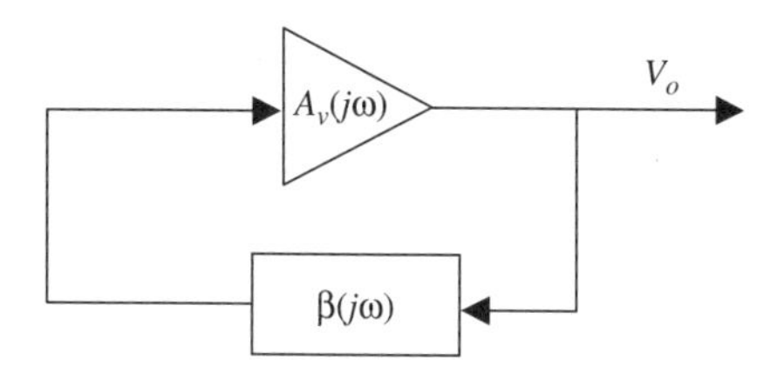

Figure P12.2

Determine the frequency of oscillation if the circuit oscillates.

12.3. (a) In the circuit shown in Figure P12.3, should the gain of the amplifier be positive or negative to create the possibility for it to oscillate?

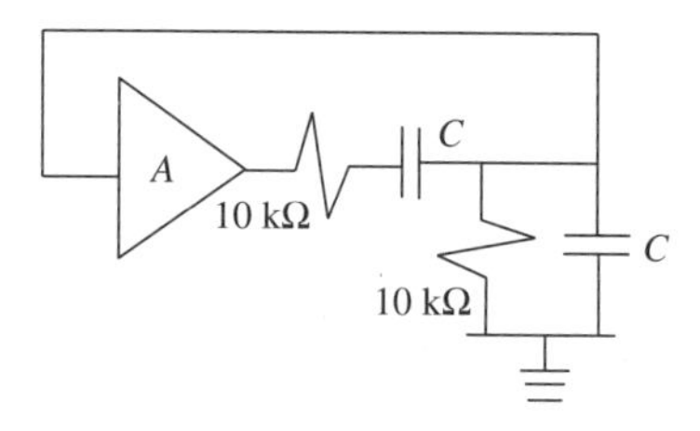

Figure P12.3

(b) What value should $|A|$ have to cause sinusoidal oscillations?

(c) Find the capacitance C for the oscillations to occur at 10 kHz.

12.4. Complete the design of the biased BJT Colpitts oscillator circuit shown in Figure P12.4 for it to oscillate at 150 MHz.

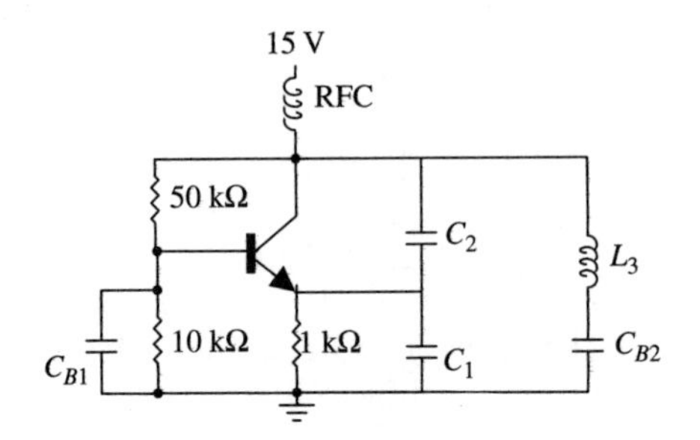

Figure P12.4

12.5. In the circuit shown in Figure P12.5, the FET is biased such that its g_m is 4.5 mS. For this circuit to oscillate at 100 MHz, find C_1 and C_2. An open-loop gain of 5 ensures the beginning of oscillations, and the unloaded Q value of a 100-nH inductor is 1000.

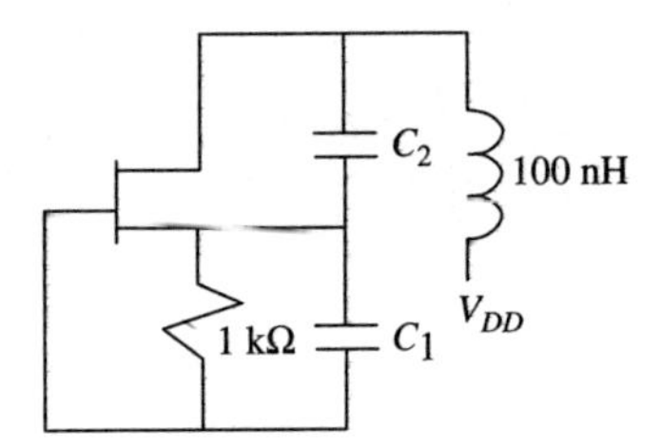

Figure P12.5

12.6. In the circuit shown in Figure P12.6 the FET is biased such that its g_m is 4.5 mS. For this circuit to oscillate at 100 MHz, find L_1 and L_2. An open-loop gain of 5 ensures the beginning of oscillations.

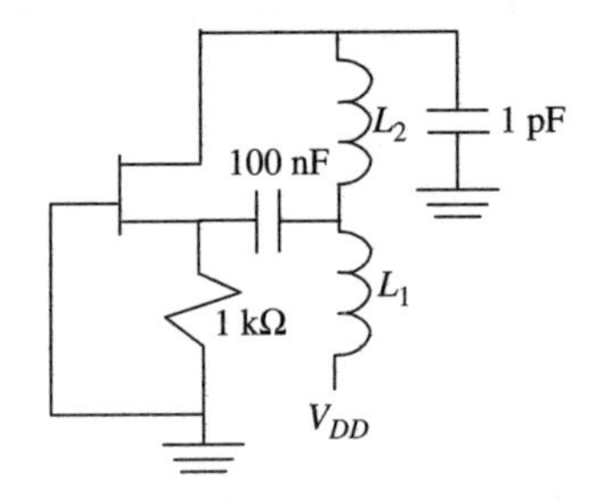

Figure P12.6

12.7. Free-running frequency and the gain factor of a voltage-controlled oscillator are given as 500 MHz and 150 kHz/V, respectively. It is being used in a PLL along with a sinusoidal phase detector with maximum output of 5 V at $\theta_e = \pi/2$ radians. If there are no amplifiers or frequency divider in the loop, what is its hold-in range?

12.8. If the phase detector in Problem 12.7 is replaced by a triangular type with $V_{e\ \max} = A = 5$ V at $\theta_{e\ \max} = \pi/2$ radians, find the new holding range.

12.9. In the PLL shown in Figure P12.9, peak amplitudes of the reference input as well as the VCO output are 1.5 V. An analog multiplier is being used as the phase detector that has 5 V output when its reference input, V_{in}, and the oscillator signal, V_{osc}, both are 2 V. The free-running frequency of the VCO is 50 MHz, which reduces to zero when its control voltage V_{cntl} is reduced by 5 V. Find the phase difference between the input and the VCO output when it is in lock with input frequency of (**a**) 55 MHz and (**b**) 45 MHz.

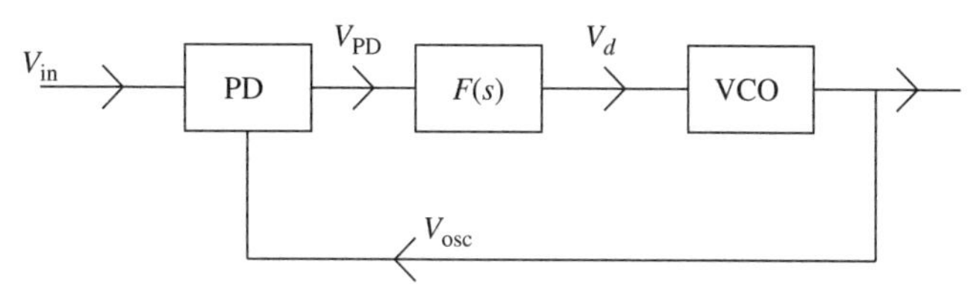

Figure P12.9

12.10. Analyze the PLL system shown in Figure P12.10. By controlling the frequency divider, its output frequency can be changed from 4 MHz to 6 MHz in steps of 200 kHz. The transfer function of the phase detector, K_d, is 0.5 V/rad. The free-running frequency and gain factor of VCO are 5 MHz and 10^7 rad/s·V, respectively. The voltage gain of the amplifier is 20. Design a passive lead–lag low-pass filter that can be used in the loop.

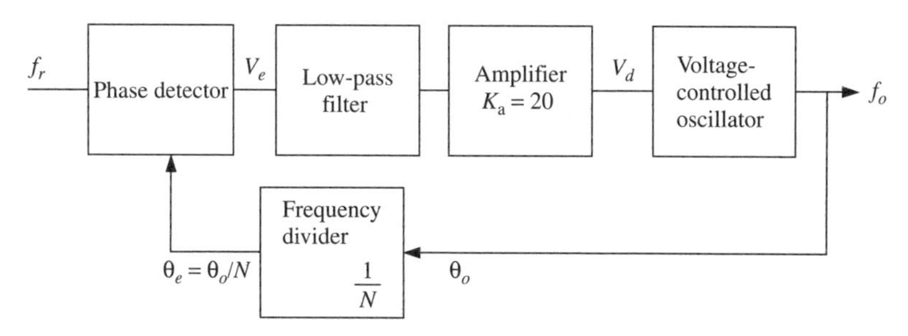

Figure P12.10

12.11. Design a multiple phase-locked-loop synthesizer to cover the frequency range 100.0 to 200.0 MHz in 100-Hz increments. The loop frequency switching time should not exceed 25 ms.

12.12. Design a multiple phase-locked-loop synthesizer to cover the frequency range 35.40 to 40.0 MHz in 10-Hz increments. The reference frequency is to be 100 kHz. No loop should operate with a reference frequency below 100 kHz.

12.13. Design a three-loop synthesizer to cover the frequency range 190 to 200 MHz in steps of 10 Hz. The loop frequency switching time should not exceed 25 ms.

12.14. A 2-GHz oscillator is to be designed using a negative resistance diode with its reflection coefficient Γ_G as $2.05\ \angle -60^\circ$ (measured with $Z_0 = 50\,\Omega$ as reference). Determine its load impedance.

12.15. The S-parameters of a GaAs MESFET are measured in the common-source mode at $V_{ds} = 6\,\text{V}$ and $I_{ds} = 50\,\text{mA}$ at 8 GHz. The results are found as follows. $S_{11} = 0.63\ \angle -150^\circ$, $S_{12} = 0.07\ \angle -30^\circ$, $S_{21} = 3.10\ \angle 40^\circ$, and $S_{22} = 0.70\ \angle -100^\circ$. Determine its S-parameters in common-gate configuration.

12.16. A silicon BJT is biased at $V_{CE} = 18\,\text{V}$ and $I_C = 30\,\text{mA}$. Its S-parameters measured at 5.0 GHz in the common-emitter configuration are found to be $S_{11} = 0.63\ \angle -150^\circ$, $S_{12} = 0.007\ \angle 30^\circ$, $S_{21} = 2.10\ \angle 40^\circ$, and $S_{22} = 0.70\ \angle -100^\circ$. Find its S-parameters in common-base configuration.

12.17. A silicon BJT is biased at $V_{CE} = 10\,\text{V}$ and $I_C = 40\,\text{mA}$. Its S-parameters measured at 8.0 GHz in the common-base configuration are found to be $S_{11} = 1.32\ \angle 88^\circ$, $S_{12} = 0.595\ \angle 99^\circ$, $S_{21} = 1.47\ \angle 172^\circ$, and $S_{22} = 1.03\ \angle -96^\circ$. Find its S-parameters in common-emitter configuration.

12.18. Design an oscillator to operate at 8 GHz using the common-gate configuration of MESFET given in Problem 12.15. Use an inductor at its gate terminal to provide positive feedback, if necessary. Determine the value of the feedback inductor.

12.19. Design an oscillator to operate at 5 GHz using the common-base configuration of BJT given in Problem 12.16. Use an inductor at its base terminal to provide positive feedback, if necessary. Determine the value of the feedback inductor.

13

DETECTORS AND MIXERS

Detector and mixer circuits employ nonlinear electronic devices to process electrical information. For example, if a transmission line is being used to send only one voice signal at a given time, it can be connected directly without a problem. On the other hand, if two different voice signals are sent over this line at the same time, no information can be retrieved at the other end due to their interference. In practice, for economic reasons, a large number of telephone subscribers may need to be connected via a single long-distance communication line. This requires some means to move each voice channel to a different frequency band. This solution is *frequency-division multiplexing* (FDM). A commonly used FDM hierarchy is illustrated in Figure 13.1. Such multiplexing systems require mixers and detectors.

Similarly, wireless systems operate at higher frequencies to achieve efficient and directional radiation from an antenna. In this case, electrical information is generally superimposed on a high-frequency carrier signal via amplitude or frequency modulation. In *amplitude modulation* (AM), the amplitude of a radio-frequency signal (the carrier) varies according to the information (the modulating) signal. In the case of *frequency modulation* (FM), the frequency of carrier signal varies according to the modulating signal. Figure 13.2 illustrates various waveforms associated with these modulations.

The chapter begins with an overview of the characteristics of these modulations and their detection schemes. Several mixer circuits are then presented that can be designed using diodes or transistors as nonlinear electronic devices.

Radio-Frequency and Microwave Communication Circuits: Analysis and Design, Second Edition,
By Devendra K. Misra
ISBN 0-471-47873-3

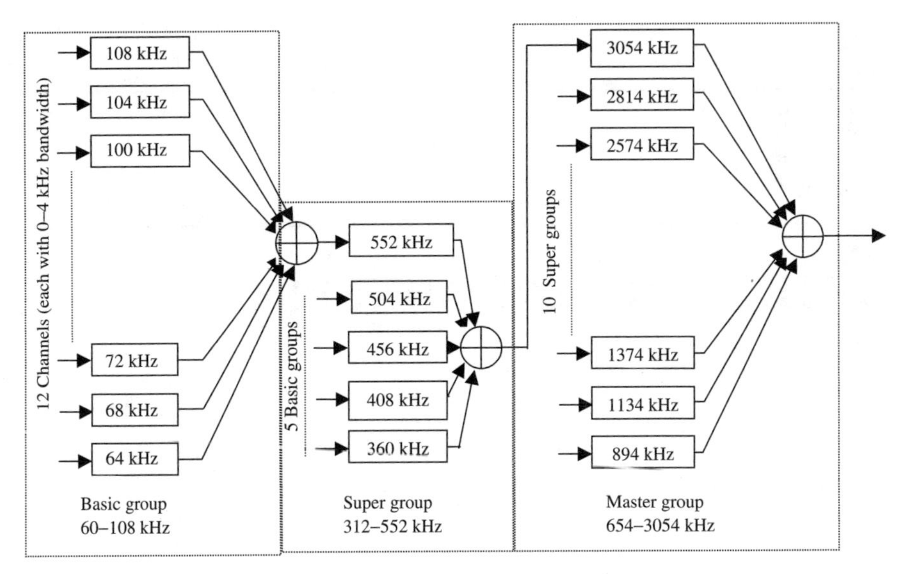

Figure 13.1 FDM hierarchy.

13.1 AMPLITUDE MODULATION

Consider the AM system illustrated in Figure 13.3. Assume that two inputs to this circuit are $s(t) = a_0 + b_0 m(t)$ and $c(t) = \cos(\omega_c t + \psi)$. The output signal $c_{AM}(t)$ is then found as

$$c_{AM}(t) = s(t)c(t) = a_0\left[1 + \frac{b_0}{a_0} m(t)\right]\cos(\omega_c t + \psi) \tag{13.1.1}$$

An analysis of this expression shows that $c_{AM}(t)$ represents an amplitude-modulated signal with $s(t)$ as the modulating signal and $c(t)$ as the carrier. Further, the ratio b_0/a_0 is called the *modulation index.*

If the modulating signal is $m(t) = \cos \omega_m t$, (13.1.1) can be simplified as follows:

$$c_{AM}(t) = a_0\left(1 + \frac{b_0}{a_0}\cos\omega_m t\right)\cos(\omega_c t + \psi)$$

or

$$c_{AM}(t) = a_0 \cos(\omega_c t + \psi) + \frac{b_0}{2}\{\cos[(\omega_c + \omega_m)t + \psi] + \cos(\omega_c - \omega_m)t + \psi\} \tag{13.1.2}$$

This can easily be graphed to verify that indeed it represents an AM signal as illustrated in Figure 13.2. Further, it shows that the output has three sinusoidal signals, each with different frequencies. Its frequency spectrum is shown in Figure 13.4. Besides the carrier frequency ω_c, there are two other frequencies as $\omega_c \pm \omega_m$. The signal spectrum $\omega_c + \omega_m$ represents the upper sideband;

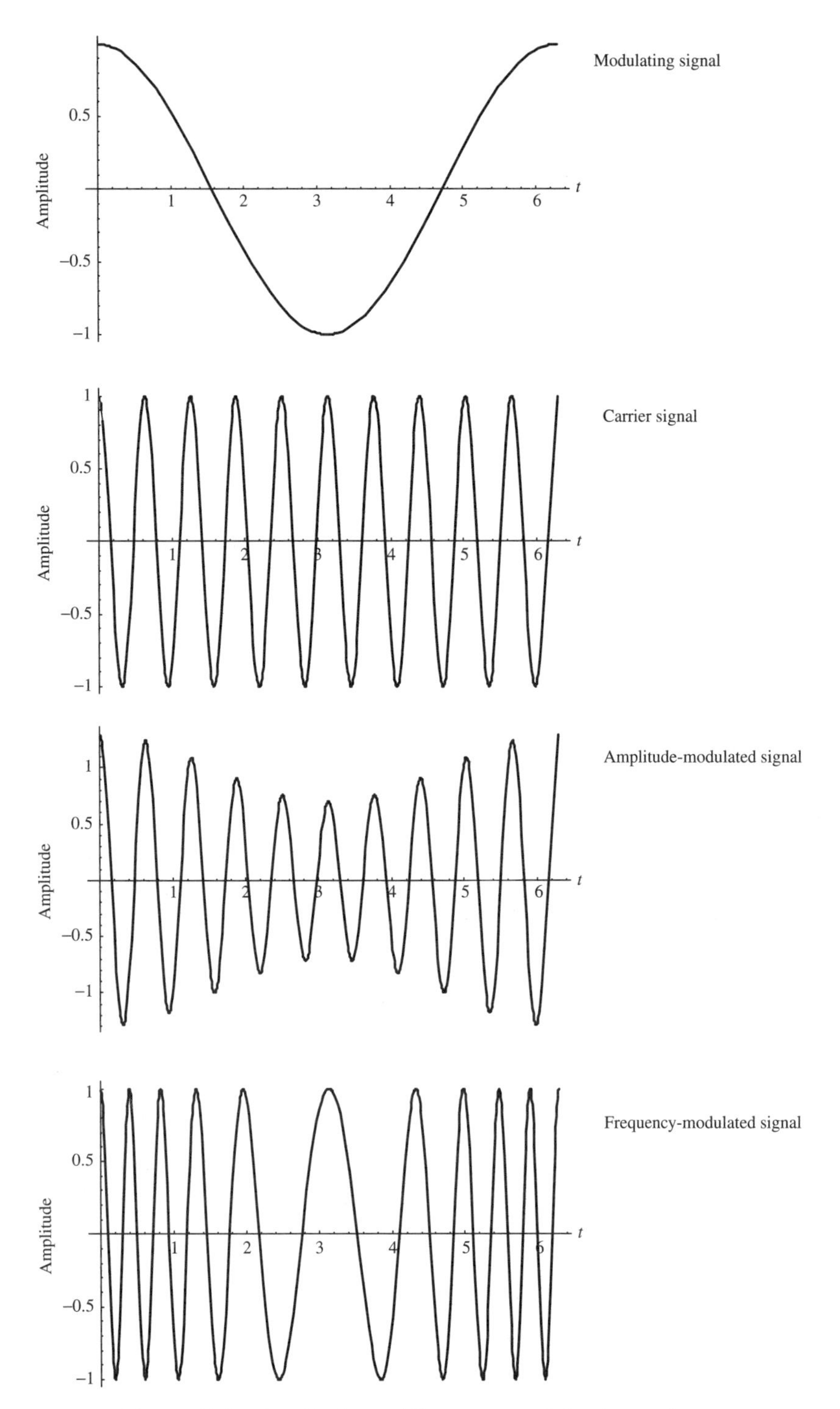

Figure 13.2 Amplitude- and frequency-modulated waveforms.

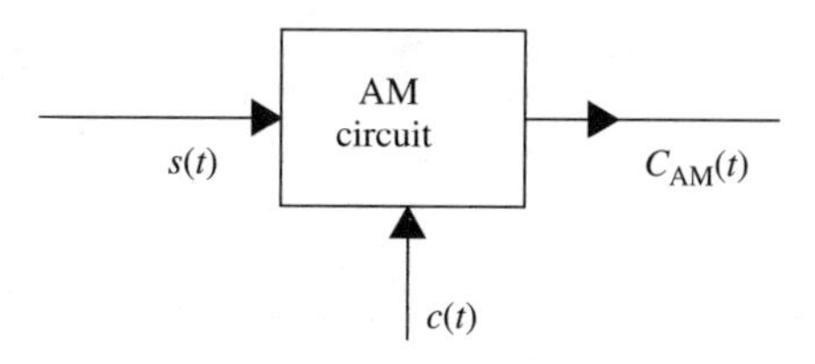

Figure 13.3 Amplitude-modulation system.

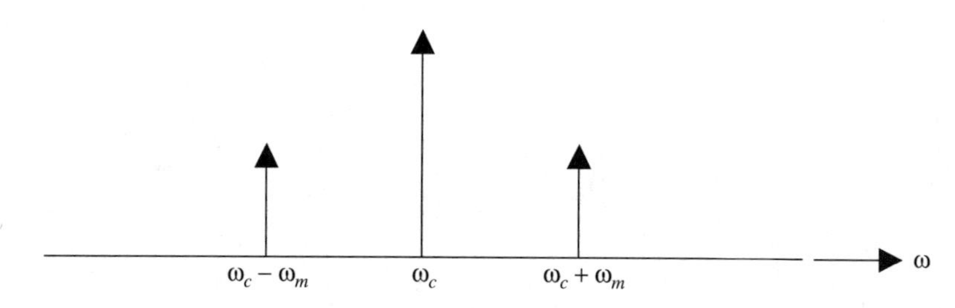

Figure 13.4 Frequency spectrum of $c_{AM}(t)$.

$\omega_c - \omega_m$ is the lower sideband. Note that each sideband includes the information signal ω_m. When only one of these signals is used to send the information, it is called *single-sideband* (SSB) *AM transmission*. On the other hand, complete $c_{AM}(t)$ represents *double-sideband* (DSB) *AM transmission*. It is called *suppressed carrier AM* if the carrier component of $c_{AM}(t)$ is filtered out before transmitting.

Example 13.1 A 100-MHz sinusoidal signal is being used as the carrier for AM transmission. Its amplitude is 1000 V and the average power is 10 kW. The modulating signal is $s(t) = 3\cos\omega_1 t + 2\cos 2\omega_1 t + \cos 3\omega_1 t$, where $\omega_1 = 4\pi \times 10^6$ rad/s. The modulation index is 0.15.

(a) Determine the frequency spectrum of the signal modulated. Find the amplitude of each sinusoidal component, the power in the carrier and each pair of sidebands, and the total bandwidth of the AM signal.

(b) Determine the peak amplitude and peak instantaneous power of this signal.

SOLUTION Since amplitude of the carrier signal is 1000 V and its power is 10 kW, the proportionality constant that relates the two can be determined. For $P = \alpha V^2$,

$$\alpha = \frac{P}{V^2} = \frac{10{,}000}{1{,}000{,}000} = 0.01$$

The AM signal can be determined from (13.1.1) as follows:

$$C_{AM}(t) = 1000[1 + 0.15(3\cos\omega_1 t + 2\cos 2\omega_1 t + \cos 3\omega_1 t)]\cos\omega_c t \qquad \text{V}$$

TABLE 13.1 Results for Example 13.1(a)

	Carrier	$\omega_c \pm \omega_1$	$\omega_c \pm 2\,\omega_1$	$\omega_c \pm 3\,\omega_1$
Frequency (MHz)	100	100 ± 2	100 ± 4	100 ± 6
Amplitude (V)	1000	225	150	75
Power (W)	10,000	1012.5	450	112.5

or

$$\begin{aligned} C_{\text{AM}}(t) = 1000 \cos \omega_c t &+ 225[\cos(\omega_c + \omega_1)t + \cos(\omega_c - \omega_1)t] \\ &+ 150[\cos(\omega_c + 2\,\omega_1)t + \cos(\omega_c - 2\,\omega_1)t] \\ &+ 75[\cos(\omega_c + 3\,\omega_1)t + \cos(\omega_c - 3\,\omega_1)t] \qquad \text{V} \end{aligned}$$

where $\omega_1 = 4\pi \times 10^6$ rad/s and $\omega_c = 2\pi \times 10^8$ rad/s.

(**a**) From the expression above, the results can be summarized as shown in Table 13.1.

(**b**)

$$\text{peak amplitude, } A = 1000 + 2 \times 225 + 2 \times 150 + 2 \times 75 = 1900\,\text{V}$$

$$\text{peak power} = \alpha A^2 = 36{,}100\,\text{W}$$

Frequency Converters

Frequency converters use nonlinear electronic devices that multiply the input signals. They can be used to generate high-frequency sinusoidal signals from a low-frequency reference. Consider the frequency converter shown in Figure 13.5. Let V_i and V_L be two sinusoidal inputs that produce V_o at its output. Mathematically,

$$V_i = a \cos \omega_1 t$$

$$V_L = b \cos \omega_2 t$$

and

$$V_o = ab \cos \omega_1 t \cos \omega_2 t$$

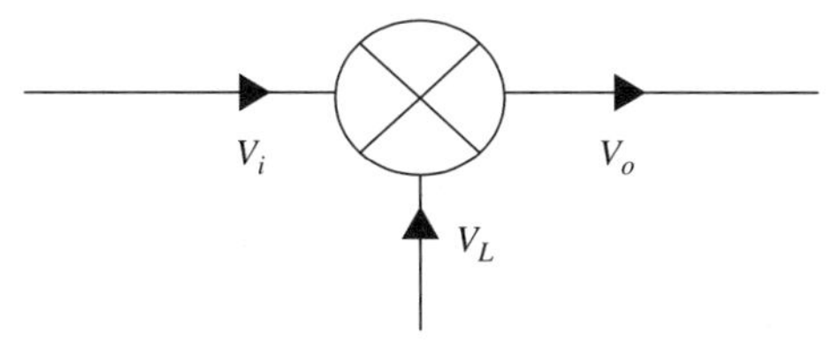

Figure 13.5 Frequency converter.

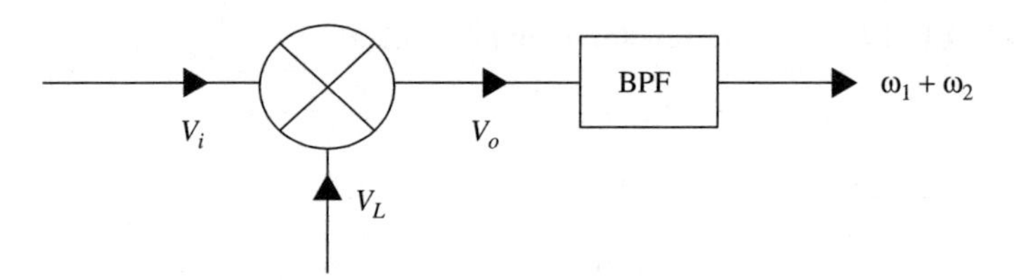

Figure 13.6 Up-converter circuit.

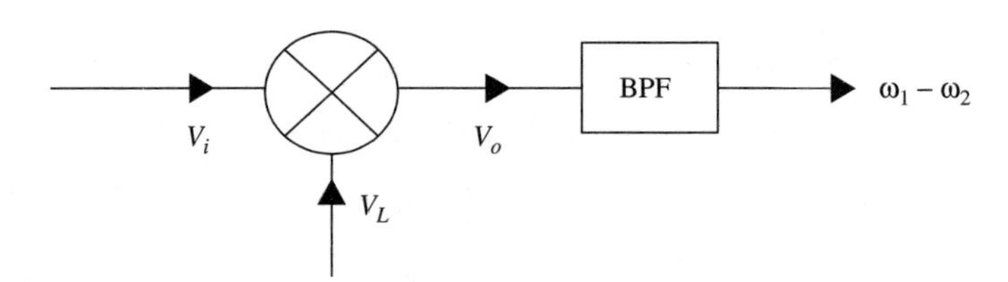

Figure 13.7 Down-converter circuit.

Output V_o can be rearranged as follows:

$$V_o = \frac{ab}{2}\{\cos[(\omega_1 + \omega_2)t] + \cos[(\omega_1 - \omega_2)t]\} \tag{13.1.3}$$

Hence, the output is a sum of two sinusoidal signals. These sum and difference frequency signals can be filtered out for up or down conversions as illustrated in Figures 13.6 and 13.7, respectively.

Single-Diode Mixer Circuit Arrangement

Consider the circuit arrangement shown in Figure 13.8. Two inputs, v_s and v_L, are added together before applying to a diode. Capacitors C_1 and C_2 are used to block the dc bias of the diode from other sides of the circuit. Thus, these capacitors are selected such that there is negligible reactance for the ac signal. Similarly, inductors L_1 and L_2 are used to block ac from short-circuiting via the dc source. Hence, these inductors should have very high reactance at ac, whereas dc passes through with negligible loss. Thus, the voltage applied to the diode has both ac and dc components.

To analyze this circuit and determine its output signal, we need to consider the actual characteristics of the diode. The $V-I$ characteristic of a typical forward-biased diode is illustrated in Figure 13.9. For the present case, bias voltage V_b is assumed high enough to keep the diode forward biased. If the current through this diode is i_d and the voltage across its terminals is v_d,

$$i_d = I_s(e^{\alpha v_d} - 1) \tag{13.1.4}$$

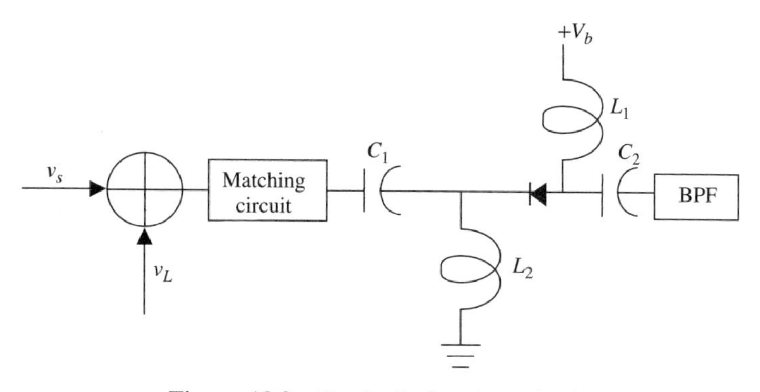

Figure 13.8 Single-diode mixer circuit.

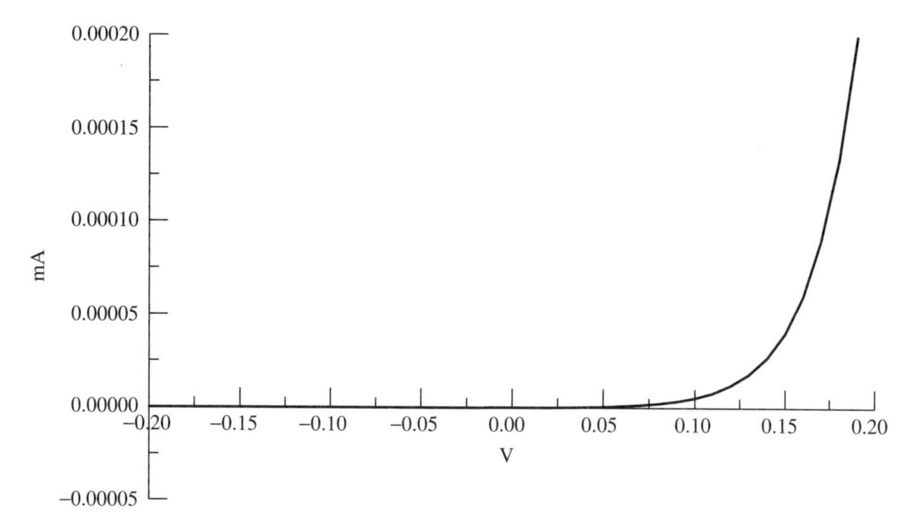

Figure 13.9 V–I characteristic of a typical diode.

where

$$\alpha = \frac{q}{nkT} \tag{13.1.5}$$

I_s is called the *reverse-saturation current* (on the order of 10^{-8} to 10^{-15} A), q is the electronic charge (1.602×10^{-19} C), n is a number ranging between 1 and 2, k is the Boltzmann constant (1.38×10^{-23} J/K), and T is the temperature in kelvin.

Since the diode has both alternating and direct voltage across it, we can write

$$v_d = V_b + v_{\text{ac}}$$

where $v_{\text{ac}} = v_s + v_L$. Therefore, current through the diode can be expressed as

$$i_d = I_s(e^{\alpha(V_b + v_{\text{ac}})} - 1) = I_s(e^{\alpha V_b} e^{\alpha v_{\text{ac}}} - 1)$$

If $v_{ac} \ll V_b$ such that αv_{ac} is a small fraction, i_d can be expanded as follows. Further, it can be approximated only by a few terms of the series:

$$i_d = I_s \left\{ e^{\alpha V_b} \left[1 + \alpha v_{ac} + \frac{(\alpha v_{ac})^2}{2} + \cdots \right] - 1 \right\}$$

After rearranging, we find that

$$i_d = I_s(e^{\alpha V_b} - 1) + I_s e^{\alpha V_b} \left[\alpha v_{ac} + \frac{(\alpha v_{ac})^2}{2} + \cdots \right] = I_b + i_{ac}$$

Thus, there are two components in the diode current: a dc (I_b) and an ac (i_{ac}). The ac component i_{ac} through the diode may be approximated as

$$i_{ac} \approx I_s e^{\alpha V_b} \left[\alpha v_{ac} + \frac{(\alpha v_{ac})^2}{2} \right] = g_d v_{ac} + g_d' \frac{v_{ac}^2}{2} \tag{13.1.6}$$

where

$$g_d = \alpha I_s e^{\alpha V_b} = \alpha(I_b + I_s) \tag{13.1.7}$$

and

$$g_d' = \alpha g_d \tag{13.1.8}$$

Parameter g_d is known as the *dynamic conductance* of the diode. It is equal to the inverse of the junction resistance. Parameter g_d' has units of A/V^2. A small-signal equivalent circuit of the diode can be drawn as shown in Figure 13.10. Its junction capacitance C_j appears in parallel with the junction resistance R_j. The lead inductance L_p and bulk resistance R_s appear in series with this combination. As illustrated, packaging capacitance C_p between the two leads appears in parallel with the circuit.

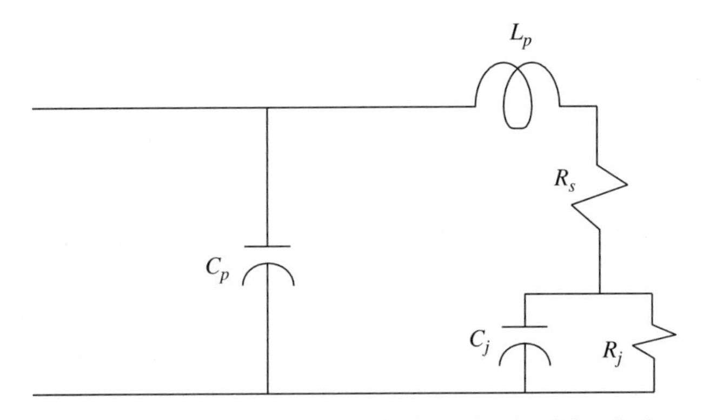

Figure 13.10 Small-signal equivalent circuit of the diode.

Example 13.2 The equivalent-circuit parameters of a diode in an axial lead package are given as follows: $C_p = 0.1\,\text{pF}$, $L_p = 2.0\,\text{nH}$, $C_j = 0.15\,\text{pF}$, $R_s = 10\,\Omega$, and $I_s = 10^{-8}$ A. Determine its junction resistance if the bias current is (**a**) zero and (**b**) $100\,\mu\text{A}$. Assume that the diode is at room temperature ($T = 290\,\text{K}$) and $n = 1$.

SOLUTION From (13.1.7),

$$R_j = \frac{1}{g_d} = \frac{1}{\alpha(I_b + I_s)}$$

and

$$\frac{kT}{q} = \frac{1.38 \times 10^{-23} \times 290}{1.602 \times 10^{-19}} = 0.025\,\text{V}$$

(**a**)

$$R_j = \frac{0.025}{10^{-8}}\,\Omega = 2500\,k\Omega$$

(**b**)

$$R_j = \frac{0.025}{(100 + 0.01) \times 10^{-6}}\,\Omega = 249.9\,\Omega$$

Radio-Frequency Detector

Diodes are used for the detection of radio-frequency signals, to convert a part of RF input to dc. They are used for monitoring the power (relative power measurement), automatic gain control circuits, and the detection of AM signals. Assume that the RF signal, $a_1 \cos \omega t$, is applied to a diode biased at V_b. Hence, the total voltage V_t applied to it is

$$V_t = V_b + a_1 \cos \omega t$$

The corresponding diode current i_d can be expressed as follows:

$$i_d \approx I_b + g_d a_1 \cos \omega t + \frac{g_d'}{2} a_1^2 \cos^2 \omega t$$

or

$$i_d \approx I_b + g_d' \frac{a_1^2}{4} + g_d a_1 \cos \omega t + \frac{g_d'}{4} a_1^2 \cos 2\,\omega t \tag{13.1.9}$$

The *current sensitivity*, β_i, of the diode is defined as a ratio of the change in output dc to that of RF input power. Hence,

$$\beta_i = \frac{g_d' a_1^2/4}{g_d a_1^2/2} = \frac{g_d'}{2 g_d} \tag{13.1.10}$$

The current sensitivity has units of A/W or V^{-1}.

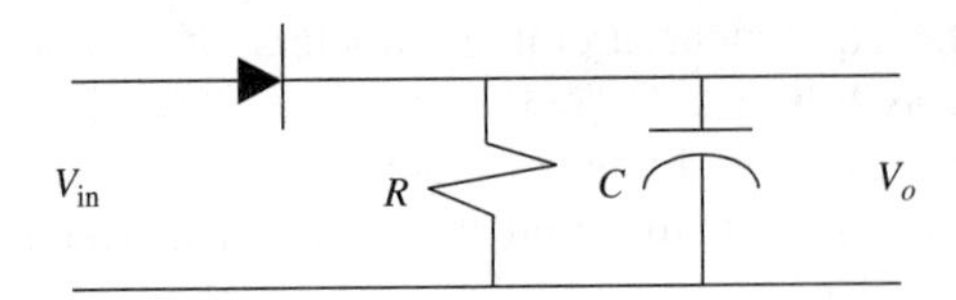

Figure 13.11 Zero-biased diode detector circuit.

An *open-circuit voltage sensitivity* β_v of the diode is defined as the ratio of change in direct voltage across the junction to that of RF input power. Hence,

$$\beta_v = \frac{g_d'}{2g_d^2} = \beta_i R_j \tag{13.1.11}$$

Voltage sensitivity has units of V/W or A^{-1}. Its typical range is 400 to 1500 mA^{-1}. Consider a diode circuit that has no dc bias voltage, as shown in Figure 13.11. If an amplitude-modulated signal V_{in} is applied to this circuit, current through the diode can be found through (13.1.6).

For $V_{in} = V_o(1 + m\cos\omega_m t)\cos\omega_c t$

$$i_{ac} \approx g_d V_o(1 + m\cos\omega_m t)\cos\omega_c t + \frac{g_d' V_o^2}{2}(1 + m\cos\omega_m t)^2\cos^2\omega_c t$$

or

$$\begin{aligned} i_{ac} \approx {} & g_d V_o\left\{\cos\omega_c t + \frac{m}{2}\cos[(\omega_m + \omega_c)t] + \frac{m}{2}\cos[(\omega_m - \omega_c)t]\right\} \\ & + \frac{g_d' V_o^2}{4}\left\{1 + \frac{m^2}{2} + 2m\cos\omega_m t + \frac{m^2}{2}\cos 2\,\omega_m t\right. \\ & + \cos 2\,\omega_c t + m\cos[(2\,\omega_c + \omega_m)t] \\ & + m\cos[(2\,\omega_c - \omega_m)t] + \frac{m^2}{2}\cos 2\,\omega_c t + \frac{m^2}{4}\cos[2(\omega_c + \omega_m)t] \\ & \left. + \frac{m^2}{4}\cos[2(\omega_c - \omega_m)t]\right\} \end{aligned} \tag{13.1.12}$$

Hence, the diode current has several frequency components, shown graphically in Figure 13.12. It includes the modulating signal frequency ω_m. If this current flows

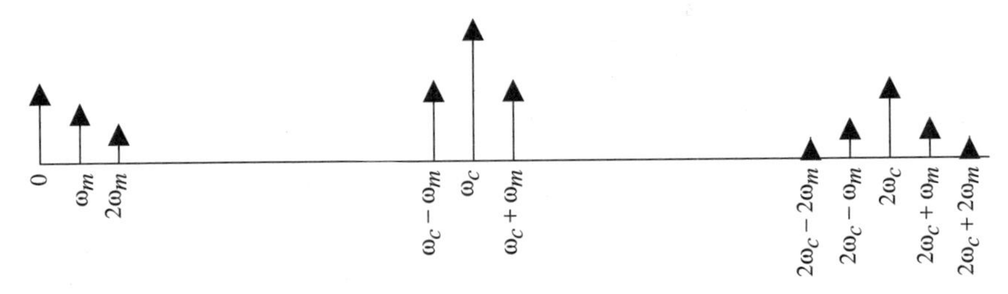

Figure 13.12 Spectrum of the diode current given by (13.1.12).

through a resistance R, the corresponding voltage contains the same spectrum as well. A low-pass filter can be used to suppress the undesired high-frequency signals. Note that there is a second harmonic $2\,\omega_m$ of the modulating signal that may sometimes be difficult to suppress. However, it may have relatively negligible amplitude, especially when the modulation index m is less than unity.

As a special case, if an unmodulated radio-frequency signal is applied to this circuit, m is zero in (13.1.12). Therefore, the expression for the diode current simplifies to

$$i_{\text{ac}} \approx \frac{g'_d V_o^2}{4} + g_d V_o \cos\omega_c t + \frac{g'_d V_o^2 m^2}{4} \cos 2\,\omega_c t$$

Therefore, a capacitor C connected across the output can easily suppress its sinusoidal components, while its first term represents a direct voltage that is proportional to the square of the RF amplitude V_o. This type of circuit is used in practice to monitor radio frequency and microwave signals.

Coherent Detection of AM

An alternative technique for detection of amplitude-modulated signals is illustrated in Figure 13.13. In this case, the amplitude-modulated signal is multiplied with a sinusoidal signal that has the same frequency as the carrier signal. Assume that

$$C_{\text{AM}}(t) = a_o[1 + m(t)]\cos(\omega_c t + \varphi)$$

The voltage $v(t)$ at the output of the multiplier is found to be

$$v(t) = C_{\text{AM}}(t) a \cos(\omega_c t + \varphi) = a a_{\text{o}}[1 + m(t)]\cos(\omega_c t + \varphi)\cos(\omega_c t + \varphi)$$

or

$$v(t) = \frac{a a_o}{2}[1 + m(t)][\cos(\theta - \varphi) + \cos(2\,\omega_c t + \theta + \varphi)]$$

If the low-pass filter has a cutoff frequency below $2\,\omega_c$, its output voltage v_o is

$$v_o(t) = \frac{a a_o}{2}[1 + m(t)]\cos(\theta - \varphi)$$

Hence, this circuit can provide a more efficient way to extract the information signal from the AM than the one considered earlier. However, the output voltage

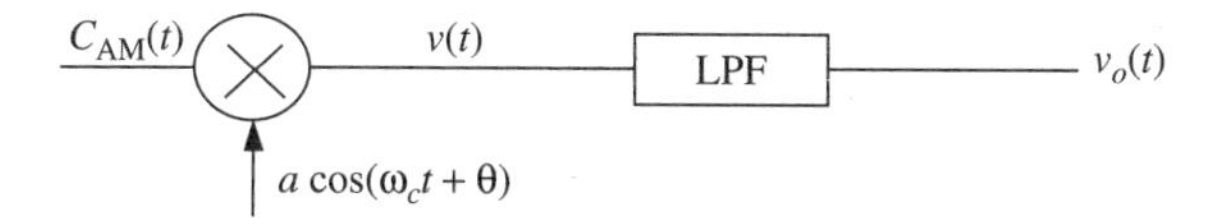

Figure 13.13 Simple coherent detector circuit.

goes to zero if the argument of the cosine function $(\theta - \varphi)$ is 0.5π (90°). To ensure maximum output voltage, this phase difference should be zero. In other words, the local oscillator signal should be phase locked to the carrier. This can be done using the PLL scheme discussed in Chapter 12. This circuit arrangement is illustrated in Figure 13.14.

Single-Sideband Generation

As illustrated in Figure 13.4, the modulating signal is included in both of the sidebands of an AM signal. The precious frequency spectrum can be economized if only one of these is used to transmit the information. For example, the FDM hierarchy shown in Figure 13.1 employs only the lower sideband. It can be achieved with a low-pass filter that passes only the desired component. On the other hand, it can be done more efficiently using the circuit in Figure 13.15. This

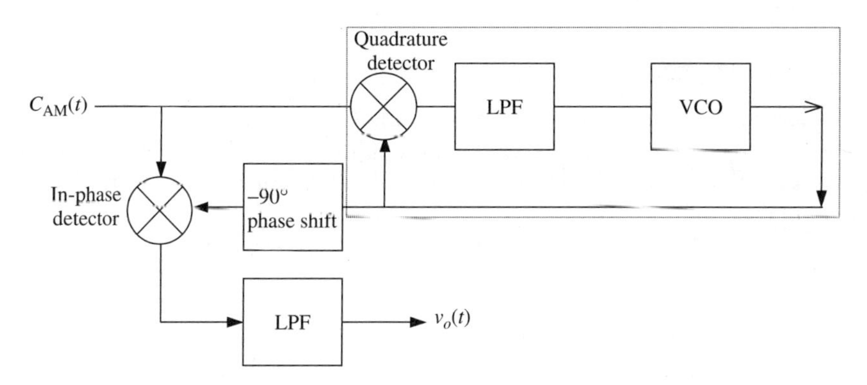

Figure 13.14 PLL-based coherent detection scheme.

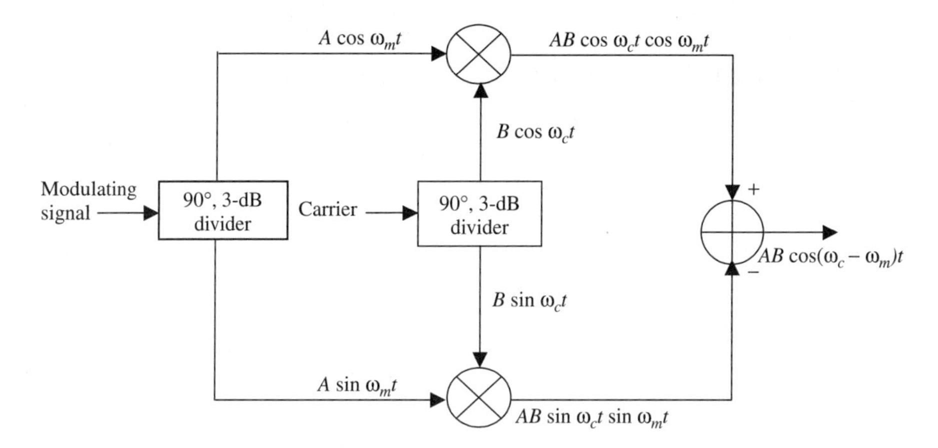

Figure 13.15 SSB generation scheme.

circuit uses two power dividers and two mixers. A power divider splits the input signal into two parts equal in amplitude but which differ in phase by 90°. Thus, a power divider divides the modulating signal into $A\cos\omega_m t$ and $A\sin\omega_m t$. Similarly, the carrier is also divided into $B\cos\omega_c t$ and $B\sin\omega_c t$. These components are applied to corresponding mixers, and the output of one is then subtracted from the other to get the single sideband.

13.2 FREQUENCY MODULATION

As illustrated in Figure 13.2, in frequency modulation the carrier frequency changes according to the modulating signal. It can be expressed mathematically as

$$c_{\mathrm{FM}}(t) = A\cos[\omega_c t + \Delta\omega \int x(t)dt + \psi] \tag{13.2.1}$$

where A is the carrier amplitude, $\Delta\omega$ the frequency deviation coefficient in rad/s·V, and $x(t)$ the modulating signal in volts. Therefore, the frequency of c_{FM}, $\omega(t)$, is

$$\omega(t) = \frac{d}{dt}[\omega_c t + \Delta\omega \int x(t)dt + \psi] = \omega_c + \Delta\omega\ x(t) \tag{13.2.2}$$

If the modulating signal is a sinusoidal wave that is given by

$$x(t) = a\cos(\omega_m t + \theta_m)$$

then

$$\int x(t)dt = \frac{a}{\omega_m}\sin(\omega_m t + \theta_m)$$

Therefore, the frequency-modulated carrier in this case will be given as

$$c_{\mathrm{FM}}(t) = A\cos\left[\omega_c t + \frac{a\ \Delta\omega}{\omega_m}\sin(\omega_m t + \theta_m) + \psi\right] \tag{13.2.3}$$

Modulation index β of the frequency-modulated signal is defined as

$$\beta = \frac{a\ \Delta\omega}{\omega_m} \tag{13.2.4}$$

Note that the modulation index is a dimensionless quantity.

In order to determine the frequency spectrum of $c_{\mathrm{FM}}(t)$, we need to find an equivalent expression for the right-hand side of (13.2.3):

$$c_{\mathrm{FM}}(t) = A\ \mathrm{Re}[e^{j\omega_c t + j\psi + j\beta\sin(\omega_m t + \theta_m)}]$$

or

$$c_{\mathrm{FM}}(t) = A\ \mathrm{Re}[e^{j(\omega_c t + \psi)}e^{j\beta\sin(\omega_m t + \theta_m)}] \tag{13.2.5}$$

From the mathematical tables,

$$e^{j\beta \sin\alpha} = \sum_{n=-\infty}^{\infty} J_n(\beta)e^{jn\alpha}$$

where $J_n(\beta)$ is the Bessel function of the first kind and order n. Therefore,

$$c_{\text{FM}}(t) = A\ \text{Re}\left[\sum_{n=-\infty}^{\infty} J_n(\beta)e^{j[\omega_c t+\psi+n(\omega_m t+\theta_m)]}\right]$$

or

$$c_{\text{FM}}(t) = A\sum_{n=-\infty}^{\infty} J_n(\beta)\cos[(\omega_c + n\omega_m)t + n\theta_m + \psi] \qquad (13.2.6)$$

where

$$J_{-n}(\beta) = (-1)^n J_n(\beta)$$

Therefore, the frequency spectrum of this FM signal extends to infinity as shown in Figure 13.16. As may be observed, the amplitudes of these harmonics are going down for the higher-order terms. Since these amplitudes include the Bessel function, a meaningful conclusion can be drawn only after understanding its characteristics. As illustrated in Figure 13.17, the magnitude of a Bessel function is much smaller than unity if its order n is much larger than its argument β. In particular,

$$J_n(\beta) \ll 1 \qquad \text{for } n \gg \beta$$

Therefore, the infinite series of (13.2.6) can be terminated at finite n. The bandwidth B_c determined this way, known as the *Carson rule bandwidth*, takes $n = \beta + 1$ terms of the infinite series into account. It is given by

$$B_c = 2(\beta + 1)\omega_m \qquad \text{rad/s} \qquad (13.2.7)$$

If β is very large, B_c occupies a bandwidth $\beta + 1$ times larger than that which would be produced by an AM signal. It is known as *wideband FM*. On the other

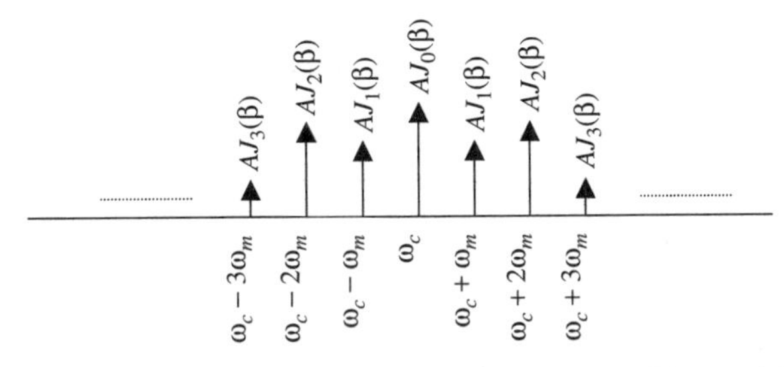

Figure 13.16 Spectrum of the FM signal with sinusoidal modulation.

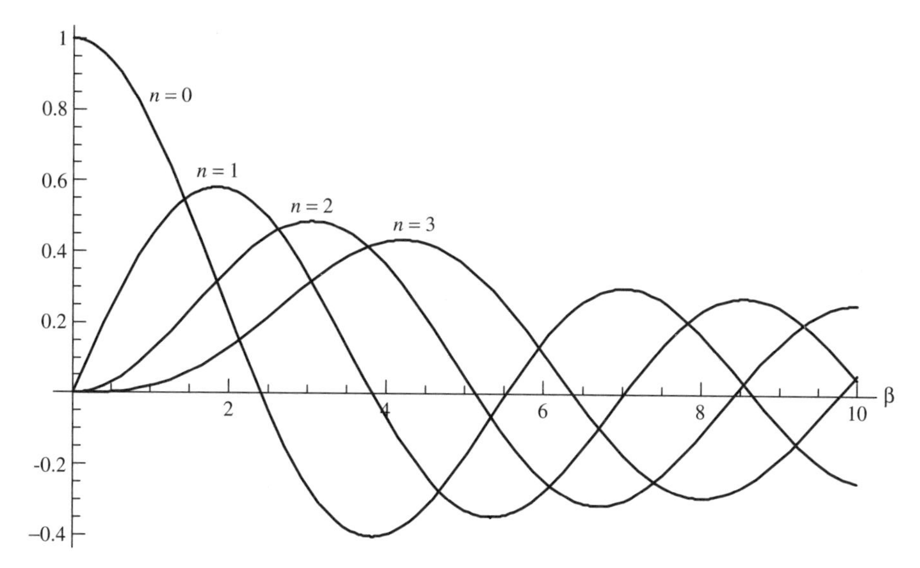

Figure 13.17 Bessel function of the first kind for $n = 0, 1, 2,$ and 3.

hand, if β is much smaller than unity, the FM bandwidth is comparable to that of an AM system and is known as *narrowband FM*.

Example 13.3 A frequency modulator is connected with a carrier, $f_c = 5\,\text{MHz}$, and an audio signal $V_m = 1\,\text{V}$ of $f_m = 1\,\text{kHz}$. It produces frequency deviation of $\pm 10\,\text{kHz}$ at the output of the modulator.

(**a**) An FM wave from the modulator is passed through a series of frequency multipliers with a total multiplication factor of 12; that is, $f_o = 12 f_{\text{in}}$. What is the frequency deviation at the output?

(**b**) The output of the modulator ($f_c = 5\,\text{MHz}, \Delta f = 10\,\text{kHz}$) is input to a mixer stage along with a 55-MHz signal from an oscillator. Find the sum frequency output f_o of the mixer and the frequency deviation Δf_o.

(**c**) What is the modulation index at the output of the modulator?

(**d**) The audio input to the modulator is changed to $V_m = 2\,\text{V}$ and $f_m = 500\,\text{Hz}$. At the modulator output, what will be the modulation index and frequency deviation Δf?

SOLUTION The problem can be solved as follows:

(**a**) Since the input frequency is multiplied by a factor of 12, the frequency deviation at the output of the multiplier will be $\pm 10\,\text{kHz} \times 12 = \pm 120\,\text{kHz}$.

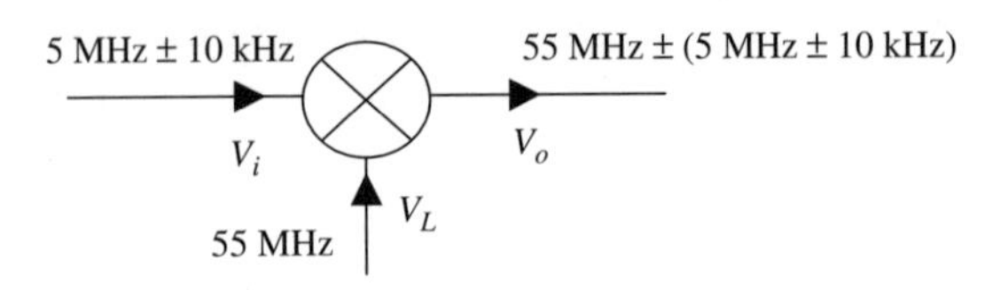

Figure 13.18 Setup for Example 13.3.

(**b**) The arrangement is illustrated in Figure 13.18. Therefore, the sum frequency in the output is given as

$$55\,\text{MHz} + 5\,\text{MHz} + 10\,\text{kHz} = 60.01\,\text{MHz}$$

The frequency deviation is $\pm$ 5.01 MHz.

(**c**) From (13.2.4),

$$\beta = \frac{a\ \Delta\omega}{\omega_m} = \frac{a\ \Delta f}{f_m} = \frac{1 \times 10\,\text{kHz}}{1\,\text{kHz}} = 10$$

(**d**) The frequency deviation in this case increases by a factor of 2. Therefore,

$$\beta = \frac{a\ \Delta\omega}{\omega_m} = \frac{a\ \Delta f}{f_m} = \frac{2 \times 10\,\text{kHz}}{500\,\text{Hz}} = 40$$

FM Detector

A PLL is frequently employed for the detection of FM signals. Consider that the reference signal $v_r(t)$ in Figure 13.19 is an FM signal:

$$v_r(t) = A_r \sin[\omega_c t + \theta_r(t)]$$

where

$$\theta_r(t) = \Delta\omega \int^t m(\tau)\, d\tau$$

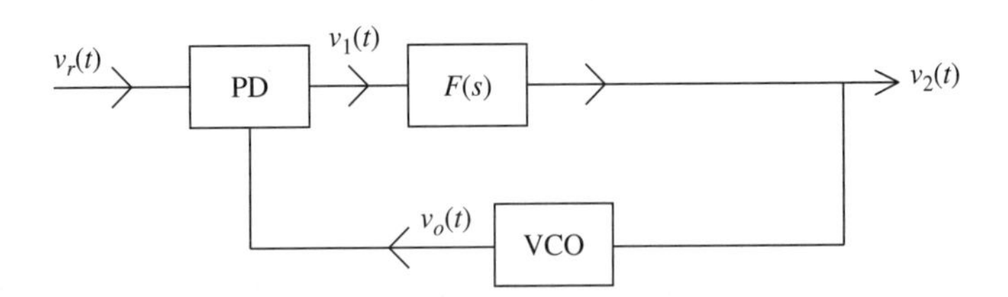

Figure 13.19 FM detection scheme using a PLL.

In the transform domain it can be expressed as

$$\theta_r(s) = \Delta\omega \frac{M(s)}{s}$$

The transfer function of the system shown in Figure 13.19 is

$$\frac{V_2(s)}{\theta_r(s)} = \frac{K_d F(s)}{1 + K_d(K_o/s)F(s)} = \frac{sK_d F(s)}{s + K_d K_o F(s)}$$

Hence, the output voltage in the frequency domain is found to be

$$V_2(s) = \frac{sK_d F(s)}{s + K_d K_o F(s)} \Delta\omega \frac{M(s)}{s} = \frac{\Delta\omega\ K_d F(s) M(s)}{s + K_d K_o F(s)}$$

Assuming that $M(s)$ has a bandwidth B and $|F(s)| = 1$ in the passband, this expression can be simplified as

$$V_2(s) = \frac{\Delta\omega\ K_d M(s)}{s + K_d K_o} = \frac{\Delta\omega\ M(s)/K_o}{1 + s/K_d K_o} \approx \frac{\Delta\omega\ M(s)}{K_o}$$

Hence, for $K_d K_o$ very large such that

$$\left| \frac{s}{K_d K_o} \right| \ll 1$$

we have

$$V_2(s) \approx \frac{\Delta\omega\, M(s)}{K_o}$$

Switching back to the time domain, we find that the output voltage will be given as

$$v_2(t) \approx \frac{\Delta\omega}{K_o} m(t)$$

Hence, the output voltage $v_2(t)$ is proportional to the information signal $m(t)$.

13.3 SWITCHING-TYPE MIXERS

It can easily be shown from (13.1.9) that the output of a single-diode mixer circuit contains both the local oscillator and RF input frequencies. Sometimes it may be hard to filter these out from the desired signal. In this section we present efficient ways to accomplish this. It is assumed for simplicity that the diodes used in these circuits are ideal. Hence, it turns the circuit on or off depending on whether it is forward or reverse biased. Because of this characteristic, these circuits are termed *switching-type mixers*.

Switching-type mixers can be divided into two categories. In *single-balanced mixers* the local oscillator or the RF input can be balanced out such that it does not appear in the output. On the other hand, the output of a *double-balanced mixer* contains neither of the input frequencies. Consider the single-balanced mixer circuit shown in Figure 13.20. The transformer and diodes are assumed ideal for simplicity. Further, it is assumed that the local oscillator voltage v_L is a square wave with a large amplitude compared with v_i. Therefore, only the local oscillator signal determines the conduction through a given diode. An equivalent of this circuit is shown in Figure 13.21. When v_L is positive, diode D_1 is on and D_2 is off. The reverse is true if v_L switches to its negative polarity. Mathematically,

$$v_o = \begin{cases} v_L + v_i & v_L > 0 \\ v_L - v_i & v_L < 0 \end{cases}$$

or

$$v_o = v_L + v_i' \tag{13.3.1}$$

where

$$v_i' = v_i s(t) \tag{13.3.2}$$

$$s(t) = \begin{cases} 1 & v_L > 0 \\ -1 & v_L < 0 \end{cases} \tag{13.3.3}$$

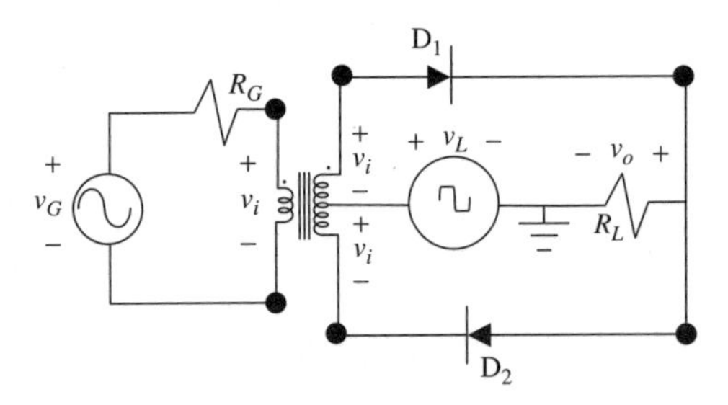

Figure 13.20 Switching type single-balanced mixer.

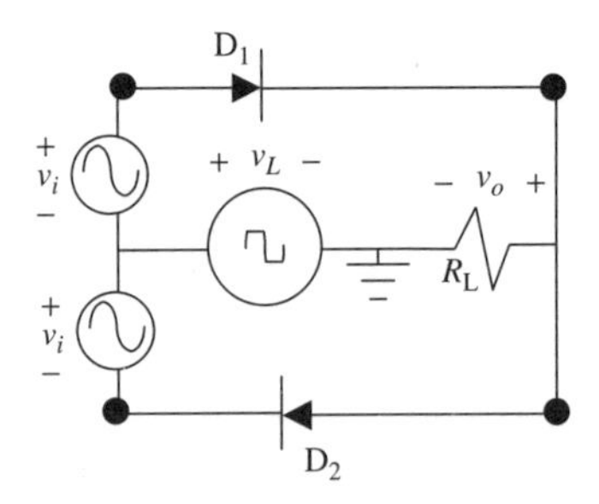

Figure 13.21 Equivalent of the circuit shown in Figure 13.20.

Hence, the switching frequency of $s(t)$ is ω_L, the same as the local oscillator frequency.

Using Fourier series representation, $s(t)$ may be expressed as follows:

$$s(t) = \frac{4}{\pi} \sum_{n=1}^{\infty} \frac{(-1)^{n+1}}{n} \sin \frac{n\pi}{2} \cos n\omega_L t \tag{13.3.4}$$

If v_i is a sinusoidal signal,

$$v_i = V \cos \omega_i t$$

then

$$v_i' = \frac{2V}{\pi} \sum_{n=1}^{\infty} \frac{(-1)^{n+1}}{n} \sin \frac{n\pi}{2} \{\cos[(n\omega_L + \omega_i)t] + \cos[(n\omega_L - \omega_i)t]\} \tag{13.3.5}$$

Therefore, the output voltage can be expressed as

$$\begin{aligned} v_o = v_L + v_i' = v_L + \frac{2V}{\pi} \sum_{n=1}^{\infty} \frac{(-1)^{n+1}}{n} \sin \frac{n\pi}{2} \{\cos[(n\omega_L + \omega_i)t] \\ + \cos[(n\omega_L - \omega_i)t]\} \end{aligned} \tag{13.3.6}$$

Thus, output consists of the local oscillator frequency ω_L and an infinite number of sum and difference frequencies of ω_i with odd multiples of ω_L. Frequency components $\omega_L + \omega_i$ and $\omega_L - \omega_i$ represent the upper and lower sidebands, respectively, each with amplitudes equal to $2V/\pi$. These components can be separated from the other higher-order terms, known as *spurious signals*.

Note that this formulation is based on the ideal switching characteristic of the diode. In other words, it assumes that the local oscillator switches the diode current instantaneously. Deviation from this characteristic increases the distortion in its output. A major disadvantage of this mixer circuit is that the local oscillator signal appears in its output. Frequency components ω_L and $\omega_L \pm \omega_i$ may be very close if the local oscillator frequency is high. In this situation, it may be difficult to separate the signals.

An alternative circuit that blocks the local oscillator signal from appearing at its output is depicted in Figure 13.22. As before, it can be analyzed easily by assuming that the diodes are ideal and that the local oscillator voltage controls its switching. A simplified equivalent of this circuit is shown in Figure 13.23. In this circuit, both diodes conduct (on) when the local oscillator voltage is positive and they do not conduct (off) otherwise. Hence, the output voltage v_o can be written as

$$v_o = v_i s(t) \tag{13.3.7}$$

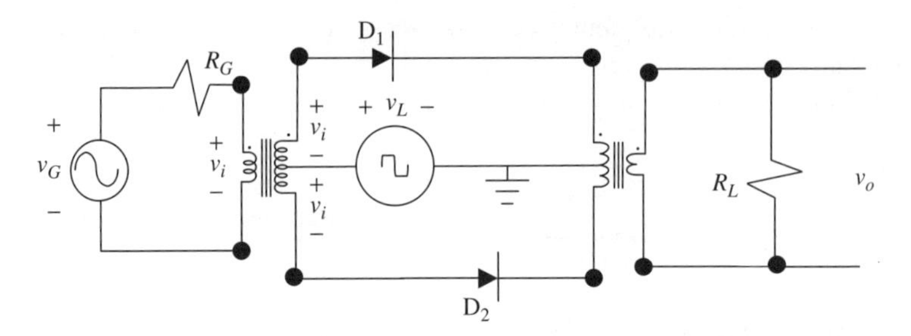

Figure 13.22 Another switching-type single-balanced mixer.

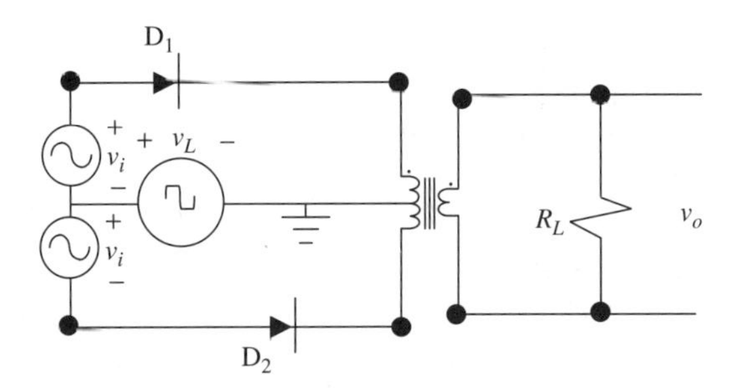

Figure 13.23 Equivalent of the single-balanced mixer shown in Figure 13.22.

where

$$s(t) = \begin{cases} 1 & v_L > 0 \\ 0 & v_L < 0 \end{cases} \tag{13.3.8}$$

This $s(t)$ can be expressed in terms of the Fourier series as

$$s(t) = \frac{1}{2} + \sum_{n=1}^{\infty} \frac{2}{n\pi} \sin\frac{n\pi}{2} \cos n\omega_L t \tag{13.3.9}$$

For $v_i = V\cos\omega_i t$, the output voltage is found to be

$$v_o = \frac{V}{2}\cos\omega_i t + \frac{V}{\pi}\sum_{n=1}^{\infty}\frac{1}{n}\sin\frac{n\pi}{2}\{\cos[(n\omega_L + \omega_i)t] + \cos[(n\omega_L - \omega_i)t]\} \tag{13.3.10}$$

Thus, the output of this mixer circuit does not contain the local oscillator signal. However, it includes the input signal frequency ω_i that appears with the sidebands and spurious signals. It can be a problem in certain applications. For example, consider a case where the input and the local oscillator signal frequencies at a receiver are 60 and 90 MHz, respectively. As illustrated in Figure 13.24, the difference signal frequency at its output is 30 MHz. If a 30-MHz signal is also

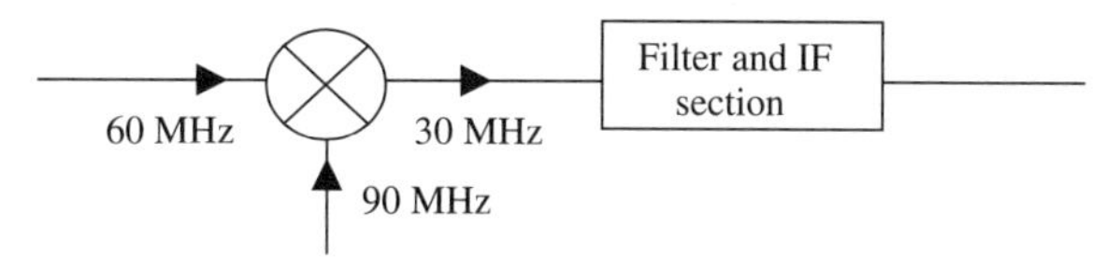

Figure 13.24 Single-balanced mixer of Figure 13.22 at the receiver.

present at the input, it will get through the mixer as well. A preselector (bandpass filter) may be used to suppress the undesired input.

Note from (13.3.10) that the amplitude of the sidebands is only V/π, whereas it was twice this in (13.3.6). It influences the sensitivity of the receiver employing this mixer. In the following section we consider another mixer circuit, which solves most of these problems.

Double-Balanced Mixer

A double-balanced mixer circuit requires four diodes and two transformers with secondary sides center-tapped, as shown in Figure 13.25. As before, it is assumed that the local oscillator voltage v_L is high compared with the input v_i. Hence, diodes D_1 and D_2 conduct when the local oscillator voltage is positive while the other two are in the off state. This situation turns the other way around when the voltage v_L reverses its polarity. Thus, diodes D_1 and D_2 conduct for the positive half cycle of v_L and D_3 and D_4 conduct for the negative half. Equivalent circuits can be drawn separately for these two states. Figure 13.26 shows one such equivalent circuit, which represents the case of positive v_L. Diode resistance in the forward-biased condition is assumed to be r_d. The equivalent circuit can be simplified further to facilitate the analysis.

Consider the simplified equivalent circuit shown in Figure 13.27. The currents in the two loops are i_1 and i_2. Using Kirchhoff's voltage law, loop equations can

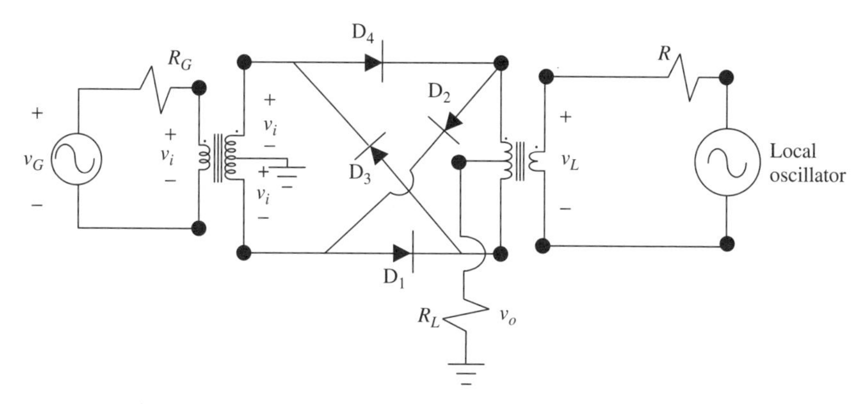

Figure 13.25 Switching-type double-balanced mixer.

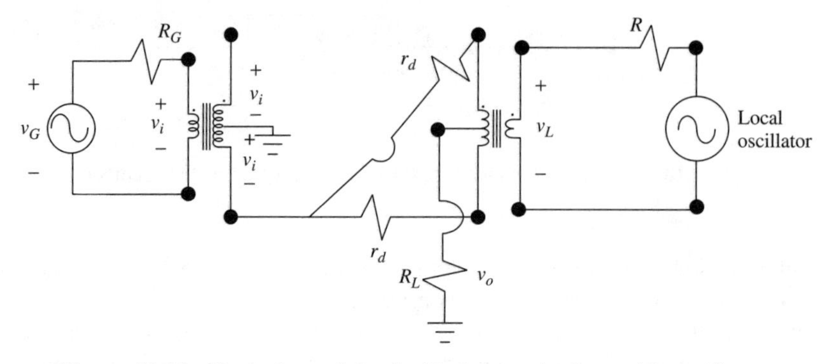

Figure 13.26 Equivalent of the double-balanced mixer with positive v_L.

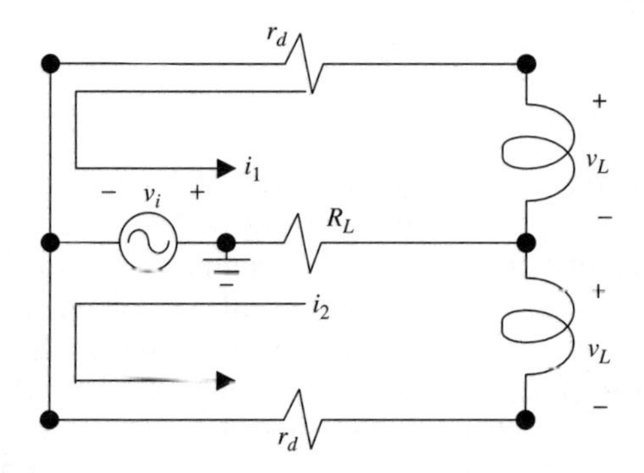

Figure 13.27 Simplified equivalent circuit of Figure 13.26.

be written as

$$v_i = R_L(i_1 - i_2) + r_d i_1 - v_L \tag{13.3.11}$$

and

$$v_i = R_L(i_1 - i_2) - r_d i_2 + v_L \tag{13.3.12}$$

Equations (13.3.11) and (13.3.12) can be solved for $i_1 - i_2$. Hence,

$$i_1 - i_2 = \frac{v_i}{R_L + r_d/2} \tag{13.3.13}$$

But from Ohm's law,

$$i_1 - i_2 = -\frac{v_o}{R_L} \tag{13.3.14}$$

Combining (13.3.13) and (13.3.14) yields

$$\frac{v_o}{v_i} = -\frac{R_L}{R_L + r_d/2} \tag{13.3.15}$$

Similarly, an equivalent circuit can be drawn for the negative half cycle of v_L. Diodes D_3 and D_4 are conducting in this case, and following an identical analysis it may easily be found that

$$\frac{v_o}{v_i} = +\frac{R_L}{R_L + r_d/2} \tag{13.3.16}$$

Equations (13.3.15) and (13.3.16) can be expressed by a single equation after using the function $s(t)$ as defined in (13.3.3). Hence,

$$v_o = \frac{R_L}{R_L + r_d/2} v_i' = \frac{R_L}{R_L + r_d/2} v_i s(t) \tag{13.3.17}$$

Now using the Fourier series representation for $s(t)$ as given in (13.3.4), (13.3.17) can be expressed as

$$v_o = \frac{R_L}{R_L + r_d/2} v_i \frac{4}{\pi} \sum_{n=1}^{\infty} \frac{(-1)^{n+1}}{n} \sin\frac{n\pi}{2} \cos n\omega_L t \tag{13.3.18}$$

In case of a sinusoidal input, $v_i = V \cos \omega_i t$, the output voltage v_o can be found as

$$v_o = \frac{R_L}{R_L + (r_d/2)} \frac{2V}{\pi} \sum_{n=1}^{\infty} \frac{(-1)^{n+1}}{n} \sin\frac{n\pi}{2} \{\cos[(n\omega_L + \omega_i)t] + \cos[(n\omega_L - \omega_i)t]\} \tag{13.3.19}$$

Hence, this mixer circuit produces the upper and lower sidebands along with an infinite number of spurious signals. However, the ω_L and ω_i signals are both isolated from the output. This analysis assumes that the diodes are perfectly matched and the transformers are ideal. Variations of these characteristics may not perfectly isolate the output from the input and the local oscillator. These double-balanced mixers are commercially available for applications in the audio through microwave frequency bands.

13.4 CONVERSION LOSS

Since mixers are used to convert the frequency of an input signal, a circuit designer would like to know if there is some change in its power as well. This information is especially important in receiver design where the signal received may be fairly weak. *Conversion loss* (or *gain*) of the mixer is defined as a ratio of the power output in one of the sidebands to the power of its input signal. It can be evaluated as follows.

Consider the double-balanced mixer circuit shown in Figure 13.24. Using its equivalent circuit illustrated in Figure 13.26, the input resistance R_{in} can be found with the help of (13.3.13) as follows:

$$R_{\text{in}} = \frac{v_i}{i_1 - i_2} = R_L + \frac{r_d}{2} \approx R_L \tag{13.4.1}$$

It is assumed here that load resistance R_L is much larger than the forward resistance r_d of the diode.

If there is maximum power transfer from the voltage source v_G at the input to the circuit, the source resistance R_G must be equal to the input resistance R_{in}. Since the input resistance is equal to the load R_L, the following condition must be satisfied:

$$R_G = R_L$$

If the sinusoidal source voltage v_G has a peak value of V_P, it divides between two equal resistances R_G and R_{in}, and only half of voltage V_P appears across the transformer input. Hence,

$$P_i = \frac{V_P^2}{8R_L} \tag{13.4.2}$$

From (13.3.19), the peak output voltage V_o of the sideband is

$$V_o|_{\omega_L \pm \omega_i} = \frac{2V}{\pi} = \frac{V_P}{\pi} \tag{13.4.3}$$

Therefore, the output power P_o is found as

$$P_o = \frac{V_P^2}{2\pi^2 R_L} \tag{13.4.4}$$

If G is defined as the conversion gain, which is a ratio of its output to input power, then from (13.4.2) and (13.4.4) we find that

$$G = \frac{P_o}{P_i} = \frac{V_P^2}{2\pi^2 R_L} \frac{8R_L}{V_P^2} = \frac{4}{\pi^2} \tag{13.4.5}$$

Since G is less than unity, it is really not a gain but a loss of power. Hence, the conversion loss, L, is

$$L = 10 \log \frac{\pi^2}{4} = 3.92\,\text{dB} \approx 4\,\text{dB} \tag{13.4.6}$$

Hence, approximately 40% of the power of the input signal is transferred to the output of an ideal double-balanced mixer. Circuit loss and mismatch reduce it further.

In the case of a single-balanced mixer where the local oscillator is isolated from the output, the sideband voltage can be found from (13.3.10). Hence, the power output in one of the sidebands is given as

$$P_o = \frac{1}{2R_L}\left(\frac{V}{\pi}\right)^2 = \frac{V_P^2}{8\pi^2 R_L} \tag{13.4.7}$$

Since the input power is still given by (13.4.2), its conversion gain may be found as

$$G = \frac{P_o}{P_i} = \frac{V_s^2}{8\pi^2 R_L} \times \frac{8R_L}{V_s^2} = \frac{1}{\pi^2} \tag{13.4.8}$$

Again, the output is less than the input power, and therefore, it represents a loss of signal power. The conversion loss for this case is

$$L = 10 \log \pi^2 = 9.943 \approx 10\,\text{dB} \tag{13.4.9}$$

Hence, the conversion loss in this single-balanced mixer is 6 dB (four times) larger than that of the double-balanced mixer considered earlier. It may be noted that the mixer circuit shown in Figure 13.19 has a conversion loss of only 4 dB.

These results show that the maximum possible power transferred from the input to the output of a diode mixer is less than 40%. Therefore, these circuits always have conversion loss. A conversion gain is possible only when a transistor (BJT or FET) circuit is used as the mixer. FET mixers are described briefly later in the chapter.

13.5 INTERMODULATION DISTORTION IN DIODE-RING MIXERS

As discussed in Chapter 2, the noise level defines the lower end of the dynamic range of a system, whereas the distortion determines its upper end. Therefore, a circuit designer needs to know the distortion introduced by the mixer and whether there is some way to limit that. We consider here the mixer circuit depicted in Figure 13.28, which has a resistance R in series with each diode.

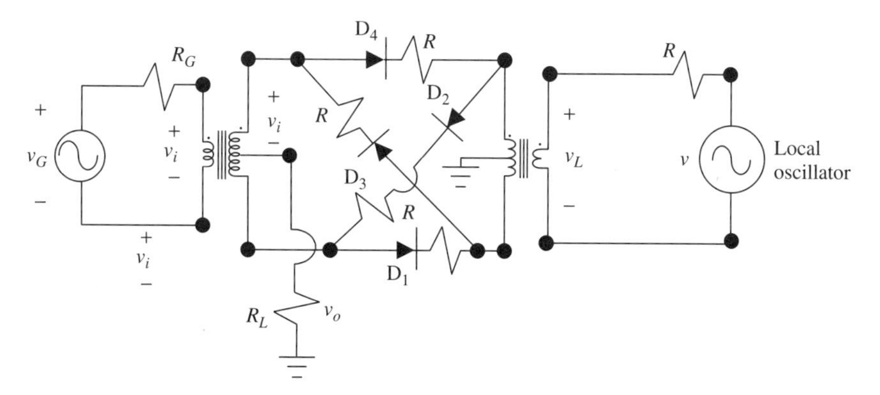

Figure 13.28 Mixer circuit with resistance R in each branch.

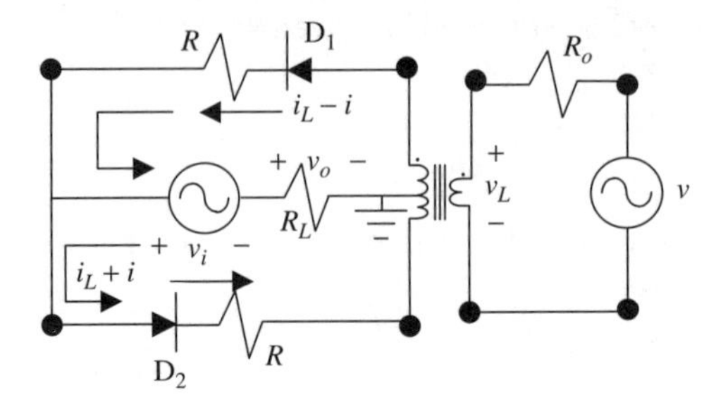

Figure 13.29 Simplified equivalent of the circuit shown in Figure 13.28 for $v_L > 0$.

Assuming that v_L is large enough that the local oscillator alone controls the conduction through the diodes, an equivalent circuit can be drawn for each half cycle. In the case of a positive half period of v_L, the simplified equivalent circuit may be drawn as illustrated in Figure 13.29. Diodes D_1 and D_2 conduct for this duration while the other two are in the off state. Assume that i_L is current due to v_L and i is due to v_i. Hence, currents through the diodes are given as follows:

$$\text{current through diode } D_1 = i_{d3} = i_L - i$$

$$\text{current through diode } D_2 = i_{d4} = i_L + i$$

Since the V–I characteristic of a diode is given as

$$i_d = I_s(e^{\alpha v_d} - 1) \approx I_s e^{\alpha v_d} \tag{13.5.1}$$

where

$$\alpha = \frac{q}{nkT} \tag{13.5.2}$$

we find that

$$v_d = \frac{1}{\alpha} \ln \frac{i_d}{I_s} \tag{13.5.3}$$

Note that net current flowing through the load resistance R_L is the difference of two loop currents: $i_L - i$ and $i_L + i$. Hence, the LO current through R_L cancels out, and two components of current due to v_i are added to make it $2i$. Loop equations for the circuit can be found to be

$$v_L = (i_L - i)R + v_{d3} + v_i - R_L \cdot 2i \tag{13.5.4}$$

and

$$v_L = R_L \cdot 2i - v_i + R(i_L + i) + v_{d4} \tag{13.5.5}$$

Subtracting (13.5.5) from (13.5.4), we have

$$0 = -2(R + 2R_L)i + 2v_i + v_{d3} - v_{d4}$$

or

$$2(R + 2R_L)i - 2v_i = \frac{1}{\alpha}\left(\ln\frac{i_L - i}{I_s} - \ln\frac{i_L + i}{I_s}\right) = -\frac{1}{\alpha}\ln\frac{i_L + i}{i_L - i}$$

or

$$v_i = (R + 2R_L)i + \frac{1}{2\alpha}\ln\frac{i_L + i}{i_L - i} \tag{13.5.6}$$

Since the LO current i_L is very large compared with i, it can be expanded as an infinite series:

$$v_i = \left(R + 2R_L + \frac{1}{\alpha i_L}\right)i + \frac{1}{3\alpha i_L^3}i^3 + \frac{1}{5\alpha i_L^5}i^5 + \cdots \tag{13.5.7}$$

An inverse series solution to this can easily be found with the help of computer software (such as Mathematica) as follows:

$$i = \frac{v_i}{R + 2R_L + 1/\alpha i_L} - \frac{v_i^3}{3\alpha i_L^3\,(R + 2R_L + 1/\alpha i_L)^4} + \cdots \tag{13.5.8}$$

Therefore, the output voltage v_o can be written as

$$v_o = 2R_L i = 2R_L\left[\frac{v_i}{R + 2R_L + 1/\alpha i_L} - \frac{v_i^3}{3\alpha i_L^3\,(R + 2R_L + 1/\alpha i_L)^4} + \cdots\right] \tag{13.5.9}$$

On comparing it with (2.5.31), we find that the term responsible for the intermodulation distortion (IMD) is

$$k_3 = -\frac{2R_L}{3\alpha i_L^3\,(R + 2R_L + 1/\alpha i_L)^4} \tag{13.5.10}$$

Since the IMD is directly proportional to the cube of k_3, (13.5.10) must be minimized to reduce the intermodulation distortion. One way to achieve this objective is to use a large local oscillator current i_L. Another possible way is to employ the resistance R in series with each diode. Sometimes an additional diode is used in place of R. This permits higher local oscillator drive levels and a corresponding reduction in IMD.

Example 13.4 In the mixer circuit shown in Figure 13.30, transformer and diodes are ideal. The transformer has 400 turns on its primary and 800 turns with a center tap on its secondary side. Find the output voltage and the value of load R_L that provides maximum power transfer from v_s.

SOLUTION For maximum power transfer, the impedance when transformed to the primary side of the transformer must be equal to 50 Ω. Since the number of turns on either side of the center tap is the same as on its primary side and only one of the diodes is conducting at a given time, the load resistance R_L must be

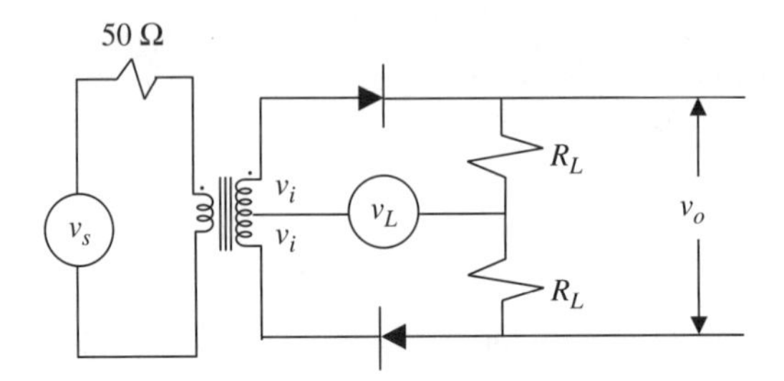

Figure 13.30 Mixer circuit for Example 13.4.

50 Ω as well. Under this condition, only half of the signal voltage appears across the primary. Hence, $v_i = v_s/2$.

Assume that v_L is large compared with v_i such that diode switching is controlled solely by the local oscillator. Therefore,

$$v_o = \begin{cases} v_L + v_i & \text{for } v_L > 0 \\ -(v_L + v_i) & \text{for } v_L < 0 \end{cases}$$

This output voltage can be expressed as follows:

$$v_o = (v_L + v_i)s(t)$$

where

$$s(t) = \begin{cases} 1 & v_L > 0 \\ -1 & v_L < 0 \end{cases}$$

We already know the Fourier series representation for this $s(t)$, given by (13.3.4) as

$$s(t) = \frac{4}{\pi}\sum_{n=1}^{\infty}\frac{(-1)^{n+1}}{n}\sin\frac{n\pi}{2}\cos n\omega_L t \tag{13.3.4}$$

If v_i and v_s are sinusoidal voltages

$$v_i = V_i \cos \omega_i t$$

and

$$v_L = V_L \cos \omega_L t$$

then the output voltage will be

$$v_o = (V_i \cos \omega_i t + V_L \cos \omega_L t)\frac{4}{\pi}\sum_{n=1}^{\infty}\frac{(-1)^{n+1}}{n}\sin\frac{n\pi}{2}\cos n\omega_L t$$

or

$$v_o = \frac{2}{\pi}\sum_{n=1}^{\infty}\frac{(-1)^{n+1}}{n}\sin\frac{n\pi}{2}\{V_i[\cos(n\omega_L+\omega_i)t+\cos(n\omega_L-\omega_i)t]$$

$$+V_L[\cos(n\omega_L+\omega_L)t+\cos(n\omega_L-\omega_L)t]\}$$

13.6 FET MIXERS

Diodes are commonly used in mixer circuits because of certain advantages, including a relatively low noise. A major disadvantage of these circuits is the conversion loss, which cannot be reduced below 4 dB. On the other hand, transistor circuits can provide a conversion gain, although they may be noisier. A well-designed FET mixer produces less distortion than do BJT circuits. Further, the range of its input voltage is generally much higher (10 times or so) with respect to BJT mixers. The switching and resistive types of mixer have been designed using the FET. In this section we present an overview of selected FET mixer circuits.

Figure 13.31 shows a FET mixer circuit. Local oscillator signal v_L is applied to the source terminal of the FET via capacitor C_1 while the input v_2 is connected to its gate. For the FET operating in saturation, we can write

$$i_D = I_{DSS}\left(1-\frac{V_{gs}}{V_p}\right)^2 \tag{13.6.1}$$

where i_D is the drain current of the FET, V_{gs} the gate-to-source voltage, V_p its pinch-off voltage, and I_{DSS} the drain current for $V_{gs}=0$. For the circuit shown in Figure 13.31, the total gate voltage V_{gs} is

$$V_{gs} = v_i - v_L + V_{GS}$$

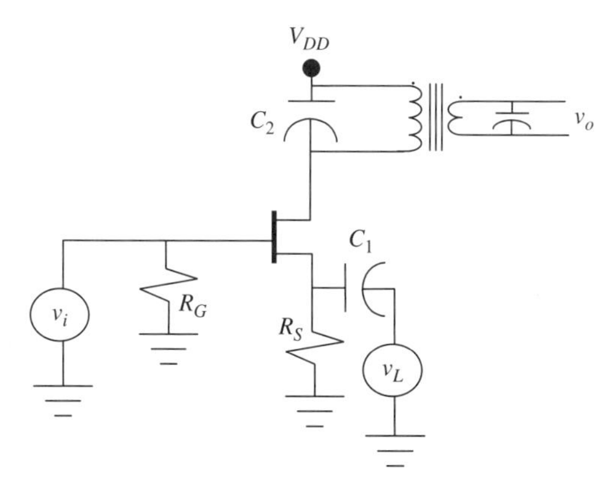

Figure 13.31 FET mixer circuit.

where V_{GS} is the gate-to-source dc bias voltage. Therefore, the drain current i_D in this case is

$$i_D = I_{DSS}\left(1 - \frac{v_i - v_L + V_{GS}}{V_p}\right)^2 \tag{13.6.2}$$

For sinusoidal v_i and v_L given as

$$v_i = V_i \cos \omega_i t$$

and

$$v_L = V_L \cos \omega_L t$$

the drain current i_D can be found from (13.6.2). Hence,

$$i_D = I_{DSS}\left[1 - \frac{V_i \cos \omega_i t - V_L \cos \omega_L t + V_{GS}}{V_p}\right]^2$$

or

$$\begin{aligned} i_D = I_{DSS}&\left\{1 - \frac{2}{V_p}[V_i \cos \omega_i t - V_L \cos \omega_L t + V_{GS}]\right\} \\ &+ \frac{I_{DSS}}{V_p^2}[V_{GS}^2 + 2V_{GS}V_i \cos \omega_i t + V_i^2 \cos^2 \omega_i t - 2V_L V_{GS} \cos \omega_L t \\ &+ V_L^2 \cos^2 \omega_L t - 2V_i V_L \cos \omega_i t \cos \omega_L t] \end{aligned} \tag{13.6.3}$$

Equation (13.6.3) can be rearranged in the following form and then the amplitudes of sum and difference frequency components can easily be identified as follows:

$$\begin{aligned} i_D = I_{DC} &+ a_1 \cos \omega_i t + a_2 \cos \omega_L t + b_1 \cos 2\,\omega_i t + 2 \cos 2\,\omega_L t \\ &- c\{\cos[(\omega_L + \omega_i)t] + \cos[(\omega_L - \omega_i)t]\} \end{aligned}$$

where

$$c = \frac{I_{DSS} V_i V_L}{V_p^2} \tag{13.6.4}$$

The ratio of amplitude c to input voltage V_i is defined as the *conversion transconductance* g_c. Hence,

$$g_c = \frac{c}{V_i} = \frac{I_{DSS} V_L}{V_p^2} \tag{13.6.5}$$

For high conversion gain, g_c must be large. It appears from (13.6.5) that the FET with high I_{DSS} should be preferable, but that is generally not the case. I_{DSS} and V_p of a FET are related such that the device with high I_{DSS} also has high V_p. Therefore, it may result in a lower g_c than that of a low I_{DSS} device. V_L

is directly related to the conversion gain. Hence, higher local oscillator voltage increases the conversion gain. Since the FET is to be operated in its saturation mode (the constant-current region), V_L must be less than the magnitude of the pinch-off voltage.

As a special case, if $V_L = |V_p|/2$, the conversion transconductance may be found as

$$g_c = \frac{I_{DSS}}{2V_p} \tag{13.6.6}$$

Since the transconductance g_m of a JFET is given as

$$g_m = \frac{\partial i_D}{\partial V_{GS}} = -\frac{2I_{DSS}}{V_p}\left(1 - \frac{V_{GS}}{V_p}\right) \tag{13.6.7}$$

and

$$g_m|_{V_{GS}=0} = -\frac{2I_{DSS}}{V_p} \tag{13.6.8}$$

the conversion transconductance is one-fourth of the small-signal transconductance evaluated at $V_{GS} = 0$ (provided that $V_L = 0.5V_p$). For a MOSFET, it can be shown that the conversion transconductance cannot exceed one-half of the small-signal transconductance.

Note from (13.6.3) that there are fundamental terms of input as well as local oscillator frequencies, their second-order harmonics, and the desired $\omega_i \pm \omega_L$ components. Unlike the diode mixer, higher-order spurious signals are not present in this output. In reality, there is some possibility for those terms to be present because of the circuit imperfections.

An alternative to the circuit shown in Figure 13.31 combines the local oscillator signal with the input before applying it to the gate of FET. The analysis of this circuit is similar to the one presented above. In either case, one of the serious problems is isolation among the three signals. There have been numerous attempts to address those problems.

Example 13.5 A given JFET has $I_{DSS} = 50\,\text{mA}$ and transconductance $g_m = 200\,\text{mS}$ when its gate voltage V_{GS} is zero. If it is being used in a mixer that has a 50-Ω load, find the conversion gain of the circuit.

SOLUTION Since

$$g_m|_{V_{GS}=0} = -\frac{2I_{DSS}}{V_p}$$

then

$$V_p = \frac{2 \times 50 \times 10^{-3}}{200 \times 10^{-3}} = \frac{1}{2} = 0.5\,\text{V}$$

If the local oscillator voltage V_L is kept at approximately 0.25 V, the conversion transconductance g_c may be found as

$$g_c = \frac{50 \times 10^{-3}}{2 \times 0.5} = 50 \times 10^{-3}\text{S} = 50\,\text{mS}$$

Therefore, the magnitude of voltage gain A_v is

$$A_v \approx g_c R_L = 50 \times 10^{-3} \times 50 = 2.5$$

Since a common-gate configuration of the FET is generally used in the mixer circuit, its current gain is close to unity. Therefore, the conversion power gain of this circuit is approximately 2.5.

A dual-gate FET is frequently employed in a mixer circuit. There are several circuit configurations and modes of operation for dual-gate FET mixers. Figure 13.32 illustrates a circuit that usually performs well in most receiver designs. The working of this circuit can be explained considering that a dual-gate FET represents two single-gate transistors connected in cascode. Hence, the upper FET works as a source follower and the lower one provides mixing. An RF signal is applied to a gate that is close to the ground while the local oscillator is connected at the other gate (one close to the drain in common-source configuration). If RF is connected to the other gate, the drain resistance of the lower transistor section appears as source resistance and thus reduces the signal.

This circuit is used as a mixer when both transistors are operating in their triode (nonsaturation) mode. A bypass circuit is needed before its output to block the input (RF) and the local oscillator signals. Further, another bypass circuit may be needed across the local oscillator to block the output frequencies. This circuit has lower gain and higher noise, but a properly designed mixer can achieve a modestly higher third-harmonic intercept compared with a single-gate mixer.

Figure 13.33 shows a FET mixer circuit that employs two transistors and transformers. A local oscillator signal is applied to both transistors in the same phase while one RF input is out of phase with respect to the other. The resulting output is a differential of the two sides, and therefore the local oscillator signal is

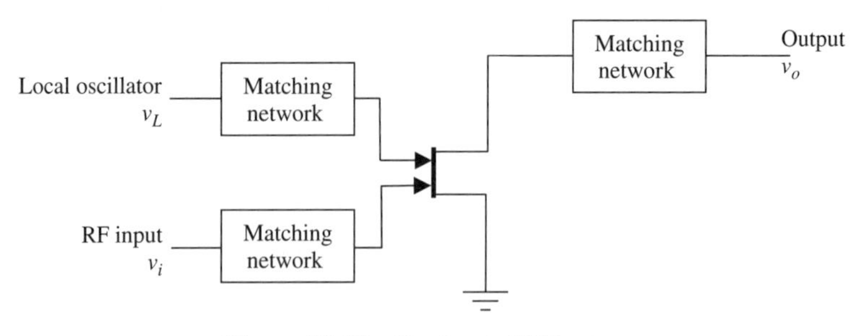

Figure 13.32 Dual-gate FET mixer.

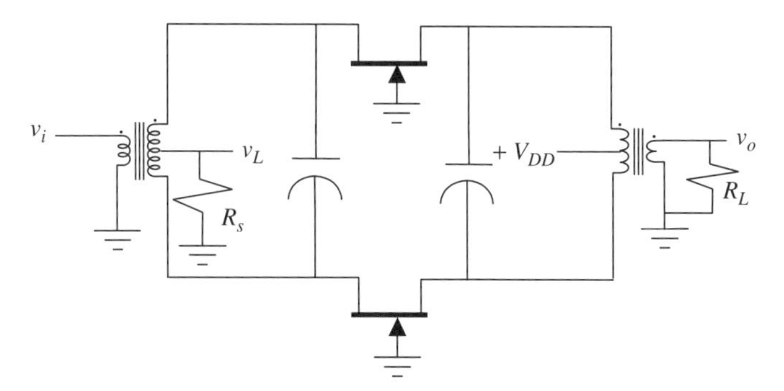

Figure 13.33 FET balanced mixer.

canceled out while the RF is added. The nonlinear characteristic of the transistors generates the upper and lower sideband signals.

If the dc bias voltage at the source is V_S, the RF input is $V_i \cos \omega_i t$, and the local oscillator voltage is $V_L \cos \omega_L t$, the output voltage of this circuit is

$$v_o = \left(\frac{4R_L V_i}{V_P} + \frac{4R_L V_i V_S}{V_P^2} \right) \cos \omega_i t + \frac{2R_L V_i V_L}{V_P^2} \times [\cos(\omega_L + \omega_i)t + \cos(\omega_L - \omega_i)t] \tag{13.6.9}$$

This shows that the output of this circuit contains the RF input frequency along with two sidebands. Asymmetry and other imperfection can add more frequency components as well.

Figure 13.34 illustrates an active double-balanced circuit, known as the *Gilbert cell mixer*, commonly employed in radio-frequency integrated circuits (RFICs). Transistors Q_1 and Q_2 are used to convert the RF voltage to current, and Q_3 to Q_6 are used for the switching. RF and local oscillator signals are applied to the circuit via the corresponding matching networks (baluns). Similarly, the IF output requires a matching network as well. A detailed analysis of the circuit is beyond the scope of this book; only a qualitative analysis is presented here.

Assume that the local oscillator signal is a large square wave that can turn on the transistors when it is high and turn them off when it is low. Note that the current i_{if1} is the sum of i_3 and i_5, and i_{if2} is the sum of i_4 and i_6. Further, the RF current i_1 is the sum of i_3 and i_4 and the sum of i_5 and i_6 is equal to i_2. The bias current I_b is the sum of i_1 and i_2. Since a transistor is on only when the local oscillator signal is high, transistors Q_4 and Q_5 will be on while Q_3 and Q_6 will be in off conditions. Therefore,

$$i_{\text{if1}} = i_5 = i_2$$

and

$$i_{\text{if2}} = i_4 = i_1$$

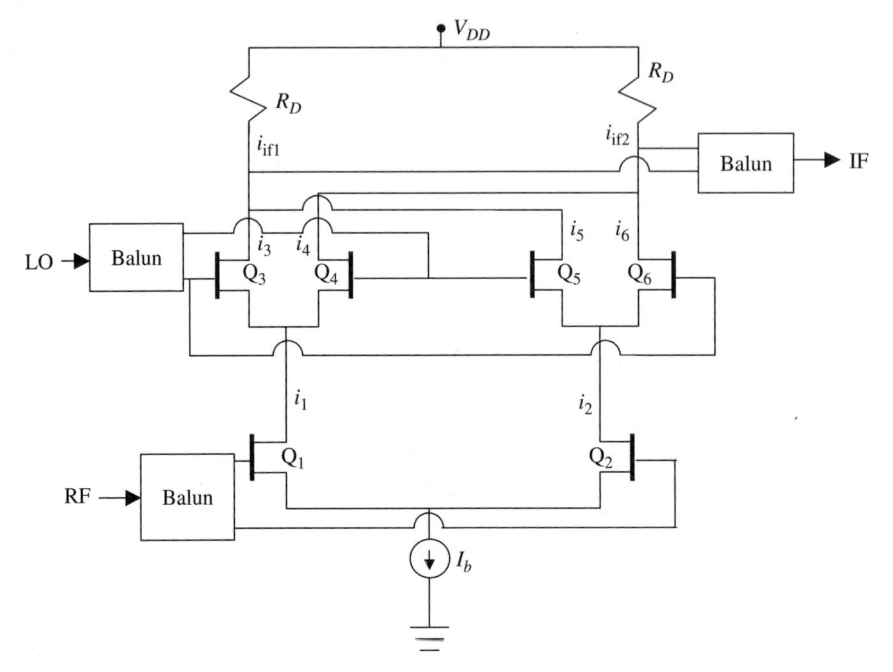

Figure 13.34 Gilbert cell mixer circuit.

Similarly, when the local oscillator signal is low, transistors Q_3 and Q_6 will be on and Q_4 and Q_5 will be off. Therefore,

$$i_{\text{if1}} = i_3 = i_1$$

and

$$i_{\text{if2}} = i_6 = i_2$$

Hence the IF output v_{if} can be found to be

$$v_{\text{if}} = \begin{cases} R_D(i_2 - i_1) & v_L > 0 \\ R_D(i_1 - i_2) & v_L < 0 \end{cases} \tag{13.6.10}$$

or

$$v_{\text{if}} = R_D(i_2 - i_1)s(t) \tag{13.6.11}$$

where $s(t)$ is defined by (13.3.3) and (13.3.4). Further, the conversion gain G_c of this circuit can be found after assuming that it is symmetrical, and therefore the transconductance g_m of Q_1 is the same as that of Q_2. It is given as

$$G_c = \frac{4}{\pi} g_m R_D \tag{13.6.12}$$

SUGGESTED READING

Bahl, I. J., and P. Bhartia, *Microwave Solid State Circuit Design*. New York: Wiley, 1988.

Collin, R. E., *Foundations for Microwave Engineering*. New York: McGraw-Hill, 1992.

Larson, L. E. (ed.), *RF and Microwave Circuit Design for Wireless Communications*. Boston: Artech House, 1996.

Leung, B., *VLSI for Wireless Communication*. Upper Saddle River, NJ: Prentice Hall, 2002.

Maas, S. A., *Microwave Mixers*. Boston: Artech House, 1993.

Pozar, D. M., *Microwave Engineering*. New York: Wiley, 1998.

Rohde, U. L., *Microwave and Wireless Synthesizers*. New York: Wiley, 1997.

Smith, J. R., *Modern Communication Circuits*. New York: McGraw-Hill, 1998.

Vendelin, G. D., A. Pavio, and U. L. Rhode, *Microwave Circuit Design Using Linear and Non-linear Techniques*. New York: Wiley, 1990.

PROBLEMS

13.1. In an AM signal, the carrier output is 1 kW. If the wave modulation is 100%, determine the power in each sideband. How much power is being transmitted?

13.2. Assume that an AM transmitter is modulated with a video signal given by $m(t) = -0.15 + 0.7 \sin \omega_1 t$, where f_1 is 4 MHz. Let the unmodulated carrier amplitude be 100 V. Evaluate and skctch the spectrum of modulated signal.

13.3. An AM broadcast station that uses 95% modulation is operating at total output power of 100 kW. Find the power transmitted in its sidebands.

13.4. With a modulation index of 80% an AM transmitter produces 15 kW. How much of this is carrier power? Find the percentage power saving if the carrier and one of the sidebands are suppressed before transmission.

13.5. When an AM station transmits an unmodulated carrier, the current through its antenna is found to be 10 A. It increases to 12 A with a modulated signal. Determine the modulation index used by the station.

13.6. In the circuit arrangement shown in Figure P13.6, $V_{out} = V_1 - V_2$ is the output signal, where $V_1 = 10V_C + 0.2V_C V_m$ and $V_2 = 6V_C - 0.2V_C V_m$.

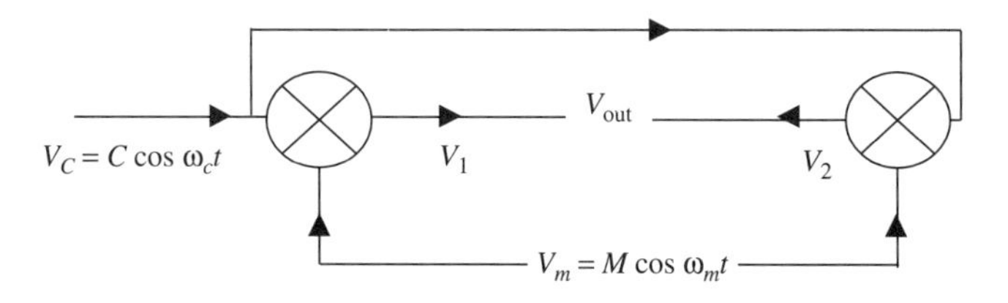

Figure P13.6

(a) Calculate the value of M that results in a modulation index of 0.8.

(b) Determine the value of C that results in an unmodulated output voltage amplitude of 8 V.

(c) If this V_{out} is applied to a 50-Ω load, find the power in the carrier and in each sideband.

13.7. The circuit shown in Figure P13.7 uses a zero-biased diode with $R_j = 500\ \Omega$, and $n = 1$ at 290 K. If the signal applied to this circuit is

$$v_{in}(t) = 0.5[1 + 0.3\cos(4\pi \times 10^4 t + 0.15)]\cos(\pi \times 10^9 t - 0.15) \qquad \text{V}$$

find the output voltage v_{out}. Also, sketch the frequency spectra of v_{in} and v_{out}.

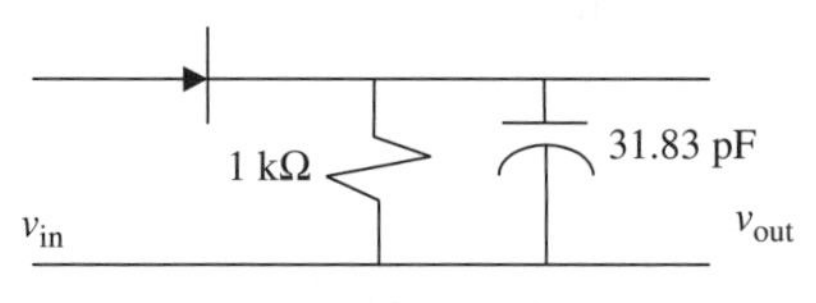

Figure P13.7

13.8. A 1-MHz carrier signal is amplitude-modulated simultaneously with 300-Hz, 800-Hz, and 2-kHz sinusoidal signals. What frequencies are present in the output?

13.9. An AM broadcast transmitter radiates 50 kW of carrier power. If the modulation index is 0.85, find the power radiated.

13.10. An FM broadcast station uses a 10-kHz audio signal of 2 V peak value to modulate the carrier. If the allowed frequency deviation is 20 kHz, determine the bandwidth requirement.

13.11. A frequency modulator is supplied with a carrier of 10 MHz and a 1.5-kHz audio signal of 1 V. The frequency deviation at its output is found to be ±9 kHz.

(a) An FM wave from the modulator is passed through a series of frequency multipliers with a total multiplication of 15 (i.e., $f_o = 15 f_c$). Determine the frequency deviation at the output of the multiplier.

(b) An output of the modulator is mixed with a 58-MHz signal. Find the sum frequency output of the mixer and its frequency deviation.

(c) Find the modulation index at the output of the modulator.

(d) The audio input to the modulator is changed to 500 Hz with its amplitude at 2 V. Determine the modulation index and frequency deviation at the output of the modulator.

13.12. An FM signal,

$$v_{FM}(t) = 1500\cos[2\pi \times 10^9 t + 2\sin(3\pi \times 10^5)] \qquad \text{V}$$

is applied to a 75-Ω antenna. Determine (**a**) the carrier frequency, (**b**) the transmitted power, (**c**) the modulation index, and (**d**) the frequency of the modulating signal.

13.13. When the modulating frequency in FM is 400 Hz and the modulating voltage is 2.4 V, the modulation index is 60. Calculate the maximum deviation in frequency. What is the modulation index when the modulating frequency is reduced to 250 Hz and the modulating voltage is simultaneously raised to 3.2 V?

13.14. The $i_D - V_{gs}$ characteristics of a FET are given by

$$i_D = 7\left(1 - \frac{2V_{gs}}{7}\right)^2 \quad \text{mA}$$

If it is being used in a mixer that has a 50-Ω load, find the conversion gain of the circuit.

13.15. Calculate the conversion loss of the double-balanced mixer shown in Figure P13.15. The diode ON resistance is much less than the load resistance R_L. What value of R_L is required for maximum power transfer if the transformer employed in the circuit has 400 turns in its primary and 800 turns with a center tap on its secondary side?

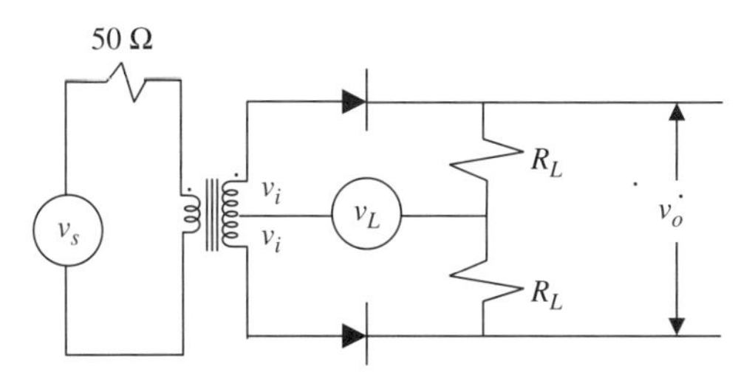

Figure P13.15

13.16. In the mixer circuit shown in Figure P13.16, the transformer and diodes are ideal. The transformer has 200 turns on each side. Find the value of R_L that provides maximum power transfer from the source.

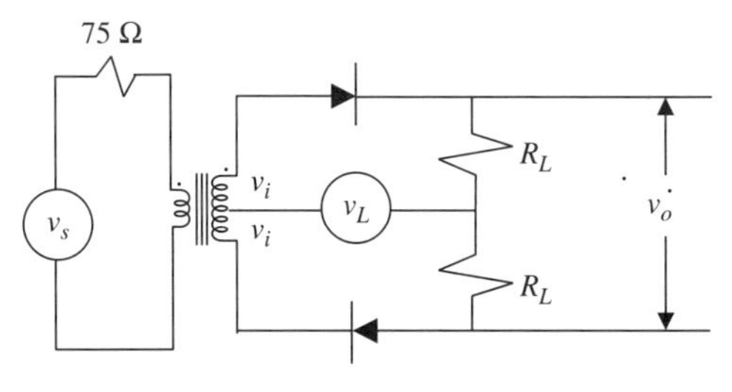

Figure P13.16

APPENDIX 1

DECIBELS AND NEPER

Consider the two-port network shown in Figure A1.1. Assume that V_1 and V_2 are the voltages at its ports. The voltage gain (or loss) G_v of this circuit is expressed in decibels as

$$G_v = 20 \log_{10} \frac{V_2}{V_1} \qquad \text{dB} \tag{A1.1}$$

Similarly, if P_1 and P_2 are the power levels at the two ports, the power gain (or loss) of this circuit in decibels is given as

$$G_p = 10 \log_{10} \frac{P_2}{P_1} \qquad \text{dB} \tag{A1.2}$$

Thus, the dB unit provides a relative level of the signal. For example, if we are asked to find P_2 in watts for G_p as 3 dB, we also need P_1. Otherwise, the only information we can deduce is that P_2 is twice P_1.

Sometimes power is expressed in logarithmic units, such as dBW and dBm. These units are defined as follows. If P is power in watts, it can be expressed in dBW as

$$G = 10 \log_{10} P \qquad \text{dBW} \tag{A1.3}$$

On the other hand, if P is in milliwatts, the corresponding dBm power is found as

$$G = 10 \log_{10} P \qquad \text{dBm} \tag{A1.4}$$

Radio-Frequency and Microwave Communication Circuits: Analysis and Design, Second Edition, By Devendra K. Misra
ISBN 0-471-47873-3

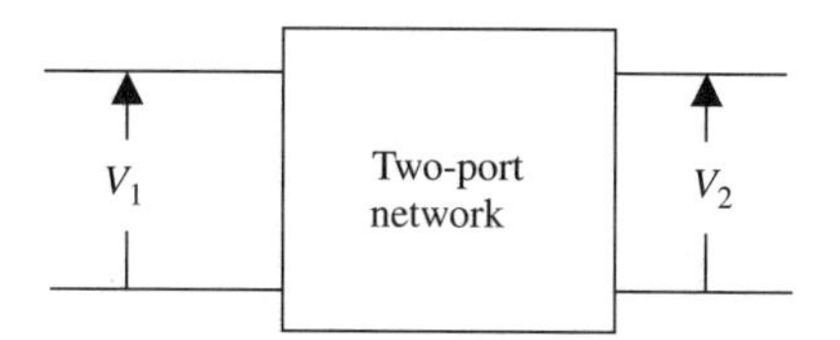

Figure A1.1 Two-port network.

Thus, the dBW and dBm units represent power relative to 1 W and 1 mW, respectively.

Another decibel unit that is commonly used to specify phase noise of an oscillator or strengths of various sidebands of a modulated signal is dBc. It specifies the signal strength relative to the carrier. Consider a 100-MHz oscillator that has an output power of -10 dBm. Suppose that its output power is -30 dBm in the frequency range 105 to 106 MHz; then power per hertz in the output spectrum is -90 dBm and the phase noise is -80 dBc.

In general, if amplitudes of the sideband and the carrier are given as V_2 and V_c, respectively, the sideband in dBc is found as

$$G_2 = 20 \log \frac{V_2}{V_c} \tag{A1.5}$$

Consider now a 1-m-long transmission line with its attenuation constant α. If V_1 is the signal voltage at its input port, the voltage V_2 at its output is given as

$$|V_2| = |V_1| e^{-\alpha} \tag{A1.6}$$

Therefore, the voltage gain G_v of this circuit in nepers is

$$G_v(\mathrm{N_P}) = \ln \frac{|V_2|}{|V_1|} = -\alpha \qquad \mathrm{N_P} \tag{A1.7}$$

On the other hand, G_v in decibels is found to be

$$G_v(\mathrm{dB}) = 20 \log \frac{|V_2|}{|V_1|} = 20 \log e^{-\alpha} = -20\alpha \log e = -8.6859\alpha \qquad \mathrm{dB}$$

Therefore,

$$1\mathrm{N_P} = 8.6859\,\mathrm{dB} \tag{A1.8}$$

The negative sign indicates that V_2 is smaller than V_1 and there is loss of signal.

APPENDIX 2

CHARACTERISTICS OF SELECTED TRANSMISSION LINES

COAXIAL LINE

Consider a coaxial line with its inner and outer conductor radii a and b, respectively, as illustrated in Figure A2.1. Further, ε_r is the dielectric constant of the insulating material. The line parameters L, R, C, and G of this coaxial line are found as follows:

$$C = \frac{55.63\varepsilon_r}{\ln(b/a)} \qquad \text{pF/m} \tag{A2.1}$$

and

$$L = 200 \ln \frac{b}{a} \qquad \text{nH/m} \tag{A2.2}$$

If the coaxial line has small losses due to imperfect conductor and insulator, its resistance and conductance parameters can be calculated as follows:

$$R \approx 10\left(\frac{1}{a} + \frac{1}{b}\right)\sqrt{\frac{f_{\text{GHz}}}{\sigma}} \qquad \Omega/\text{m} \tag{A2.3}$$

and

$$G = \frac{0.3495\varepsilon_r f_{\text{GHz}} \tan\delta}{\ln(b/a)} \qquad \text{S/m} \tag{A2.4}$$

Radio-Frequency and Microwave Communication Circuits: Analysis and Design, Second Edition,
By Devendra K. Misra
ISBN 0-471-47873-3

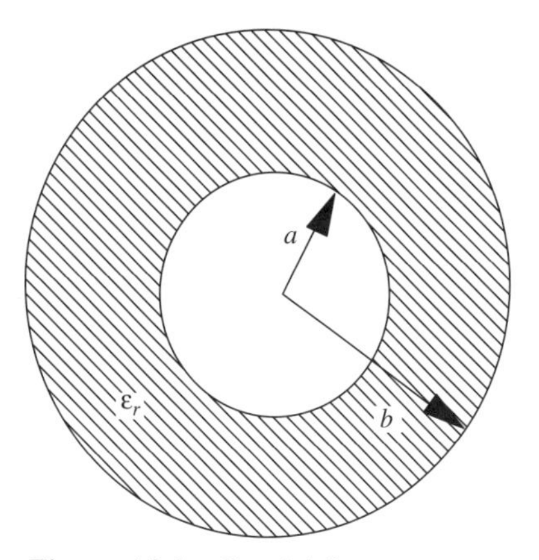

Figure A2.1 Coaxial line geometry.

where $\tan\delta$ is the loss tangent of the dielectric material, σ the conductivity of the conductors in S/m, and f_{GHz} the signal frequency in GHz. The characteristic impedance and propagation constant of the coaxial line can be calculated easily using the formulas given in Chapter 3.

Attenuation constants α_c and α_d due to conductor and dielectric losses, respectively, may be determined from

$$\alpha_c = \frac{R_s}{2\sqrt{\mu_0/\varepsilon_0\varepsilon_r}\,\ln(b/a)}\left(\frac{1}{a}+\frac{1}{b}\right) \tag{A2.5}$$

$$\alpha_d = \frac{\omega}{2}\sqrt{\mu_0\varepsilon_0\varepsilon_r}\,\tan\delta \tag{A2.6}$$

where

$$R_s = \sqrt{\frac{\omega\mu_0}{2\sigma}} \tag{A2.7}$$

STRIP LINE

Strip line geometry is illustrated in Figure A2.2. Insulating material of thickness h has a dielectric constant ε_r. The width and thickness of the central conducting strip are w and t, respectively. For the case of $t = 0$, its characteristic impedance can be found as follows:

$$Z_0 = \frac{1}{\sqrt{\varepsilon_r}}\begin{cases} 296.1\left(0.6931+\ln\dfrac{1+\sqrt{x'}}{1-\sqrt{x'}}\right)^{-1} & 0 < x \le 0.7 \\ 30\left(0.6931+\ln\dfrac{1+\sqrt{x}}{1-\sqrt{x}}\right) & 0.7 \le x < 1 \end{cases} \tag{A2.8}$$

$$x' = \tanh\frac{\pi w}{2h} \tag{A2.9}$$

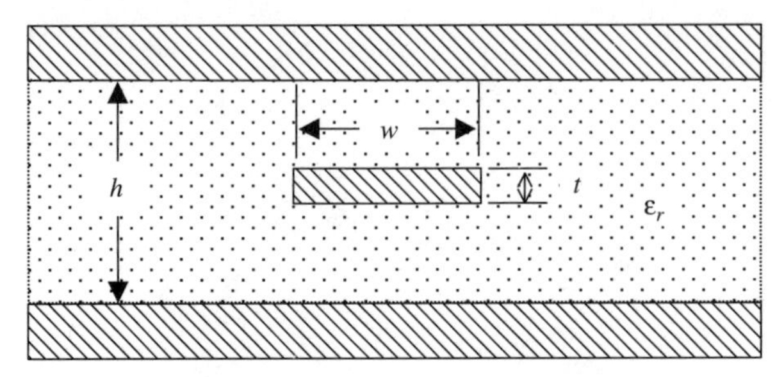

Figure A2.2 Strip line geometry.

and

$$x = \sqrt{1 - x'^2} \tag{A2.10}$$

For the design of a strip line, the following convenient formulas can be used:

$$\frac{w}{h} = 0.6366 \tanh^{-1} \sqrt{k} \tag{A2.11}$$

where

$$k = \begin{cases} [(e^{\pi/y} - 2)/(e^{\pi/y} + 2)]^2 & 0 \le y \le 1 \\ \sqrt{1 - [(e^{\pi y} - 2)/(e^{\pi y} + 2)]^4} & 1 \le y \end{cases} \tag{A2.12}$$

$$y = Z_0 \frac{\sqrt{\varepsilon_r}}{94.18} \tag{A2.13}$$

For $t \neq 0$,

$$\frac{w}{h} = \Delta_1 - \Delta_2 \tag{A2.14}$$

$$\Delta_1 = 2.5465(1 - \Lambda) \frac{\sqrt{e^y + 0.568}}{e^y - 1} \tag{A2.15}$$

and

$$\Delta_2 = \frac{\Lambda}{\pi} \left\{ 1 - 0.5 \ln \left[\left(\frac{\Lambda}{2 - \Lambda} \right)^2 + \left(\frac{0.0796\Lambda}{\Delta_1 - 0.26\Lambda} \right)^\delta \right] \right\} \tag{A2.16}$$

where

$$\Lambda = \frac{t}{h} \tag{A2.17}$$

$$\delta = \frac{2}{1 + \frac{2}{3}[\Lambda/(1 - \Lambda)]} \tag{A2.18}$$

and y is as defined in (A2.13).

MICROSTRIP LINE

The geometry of a microstrip line is illustrated in Figure A2.3. Dielectric substrate on the conducting ground is h meters high and its dielectric constant is ε_r. The width and thickness of the conducting strip on its top are w and t, respectively. The characteristic impedance of this line can be found as follows:

$$Z_0 = \begin{cases} \dfrac{60}{\sqrt{\varepsilon_{re}}} \ln\left(\dfrac{8h}{w_e} + \dfrac{w_e}{4h}\right) & \text{for } \dfrac{w_e}{h} \leq 1 \\ \dfrac{376.7}{\sqrt{\varepsilon_{re}}} \left[\dfrac{w_e}{h} + 1.393 + 0.667 \ln\left(\dfrac{w_e}{h} + 1.444\right)\right]^{-1} & \text{for } \dfrac{w_e}{h} \geq 1 \end{cases} \tag{A2.19}$$

where

$$\frac{w_e}{h} = \begin{cases} \dfrac{w}{h} + 0.3979\dfrac{t}{h}\left[1 + \ln\left(12.5664\dfrac{w}{t}\right)\right] & \dfrac{w}{h} \leq \dfrac{1}{2\pi} \\ \dfrac{w}{h} + 0.3979\dfrac{t}{h}\left[1 + \ln\left(2\dfrac{h}{t}\right)\right] & \dfrac{w}{h} \geq \dfrac{1}{2\pi} \end{cases} \tag{A2.20}$$

The effective dielectric constant ε_{re} of a microstrip line ranges between ε_r and 1 because of its propagation characteristics. If the signal frequency is low such that the dispersion is not a problem, it can be determined as follows:

$$\varepsilon_{re} = 0.5\left[\varepsilon_r + 1 + (\varepsilon_r - 1)F\left(\frac{w}{h}\right)\right] - \frac{\varepsilon_r - 1}{4.6}\frac{t}{h}\sqrt{\frac{h}{w}} \tag{A2.21}$$

where

$$F\left(\frac{w}{h}\right) = \begin{cases} \left(1 + 12\dfrac{h}{w}\right)^{-0.5} + 0.04\left(1 - \dfrac{w}{h}\right)^2 & \dfrac{w}{h} \leq 1 \\ \left(1 + 12\dfrac{h}{w}\right)^{-0.5} & \dfrac{w}{h} \geq 1 \end{cases} \tag{A2.22}$$

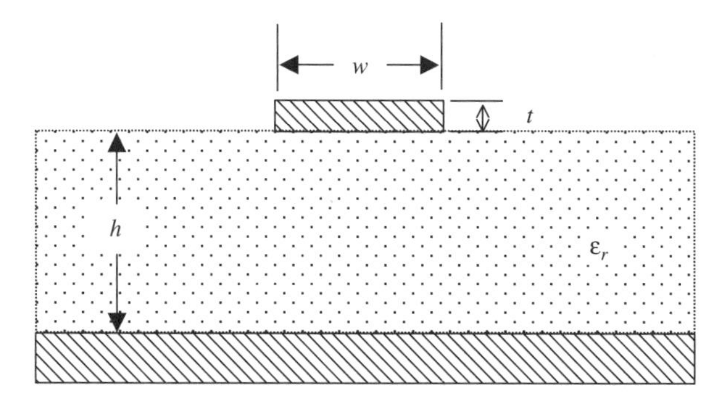

Figure A2.3 Microstrip line geometry.

If dispersion cannot be ignored, the effective dielectric constant may be found as follows:

$$\varepsilon_{re}(f) = \left(\frac{\sqrt{\varepsilon_r} - \sqrt{\varepsilon_{re}}}{1 + 4F^{-1.5}} + \sqrt{\varepsilon_{re}} \right)^2 \tag{A2.23}$$

where

$$F = \frac{40}{3} f_{\mathrm{GHz}} h \sqrt{\varepsilon_r - 1} \left\{ 0.5 + \left[1 + 2 \log \left(1 + \frac{w}{h} \right) \right]^2 \right\} \tag{A2.24}$$

The corresponding characteristic impedance is determined from the following formula:

$$Z_0(f) = Z_0 \frac{\varepsilon_{re}(f) - 1}{\varepsilon_{re} - 1} \sqrt{\frac{\varepsilon_{re}}{\varepsilon_{re}(f)}} \tag{A2.25}$$

Attenuation constants α_c and α_d for the conductor and dielectric losses, respectively, are determined as follows:

$$\alpha_c = \begin{cases} 9.9825 \dfrac{\zeta}{h Z_0} \sqrt{\dfrac{f_{\mathrm{GHz}}}{\sigma}} \dfrac{32 - (w_e/h)^2}{32 + (w_e/h)^2} \ \text{Np/m} & \dfrac{w}{h} \le 1 \\ 44.1255 \times 10^{-5} \dfrac{\zeta Z_0 \varepsilon_{re}}{h} \sqrt{\dfrac{f_{\mathrm{GHz}}}{\sigma}} \left[\dfrac{w_e}{h} + \dfrac{0.667(w_e/h)}{(w_e/h) + 1.444} \right] \ \text{Np/m} & \dfrac{w}{h} \ge 1 \end{cases} \tag{A2.26}$$

where

$$\zeta = \begin{cases} 1 + \dfrac{h}{w_e} \left[1 + \dfrac{1.25t}{\pi w} + \dfrac{1.25}{\pi} \ln \left(4\pi \dfrac{w}{t} \right) \right] & \dfrac{w}{h} \le \dfrac{1}{2\pi} \\ 1 + \dfrac{h}{w_e} \left[1 - \dfrac{1.25t}{\pi h} + \dfrac{1.25}{\pi} \ln \left(2 \dfrac{h}{t} \right) \right] & \dfrac{w}{h} \ge \dfrac{1}{2\pi} \end{cases} \tag{A2.27}$$

σ is conductivity of the conductor and f_{GHz} is the signal frequency in GHz.

$$\alpha_d = 10.4766 \frac{\varepsilon_r}{\varepsilon_r - 1} \frac{\varepsilon_{re} - 1}{\sqrt{\varepsilon_{re}}} f_{\mathrm{GHz}} \tan \delta \qquad \text{Np/m} \tag{A2.28}$$

For the design of a microstrip line that has negligible dispersion, the following formulas may be more convenient. For $A > 1.52$,

$$\frac{w}{h} = \frac{8e^A}{e^{2A} - 2} \tag{A2.29}$$

For $A \le 1.52$,

$$\frac{w}{h} = 0.6366 \left\{ B - 1 - \ln(2B - 1) + \frac{\varepsilon_r - 1}{2\varepsilon_r} \left[\ln(B - 1) + 0.39 - \frac{0.61}{\varepsilon_r} \right] \right\} \tag{A2.30}$$

where

$$A = \frac{Z_0}{84.8528}\sqrt{\varepsilon_r + 1} + \frac{\varepsilon_r - 1}{\varepsilon_r + 1}\left(0.23 + \frac{0.11}{\varepsilon_r}\right) \tag{A2.31}$$

and

$$B = \frac{592.2}{Z_0\sqrt{\varepsilon_r}} \tag{A2.32}$$

Experimental verification indicates that (A2.29) and (A2.30) are fairly accurate as long as $t/h \leq 0.005$.

APPENDIX 3

SPECIFICATIONS OF SELECTED COAXIAL LINES AND WAVEGUIDES

TABLE A3.1 Selected Computer, Instrumentation, and Broadcast Cables

Cable Type RG(−)	Insulation	Core O.D. (in.)	Cable O.D. (in.)	Z_0 (Ω)	Capacitance (pF/ft)	Attenuation (dB/100 ft) at 400 MHz
8/U	Polyethylene	0.285	0.405	50	26.0	3.8
9/U	Polyethylene	0.280	0.420	51	30.0	4.1
11/U	Polyethylene	0.285	0.405	75	20.5	4.2
58/U	Polyethylene	0.116	0.195	53.5	28.5	9.5
59/U	Polyethylene	0.146	0.242	75	17.3	5.6
122/U	Polyethylene	0.096	0.160	50	30.8	16.5
141A/U	Teflon	0.116	0.190	50	29.0	9.0
142B/U	Teflon	0.116	0.195	50	29.0	9.0
174/U	Teflon	0.060	0.100	50	30.8	20.0
178B/U	Teflon	0.034	0.072	50	29.0	28.0
179B/U	Teflon	0.063	0.100	75	19.5	21.0
180B/U	Teflon	0.102	0.140	95	15.0	17.0
213/U	Polyethylene	0.285	0.405	50	30.8	4.7
214/U	Polyethylene	0.285	0.425	50	30.8	4.7
223/U	Polyethylene	0.116	0.206	50	30.8	10.0
316/U	Teflon	0.060	0.098	50	29.0	20.0

Radio-Frequency and Microwave Communication Circuits: Analysis and Design, Second Edition,
By Devendra K. Misra
ISBN 0-471-47873-3

TABLE A3.2 Selected Semirigid Coaxial Lines

MIL-C-17F Designation	Nominal Impedance (Ω)	Dielectric Diameter (in.)	Center Conductor Diameter (in.)	Maximum Operating Frequency (GHz)	Capacitance (pF/ft)	Attenuation at 1 GHz (dB/100 ft)
129	50	0.209	0.0641	18	29.6	7.5
130	50	0.1175	0.0362	20	29.9	12
133	50	0.066	0.0201	20	32	22
151	50	0.037	0.0113	20	32	40
154	50	0.026	0.008	20	32	60

TABLE A3.3 Standard Rectangular Waveguides

EIA Nomenclature WR (−)	Inside Dimension (in.)	TE_{10} Mode Cutoff Frequency (GHz)	Recommended Frequency Band for TE_{10} Mode
2300	23.0 × 11.5	0.2565046	0.32–0.49
2100	21.0 × 10.5	0.2809343	0.35–0.53
1800	18.0 × 9.0	0.3277583	0.41–0.62
1500	15.0 × 7.5	0.3933131	0.49–0.75
1150	11.5 × 5.75	0.5130267	0.64–0.98
975	9.75 × 4.875	0.6051054	0.76–1.15
770	7.7 × 3.85	0.7662235	0.96–1.46
650	6.5 × 3.25	0.9077035	1.14–1.73
510	5.1 × 2.55	1.1569429	1.45–2.2
430	4.3 × 2.15	1.3722704	1.72–2.61
340	3.4 × 1.7	1.7357340	2.17–3.30
284	2.84 × 1.34	2.0782336	2.60–3.95
229	2.29 × 1.145	2.5779246	3.22–4.90
187	1.87 × 0.872	3.1530286	3.94–5.99
159	1.59 × 0.795	3.7125356	4.64–7.05
137	1.372 × 0.622	4.3041025	5.38–8.17
112	1.122 × 0.497	5.2660611	6.57–9.99
90	0.9 × 0.4	6.5705860	8.20–12.5
75	0.75 × 0.375	7.8899412	9.84–15.0
62	0.622 × 0.311	9.4951201	11.9–18.0
51	0.51 × 0.255	11.586691	14.5–22.0
42	0.42 × 0.17	14.088529	17.6–26.7
34	0.34 × 0.17	17.415732	21.7–33.0
28	0.28 × 0.14	21.184834	26.4–40.0
22	0.244 × 0.112	26.461666	32.9–50.1
19	0.188 × 0.094	31.595916	39.2–59.6
15	0.148 × 0.074	40.058509	49.8–75.8
12	0.122 × 0.061	48.54910	60.5–91.9
10	0.1 × 0.05	59.35075	73.8–112
8	0.08 × 0.04	74.44066	92.2–140
7	0.065 × 0.0325	91.22728	114–173
5	0.051 × 0.0255	116.47552	145–220
4	0.043 × 0.0215	137.93866	172–261
3	0.034 × 0.017	174.43849	217–330

APPENDIX 4

SOME MATHEMATICAL FORMULAS

$$\sin(A + B) = \sin A \cos B + \cos A \sin B$$

$$\sin(A - B) = \sin A \cos B - \cos A \sin B$$

$$\cos(A + B) = \cos A \cos B - \sin A \sin B$$

$$\cos(A - B) = \cos A \cos B + \sin A \sin B$$

$$\tan(A + B) = \frac{\tan A + \tan B}{1 - \tan A \tan B}$$

$$\tan(A - B) = \frac{\tan A - \tan B}{1 + \tan A \tan B}$$

$$\sin^2 A + \cos^2 A = 1$$

$$\tan^2 A + 1 = \sec^2 A$$

$$\cot^2 A + 1 = \csc^2 A$$

$$\sin A + \sin B = 2 \sin \frac{A + B}{2} \cos \frac{A - B}{2}$$

$$\sin A - \sin B = 2 \cos \frac{A + B}{2} \sin \frac{A - B}{2}$$

$$\cos A + \cos B = 2 \cos \frac{A + B}{2} \cos \frac{A - B}{2}$$

Radio-Frequency and Microwave Communication Circuits: Analysis and Design, Second Edition,
By Devendra K. Misra
ISBN 0-471-47873-3

$$\cos A - \cos B = 2\sin\frac{A+B}{2}\sin\frac{B-A}{2}$$

$$2\sin A\cos B = \sin(A+B) + \sin(A-B)$$

$$2\cos A\sin B = \sin(A+B) - \sin(A-B)$$

$$2\cos A\cos B = \cos(A+B) + \cos(A-B)$$

$$2\sin A\sin B = \cos(A-B) - \cos(A+B)$$

$$\sin\frac{A}{2} = \pm\sqrt{\frac{1-\cos A}{2}}$$

$$\sin 2A = 2\sin A\cos A$$

$$\cos 2A = \cos^2 A - \sin^2 A = 2\cos^2 A - 1 = 1 - 2\sin^2 A$$

$$\tan 2A = \frac{2\tan A}{1-\tan^2 A}$$

$$\cos\frac{A}{2} = \pm\sqrt{\frac{1+\cos A}{2}}$$

$$\tan\frac{A}{2} = \pm\sqrt{\frac{1-\cos A}{1+\cos A}} = \frac{\sin A}{1+\cos A} = \frac{1-\cos A}{\sin A}$$

$$e^{jA} = \cos A + j\sin A$$

$$e^{-jA} = \cos A - j\sin A$$

$$\sin A = \frac{e^{jA}-e^{-jA}}{2j} = A - \frac{A^3}{3!} + \frac{A^5}{5!} - \frac{A^7}{7!} + \cdots$$

$$\cos A = \frac{e^{jA}+e^{-jA}}{2} = 1 - \frac{A^2}{2!} + \frac{A^4}{4!} - \frac{A^6}{6!} + \cdots$$

$$\tan A = j\frac{e^{-jA}-e^{jA}}{e^{jA}+e^{-jA}} = A + \frac{A^3}{3} + 2\frac{A^5}{15} + 17\frac{A^7}{315} + \cdots$$

$$\sinh A = \frac{e^{A}-e^{-A}}{2}$$

$$\cosh A = \frac{e^{A}+e^{-A}}{2}$$

$$\tanh A = \frac{\sinh A}{\cos A} = \frac{e^{A}-e^{-A}}{e^{A}+e^{-A}}$$

$$\coth A = \frac{\cosh A}{\sinh A} = \frac{e^{A}+e^{-A}}{e^{A}-e^{-A}}$$

$$\sinh(A+B) = \sinh A\cosh B + \cosh A\sinh B$$

$$\cosh(A+B) = \cosh A\cosh B + \sinh A\sinh B$$

$$\sinh(A - B) = \sinh A \cosh B - \cosh A \sinh B$$

$$\cosh(A - B) = \cosh A \cosh B - \sinh A \sinh B$$

$$\tanh(A + B) = \frac{\tanh A + \tanh B}{1 + \tanh A \tanh B}$$

$$\tanh(A - B) = \frac{\tanh A - \tanh B}{1 - \tanh A \tanh B}$$

$$\cos jA = \cosh A$$

$$\sin jA = j \sinh A$$

$$\cosh^2 A - \sinh^2 A = 1$$

$$\sinh jA = j \sin A$$

$$\cosh jA = \cos A$$

$$\tanh jA = j \tan A$$

$$e^A = 1 + A + \frac{A^2}{2!} + \frac{A^3}{3!} + \frac{A^4}{4} + \cdots$$

$$e^{-A} = 1 - A + \frac{A^2}{2} - \frac{A^3}{3} + \frac{A^4}{4!} - \cdots$$

$$\sinh A = \frac{e^A - e^{-A}}{2} = A + \frac{A^3}{3!} + \frac{A^5}{5!} + \frac{A^7}{7!} + \cdots$$

$$\cosh A = \frac{e^A + e^{-A}}{2} = 1 + \frac{A^2}{2!} + \frac{A^4}{4!} + \frac{A^6}{6!} + \cdots$$

$$\log_a xy = \log_a x + \log_a y$$

$$\log_a \frac{x}{y} = \log_a x - \log_a y$$

$$\log_a x^y = y \log_a x$$

$$\log_a x = \log_b x \times \log_a b = \frac{\log_b x}{\log_b a}$$

$$\ln x = \log_{10} x \times \ln 10 = 2.302585 \times \log_{10} x$$

$$\log_{10} x = \ln x \times \log_{10} e = 0.434294 \times \ln x$$

$$e = 2.718281828$$

APPENDIX 5

VECTOR IDENTITIES

1. $\mathbf{A} \cdot (\mathbf{B} \times \mathbf{C}) = \mathbf{B} \cdot (\mathbf{C} \times \mathbf{A}) = \mathbf{C} \cdot (\mathbf{A} \times \mathbf{B})$
2. $\mathbf{A} \times (\mathbf{B} \times \mathbf{C}) = \mathbf{B}(\mathbf{A} \cdot \mathbf{C}) - \mathbf{C}(\mathbf{A} \cdot \mathbf{B})$
3. $\nabla(\phi_1\phi_2) = \phi_1\nabla\phi_2 + \phi_2\nabla\phi_1$
4. $\nabla \cdot (\mathbf{A} \times \mathbf{B}) = \mathbf{B} \cdot (\nabla \times \mathbf{A}) - \mathbf{A} \cdot (\nabla \times \mathbf{B})$
5. $\nabla \times \phi\mathbf{A} = \nabla\phi \times \mathbf{A} + \phi(\nabla \times \mathbf{A})$
6. $\nabla \cdot \phi\mathbf{A} = \nabla\phi \cdot \mathbf{A} + \phi(\nabla \cdot \mathbf{A})$
7. $\nabla \cdot (\nabla \times \mathbf{A}) = 0$
8. $\nabla \times (\nabla\phi) = 0$
9. $\nabla \times \nabla \times \mathbf{A} = \nabla(\nabla \cdot \mathbf{A}) - \nabla^2\mathbf{A}$
10. **Stokes's Theorem**. The circulation around a simple closed curve is equal to the integral over any simple surface spanning the curve, of the normal component of the curl, the positive sense on the curve being counterclockwise as seen from the side of the surface toward which the positive normal points. Mathematically,

$$\oint_C \mathbf{A} \cdot d\mathbf{l} = \int_S (\nabla \times \mathbf{A}) \cdot d\mathbf{s}$$

11. **Divergence Theorem** (also known as *Gauss's theorem*). The integral of the divergence of a vector field over a region of space is equal to the

Radio-Frequency and Microwave Communication Circuits: Analysis and Design, Second Edition,
By Devendra K. Misra
ISBN 0-471-47873-3

integral over the surface of that region of the component of the field in the direction of the outward directed normal to the surface. Mathematically,

$$\int_V (\nabla \cdot \mathbf{A}) d\mathbf{v} = \oint_S \mathbf{A} \cdot d\mathbf{s}$$

12. **Helmholtz's Theorem**. A vector is specified uniquely if its divergence and curl are given within a region and its normal component is given over the boundary.

Cartesian Coordinates

$$\nabla f = \hat{x}\frac{\partial f}{\partial x} + \hat{y}\frac{\partial f}{\partial y} + \hat{z}\frac{\partial f}{\partial z}$$

$$\nabla^2 f = \frac{\partial^2 f}{\partial x^2} + \frac{\partial^2 f}{\partial y^2} + \frac{\partial^2 f}{\partial z^2}$$

$$\nabla \cdot \mathbf{A} = \frac{\partial A_x}{\partial x} + \frac{\partial A_y}{\partial y} + \frac{\partial A_z}{\partial z}$$

$$\nabla \times \mathbf{A} = \begin{vmatrix} \hat{x} & \hat{y} & \hat{z} \\ \dfrac{\partial}{\partial x} & \dfrac{\partial}{\partial y} & \dfrac{\partial}{\partial z} \\ A_x & A_y & A_z \end{vmatrix}$$

Cylindrical Coordinates

$$\nabla f = \hat{\rho}\frac{\partial f}{\partial \rho} + \hat{\phi}\frac{1}{\rho}\frac{\partial f}{\partial \phi} + \hat{z}\frac{\partial f}{\partial z}$$

$$\nabla^2 f = \frac{1}{\rho}\frac{\partial}{\partial \rho}\left(\rho\frac{\partial f}{\partial \rho}\right) + \frac{1}{\rho^2}\frac{\partial^2 f}{\partial \phi^2} + \frac{\partial^2 f}{\partial z^2}$$

$$\nabla \cdot \mathbf{A} = \frac{1}{\rho}\frac{\partial(\rho A_\rho)}{\partial \rho} + \frac{1}{\rho}\frac{\partial A_\phi}{\partial \phi} + \frac{\partial A_z}{\partial z}$$

$$\nabla \times \mathbf{A} = \begin{vmatrix} \hat{\rho} & \rho\hat{\phi} & \hat{z} \\ \dfrac{\partial}{\partial \rho} & \dfrac{\partial}{\partial \phi} & \dfrac{\partial}{\partial z} \\ A_\rho & \rho A_\phi & A_z \end{vmatrix}$$

Spherical Coordinates

$$\nabla f = \hat{r}\frac{\partial f}{\partial r} + \hat{\theta}\frac{1}{r}\frac{\partial f}{\partial \theta} + \hat{\phi}\frac{1}{r\sin\theta}\frac{\partial f}{\partial \phi}$$

$$\nabla^2 f = \frac{1}{r^2\sin\theta}\left[\sin\theta\frac{\partial}{\partial r}\left(r^2\frac{\partial f}{\partial r}\right) + \frac{\partial}{\partial \theta}\left(\sin\theta\frac{\partial f}{\partial \theta}\right) + \frac{1}{\sin\theta}\frac{\partial^2 f}{\partial \phi^2}\right]$$

$$\nabla \cdot \mathbf{A} = \frac{1}{r^2 \sin\theta}\left[\sin\theta \frac{\partial}{\partial r}(r^2 A_r) + r\frac{\partial}{\partial\theta}(\sin\theta A_\theta) + r\frac{\partial A_\phi}{\partial\phi}\right]$$

$$\nabla \times \mathbf{A} = \frac{1}{r^2 \sin\theta}\begin{vmatrix} \hat{r} & r\hat{\theta} & r\sin\theta\hat{\phi} \\ \dfrac{\partial}{\partial r} & \dfrac{\partial}{\partial\theta} & \dfrac{\partial}{\partial\phi} \\ A_r & rA_\theta & r\sin\theta A_\phi \end{vmatrix}$$

APPENDIX 6

SOME USEFUL NETWORK TRANSFORMATIONS

This appendix contains transformations of networks that are commonly encountered. These transformations can save time of analysis and design. Figure A6.1 shows a series-connected $R_s - L_s$ circuit transformed to a parallel-connected $R_p - L_p$ circuit. Resistors in the two circuits may represent loss associated with the inductor. These two circuits are equivalent if their impedance (as well as their admittance) values are equal. This equality gives

$$R_p = R_s + \frac{(\omega L_s)^2}{R_s} \tag{A6.1}$$

and

$$L_p = L_s + \frac{R_s^2}{\omega^2 L_s} \tag{A6.2}$$

Transforming the other way gives

$$R_s = \frac{R_p(\omega L_p)^2}{R_p^2 + (\omega L_p)^2} \tag{A6.3}$$

and

$$L_s = \frac{L_p(R_p)^2}{R_p^2 + (\omega L_p)^2} \tag{A6.4}$$

Radio-Frequency and Microwave Communication Circuits: Analysis and Design, Second Edition,
By Devendra K. Misra
ISBN 0-471-47873-3

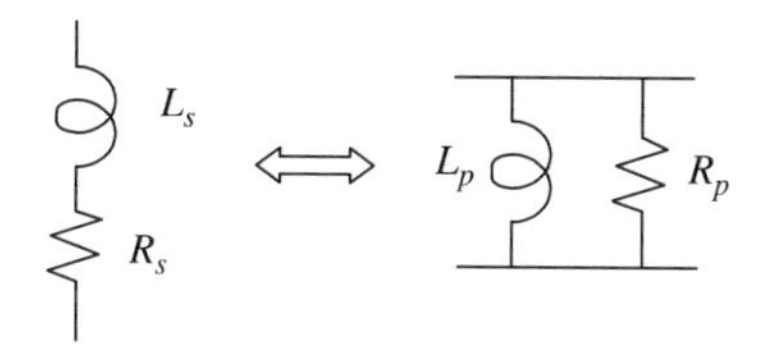

Figure A6.1 Transformation of a series R–L to a shunt R–L circuit, or vice versa.

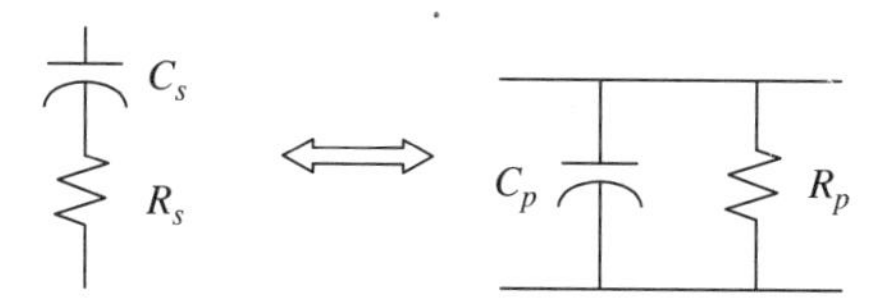

Figure A6.2 Transformation of a series R–C to a shunt R–C circuit, or vice versa.

Similarly, a series R_s–C_s circuit can be transformed to a parallel R_p–C_p circuit, or vice versa, as illustrated in Figure A6.2. By equating their impedance (or admittance) expression, following relations may be found:

$$R_p = \frac{1 + (\omega C_s R_s)^2}{R_s(\omega C_s)^2} \tag{A6.5}$$

and

$$C_p = \frac{C_s}{1 + (\omega C_s R_s)^2} \tag{A6.6}$$

Transforming other way gives

$$R_s = \frac{R_p}{1 + (\omega C_p R_p)^2} \tag{A6.7}$$

and

$$C_s = \frac{1 + (\omega C_p R_p)^2}{\omega^2 C_p R_p^2} \tag{A6.8}$$

The inductor-tapped circuit shown in Figure A6.3 may be transformed to a parallel R_p–L_p circuit. This type of transformation facilitates the analysis of a Hartley oscillator. Again, the impedance of an inductor-tapped circuit must be equal to that of a parallel R_p–L_p circuit. Therefore,

$$R_p = \frac{(\omega L_1 L_2)^2 + R^2(L_1 + L_2)^2}{R L_2^2} \tag{A6.9}$$

and

$$L_p = \frac{(\omega L_1 L_2)^2 + R^2(L_1 + L_2)^2}{\omega^2 L_1 L_2^2 + R^2(L_1 + L_2)} \tag{A6.10}$$

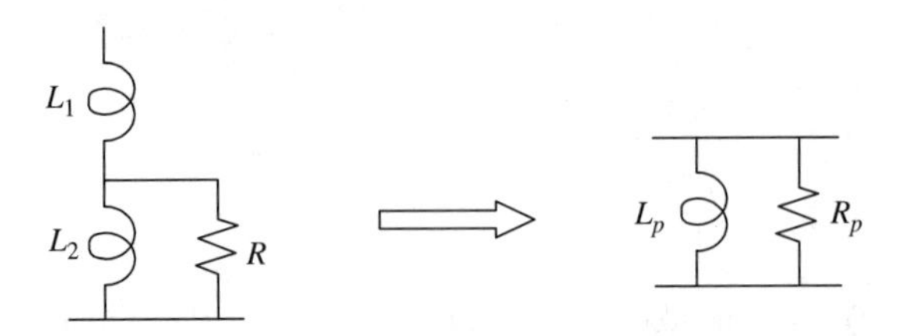

Figure A6.3 Transformation of an inductor-tapped circuit.

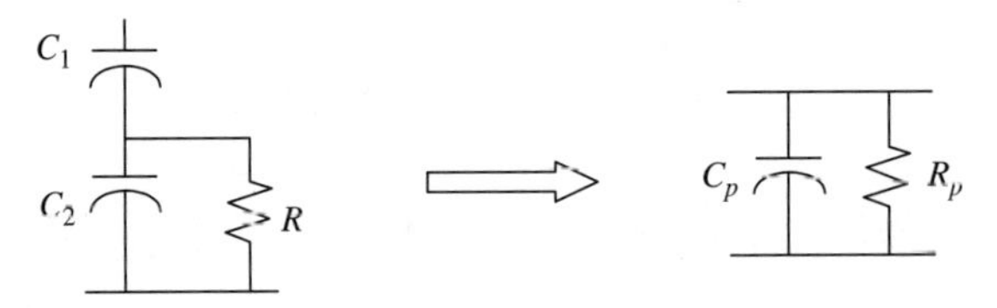

Figure A6.4 Transformation of a capacitor-tapped circuit.

Similarly, a capacitor-tapped circuit can be transformed to a shunt R_p–C_p circuit as illustrated in Figure A6.4. In this case, the following relations must hold if the two circuits are equal:

$$R_p = \frac{1 + \omega^2 R^2 (C_1 + C_2)^2}{\omega^2 R C_1^2} \tag{A6.11}$$

and

$$C_p = \frac{C_1[1 + \omega^2 R^2 C_2 (C_1 + C_2)]}{1 + \omega^2 R^2 (C_1 + C_2)^2} \tag{A6.12}$$

APPENDIX 7

PROPERTIES OF SOME MATERIALS

TABLE A7.1 Conductivity of Some Materials

Material	Conductivity (S/m)
Aluminum	3.54×10^7
Brass	1.5×10^7
Bronze	1.0×10^7
Copper	5.813×10^7
Gold	4.1×10^7
Iron	1.04×10^7
Nichrome	0.09×10^7
Nickel	1.15×10^7
Silver	6.12×10^7
Stainless steel	0.11×10^7

Radio-Frequency and Microwave Communication Circuits: Analysis and Design, Second Edition,
By Devendra K. Misra
ISBN 0-471-47873-3

TABLE A7.2 Dielectric Constant and Loss Tangent of Some Materials at 3 GHz

Material	Dielectric Constant, ε_r	Loss Tangent, tan δ
Alumina	9.6	0.0001
Glass (Pyrex)	4.82	0.0054
Mica (ruby)	5.4	0.0003
Nylon (610)	2.84	0.012
Polystyrene	2.55	0.0003
Plexiglas	2.60	0.0057
Quartz (fused)	3.8	0.00006
Rexolite (1422)	2.54	0.00048
Styrofoam (103.7)	1.03	0.0001
Teflon	2.1	0.00015
Water (distilled)	77	0.157

APPENDIX 8

COMMON ABBREVIATIONS

AAU	analog audio
ABCMOS	advanced bipolar CMOS
ABR	available bit rate
ACI	adjacent channel interference
ACTS	advanced communication technology satellite
ADC	analog-to-digital converter
ADPCM	adaptive differential pulse code modulation
ADSL	asymmetric digital subscriber line
AFC	automatic frequency control
AGC	automatic gain control
AM	amplitude modulation
AMPS	advanced mobile phone service
ANSI	American National Standards Institute
ARDIS	advanced radio data information service
ASCII	American Standard Code for Information Interchange
ASIC	application-specific integrated circuit
ASK	amplitude-shift keying
ATM	asynchronous transfer mode
AWGN	additive white Gaussian noise
B-CDMA	broadcast CDMA
BER	bit error rate
BIFET	bipolar field-effect transistor

Radio-Frequency and Microwave Communication Circuits: Analysis and Design, Second Edition,
By Devendra K. Misra
ISBN 0-471-47873-3

BIMOS	bipolar metal-oxide semiconductor
BIOS	basic input–output system
BJT	bipolar junction transistor
BONDING	bandwidth on demand interoperability
BPSK	biphase-shift keying
CAD	computer-aided design
CBR	continuous bit rate
CCD	charge-coupled device
CCIR	International Radio Consultative Committee
CDMA	code-division multiple access
CDPD	cellular digital packet data
CIR	carrier-over-interference ratio
CMOS	complementary metal-oxide semiconductor
CNR	carrier-over-noise ratio
COB	chip on board
Codec	coder–decoder
CRC	cyclic redundancy check
CSDN	circuit-switched digital network
CSMA/CD	carrier sense multiple access with collision detection
CT	cordless telephone
CW	continuous wave
DAB	digital audio broadcast
DAC	digital-to-analog converter
DAS	data acquisition system
DBS	direct broadcast satellite
DCE	data communication equipment
DCP	data communication protocol
DCPSK	differentially coherent phase-shift keying
DCS-1800	digital communication system 1800
DDS	direct digital synthesis
DECT	digital European cordless telecommunications
DFT	discrete Fourier transforms
DIP	dual-in-line package
DMR	digital mobile radio
DPCM	differential pulse code modulation
DPSK	differential phase-shift keying
DQPSK	differential quadrature phase-shift keying
DRAM	dynamic random access memory
DRO	dielectric resonator oscillator
DSMA	digital sensed multiple access
DSO	digital sampling oscilloscope
DSP	digital signal processing
DUT	device under test
DWT	discrete wavelet transforms

EBS	emergency broadcast system
EDA	electronic design automation
EIA	Electronics Industries Association
EIRP	effective isotropic radiated power
EMC	electromagnetic compatibility
EMI	electromagnetic interference
ERC	European Radio Commission
ESD	electrostatic discharge
ESDI	enhanced small device interface
ETN	electronic tandem network
ETSI	European Telecommunication Standard Institute
FCC	Federal Communications Commission
FDD	frequency-division duplex
FDDI	fiber-distributed data interface
FDMA	frequency-division multiple access
FET	field-effect transistor
FFT	fast Fourier transforms
FH	frequency hopping
FIFO	first-in first-out system
FIR	finite impulse response filter
FM	frequency modulation
FPGA	field-programmable gate array
FSK	frequency-shift keying
GEO	geosynchronous satellite
GFSK	Gaussian frequency shift keying
GIGO	garbage in, garbage out
GMSK	Gaussian minimum shift keying
GPIB	general-purpose interface bus
GPS	global positioning system
GSM	(Groupe Special Mobile) global system for mobile communication
GUI	graphical user interface
HBT	heterojunction bipolar transistor
HDTV	high-definition television
HEMT	high-electron mobility transistor
HF	high frequency
HIPERLAN	high-performance radio LAN
I and Q	in-phase and quadrature phase
IC	integrated circuits
IDE	integrated device electronics
IF	intermediate frequency
IGBT	insulated gate bipolar transistor
IIT	infinite impulse response filter

IMD	intermodulation distortion
IMPATT	impact ionization avalanche transit time diode
INTELSAT	International Telecommunication Satellite Consortium
IP	intermodulation product
IP	Internet protocol
IP2	second-order intercept
IP3	third-order intercept
ISDN	integrated services digital network
ISI	intersymbol interference
ISM	industrial, scientific, and medical bands
ITU	International Telecommunication Union
JDC	Japanese digital cellular standard
JFET	junction field-effect transistor
JPEG	Joint Photographic Experts Group
LAN	local area network
LATA	local access and transport area
LCD	liquid-crystal display
LED	light-emitting diode
LEOs	low-Earth-orbit satellite
LHCP	left-hand circular polarization
LNA	low-noise amplifier
LO	local oscillator
LOS	line of sight
LUT	lookup table
MAC	medium access control
MAN	metropolitan area network
MAU	multistation access unit
MBE	molecular beam epitaxy
MDS	minimum detectable signal
MDSL	medium-bit-rate digital subscriber line
MEOs	medium-Earth-orbit satellite
MESFET	metal-semiconductor field-effect transistor
MIDI	musical instrument digital interface
MIPS	million instructions per second
MLC	multilayer ceramics
MMI	man–machine interface
MMIC	monolithic microwave integrated circuit
MMSI	multimode stereoscopic imaging
MODFET	modulation doped field-effect transistor
MOSFET	metal-oxide-semiconductor field-effect transistor
MPEG	Motion Picture Experts Group
MPP	massively parallel processing
MPSK	minimum phase-shift keying

MSK minimum shift keying
MSP mixed signal processor
MSS mobile satellite service
MTI moving-target indicator
MTSO mobile telephone switching office
MUX multiplexing

NADC North American Digital Cellular/Cordless
NF noise figure
NMT Nordic Mobile Telephone
NTSC National Television Standards Committee
NTT Nippon Telephone and Telegraph

OFDM orthogonal frequency-division multiplexing
OOK on–off keying
OOP object-oriented programming

PAD packet assembler and disassembler
PAL phase alternating line
PAM pulse amplitude modulation
PAMR public access mobile radio
PBX private branch exchange
PCB printed circuit board
PCM pulse code modulation
PCMCIA Personal Computer Memory Card International Association
PCN personal communication network
PCS personal communication service
PDA personal digital assistant
PDC personal digital cellular
PDIP plastic dual-in-line package
PGA programmable gain amplifier; *also* pin grid array
PHP personal handy phone
PHS personal handyphone system (formerly PHP)
PID proportional integrating differentiating
PIN p-type, intrinsic, n-type diode
PLC programmable logic controller
PLL phase-locked loop
PM pulse modulation
PMR private mobile radio
POTS plain old telephone service
PPGA plastic pin grid array
PQFP plastic quad flat pack
PSK phase-shift keying
PSTN public switched telephone network
PVC permanent virtual connection

PWB	printed wire board
PWM	pulse width modulation
QAM	quadrature amplitude modulation
QCELP	Qualcomm coded excited linear predictive coding
QDM	quadrature demodulator
QMFSK	quadrature modulation frequency-shift keying
QPSK	quadrature phase-shift keying
RADAR	radio detection and ranging
RAID	redundant array of independent disks
RBDS	radio broadcast data system
RCS	radar cross section
RES	Radio Expert Systems Group
RF	radio frequency
RFI	radio-frequency interference
RHCP	right-handed circular polarization
RMS	root mean square
RPE-LTP	regular pulse excitation long-term predictor
RTD	resistive thermal detector
RTMS	radio telephone mobile system
SAW	surface acoustic wave
SCSI	small computer systems interface
SDH	synchronous digital hierarchy
SDIP	shrink dual-in-line package
SDN	switched digital network
SFDR	spur-free dynamic range
SINAD	signal-over-noise and distortion
SIP	single-in-line package
SLICs	standard linear integrated circuits
SMD	surface-mount device
SMR	specialized mobile radio
SNR	signal-to-noise ratio
SOI	silicon on insulator
SOIC	small outline integrated circuit
SONET	synchronous optical network
SOT	small outline transistor
SPLIC	special-purpose linear integrated circuit
SQPSK	staggered quadriphase-shift keying
SQUID	superconducting quantum interface
SSB	single sideband
STP	shielded twisted pair
SVD	simultaneous voice and data
SWR	standing wave ratio

TACS total access communication system
TCP transmission control protocol
T-DAB terrestrial-digital audio broadcasting
TDD time-division duplex
TDMA time-division multiple access
TDR time-domain reflectometry
TE transverse electric
TEM transverse electromagnetic wave
TETRA trans-European trunked radio system
THD+N total harmonic distortion plus noise
TIA Telecommunications Industry Association
TM transverse magnetic
TNC threaded Neil-Councilman connector
TSAPI telephony services–oriented application programming interface
TTL transistor–transistor logic
TVG time-variable gain
TVRO TV receive-only
TVS transient voltage suppression
TWTA traveling wave tube amplifier
Tx transmitter

UART universal asynchronous receiver/transmitter
UDPC universal digital personal communications
UHF ultrahigh frequency
UNC universal control network
UNI user-to-network interface
UPS uninterruptible power supply
USART universal synchronous–asynchronous receiver transmitter
USB universal serial bus
UTC universal time code
UTP unshielded twisted pair

VCO voltage-controlled oscillator
VDT video display terminal
VHDL very high definition language
VHF very high frequency
VLF very low frequency
VLSI very-large-scale integration
VPSK variable phase-shift keying
VSAT very small aperture (satellite ground) terminal
VSELP vector sum excited linear predictive coding
VSWR voltage standing wave ratio
VXCO voltage-controlled crystal oscillator

WAN wide area network
WATS wide area telecommunication network

WDM	wavelength-division multiplexing
WER	word error rate
WFAU	wireless fixed access unit
WLAN	wireless local area network
WLL	wireless local loop
WORM	write once, read many
WPABX	wireless private branch exchange

APPENDIX 9

PHYSICAL CONSTANTS

Permittivity of free space, ε_0	8.8542×10^{-12} F/m
Permeability of free space, μ_0	$4\pi \times 10^{-7}$ H/m
Impedance of free space, η_0	$376.7\ \Omega$
Velocity of light in free space, c	2.997925×10^{8} m/s
Charge of electron, q_e	1.60210×10^{-19} C
Mass of electron, m_e	9.1091×10^{-31} kg
Boltzmann's constant, k	1.38×10^{-23} J/K
Planck's constant, h	6.6256×10^{-34} J·s

Radio-Frequency and Microwave Communication Circuits: Analysis and Design, Second Edition,
By Devendra K. Misra
ISBN 0-471-47873-3

INDEX

Radio-Frequency and Microwave Communication Circuits: Analysis and Design, Second Edition,
By Devendra K. Misra
ISBN 0-471-47873-3